DIPLOMA PROG[RAMME]

Biology

Course Companion

Andrew Allott
David Mindorff

OXFORD

This book is dedicated to our families, who have supported us during its writing:
Alison and William Allott.
Michèle, Nicholas, Michael, Benjamin and Jonathan Mindorff
It is also dedicated to all the IB Biology teachers and students throughout the world, who use it.

Great Clarendon Street, Oxford OX2 6DP

Oxford University Press is a department of the University of Oxford.
It furthers the University's objective of excellence in research, scholarship, and education by publishing worldwide in

Oxford New York

Auckland Cape Town Dar es Salaam Hong Kong Karachi
Kuala Lumpur Madrid Melbourne Mexico City Nairobi
New Delhi Shanghai Taipei Toronto

With offices in

Argentina Austria Brazil Chile Czech Republic France Greece
Guatemala Hungary Italy Japan Poland Portugal Singapore
South Korea Switzerland Thailand Turkey Ukraine Vietnam

First published 2007

British Library Cataloguing in Publication Data

Data available

ISBN 978 0 19 915145 5

10 9 8 7 6 5 4 3 2

Printed in Great Britain by Bell and Bain Ltd, Glasgow

Authors

Andrew Allott has taught for over 25 years and is currently Head of Biology at Shrewsbury School. He is a former chief examiner for IB Biology and is currently involved in examining and curriculum review.

David Mindorff is deputy chief examiner, internal assessment moderator, biology faculty member for the IB online curriculum centre, workshop leader and an IB diploma coordinator at Branksome Hall. He has been a teacher of TOK and biology for 15 years.

Dedication

This book is dedicated to our families who have supported us during it's writing: Alison and William Allott. Michèle, Nicholas, Michael, Benjamin and Jonathan Mindorff. It is also dedicated to all IB Biology teachers and students throughout the world who use it, and The Branksome Hall SBI 4U class of 2006/07.

Acknowledgements

We are grateful for permission to reprint the following copyright material:
P7 Dr. Jeremy Burgess/Science Photo Library; p13br Per-Anders Jansson; p16t Roland Birke/Phototake Inc/Photolibrary; p16m Science Photo Library/Photolibrary; p17 Science Photo Library; p19 Biophoto Associates/Photo Researchers, Inc/Photolibrary; p20 reproduced from Carolyn Strange: 'Biological ties that bind', *Bioscience* 47 (1), 1997, copyright © American Institute of Biological Sciences, by permission of the AIBS.
p24b Dr. Gopal Murti/Science Photo Library; p30 J.C. Revy/Science Photo Library; p36 reproduced from figure 3 published online at www.jpchemnet.de/dnaback.html, by permission of Dr Frank J Jochem.
P52; reproduced from fig 10.18 in M Cummings: *Human Heredity: Principles and Issues* (4e, 1997), by permission of Brooks/Cole, a division of Thomson Learning: www.thomsonrights.com. Fax 800 730-2215
p55 Science Photo Library; p83 Photo Researchers, Inc/Photolibrary; p93 reproduced with permission of Dr Carmen Manella, Wadsworth Center, State of New York Department of Health.
p110 Ed Reschke/Peter Arnold Images Inc/Photolibrary; p115 Dr. Jeremy Burgess/Science Photo Library; p126l reproduced from Tse and Chan: *Journal of Biological Education* 35 (2001) by permission of the Institute of Biology
p130br Science Photo Library/Photolibrary; p134tl Bettmann/Corbis; p134tm Zigmund Leszczynski/Animals Animals / Earth Scenes/Photolibrary; p134tr Joseph T Collins/Photo Researchers, Inc/Photolibrary; p134bl John Cancalosi/Peter Arnold Images Inc/Photolibrary; p134bm Gary Nafis; p134br Jose Luis Gomez de Francisco /Naturepl.com; p 158 Courtesy of Affymetrix; p163 reproduced from M Krings et al: 'Neanderthal DNA sequences and the original of modern humans', *Cell* 90 (1997), copyright © 1997, by permission of Elsevier.p168 (lower l) reproduced from Daniel Pauly et al: 'Towards sustainability in world fisheries', *Nature* 418, copyright © 2001, by permission of Macmillan Publishers Ltd.
p171bl Laurent Geslin /Naturepl; p 172 (t) reproduced from fig 50.12 in Purves et al: *Life: The Science of Biology* 4e (Sinauer, 1995)
by permission of Sinauer Associates, Inc.
p 173t reproduced from fig 1 in Scheffer: *Scientific Monthly* 73 , 1951, by permission of AAAS.
p 180t Wang, M., and J.E. Overland (2004): Detecting Arctic Climate change using Köppen classification. Clim. Change, 67(1), 43-62. **p184t Omikron/Photo Researchers Inc/Photolibrary; p187 The Print Collector / Alamy; p197(l)** reproduced from Ray Gibson, B Hextall & A Rogers: *Photographic Guide to Sea and Seashore Life* (OUP, 2001), by permission of Oxford University Press.
P 198; reproduced from P Bang and P Dahlstrom: *Animal Tracks and Signs* (OUP, 2001), by permission of Oxford University Press. **p200 (t)** reproduced from Ray Gibson, B Hextall & A Rogers: *Photographic Guide to Sea and Seashore Life* (OUP, 2001), by permission of Oxford University Press.
P 202 reproduced from *Oxford Book of Wild Flowers* (OUP, 1976), by permission of Oxford University Press and *Oxford Book of Flowerless Plants* (OUP, 1966), by permission of Oxford University Press.
p205 Biophoto Associates/Science Photo Library; p209 reproduced from T Netherwood et al: 'Assessing the survival of transgenic plant DNA in the human gastrointestinal tract', *Nature Biotechnology* 22, copyright © 2004, by permission of Macmillan Publishers Ltd. **p216 Roger A. Gorski/Brain Research Institute, UCLA; p223 (t)** reproduced from H Honda et al 'No effect of MMR withdrawal on the incidence of autism: a total population study', *Journal of Child Psychology and Psychiatry*, 2005, by permission of Blackwell Publishing.
p224 Eye of Science/Science Photo Library/Photolibrary; p226 Steve Gschmeissner/Science Photo Library; p228l Astrid & Hanns-Frieder Michler/Science Photo Library; p228r Manfred Kage/Science Photo Library; p234 Astrid & Hanns-Frieder Michler/Science Photo Library; p260 Andrew Syred/Science Photo Library; p265 Astrid & Hanns-Frieder Michler/Science Photo Library; p266 Biophoto Associates/Science Photo Library; p261 reproduced from Karen Birch: *British Medical Journal* 330 (2005), by permission of BMJ Publishing Group.
p277 Library of Congress; p314 reproduced from Krebs and Davies: *An Introduction to Behavioural Ecology* (3e, Blackwell, 1993) , by permission of Blackwell Publishing.
p 322 "LMP-420, a small-molecule inhibitor of TNF-alpha, reduces replication of HIV-1 and *Mycobacterium tuberculosis in human cells*", Haraguchi et al. *AIDS Research and Therapy* 2006, **3**:8 p302 drawings of hominid skulls by Cornelia Blik for diagram in W Leonard: 'Food for Thought', *Scientific American*, 13.12.2002, reproduced with her permission.
p326 John Durham/Science Photo Library/Photolibrary; p328 Courtesy of Isle of Wight Natural History & Archaeological Society. Image taken from http://www.medinavalleycentre.org.uk.
p330 Yann Arthus-Bertrand/Corbis; p333 reproduced from W C Henson: 'Cesium-137 in Alaska Lichens, Caribou and Inuit', *Health Physics* 13 (1967) pp 383-389, by permission from the Health Physics Society.
p334b Biofoto/Heather Angel; p341 Steve Gschmeissner/Science Photo Library; p344 Innerspace Imaging/Photolibrary

Extracts

David Appell: abbreviated version of 'The New Uncertainty Principle', *Scientific American*, January 2001, copyright © 2001 by Scientific American, Inc, reprinted by permission of Scientific American, Inc. All rights reserved.
Thomas Barlow: extract from 'Settling the stomach, not lining the pocket', *Financial Times* 21 Oct 2000, reprinted by permission of the author.
Timothy M Beardsley: extract from 'Framing Biology', *Bioscience* 56 (7), 2006, copyright © American Institute of Biological Sciences, reprinted by permission of the AIBS.
Larry Cahill: extracts from 'His Brain, Her Brain', *Scientific American*, May 2005, copyright © 2005 by Scientific American, Inc. reprinted by permission of Scientific American, Inc. All rights reserved.
Jeff Collin: extract from article on tobacco control (Institute for Development Studies), reprinted by permission of the author.
Marcus W Feldman, Richard C Lewontin and Mary-Claire King: extracts from 'Race: A Genetic melting-pot', *Nature* 424, copyright © 2003, by permission of Macmillan Publishers Ltd.
Watson Fuller: extract from '"Who said 'Helix'?" Right and wrong in the story of how the structure of DNA was discovered', *Nature* 424, copyright © 2003, by permission of Macmillan Publishers Ltd.
N Eldredge and S J Gould: extract from 'Punctuated equilibria: an alternative to phyletic gradualism' in T J M Schopf (ed.): *Models in Paleobiology* (Freeman Cooper Co, 1972), reprinted by permission of Niles Eldredge.
Gerald L Geison: extracts from *The Private Science of Louis Pasteur* (Princeton, 1995), copyright © Princeton University Press 1995, reprinted by permission of Princeton University Press.
W Wayt Gibbs: extracts from 'The Unseen Genome: Gems among the Junk', *Scientific American*, November 2003, copyright © 2003 by Scientific American, Inc, reprinted by permission of Scientific American, Inc. All rights reserved.
Ruth Hubbard: extract from 'Science, Power, Gender: How DNA Became the Book of Life', *Signs: Journal of Women in Culture and Society*, 28:3 (2003), reprinted by permission of the University of Chicago Press.
Ruth Hubbard and Elijah Wald: extract from *Exploding the Gene Myth* (Beacon Press, 1993), copyright © 10997, 1993 by Ruth Hubbard and Elijah Wald, reprinted by permission of Beacon Press, Boston.
Lisa Jardine: extract from interview on BBC Radio 4 reprinted by permission of Professor Jardine.
Tedd Jesdale: interview reprinted by permission of Tedd Jesdale.
Hans Krebs: extract from Nobel Prize Lecture 1953, copyright © The Nobel Foundation 1953, reprinted from www.nobelprize.org, by permission of The Nobel Foundation.
Elaine Larson: extract from 'Innovations in Health Care: Antisepsis as a Case Study', *American Journal of Public Health*, Jan 1989, 79 (1), by permission of the American Public Health Association.
Dr Carmen Mannella: comments on the development of the understanding of mitochondria, reprinted by permission of Dr Carmen and the State of New York Department of Health.
Jean Marx: extract from 'Unravelling the causes of Diabetes', *Science* 296, 2002, reprinted by permission of AAAS.
Ernst Mayer: extract from 'The Evolution of Ernst: Interview with Ernst Mayer', www.sciam.com. 'In Focus', July 6, 2004, copyright © 2006 by Scientific American, Inc. reprinted by permission of Scientific American, Inc. All rights reserved.
Peter Mitchell: extract from Nobel Prize Banquet Speech 1978, copyright © The Nobel Foundation 1978, reprinted from www.nobelprize.org, by permission of The Nobel Foundation.
National Biothetics Advisory Commission: extracts from testimonies in *Ethical Issues in Human Stem Research, Volume III: Religious Persepective* (NBAC, June 2000), reprinted by permission of Eric Meslin, former Executive Director of NBAC.
David S Newberger: extract from 'Down Syndrome: Prenatal Risk Assessment and Diagnosis', *American Family Physician* 62 (4), 2000, p 825, copyright © 2000 American Academy of Family Physicians, reprinted by permission of the AAFP.
New Scientist: extract from '13 things that do not make sense', *The New Scientist*, March 2005, reprinted by permission of New Scientist.
Marshall Nirenberg: extract from an interview in 1968, copyright © The Nobel Foundation 1968, reprinted from www.nobelprize.org, by permission of The Nobel Foundation.
Carolyn Raffensperger and Katherine Barrett: extract from 'In defense of the Precautionary Principle', *Nature Biotechnology* 19 (9), copyright © 2001, by permission of Macmillan Publishers Ltd.
T Rahemtulla and R Bhopal: extract from 'Pharmacogenetics and ethnically targeted therapies', *British Medical Journal* 330, May 2005, reprinted by permission of BMJ Publishing Group.
Richard Smith: extract from 'The truth about juice', *The Guardian*, 5 Sept 2006, copyright © Guardian News & Media Ltd 2006, by permission of the Guardian Media Group.
Carolyn Strange: extract from 'Biological ties that bind', *Bioscience* 47 (1), 1997, copyright © American Institute of Biological Sciences, reprinted by permission of the AIBS.
Dr. Swezey: extract from 'Arthritis and Rheumatism', *Arthritis Rheumatism* 41 (5), 1998, by permission of John Wiley and Sons, Inc.
Simon Winchester: extract from *The Map that Changed the World: A Tale of Rocks, Ruin and Redemption* (Viking, 2001), copyright © Simon Winchester 2001, reprinted by permission of Penguin Books Ltd
World Health Organization: extract from communique on World Tobacco Day 2006 by Catherine Le Gales-Camus, WHO Assistant Director-General, Non-communicable Diseases & Mental Health, reprinted by permission of WHO.

We have tried to trace and contact all copyright holders before publication. If notified the publishers will be pleased to rectify any errors or omissions at the earliest opportunity.

Course Companion definition

The IB Diploma Programme Course Companions are resource materials designed to provide students with extra support through their two-year course of study. These books will help students gain an understanding of what is expected from the study of an IB Diploma Programme subject.

The Course Companions reflect the philosophy and approach of the IB Diploma Programme and present content in a way that illustrates the purpose and aims of the IB. They encourage a deep understanding of each subject by making connections to wider issues and providing opportunities for critical thinking.

These Course Companions, therefore, may or may not contain all of the curriculum content required in each IB Diploma Programme subject, and so are not designed to be complete and prescriptive textbooks. Each book will try to ensure that areas of curriculum that are unique to the IB or to a new course revision are thoroughly covered. These books mirror the IB philosophy of viewing the curriculum in terms of a whole-course approach; the use of a wide range of resources; international-mindedness; the IB learner profile and the IB Diploma Programme core requirements; theory of knowledge; the extended essay; and creativity, action, service (CAS).

In addition, the Course Companions provide advice and guidance on the specific course assessment requirements and also on academic honesty protocol.

IB mission statement

The International Baccalaureate aims to develop inquiring, knowledgeable, and caring young people who help to create a better and more peaceful world through intercultural understanding and respect.

To this end the IB works with schools, governments, and international organizations to develop challenging programmes of international education and rigorous assessment.

These programmes encourage students across the world to become active, compassionate, and lifelong learners who understand that other people, with their differences, can also be right.

The IB learner profile

The International Baccalaureate aims to develop internationally minded people who, recognizing their common humanity and shared guardianship of the planet, help to create a better and more peaceful world. IB learners strive to be:

Inquirers They develop their natural curiosity. They acquire the skills necessary to conduct inquiry and research and show independence in learning. They actively enjoy learning and this love of learning will be sustained throughout their lives.

Knowledgeable They explore concepts, ideas, and issues that have local and global significance. In so doing, they acquire in-depth knowledge and develop understanding across a broad and balanced [illegible] isciplines.

Thinkers They exercise initiative in applying thinking skills critically and creatively to recognize and approach complex problems, and make reasoned, ethical decisions.

Communicators They understand and express ideas and information confidently and creatively in more than one language and in a variety of modes of communication. They work effectively and willingly in collaboration with others.

Principled They act with integrity and honesty, with a strong sense of fairness, justice, and respect for the dignity of the individual, groups, and communities. They take responsibility for their own actions and the consequences that accompany them.

Open-minded They understand and appreciate their own cultures and personal histories, and are open to the perspectives, values, and traditions of other individuals and communities. They are accustomed to seeking and evaluating a range of points of view, and are willing to grow from the experience.

Caring They show empathy, compassion, and respect towards the needs and feelings of others. They have a personal commitment to service, and act to make a positive difference to the lives of others and to the environment.

Risk-takers They approach unfamiliar situations and uncertainty with courage and forethought, and have the independence of spirit to explore new roles, ideas, and strategies. They are brave and articulate in defending their beliefs.

Balanced They understand the importance of intellectual, physical, and emotional balance to achieve personal well-being for themselves and others.

Reflective They give thoughtful consideration to their own learning and experience. They are able to assess and understand their strengths and limitations in order to support their learning and personal development.

A note on academic honesty

It is of vital importance to acknowledge and appropriately credit the owners of information when that information is used in your work. After all, owners of ideas (intellectual property) have property rights. To have an authentic piece of work, it must be based on your individual and original ideas with the work of others fully acknowledged. Therefore, all assignments, written or oral, completed for assessment must use your own language and expression. Where sources are used or referred to, whether in the form of direct quotation or paraphrase, such sources must be appropriately acknowledged through the use of footnotes and bibliographies.

Footnotes (placed at the bottom of a page) or endnotes (placed at the end of a document) are to be provided when you quote or paraphrase from another document, or closely summarize the information provided in another document. You do not need to provide a footnote for information that is part of a "body of knowledge". That is, definitions do not need to be footnoted as they are part of the assumed knowledge.

Bibliographies should include a formal list of the resources that you used in your work. "Formal" means that you should use one of the several accepted forms of presentation. This usually involves separating the resources that you use into different categories (e.g. books, magazines, newspaper articles, Internet-based resources, CDs, and works of art) and providing full information as to how a reader or viewer of your work can find the same information. A bibliography is compulsory in the extended essay.

What constitutes malpractice?

Malpractice is behaviour that results in, or may result in, you or any student gaining an unfair advantage in one or more assessment component. Malpractice includes plagiarism and collusion.

Plagiarism is defined as the representation of the ideas or work of another person as your own. The following are some of the ways to avoid plagiarism:

- words and ideas of another person to support one's arguments must be acknowledged
- passages that are quoted verbatim must be enclosed within quotation marks and acknowledged
- CD-ROMs, email messages, web sites on the Internet, and any other electronic media must be treated in the same way as books and journals
- the sources of all photographs, maps, illustrations, computer programs, data, graphs, audio-visual, and similar material must be acknowledged if they are not your own work
- copying works of art, whether music, film, dance, theatre arts, or visual arts, and where the creative use of a part of a work takes place, must be acknowledged.

Collusion is defined as supporting malpractice by another student. This includes:

- allowing your work to be copied or submitted for assessment by another student
- duplicating work for different assessment components and/or diploma requirements.

Contents

Ode to Scientific Integrity

– a poem written by IB Biology Student Kelly M[c]Elroy

I have my principles:
measure twice,

label carefully,
sterilise, sterilise, sterilise.

Where has the poet found
a scientific conscience?
She will never reuse pippettor tips
or shake test tubes with malevolence.
She will, however, always
carefully assess procedure,
calculate error,
and question.

A question tested,
with time can build a theory,
from lots of little affirmations.

There is no truth in science,
only theory, observation,
and clean laboratory hope.

And, for all scientific rationalism,
a body is still built of cells,
(specialised, multiplying, shuffling)
commanded by chromosomes
and filled with melancholy humors.

Introduction

The book has been written to be a companion for students of Biology in the International Baccalaureate Diploma Programme. Biology is a vital area of study for the 21st century. In a world where rapid change is threatening the survival of many species, you need to be both mindful of existing challenges and capable of acting on those challenges. The approach that the authors of the IB Biology Course Companion have tried to follow is perfectly summed up in a quote from Alice in Wonderland:

> *"Why"*, said the dodo, *"the best way to explain it is to do it."*

We believe that the optimal approach to learning is to be active – the more that students do for themselves, guided by their teacher, the better they will learn. During the two years of an IB Diploma Programme Biology Course students are encouraged to develop both a knowledge of Biology and an undersanding of the nature of scientific enquiry.

Core topics

Information needed for a comprehensive understanding of the IB Biology Core and selected topics in Options A to H.

Advanced topics

Information needed for IB Biology Additional Higher Level topics and for selected topics in Options D to H.

Questions

At the end of each chapter is a page of questions, including both past IB Biology exam questions and new questions. Multiple choice questions, short answer questions and extended response questions are all included, with mark allocations, following the pattern that has been established in IB Biology exams.

Inquiry

Short topics for students to research themselves, using information readily available in textbooks or on the Internet. Clear instructions or a series of questions are given to guide the inquiry. The aim is to promote an independent approach to learning.

Working with data

Instructions for handling data from experiments as well as in the use of Information and Communication Technology. This is assessed in Internal Assessment (IA) in the Data Collection and Processing, and in the Conclusion and Evaluation sections.

Rediscovering Biology

Ideas for experimental work, usually with known outcomes, aimed at promoting understanding through "doing and seeing." Another aim is the development of practical techniques. These are assessed in the Manipulative Skills section of Internal Assessment (IA). I

Investigating Biology

Ideas for experimental work, with unknown outcomes, where the student can define the problem and the methods. These are skills needed for the Design section of Internal Assessment (IA).

Critical Considerations

These are tasks involving discussion of issues in science, technology and society, including ethical issues and internationalism.

Thinking about science

Tasks involving discussion of the scientific methods and issues of knowledge in science. These are issues involved in the Theory Of Knowledge – a core element in the IB Diploma Programme model.

Data based questions

Questions that involve studying and analysing data from biological research – this type of question appears in both Paper 2 and Paper 3 for SL and HL IB Biology.

1 Cells

The invention of the microscope in the 17th century quickly led to the discovery of cells. Robert Hooke was the first to use the term **cell** in 1665, after examining cork and other plant tissues under the microscope. In the centuries that followed, the **cell theory** was developed.

- Cells are the smallest unit of life and nothing smaller can survive independently.
- All living things consist of cells, although the smallest organisms may consist of one cell only.
- All cells come from other pre-existing cells, by division, and therefore new cells cannot be constructed from non-living chemical substances.

Microscopes have been improved considerably since the 17th century. The parts of a microscope are shown in Figure 1.

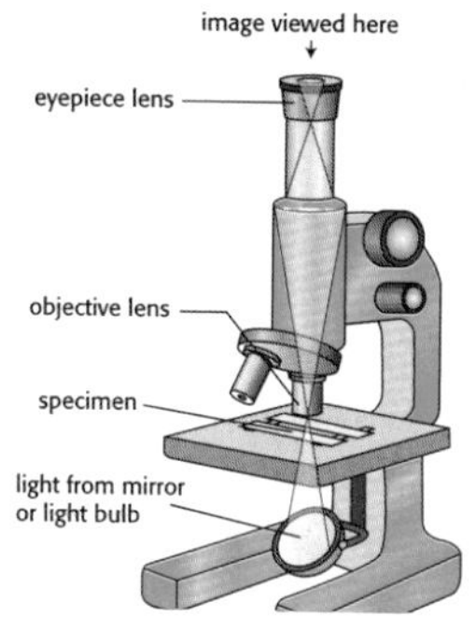

Figure 1 Compound light microscope

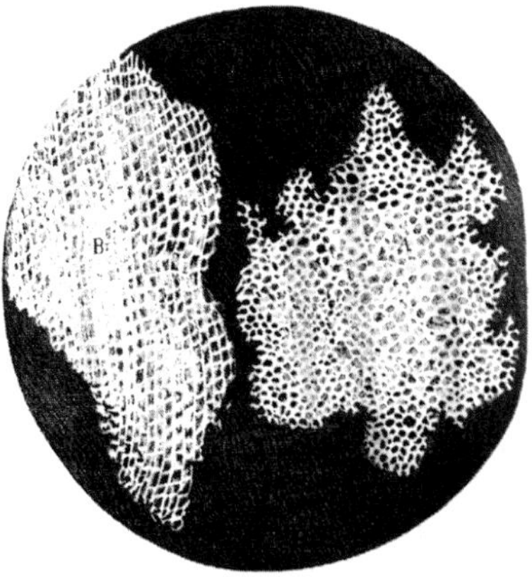

Figure 2 Robert Hooke's drawing of cork cells

Rediscovering biology: cells in cork tissue

Robert Hooke's discovery of cells can easily be repeated.

1. Use a very sharp blade to cut a very thin slice of cork.
2. Place this on the centre of a microscope slide. No water or cover slip is necessary.
3. Examine the cork using the low power lens of a microscope. Move the slide until you find the thinnest part of the cork, probably near an edge. Move the slide until this part is in the centre of the field of view.
4. Examine the thinnest area using the high power lens of the microscope. Draw a small group of cells.

Working with data: instructions for drawings

Careful drawings are a useful way of recording the structure of cells or other biological structures. Usually the lines on the drawing represent the edges of structures. Do not show unnecessary detail and only use faint shading. Drawings of structures seen using a microscope will be larger than the structures actually are – the drawing shows them magnified. On page 12 the method for calculating the magnification of a drawing is explained. Everything on a drawing should be shown to the same magnification.

(a) Use a sharp pencil with a hard lead to draw single sharp lines.

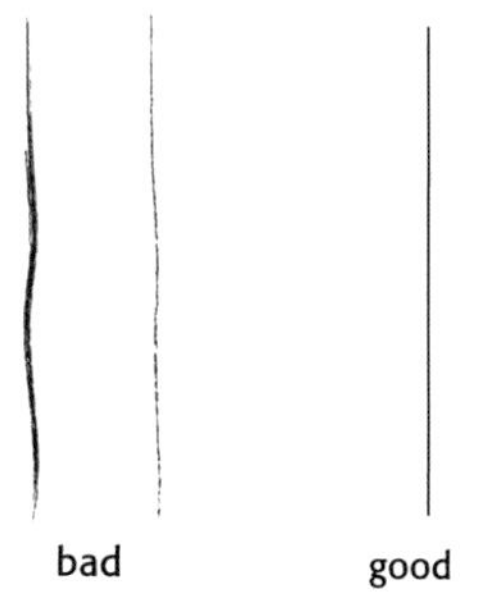

(b) Join up lines carefully to form continuous structures such as cells.

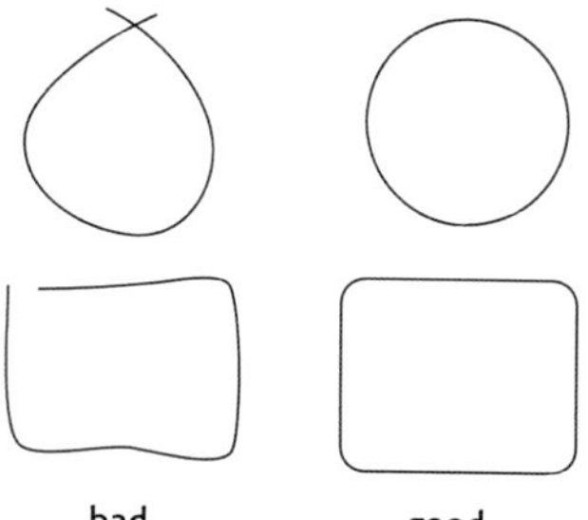

(c) Draw lines freehand, but use a ruler for labelling lines.

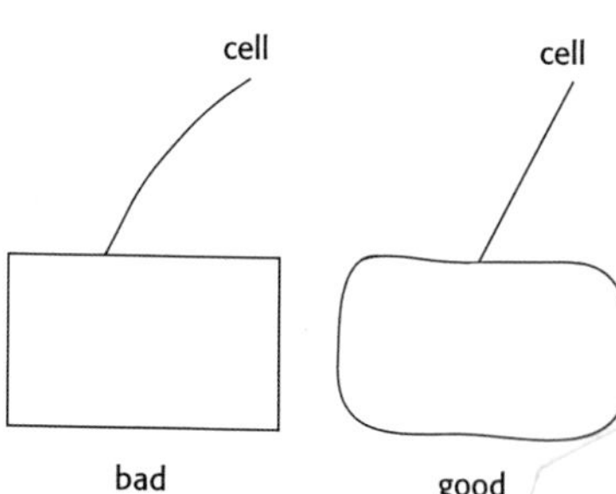

Figure 3 Examples of drawing styles

Plant and animal cells

There are many similarities in the structure of plant and animal cells (Figure 4). A nucleus is usually present, with cytoplasm around it. The nucleus is colourless but it can be stained to make it easily visible. The cytoplasm is enclosed by a plasma membrane, which is too thin to see with a light microscope, but its position can be deduced by the sharp edge of the cytoplasm.

There are also some differences between plant and animal cells.

- Plant cells have a cellulose cell wall outside their plasma membrane, which animal cells never have. The cellulose is in bundles of molecules called microfibrils, which have high tensile strength. The cell wall also maintains a plant cell in a fixed shape unless the cell is actively growing.
- Plant cells often have a large permanent vacuole, which contains a solution of mineral ions such as potassium and other solutes. It is bounded by a vacuole membrane. Because there are no proteins or other macromolecules, it does not stain densely and so is difficult to see with a light microscope. Animal cells usually only have small fluid filled sacs in their cytoplasm, called vesicles.
- Many plant cells in stems and leaves have chloroplasts in their cytoplasm. These are green in colour because of the chlorophyll that they contain, and they can also store starch. Animal cells never contain chloroplasts unless they have taken in algal cells that have them, as in the cells of some corals. Also, animal cells do not synthesize or store starch.

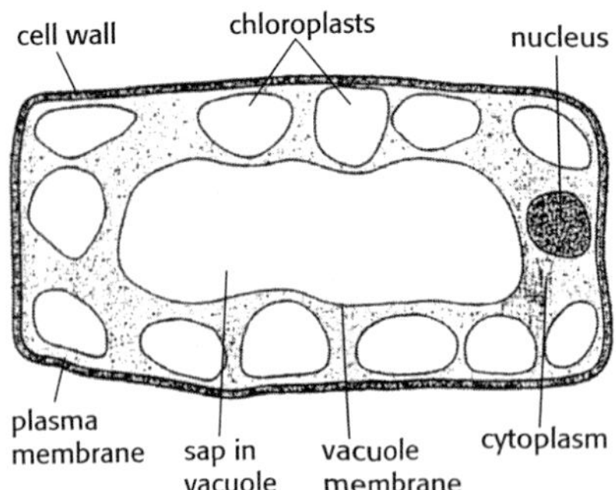

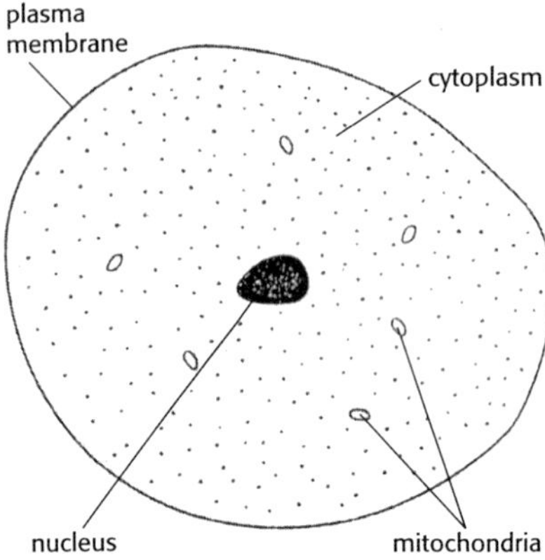

Figure 4 Drawings of a moss leaf cell (above) and a human cheek epithelium cell (below)

Thinking about science: evolution and cell structure

Theodore Dobzhansky, a Russian geneticist who moved to the United States, famously said, "Nothing in biology makes sense, except in the light of evolution." Evolution by natural selection helps to explain differences in cell structure: if some members of a population can adapt to take advantage of an unexploited resource, then they may eventually evolve into a new species. Organisms therefore tend to become different from each other. Table 1 shows how the cell structure of plants and animals allows them to exploit different resources. Fungi are in a different kingdom from plants and animals. They have cell walls and vacuoles, but no chloroplasts. Figure 5 shows a yeast cell. Yeast is a fungus. Discuss whether fungi's cell structure will cause them to exploit different resources from plants or animals, or the same resources.

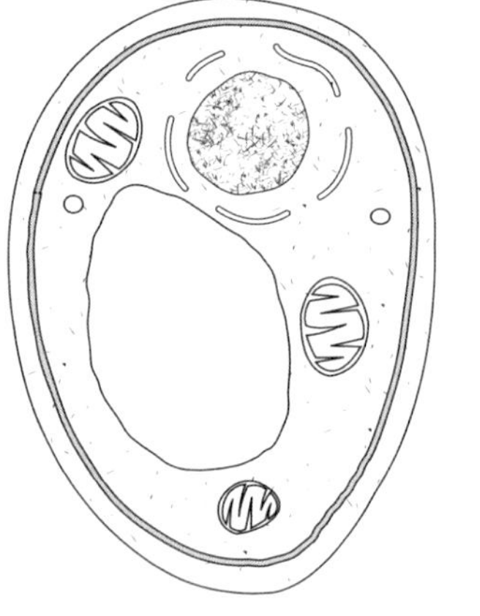

Figure 5 A drawing of a yeast cell

Table 1

	Chloroplasts	Cell walls	Vacuoles
Plants	Chloroplasts make food for the cell from simple inorganic molecules absorbed from the environment.	The cell wall makes it impossible to take in solids such as food, but supports the cell and prevents excess water entering by osmosis.	The vacuole can expand quickly by absorbing mineral ions and water and therefore allows rapid growth of plant cells.
Animals	Animal cells must absorb food made by other organisms. They can either take in undigested solids, or organic compounds that have been digested.	Solids can be taken in by vesicle formation, but excess water may enter by osmosis, unless solute concentrations are the same inside and outside the cell.	Lack of a vacuole reduces the mass of the cell and makes movement easier, so animals have much greater range of locomotion methods, including flying.

Rediscovering biology: comparing plant and animal cells

Cells are fascinating, beautiful structures when viewed with a light microscope, especially when they are still living. The same general method can be used for preparing a temporary mount of cells on a microscope slide.

- Place the cells on the slide in a layer not more than one cell thick.
- Add a drop of water or stain.
- Carefully lower a cover slip onto the drop. Try to avoid trapping any air bubbles.
- Remove excess fluid or stain by putting the slide inside a folded piece of paper towel and pressing lightly on the cover slip.

It is best to examine the slide first using low power. Move the slide to get the most promising areas in the middle of the field of view and then move up to high power. Draw a few cells, so you remember their structure.

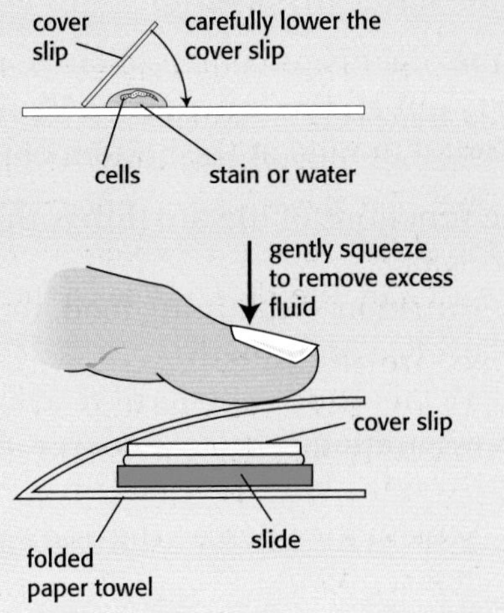

Figure 6 Making a temporary mount

1 *Moss leaf*

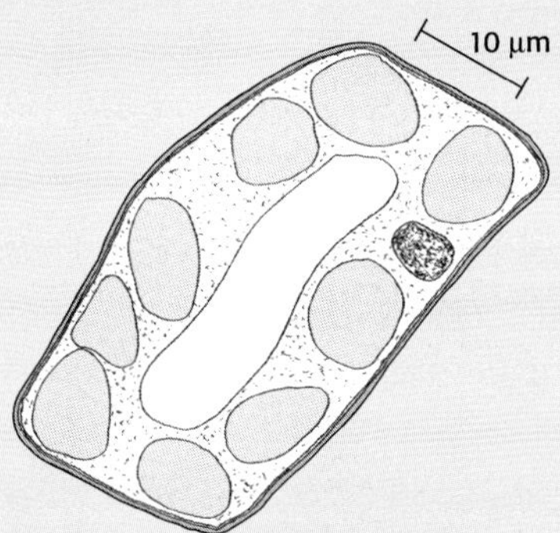

Use a moss plant with very thin leaves. Mount a single leaf in a drop of water or methylene blue stain.

2 *Banana fruit cell*

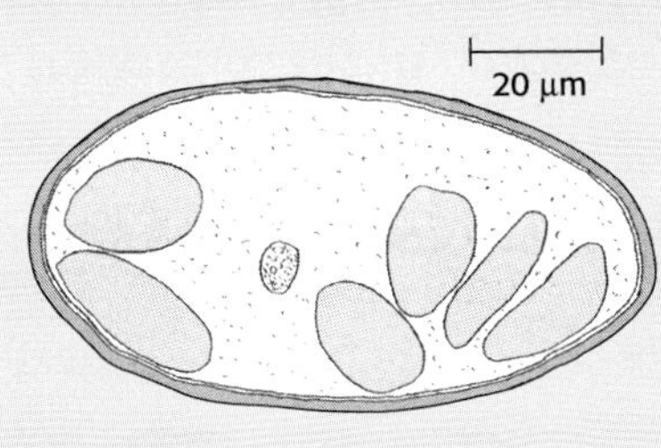

Scrape a small amount of the soft tissue from a banana and place on a slide. Mount in a drop of iodine solution.

3 *Mammalian liver cell*

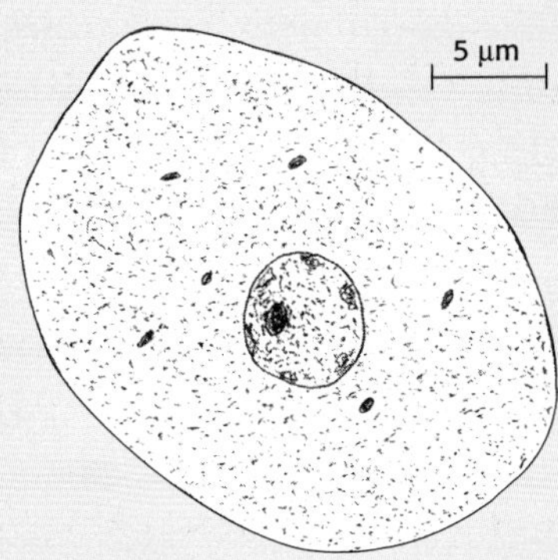

Scrape cells from a freshly cut surface of liver (not previously frozen). Smear onto a slide and add methylene blue to stain.

4 *Leaf lower epidermis*

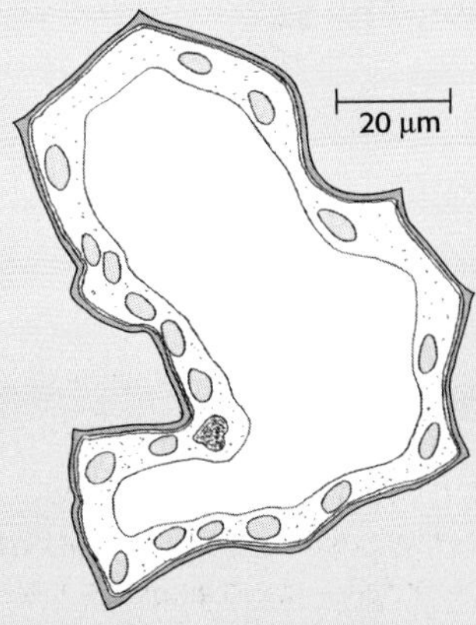

Peel the lower epidermis off a leaf. The cell drawn here was from *Valeriana*. Mount in water or in methylene blue.

5 *Human cheek cell*

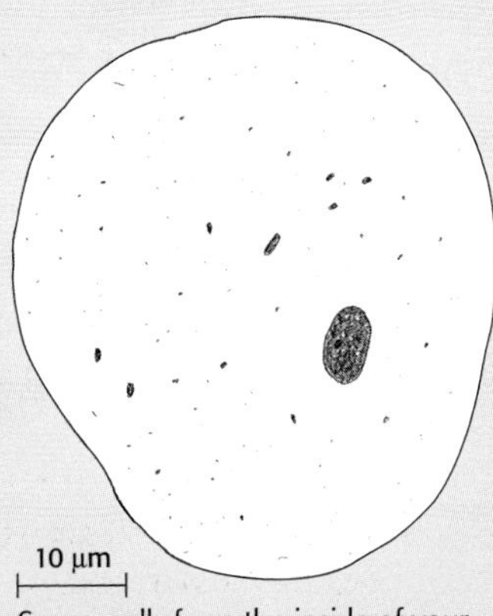

Scrape cells from the inside of your cheek with a cotton bud. Smear them on a slide and add methylene blue to stain.

6 *White blood cell*

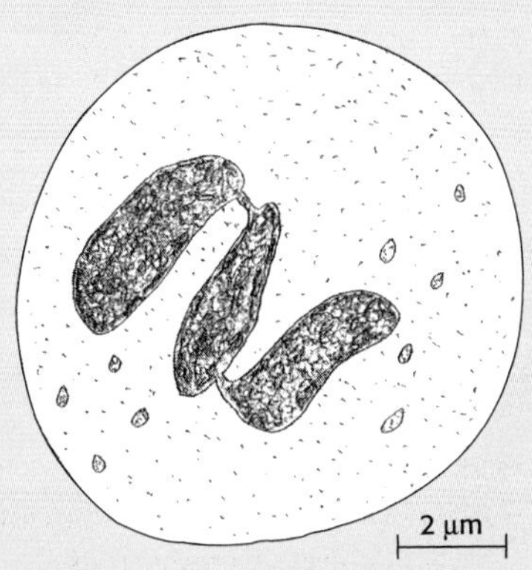

A thin layer of mammalian blood can be smeared over a slide and stained with Leishman's stain.

Figure 7 Plant and animal cell drawings

Unicellular organisms

Some organisms consist of only one cell. This cell therefore has to carry out all the functions of life. Figure 8 shows an amoeba, which is found in mud at the bottom of ponds.

The functions of life are things that all organisms must do to ensure survival.

- Nutrition – obtaining food, to provide energy and the materials needed for growth.
- Metabolism – chemical reactions inside the cell, including cell respiration to release energy.
- Growth – an irreversible increase in size.
- Sensitivity – perceiving and responding to changes in the environment.
- Homeostasis – keeping conditions inside the organism within tolerable limits.
- Reproduction – producing offspring either sexually or asexually.

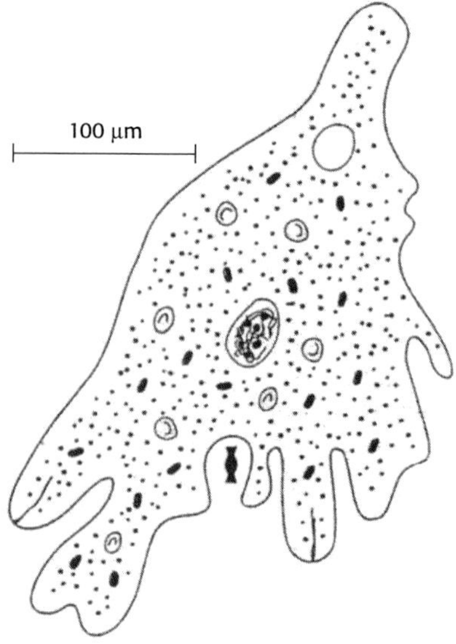

Figure 8 *Amoeba*

The structure of the single cells of unicellular organisms is therefore more complex than most of the cells of multicellular organisms. Figure 9 shows the structure of *Paramecium*, an organism that lives in the freshwater of ponds and rivers, feeding on smaller organisms by ingesting them.

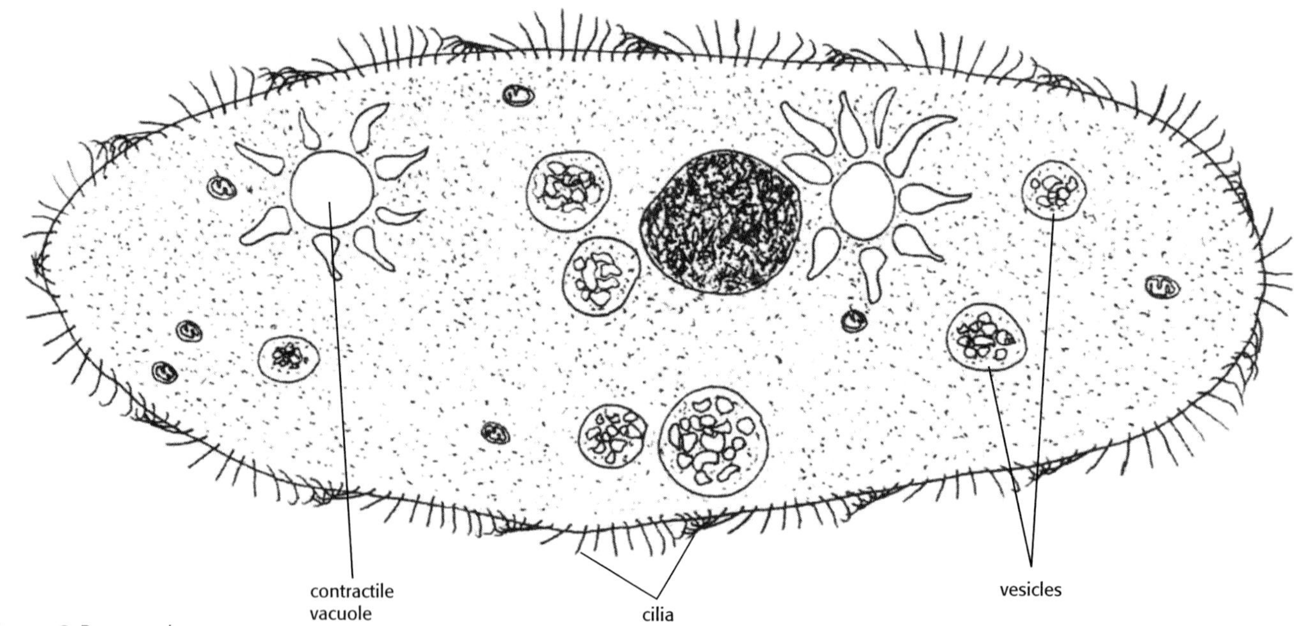

Figure 9 *Paramecium*

1 Vesicles contain smaller organisms that the *Paramecium* has consumed. These are gradually digested. The undigested remains are then expelled.

Which life process is this? Which other life processes depend on the resources obtained by it?

2 Beating of the cilia moves the *Paramecium* through the water. Energy is used to do this. Stimuli sometimes cause the direction of movement to change.

Which life processes are involved? Which life process is needed to provide energy for the beating of the cilia?

3 The contractile vacuoles at each end of the cell fill up with water and then expel it through the plasma membrane of the cell. This takes energy.

Which life processes are involved? Why does *Paramecium* expel water, despite having to use energy to do it?

Rediscovering biology: observing unicellular organisms

Living unicellular organisms can be found in many places including water from ponds. A centrifuge can be used to concentrate them in water. Mud or soil, shaken with water, can also be a useful source.

Unicells can then be examined using a microscope, using the method described below.

Make a ring or square of vaseline or methyl cellulose on a microscope slide. This should be the same shape and size as the cover slip, so that the cover slip will be raised up about one millimetre when it is placed on the slide. An alternative is to use a cavity slide.

Place one drop of the liquid containing the unicells inside the Vaseline/methyl cellulose or the depression of a cavity slide, then add a cover slip.

Examine using the low power lens of your microscope and then the high power lens when you find something interesting.

If you can obtain a culture of *Paramecium*, you can try feeding it with yeast cells stained with Congo red.

Dip the tip of a pin or needle into the stained yeast and stir it into the *Paramecium* on a slide. Add a cover slip and examine.

?

1 How many functions of life can you observe in the unicellular organisms that you can see?
2 How does *Paramecium* eat yeast cells?
3 Congo red is red in acid conditions and blue in neutral or alkaline conditions. After about an hour inside *Paramecium's* vesicles the red stained yeast have usually turned blue. How could the *Paramecium* change the pH inside the vesicles and how might this help in the digestion of the yeast cells?
4 In the *Paramecium* culture there may be both small cells and large adult cells. What two life processes does this suggest are occurring in the culture?

Instructions for staining yeast with Congo red

Mix 3g of yeast with 25ml water. Add 0.3g of Congo red and boil for 5 minutes.

Thinking about science: evidence for the cell theory

Scientists base their understanding of the natural world on the evidence that they obtain from observations and experiments. They also use observations and experiments to test hypotheses and theories.

1 How can we obtain evidence for the part of the cell theory that states that all living things consist of cells?
  Can we *prove* that this statement is true?
2 The statement that cells can only come from pre-existing cells implies that life has always existed. Is this possible, or if it is impossible, must we discard this part of the cell theory?
3 Discuss the cell theory in relation to the following observations:
  (a) Some fungi consist of long narrow tubular hyphae, containing cytoplasm and many nuclei, not separated by walls or membranes.
  (b) Muscle fibres are large structures, surrounded by a single membrane and containing many nuclei.
  (c) Bone tissue contains a few scattered cells separated by large amounts of acellular or extracellular matrix, made of proteins and semi-crystalline minerals.
  (d) Some organisms are large and yet are not divided up into small cells. Giant algae, for example, can be 100 mm or more long and contain many nuclei.
4 (a) If a civilized planet were discovered in another part of the Universe, state what the cell theory would predict about the life found there.
  (b) If the organisms found there were composed of cells, would this prove the cell theory?

A **hypothesis** is a specific prediction that can be tested. A **theory** is a more general system of ideas, which is used to explain or interpret observations. In addition a theory generally provides scientists with an element of predictive power. Sometimes evidence is obtained that cannot be explained using the existing theory. This theory must then either be modified or replaced with a new theory. The "truth" of a scientific theory is therefore based more on whether it has explanatory power, than if it corresponds to reality.

Sizes of cells

The organisms that we study in biology vary hugely in size from redwood trees over 110 metres tall, to a satellite tobacco mosaic virus that is 7 billion times smaller! To measure the sizes of these structures, a range of units is needed. Scientists throughout the world use SI units. There is a series of SI units for size, with each unit one thousand times smaller than the previous one in the series: metres (m), millimetres (mm), micrometres (μm) and nanometres (nm).

A DNA molecule in a human chromosome is 2 nm wide, but it is an incredible 70,000,000 nm (70 mm) long on average. A phospholipid molecule is about 5 nm long, so a membrane in a cell, made of two layers of phospholipids arranged toe-to-toe, is about 10 nm thick.

A virus would be large at 100 nm. Most prokaryotic cells are within the range of 1 to 10 μm. Most animal cells are in the range 10 to 30 μm and most plant cells are between 10 and 100 μm. Table 2 shows some cells that are surprising for their size! Some multinucleate fungal "cells" grow even larger.

Table 2	
Mycoplasma pneumoniae, (a very small bacterium)	200 nm
Mimivirus (a giant virus found in amoebae)	400 nm
Megakaryocyte (a cell that makes blood platelets)	160 μm
Thiomargarita namibiensis (largest known bacterium)	0.7 mm
Diameter of the nerve fibre of a squid giant nerve cell	1 mm
Diameter of the yolk cell of an ostrich egg	120 mm

Critical considerations: investigating a knowledge claim

"In the human body, for every one of our own cells, there are ten prokaryote cells resident in us." (Black J., *Microbiology*) Does this seem a reasonable knowledge claim?

Eukaryotic cells are, on average, ten times larger, in a given dimension, than prokaryotic cells.

1 Obtain some modelling clay.
2 Construct a model of a prokaryotic cell that is 10 mm × 5 mm × 5 mm.
3 Use the clay to construct a model of a eukaryotic cell that is 10 times larger in every dimension, i.e. 100 mm × 50 mm × 50 mm.
4 Compare the two models. Does the claim seem reasonable?

Working with data: magnification and scale bars

Biologists often need to do calculations involving the magnification and sizes of images and specimens. The size of a specimen is how large something actually is. The size of an image is how large the specimen appears in a drawing or a photograph. Magnification is how much larger the image is than the actual size. This formula is used for these calculations:

$$\text{Magnification} = \frac{\text{size of image}}{\text{actual size of specimen}}$$

The formula can either be memorized or it can be derived intuitively, as we expect images that show enlargement to have a magnification greater than one. In these calculations it is important to use the same metric prefixes for the sizes of the image and the specimen, for example micrometres for both or millimetres for both, not micrometres for one and millimetres for the other.

Scale bars are sometimes used on drawings or micrographs. A micrograph is a photograph of an image under the microscope. The scale bar consists of a line, with the actual size that the line represents. For example if there was a 10 mm long scale bar on a micrograph with a magnification of 10,000 X, the scale bar would have a label saying "1 μm" above the scale bar.

Example:

The length of an image is 30 mm. It represents a structure that has an actual size of 3 μm. Determine the magnification of the image.

Either:

$30 \text{ mm} = 30 \times 10^{-3} \text{ m}$

$3 \text{ μm} = 3 \times 10^{-6} \text{ m}$

$$\text{Magnification} = \frac{30 \times 10^{-3}}{3 \times 10^{-6}}$$

= 10 000 X

Or:

30 mm = 30000 μm

$$\text{Magnification} = \frac{30000}{3}$$

= 10 000 X

Working with data: measuring size with a graticule

One method of making precise measurements of an object's size when viewed under the microscope is to use an eyepiece graticule. This is a scale that is part of the eyepiece. Different graticules vary in the width of their divisions. It is therefore necessary to calibrate the graticule.

In order to calibrate the graticule, place a graduated slide or a clear plastic ruler with divisions of known size on the stage. Focus on the slide with the 10 X objective lens. Move the slide or the eyepiece containing the graticule until the two scales are lined up. Align the divisions on the two scales by moving the slide. The size of the graticule units can now be measured. This calibrates the graticule for the 10 X objective lens. The other objective lenses can either be calibrated by repeating this procedure or by calculation. For example, with a 40 X objective lens the magnification is four times greater, so the graticule units are one quarter of those for the 10 X lens.

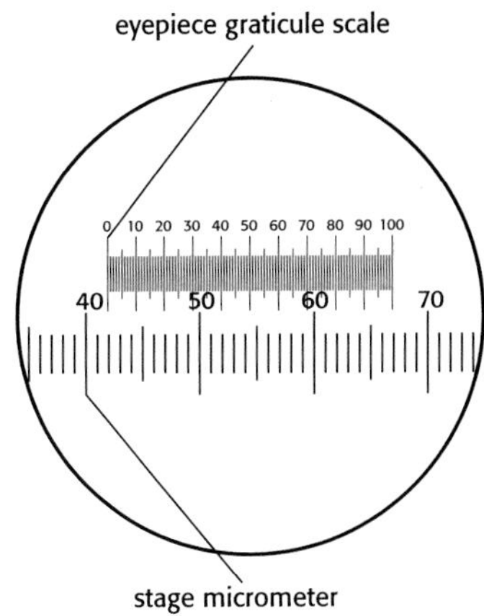

Figure 10 Calibrating a graticule

Data-based questions: size and magnification calculations

1 (a) Determine the magnification of the string of *Thiomargarita* cells in Figure 11, if the scale bar represents 0.2 mm [3]

(b) Determine the width of the string of cells. [2]

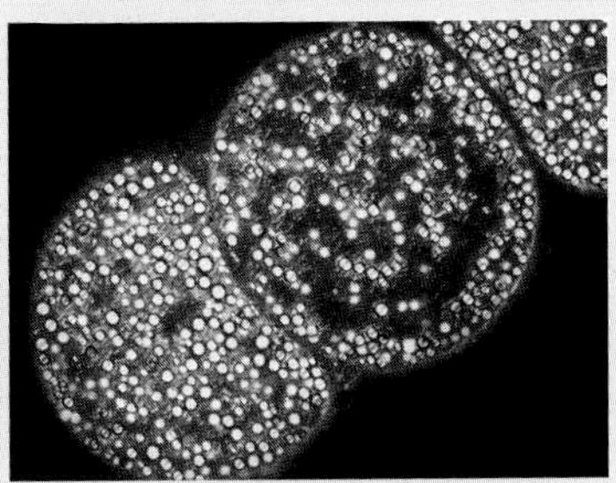

Figure 11 *Thiomargarita*

2 In Figure 12 the actual length of the mitochondrion is 8 μm.

(a) Determine the magnification of this electron micrograph. [2]

(b) Calculate how long a 5 μm scale bar would be on this electron micrograph. [2]

(c) Determine the width of the mitochondrion. [1]

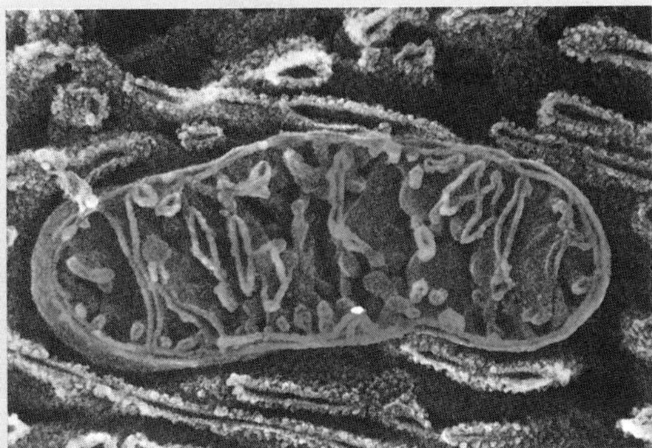

Figure 12 Mitochondrion

3 The magnification of the human cheek cells from a compound microscope (Figure 13) is 2000 X.

(a) Calculate how long a 20 μm scale bar would be on the image. [2]

(b) Determine the average length of the cheek cells. [3]

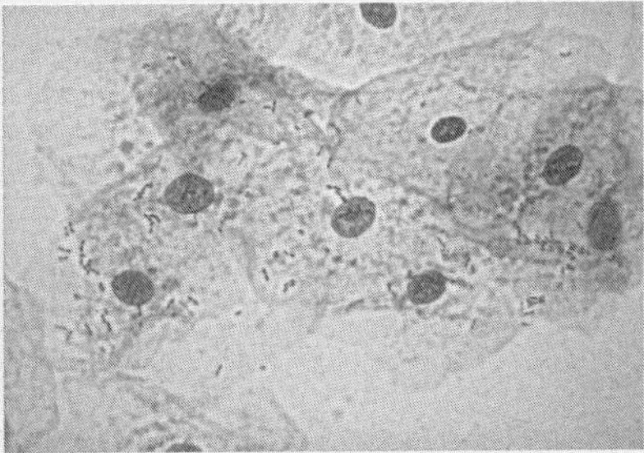

Figure 13 Human cheek cells

4 (a) Using the width of the hen's egg as a guide, estimate the actual length of the ostrich egg (Figure 14). [2]

(b) Estimate the magnification of the image. [2]

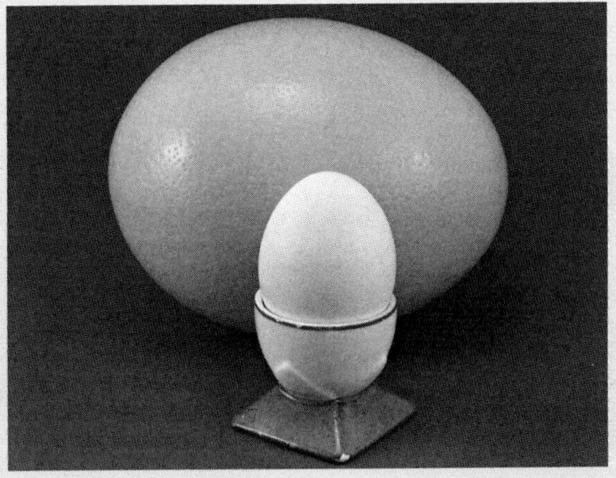

Figure 14 Ostrich egg

Surface area to volume ratios in cells

In the cytoplasm of cells, large numbers of chemical reactions take place. These reactions are known collectively as the **metabolism** of the cell. The rate of these reactions (the metabolic rate of the cell) is proportional to the volume of the cell.

For metabolism to continue, substances used in the reactions must be absorbed by the cell and waste products must be removed. Substances move into and out of cells through the plasma membrane at the surface of the cell. The rate at which substances cross this membrane depends on its surface area.

The surface area to volume ratio of a cell is therefore very important. If the ratio is too small then substances will not enter the cell as quickly as they are required and waste products will accumulate because they are produced more rapidly than they can be excreted.

Surface area to volume ratio is also important in relation to heat production and loss. If the ratio is too small then cells may overheat because the metabolism produces heat faster than it is lost over the cell's surface.

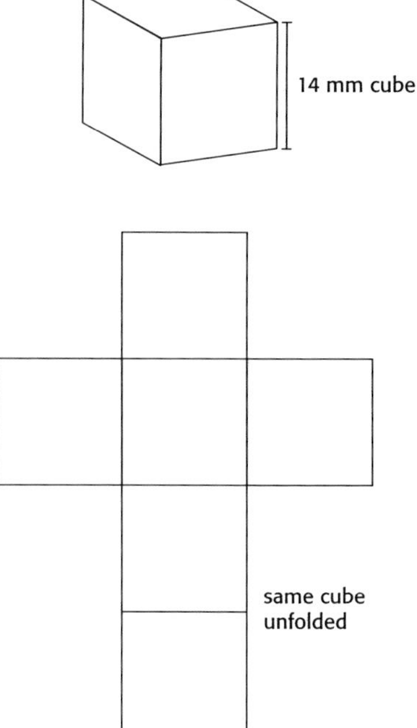

Figure 15 Volume and surface area of a cube

Rediscovering biology: cell size and surface area to volume ratio

Consider a cube with 1 mm wide sides.

Calculate its volume in mm^3 and its surface area in mm^2. Remember that it has six faces in total. Calculate the surface area to volume ratio by dividing the surface area by the volume.

Repeat this calculation for a cube with sides of 10 mm and one with sides of 100 mm.

1 What is the relationship between the size of a cube and its surface area to volume ratio?

The calculation of surface area to volume ratios for structures with the same shape but different size can be repeated for other shapes, for example for spheres.

The volume of a sphere is $\frac{4}{3}\pi r^3$. The surface area of a sphere is $4\pi r^2$.

2 Is the relationship between size and surface area to volume ratio the same as with cubes?

Cells tend to grow to a maximum size and then divide.

3 Explain clearly what you have learned about limits to cell size from these surface area to volume ratio calculations.

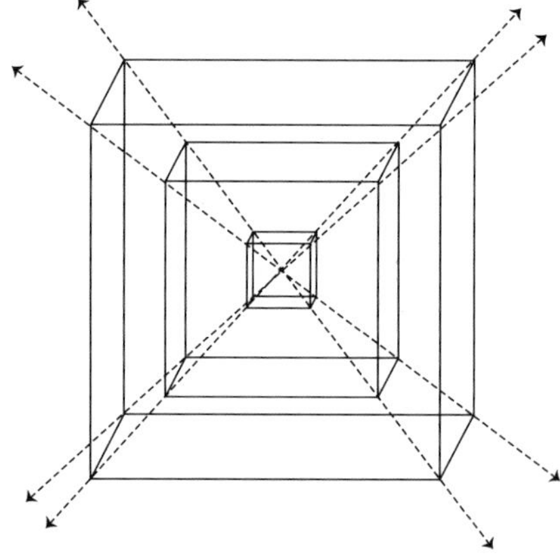

Figure 16 An expanding cube

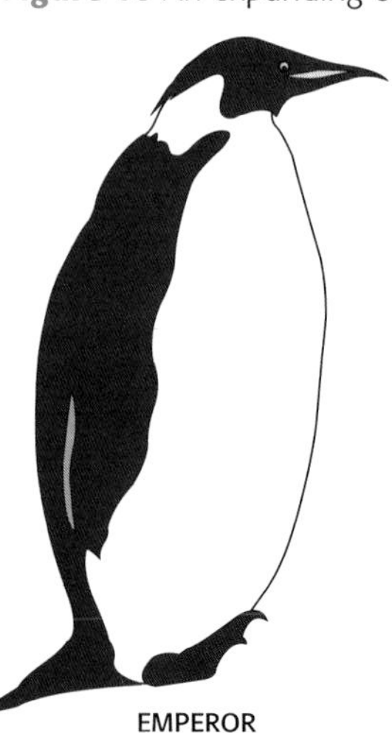

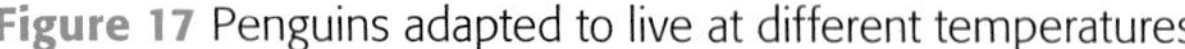

Figure 17 Penguins adapted to live at different temperatures

Rediscovering biology: size and rates of heat loss

It would be difficult to investigate heat loss from real cells. Instead, beakers can be used to represent cells of different size. At the start of the experiment all the beakers are filled with hot water. They are then left to cool, as heat is lost from them. At the end of the experiment the temperature of water in each beaker is measured to find out how much it has fallen. The only factor that should be varied in this experiment is the size of the beakers. Other factors that are kept constant for all of the beakers are known as control factors.

What are the control factors – the factors that should be the same for each beaker?

The results of this experiment can be plotted on a line graph. Instructions for drawing line graphs are shown on the next page.

Working with data: plotting graphs in biology

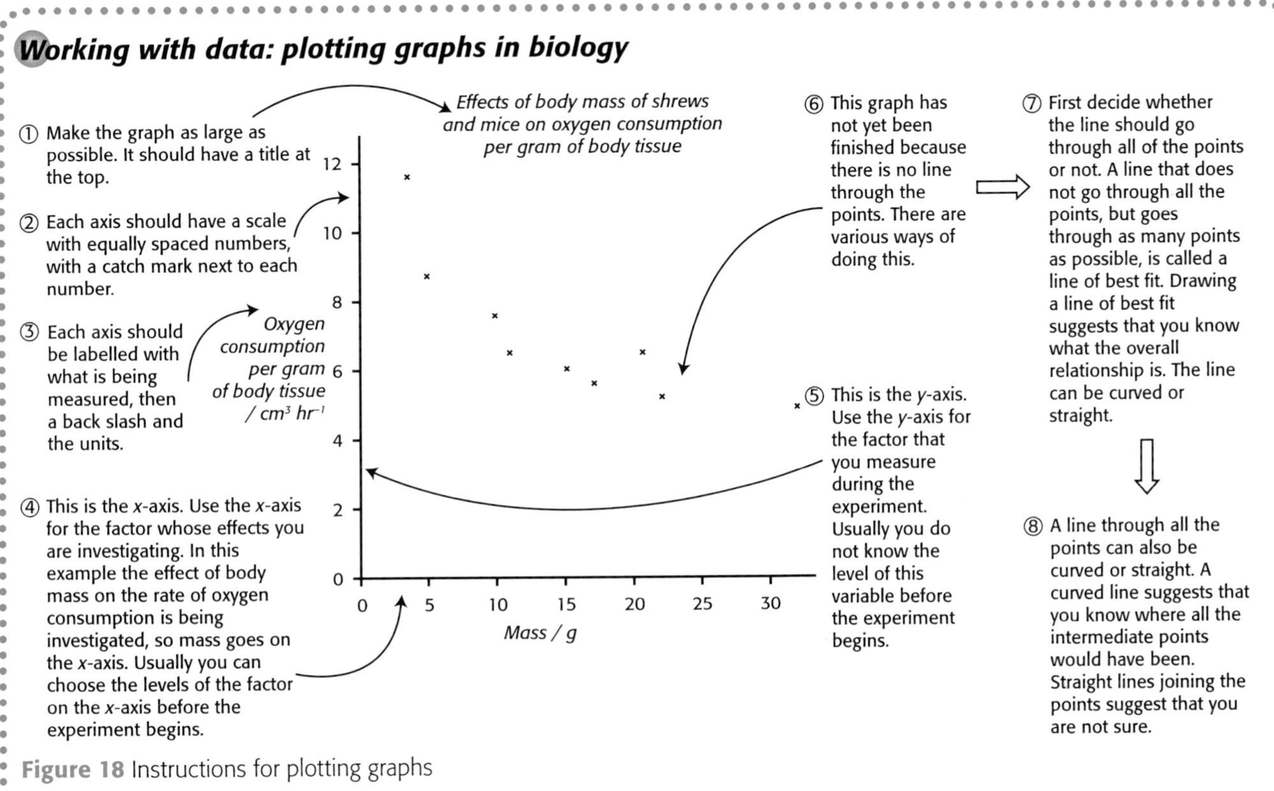

Figure 18 Instructions for plotting graphs

Data-based question: size and surface area of small mammals

Table 3 shows the surface area of skin and mass of five small mammals.

Table 3		
Mammal	**Mass (g)**	**Surface area (mm²)**
Etruscan shrew	2	12
Gerbil	74	90
Mole	76	83
Mouse	29	72
Pygmy shrew	5	21

1 Plot a graph of the data in the table, with scales and a label on each axis and the points plotted with small crosses. [5]
2 Discuss what type of line should be plotted through the points. [3]
3 Explain reasons for the relationship between mass and surface area, shown by the graph. [2]

Multicellular organisms and cell differentiation

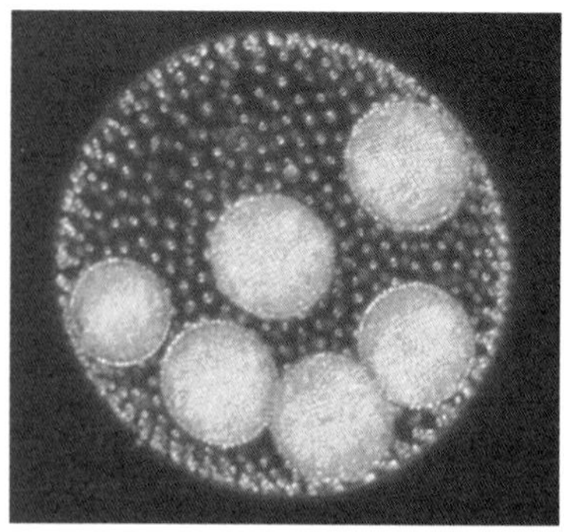

Figure 19 *Volvox* colony

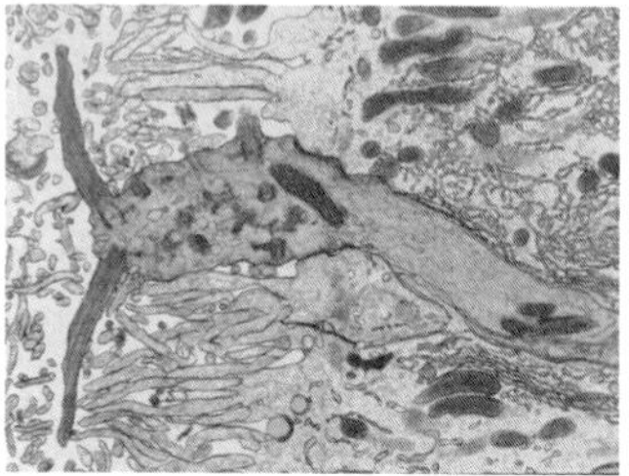

Figure 20 Olfactory receptor cell

benign | malignant

Figure 21 Microarray data

Some unicellular organisms live together in colonies, for example a type of alga called *Volvox aureus*. Each colony consists of a ball made of a protein gel, with 500 or more identical cells attached to its surface. Figure 19 shows a colony, with daughter colonies forming inside. Although the cells are co-operating they are not fused to form a single cell mass and so do not form a single organism.

Organisms consisting of a single mass of cells fused together are multicellular organisms. The cells in these organisms become specialized for specific functions. For example a red blood cell carries oxygen around the body. A rod cell in the retina of the eye absorbs light and transmits a nerve impulse to the brain. To carry out their functions, each cell type develops its own specialized structure.

The development of cells in different ways to perform different functions is called **differentiation**. This involves each cell type using some of some of the genes in its nucleus, but not others. When a gene is being used in a cell, we say that the gene is being **expressed**. In simple terms, the gene is switched on and the information in it is used to make a protein or other gene product.

An extreme example of differentiation involves a large family of genes in humans, which carry the information for making receptors that detect odorant chemicals (smells). These genes are only expressed in cells in the skin inside the nose, called olfactory receptor cells. Each of these cells expresses just one of the genes and so makes one type of receptor to detect one type of odorant. This is how we can distinguish between so many different smells. Richard Axel and Linda Buck were given the Nobel Prize for Medicine in 2004 for their work on this system. The olfactory receptor cell in Figure 20 has two projections visible on the left that contain the odorant receptors.

Data-based question: gene expression in thyroid tumour cells

A microarray consists of a large number of gene probes, arranged in precise positions on a small chip of glass. Each gene probe is a single-stranded piece of DNA, that is a copy of part of a gene. When a gene is expressed, a copy of it is made called mRNA. In the microarray procedure, mRNAs are extracted from cells and converted to single-stranded DNA (cDNA) which can then bind to gene probes on the microarray. Before placing on the microarray, coloured markers are added to the cDNA. This allows the genes that are being expressed to be recognised on the microarray.

Figure 21 shows the results of an investigation of thyroid tumours from 13 cancer patients. It shows the relative level of expression of 1139 genes in these tumours. White bands indicate increased gene expression and orange bands decreased expression. The bands and the results for different patients have been grouped by computer analysis to emphasis similarities in gene expression.

1 State how many patients had benign tumours and how many had malignant ones. [1]
2 Compare the amount of variation in gene expression between the benign and the malignant tumours with the variation within these groups. [2]
3 Suggest how the research into gene expression in thyroid tumours could be useful. [2]
4 Discuss the reasons for differences in gene expression between benign and malignant thyroid tumours. [3]

Stem cells

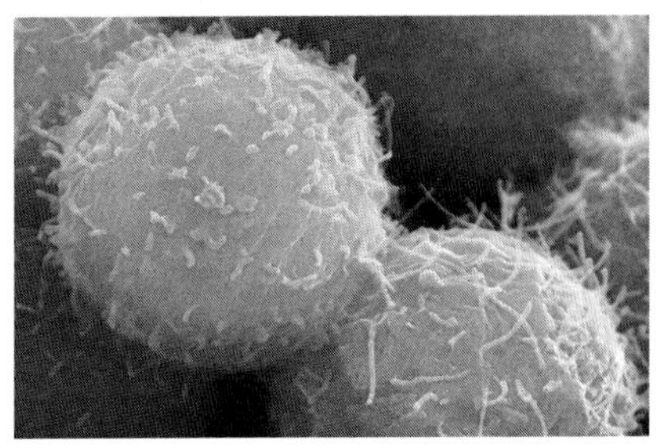

Figure 22 Mouse embryonic stem cells

Stem cells are defined as cells that have the capacity to self-renew by cell division and to differentiate. At an early stage the whole of a human embryo consists of stem cells, but gradually the cells in the embryo become committed to differentiating in a particular way. Once committed, a cell may still be able to divide about six more times. However, all the cells produced will differentiate in the same way and so they are no longer stem cells. Small numbers of cells seem to remain as stem cells, however and they are still present in the adult body. They are present in many human tissues, including bone marrow, skin and liver. They give some human tissues considerable powers of regeneration and repair. The stem cells in other tissues – for example brain, kidney and heart – only allow limited repair.

There has been great interest in stem cells because of their potential for tissue repair and for treating a variety of degenerative conditions. For example Parkinson's disease, multiple sclerosis and strokes are all caused by the loss of neurons or other cells in the nervous system. Although largely experimental, there is the potential to use stem cells to replace these cells. Embryonic stem cells and adult stem cells both have advantages and these are shown in Table 4.

Table 4

Embryonic stem cells	Adult stem cells
Easier to obtain than adult stem cells. Almost unlimited growth potential – much greater than for adult stem cells. Less chance of genetic damage than with adult stem cells. Greater capacity to differentiate into different cell types than adult stem cells.	No embryo needs to be destroyed, which usually happens when embryonic stem cells are obtained. The cells are fully compatible with the tissues of the adult, so no rejection problems occur, whereas embryonic tissue is genetically different from the patient receiving the tissue. Less chance of malignant tumours developing than with embryonic stem cells.

Critical considerations: ethics of use of embryonic stem cells

The following two extracts were parts of testimonies to the US National Bioethics Advisory Commission. They represent the views of members of the Islamic and Roman Catholic faiths.

Abdulaziz Sachedina, University of Virginia

... it is possible to propose the following as acceptable to all schools of thought in Islam.

1 *The Koran and the Tradition regard perceivable human life as possible at the* later *stages of the biological development of the embryo.*
2 *The fetus is accorded the status of a legal person only at the later stages of its development, when perceptible form and voluntary movement are demonstrated. Hence, in earlier stages, such as when it lodges itself in the uterus and begins its journey to personhood, the embryo cannot be considered as possessing moral status.*
3 *The silence of the Koran over a criterion for moral status (i.e. when the ensoulment occurs) of the fetus allows a distinction between a biological and a moral person, placing the latter stage after, at least, the first trimester of pregnancy.*

continued overleaf

In Islam, research on stem cells made possible by biotechnical intervention in the early stages of life is regarded as an act of faith in the ultimate will of God as the Giver of all life, as long as such an intervention is undertaken with the purpose of improving human health.

Kevin Wildes, Georgetown University

...I do not think one can argue that there is, in Roman Catholic thought, opposition to stem cell research itself. The crucial moral issues and stumbling blocks are the problems of the derivation of the stem cells used in the research itself. That is, the destruction of embryos or the use of fetal tissue from abortion are the key moral problems. If you think that embryos should be treated as human persons, then it makes sense to argue that they should not be destroyed for purposes of research. However, if there were a way to conduct stem cell research without destroying human life, either embryonic or fetal, I do not think the Roman Catholic tradition would have a principled opposition to such research. It is important to point out, however, that there is no single Roman Catholic "position" on this topic or many moral topics. In addition, if one thinks with a Roman Catholic imagination, one must also ask about the questions of justice in devoting resources, especially national resources, to such research when there are so many other basic medical and health needs that are unmet.

?

1. To what extent are these positions similar and how do they differ?
2. What might be the views of other stakeholders in the debate – patients, doctors, parents, research companies?

Therapeutic use of stem cells

Most of the therapeutic uses of stem cells are at a very early stage of research.

- Juvenile-onset diabetes can be treated by transplanting pancreas tissue from donors after their death, but the supply of this tissue is insufficient. A reliable method for stimulating embryonic stem cells to become insulin-secreting cells might allow more patients with diabetes to be treated.
- Stem cells have been transplanted into injured spinal cords of experimental animals and there has been some recovery of mobility. This was due to the insulating myelin sheaths around neurons being repaired. However, side-effects have been observed, including stimuli that had previously been painless causing pain to the animal.

The greatest success so far in the therapeutic use of stem cells involves bone marrow transplants. The cells needed are hematopoietic stem cells – HS cells. They are normally found in the bone marrow and divide continually to produce new cells that differentiate into red and white blood cells. HS cells have huge growth potential – just 100 of them can completely replace the blood system of mice when all cells in the bone marrow have been destroyed by radiation. HS cells are used in the treatment of a variety of blood disorders including acute leukemia, SCID (severe combined immune deficiency), multiple myeloma and lymphoma. The treatment of lymphomas is described here.

Lymphomas are cancers of the lymphatic system. The lymph nodes swell up, the patient feels lacking in energy and loses body mass.

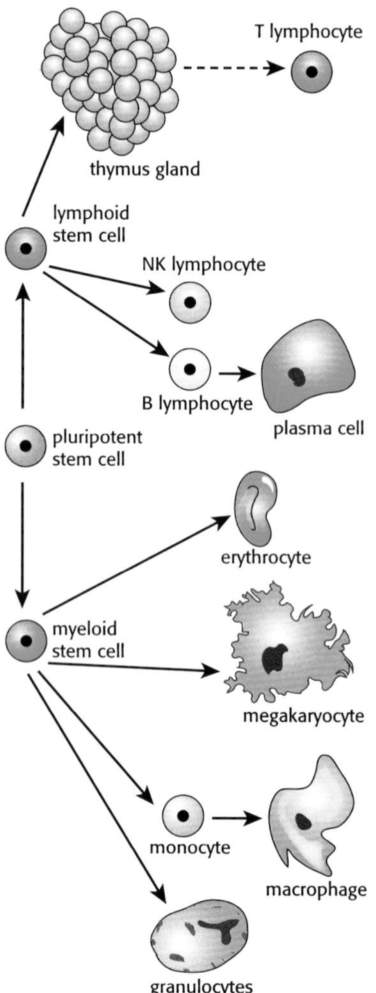

Figure 23 Origins of blood cells

Some forms of lymphoma can be treated by the following procedure.

- Cells are removed from the bone marrow of the patient.
- The patient is given high doses of chemotherapy drugs to kill dividing cells in their body. Both the cancer cells and normal cells in the bone marrow are killed.
- HS cells from the bone marrow are then transplanted back into the patient. The HS cells can fully restore healthy production of blood cells in the bone marrow.

Thinking about science: risks of stem cell therapies

Stem cells can form tumours called teratocarcinomas, in which rapid cell division occurs and up to 30 different tissue types are formed by cell differentiation. Teratocarcinomas develop naturally from stem cells in the human body. Among young adult men one of the commonest forms of tumour is formed from stem cells in the testis.

Single cells have been removed from teratocarcinomas and have been transplanted into experimental organisms, where new tumours grew, with all the tissue types of the original tumour. Although the tissues were mostly disorganized, in some cases structures resembling parts of organs have been found. This raises the possibility of growing replacement organs for transplantation, using embryonic stem cells. Research in this field is encouraged in some countries but in other countries it is banned by the national government.

Even where research is allowed, there is some concern that the ability of embryonic stem cells to divide and grow rapidly is not just an opportunity to grow replacement tissues or organs, but also a risk. It would only take one cell starting to divide out of control for a tumour to form, which would threaten the life of the patient rather than help to preserve it.

1 Which groups of people do scientists need to inform about their plans for research into growing replacement tissues or organs?
2 Do scientists need to reveal the risks as well as the opportunities, or should they assume that the risks can be avoided?
3 Why do national governments vary in their attitude to this research?
4 Should all scientists be governed by the same rules and if so, who should decide them?

Extracellular components

Cells sometimes make structures outside their plasma membranes. These are called extracellular components. One example is the plant cell wall. Cellulose microfibrils are assembled inside the cell and pass out through the plasma membrane to add to the thickness of the wall. They are visible in Figure 24. When a plant cell grows, the wall becomes attenuated or thinner and so more cellulose must be added to maintain its thickness. The wall maintains the shape of the cell and supports the plasma membrane. When water enters the plant cell by osmosis the wall prevents the expansion of the cell contents. Instead of this pressure builds up inside the cell. This eventually prevents more water from entering. The pressure also makes the cell almost rigid, so that it helps to hold the whole plant up against the force of gravity.

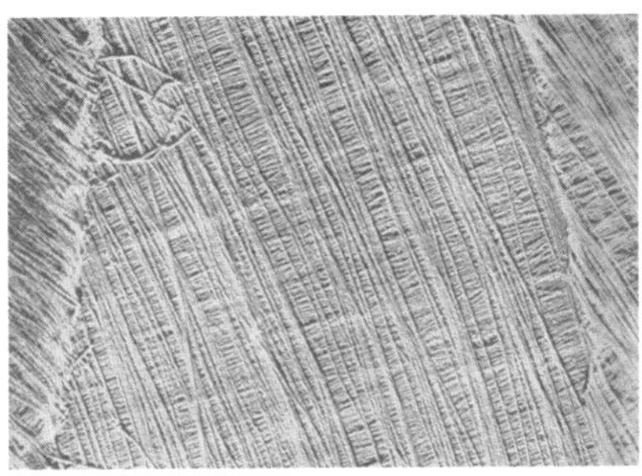

Figure 24 Scanning electron micrograph of part of a plant cell wall showing cellulose microfibrils

A second example is the extracellular matrix of animal cells, which is described in the article on page 20, written by Carolyn Strange.

Thinking about science: the extracellular matrix

Consider a slab of meat – or an intact animal, for that matter. Why don't their cells and tissues slip past each other, flowing into a puddle a few cells deep? Cells in a tissue don't only stick together – they work together, and therefore must communicate. In multicellular animals, cells are intricately connected to each other directly, and also via the extracellular matrix surrounding all cells – the ECM.

The ECM is not merely a passive scaffolding. Biologists are discovering that ECM molecules have striking effects on cell behaviour. They influence shape, orientation and polarity, movement, metabolism and differentiation. "Half the secret of life is outside the cell – the extracellular matrix is what tells the cell where it is in time and space", according to Zena Werb, an anatomy professor at the University of California San Francisco.

"What really tells the cells to remember who they are? Why is your nose your nose, and your elbow your elbow, and why don't they turn into each other?" asks Mina Bissell, director of the Life Sciences Division at Berkeley. She hypothesized more than 15 years ago that ECM possesses information crucial to a cell's ability to function properly. Although not widely embraced at the time, the idea is finally catching on as evidence mounts to support it. Frustrated with reductionist approaches, some researchers say it is time to wade in and explore the complexity. "In the traditional view, the cell was this unit by itself and the matrix was something else," says Frederick Grinnell, of the University of Texas Southwestern Medical Center. "It's actually quite difficult to say where the edge of the cell is." The nature, composition, and amount of ECM is tailored to the specific tissue. Tissues like bone and cartilage contain more matrix than cells. One common ECM protein is collagen, which accounts for approximately a third of a vertebrate's dry weight.

ECM is made and oriented by the cells within it and takes two general forms. Interstitial matrix is a three-dimensional gel that surrounds cells and fills space. The other form, basement membrane, is a mesh-like sheet formed at the base of epithelial tissues, the thin layers of cells that cover internal and external surfaces of the body and that perform protective, secretory, or other functions. Basement membrane is a remarkable cellular organizer. In culture on plastic, the cells just sit in a layer, but when you put them on basement membrane they differentiate. The cells that line blood vessels form capillary-like tubes all over the culture dish. Neuronal cells send out long, thin extensions. Salivary gland cells join into little balls and begin producing secretory proteins. Such behaviour is characteristic of normal cells. They do not grow unless properly anchored to the matrix.

(a) 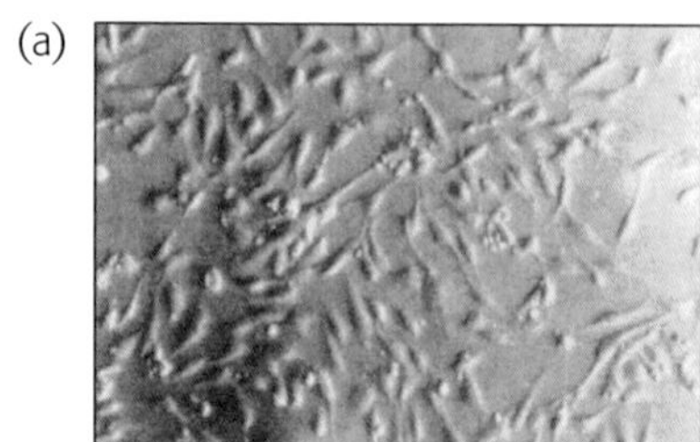

(b) 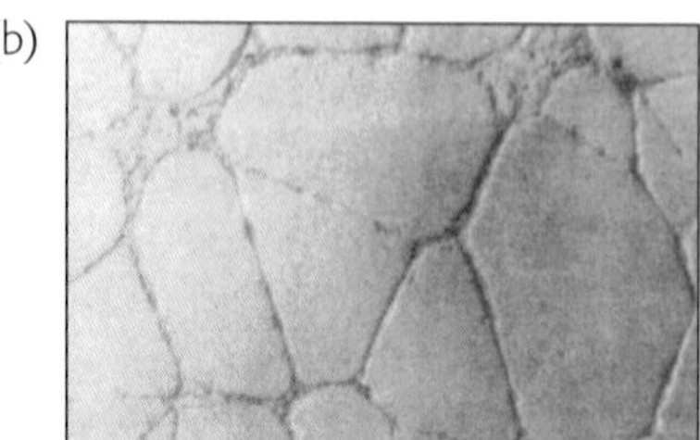

Figure 25 Basement membrane stimulates differentiation in certain cell types. Blood vessel cells placed on plastic (a) simply spread into a monolayer, but when they are put on basement membrane (b), they form capillary-like tubes.

1 Major changes in the way that scientists understand the natural world are sometimes called paradigm shifts. Explain the paradigm shift that has taken place in our understanding of the ECM.
2 Why do paradigm shifts take decades to be accepted by all of the scientific community?
3 Explain what is meant by "reductionist approaches".
4 What other possible approach is there?
5 What is the best approach to investigation for a biologist?

Life as an emergent property

Emergent properties are those that arise from the interaction of component parts, i.e. the whole is greater than the sum of its parts. For example surface tension is an emergent property of the interaction of water molecules, while consciousness is a property that emerges from the interaction of nerve cells in the brain. Life itself is an emergent property.

"Life is not inherent in any single element constituting the living cell. DNA is not alive, neither are proteins, carbohydrates or lipids. Indeed, for a single short moment, a living cell and a dead cell may, upon analysis, be found to contain precisely the same catalogue of 'dead' chemicals in identical concentrations... What distinguishes the living from the dead? Nothing more than actions and interactions. Life emerges from inert matter as a consequence of metabolism, the continuous transfer of energy and information systematically packaged in cells in a way that leads to self-perpetuation. The complexity of dynamic behavior that generates metabolism, growth and genetic inheritance is what we call life."

Excerpt from Tending Adam's Garden: Evolving the Cognitive Immune Self *by Irun Cohen*

Investigating biology: Is it alive?

Design a procedure to test whether a sample of material is alive. Suggested samples include a bacterial colony, a tube of yeast culture, the Resurrection plant *(Selaginella lepidophylla)* and pond scum.

You will need to provide evidence of some of the following: the use of energy (i.e. metabolism); growth; development; reproduction; homeostasis; response to the environment; evolutionary adaptation; composition of one or more cells.

Working with data: the normal distribution and standard deviation

Most biological samples show variability. The cross-section of a root (Figure 26) shows variations in the size of the cells. If the width of all of the cells were measured (using a graticule, or the picture-processing tool in software such as Logger Pro©), the data could be organized into a tally chart and then reorganized as a frequency distribution.

A common type of frequency distribution is called the normal distribution. It is sometimes called the bell curve because it has a bell shape, i.e. it is symmetrical around the mean. Values tend to be concentrated near the mean and decrease in frequency as the distance from the mean increases. A further requirement of a normal distribution is that the variable should be continuous, i.e. all values across the range are possible.

Standard deviation is a measure of the range of variation from an average of a group of measurements. For a normal distribution, 68% of all measurements fall within one standard deviation of the average, or mean. 95% of all measurements fall within two standard deviations of the mean. Standard deviation is explained in greater detail in an activity on page 132.

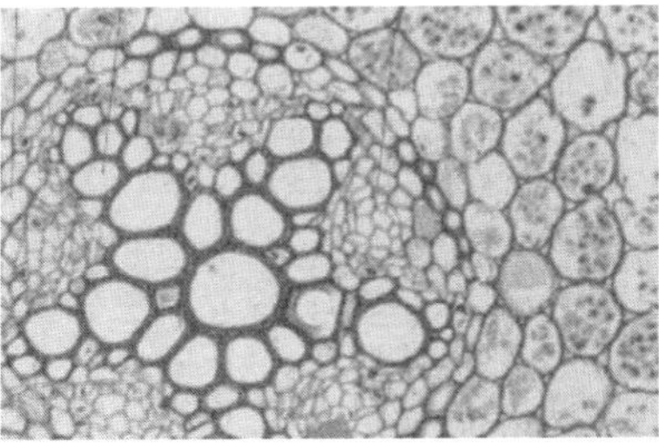

Figure 26 Transverse section of the centre of a root

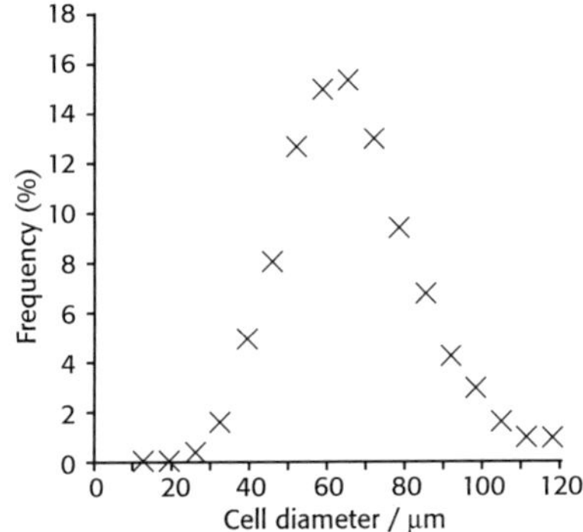

Figure 27 Frequency distribution of cell diameter in a sample of root cells

Prokaryotic cells

Prokaryotes were the first organisms to evolve on Earth and they still have the simplest cell structure. Bacteria are prokaryotes. They are mostly small in size, unicellular and are found almost everywhere – in soil, in water, on our skin, in our intestines and even in pools of hot water in volcanic areas.

The electron micrograph below shows a cell of *Escherichia coli* (*E. coli*), a bacterium found in the human intestines. Most strains of *E. coli* are harmless, but some cause food poisoning.

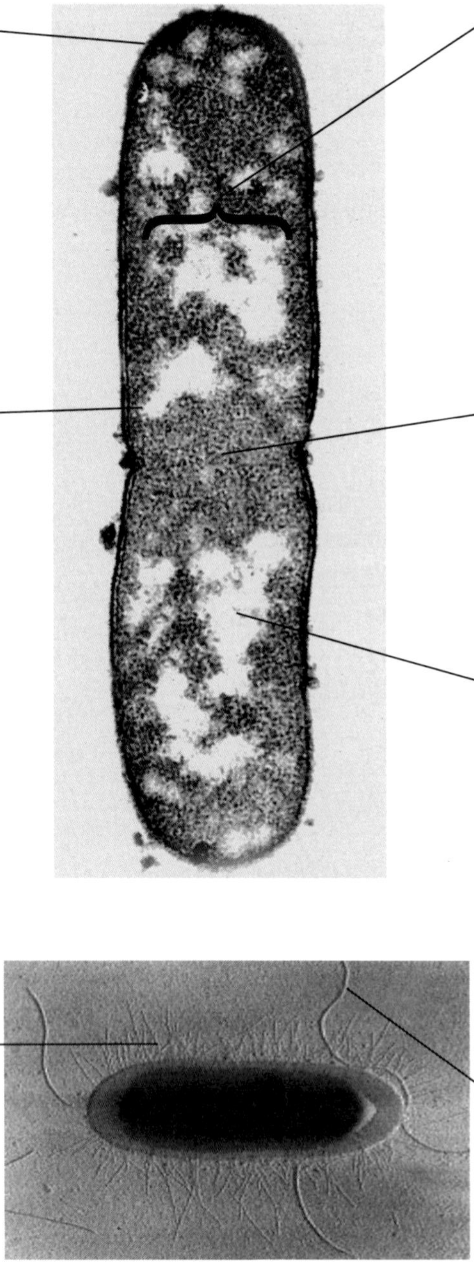

Cell wall
- always present
- composed of peptidoglycan
- protects the cell
- maintains its shape
- prevents cell from bursting

Cytoplasm
- fluid filling the space inside the plasma membrane
- water with many dissolved substances
- contains many enzymes
- contains ribosomes
- does not contain any membrane-bound organelles
- carries out the chemical reactions of metabolism

Plasma membrane
- thin layer mainly composed of phospholipids, pushed up against the inside of the cell wall in healthy cells
- partially permeable
- controls entry and exit of substances
- can also pump substances in or out by active transport
- produces ATP by aerobic cell respiration

Ribosomes
- small granular structures (70S)
- smaller than eukaryotic ribosomes which are 80S
- synthesizes proteins

Nucleoid
- region of cytoplasm containing the genetic material (usually one molecule of DNA)
- DNA molecule is circular and naked (not associated with protein)
- total amount of DNA is much smaller than in eukaryotes
- the nucleoid is stained less densely than the rest of the cytoplasm because there are fewer ribosomes in it and less protein

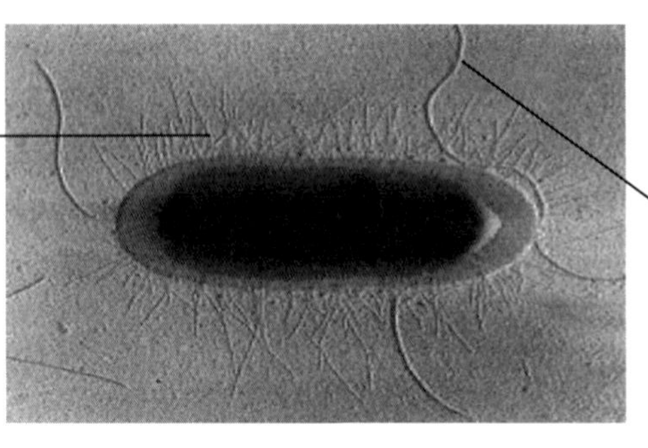

Pili
- protein filaments protruding from the cell wall
- can be pulled in or pushed out by a ratchet mechanism
- used for cell to cell adhesion
- used when bacteria stick together to form aggregations of cells
- used when two cells are exchanging DNA during a process called conjugation

Flagella
- structures protruding from the cell wall with a corkscrew shape
- base is embedded in the cell wall
- using energy they can be rotated, to propel the cell from one area to another
- unlike eukaryotic flagella they are solid and inflexible

Figure 28 Electron micrographs of *E. coli*

Eukaryotic cells

Eukaryotic cells have a much more complicated internal structure than prokaryotic cells. They have a nucleus and organelles in the cytoplasm with single or double membranes. Each organelle has a distinctive structure and function. Six types are described here.

> ? How many of each type of organelle are visible in the electron micrograph of liver tissue?

Nucleus

The nuclear membrane is double and has pores through it. Uncoiled chromosomes are spread through the nucleus and are called chromatin. There are often densely staining areas of chromatin around the edge of the nucleus. The nucleus stores almost all the genetic material of the cell. It is where DNA is replicated and transcribed, and where mRNA is modified before export to the cytoplasm.

Rough endoplasmic reticulum (rER)

The rER consists of flattened membrane sacs called cisternae. Attached to the outside of these cisternae are ribosomes. The main function of the rER is to synthesise protein for secretion from the cell. Protein synthesised by the ribosomes of the rER passes into the cisternae and is then carried by vesicles (small membrane sacs), which bud off and are moved to the Golgi apparatus.

Golgi apparatus

This organelle consists of flattened membrane sacs called cisternae, like rER. However the cisternae are not as long, are often curved, do not have ribosomes attached and have many vesicles nearby. The Golgi apparatus processes proteins brought in vesicles from the rER. Most of these proteins are then carried in vesicles to the plasma membrane for secretion.

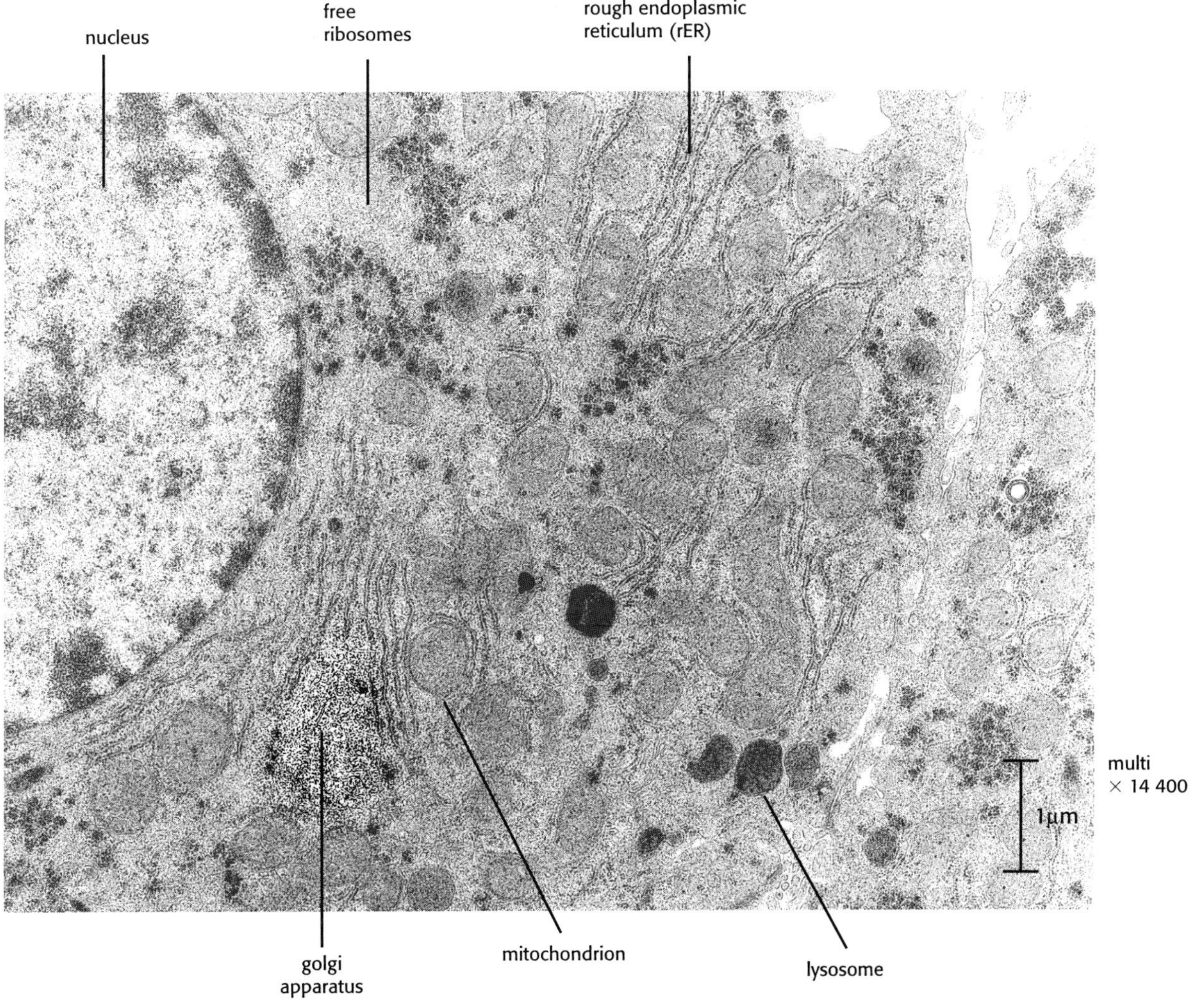

Figure 29 Electron micrograph of part of a liver cell

Lysosomes

These are approximately spherical with a single membrane. They are formed from Golgi vesicles. Lysosomes contain high concentrations of protein, which makes them densely staining in electron micrographs. They contain digestive enzymes, which can be used to break down ingested food in vesicles or break down organelles in the cell or even the whole cell.

Mitochondria

A double membrane surrounds mitochondria, with the inner of these membranes invaginated to form structures called cristae. The fluid inside is called the matrix. The shape of mitochondria is variable but is usually spherical or ovoid. They produce ATP for the cell by aerobic cell respiration. Fat is digested here if it is being used as an energy source in the cell.

Free ribosomes

These appear as dark granules in the cytoplasm and are not surrounded by a membrane. They are the same size as ribosomes attached to the rER – about 20 nm in diameter. Free ribosomes synthesize protein, releasing it to work in the cytoplasm, as enzymes, or in other ways. Ribosomes are constructed in a region of the nucleus called the nucleolus.

Chapter 1 questions

1 Figure 30 represents a cell from a multicellular organism.

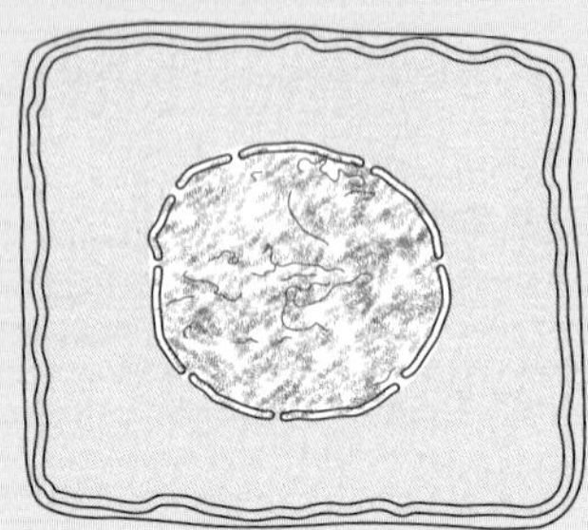

Figure 30

(a) Identify, with a reason, whether the cell is
  (i) prokaryotic or eukaryotic; [1]
  (ii) part of a root tip or a finger tip; [1]
  (iii) in a phase of mitosis or in interphase. [1]

(b) The magnification of the drawing is 2500 ×.
  (i) Calculate the actual size of the cell. [2]
  (ii) Calculate how long a 5 μm scale bar should be if it was added to the drawing. [1]

(c) Predict what would happen to the cell if it was placed in a concentrated salt solution for one hour. Include reasons for your answer. [3]

2 The electron micrograph in Figure 31 shows part of an animal cell.

(a) Identify the labelled structures. [3]

(b) The structure indicated by the first label is 1.5 μm long. Calculate the magnification of the micrograph. [2]

(c) Determine how long a 10 μm scale bar would be on the micrograph. [2]

(d) Calculate the length of the structure indicated by label III. [3]

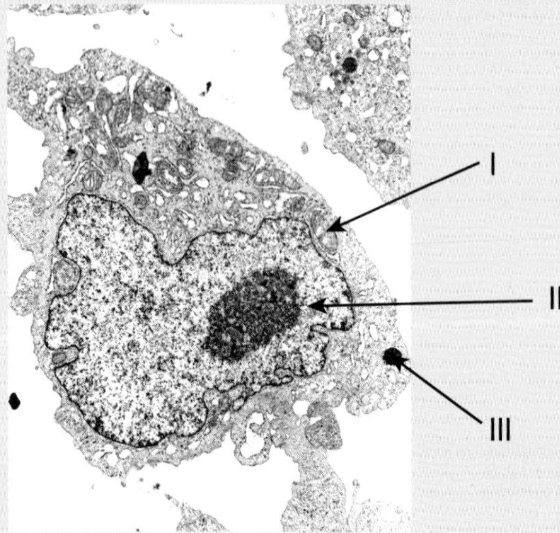

Figure 31

3 Siphonous green algae are marine organisms, found on many coral reefs. They are ecologically very successful and some species have even caused problems when accidentally introduced to new areas. *Codium fragile* for example has damaged shellfish industries after spreading off the north-west coast of the United States. *Bryopsis pennata* has become a pest species in aquaria, after accidentally being introduced on coral rock.

Figure 32 is a photograph of part of an individual of *Bryopsis pennata*. It can be 100 mm tall overall and consists of branched structures called siphons (scale bar = 0.6 mm).

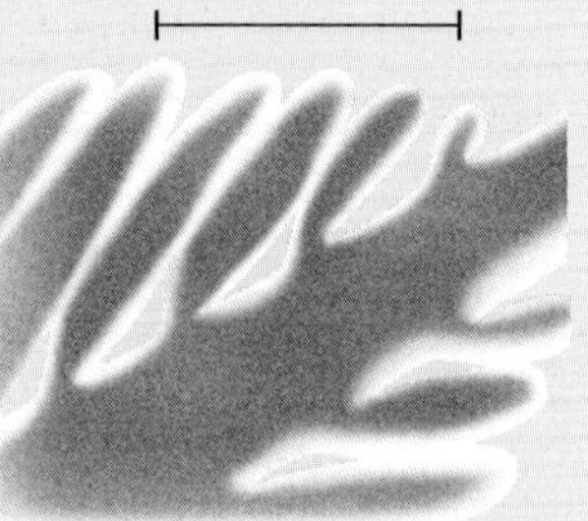

Figure 32

(a) Calculate the length of the smallest branch of the siphon, visible in the photograph. Give your answer in micrometres. [2]

Figure 33 is a diagram of part of one siphon. (magnification = 180 ×)

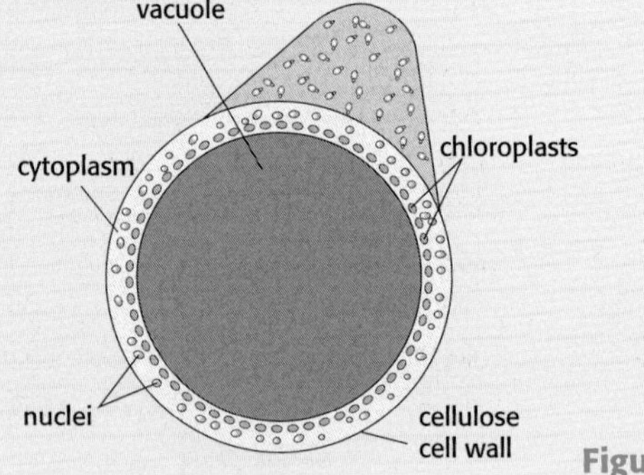

Figure 33

(b) Calculate the actual diameter of the siphon. [3]

(c) The structure of corals shows that they are animals. Deduce whether *Bryopsis pennata* is an animal, from the structure of its siphon. [2]

(d) According to the cell theory, living organisms are composed of cells. Discuss whether *Bryopsis pennata* should be described as multicellular, unicellular or acellular. [4]

(e) The vacuoles in the branched siphons are all interconnected and the fluid inside them is under pressure Suggest one advantage and one disadvantage of having interconnected, pressurised vacuoles. [2]

(f) The aquaria in which this species has become a pest contain water with salt dissolved, like the sea. Predict the effect of transferring *Bryopsis pennata* from

Membranes

Phospholipids are essential components of membranes – both the plasma membrane that surrounds the cytoplasm and all the membranes inside cells. Phospholipids are remarkable in their structure. They have two regions, with very different properties:

- two hydrocarbon tails which are not attracted to water (they are **hydrophobic**)
- a phosphate head, that is negatively charged to which water is attracted, so it is **hydrophilic**.

Figure 1 shows the molecular structure of a phospholipid.

When phospholipids are mixed with water they become arranged in double layers, with the heads facing outwards and the tails inwards. This arrangement, called a phospholipid bilayer, is the basis of the membranes of cells. It is a very stable structure, because bonds form between the phosphate heads and the surrounding water, and the tails form hydrophobic interactions – a weak type of intermolecular interaction, but collectively strong because of the huge numbers of tails in a membrane. Membranes are continuous structures, without edges, and they rarely become perforated or torn accidentally because many interactions would have to be broken for this to happen.

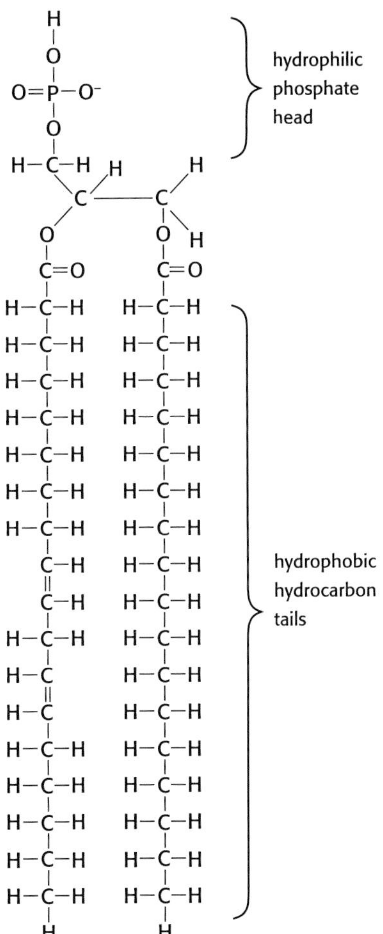

Figure 1 The molecular structure of a phospholipid. The phosphate often has other hydrophilic groups attached to it, but these are not shown in this diagram.

Rediscovering biology: lipid monolayers

In a famous letter of 1773, Benjamin Franklin described the effect of putting a small volume of olive oil onto the surface of a pond in an English village. He wanted to test an ancient observation that pouring oil on stormy water calms it. He was impressed by "…the sudden, wide and forcible spreading of a drop of oil on the face of the water…" He observed that whereas a drop of oil placed on a horizontal mirror remains in place and spreads very little, "…when put on water, it spreads instantly, many feet round, becoming so thin as to produce the prismatic colours, for a considerable space, and beyond them, so much thinner as to be invisible, except in its effect of smoothing the waves at a much greater distance."

The molecular structure of olive oil and other similar lipids suggests that they will spread to form a layer one molecule thick on the surface of water. These lipids consist of three hydrophobic hydrocarbon tails linked to a head composed of glycerol. A monolayer of lipids is shown in Figure 2. The length of the lipid molecules is approximately 2 nm, so this should be the thickness of the monolayer. This hypothesis can be tested using the apparatus in Figure 3.

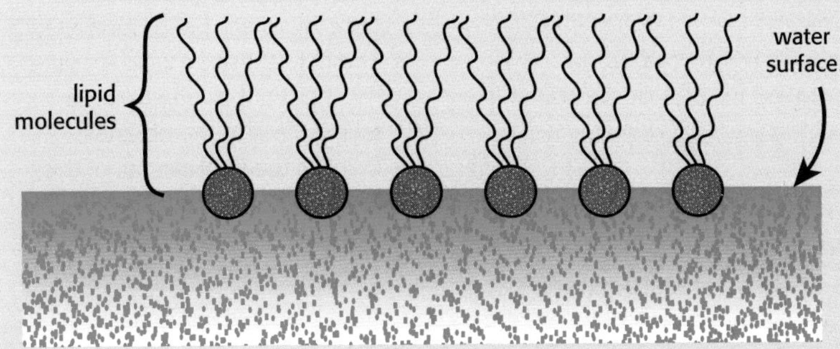

Figure 3 Apparatus to measure thickness of an oil layer

Membrane proteins

Membranes are not composed entirely of phospholipids – they also contain proteins. Some of these are embedded in the phospholipid and are called integral proteins. Others are more loosely attached to the surface of the membrane and are called peripheral proteins. The arrangement of phospholipids and proteins in membranes is shown in Figure 4. The diagram shows how membranes are asymmetrical, with differences between the two faces of the membrane in terms of the proteins that are exposed. For example, some proteins in plasma membranes have short chains of sugar molecules attached on the outside of the membrane, but not to the side facing the cytoplasm. These are called glycoproteins.

The many different types of protein in the membrane have a wide range of functions. Six examples are listed in Table 1.

Table 1

- **Hormone binding sites** (also called hormone receptors), for example the insulin receptor.
- **Immobilized enzymes** with the active site on the outside, for example in the small intestine.
- **Cell adhesion** to form tight junctions between groups of cells in tissues and organs.
- **Cell-to-cell communication,** for example receptors for neurotransmitters at synapses.
- **Channels for passive transport** to allow hydrophilic particles across by facilitated diffusion.
- **Pumps for active transport** which use ATP to move particles across the membrane.

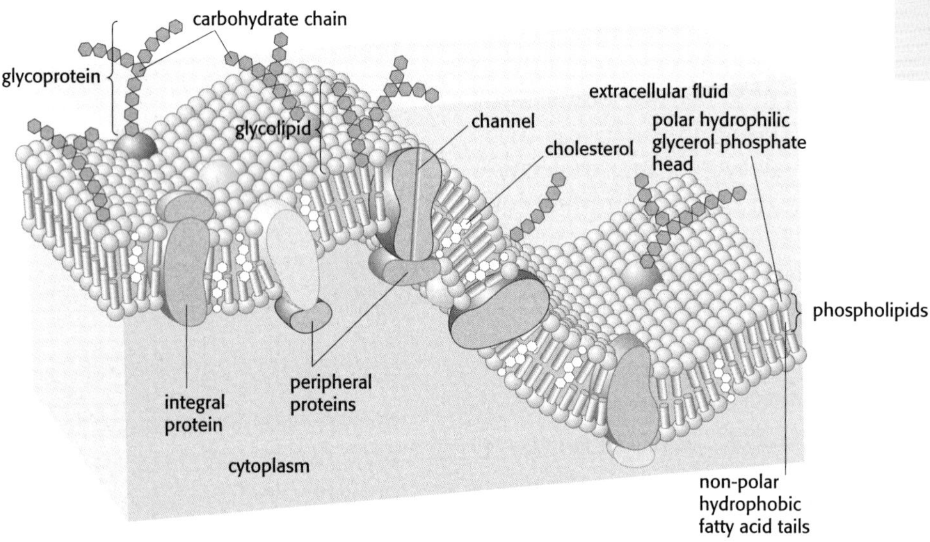

Figure 4 Fluid mosaic model of membrane structure

Thinking about science: models of membrane structure

Figure 4 shows a model of membrane structure that was developed in the early 1970s. It is called the Singer–Nicolson fluid mosaic model. According to the model, the phospholipids and proteins are like pieces of a mosaic that can rotate and move sideways, as in a fluid. They rarely swap from one half of the phospholipid bilayer to the other. Evidence for the model is given in the data-based question at the end of this chapter.

The fluid mosaic membrane model explains most, but not all, of the research findings about membranes. The following findings do not fit with it.

- Some of the lipids in the membrane are grouped in rafts or domains.
- Some membrane proteins are associated with particular domains of lipids in the membrane.
- These proteins can only move freely within their lipid domain, with the edge of the domain forming a diffusion barrier.
- Some membrane proteins are anchored to other membrane proteins or to proteins inside the cell, so are not free to move.
- Some membrane proteins that function as receptors are arranged in non-random patterns.

1 In this case, the word model is used to mean a type of hypothesis. Explain the advantage to scientists of developing models or hypotheses.
2 Suggest what happens in the period after a model or hypothesis has been developed.
3 The fluid mosaic model explains most but not all phenomena taking place in membranes. Discuss the implications of this.

Diffusion

Liquids and gases are fluids, because the particles in them can flow together in a mass. A river flowing to the sea is an example of this. The process is known as mass flow. Within fluids, individual particles also move independently and the direction of their movement is random. These movements of particles result in a process called diffusion. More particles move randomly from an area of high concentration to an area of low concentration than vice versa. There is therefore a *net* movement from the higher to the lower concentration. This diffusion is an important process in all organisms. For example, oxygen and carbon dioxide move into and out of cells by diffusion.

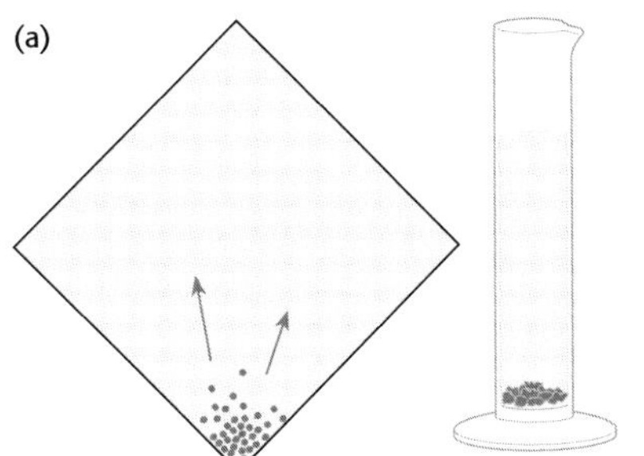

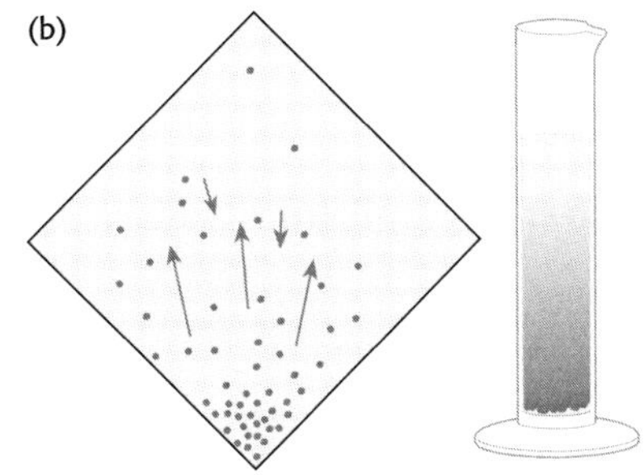

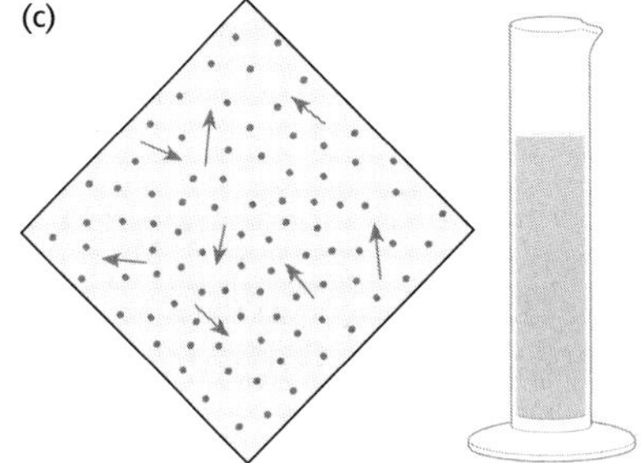

Figure 5 Model of diffusion with dots representing particles. The corresponding appearance of dye diffusing in cylinders of water is also shown.

Investigating biology: factors affecting diffusion rates

Gels are useful for investigating diffusion, because the water in them is liquid but the lattice of proteins that forms the gel is solid and prevents mass flow. Gels can be made up containing alkali and pH indicator. The rate of diffusion of acid into the gel can be followed by the gradual change in colour of the indicator, as the acid penetrates and neutralizes the alkali.

The tube (right) has gel set in the bottom with acid above it.

How long do you predict it will take for the acid to cause the indicator to change colour to the bottom of the tube?

These are the materials that are needed to test your prediction:

- Test tubes contain a 30 mm depth of gel, set in the bottom of the tube. The gel is made by dissolving 40 g of gelatin in 200 ml of boiling water, with 10 ml of 1% cresol red and 4 ml of 2 mol dm^{-3} ammonium hydroxide added.
- 1.0 mol dm^{-3} hydrochloric acid.
- 5 ml syringe.

Was the diffusion of the acid as rapid as you predicted, more rapid or less rapid?

Designing an experiment

1. Discuss with your partners which factors are likely to affect the rate at which the indicator changes colour, due to diffusion of acid.
2. Choose one of these factors and discuss how you could vary it in an experiment, to test whether it affects the rate of diffusion.
3. Decide on the details of your experiment, including how you can make sure that all factors will be kept constant, apart from the one that you are deliberately varying.

In this sort of experiment there should only be two things that vary – the independent variable and the dependent variable. The difference between dependent and independent variables is described on page 29.

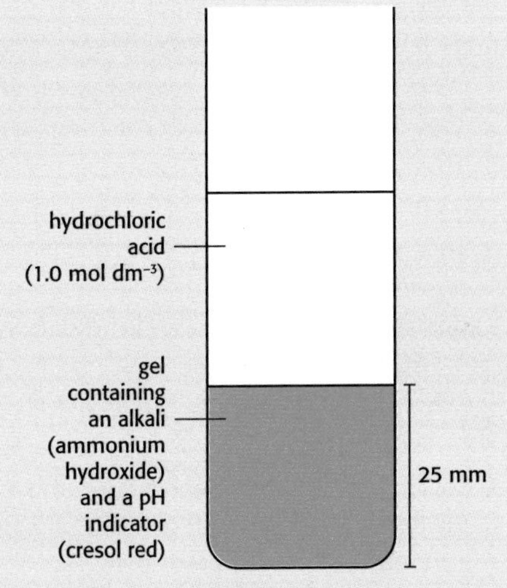

Figure 6 Test tube at start of experiment

1. In your investigation of diffusion what was the independent variable and what was the dependent variable?
2. Did your results show that the independent variable affected the level of the dependent variable?

Simple diffusion across membranes

Simple diffusion across membranes involves particles passing between the phospholipids in the membrane. It can only happen if the phospholipid bilayer is permeable to the particles. The centre of the membrane is hydrophobic, so ions with positive or negative charges cannot easily pass through. Polar molecules, which have partial positive and negative charges over their surface, can only diffuse at low rates between the phospholipids of the membrane. Small particles can pass through more easily than large particles.

Simple diffusion also only happens if the concentration of the particle is higher on one side of the membrane than the other. This is called a concentration gradient. Figure 7 illustrates simple diffusion across a membrane.

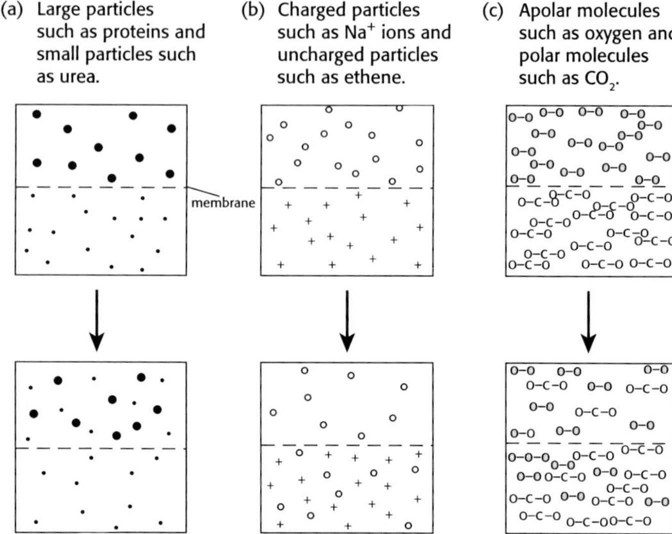

Figure 7 Diffusion across a phospholipid membrane

Rediscovering biology: membrane permeability

It would be difficult to carry out experiments on membrane permeability using real cell membranes, but artificial membranes can be used as models of cell membranes. Tubing made for use in kidney dialysis machines can be used.

Figure 8 shows apparatus set up to test whether substances in cola drink can diffuse through a membrane. Study the diagram and then answer the questions below.
If you are able to do this experiment, to test your predictions, you can use the following procedures to find out if the three substances have diffused through the membrane:

(a) Each time you are going to do the tests, lift the bag up and down a few times to mix the water in the tube. Use the dropping pipette to remove a few drops of water. Test with a drop of narrow-range pH indicator. Use a colour chart to work out the pH.
(b) Dip a glucose test strip into the water and record the colour that it turns. Instructions vary for these test strips. Follow the instructions and work out the glucose concentration in the water.
(c) Look carefully at the colour of the water to see whether it is still clear or if it is brown, like the cola.

You will need to note your results carefully during the experiment. Instructions for drawing up results tables are on the next page.

1. Cola contains glucose, phosphoric acid and caramel, a brown colouring substance. Predict which of these substances will diffuse out of the bag, with reasons for your predictions.
2. Before being put into the tube of water, the outside of the bag should be rinsed. Explain why this is a useful precaution.
3. If the mass of the bag is found before it is placed in the water and after it has been in the water for one hour, predict whether the mass of the bag will increase or decrease.

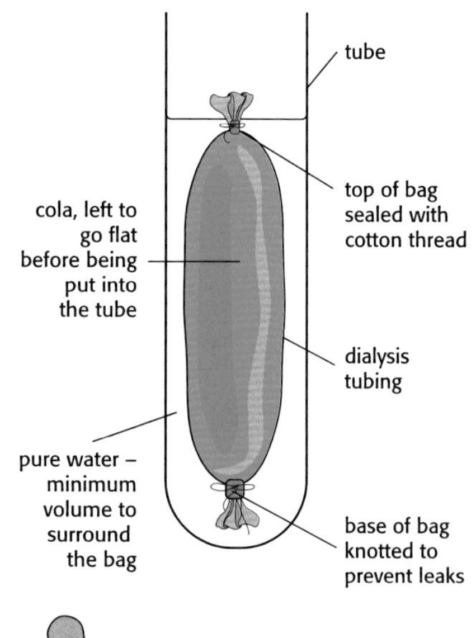

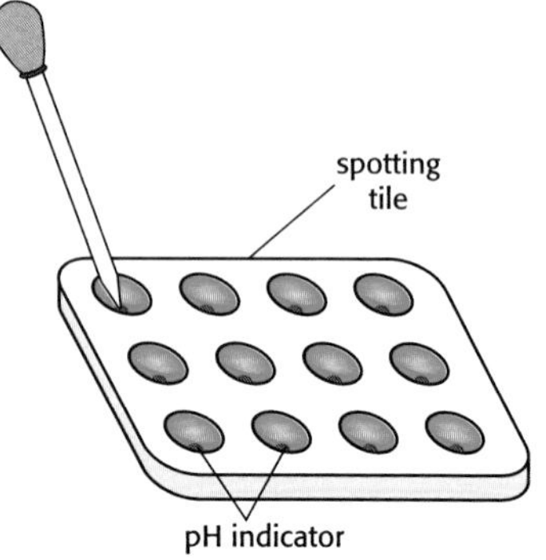

Figure 8 Apparatus for membrane permeability experiment

Working with data: results tables

- Results tables (see Table 2) can be drawn by hand or by computer, but they must be designed clearly.
- Always use a grid for results tables.
- Every row and column must have a heading saying what results are recorded in it: time, or colour of pH indicator, for example.
- If the results are numerical, you will nearly always need to show the units in the row or column heading. The units are shown after a back slash, or sometimes in parentheses. Examples of units are grams and minutes.
- Negative indices are sometimes used in the units, for example mol dm^{-3}. This means moles per dm^3, or the number of moles divided by the number of cubic decimetres. Cubic decimetres are also known as litres.
- Refer to Chapter 35 for instructions about uncertainty.

Table 2 A results table for the membrane permeability experiment on page 28.

	Time / minutes				
	0	4	8	16	32
Colour of glucose test					
Glucose conc. / mol dm^{-3}					
Colour of pH indicator					
pH					
Colour of water					
Mass of bag /g		-	-	-	

Working with data: independent and dependent variables

Many experiments in biology involve deliberately varying one factor. This is called the **independent variable**, because you can choose what the levels of the factor will be. The level of something else is measured during the experiment, which may or may not be affected by the independent variable. This is therefore called the **dependent variable**. You do not know what the levels of the dependent variable will be beforehand and you cannot choose them.

When plotting graphs, the independent variable almost always goes on the *x*-axis and the dependent variable on the *y*-axis.

Working with data: percentage calculations

If you want to express one number as a percentage of a second number, you should use this formula:

$$\frac{\text{first number}}{\text{second number}} \times 100\%$$

This formula tells you to divide the first number by the second number and then multiply by 100. The % symbol is written next to the answer.

Example 1

A mouse has a body mass of 31 g and the mass of its heart is 144 mg. What percentage of the body mass is the mass of the heart?

The mass units must be the same, so they cancel out when the percentage is calculated. 144 mg = 0.144 g

$$\frac{\text{heart mass}}{\text{body mass}} \times 100\% = \frac{0.144\text{g}}{31\text{ g}} \times 100\% = 0.46\%$$

Example 2

The mass of a sample of leaves falls from 12 g to 1.92 g when it is heated to evaporate the water in the leaves. What is the percentage mass change and therefore the percentage water content of the leaves?

The mass change is calculated by subtracting the initial mass from the final mass.

$$\text{mass change} = 1.92\text{ g} - 12.00\text{ g} = -10.08\text{ g}$$

$$\%\text{ mass change} = \frac{\text{mass change}}{\text{initial mass}} \times 100\% = \frac{-10.08\text{g}}{12.00\text{g}} \times 100\% = -84\%$$

Note that the percentage is negative, because the mass has fallen. The percentage of water in the leaves was 84%.

1 During each menstrual cycle in women, a large cell in the ovary called a primary oocyte divides to produce one cell called a secondary oocyte and another cell called a polar body. If a secondary oocyte has a volume of 0.6163 mm^3 and the polar body produced at the same time has a volume of 0.0015 mm^3, what percentage of the volume of the primary oocyte did each cell receive?

2 If a spherical human zygote, with a volume of 0.62 mm^3, divides to produce two spherical embryo cells, each with a volume of 0.31 mm^3, the total area of plasma membrane needed to surround the cells changes from 3.52 mm^2 for the zygote to 2.22 mm^2 for each of the two embryo cells. What is the percentage increase in the total amount of plasma membrane required when the zygote divides?

Facilitated diffusion across membranes

Ions and other particles, which cannot diffuse between phospholipids, can pass into or out of cells if there are channels for them through the plasma membrane. These channels are holes with a very narrow diameter. The walls of the channel consist either of a single protein molecule, or a group of proteins. The diameter and chemical properties of the channel ensure that only one type of particle passes through, for example sodium ions, or potassium ions, but not both. Because these channels help particles to pass through the membrane, from a higher concentration to a lower concentration, the process is called **facilitated diffusion**. Cells can control which types of channel are synthesized and placed in the plasma membrane and in this way they can control which substances diffuse in and out of the cell. Figure 9 shows the structure of a channel for magnesium ions, viewed from the side and from the outside of the membrane. The structure of the proteins making up the channel ensures that only magnesium ions are able to pass through the hole in the centre.

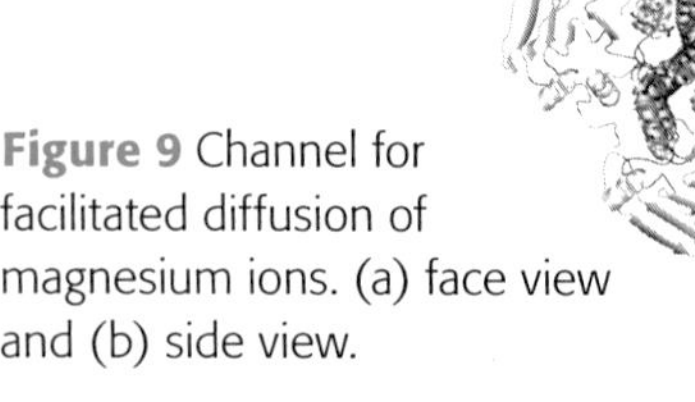

Figure 9 Channel for facilitated diffusion of magnesium ions. (a) face view and (b) side view.

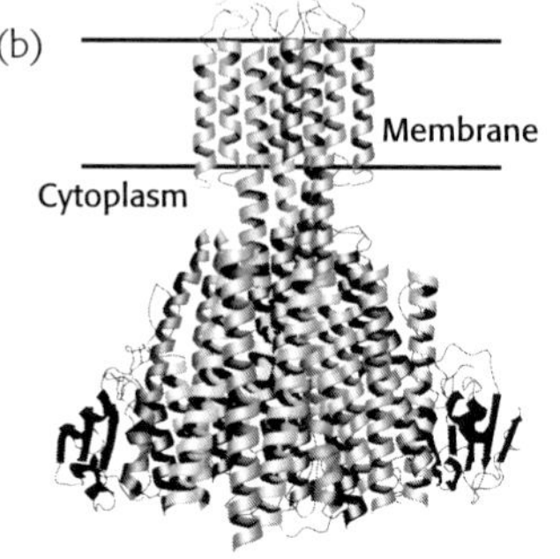

Data-based question: patch clamp analysis

The Nobel Prize for Physiology was awarded in 1991 to the German scientists Erwin Neher and Bert Sakmann for their work on ion channels in biological membranes. The technique that they developed is called patch clamp analysis (Figure 10). They used pipettes with extremely narrow tips (about 1 μm across) and placed them on the plasma membrane at the surface of cells. They found that by applying slight suction to the pipette and pulling it away from the cell, they could seal the end of the pipette with a patch of plasma membrane. Neher and Sakmann could then vary ion concentrations on the two sides of the membrane, inside and outside the pipette, and investigate the movement of ions through channels in the patch of membrane. Because ions are charged, their movement through a channel causes an electrical current, which can be measured. The results from patch clamp analysis therefore consist of trace recordings of electrical currents.

Trace (i) is from an area of plasma membrane from a muscle fibre, where the terminal of a neuron formed a synapse with the muscle fibre.

Trace (i) was obtained with a low concentration of acetylcholine (ACh) on the outside of the membrane. The experiment was then repeated at higher ACh concentrations. Higher concentrations increased the frequency of opening of the channels but not the average time that the channels remained open.

Trace (ii) was obtained using a patch of membrane from mouse spinal neurone. Channels in this membrane are activated by glycine, an amino acid.

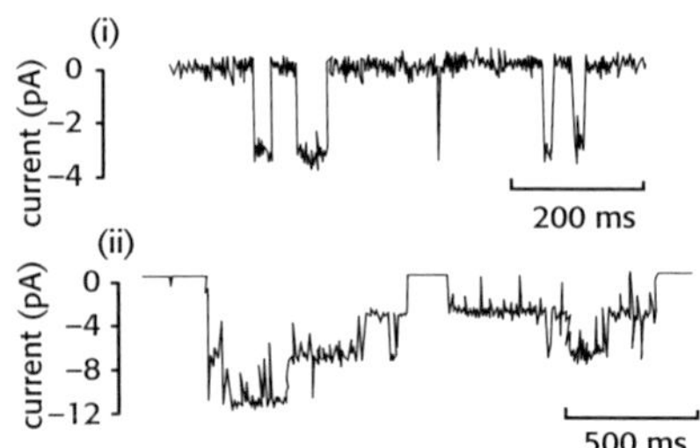

Figure 10 Results of patch clamp analysis

1 Looking at Trace (i):
   (a) State the size of current flowing through the membrane when the ion channels are closed. [1]
   (b) State how many picoamps there are in one amp. [1]
2 Trace (i) shows five occasions when an ion channel opens and then closes.
   (a) State the maximum length of time that a channel remains open. [1]
   (b) State the size of the current flowing when an ion channel is open. [1]
3 ACh binds to the ion channels in this membrane at special binding sites, causing the channels to open. The resulting flow of ions leads to contraction of the muscle fibre. Suggest what the natural source of ACh is in living muscle fibres. [1]
4 Look at Trace (i). Explain why higher ACh concentrations:
   (a) increased the frequency of opening of channels [2]
   (b) did not increase the average time that the channels remained open. [2]
5 (a) Compare Trace (ii) with Trace (i). [3]
   (b) Suggest reasons for the differences between the two traces. [3]

Osmosis

In most living cells the movement of solutes through the plasma membrane is restricted but water is able to move in and out of the cell freely – the membrane is partially permeable. Sometimes the number of water molecules moving in and out are the same – there is no net movement. Often more molecules move in or out and this net movement is osmosis. Osmosis only occurs when there are substances dissolved in water – solutes. Substances that dissolve form intermolecular bonds with the water molecules. The water molecules that bond to solutes cannot move freely, reducing the effective concentration of water. Regions with a higher solute concentration therefore have a lower water concentration than regions with a lower solute concentration. This also tends to cause water to move from regions of lower to higher solute concentration. No energy has to be expended directly to make the water molecules move – it is a passive process.

osmosis *the passive movement of water molecules, across a partially permeable membrane, from a region of lower solute concentration to a region of higher solute concentration.*

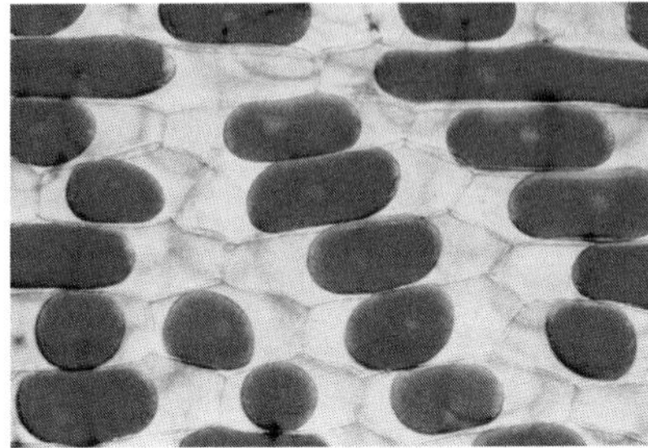

Figure 11 Micrograph of red onion cells placed in salt solution

Rediscovering biology: plasmolysis in red onion cells

Some varieties of onion have cells in the epidermis of their bulb scales with red cytoplasm, due to the presence of anthocyanin pigments. It is very easy to see the area occupied by the cytoplasm of these cells, so they are useful in osmosis experiments.

Method

1 Peel off some red epidermis from the scale of an onion bulb.
2 Cut out a sample of it, about 5 × 5 mm.
3 Mount the sample in a drop of distilled water on a microscope slide, with a cover slip.
4 Draw a diagram of a typical cell, labelling the cell wall and cytoplasm and deducing the position of the plasma membrane.
5 Put a drop of 2% sodium chloride solution next to the cover slip and draw it under the cover slip by touching the opposite side with paper towel, to cause capillarity.
6 Observe the onion cells for a few minutes, and then draw a cell that has altered in appearance. Again label the cell wall, plasma membrane and cytoplasm.
7 Repeat the procedure in **5** using distilled water to replace the sodium chloride solution.

?

1 In the cells in distilled water:
  (a) Deduce the location of the plasma membrane.
  (b) Deduce whether water moved in or out by osmosis.
  (c) Deduce whether the pressure is higher or lower than in the atmosphere.
2 Explain, using your knowledge of the process of osmosis, the effect of the sodium chloride solution on:
  (a) the volume of cytoplasm of the red onion cells
  (b) the position of the plasma membrane inside the cells.
3 Discuss whether the observed changes are reversible.

Data-based question: albumin in the blood

Figure 12 shows the concentration of albumin in the blood of children in northern Thailand. Albumin is a solute in blood. It causes 70 per cent of the reabsorption of water into blood capillaries that occurs in healthy tissues. Kwashiorkor is caused by a lack of proteins in the diet and marasmus is caused by a general lack of food.

1 (a) Compare the levels of albumin in the three groups of children. [3]
  (b) Explain the differences in the levels of albumin. [2]
2 Explain how albumin dissolved in blood plasma causes reabsorption of water into blood. [3]
3 When water is not reabsorbed into blood capillaries, tissues become swollen by fluid retention. This condition is called edema. Deduce which of the groups of children were most at risk of edema. [2]

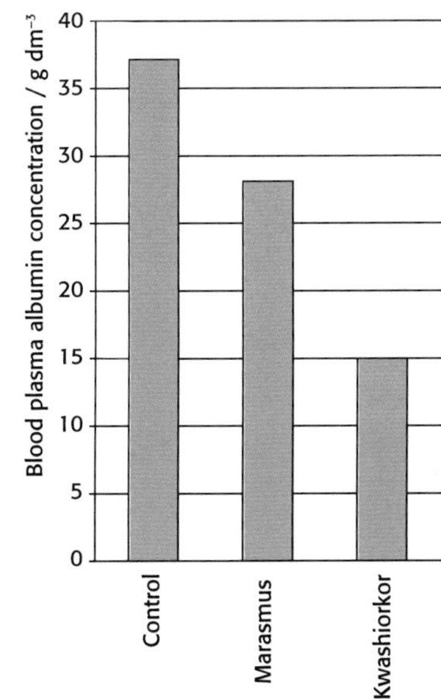

Figure 12 Albumin concentrations

Data-based question: osmosis in plant tissues

If samples of plant tissue are bathed in salt or sugar solutions for a short time, any increase or decrease in mass is due almost entirely to water entering or leaving the cells by osmosis. Figure 13 shows the percentage mass change of four tissues, when they were bathed in salt solutions of different concentrations.

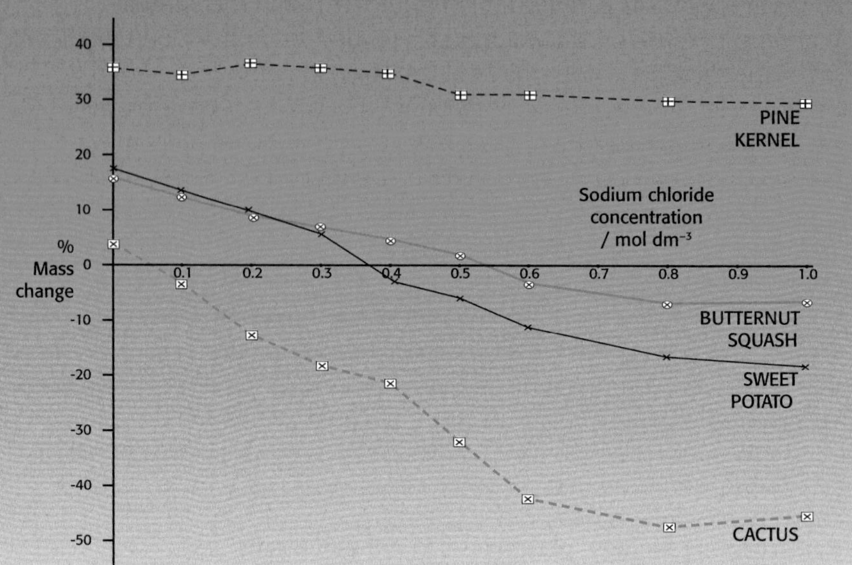

Figure 13 Mass changes in plant tissues bathed in salt solutions

1 (a) State whether water moved into or out of the tissues at 0.0 mol dm^{-3} sodium chloride solution. [1]

(b) State whether water moved into or out of the tissues at 1.0 mol dm^{-3} sodium chloride solution. [2]

2 Deduce which tissue had the lowest solute concentration in its cytoplasm. Include how you reached your conclusion in your answer. [2]

3 Suggest reasons for the differences in solute concentration between the tissues. [3]

4 Explain the reasons for using percentage mass change rather than the actual mass change in grams in this type of experiment. [2]

Rediscovering biology: osmosis in potato tissue

The experiment described above can be repeated using potato tubers, or any other plant tissue that is homogeneous and tough enough to be handled without disintegrating.

If this experiment is carried out, percentage mass change should be calculated.

Discuss with a partner or group how you could do the following things.

1 Dilute a 1 mol dm^{-3} salt solution, to obtain the salt concentrations shown on the graph.
2 Obtain samples of a plant tissue that are similar enough to each other to give comparable results.
3 Ensure that the surface of the tissue samples is dry when finding their mass, both at the start and end of the experiment.
4 Ensure that all variables are kept constant, apart from salt concentration of the bathing solution.
5 Leave the tissue in the solutions for long enough to get a significant mass change, but not so long that other factors affect the mass, such as decomposition!
6 You might choose to be more inventive in your experimental approach. Figure 14 gives one idea for measuring changes to the turgidity of the tissue, but other methods could be used.

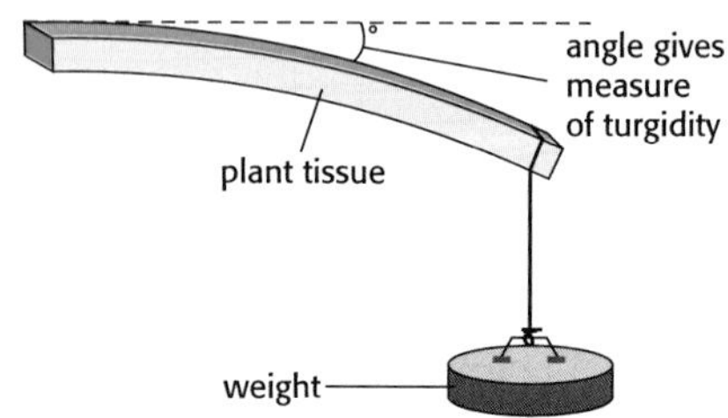

Figure 14 Method of assessing turgidity of plant tissue

Active transport

Cells sometimes take in substances even when there is a higher concentration of the substance inside the cell than outside it. The substance is absorbed against the concentration gradient. Less commonly, cells pump substances out, even though there is already a larger concentration outside. Various conclusions can be drawn – this type of movement across membranes is not diffusion, and energy must be necessary for it to happen. It is therefore called active transport. A substance called ATP supplies the energy. Every cell produces its own supply of ATP by cell respiration.

Globular proteins in membranes, usually called pump proteins or transporter proteins, carry out active transport. The membranes of cells contain many different transporter or pump proteins, allowing the cell to control the content of its cytoplasm precisely.

There are many example of active transport. For instance:

- Plant roots absorb potassium and other ions by active transport.
- Neurons pump potassium ions in and sodium ions out by active transport, to build up a store of potential energy, or electrical potential, that is used to transmit a nerve impulse.

Figure 15 illustrates how a pump protein works. The molecule or ion enters the pump protein and can reach as far as a central chamber. A conformational change to the protein takes place using energy from ATP. After this, the ion or molecule can pass to the opposite side of the membrane and the pump protein returns to its original conformation. The pump protein shown transports Vitamin B_{12} into *E. coli.*

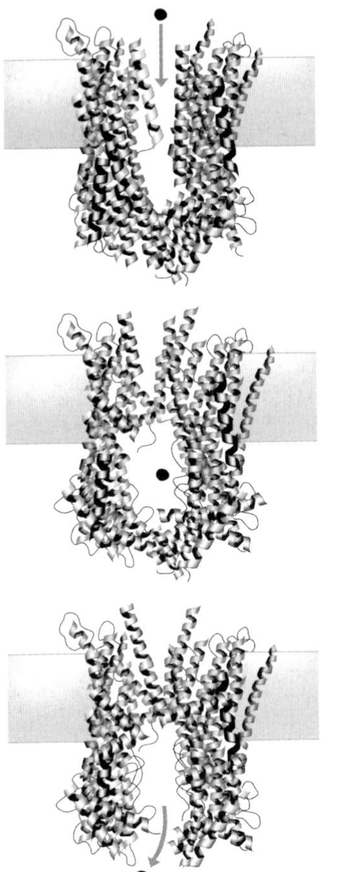

Figure 15 Action of a pump protein

Data-based question: phosphate absorption in barley roots

Roots were cut off from barley plants and were used to investigate phosphate absorption. Roots were placed in phosphate solutions and air was bubbled through. The phosphate concentration was the same in each case, but the percentage of oxygen and nitrogen was varied in the air bubbled through. The rate of phosphate absorption was measured. Table 3 shows the results.

Table 3

Oxygen /%	Nitrogen /%	Phosphate absorption /μmol g^{-1} h^{-1}
0.1	99.9	0.07
0.3	99.7	0.15
0.9	99.1	0.27
2.1	97.1	0.32
21.0	79.0	0.33

1 Describe the effect of reducing the oxygen concentration below 21.0% on the rate of phosphate absorption by roots. You should only use information from the table in your answer. [3]

2 Explain the effect of reducing the oxygen percentage from 21.0 to 0.1 on phosphate absorption. In your answer you should use as much biological understanding as possible of how cells absorb mineral ions. [3]

An experiment was done to test which method of membrane transport was used by the roots to absorb phosphate. Roots were placed in the phosphate solution as before, with 21.0% oxygen bubbling through. Varying concentrations of a substance called DNP were added. DNP blocks the production of ATP by aerobic cell respiration. Figure 16 shows the results of the experiment.

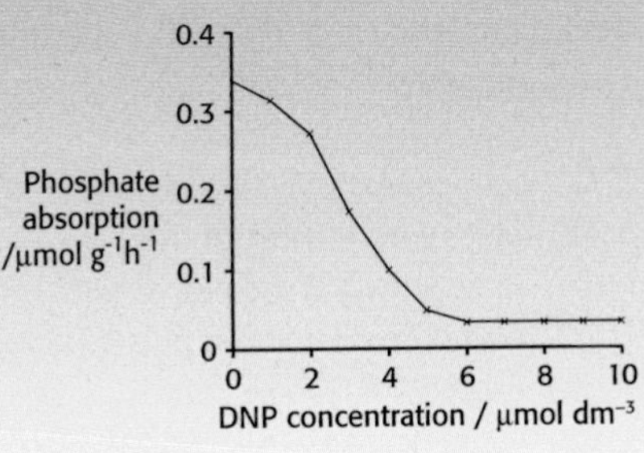

Figure 16 Effect of DNP concentration on phosphate absorption

3 Deduce, with a reason, whether the roots absorbed the phosphate by diffusion or active transport. [2]

4 Discuss the conclusions that can be drawn from the data in the graph about the method of membrane transport used by the roots to absorb phosphate. [2]

Vesicles – formation, breakage and movement

A vesicle is a small sac of membrane with a droplet of fluid inside. Vesicles are usually spherical and are present in most eukaryotic cells. They are formed from larger pieces of membrane. Membranes are fluid, so they can change shape and move. To form a vesicle, a small region of a membrane is pulled from the rest of the membrane and is pinched off. Proteins in the membrane carry out this process, using energy from ATP. Vesicles often form at the inner surface of the plasma membrane of a cell, trapping fluid from outside the cell. This process is called **endocytosis** (Figure 17). The fluid may contain larger solids in it, for example when unicellular organisms ingest their food.

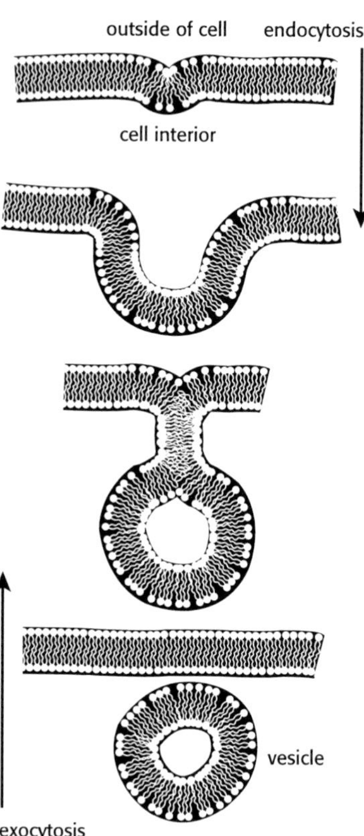

Figure 17 Exocytosis and endocytosis

Vesicles are used to move materials around inside cells. An example of this occurs in **secretory cells**. Protein is synthesized by ribosomes on the rER and is stored in the cisternae of the rER. Vesicles bud off the cisternae and carry the protein to the Golgi apparatus. Here the vesicles fuse with the cisternae of the Golgi apparatus, so that the protein can be processed, for example by adding carbohydrate to polypeptides to form glycoprotein. The protein gradually moves through the stack of cisternae, until it is finally released in more vesicles. These move to the plasma membrane and fuse with it, releasing protein from the cell. This is called **exocytosis**. Exocytosis can be used in other ways, for example to remove excess water from the cells of unicellular organisms.

Data-based question: autoradiography

In 1974, Albert Claude and Christian de Duve of Belgium and George Palade of the United States shared the Nobel Prize for Medicine, "for their discoveries concerning the structural and functional organization of the cell". They were pioneers in the field of electron microscopy and investigation of the functions of the organelles of cells. George Palade developed a technique called autoradiography. This involves living tissue that is actively synthesizing protein from amino acids. The tissue is given radioactive amino acids for a period of time called the pulse, followed by non-radioactive amino acids for a period called the chase. The tissue is then fixed, sectioned and examined using an electron microscope. Before it is examined, a thin film of photographic emulsion is painted onto the section and after a time is developed. Where there are radioactive amino acids in the section of tissue, black grains are formed in the photographic emulsion and these are visible in the electron microscope. It is therefore possible to follow the movements of amino acids and proteins in cells, by tracing the location of the radioactivity, at varying lengths of time after it entered the cell. Table 4 gives the results used in George Palade's Nobel Lecture in 1974.

Table 4

Cell compartments in guinea pig pancreas cells	% autoradiographic grains			
	Pulse 3 min	Chase minutes 7	37	117
Rough endoplasmic reticulum	86.3	43.7	24.3	20.0
Vesicles between rER and Golgi apparatus	2.7	43.0	14.9	3.6
Vesicles released by Golgi apparatus	1.0	3.8	48.5	7.5
Large storage vesicles near plasma membrane	3.0	4.6	11.3	58.6
Extracellular space –duct adjacent to the cells	0.0	0.0	0.0	7.1

1 (a) State where most of the radioactivity is located at the end of the 3-minute pulse. [1]

(b) Explain the reasons why most of the radioactivity is located in that part of the cell at that time. [2]

2 Deduce from the data how proteins are carried from the rough endoplasmic reticulum to the Golgi apparatus, giving reasons for your answer. [3]

3 Discuss, using the evidence from the table, how long it takes for a protein to be synthesized and secreted from a pancreas cell. [3]

4 Suggest reasons for the amounts of radioactivity in the duct adjacent to the pancreas cell not rising above 7.1%. [2]

Data-based question: diffusion of proteins in membranes

L. D. Frye and M. Edidin (1970) used an elegant technique to obtain evidence for the fluid nature of membranes. They attached fluorescent markers to membrane proteins – green markers to mouse cells and red markers to human cells. In both cases, spherical cells growing in tissue culture were used. The marked mouse and human cells were then fused together. At first the fused cells had one green hemisphere and one red one, but over the minutes following fusion, the red and green markers gradually merged, until they were completely mixed throughout the whole of the cell membrane. Inhibition of protein synthesis and blocking of ATP production did not prevent this mixing. (Remember ATP supplies energy for active processes in the cell.) Figure 18 shows the effect of temperature on the rate of mixing.

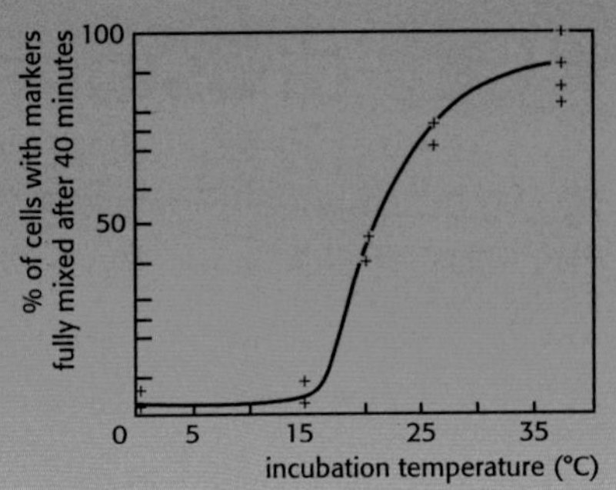

Figure 18 Effect of temperature on the rate of diffusion of fluorescent markers in membranes

1 Explain what conclusion can be drawn from each of these pieces of experimental evidence:
   (a) When the cells were kept at normal body temperatures for mouse and human cells, the red and the green markers became mixed. [3]
   (b) Blocking ATP synthesis in the cells did not prevent the mixing of the red and green markers. [2]
   (c) Inhibition of protein synthesis in the cells did not prevent mixing of the markers. [2]

2 Explain the trends shown in the graph
   (a) for temperatures between 15 and 35°C. [3]
   (b) for temperatures below 15°C. [2]

3 Predict, with reasons, the results of the experiment if it was repeated using cells from Arctic fish rather than from mice or humans. [3]

Chapter 2 questions

1 Particles can enter or leave cells by diffusion, active transport, osmosis, endocytosis or exocytosis. Identify which method is used in each of these examples.
   (a) water entering a root cell from the soil [1]
   (b) antibodies being absorbed from milk in the stomach of a newborn baby [1]
   (c) oxygen entering a cell in the gill of a salmon [1]
   (d) proteins being secreted from gland cells in the pancreas [1]
   (e) glucose from digested foods entering cells in the lining of the small intestine. [1]

2 In human secretory cells, for example in the lung and the pancreas, positively charged ions are pumped out, and chloride ions follow passively through chloride channels. Water also moves from the cells into the liquid that has been secreted.
   In the genetic disease cystic fibrosis, the chloride channels malfunction and too few ions move out of the cells. The liquid secreted by the cells becomes thick and viscous, with associated health problems.
   (a) State the names of the processes that:
      (i) move positively charged ions out of the secretory cells [1]
      (ii) move chloride ions out of the secretory cells. [1]
      (iii) move water out of the secretory cells. [1]
   (b) Explain why the fluid secreted by people with cystic fibrosis is thick and viscous. [4]

3 Table 5 shows the area of membranes in a rat liver cell.

Table 5

Membrane component	Area (μm²)
Plasma membrane	1780
Rough endoplasmic reticulum	30400
Mitochondrial outer membrane	7470
Mitochondrial inner membrane	39600
Nucleus	280
Lysosomes	100
Other components	18500

   (a) Calculate the total area of membranes in the liver cell. [2]
   (b) Calculate the area of plasma membrane as a percentage of the total area of membranes in the cell. Show your working. [3]
   (c) Explain the difference in area of the inner and outer mitochondrial membranes. [3]
   (d) Using the data in the table, identify two of the main activities of liver cells. [2]

4 (a) Draw and label a diagram to show the structure of membranes. [6]
   (b) Outline ways in which proteins help cells receive and send messages. [4]
   (c) Explain how cells use vesicles to transport materials. [8]

3 Cell division

Cell cycle

Growth, asexual reproduction, tissue repair and maintenance are examples of processes that require the creation of new cells.

In eukaryotic cells, division of the nucleus to form two genetically identical nuclei is termed mitosis. Division of the cytoplasm to form two cells is called cytokinesis.

Prokaryotic cells reproduce by a process called binary fission. This involves replication of the single circular chromosome. The two copies of the chromosome move to opposite ends of the cell, and cytokinesis quickly follows.

The life of a cell can be thought of as an ordered sequence of events, called the cell cycle. The cell cycle refers to the events between one cell division and the next in a eukaryotic cell. It can be roughly divided into interphase and cell division. Interphase is an active period in the life of a cell when many metabolic reactions occur, including protein synthesis, DNA replication and an increase in the number of mitochondria and/or chloroplasts. It is not necessarily a period of preparation for mitosis, as a cell can remain in interphase indefinitely.

Interphase consists of three phases, the G1 phase, the S phase and the G2 phase. During the S phase the cell copies all genetic material, so that after mitosis both new cells have a complete set of genes.

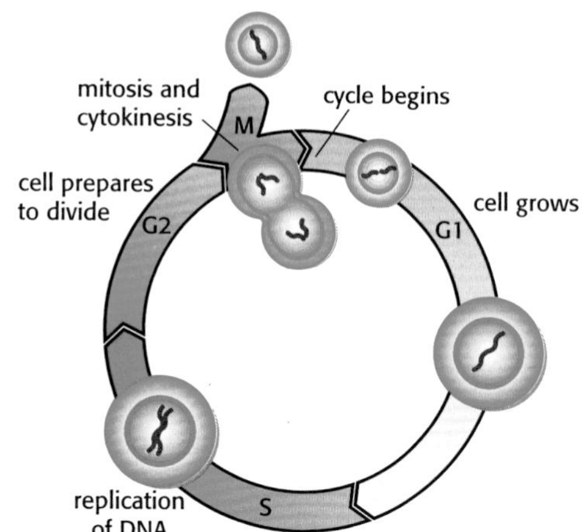

Figure 1 The cell cycle. Note that during the S phase, the chromosome in the model cell is duplicated through the process of replication.

Data-based question: cell size and the cell cycle

Figure 2 shows the daily life cycle pattern of *Emiliania huxleyi* (a species of phytoplankton) as observed under laboratory conditions. The hypothsis is that the cell cycle appears to be timed so that the light period can be used for photosynthesis linked to growth whereas energy consuming processes can occur in the dark, the daughter cells being prepared for photosynthesis by the onset of the next day.

1 State the time of day when:
   (a) most DNA replication occurs
   (b) when mitosis is most likely to occur. [2]
2 Identify the cell cycle stage when most of the increase in cell size is occurring. [1]
3 Evaluate the claim that the timing of the cell cycle in *Emiliania huxleyi* is an adaptation to take advantage of light resources. [3]

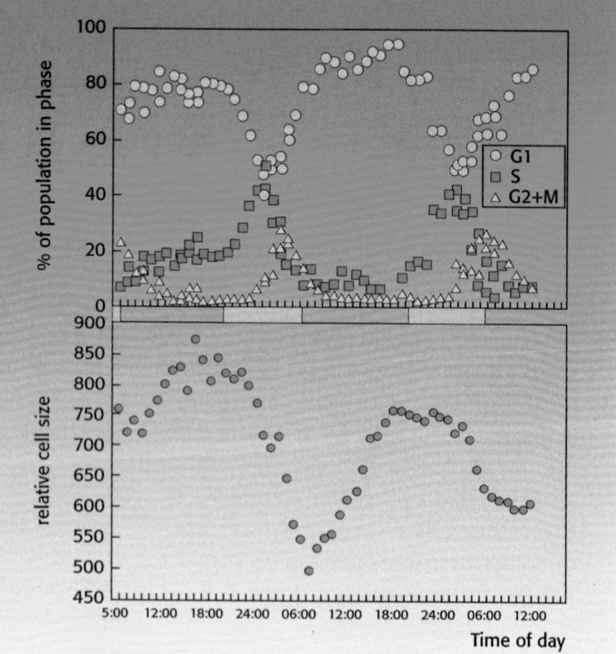

Figure 2 The cell cycle in *Emiliania huxleyi* follows a daily pattern.

The four phases of mitosis

Mitosis is the division of a eukaryotic nucleus into two genetically identical nuclei. Before mitosis can occur, two copies of each chromosome are needed. Each chromosome initially consists of a single DNA molecule. This has to be replicated before mitosis, and it then consists of two identical DNA molecules, called sister chromatids. Although it is a continuous process, cytologists have divided the events of mitosis into four phases: prophase, metaphase, anaphase and telophase. The events that occur during these stages in an animal cell are summarized here.

Prophase

The chromosomes become shorter and fatter by coiling (Figure 3a). To become short enough they have to coil repeatedly. This is called supercoiling. At the end of prophase the nuclear membrane breaks down.

Microtubules grow from the poles of the cell from a structure called the microtubule organizing centre (MTOC) to the chromosomes (Figure 3b). These microtubules form a spindle shape and so the MTOCs together with the microtubules are referred to as the mitotic spindle.

Metaphase

Spindle microtubules attach to the centromeres. Chromosomes are moved to the equator of the cell (Figure 3c), with a spindle microtubule attached to one of the sister chromatids from one pole and another spindle microtubule attached to the opposite sister chromatid from the other pole.

Anaphase

At the start of anaphase, the pairs of sister chromatids separate and the spindle microtubules pull them towards the poles of the cell (Figure 3d). Until then the centromeres had held them together. Mitosis produces two genetically identical nuclei because sister chromatids are pulled to opposite poles. To ensure this, the centromeres of sister chromatids must be attached in metaphase to spindle microtubules from different poles.

Telophase

Nuclear membranes reform around the chromatids, now called chromosomes, at each pole (Figure 3e). The chromosomes uncoil, the cell divides and the two daughter cells enter interphase again.

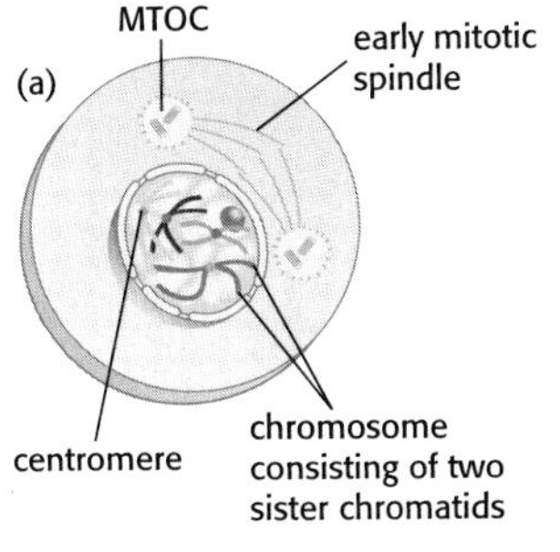

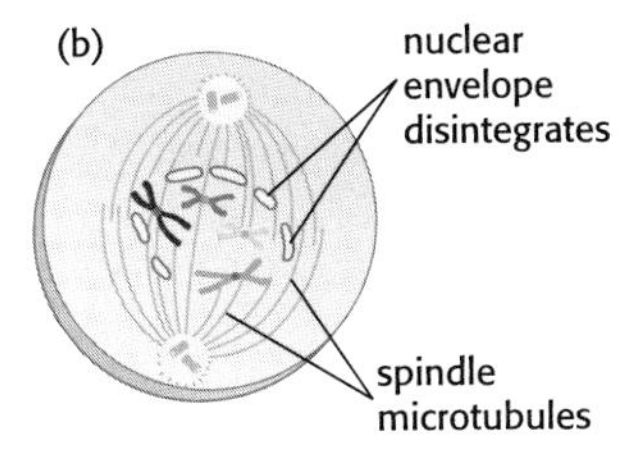

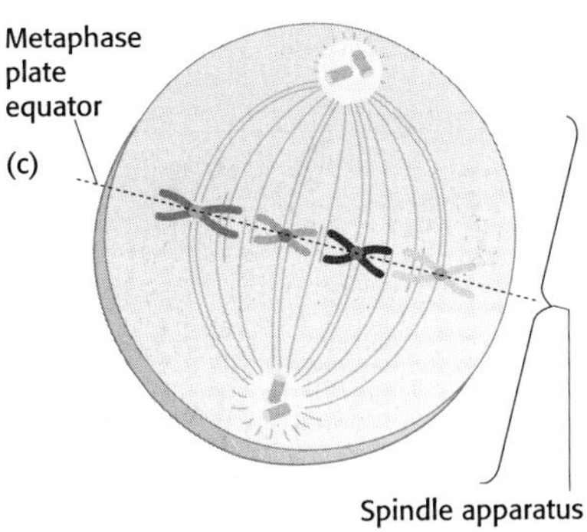

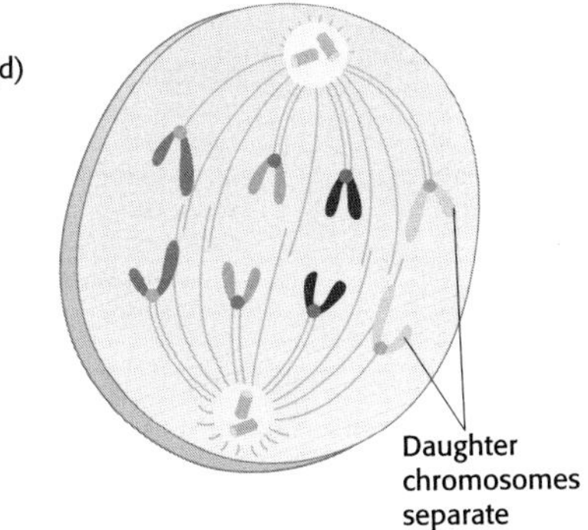

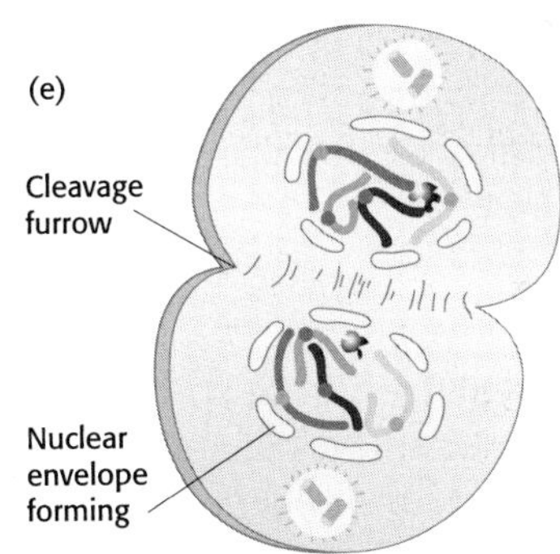

Figure 3 The stages of mitosis: (a) early and (b) late prophase. (c) metaphase. (d) anaphase. (e) telophase.

Rediscovering biology: determining the mitotic index

- Obtain a prepared slide of an onion root tip. Find and examine the meristematic region, i.e. a region of rapid cell division.
- Create a tally chart. Classify each of about 100 cells in this region as being either in interphase or in any of the stages of mitosis.
- Use this data to calculate the mitotic index; i.e. the fraction of cells undergoing mitosis.

The mitotic index is an important diagnostic signal in the classification of tumours.

Working with data: using Excel to construct a pie graph

1 To estimate the relative length of each stage of the cell cycle, the number of cells in each stage can be counted, as in the sample data in Table 1.
2 To construct a pie graph of the data, open an Excel spreadsheet.
3 Enter the names of the phases of the cell cycle in the first column.
4 Enter the total number of cells in the corresponding phase in the second column. Your table should look like Table 1. Enter the total number of cells in the corresponding phase in the second column.
5 Highlight both columns and then click on the chart wizard toolbar icon, which looks like this: 
6 For a colour pie chart (Figure 4), choose "Pie" from the "Chart type" list. For black and white, click on the "Custom type" tab and choose "B&W Pie".
7 Click "Next" and then from the "Data Range" window, press "Next" again.
8 Give the chart a title by choosing the "Title" tab and typing a title in the "Chart title" field. A suitable title might be "Fraction of total cells in each of the phases of the cell cycle".
9 Choose the "Legend" tab, click "Show legend" and for the "Placement" choose "Right". Alternatively you can have labels rather than a key, as in the diagram below. For this option, choose the "Data labels" tab and click on "percentage". Choose "Category name" as well.

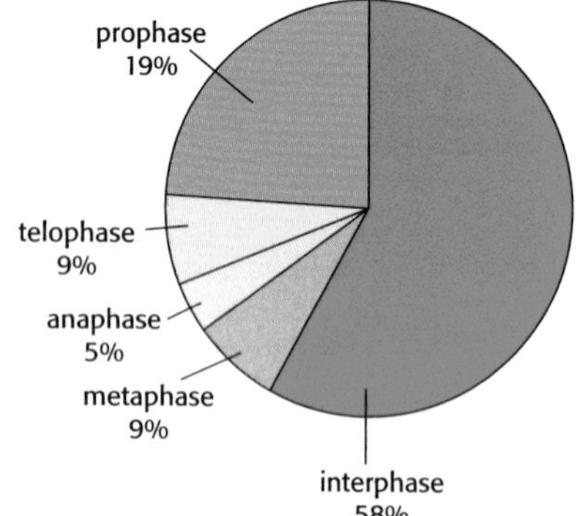

Table 1	
metaphase	20
anaphase	10
telophase	20
prophase	40
interphase	125

Figure 4 Fraction of total cells in each of the phases of the cell cycle.

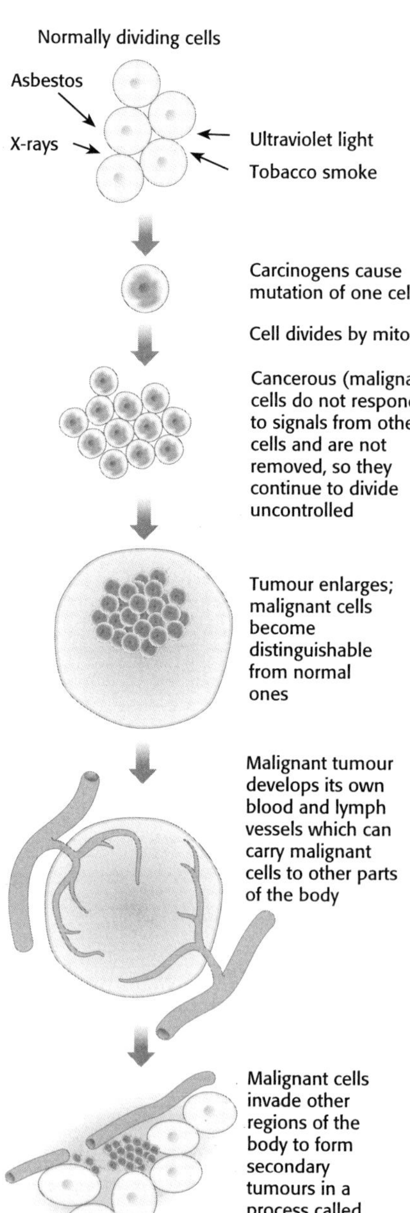

Figure 5 The development of a malignant tumour.

Cancer

The cell division cycle is regulated in a number of ways. Certain agents can damage these regulatory elements. Examples of such agents include ultraviolet light, chemicals called carcinogens or viruses. When this happens, uncontrolled cell division can occur resulting in the formation of tumours.

Thinking about science: the median is not the message

Abridged from "The Median is not the Message" by Steven Jay Gould, *Discover Magazine*, 1985.

This is a personal story of statistics, properly interpreted, as profoundly nurturant and life-giving. It declares war on the downgrading of intellect by telling a small story about the utility of dry, academic knowledge about science. Heart and head are focal points of one body, one personality.

In July 1982, I learned that I was suffering from abdominal mesothelioma, a rare and serious cancer usually associated with exposure to asbestos. When I revived after surgery, I asked my first question of my doctor and chemotherapist: "What is the best technical literature about mesothelioma?" She replied, with a touch of diplomacy... that the medical literature contained nothing really worth reading.

As soon as I could walk, I made a beeline for Harvard's Countway medical library and punched mesothelioma into the computer's bibliographic search program. An hour later, surrounded by the latest literature on abdominal mesothelioma, I realized with a gulp why my doctor had offered that humane advice. The literature couldn't have been more brutally clear: mesothelioma is incurable, with a median mortality of only eight months after discovery. I sat stunned for about fifteen minutes, then smiled and said to myself: so that's why they didn't give me anything to read.

If a little learning could ever be a dangerous thing, I had encountered a classic example. Attitude clearly matters in fighting cancer... Match people with the same cancer for age, class, health, socioeconomic status, and, in general, those with positive attitudes, with a strong will and purpose for living, with commitment to struggle, with an active response to aiding their own treatment and not just a passive acceptance of anything doctors say, tend to live longer.

Hence the dilemma for humane doctors: since attitude matters so critically, should such a sombre conclusion be advertised, especially since few people have sufficient understanding of statistics to evaluate what the statements really mean? From years of experience with the small-scale evolution of Bahamian land snails treated quantitatively, I have developed this technical knowledge – and I am convinced that it played a major role in saving my life. Knowledge is indeed power, in Bacon's proverb.

The problem may be briefly stated: What does "median mortality of eight months" signify in our vernacular? I suspect that most people, without training in statistics, would read such a statement as "I will probably be dead in eight months" – the very conclusion that must be avoided, since it isn't so, and since attitude matters so much...

When I learned about the eight-month median, my first intellectual reaction was: fine, half the people will live longer; now what are my chances of being in that half. I read for a furious and nervous hour and concluded, with relief: damned good. I possessed every one of the characteristics conferring a probability of longer life: I was young; my disease had been recognized in a relatively early stage; I would receive the nation's best medical treatment; I had the world to live for; I knew how to read the data properly and not despair.

Another technical point then added even more solace. I immediately recognized that the distribution of variation about the eight-month median would almost surely be what statisticians call "right skewed." The distribution of variation had to be right skewed, I reasoned. After all, the left of the distribution contains an irrevocable lower boundary of zero (since mesothelioma can only be identified at death or before). Thus, there isn't much room for the distribution's lower (or left) half – it must be scrunched up between zero and eight months. But the upper (or right) half can extend out for years and years, even if nobody ultimately survives.

The distribution was indeed, strongly right skewed, with a long tail (however small) that extended for several years above the eight month median. I saw no reason why I shouldn't be in that small tail, and I breathed a very long sigh of relief. My technical knowledge had helped. I had read the graph correctly. I had asked the right question and found the answers. I had obtained, in all probability, the most precious of all possible gifts in the circumstances - substantial time. I didn't have to stop and immediately follow Isaiah's injunction to Hezekiah – set thine house in order for thou shalt die, and not live. I would have time to think, to plan, and to fight.

1 What evidence does Gould offer that the separation of heart and mind is an absurd dichotomy?
2 What can be meant by the statement, "Knowledge is power"?
3 What is the value of measures of central tendency?

Chapter 3 questions

1 Images A to H (Figure 6) represent light micrographs of cells in the various stages of mitosis. The two bright regions represent telomeres (which are the ends of the chromosomes) and centromeres.
   (a) Determine the correct sequence in which the events occurred.
   (b) Identify the stage depicted in each micrograph.
   (c) Outline the patterns seen in the movement of the telomeres.

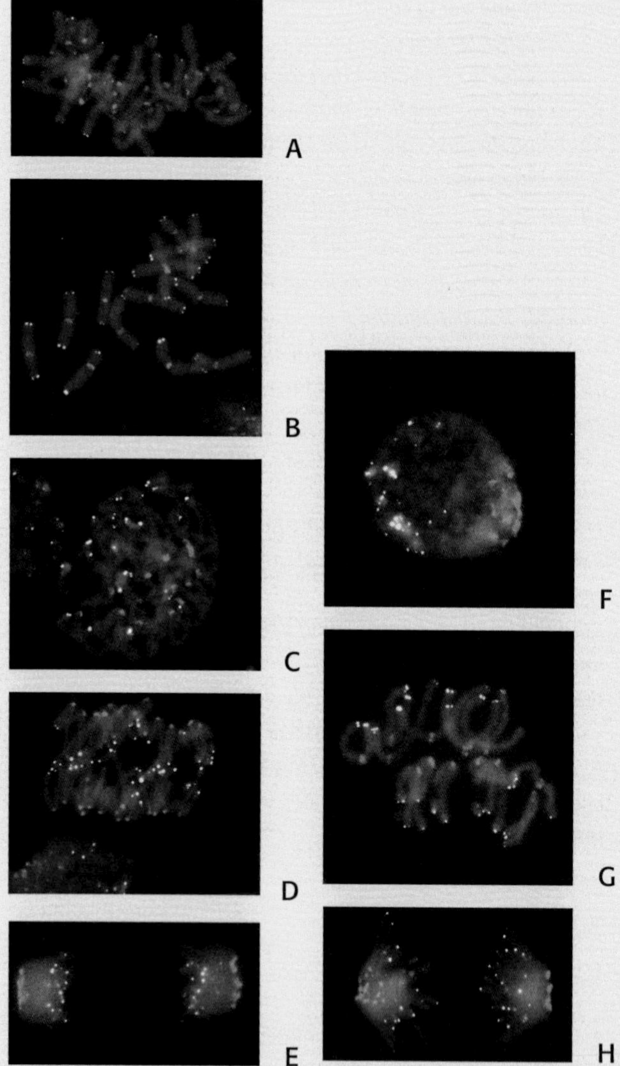

Figure 6

2 In each of the following combinations of words or phrases, one word or phrase does not belong. Identify which it is and explain why it does not belong.
   (a) spindle microtubule formation, replication, growth
   (b) spindle microtubule formation, supercoiling of chromosomes, attachment of microtubules to centromeres
   (c) maintenance, repair, formation of gametes.

3 The amount of DNA present in each cell nucleus was measured in a large number of cells taken from two different cultures of human bone marrow (Figure 7).

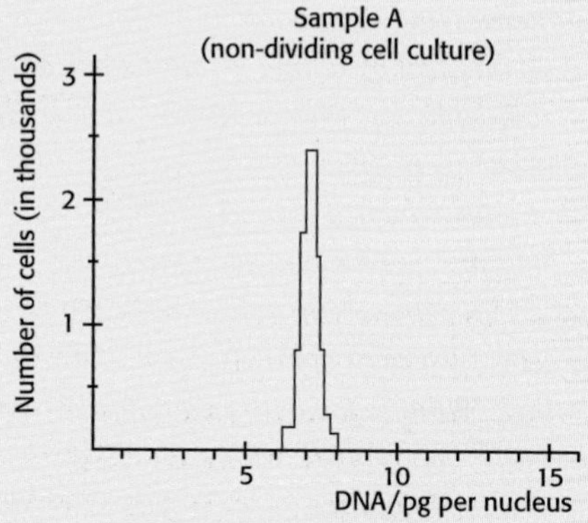

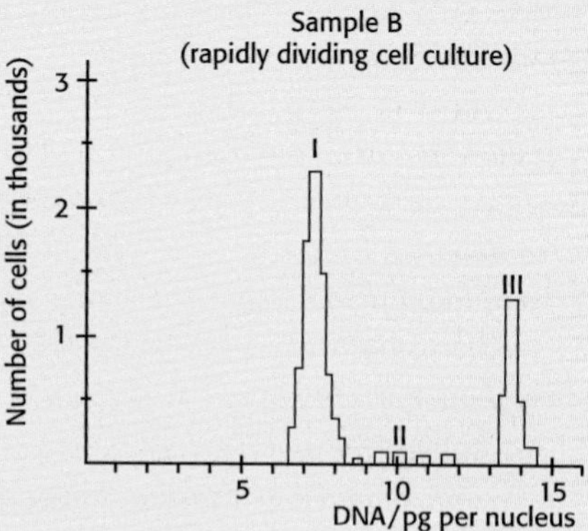

Figure 7

   (a) For each label (I, II and III) in the Sample B graph, deduce which phase of the cell cycle the cells could be in; i.e. G1, G2 or S. [3]
   (b) Estimate the approximate amount of DNA per nucleus that would be expected in the following human cell types:
      (i) bone marrow at prophase
      (ii) bone marrow at telophase. [2]

4 Define the following terms:
   (a) diploid (b) cytokinesis (c) S phase. [3]

5 Draw diagrams to show the four stages of mitosis in an animal cell with four chromosomes. [5]

6 Which phases of mitosis are shown in Figure 8? [2]

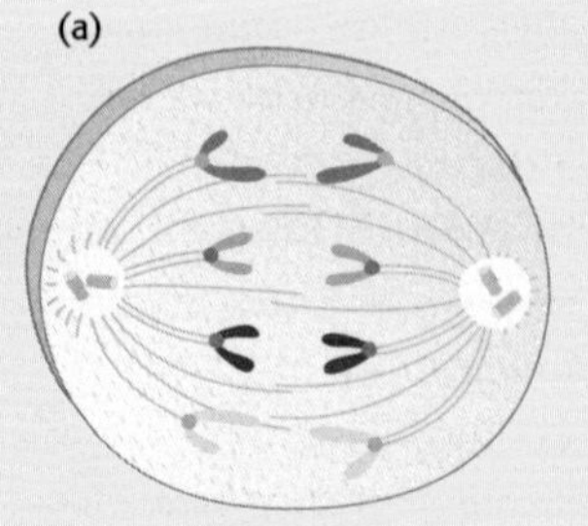

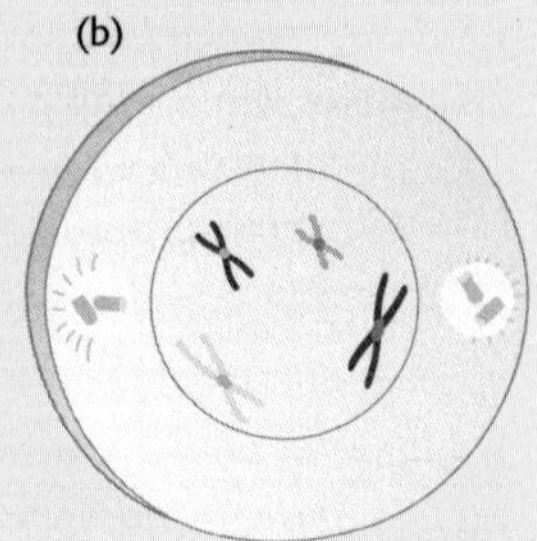

Figure 8

4 Chemicals of life

The elements

Elements are pure substances. An element is a material that is made up of just one kind of atom. Each element has its own unique chemical and physical properties. The most common elements found in living things are carbon, hydrogen, oxygen and nitrogen.

A very large range of elements serve roles in animals, plants and prokaryotes. Some examples are:

- Sulfur is a component of amino acids.
- Phosphorus is a component of nucleotides in DNA and RNA.
- Iron serves a role in oxygen transport in animal blood and is a component of cytochrome proteins involved in electron transport in plants, animals and prokaryotes.
- Sodium plays a role in nerve impulses in animals and in some halophilic bacteria it is essential for maintaining metabolic activities. Sodium is not essential in most plants, though for some, it can play a role in maintaining osmotic balance.

Two or more atoms can chemically combine to form a molecule through the formation of a covalent bond. If an atom or a molecule has an electric charge, it is referred to as an ion. Table 1 shows some important ions in biological systems.

? Obtain a copy of the periodic table. What fraction of the elements in the periodic table of elements have a role in living things?

Table 1 Some important ions in biological systems

Name	Symbol
iron	Fe^{2+}, Fe^{3+}
sodium	Na^+
chlorine	Cl^-
potassium	K^+
calcium	Ca^{2+}
hydrogen	H^+
magnesium	Mg^{2+}
carbonate	CO_3^{2-}
nitrate	NO_3^-
phosphate	PO_4^{3-}

Water

Water is a molecule formed by covalent bonds between an oxygen atom and two hydrogen atoms. The bond between hydrogen and oxygen involves unequal sharing of electrons, i.e. it is a polar covalent bond. As a result, the water molecule has an unequal charge distribution (Figure 1).

The unequal charge distribution causes hydrogen to have a partial positive charge and oxygen to have a partial negative charge. The consequence is that water molecules tend to stick together because of the formation of "hydrogen bonds".

A hydrogen bond is a type of intermolecular force that forms when a hydrogen in one polar covalent molecule is attracted to the slightly negative atom of another polar covalent molecule. The term "bond" is misleading in some ways, because:

- It is an interaction between molecules, not a "bond" within a molecule like a covalent bond.
- It does not lead to a change in the chemical properties of its constituent particles the way a covalent bond would.
- It is significantly weaker than a covalent bond or an ionic bond.

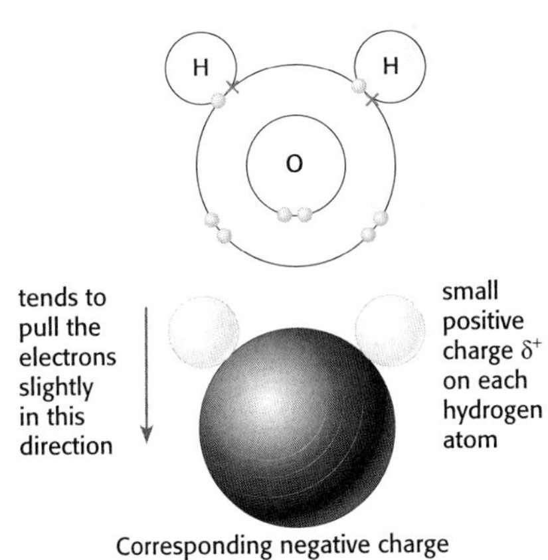

Figure 1 In a water molecule, the oxygen atom tends to have a partial negative charge and the hydrogen atom a partial positive charge.

? Explain why the hydrogen ion is sometimes referred to as a proton.

The hydrogen bond (Figure 2) gives water its unique properties and these properties are, in turn, of immense importance to living things. Cohesion refers to the binding together of two molecules of the same type, for instance two water molecules. Adhesion refers to the binding between different polar molecules, for example water binding to the mineral fragments in soil.

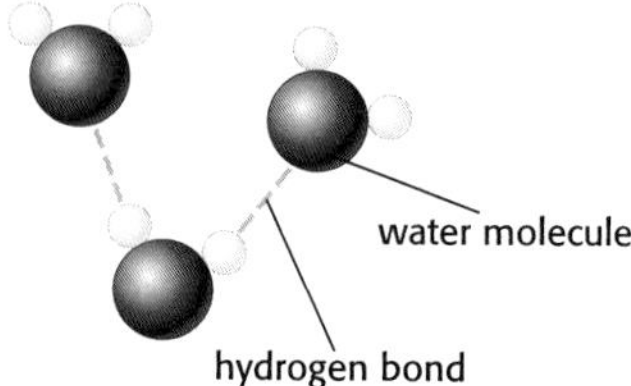

Figure 2 The dotted line indicates the presence of an intermolecular force between the molecules. This is called a hydrogen bond.

Rediscovering biology: the strength of the hydrogen bond

Despite being weaker than a covalent bond, the strength of the hydrogen bond is still considerable.

Demonstration A (Figure 3)

Take one large test tube and one small test tube, e.g. 10 cm^3 and 20 cm^3. The small test tube must just fit into the large test tube. Fill the large tube to the top with water. Immerse the small tube in water so that the outside is wet. Place the empty small tube into the large tube until it sinks and a skin of water climbs up the side of the small tube. Invert the pair while holding on to the larger tube. (you might want to test this once with a hand in place to catch the small tube in case you don't have the right combination of test tubes). If you wait patiently, you should see the small tube rise into the large tube. Explain your observations.

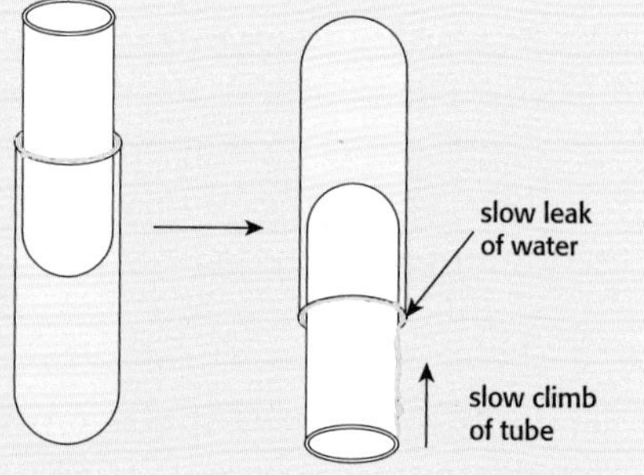

Figure 3 The set-up for demonstration A

Demonstration B

Immerse two microscope slides in water. Remove them from the water and place them together lengthways so that the bottom slide projects to form a lip. Hold the slides above a towel and begin to pile small coins on the lip. How many coins can the seal hold before it breaks? Take care – the slide may break before the seal does!

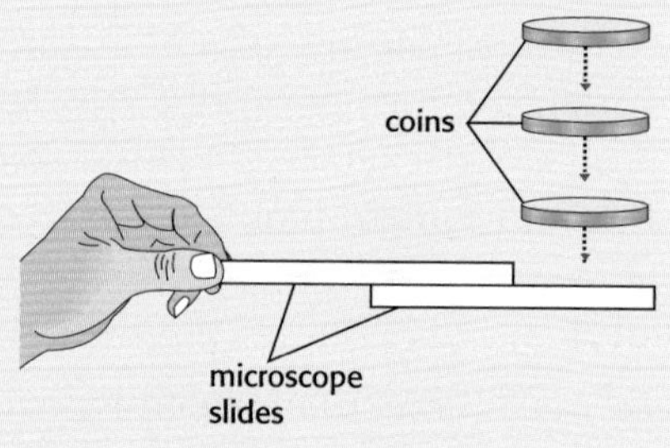

Figure 4 The set-up for demonstration B

Thinking about science: outlines and explanations

The first of the following two sections presents an *outline* of the thermal properties of water, while the second section represents an *explanation* of the thermal properties of water.

Outline: Water has a high specific heat capacity. This means that it resists changes to its temperature. Its temperature remains relatively stable in comparison to air or land. Water is most dense at 4 °C and as a result, ice floats, allowing aquatic organisms to remain alive below the ice. It is a liquid over a broad range of temperatures – from 0 °C to 100 °C which happens to be a temperature range found in most habitats on Earth. For it to vaporize, considerable quantities of heat are required. This makes it a good evaporative coolant.

"Explain" and "Outline" are command verbs, used in IB exam questions. Distinguish between an outline and an explanation.

Explanation: Water molecules interact through hydrogen bonding. This intermolecular force restricts the motion of the molecules. In order to achieve changes of state, energy must be provided to increase the motion of the molecules and to break their hydrogen bonds. For this reason, water resists changes to its temperature. This also explains why sweat dissipates so much heat as it evaporates when we are sweating. Water molecules form orderly structures involving hydrogen bonds. The most orderly structure occurs at a temperature of 4 °C. Below this temperature, the orderly array becomes distorted, resulting in a higher volume. This explains why ice can float.

The solvent properties of water

The polar nature of the water molecule means that it forms shells around charged and polar molecules, preventing them from clumping together and therefore keeping them in solution (Figure 5). Because such a broad range of materials dissolve in water, it is often referred to as the universal solvent.

Life depends on the solvent properties of water. Water is the medium in which most of the chemical reactions in organisms occur. Blood and the sap that flows through the vascular system of a plant are water-based transport media.

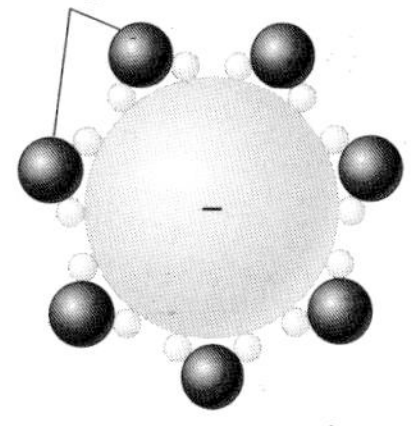

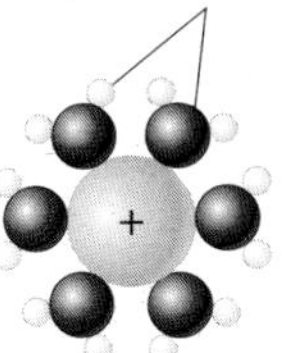

Figure 5 When table salt (NaCl) is added to water, the negative ends of the water molecules form a layer around the Na^+ ions and the positive ends of water molecules from a layer around the Cl^- ions, keeping the ions in solution

Thinking about science: homeopathy and the memory of water

In homeopathic medicine, extremely dilute solutions of chemicals are given as cures. The dilution factor is so high that it is possible that not a single molecule of the chemical remains. It is claimed that the water has a "memory" of the chemical. The following is an extract from the article "13 Things that do Not Make Sense", from *New Scientist* March 2005.

Madeleine Ennis, a pharmacologist at Queen's University, Belfast, was the scourge of homeopathy. She railed against its claims that a chemical remedy could be diluted to the point where a sample was unlikely to contain a single molecule of anything but water, and yet still have a healing effect. Until, that is, she set out to prove once and for all that homeopathy was bunkum…

Ennis describes how her team looked at the effects of ultra-dilute solutions of histamine on human white blood cells involved in inflammation. These 'basophils' release histamine when the cells are under attack. Once released, the histamine stops them releasing any more. The study, replicated in four different labs, found that homeopathic solutions – so dilute that they probably didn't contain a single histamine molecule – worked just like histamine. Ennis might not be happy with the homeopaths' claims, but she admits that an effect cannot be ruled out.

So how could it happen? Homeopaths prepare their remedies by dissolving things like charcoal, deadly nightshade or spider venom in ethanol, and then diluting this "mother tincture" in water again and again. No matter what the level of dilution, homeopaths claim, the original remedy leaves some kind of imprint on the water molecules. Thus, however dilute the solution becomes, it is still imbued with the properties of the remedy.

*You can understand why Ennis is sceptical. And it remains true that no homeopathic remedy has ever been shown to work in a large randomized, placebo-controlled clinical trial. But the Belfast study (*Inflammation Research*, Vol 53, p 181) suggests that something is going on. "We are", Ennis says in her paper, "unable to explain our findings and are reporting them to encourage others to investigate this phenomenon". If the results turn out to be real, she says, the implications are profound: we may have to rewrite physics and chemistry.*

1 Explain what is meant by the term "pseudoscience".
2 What features allow you to categorize a knowledge claim as "pseudoscientific"?
3 Is homeopathy "bunkum"? Conduct a survey to determine the scientific consensus on homeopathy.

Organic molecules

The chemistry of living things is based on the element carbon. Any molecule which contains carbon is defined as organic. There are a few exceptions: the oxides of carbon, i.e. CO, CO_2, CO_3^{2-} and HCO_3^- are not categorized as organic.

Originally, "organic" was synonymous with "living", and the term is often still used in this way. However, as has been pointed out earlier, a large number of inorganic substances are important to life. Also, many organic chemicals, such as plastics and petrol, are not found in living things.

Structural formulas show all of the atoms found in a compound and the way in which the atoms are bound to each other (Figure 6). A covalent bond is represented by a straight line. In a condensed structural formula, the bonds between hydrogen and other atoms are often omitted because their positions are easy to predict.

Macromolecules are large molecules that results from assembling long chains of repeating subunits, called monomers. For example, the nucleic acid RNA is made up of a long chain of individual nucleotides, while a polypeptide is made up of a long chain of amino acids.

(a)

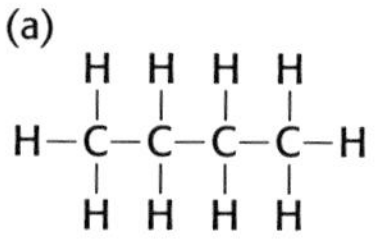

H H H H
H–C–C–C–C–H
H H H H

(b)

$CH_3–CH_2–CH_2–CH_3$

Figure 6 The organic compound butane is a gas used in cigarette lighters. (a) The structural formula for butane (b) The condensed structural formula

Carbohydrates

Carbohydrates are characterized by their composition. They are composed of carbon, hydrogen and oxygen with hydrogen and oxygen being present in the ratio of 2H : 1O; thus the name carbo-hydrate. The sub-units, or monomers, of carbohydrates are monosaccharides. Some examples of monosaccharides are ribose, fructose and glucose. Maltose is a disaccharide formed from two glucose molecules, sucrose is formed from one glucose molecule and one fructose molecule, and lactose is formed from one glucose molecule and one galactose molecule. Carbohydrates with many monomers are called polysaccharides (see Table 2).

Table 2 Some functions of common carbohydrates

Example	Type of carbohydrate	Function
glucose	monosaccharide	Energy source in animal cells.
fructose	monosaccharide	Component of flower nectar.
lactose	disaccharide	Sugar found in milk.
sucrose	disaccharide	Transport form of carbohydrates in phloem.
glycogen	polysaccharide	Carbohydrate storge in liver of animals.
cellulose	polysaccharide	Component of plant cell walls.

Rediscovering biology: investigating the structure of carbohydrates

Monosaccharides can be classified by the number of carbons they possess. For example three-carbon sugars are called trioses, five-carbon sugars are pentoses, and six-carbon sugars are hexoses. Pentoses and hexoses are unique in that they can exist in straight-chain form but they can also fold into rings.

1 Using a molecular model kit, construct a model of structure A.
2 Twist the model so that carbon 1 comes near the oxygen attached to carbon 5 as shown in structure B.
3 Break the double bond on carbon 1 and remove the hydrogen attached to the oxygen attached to carbon 5.
4 Attach carbon 1 to the oxygen on carbon 5 and reposition the detached hydrogen as shown in structure C.

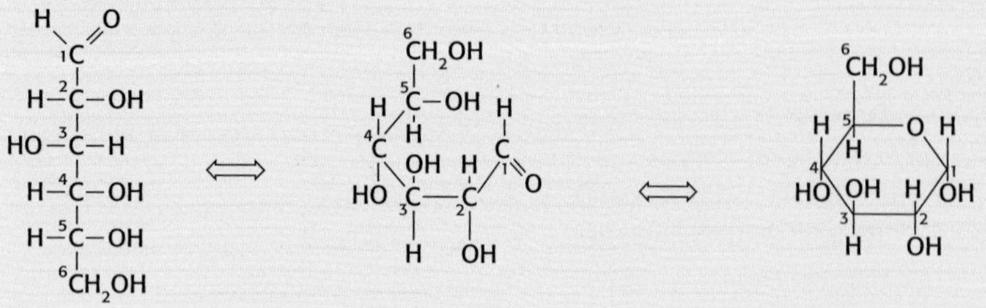

Figure 7 Converting the straight chain form of a hexose sugar into the ring form

Rediscovering biology: reducing sugars

Reducing sugars react with Benedict's reagent to produce a precipitate, often red-brown in colour. All sugars that can exist in the open form including all monosaccharides, are reducing sugars. In order for monosaccharides to form disaccharides, they must be in the ring form. In some disaccharides, one of the rings can open and so some disaccharides, such as maltose and lactose, are reducing sugars. However, in the acidic conditions of Barfoed's reagent, only monosaccharides will produce a positive result. A positive Barfoed's test appears similar to a positive Benedict's test.

1 Design a data table to record your results. You will need to make clear and descriptive observations of both the colour and amount of precipitate
2 Create a solution of 1.0 g in 50 cm^3 of all of the sugars you have available. Your teacher will also supply you with a solution of an unknown carbohydrate.
3 Fill a test tube for every test required with 3 cm^3 of Benedict's reagent. Add 1 cm^3 of the sugar solution to a tube containing the Benedict's reagent. Repeat for all sugar solutions to be tested.
4 Set the test tubes in a boiling water bath for three minutes.
5 Remove the test tubes from the boiling water bath. Determine the relative amount of precipitate formed and the colour.
6 Repeat the procedure using Barfoed's reagent in place of Benedict's reagent in a separate set of fresh test tubes. You may need to boil the tubes for considerably longer to get a proper result.
7 Using the information in your table, identify the unknown carbohydrate.
8 Which carbohydrates will be largely indistinguishable from the results of these two tests?

Working with data: molecular visualization of molecules

Many software applications allow manipulation of images of molecules. Applications such as JMol are available for free download. Image libraries are also available. For example, Protein Data Bank files can be searched for on the internet by adding ".pdb" after the name of the molecule, for example "ribose.pdb" or "maltose.pdb".

1 Use molecular visualization software to compare ribose, glucose and maltose (Figure 8).
2 Use molecular visualization software to compare cellulose to glycogen (Figure 8).

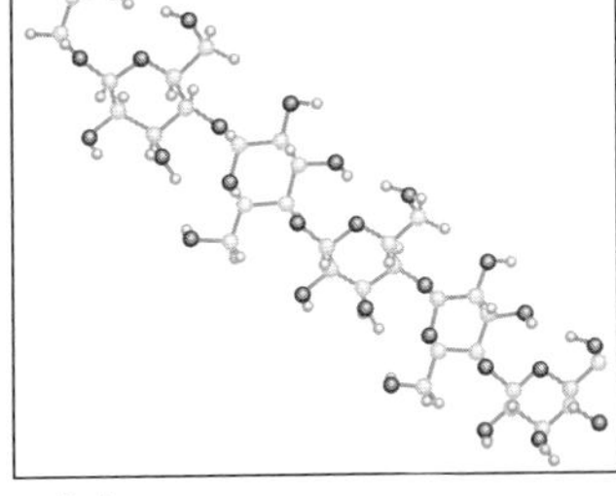

cellulose

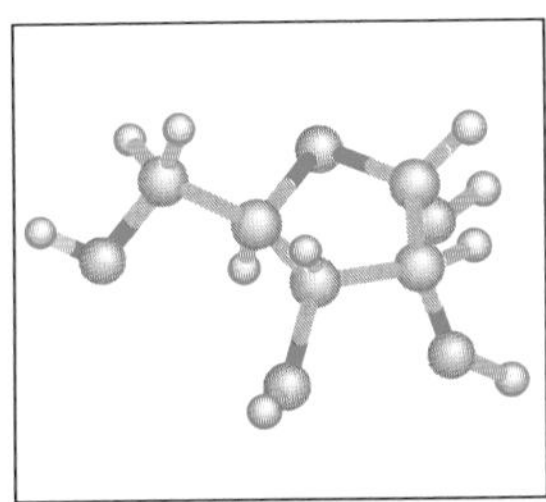

ribose

glucose

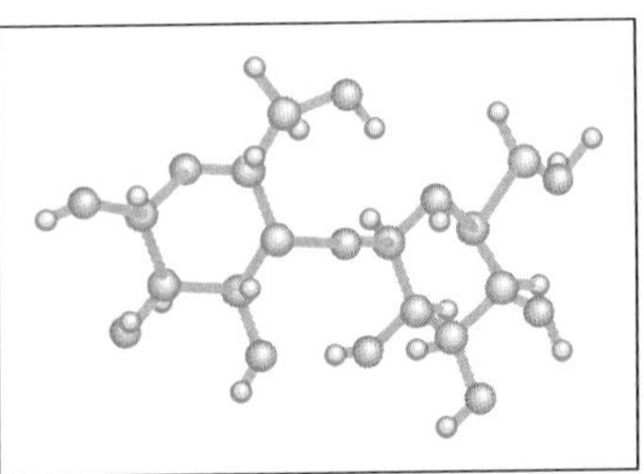

maltose

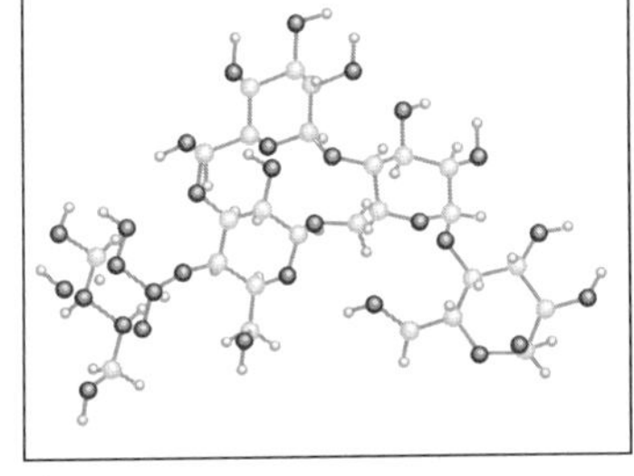

glycogen

Figure 8 Carbohydrates as viewed through molecular visualization software

Condensation and hydrolysis

Monosaccharides, amino acids and nucleotides are the subunits of polysaccharides, polypeptides and nucleic acids respectively. When these subunits combine, they do so by a process called condensation (Figure 9). This involves the loss of an –OH from one molecule and an –H from another molecule, which together form H_2O. Thus, condensation involves the combination of subunits and yields water.

To split the bonds formed by condensation, water is added in a reaction called hydrolysis. For example, a disaccharide can be split, using water, into two disaccharides.

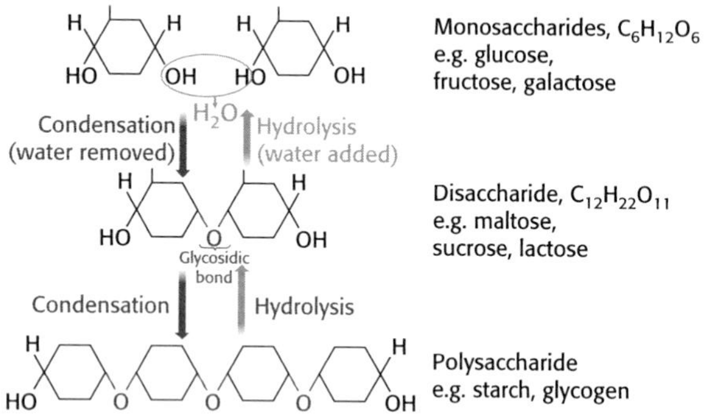

Monosaccharides, $C_6H_{12}O_6$ e.g. glucose, fructose, galactose

Condensation (water removed) H_2O Hydrolysis (water added)

Disaccharide, $C_{12}H_{22}O_{11}$ e.g. maltose, sucrose, lactose

Glycosidic bond

Condensation Hydrolysis

Polysaccharide e.g. starch, glycogen

Figure 9 Condensation and hydrolysis reactions between monosaccharides and disaccharides

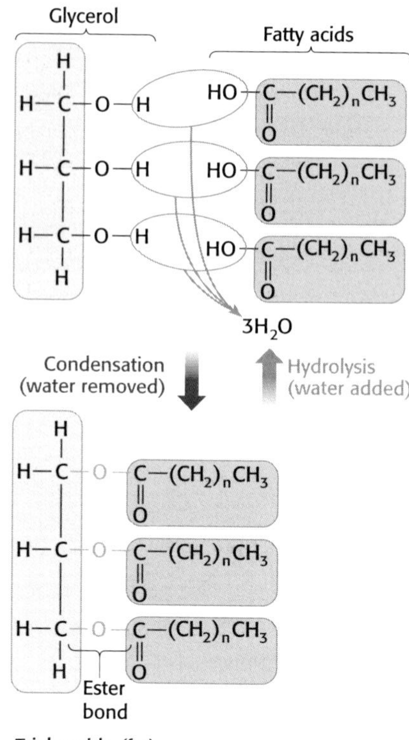

Figure 10 Formation of a triglyceride from glycerol and three fatty acids

Lipids

Lipids are a broad category of molecules that include steroids, waxes, fatty acids and triglycerides. In common language, triglycerides are fats if they are solid at room temperature or oils if they are liquid at room temperature. Three fatty acids combine with one molecule of glycerol to form a triglyceride (see Figure 10). When this happens, three water molecules are produced. This is also a condensation reaction. Note that the breakdown of triglycerides occurs by hydrolysis. Triglycerides can serve as insulation, for example in the blubber of Arctic marine mammals. They also have a role in energy storage.

Steroids act as hormones, waxes serve as hydrophobic barriers on the upper surface of leaves, and fatty acids serve as structural components of cell membranes. Figure 11 shows the structure of two fatty acids.

Lipids and carbohydrates are both used for energy storage, but there are several differences. Lipids are normally used for long-term energy storage. When an animal is starved, it is likely to use stores of glycogen before it calls on lipid stores. Lipids contain twice as much energy per gram as carbohydrates. Carbohydrates are more soluble and so are easier to transport. However, they also tend to have a stronger impact on osmotic balance.

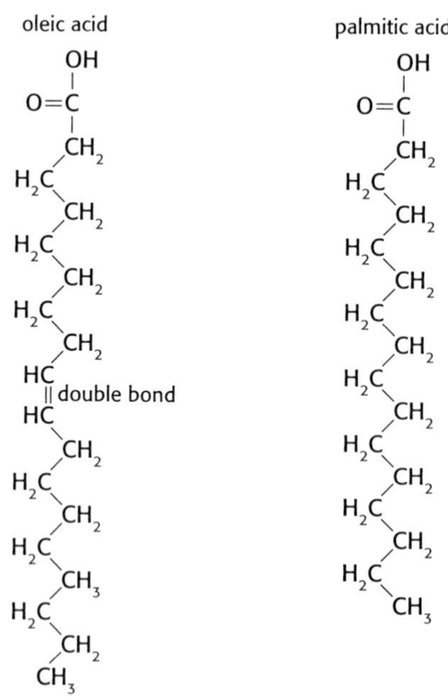

Figure 11 Oleic acid and palmitic acid are two examples of fatty acids

?

Lipase is a digestive enzyme that accelerates the breakdown of triglycerides in the small intestine. In the laboratory the rate of activity of lipase can be detected by a decline in pH. What causes the pH to decline?

Amino acids and proteins

Amino acids have the general formula shown in Figure 12. The "R" group of the amino acid is variable and can range from non-polar through polar to positively or negatively charged. Four examples of the twenty commonly occurring amino acids are shown in Figure 13.

The bond between the two amino acids in a dipeptide is called a peptide bond. Amino acids join together through condensation reactions, and separate via hydrolysis.
Figure 14 shows these reactions.

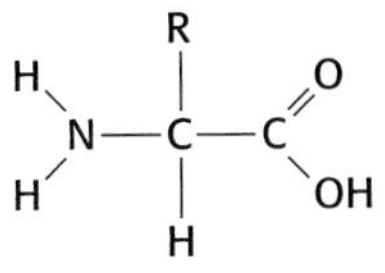

R
H–N(H)–C(H)–COOH

Figure 12 A generalized amino acid

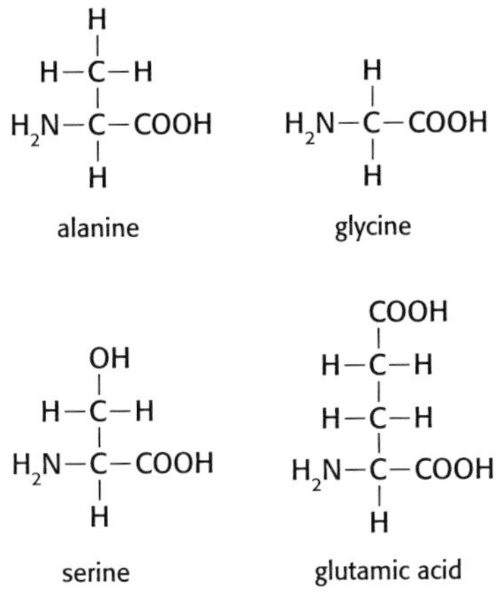

carboxyl group, amino group, amino acid, amino acid, H_2O, condensation (water removed), hydrolysis (water added), peptide bond

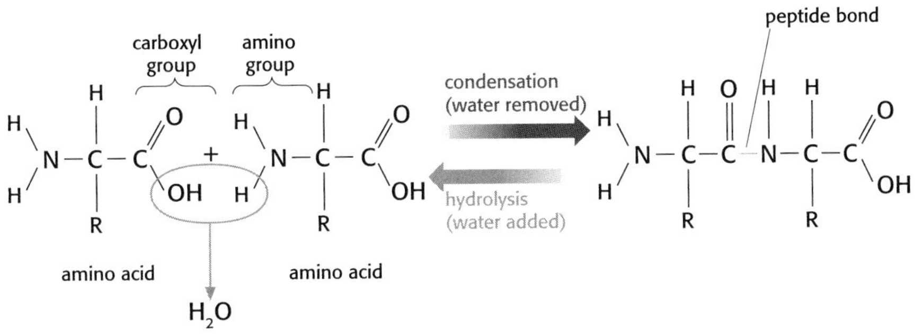

Figure 14 Condensation joins two amino acids and hydrolysis splits a dipeptide into two amino acids.

alanine, glycine, serine, glutamic acid

Figure 13 Some common amino acids

Rediscovering biology: paper chromatography of an amino acid

1. Obtain approximately 30 ml of chromatography solution from your teacher, pour it into a chromatography jar or a tall, narrow beaker (it will need to be taller than 10 cm, but not too wide, as 30 ml of solution may not cover the bottom of the beaker). Cover it and place it in the fume hood. The recipe for the solvent is 8 parts ethanol:1 part water:1 part concentrated ammonia solution.
2. Obtain a piece of rectangular filter paper approximately 10 cm x 10 cm. *It is important to avoid contaminating the paper with amino acids from your fingers, surfaces or your ruler.*
3. Draw a pencil line with a ruler approximately 15 mm from the bottom of the paper and mark four equally spaced marks on the line.
4. Your teacher will supply you with four amino acid solutions: a polar amino acid, a hydrophobic amino acid and an electrically charged amino acid. The fourth amino acid will be unknown to you but will be one of the other three.
5. Taking care to use a different capillary tube each time, spot each amino acid solution on a different mark on the paper so that a small circle of solution is visible.
6. Fold the paper so that it will fit into the beaker. Replace the cover.
7. Monitor the jar – watch how high that the solvent has climbed.
8. Remove the paper from the solvent just before the solvent front reaches the top of the paper.
9. Quickly mark with a pencil where the solvent front reached. Use a clothes peg or a paper clip to hang the paper from a string to dry.
10. When it is dry, your teacher will spray the paper with ninhydrin solution and heat it in a drying oven until purple spots marking the location of amino acids become visible.
11. Determine the R_f values for each spot using the formula:

$$R_f = \frac{\text{distance migrated by solute}}{\text{distance migrated by solvent front}}$$

12. Using R_f values, determine the identity of the unknown amino acid.

? Identify which part of each amino acid in Figure 13 represents the "R" group.

(a) alpha-helix

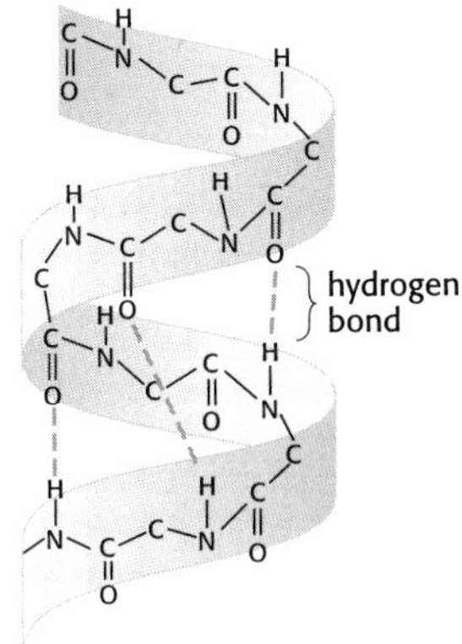

(b) beta-pleated sheet

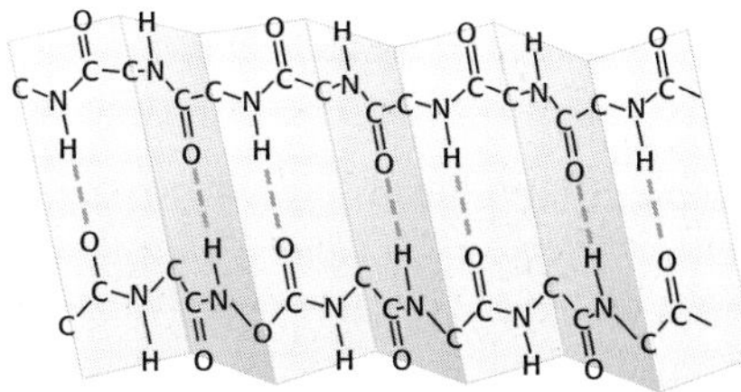

Figure 15 Two examples of protein secondary structure (see page 48)

HL

Protein structure

A chain of amino acids is called a polypeptide. Given that the 20 commonly occurring amino acids can be combined in any sequence, it should not be surprising that there is a huge diversity of proteins.

The sequence of amino acids in a polypeptide is termed its **primary structure**.

Because the chain of amino acids in a polypeptide has polar covalent bonds within its backbone, it tends to fold in such a way that hydrogen bonds form between the carboxyl (C═O) group of one residue and the amino group (N—H) group of an amino acid in another part of the chain. This results in the formation of patterns within the polypeptide called **secondary structures**. The α-helix and the β-pleated sheet are examples of secondary structures (see Figure 15 on page 47).

Tertiary structure refers to the overall three-dimensional shape of the protein (Figure 16). This shape is a consequence of the interaction of R-groups with one another and with the surrounding water medium. There are several different types of interaction.

- Positively charged R-groups will interact with negatively charged R-groups.
- Hydrophobic amino acids will orientate themselves toward the centre of the polypeptide to avoid contact with water, while hydrophilic amino acids will orientate themselves outward.
- Polar R-groups will form hydrogen bonds with other polar R-groups.
- The R-group of the amino acid cysteine can form a covalent bond with the R-group of another cysteine forming what is called a disulfide bridge.

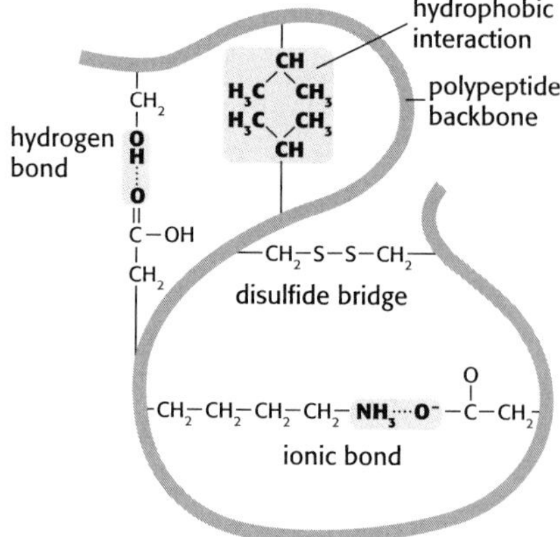

Figure 16 R-group interactions contribute to tertiary structure.

Proteins can be formed from a single polypeptide chain or they can be formed from more than one polypeptide chain. **Lysozyme** is composed of a single chain. So lysozyme is both a polypeptide and a protein. **Insulin** is formed from two polypeptides, and hemoglobin is made up of four chains. **Quaternary structure** refers to the way polypeptides fit together when there is more than one chain.

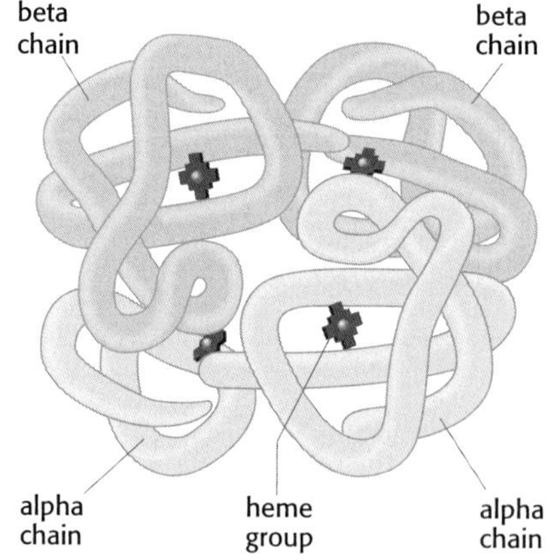

Figure 17 Hemoglobin

The quaternary structure of hemoglobin in adults consists of four chains: two α–chains and two β–chains (Figure 17). Each subunit contains a molecule called a heme group.

The biological activity of a molecule is related to its structure. Certain treatments such as exposure to high temperatures, or changes in pH can cause alterations in the structure of a molecule and therefore disrupt its biological activity. When a protein has permanently lost its structure it is said to be denatured.

Molecular visualization software often contains a function where the different types of secondary structure can be visualized. Figure 18 is an image of insulin showing three areas where the α-helix can be seen. It also shows the quaternary structure of insulin, i.e. the relative positions of the two polypeptides.

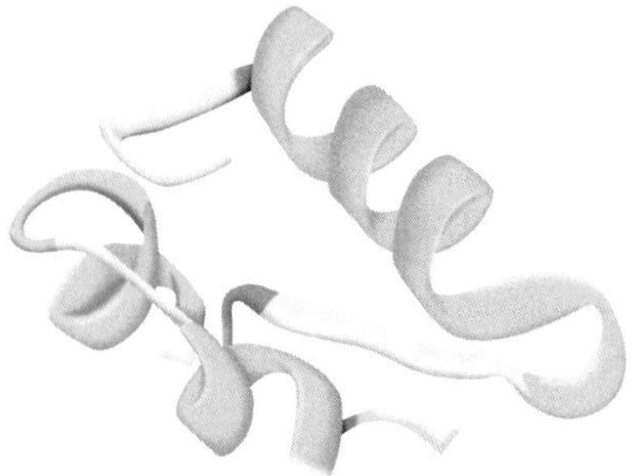

Figure 18 The structure of insulin

Rediscovering biology: modelling tertiary structure

Obtain 5 different colours of plasticine. You can use it to model primary and tertiary structure. The colours you use may be different.

1 Form 11 blue plasticine spheres. These represent polar amino acids. An example of a polar amino acid is serine.
2 Make 1 yellow plasticine sphere. This represents a positively charged amino acid. An example of a positively charged amino acid is lysine.
3 Make 1 green plasticine sphere. This represents a negatively charged amino acid. An example of a negatively charged amino acid is aspartic acid.
4 Make 2 pink plasticine spheres. These represent cysteine.
5 Make 2 red plasticine spheres. These represent non-polar amino acids. An example of a non-polar amino acid is valine
6 Connect the spheres in the following sequence:

blue-blue-blue-blue-blue-pink-red-red-yellow-blue-blue-green-blue-blue-pink-blue

(a) Assuming the spheres represent the examples of amino acids given in steps 1 through 5, state the primary structure of your polypeptide.
(b) Assume the polypeptide was placed into water. It would fold according to the following rules:
(i) polar groups on the outside
(ii) non-polar groups on the inside
(iii) positive near negative
(iv) cysteine beside the other cysteine.
Try to fold the polypeptide in a way that complies with these rules. (Hint: it should have an S-shape.)

Note that the folding rules (i–iv) set limits to the stability of certain proteins, so there are more possible primary sequences than there are polypeptides occuring in nature.

Fibrous and globular proteins

Protein shapes can be broadly categorized into two categories: fibrous and globular. Fibrous proteins have elongated shapes, tend to be insoluble in water and physically tough. Examples of fibrous proteins include collagen in skin and keratin found in hair and fingernails. Globular proteins tend to be compact and rounded and they are usually water-soluble. Enzymes and other functional proteins such as antibodies are globular proteins.

Proteins have many functions. Some are shown in Table 3, but the list is not exhaustive.

Investigating biology: albumin and gelatin

Egg albumin is a globular protein and gelatin is a fibrous protein. Fibrous and globular proteins differ in their 3-D structure. Make a solution of each. Do the solutions differ in their response to denaturing agents? How do they respond when heated? Compare how their viscosity changes as temperature changes. What is the effect of adding acid to each protein?

Table 3 The shapes of some common proteins

Example	Function	Shape
Sucrase	Enzyme that breaks down sucrose into glucose and fructose	globular
Elastin	Component of connective tissue that helps skin recover its shape	fibrous
Insulin	Hormone involved in the regulation of blood sugar	globular
Immunoglobulins	Involved in the defence against disease	globular
Na^+/K^+ pump	A transport protein in the membrane of nerve cells	globular
Collagen	Protein strengthening tissues	fibrous

Rediscovering biology: measuring the protein content in milk

1 Casein is a group of proteins that are found in milk. Obtain 125 ml of milk and 125 ml of soya milk. Gather information from the nutrition panels about the protein content in each type of milk.
2 Warm the milk to room temperature. Determine the mass of the milk.
3 Add 10 ml of dilute acid and stir continuously for about 90 s.
4 Allow the milk to settle for about 5 minutes so that the casein can precipitate out of the milk.
5 Filter the precipitate through cheese cloth. Carefully run water through the casein to rinse off the acid.
6 Squeeze the solid to remove as much liquid as possible and then lay the cheesecloth and casein on to an absorbent paper to dry.
7 Determine the mass of the soya milk.
8 Heat the soya milk until it begins to boil. Remove it from the heat and stir in 2 g of solid magnesium sulfate.
9 Wait for the solution to sufficiently cool.
10 Filter the precipitated protein through cheese cloth. Run water through the protein as a control.
11 Squeeze the solid to remove as much liquid as possible and then lay the cheese cloth and protein on paper to dry.
12 Compare the protein yields from the two types of milk. Draw a conclusion. Discuss limitations and possible modifications to this procedure.

Data-based question: function of the protein metallothionein

The level of blood urea nitrogen (BUN) is a measure of kidney function. If BUN rises, then kidney function is impaired. At high levels of exposure, mercury can impair kidney function.

Two groups of mice were exposed to increasing concentrations of mercury chloride ($HgCl_2$). Mice that lacked the ability to produce the protein metallothionein were compared to mice that possessed the wild-type genetic make-up; that is, the genetic make-up most commonly seen in mice. The results are shown in Figure 19.

1 Outline the relationship between the concentration of $HgCl_2$ and BUN in wild-type mice. [2]
2 (a) Using the data in the bar chart, deduce the effect of metallothionein in the wild-type mice. [4]
(b) Suggest a way that the protein could have these effects. [1]
3 Discuss the ethics of conducting this experiment. Is any additional information required to make a judgement? [3]

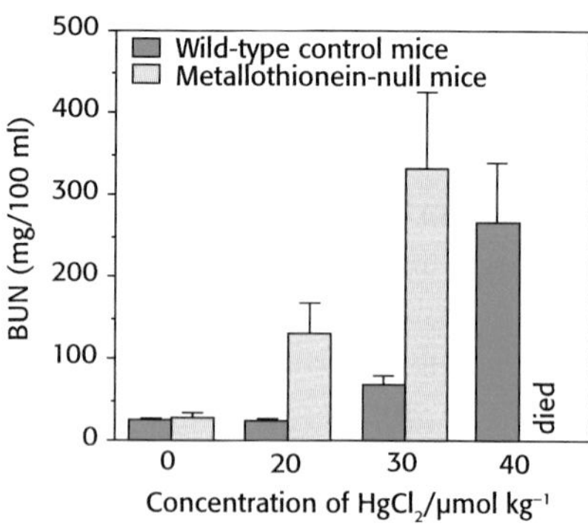

Figure 19 The effect of exposure to $HgCl_2$ on blood urea nitrogen

Investigating biology: the breaking force of hair

Hair is composed largely of keratin, a fibrous protein. Fibrous proteins are often characterized by strength. However, chemical or physical alteration of the hair protein can affect its strength.

The following procedure can be used to assess the breaking force of hair.

Materials

- digital force metre
- computer
- hair strands of fixed lengths

The set-up for the experiment is shown in Figure 20.

Procedure

1. Calibrate the sensor so that when the hook on the sensor hangs down, it reads zero with nothing hanging on it.
2. Determine a method of suspending the hair from the sensor. One method would be to loop it around the sensor hook and pull on both ends of the hair at once. Another would be to wind it around the hook several times.
3. Set data collection to be time-based, with a maximum force of 10 N.
4. Begin collecting data by slowly and gently pulling on the hair until it snaps. Determine the maximum force by using the statistics function of the software or by reading through the data to find the highest value.
5. Possible independent variables include comparing wet and dry; comparing hair/fur across species; chemically treated versus not chemically treated hair.
6. Within one person, there will be significant variation, all else being controlled. The "sufficient relevant data" aspect will be important here.
7. You might consider using the t-test described on page 52 for determining whether differences in mean breaking force are statistically significant.

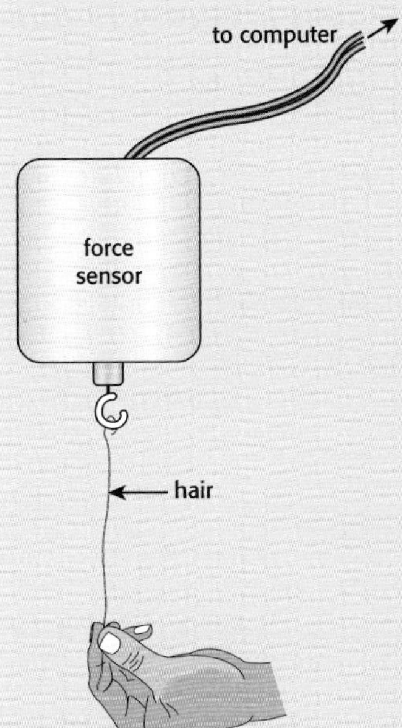

Figure 20 Apparatus for measuring the breaking force of hair

Table 4 Breaking force of 50 samples each of non-blow-dried hair (NBD) and blow-dried hair (BD)

NBD hair breaking force /N ± .01 N			BD hair breaking force /N ± .01 N		
1.794	0.854	1.779	0.634	2.434	1.104
1.027	1.784	2.503	1.811	1.403	1.227
3.472	2.787	1.166	2.142	1.84	0.856
2.884	2.454	1.196	1.967	1.932	0.859
2.111	1.963	3.213	0.155	1.994	0.061
1.956	1.819	2.872	2.024	1.626	0.031
1.413	2.607	2.616	1.563	1.932	1.072
1.782	2.055	3.421	2.076	1.714	2.001
2.186	1.411	3.011	2.065	0.061	1.721
1.738	2.362	2.704	3.881	2.454	1.231
1.765	2.276	1.986	2.399	2.856	2.172
1.897	2.546	2.721	2.481	2.515	1.982
2.261	2.669	3.002	2.434	2.853	2.21
1.309	2.393	2.947	2.445	1.012	1.70
1.165	1.334		2.787	1.258	
1.534	1.963		1.188	1.012	
1.251	1.503		2.474	1.258	
0.854	1.626		2.007	0.031	
$\bar{x}$=2.10 N s = 0.66 N			$\bar{x}$=1.70 N s = 0.84 N		

Working with data: using the Ti-83 graphic calculator to perform the t-test

Sample results for a comparison between breaking force of blow-dried and non-blow-dried hair are shown in Table 4 (see page 51). The mean breaking force appears to be different, but can we conclude from this data that the treatment is really different or is different due to random variation within the two samples of hair? Biologists often use a t-test to determine whether *differences in means* are significant or just due to chance. The use of the *t*-test is discussed more fully in the chapter covering the extended essay.

To perform the test using the graphic calculator, first you need to enter your data. The instructions that follow are for the Texas Instruments Ti-83 calculator.

1 Press the **STAT** button. The **Edit** menu is highlighted by default. Press **ENTER**.
2 Now specify which list you want to enter the data into by pressing the button "2nd" (top left corner) and then $\mathbf{L_1}$ (above the number 1 key).
3 Enter the data.
4 Enter the second set of data into $\mathbf{L_2}$ following the same procedure.

To perform the t-test:

5 Push the **STAT** button.
6 Move the cursor to the **TESTS** menu.
7 Press 4; the default should be **Data** (it will be flashing).* Pooled should be highlighted as **Yes**.
8 Move the cursor down to the last row off the screen until you see the flashing **Calculate**.
9 Press **ENTER**. A ***t*** value of 2.65 with a (p) value of 0.009 shows that the difference in means is indeed statistically significant.

*An alternative at stage (g) is to enter the statistics (STATS) if they are known. Table 4 tells you the values of $\bar{x}$ = and s for both data sets, so these could be entered instead of using the data in the lists.

The standard deviation for the non-blow-dried hair is higher than the standard deviation for the blow-dried hair.

1 Explain the concept of standard deviation. [2]
2 Discuss the observed differences in standard deviation. [1]

Chapter 4 questions

1 Distinguish between the following pairs of words.
   (a) protein and polypeptide [1]
   (b) fat and oil [1]
   (c) starch and glycogen [1]
   (d) condensation and hydrolysis [1]
   (e) hydrophobic and hydrophilic. [1]

2 Write a word equation for each of the following.
   (a) hydrolysis of maltose [1]
   (b) the condensation reaction that forms a triglyceride [1]
   (c) hydrolysis of starch to remove a single molecule. [1]

3 Explain the importance of the transparency of water to life. [3]

4 Hemoglobin is a protein composed of two pairs of globin subunits. During the process of development from conception through to 6 months after birth, human hemoglobin changes in composition. Adult hemoglobin consists of two alpha- and two beta-globin subunits. Four other polypeptides are found during development: zeta, delta, epsilon and gamma.

Figure 21 illustrates the changes in hemoglobin composition during gestation and after birth in a human.

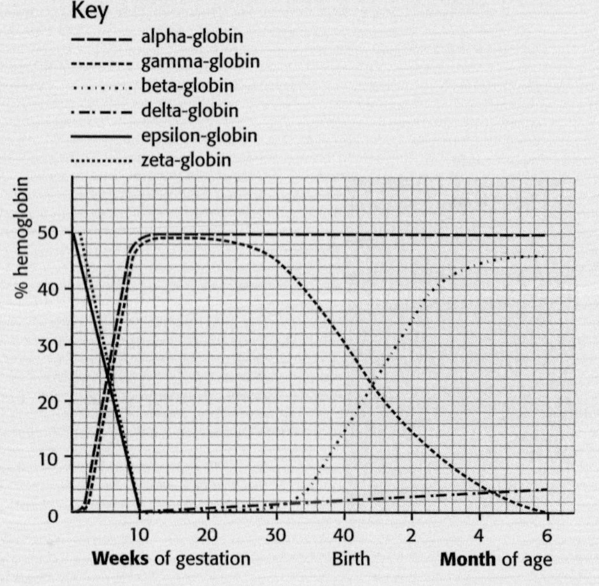

Figure 21

(a) State which two subunits are present in highest amounts early in gestation. [1]
(b) Compare changes in the amount of the gamma-globin gene with beta-globin. [3]
(c) Determine the composition of the hemoglobin at 10 weeks of gestation and at 6 months of age. [2]
(d) State the source of oxygen for the fetus. [1]
(e) The different types of hemoglobin have different affinities for oxygen. Suggest reasons for the changes in hemoglobin type during development and after birth. [3]

5 During the Antarctic winter female Emperor penguins live and feed at sea, but males have to stay on the ice to incubate the single egg the female has laid. Throughout this time the males eat no food. After 16 weeks the eggs hatch and the females return. While the males are insulating the eggs they stand in tightly packed groups of about 3000 birds. To investigate the reasons for standing in groups, 10 male birds were taken from a colony at Pointe Geologie in Antarctica. They had already survived 4 weeks without food. They were kept for 14 more weeks without food in fenced enclosures where they could not form groups. All other conditions were kept the same as in the wild colony. The mean air temperature was –16.4 °C. The composition of the captive and the wild birds' bodies was measured before and after the 14-week period of the experiment. The results in kilograms are shown in Figure 22.

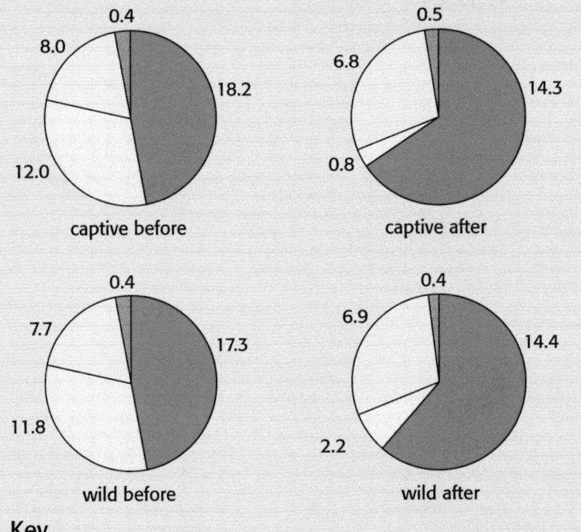

Figure 22

(a) Calculate the total mass loss for each group of birds. [2]
(i) wild
(ii) captive
(b) Compare the changes in lipid content of the captive birds with those of the birds living free in the colony. [2]
(c) Besides being used as an energy source, state another function of lipid which might be important for penguin survival. [1]

Higher-level questions

6 (a) State **one** type of secondary structure of a protein. [1]
(b) Outline the differences between globular and fibrous proteins, giving a named example of each. [3]
(c) Explain the significance of polar amino acids for membrane proteins. [2]
(d) Distinguish between a secondary and tertiary structure using examples. [4]

7 The protein content of harvested wheat grain depends on the water content in the soil at the time of sowing. In an experiment carried out in semi-arid soil in Queensland, Australia, over several years, researchers measured the protein obtained from wheat sown in different soil conditions.

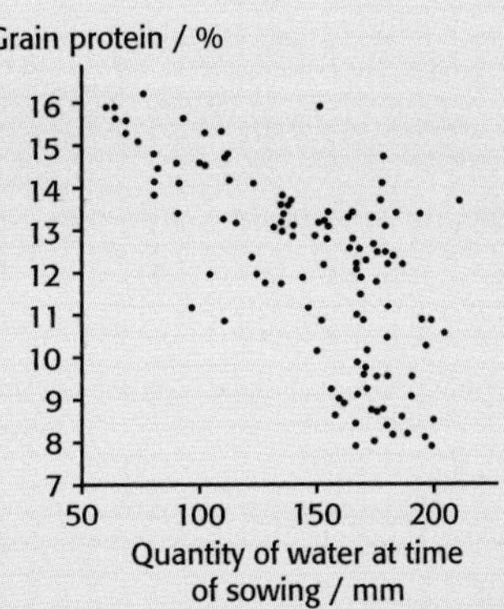

Figure 23

(a) Outline the relationship between water content at planting time and protein content. [1]
(b) Suggest why this relationship exists. [3]

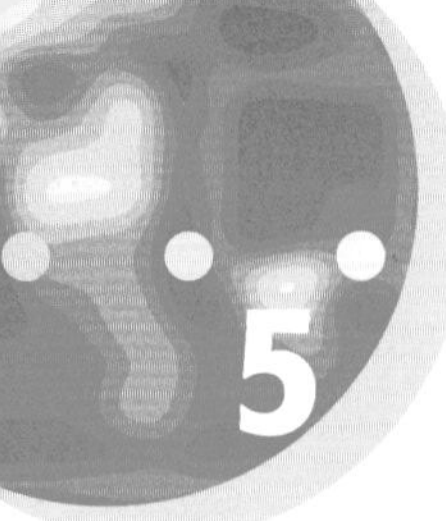

DNA structure and replication

Nucleotides and nucleic acids

DNA is a polymer of nucleotides. Nucleotides are molecules that consist of three parts: a sugar, a phosphate group and a nitrogen-containing ring structure called a nitrogenous base. Figure 1 shows these parts. A polymer of nucleotides is called a nucleic acid. In the polymer a covalent bond links the phosphate of one nucleotide to the sugar of the next nucleotide. One example of a nucleic acid is ribonucleic acid or RNA. Another example is deoxyribonucleic acid, or DNA. The sugar within DNA is deoxyribose and the sugar in RNA is ribose (Figure 2).

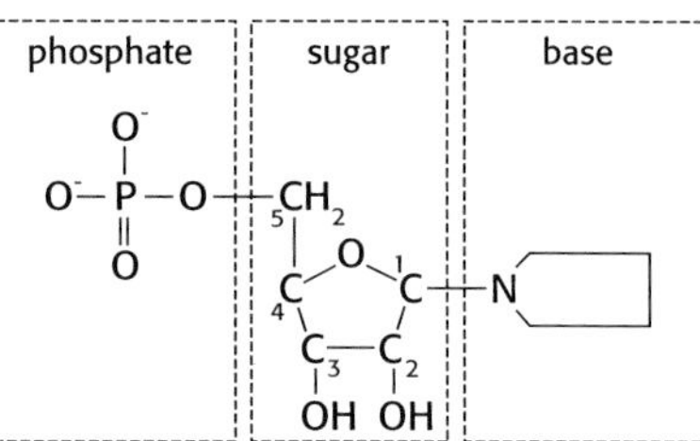

Figure 1 The parts of a nucleotide

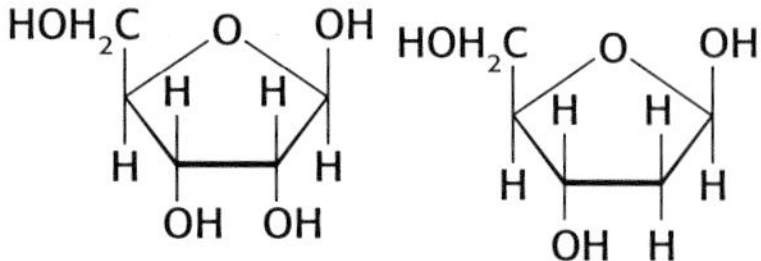

Figure 2 The sugar on the left is ribose and the sugar on the right is deoxyribose.

HL

Nucleotides

The carbons of the sugar molecule are assigned numbers as shown in Figure 1. These numbers can be used to indicate points of attachment. The nitrogenous bases in DNA are of two types: the **purines** adenine and guanine and the **pyrimidines** cytosine and thymine. Uracil is a pyrimidine found in RNA.

Data-based question: Chargaff's data

DNA samples from multiple species were analysed in terms of the their nucleotide composition by Erwin Chargaff, an Austrian biochemist. The data is presented in Table 1 in the form of ratios.

Table 1

Source	Adenine to guanine	Thymine to cytosine	Adenine to thymine	Guanine to cytosine	Purines to pyrimidines
Ox	1.29	1.43	1.04	1.00	1.1
Human	1.56	1.75	1.00	1.00	1.0
Hen	1.45	1.29	1.06	0.91	0.99
Salmon	1.43	1.43	1.02	1.02	1.02
Wheat	1.22	1.18	1.00	0.97	0.99
Yeast	1.67	1.92	1.03	1.20	1.0
Hemophilus influenzae	1.74	1.54	1.07	0.91	1.0
E. coli K2	1.05	0.95	1.09	0.99	1.0
Avian tubercle bacillus	0.4	0.4	1.09	1.08	1.1
Seratia marcescens	0.7	0.70	0.95	0.86	0.9
Bacillus schatz	0.7	0.6	1.12	0.89	1.0

Source: After E. Chargaff *et al., J. Biol Chem.* 177 (1949).

1 Compare the ratios in humans with those in wheat. [2]
2 Evaluate the following claims, based on the data:
   (a) the ratio of guanine to cytosine is 1:1 in all species [2]
   (b) the ratio of pyrimdines to purines is 1:1 [2]
   (c) within experimental error, the ratio of adenine to thymine is 1:1. [2]

Thinking about science: co-operation in science

When X-rays pass through a strand of DNA they bend, or diffract, like water waves passing through a space between two rocks. The famed X-ray crystallograph shown in Figure 3 was produced by Rosalind Franklin's working group at King's College, London. It took great skill and tenacity to achieve such a clear image. It provided crucial evidence that DNA was a double helix.

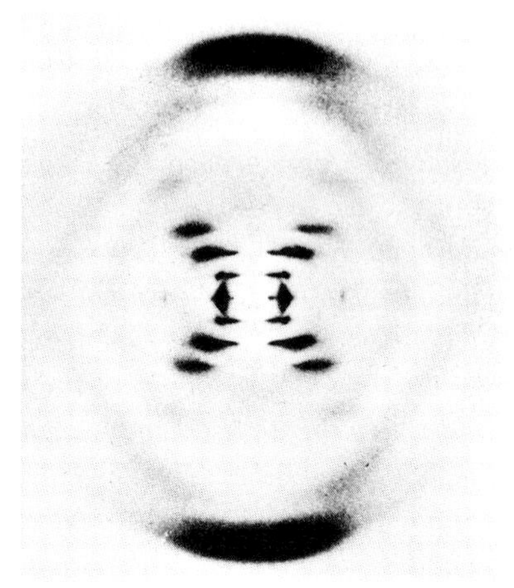

Figure 3 Rosalind Franklin's X-ray diffraction photograph of DNA

James Watson's account of the elucidation of the structure of DNA, in his book *The Double Helix,* describes a competition between research groups at King's College, London, where Franklin and her colleague Maurice Wilkins worked; the Cavendish laboratory at Cambridge University where Watson and his colleague Francis Crick worked; and another group at the California Institute of Technology. Ultimately James Watson, Frances Crick and Maurice Wilkins shared the Nobel Prize for the discovery.

Much has been written about the fact that Watson and Crick used data obtained by Rosalind Franklin at King's College, as well as further research information found in a report by King's researchers, without her permission. The following four sources are relevant to a consideration of the interaction of these researchers.

Source A:

Excerpt from the speech to the Association of Los Alamos Scientists by J. Robert Oppenheimer Los Alamos, NM November 2, 1945.

It is not possible to be a scientist unless you believe... it is of the highest value to share your knowledge, to share it with anyone who is interested.

Source B:

Hubbard, Ruth, 2003 "Science, Power, Gender: How DNA Became the Book of Life", ***Journal of Women in Culture and Society,*** **vol. 28, no. 3. University of Chicago Press.**

Rosalind Franklin accepted a fellowship in the biophysics unit at King's College, London, directed by Professor John Randall. The unit was working on the structure of DNA, and Randall asked Franklin to build a high-resolution camera with which to make more detailed measurements of the X-ray diffraction patterns of DNA than had previously been possible. Also, intentionally or not, Randall put Franklin into a highly ambiguous situation by leading Maurice Wilkins, the unit's assistant director, to believe that Franklin and he would be working on DNA together, while telling Franklin she would be doing the X-ray diffraction studies on her own. When personality conflicts began to develop between Franklin and Wilkins, she decided they would not be able to work together. And before long, she obtained the sharpest X-ray diffraction image of DNA in existence that clearly showed that DNA can form a helix. Unbenownst to Franklin, two crucial things happened. One was that Wilkins showed Watson Franklin's best X-ray diffraction image, which clearly indicated that DNA forms a helix. The other was that Max Perutz, a senior researcher at Cambridge, received a research report the King's group had submitted to their funders. Knowing of Watson's and Crick's interest in DNA, he showed them the report, which included the conclusions Franklin had drawn on the basis of her X-ray image, conclusions that specified all the critical dimensions of the DNA helix.

At just about this time, Watson and Crick found out that the famous U.S. chemist Linus Pauling was about to propose a transparently incorrect structure for DNA. With that, they decided they no longer needed to consider DNA the property of King's. Armed with Franklin's calculations against which to check possible models, they went into a frenzy of model building and, within about six weeks, came up with the now-famous double helix. The first time the group at King's realized that Watson and Crick had gone back to working on DNA was when Wilkins received in the mail a copy of the note Watson and Crick were submitting to Nature. *He promptly decided to write an accompanying note with his coworkers Stokes and Wilson, and so did Franklin with her coworker Gosling. Franklin framed her note as though her data confirmed the Watson-Crick structure, since she had no idea that those data had been in their hands while they puzzled out the structure. And she never realized it, because five years later, in 1958, at thirty-seven years old, she died of cancer.*

Source C:

Extract from Fuller, Watson, 21 August 2003 "Who said 'helix?' Right and wrong in the story of how the structure of DNA was discovered", *Nature* Vol. 424.

Crucial for the success of Watson and Crick's approach had been the use of molecular models.

Franklin's insistence on direct analysis of the X-ray data …and her dismissal of model-building, so successful in the discovery of the α-helix [protein structure], were serious errors of judgment. Crick especially had a strong commitment to model-building. It is important to emphasize, in view of the attention that has been given to the exploitation of King's data by the Cambridge workers, that before the discovery of the double helix, Crick encouraged Wilkins and Franklin to build models and provided them with jigs developed in Cambridge for constructing the atomic components.

The history of the discovery of DNA is too often presented in popular accounts in terms of results 'stolen' by Watson and Crick with Franklin as the victim. Yet in the complex interactions in and between the two laboratories, it is not sustainable to view Franklin merely as a victim of other people's actions.

It was to the sad detriment not only of herself but also of the King's laboratory as a whole that Franklin chose to work in isolation on a problem, the solution of which depended on confluent results from several workers using different techniques. This is particularly a matter for regret because the experimental work that Franklin performed at King's was of the highest quality. Franklin's approach contrasted markedly with that of Wilkins, who made his results freely known.

Source D:

Letter from Wilkins to Crick after he had visited Cambridge to see the Watson and Crick model of the double-helix. Quoted in "Who said 'helix?'…" (see Source C).

Dear Francis: I think you're a couple of old rogues, but you may well have something. I like the idea…. I was a bit peeved because I was convinced that the 1:1 ratio was significant and had a 4 planar group sketched and was going to look into it…. I might, given a little time, have got it. But there is no good grousing – I think it's a very exciting notion and who the hell got it isn't what matters.

1 IB students must undertake an interdisciplinary collaborative project with students studying other sciences. The group project assessment criteria defines behaviors that a student should display when working as a scientist:
- self-direction
- perseverance
- collaboration
- communication
- open-mindedness
- reflectiveness.

(a) Are all of these criteria equally important to science?
(b) Discuss the relevance of these criteria to the discovery of the structure of DNA.
(c) Which of these criteria are consistent with features Oppenheimer is claiming are essential for science?

2 Accounts differ, but evidence suggests that without the flow of information from Kings to Cambridge, Franklin's contribution would have achieved earlier prominence. To what extent do you agree with Wilkin's statement, "…who the hell got it isn't what matters"?

DNA structure

Figure 4 shows a simple diagram of the structure of DNA. Note the following features in the diagram.

- The nucleotides within each strand are joined by covalent bonds.
- Alternating sugar and phosphate molecules form a backbone for each strand.
- The two strands are held together by hydrogen bonds between the nitrogenous bases.
- The two strands are arranged in an anti-parallel fashion.
- There are four different nitrogenous bases: adenine (A), thymine (T), guanine (G) and cytosine (C).
- Adenine (A) is always paired with thymine (T) and guanine (G) is always found paired with cytosine (C).

The two strands of DNA are said to be complementary, i.e. the sequence on one strand determines the sequence on the other strand. It is this feature of DNA that makes it capable of self replication and allows it to serve as a guide for the production of RNA.

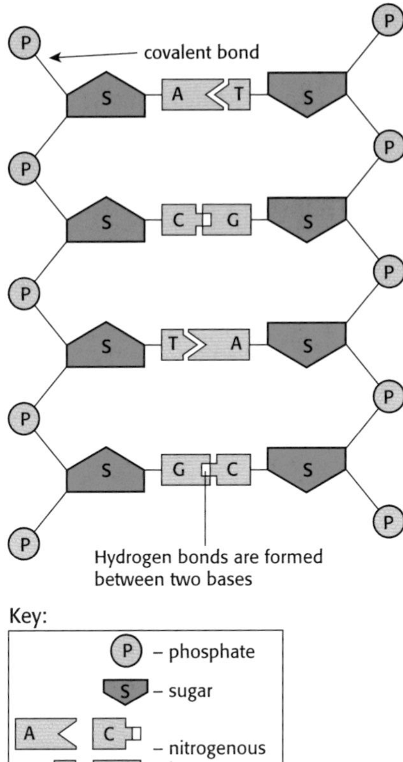

Figure 4 Simplified diagram of DNA

HL Figure 5 provides further details regarding DNA structure.

- One strand is oriented in the direction 5′ to 3′ and the other is oriented in the direction 3′ to 5′.
- It is composed of two polymers each coiled around the other in a structure called a double helix.
- A pyrimidine is always opposite a purine.
- There are three hydrogen bonds between G and C whereas there are only two between A and T.

Central dogma of genetics

Reproduction, even at the cellular level, requires the transmission of information from parent to offspring. This transmission is termed heredity. A gene is a finite or discrete unit of heredity that is coded for by a sequence of DNA bases.

The function of most genes is to specify the sequence of amino acids in a particular polypeptide. A protein is composed of one or more polypeptides. Proteins often directly or indirectly determine the observable characteristics of an individual.

The DNA sequence within a gene can be used to create an RNA molecule in a process called transcription. These RNA molecules are in turn decoded into the amino acid sequence of a polypeptide through the process of translation. Figure 6 shows an RNA strand beside one of the strands of a DNA molecule.

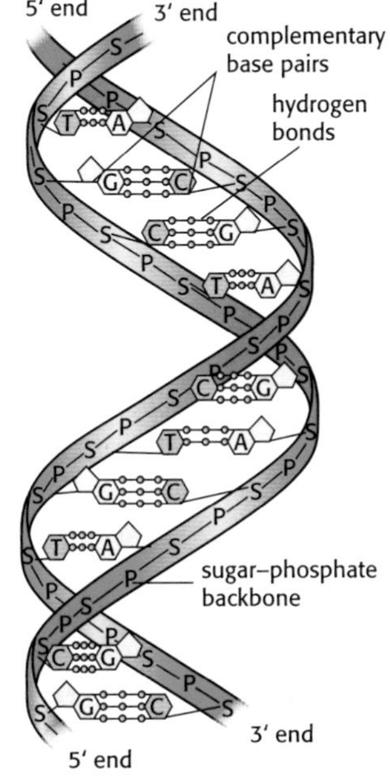

Figure 5 The two strands of DNA form a double helix.

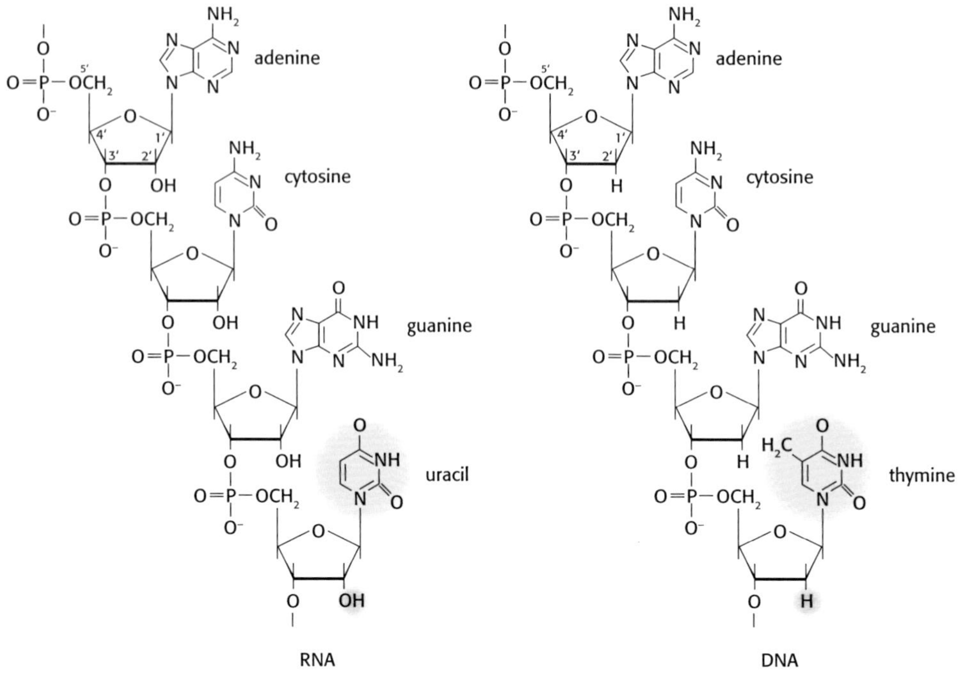

RNA

DNA

Figure 6 Structure of DNA and RNA strands

Rediscovering biology: isolating DNA

Materials

- Cold ethanol
- Detergent
- Beaker
- Salt (sodium chloride)
- Distilled water
- Glass stirring rod

1. Put 1 cm^3 of an 8% salt solution into a test tube.
2. Put about 10 cm^3 of water into a cup and swirl this water in your mouth vigorously for about 30 seconds. The swirling should be vigorous enough to remove cells from the lining of your cheeks. Spit out the solution back into the cup and then transfer the solution into the test tube. You should have a cloudy solution.
3. Add 1 cm^3 of detergent in to the test tube.
4. Stopper the tube and gently invert it a few times until the solution clears.
   Optional: add a drop of protease such as proteinase K to disrupt histone binding and increase yields.
5. Obtain a test tube of cold alcohol. Add about 5 cm^3 to the cell extract by gently pouring it down the side of the test tube.
6. Place the test tube with the alcohol and cells extract in a beaker or test tube holder. Let it sit undisturbed for about four minutes. Do not shake. The white material coming out of solution as a precipitate is DNA.
7. Dip the glass rod into the tube, slowly rotating it to spool out DNA.

1 Suggest the function of the following ingredients in the experiment:
(a) the detergent
(b) the salt solution
(c) the alcohol
(d) the protease.

HL

The nucleosome

In bacteria, DNA is naked but in eukaryotes, the DNA is associated with proteins. The most common proteins are histones. Histone proteins serve a dual purpose:

- they help to supercoil the DNA to package up the chromatin
- they also help to regulate the expression of genes.

A nucleosome is a group of eight histone proteins arranged to form an octomer. An additional protein is associated with the DNA to link one nucleosome to the next.

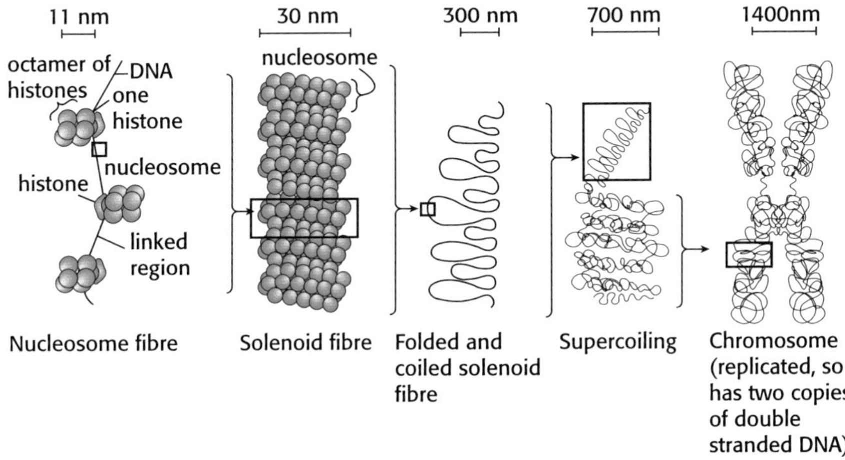

Figure 7 Levels of organization within the chromosome

Rediscovering biology: supercoiling

Obtain a long elastic band. Hold one end and turn the other, until tension in the rubber band causes twists to form. The tighter the band is wound, the more twists the band forms. This twisting is referred to as supercoiling.

1 What effect does supercoiling have on the total space occupied by the rubber band?
2 Suggest when it would be useful for the DNA in a cell to supercoil.
3 Use molecular visualization software such as JMol to view a rendering of a nucleosome. Note that the DNA coils around the octomer, but that a tail-like extension of amino acids projects to the outside. This is the histone tail. It is a chemical modification to these tails that is thought to play a role in the regulation of gene expression.

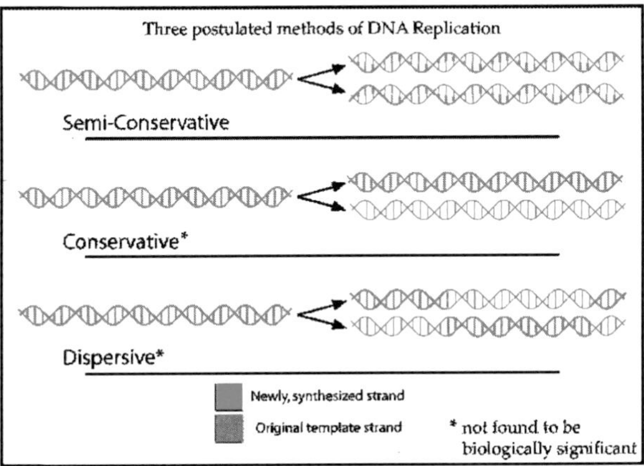

Figure 8 Three possible mechanisms for DNA replication

Data-based question: the Meselson–Stahl experiment

"It has not escaped our notice that the specific pairing we have postulated immediately suggests a possible copying mechanism for the genetic material."

James Watson and Francis Crick, *Nature,* 1953.

In order for cell division to occur, DNA must be duplicated to ensure that progeny cells have the same genetic information as the parent cells. The process of duplicating DNA is termed replication. The Meselson–Stahl experiment sought to understand the mechanism of replication. Did it occur in a conservative fashion, a semi-conservative fashion or in a dispersive fashion (see Figure 8 on page 59)?

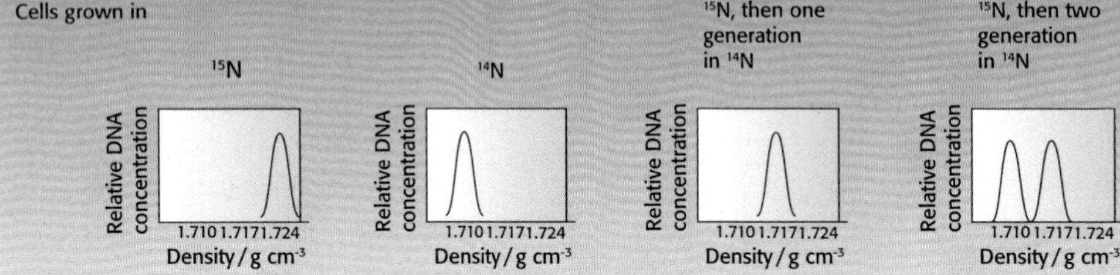

Figure 9 Evidence that DNA replication is semi-conservative

Meselson and Stahl grew *E. coli* in a medium containing "heavy" nitrogen (^{15}N) for a number of generations. They then transferred the bacteria to a ^{14}N medium. Samples of the bacteria were taken over a period of time and separated by density gradient centrifugation, a method in which heavier molecules settle further down in a centrifuge tube than lighter ones.

Explain why the data in Figure 9 suggests that replication is semi-conservative and why the other models are not supported. [5]

Semi-conservative replication

When a cell prepares to divide, the two strands of the double helix separate (see Figure 10). Each of these original strands serves as guides, or templates, for the creation of a new strand. The result will be two DNA molecules, both composed of an original strand and a newly synthesized strand. For this reason, DNA replication is referred to as being semi-conservative.

The sequence on one DNA strand determines the sequence on the other strand, i.e. one strand is complementary to the other.

DNA replication proceeds as follows:

- A large number of enzymes are involved in the process of replication. For these enzymes to gain access to the DNA the helix has to be unwound and separated by the enzyme DNA helicase.
- Nucleotides hydrogen bond to their complementary bases within the template.
- DNA polymerase links the phosphate of the newest nucleotide to the sugar of the nucleotide before it by a covalent bond.

Parental DNA

Replication fork

Parental strand | New strand | New strand | Parental strand

Figure 10 Semi-conservative replication

HL

DNA replication

There are a differences in the details of DNA replication in prokaryotes and eukaryotes, but in outline the processes are very similar. Replication begins at sites called origins of replication.

In prokaryotic cells there is one origin and in eukaryotic cells there are many. Replication proceeds in both directions from the origin and on both strands.

During replication, each new unit added to the growing nucleic acid is a nucleoside triphosphate. A nucleoside is a sugar and a base. A synonym for a nucleotide is "nucleoside monophosphate". A nucleoside triphosphate hydrogen bonds to its complementary base in the exposed DNA molecule. The hydrolysis of two molecules of phosphate occurs to convert the nucleoside triphospate to a nucleotide. The hydrolysis of the two molecules of phosphate provides the energy necessary to add the nucleotide to the growing polymer. Replication is catalysed by the enzyme DNA polymerase III. It requires a short sequence called a primer to start the process. The primer is made of RNA and is made by an enzyme called primase. DNA polymerase I, is the enzyme that later digests away the RNA primer and replaces it with DNA.

The nucleoside triphosphates can only be added at the 3′ end of the nucleotide before it. Because the DNA strands are organized in an anti-parallel fashion, the result is that DNA must be synthesized either in a continuous fashion, creating a leading strand, or in fragments, creating a lagging strand.

These fragments on the lagging strand are called Okazaki fragments (see Figure 11). The fragments are later joined together by the enzyme DNA ligase.

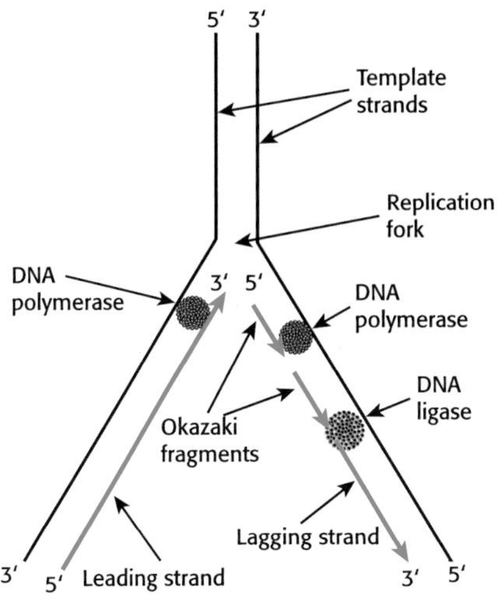

Figure 11 Because new nucleotides are always added at the 3′ end of the molecule, replication is continuous on the leading strand but discontinuous on the lagging strand.

Data-based question: evidence for discontinuous DNA synthesis

Different samples of bacteria were supplied with radioactive nucleoside triphosphates for a series of times (5, 10, 30 or 60 seconds). This was the "pulse" period. This was followed by adding a large excess of non-radioactive nucleoside triphosphates for a longer period of time. This is called the "chase" period. The appearance of radioactive nucleotides (incorporated during the pulse) in parts of the product DNA give an indication of the process of converting intermediates to final products.

DNA was isolated from the bacterial cells, denatured (separated into two strands by heat) and centrifuged to separate molecules by size. The closer to the top of the centrifuge tube, the smaller the molecule.

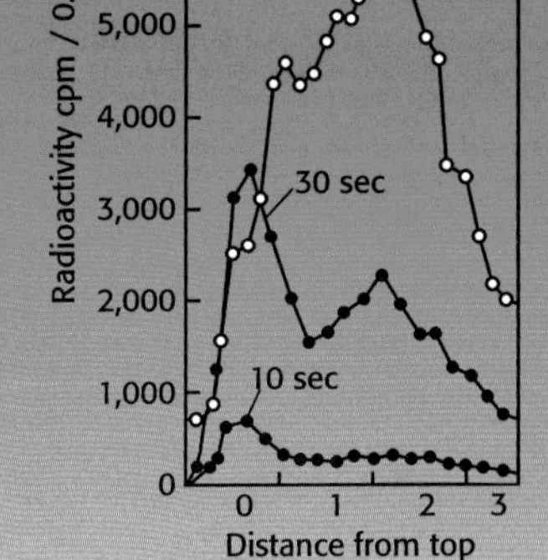

Figure 12

1 Compare the sample that was pulsed for 10 seconds with the sample that was pulsed for 30 seconds. [2]
2 Explain why the sample that was pulsed for 30 seconds provides evidence for the presence of both a leading strand and many lagging strands.
3 Explain why the sample that was pulsed for 60 seconds provides evidence for the activity of DNA ligase.

Chapter 5 questions

1 Give a synonym for the term "nucleoside monophosphate". [1]

2 State the name of the enzyme with each of these functions.

(a) Creates an RNA primer during replication. [1]
(b) Removes the primer at the end of replication. [1]
(c) Seals gaps between fragments on the lagging strand. [1]
(d) Unwinds DNA prior to replication. [1]

3 Refer to Figure 13 when answering the following questions.

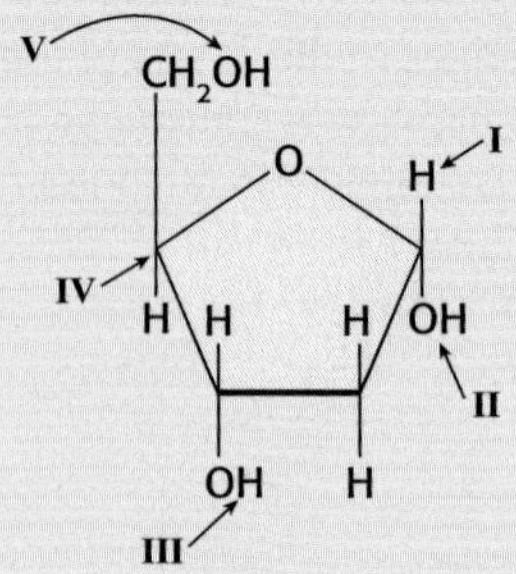

Figure 13

(a) State what molecule is represented. [1]
(b) State whether the molecule would be found in DNA or RNA. [1]
(c) State the part of the molecule to which phosphates bind. [1]
(d) Identify the part of the molecule that refers to the 3' end. [1]

4 (a) Outline the steps involved in DNA replication. [5]
(b) Describe the structure and function of nucleosomes. [5]
(c) Compare RNA and DNA. [5]

5 State two functions of nucleosomes. [2]

6 With reference to Figure 14, answer the following questions.

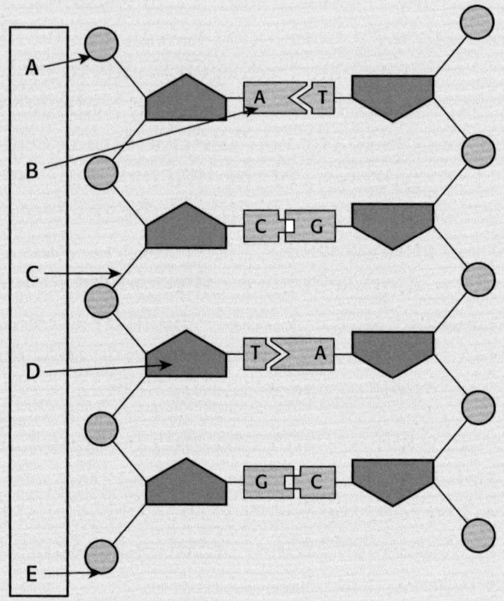

Figure 14

(a) What part of the nucleotide is labelled A? [1]
(b) What kind of bond forms between the structures labelled B? [1]
(c) What kind of bond is indicated by label C? [1]
(d) What sub-unit is indicated by label D? [1]
(e) Which terminal of the molecule is indicated by label E? [1]

7 Figure 15 shows a short section of DNA before and after replication. If the nucleotides used to replicate the DNA were radioactive, which strands in the replicated molecules would be radioactive?

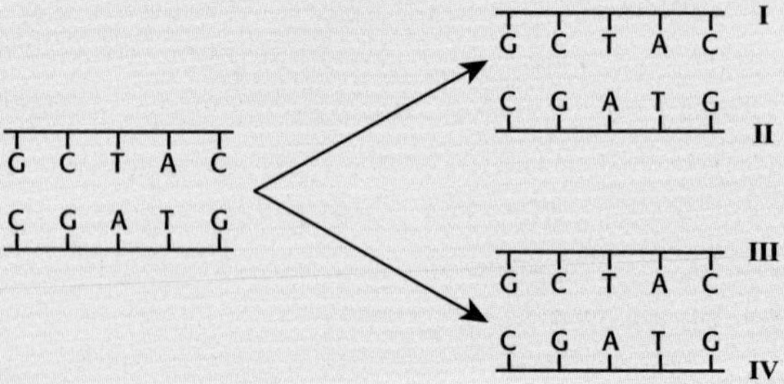

Figure 15

6 Transcription and translation

The production of polypeptides (protein chains) is under hereditary control and changes depending on cell needs and environmental conditions. The gene that codes for the production of a particular polypeptide needs to be consulted every time the cell needs to make that polypeptide. When making a polypeptide, the stable DNA molecule is used as a template or guide to make a more short-lived copy called messenger RNA or mRNA.

In a paper published in 1970, Francis Crick defined the central dogma of genetics:

> **The central dogma** *of molecular biology deals with the detailed residue-by-residue transfer of sequential information. It states that such information cannot be transferred from protein to either protein or nucleic acid.*

Crick chose the word dogma because he thought of it as a grand hypothesis that, however plausible, had little direct experimental support.

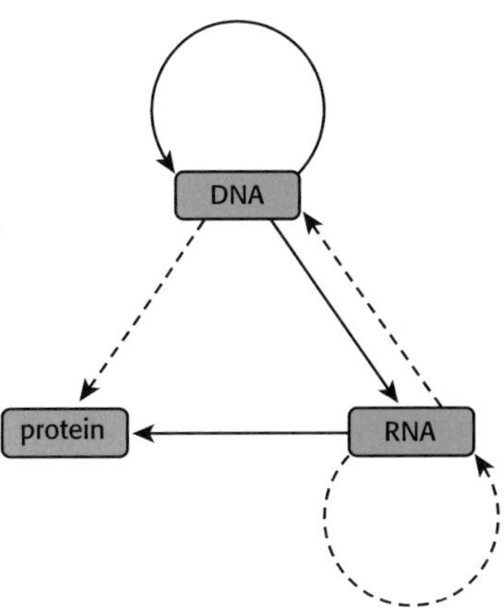

Figure 1 The central dogma of genetics. The arrows represent the flow of information. Solid arrows show what Crick thought to be likely pathways of information flow, while the dotted arrows show what were thought to be possible pathways of information flow.

Transcription

mRNA is made in a process called transcription. There are other types of RNA such as transfer RNA (tRNA), which has a role in translation, and ribosomal RNA (rRNA) which is a structural and functional component of ribosomes. The production of these other forms of RNA also occurs through the process of transcription.

Because RNA is single-stranded, transcription occurs along one strand only. The RNA that is produced has a sequence that is complementary to the DNA that is used as a template for transcription.

A dogma is something which is held to be true without proof. To what extent is dogma unscientific?

What follows is an outline of transcription:

- The enzyme RNA polymerase binds to a site called the promoter on the DNA.
- The DNA to be transcribed is separated by the RNA polymerase in the region of the gene to be transcribed.
- RNA nucleotides pair with their complementary bases on one strand of the DNA only. There is no thymine in RNA, so uracil pairs in a complementary fashion with adenine.
- RNA polymerase forms covalent bonds between the nucleotides.
- The RNA separates from the DNA and the double helix reforms.

The process is summarized in Figure 2.

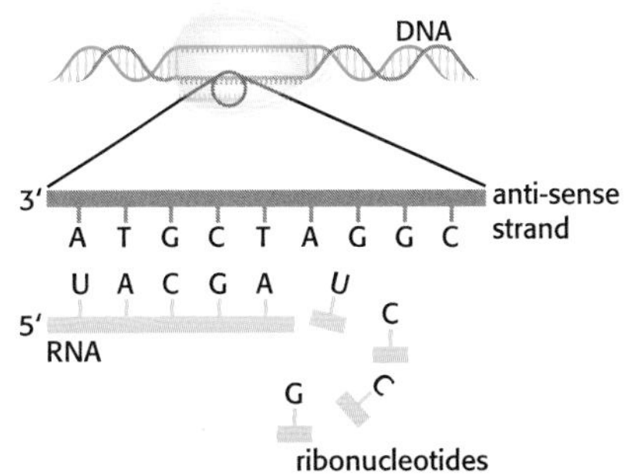

Figure 2 Transcription

HL

Details of transcription

The two sections of DNA involved in transcription have complementary sequences. One sequence is referred to as the "sense" sequence and the other the "anti-sense" sequence. The mRNA that is created by transcription has the same sequence as the sense strand of the DNA. The template DNA strand has the anti-sense sequence.

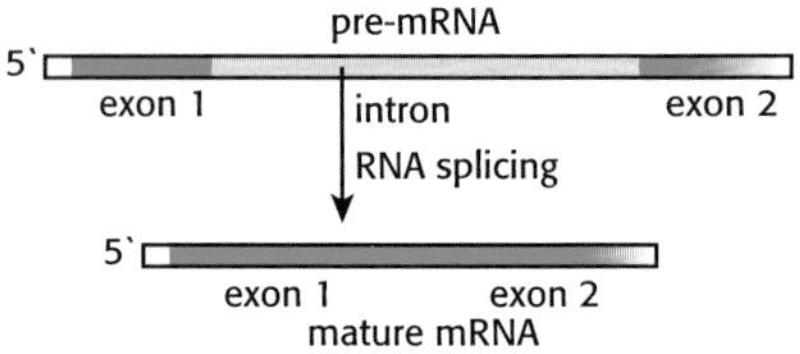

Figure 3 RNA splicing is a form of post-transcriptional modification.

In eukaryotes, the immediate product of mRNA transcription is referred to as pre-mRNA, as it must go through several stages of post-transcriptional modification to become mature mRNA. One of these stages is called RNA splicing, shown in Figure 3. Interspersed throughout the mRNA are sequences that will not contribute to the formation of the polypeptide. They are referred to as intervening sequences, or introns. These introns must be removed. The remaining coding portions of the mRNA are called exons. These will be spliced together to form the mature mRNA.

Figure 4 gives an outline of the main stages involved in transcription.

(a) RNA polymerase binds at a site called the promoter. In addition to determining where transcription starts, the promoter determines which of the two DNA strands will be transcribed.

(b) The RNA polymerase unwinds the DNA and proceeds to catalyze the formation of covalent bonds in the growing RNA transcript.

(c) RNA nucleoside triphosphates can only be added at the 3′ so the RNA transcript grows in the 5′ to 3′ end. As the RNA polymerase moves along the DNA, it continues to unwind the helix while the helix re-winds after the RNA polymerase passes and transcription is completed over that segment.

(d) When the RNA polymerase reaches a termination sequence, it detaches from the DNA and the newly formed RNA is released, ready to be used.

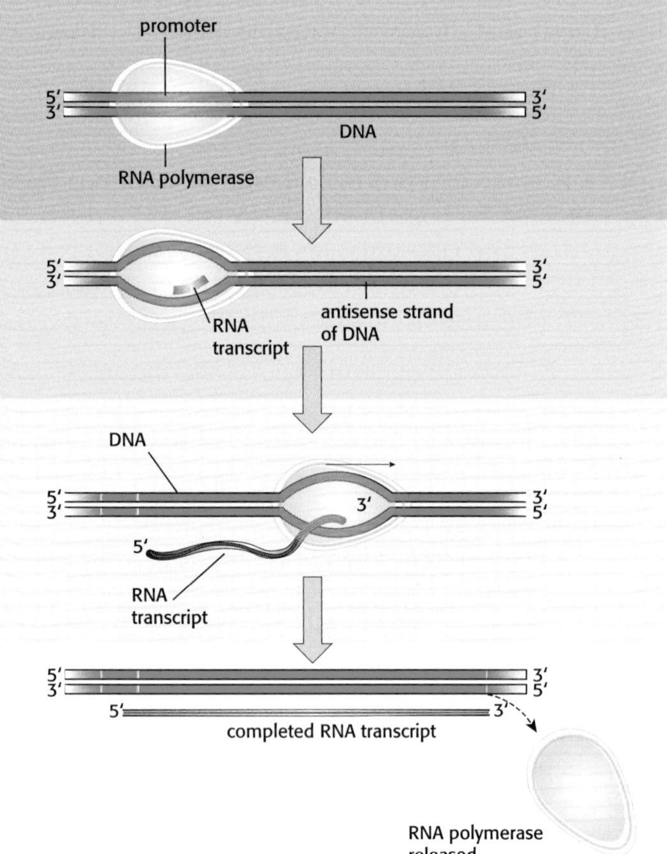

Figure 4

Translation

The base sequence in the mRNA molecule is used as a guide for assembling the sequence of amino acids that will be a polypeptide. The process of protein production using mRNA as a guide is called translation.

The "translation dictionary" that enables the cellular machinery to convert the base sequence on the mRNA into an amino acid sequence is called the genetic code.

A sequence of three bases on the mRNA is called a codon. Each codon codes for a specific amino acid to be added to the polypeptide. Amino acids are carried on another kind of RNA, called tRNA. Each amino acid is carried by a specific tRNA, which has a three-base anti-codon complementary to the mRNA codon for that particular amino acid. A tRNA with the correct anti-codon attaches to the codon on the mRNA.

Table 1 lists all of the 64 possible codons. The three bases of an mRNA codon are designated in the table as first, second and third positions. Note that different codons can code for the same amino acid. For example the codons GUU and GUC both code for the amino acid valine. For this reason, the code is said to be "degenerate". Note also that three codons are "stop" codons that code for the end of translation.

The genetic code is universal in that it operates in the same way in nearly all life on Earth. There are some very rare exceptions. For example, in some cases, the stop codons are used to code for non-standard amino acids.

Translation takes place on cell structures known as ribosomes. These are in the cytoplasm, outside the cell nucleus. Each ribosome comprises a small and a large subunit. An outline of translation follows.

- An mRNA binds to the small subunit of the ribosome.
- tRNA molecules are present, each one carrying the specific amino acid corresponding to its anti-codon. Figure 5 shows a tRNA with the anti-codon for phenylalanine.
- The tRNA binds to the ribosome at the site where its anti-codon matches the codon on the mRNA.
- Two tRNAs bind at once and the first one in transfers the growing polypeptide chain to the second one in.
- The ribosome moves along the mRNA and the process continues until a stop codon is reached when the completed polypeptide is released.

Table 1

First position (5′ end)	Second position U	C	A	G	third position (3′ end)
U	Phe	Ser	Tyr	Cys	U
	Phe	Ser	Tyr	Cys	C
	Leu	Ser	Stop	Stop	A
	Leu	Ser	Stop	Trp	G
C	Leu	Pro	His	Arg	U
	Leu	Pro	His	Arg	C
	Leu	Pro	Gin	Arg	A
	Leu	Pro	Gin	Arg	G
A	Ile	Thr	Asn	Ser	U
	Ile	Thr	Asn	Ser	C
	Ile	Thr	Lys	Arg	A
	Met	Thr	Lys	Agr	G
G	Val	Ala	Asp	Gly	U
	Val	Ala	Asp	Gly	C
	Val	Ala	Glu	Gly	A
	Val	Ala	Glu	Gly	G

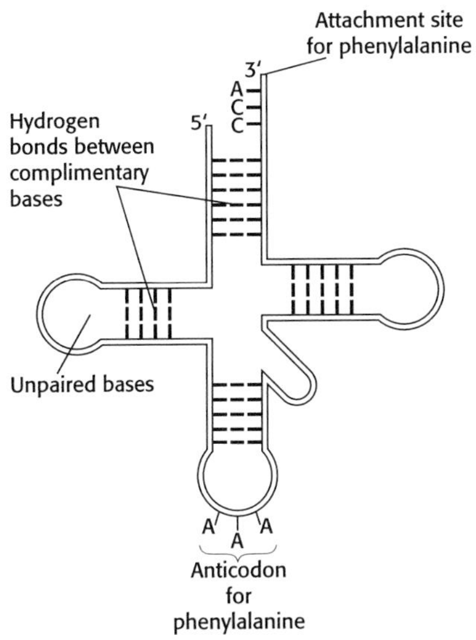

Figure 5 A tRNA molecule

HL Details of translation

Each tRNA molecule is recognized by a tRNA-activating enzyme that binds a specific amino acid to the tRNA, using ATP for energy. The amino acid attaches at the 3′ end of the tRNA. The 3′ end of the tRNA terminates with the nucleotide sequence CCA.

Ribosomes are composed of ribosomal RNA, or rRNA and a large variety of individual proteins. Each ribosome is composed of two subunits, one larger than the other. Each ribosome has three tRNA binding sites – the "E" or exit site, the "P" or peptidyl site and the "A" or aminoacyl site.

Often, more than one ribosome can be actively translating the same mRNA molecule at the same time. The resulting complex of ribosomes along a single mRNA is called a polyribosome or polysome (Figure 6).

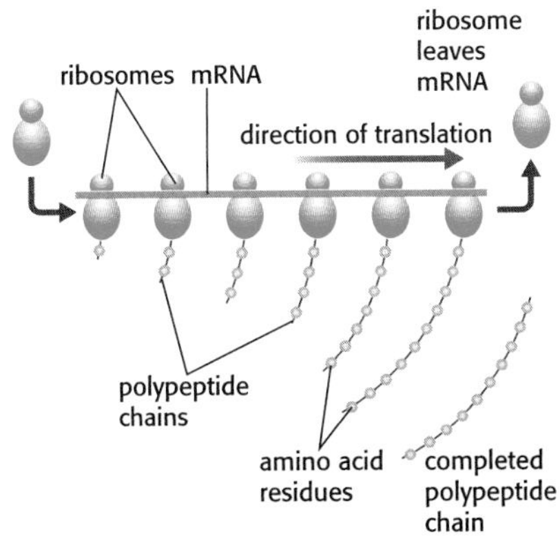

Figure 6 A polysome

Translation always begins in the **cytoplasm**. If the proteins are destined eventually for **lysosomes** or for export, then the ribosomes bind to the endoplasmic reticulum and complete the process of translation while bound. Proteins that are for use within the cytosol are synthesized by unbound ribosomes.

The stages involved in translation are shown in detail in Figure 7.

> **?**
> 1 Identify the amino acid coded for by the codon
> (a) UCC
> (b) CAU.
> 2 Identify all the codons that code for arginine.

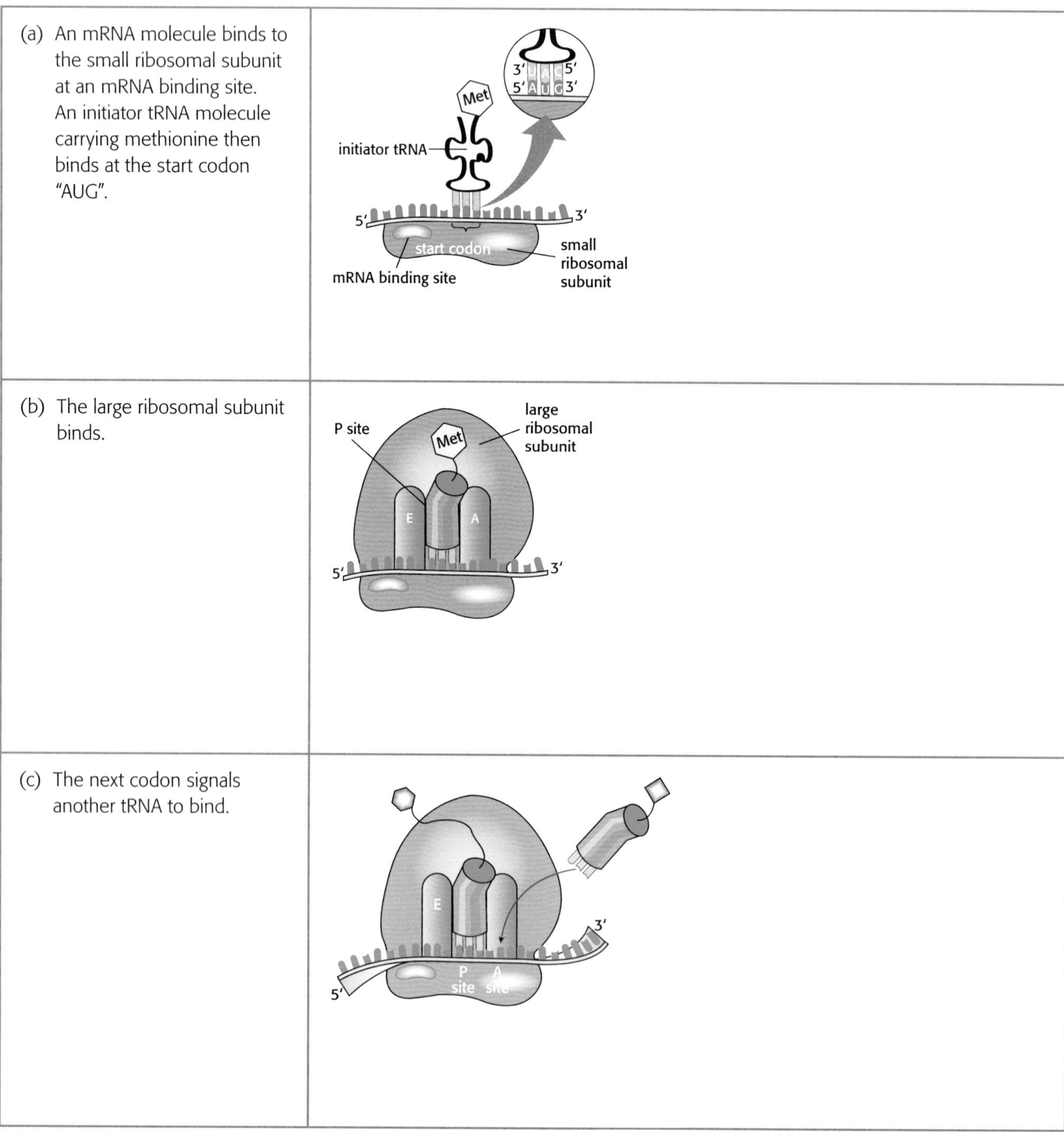

Figure 7

(d) A peptide bond is formed between the amino acid in the "A" site and the amino acid in the "P" site.	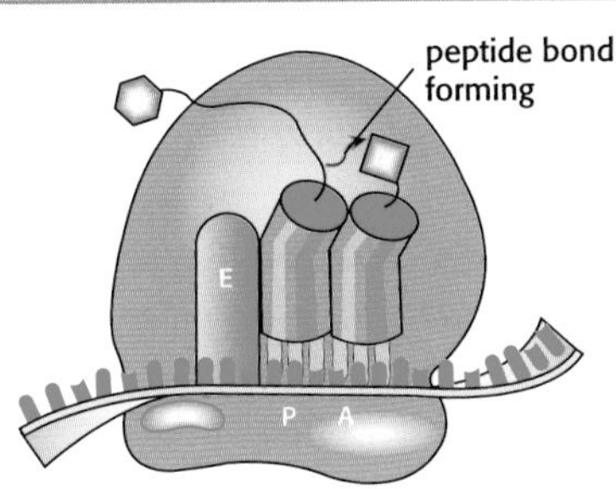
(e) The ribosome translocates, moving the first tRNA into the "E" site freeing the first tRNA and opening up the next codon to the appropriate tRNA.	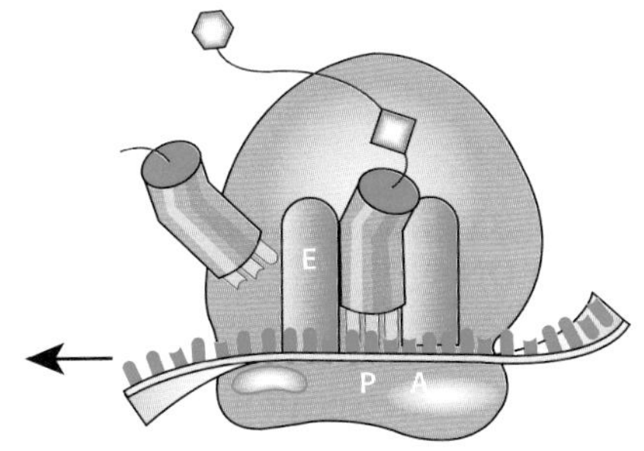
(f) The process continues until a stop codon is reached when the free polypeptide is released. Note the direction of movement along the mRNA is from the 5′ to the 3′ end of the mRNA.	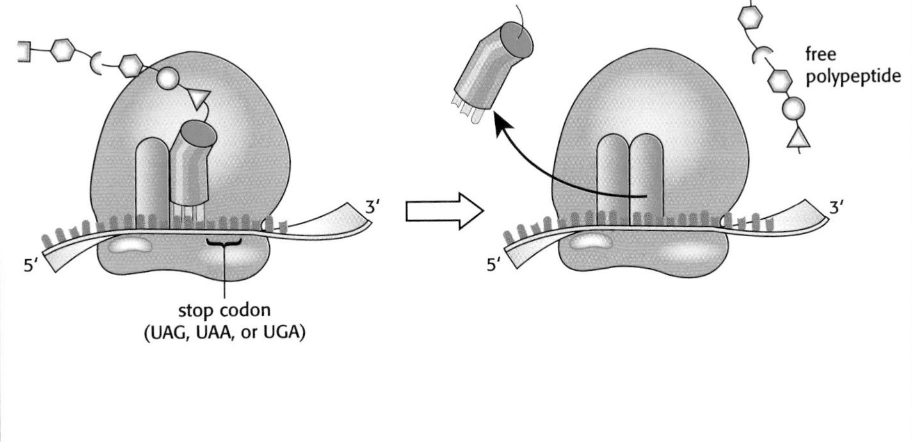

Figure 7 (cont)

Thinking about science: language

Excerpt from Hubbard, Ruth and Wald, Elijah, 1993, *Exploding the Gene Myth*, Beacon Press.

The language that geneticists use often carries considerable ideological baggage. Molecular biologists as well as the press, use verbs like "control", "program" or "determine" when speaking about what genes or DNA do. These are all inappropriate because they assign far too active a role to DNA. The fact is that DNA doesn't "do" anything; it is a remarkably inert molecule. It just sits in our cells and waits for other molecules to interact with it.

In a way, the DNA in our cells is like a cookbook. We need a cookbook if we want to make a complex dish, but it does not make the dish, nor can it determine which dish to make or whether the dish will come out right. The cook and the ingredients will determine whether and how a recipe is used, whether we end up eating soup or cake and how the food tastes… if cells and organisms are the cooks in this metaphor, many ingredients, among them genes and environmental factors, combine to produce a "dish" that could not have been predicted by looking at the ingredients separately.

1 What are the strengths and weaknesses of using metaphors to explain a concept?
2 To what extent is the cookbook an appropriate metaphor for DNA?
3 Imagine you are a scientist who is in the position of having to create a term to describe a particular process. Should your word choice provide an element of interpretation or should it be merely descriptive?

HL

Non-coding sequences

The **Human Genome Project** has led us to understand that there are a number of recognizable patterns observed in DNA.

It has been estimated that there are approximately 25,000 protein-coding genes in the human genome. In addition, some sequences are transcribed to produce other forms of RNA other than mRNA.

Originally, estimates for the number of genes were much higher. This prediction was based on observed diversity of phenotypes. It has been found that non-gene factors can have an influence on phenotype and on gene expression, and these non-gene factors may be one source of the diversity.

Most of the genome is not transcribed. Originally called "junk DNA", it is being increasingly recognized that elements of this "junk" play roles in gene expression. Within this "junk" region, there are elements that affect gene expression as well as highly repetitive sequences (satellite DNA). The latter can form between 5 and 45 per cent of the genome. The repeating sequences are typically between 5 and 300 base pairs long. A repeated sequence may be duplicated as many as 10^5 times per genome.

Rediscovering biology: influence of the environment on gene expression in *S. marcescens*

The phenotype of *S. marcescens* can change markedly with temperature. *S. marcescens* is a common contaminant of bacterial plates or a pure culture can be obtained from a biological supply company. Create multiple streak plates from the same clone. Expose the streak plates to different temperatures and observe the difference in pigmentation.

Research the possible adaptive advantage of differential gene expression. For example, is the pigment expressed or not expressed at the temperature of the human body? How might this affect the immune response of the body and therefore the survival of the *S. marcescens* if its habitiat is the human body.

Thinking about science: junk DNA

Extracts from Wayt Gibbs, W., Nov 2003 *The Unseen Genome: Gems among the Junk, Scientific American*, Vol. 289 Issue 5, pp 46–53.

Journals and conferences have been buzzing with new evidence that contradicts conventional notions that genes, those sections of DNA that encode proteins, are the sole mainspring of heredity and the complete blueprint for all life. Much as dark matter influences the fate of galaxies, dark parts of the genome exert control over the development and distinctive traits of all organisms, from bacteria to humans. The genome is home to many more actors than just the protein-coding genes.

Some scientists now suspect that much of what makes one person, and one species, different from the next are variations in the gems hidden within our "junk" DNA.

It will take years, perhaps decades, to construct a detailed theory that explains how DNA, RNA and the epigenetic machinery all fit into an interlocking, self-regulating system. But there is no longer any doubt that a new theory is needed to replace the central dogma that has been the foundation of molecular genetics and biotechnology since the 1950s.

The central dogma, as usually stated, is quite simple: DNA makes RNA, RNA makes protein, and proteins do almost all the real work of biology. The idea is that information is stored in the twisted ladder of DNA, specifically in the chemical bases (commonly labeled A, T, G and C) that pair up to form the rungs of the ladders. A gene is just a particular sequence of bases on one side of the ladder that specifies a protein.

There is no clear correspondence between the complexity of a species and the number of genes in its genome. "Fruit flies have fewer coding genes than roundworms, and rice plants have more than humans," notes John S. Mattick, director of the Institute for Molecular Bioscience at the University of Queensland in Brisbane, Australia. "The amount of noncoding DNA, however, does seem to scale with complexity."

1 The term "junk DNA" was applied because these non-coding sequences were believed to be an artifact of evolution. Which observation described in the article supports this assumption?

2 Which observation mentioned in the article is not accounted for by the central dogma?

Data-based question: exons and species complexity

Exons are the DNA sequences that code for proteins while introns are the intervening sequences that have to be removed. Figure 8 shows the number of exons found in genes for three different groups of eukaryotes.

1 Determine the percentage of genes in yeast that have only 1 exon. [1]

2 State the most common number of exons found in mammal genes. [1]

3 With reference to the data, discuss the statement "The amount of noncoding DNA does seem to scale with complexity." [3]

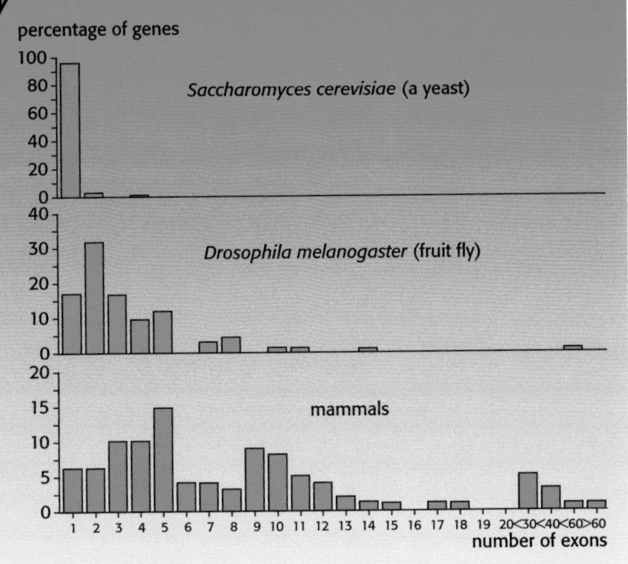

Figure 8 Graphs of the number of exons in three different species

Chapter 6 questions

1 In which process is transfer RNA (tRNA) involved?
(a) recombination
(b) replication
(c) transcription
(d) translation.

2 Which two processes involve the unwinding (uncoiling) of the DNA double helix and its separation into two strands of nucleotides?
(a) replication and telophase of mitosis
(b) telophase of mitosis and translation
(c) translation and transcription
(d) transcription and replication.

3 Where is an intron found?
(a) DNA
(b) RNA
(c) genes
(d) polypeptides.

4 A certain gene codes for a polypeptide that is 120 amino acids long. Approximately how many nucleotides long is the mRNA that codes for this polypeptide likely to be?
(a) 30
(b) 40
(c) 360
(d) 480.

5 A concept map involves drawing lines between associated concepts, and labelling the lines to show the relationships between the connected concepts. Create a concept map that links the following terms: DNA, nucleotide, mRNA, tRNA, rRNA, protein, amino acid, ribosome.

6 Explain the terms universal and degenerate as they relate to the genetic code. [2]

7 Given a certain amino acid sequence, to what extent can the following be predicted?
(a) The sequence of the mRNA that coded for it.
(b) The sequence of the gene that coded for it. [3]

8 A nucleic acid is composed of nucleotides. What is the composition of each of the following?
(a) a gene
(b) a codon
(c) a nucleotide
(d) a polypeptide.

9 Compare transcription and translation. [5]

10 Describe the genetic code. [5]

11 Determine two possible DNA sequences that could code for: Phe-Ile-Val-Leu. [4]

Higher-level questions

12 Outline how the physical structure of tRNA is related to its function. [3]

13 Compare the role of enzymes in the processes of transcription and replication. [4]

14 Place the following events of translation in the correct sequence:
(a) joining of the small and large ribosome subunits
(b) binding of the met-tRNA to the start (AUG) codon
(c) covalent bonding between two amino acids
(d) formation of polysomes
(e) binding of a second tRNA to the A-site. [2]

7 Enzymes

Enzymes are globular proteins that work as catalysts – they speed up chemical reactions without being altered themselves. Enzymes are often called biological catalysts because they are made by living cells and speed up biochemical reactions. The substances that enzymes convert into products in these reactions are called **substrates**. A general equation for an enzyme-catalysed reaction is:

$$\text{substrate} \xrightarrow{\text{enzyme}} \text{product}$$

Enzymes are found in all living cells and are also secreted by some cells to work outside. Living organisms produce many different enzymes – literally thousands of them. This is because most enzymes only catalyze one biochemical reaction and as thousands of reactions take place in cells, nearly all of which need to be catalyzed, many different enzymes are needed. A typical enzyme will therefore only work on the substrates used in one reaction. This property of enzymes is called **enzyme–substrate specificity**.

To be able to explain enzyme–substrate specificity, we must look at the mechanism by which enzymes catalyze reactions. This involves the substrate, or substrates binding to a special region on the surface of the enzyme called the **active site** (see Figure 1). The shape and chemical properties of the active site and the substrate match each other. This allows the substrate to bind, but not other substances. Substrates are converted into products while they are bound to the active site and the products are then released, freeing the active site to catalyze another reaction. The match between the enzyme's active site and the substrate has been likened to the fitting of a key into a lock – other keys will not fit. This **lock and key model** explains the substrate specificity of enzymes.

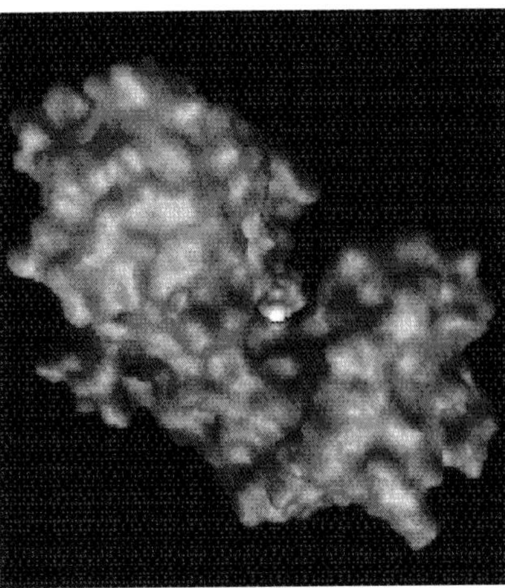

Figure 1 Computer-generated image of the enzyme hexokinase, with a molecule of its substrate glucose bound to the active site. The enzyme bonds a second substrate, phosphate, to the glucose, to make glucose phosphate.

In 1833, the French scientists Anselme Payen and Jean-François Persoz discovered a substance in germinated barley seeds that converted starch into the sugar maltose. They devised a method of purifying it to form a white powder. It was the first enzyme to be isolated and investigated, although the name enzyme wasn't used until 1876.

Rediscovering biology: diastase in barley seeds

The activity of the enzyme diastase can be investigated microscopically, using foods as test substrates.

1 (a) Smear the freshly cut surface of a piece of potato tuber over the centre of two microscope slides, leaving behind many starch granules (see Figure 2).
  (b) Smear cottage cheese over the centre of two other slides, leaving many small lumps of protein.
  (c) Place a small drop of cream or creamy milk (not homogenized) on the centre of two other slides, leaving many droplets of fat.
2 Heat all the slides on a hot plate at 100°C for one minute and then allow them to cool.
3 With each pair of slides, place a drop of water over the food on one slide and a drop of 1% diastase solution on the other.
4 Cover the slides to prevent drying and leave them for a few hours.
5 Stain the foods – use iodine solution for the starch, eosin for the protein and Sudan III for the fat. Examine the slides under a microscope. Compare the pairs of slides to see whether the grains or droplets of food are still visible or have been digested by the diastase.

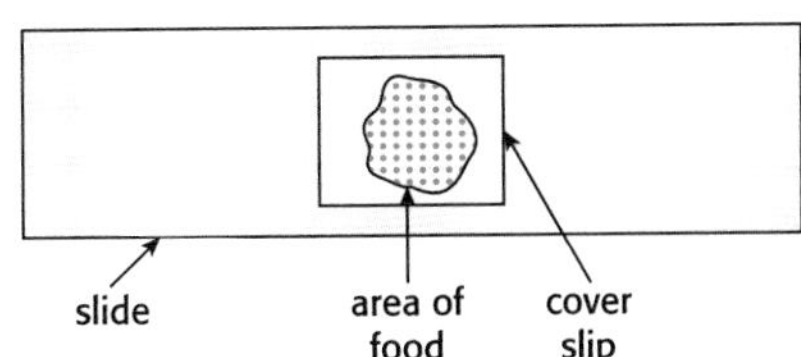

Figure 2 Slide used to investigate the substrate specificity of amylase

1 (a) Predict which foods diastase will digest. [2]
  (b) State the property of enzymes this would demonstrate. [1]
2 Explain why stains are needed to give clear results in this experiment. [2]
3 Explain why a slide with no enzyme is needed for each of the foods. [3]
4 Explain why the slides must be left for several hours before looking at the results. [2]

Effect of pH on enzyme activity

The acidity or alkalinity of a solution is measured using the pH scale. The lower the pH, the more acid or the less alkaline a solution is. Acidity is due to the presence of hydrogen ions, so the lower the pH, the higher the hydrogen ion concentration. The pH scale is logarithmic. This means that reducing the pH by one unit makes a solution ten times more acidic. A solution at pH 7 is neutral. A solution at pH 6 is slightly acidic; pH 5 is ten times more acidic than pH 6, pH 4 is one hundred times more acidic than pH 6 and so on.

Enzymes are sensitive to pH. Most enzymes have an optimum pH at which their activity is highest. If the pH is increased or decreased from the optimum, enzyme activity decreases and eventually stops altogether. When the hydrogen ion concentration is higher or lower than the level that the enzyme naturally works at, the structure of the enzyme is altered, including the active site. Beyond a certain pH the structure of the enzyme is irreversibly altered. This is called **denaturation**. Enzymes do not all have the same pH optimum – in fact, there is a wide range. This reflects the wide range of pH environments in which enzymes work. Figure 3 shows the pH range of some of the places where enzymes work.

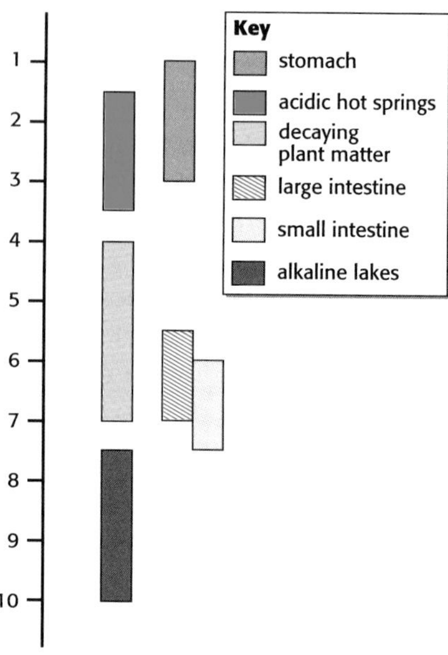

Figure 3 Chart showing the range of pH environments in which enzymes work

Rediscovering biology: the effect of pH on proteases

Proteases are enzymes that digest protein. Although all proteins have this function, they are a varied group of enzymes in other ways. The activity of proteases can be investigated using cubes of the protein gelatine. Either pure gelatine can be used, or sugar-free jello, which is more than 50 per cent gelatine.

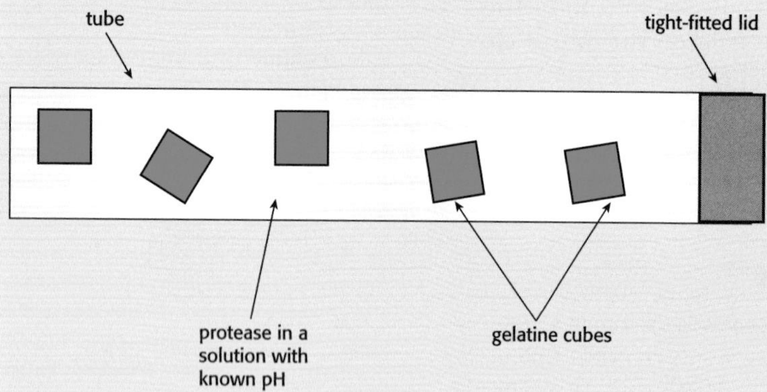

Figure 4 Tube used to investigate the rate of digestion of gelatine

The rate of digestion of the gelatine can be measured by finding the mass of the gelatine cubes before and after the digestion period. If sugar-free jello cubes are used, the colouring that they contain is released as the gelatine in the cube is digested. The concentration of the pigment can be measured using a colorimeter. This gives an alternative way of assessing enzyme activity.

Table 1 gives results that were obtained using a protease called papain, extracted from the flesh of fresh pineapples and cubes of gelatine.

Table 1

pH	Mass decrease (mg)
2	80
3	127
4	163
5	177
6	213
7	167
8	157
9	142

1 Draw a graph of the results in the table. [5]
2 Describe the relationship between pH and papain activity, following the instructions in graph analysis part 1 on page 73. [3]
3 Using the graph and the data in the table, discuss the conclusions that can be drawn about the precise pH optimum of papain. [2]

Dealing with data: graph analysis part 1

Why do biologists use graphs so often? A simple answer is that they show you, in a strongly visual way, what happens to the level of something if you change the level of something else. In other words, graphs display the relationship between two variables. There are three main types of relationship.

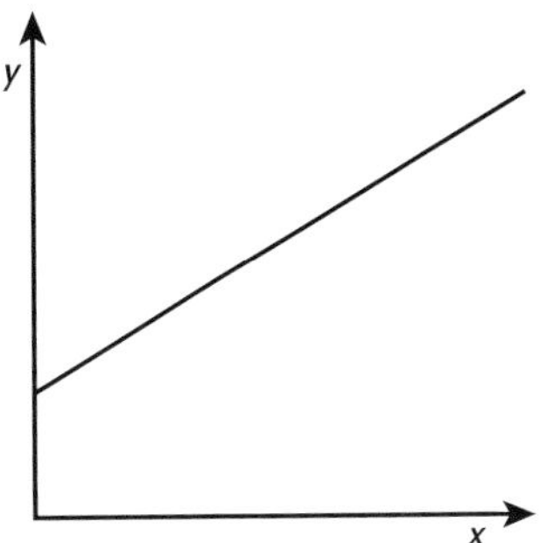

The level of *y* increases as the level of *x* is increased.

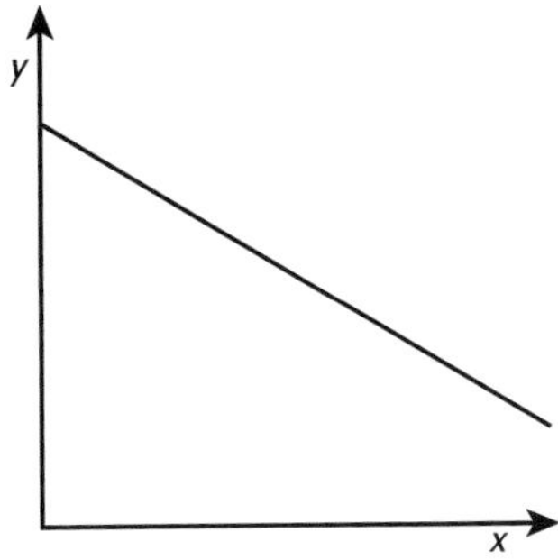

The level of *y* decreases as the level of *x* is increased.

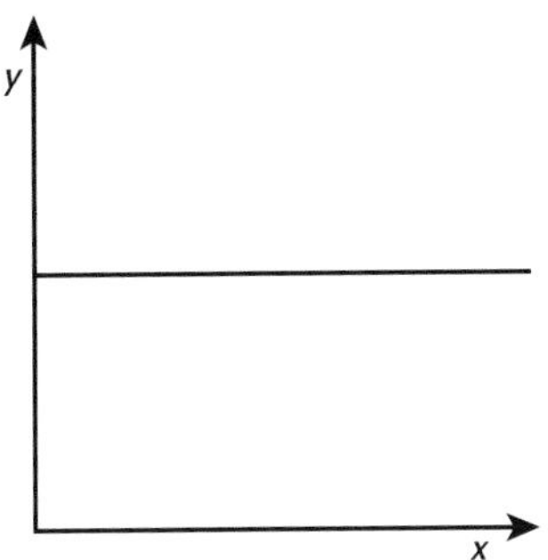

There is no relationship between *x* and *y* (when *x* is increased *y* doesn't change).

Figure 5 Graphs showing relationships between variables

When we look at graphs in biology, they are often more complicated than the three shown above. The relationship between *x* and *y* changes as *x* is increased. When you describe graphs like this, you should divide the graph up into different ranges of *x* and say what the relationship is in each range. You should also say at what level of *x* the level of *y* reaches a maximum or a minimum.

For example, in Figure 6, as *x* is increased from 0 to 4.5 units, *y* decreases, reaching a minimum when $x = 4.5$ units. As *x* is increased from 4.5 to 9 units, *y* increases.

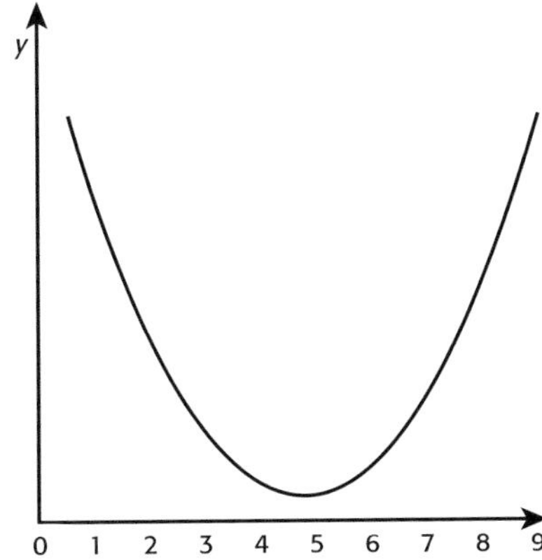

Figure 6 Graph showing a curve

Describe the relationship between *x* and *y* shown in the graphs in Figure 7.

(a)

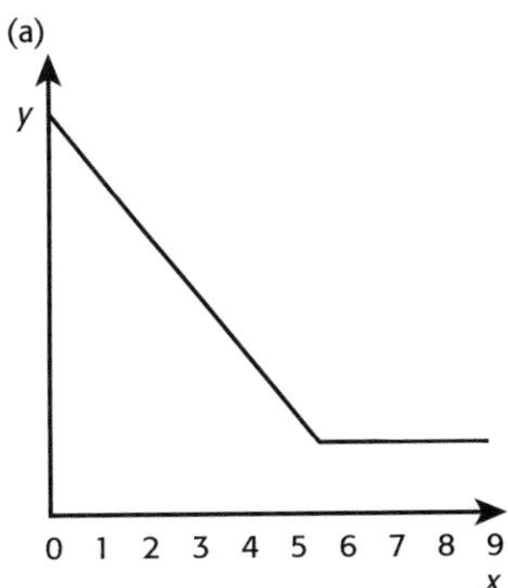

(b)

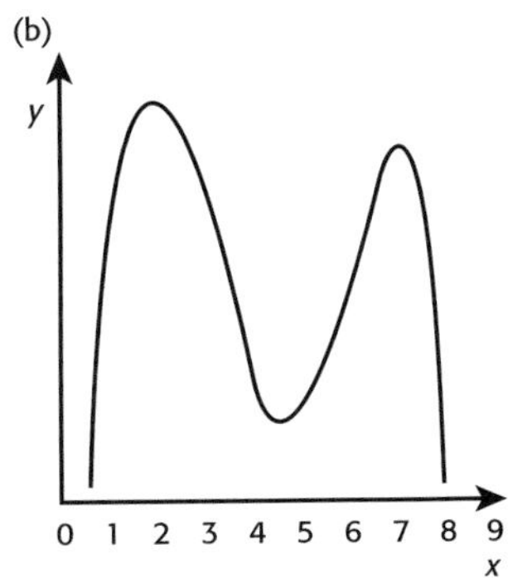

(c)

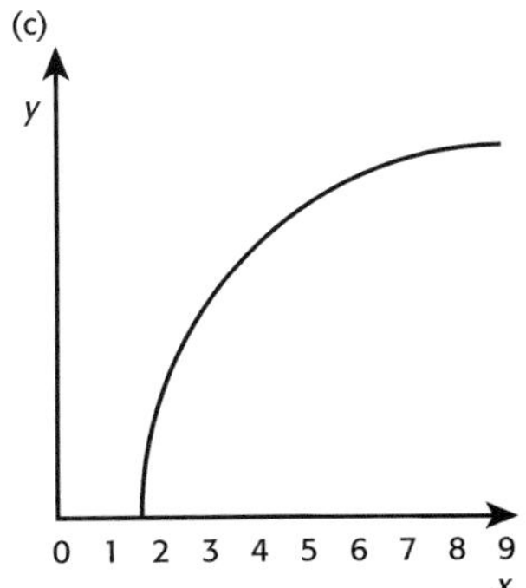

Figure 7 Examples of graphs from Biology experiments

Effect of temperature on enzyme activity

Temperature affects enzyme activity in two ways:

- In liquids, the particles are in continual random motion. When a liquid is heated, the particles in it are given more kinetic energy. Both enzyme and substrate molecules therefore move around faster at higher temperatures and the chance of a substrate molecule colliding with the active site of the enzyme is increased. Enzyme activity therefore increases.
- When enzymes are heated, bonds in the enzyme vibrate more and the chance of the bonds breaking is increased. When bonds in the enzyme break, the structure of the enzyme changes, including the active site. This change is permanent and is called **denaturation**. When an enzyme molecule has been denatured, it is no longer able to catalyze reactions. As more and more enzyme molecules in a solution become denatured, enzyme activity falls. Eventually it stops altogether, when the enzyme has been completely denatured.

So, as temperature rises there are reasons for both increases and decreases in enzyme activity. Figure 8 shows the effects of temperature on a typical enzyme.

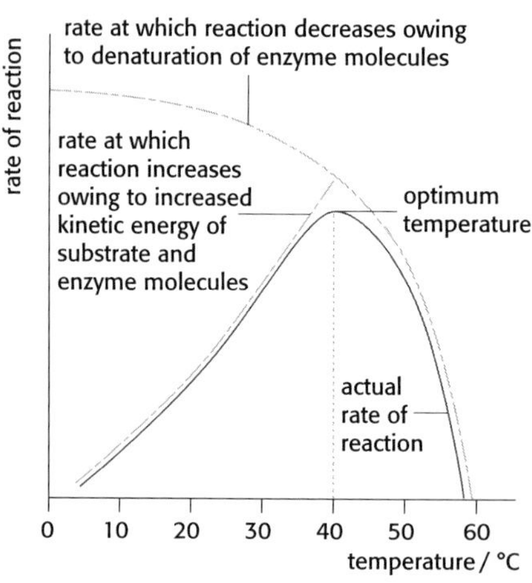

Figure 8 Temperature and enzyme activity

Rediscovering biology: the effect of temperature on lipase

Lipase converts fats into fatty acids and glycerol. It therefore causes a decrease in pH. This pH change can be used to measure the activity of lipase. Figure 9 shows suitable apparatus.

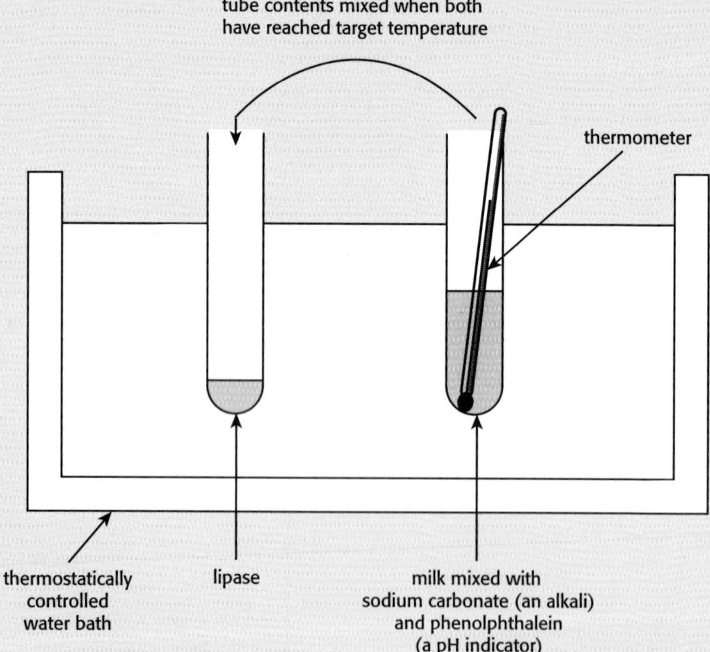

Figure 9 Apparatus for investigating the activity of lipase

Phenolphthalein is pink in alkaline conditions, but becomes colourless when the pH drops to 7. The time taken for this colour change can be used to measure the activity of lipase at different temperatures. Alternatively, pH changes could be followed using a pH probe and data-logging software.

1 Suggest a reason for:
   (a) milk being used to provide a source of fat in this experiment rather than vegetable oil [1]
   (b) the thermometer being placed in the tube containing the larger, rather that the smaller volume of liquid [1]
   (c) the substrate being added to the enzyme, rather than the enzyme to the substrate. [1]

2 State the factors that should be kept constant in this experiment. [2]

3 Explain whether lipase from the human pancreas or from germinating castor oil seeds would be expected to have the higher optimum temperature. [2]

4 Sketch the shape of the graph that you would expect from this experiment, with a temperature range from 0°C to 80°C on the x axis and time taken for the indicator to change colour on the y axis. [3]

Enzymes and substrate concentrations

Enzymes cannot catalyze reactions until the substrate binds to the active site. This happens because of the random movement of molecules in liquids that result in collisions between substrates and active sites. If the concentration of substrates is increased in a solution with a fixed concentration of enzyme, substrate–active site collisions will take place more frequently and the rate at which the enzyme catalyzes its reaction (enzyme activity) increases.

However, there is another trend that needs to be considered. After the binding of a substrate to an active site, the active site is occupied and unavailable to other substrate molecules until products have been formed and released from the active site. As the substrate concentration rises, more and more of the active sites are occupied at any moment. A greater and greater proportion of substrate–active site collisions are therefore blocked. For this reason, the increases in the rate at which enzymes catalyse reactions get smaller and smaller as substrate concentration rises. If the relationship between substrate concentration and enzyme activity is plotted on a graph, a distinctive curve is seen (Figure 10), rising less and less steeply, but never quite reaching a maximum.

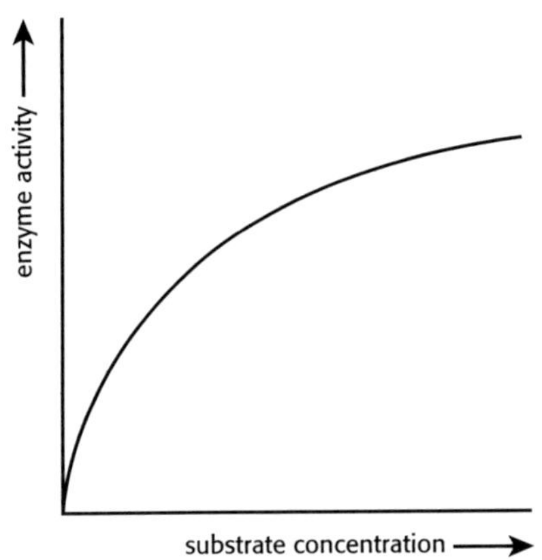

Figure 10 The effect of substrate concentration on enzyme activity. Part 2 of the graph analysis on page 76 explains how to describe this sort of curve.

Investigating biology: the effect of substrate concentration on catalase

Catalase is one of the most widespread enzymes. It catalyses the conversion of hydrogen peroxide, a toxic by-product of metabolism, into water and oxygen. The apparatus shown in Figure 11 can be used to investigate the activity of catalase in yeast.

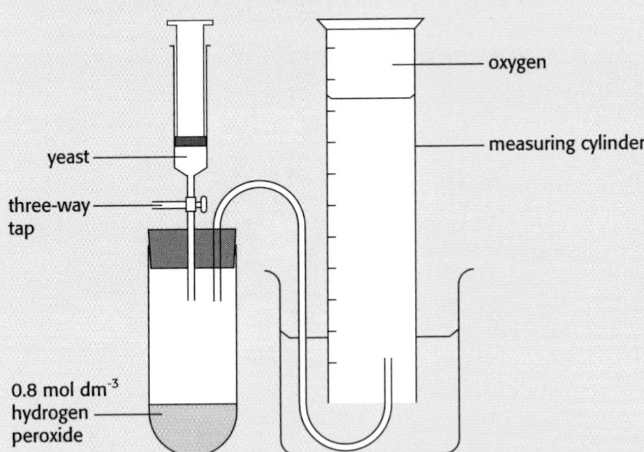

Figure 11 Apparatus for measuring catalase activity

The experiment could be repeated using the same concentration of yeast, but different hydrogen peroxide concentrations. Another possible investigation would be to assess the catalase concentrations in other cell types, such as liver, kidney or germinating seeds. These tissues would have to be macerated and then mixed with water at the same concentration as the yeast.

?

1 Describe how the activity of the enzyme catalase could be measured using the apparatus shown in Figure 11. [2]
2 Explain why a yeast suspension must always be thoroughly stirred before a sample of it is taken for use in an experiment. [2]
3 State two factors, apart from enzyme concentration, that should be kept constant if investigating the effect of substrate concentration. [2]
4 Predict whether the enzyme activity will change more if substrate concentration is increased by 0.2 mol dm^{-3} or if it is decreased by the same amount. [2]
5 Explain why tissues such as liver must be macerated before investigating catalase activity in them. [2]

Safety goggles must be worn if this experiment is performed. Care should be taken not to get hydrogen peroxide on the skin.

Dealing with data: graph analysis part 2

In Part 1 of graph analysis, the relationship where one variable increases when another variable is increased was described. The increase can take various different forms. If it takes the form of a straight line on the graph, it is called a linear increase. Figure 12 shows three special relationships.

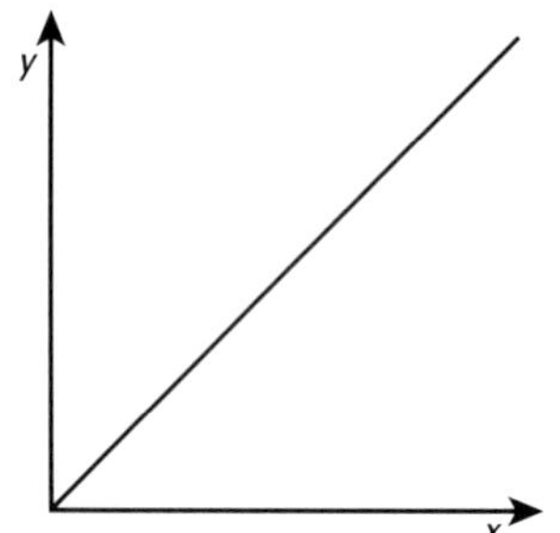

x is **directly proportional** to *y*, if there is a linear increase in y as x is increased and when *x* is zero, *y* is also zero.

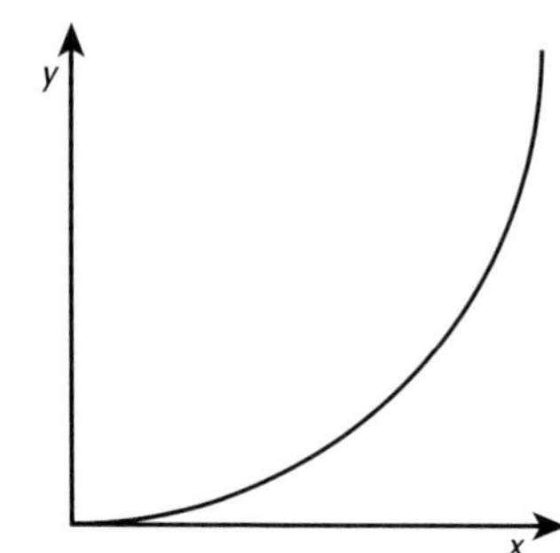

An **exponential increase** happens if the increase in *y* is directly proportional to the size of *y*, so that as *x* increases, the increases in *y* become larger and larger.

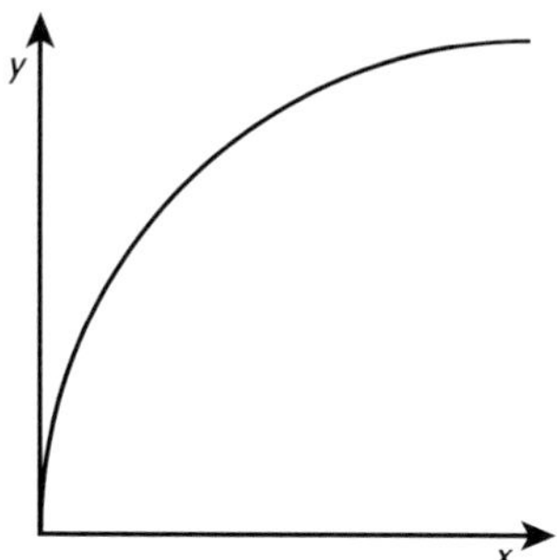

A **decelerating increase** happens if the increases in *y* become smaller and smaller, as *x* increases. This is because the level of *y* is approaching a maximum or saturation level.

Figure 12

1 Sketch a graph to show the following relationships:
   (a) a linear decrease in y as x is increased
   (b) an exponential decrease in y as x is increased
   (c) an accelerating decrease in y as x is increased.
2 Suggest a biological example of each of the graphs in question 1.

Data analysis questions: biosynthesis of glycogen

The Nobel prize for medicine was won in 1947 by Gerty Cori and her husband Carl. They isolated two enzymes that convert glucose phosphate into glycogen. Glycogen is a polysaccharide, composed of glucose molecules bonded together in two ways, called 1"4 and 1"6 bonds (see Figure 13).

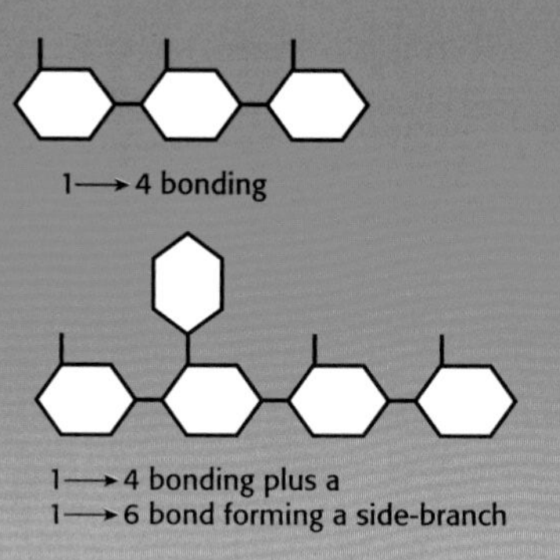

Figure 13 Bonding in glycogen

1 Explain why two different enzymes are needed for the synthesis of glycogen from glucose phosphate. [2]
2 The formation of side-branches increases the rate at which glucose phosphate molecules can be linked on to a growing glycogen molecule. Explain the reason for this.

Figure 14 shows the percentage conversion of glucose phosphate to glycogen by the two enzymes, over a 50-minute period.

3 Curve A was obtained using heat-treated enzymes. Explain the shape of curve A. [2]
4 Curve B was obtained using enzymes that had not been heat-treated.
   (a) Describe the shape of Curve B. [2]
   (b) Explain the shape of Curve B. [2]

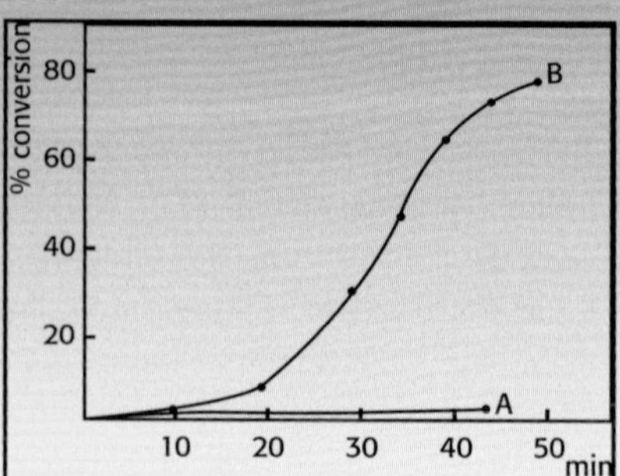

Figure 14 % conversion of glucose phosphate to glycogen

Thinking about science: using enzymes for fermentation

After the discovery in the 19th century of the conversion of sugar into alcohol by yeast, a dispute developed between two scientists, Justus von Liebig and Louis Pasteur. In 1860 Pasteur argued that this process, called fermentation, could not occur unless live yeast cells were present. Liebig claimed that the process was chemical and that living cells were not needed. Pasteur's view reflected the vitalistic dogma – that the substances in animals and plants could only be made under the influence of a "vital spirit" or "vital force". These contrasting views were as much influenced by political and religious factors as by scientific evidence. The dispute was only resolved after the death of both men. In 1897 the Buchner brothers, Hans and Eduard, showed that an extract of yeast, containing no yeast cells, did indeed convert sucrose into alcohol. The vitalistic dogma was overthrown and the door was opened to the use of enzymes to catalyse chemical processes outside living cells.

1. What is the difference between dogma and theory?
2. Is dogma ever acceptable in science or in other academic disciplines?
3. How were the Buchner brothers successful in ending the long-standing dispute about fermentation?
4. Should the vitalistic dogma be completely abandoned? Can the properties of living organisms be explained in purely mechanistic terms?
5. Mechanism implies that the whole of life can be explained in physical terms and that if physical explanations have not yet been found for aspects of life, they will ultimately be found. Is this also a dogma?

Using lactase to produce lactose-reduced milk

Lactose is the sugar that is naturally present in milk. It can be converted into glucose and galactose by the enzyme lactase:

$$\text{lactose} \xrightarrow{\text{lactase}} \text{glucose + glactose}$$

Lactase is obtained from *Kluveromyces lactis*, a type of yeast that grows naturally in milk. Biotechnology companies culture the yeast, extract the lactase from the yeast and purify it for sale to food manufacturing companies. There are several reasons for using lactase in food processing:

- Some people are lactose-intolerant and cannot drink more than about 250 ml of milk per day, unless it is lactose-reduced (see Figure 15).
- Galactose and glucose are sweeter than lactose, so less sugar needs to be added to sweet foods containing milk, such as milk shakes or fruit yoghurt.
- Lactose tends to crystallise during the production of ice cream, giving a gritty texture. Because glucose and galactose are more soluble than lactose they remain dissolved, giving a smoother texture.
- Bacteria ferment glucose and galactose more quickly than lactose, so the production of yoghurt and cottage cheese is faster.

Figure 15 Reduced-lactose milk

Rediscovering biology: immobilizing lactase

During food processing, lactase can either be added to milk, or it can be immobilized on a surface or in beads of a porous material. This latter process avoids contamination of the product with lactase. Lactase can be immobilized using the following method:

1. 2 ml of lactase solution is mixed thoroughly with 8 ml of 2% sodium alginate solution.
2. This mixture is drawn up into a syringe and is then dripped into a 2% calcium chloride solution. The drops of sodium alginate become gel beads, with the lactase immobilized inside.
3. The gel beads are strained off and rinsed and are placed in a syringe with a screw-gate clip.
4. The clip is closed and the syringe is filled up with milk.
5. Drips of milk are removed immediately and then 1, 2, 5, 10 and 20 minutes later. The glucose content of the milk samples is tested using glucose test strips.

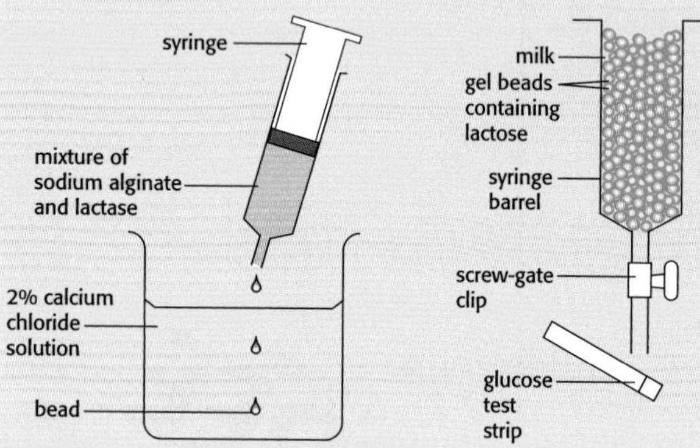

Figure 16 Apparatus for immobilized lactase experiment

1. Sketch a graph to predict the results of the experiment, assuming that the twenty-minute period was long enough for all or nearly all of the lactose to be converted. Consider carefully which axis should be used for time and which for relative glucose concentration. Units for glucose concentration are not needed. [5]
2. Discuss the advantages of using immobilised lactose for production of reduced lactose milk. [5]

Critical considerations: attitudes to lactose intolerance

In some adult humans, levels of lactase are too low to digest lactose in milk adequately. Instead, lactose passes through the small intestine into the large intestine, where bacteria feed on it, producing carbon dioxide, hydrogen and methane. These gases cause some unpleasant symptoms, discouraging consumption of milk. The condition is known as lactose intolerance. It has sometimes in the past been regarded as an abnormal condition, or even as a disease, but it could be argued that lactose intolerance is the normal human condition.

The first argument for this view is a biological one. Female mammals produce milk to feed their young offspring. When a young mammal is weaned, solid foods replace milk and lactase secretion declines. Humans who continue to consume milk into adulthood are therefore unusual. Inability to consume milk because of lactose intolerance should not therefore be regarded as abnormal.

The second argument is a simple mathematical one: 90–95% of humans are lactose intolerant (see Figure 17).

The third argument is evolutionary. Our ancestors were almost certainly all lactose intolerant, so this is the natural or normal state. Lactose tolerance appears to have evolved separately in at least three centres: Northern Europe, parts of Arabia, the Sahara and eastern Sudan, and parts of East Africa inhabited by the Tutsi and Maasai peoples. Elsewhere, tolerance is probably due to migration from these centres.

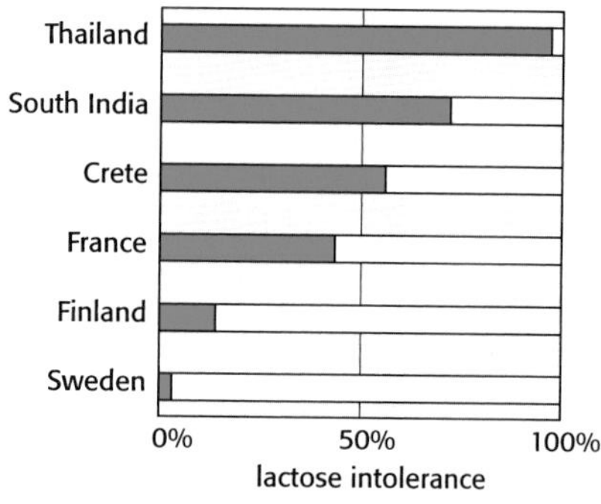

Figure 17 Rates of lactose intolerance

1. Explain why lactose tolerance has been thought of as normal, rather than lactose intolerance.
2. Discuss whether the terms "normal" and "abnormal" are ever appropriate in descriptions of people.
3. Lactose tolerance is correlated with traditional consumption of milk. Is either of the following statements correct?
   (a) Lactose tolerance caused milk consumption in adulthood to begin.
   (b) Milk consumption in adulthood caused lactose tolerance to develop.

HL

Activation energy and enzymes

Chemical reactions are not single-step processes. Substrates have to pass through a transition state before they are converted into products. Energy is required to reach the transition state, and although energy is released in going from the transition state to the product, some energy must be put in to reach the transition state. This is called the activation energy. The activation energy is used to break or weaken bonds in the substrates. Figure 18a shows these energy changes for an exothermic reaction that is not catalyzed by an enzyme.

When an enzyme catalyzes a reaction, the substrate binds to the active site and is altered to reach the transition state. It is then converted into the products, which separate from the active site. The binding lowers the overall energy level of the transition state. The activation energy of the reaction is therefore reduced (Figure 18b). The net amount of energy released by the reaction is unchanged by the involvement of the enzyme. However the activation energy is reduced and the rate of the reaction is greatly increased, typically by a factor of a million or more.

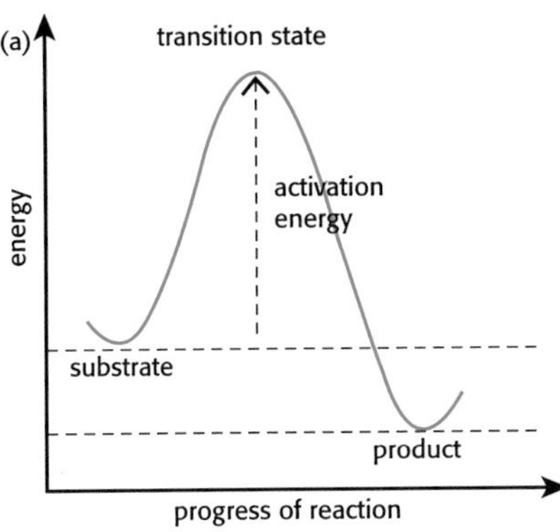

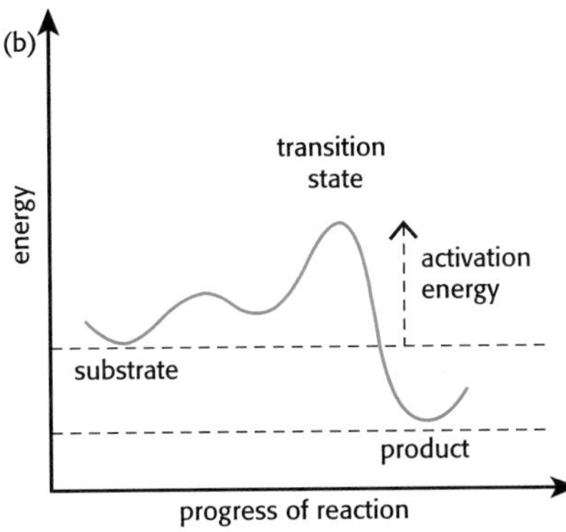

Figure 18 Graphs showing activation energy (a) without an enzyme and (b) with an enzyme

The induced-fit model

Detailed studies of the structure and shape of active sites and their substrates have shown that although they are complementary to each other, they do not exactly fit. Binding nonetheless does happen. The *induced-fit model* provides an explanation for how this occurs. According to this model of enzyme activity, when substrates bind, they cause the structure of the active site to change, so that it fits the substrates (see Figure 19). Of course, the structure of the substrates is also altered, as they are changed into the transition state. So, both substrates and active site undergo conformational changes during binding and these changes help to weaken or break bonds in the substrate and lower the activation energy and catalyze the reaction.

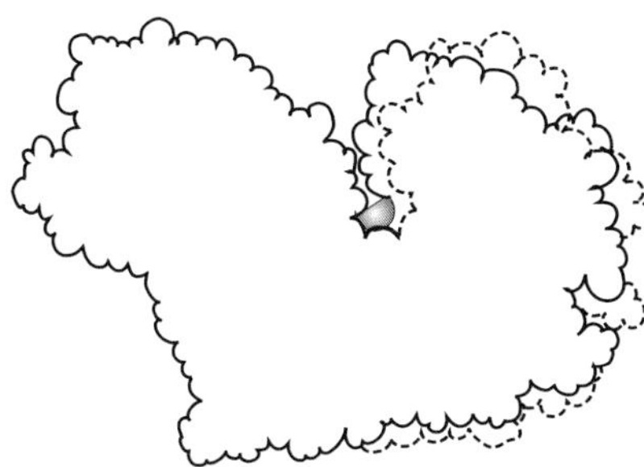

Figure 19 Induced fit in the enzyme hexokinase. The solid line shows the shape of the hexokinase molecule with its substrate (glucose) in place. The dotted line shows the shape of the enzyme without the substrate.

Thinking about science: making and testing theoretical models

The lock and key model (page 71) and the induced-fit model were both developed to help to explain enzyme activity. Models like these are simplified descriptions, which can be used to make predictions. Scientists test these predictions, usually by performing experiments. If the results agree with the predictions, then the model is retained; if not then the model is modified or replaced.

The German scientist Emil Fischer introduced the lock and key model in 1890. Daniel Koshland suggested the induced-fit model in 1959 in the United States. The conformational changes predicted by Koshland's model were subsequently observed using high-resolution X-ray analysis of enzymes and other newly developed techniques. Although much experimental evidence has accumulated confirming predictions based on the induced-fit model, it is still just viewed as a model of enzyme activity.

?

1 Why has the lock and key model been superseded by the induced-fit model?
2 Could there ever be a situation in which two models are both retained?
3 Why is the induced-fit model not now called the Law of Induced Fit?

HL

Enzyme inhibition

Some chemical substances bind to enzymes and reduce the activity of the enzyme. They are therefore known as inhibitors. The two main types of inhibitor are competitive and non-competitive inhibitors. Table 2 shows examples of each type.

Table 2 Examples of competitive and non-competitive inhibitors

	Competitive	Non-competitive
Enzyme	dihydropteroate sythetase	phosphofructokinase
Inhibitor	sulfadiazine	xylitol-5-phosphate
Substrate	para-aminobenzoate	fructose-6-phosphate
Binding	The inhibitor binds reversibly to the enzyme's active site. While it remains bound, substrates cannot bind.	The inhibitor binds reversibly to a site away from the active site. While it remains bound, the active site is distorted and substrate cannot bind.

?

1 State two similarities between non-competitive and competitive inhibitors. [2]
2 Explain why the two types of inhibitor bind at different sites on enzymes. [4]
3 Explain why the terms competitive and non-competitive are used. [4]

Investigating biology: catechol oxidase in bananas

When some fruits are cut open, the surface that is exposed to oxygen in the air turns brown. This is due to a two-stage process:

- A substance called catechol is converted to a yellow pigment by the enzyme catechol oxidase.
- The yellow pigment reacts with oxygen to become the brown pigment melanin.

Catechol oxidase can be extracted from bananas by grinding slices of banana with water and then straining the mixture through muslin cloth. A clear liquid containing the enzyme is produced. This can be used to investigate any of the factors affecting enzyme activity described on the previous pages. Figure 20 shows the basic apparatus that is needed. The intensity of the yellow pigment could be measured using a colorimeter with a blue filter (440 nm). A suitable control should also be prepared, to show that the enzyme really has increased the amount of yellow pigment.

One factor that could be investigated is the concentration of an inhibitor. Lead inhibits many enzymes and can be added to the reaction mixture as lead ethanoate solution. Care is needed to ensure that other factors are not varied at the same time.

Lead should not be disposed of down drains for waste water as it may pass into watercourses.

1 List five factors that affect the activity of catechol oxidase. [2]
2 Describe how the control experiment should be performed. [2]
3 Describe the contents of two tubes that could be set up to show that lead inhibits catechol oxidase. [2]
4 Explain why a blue filter, rather than a yellow filter, is used to measure the concentration of a yellow pigment. [2]
5 Explain why the concentration of yellow pigment must be measured immediately, rather than on the next day. [2]

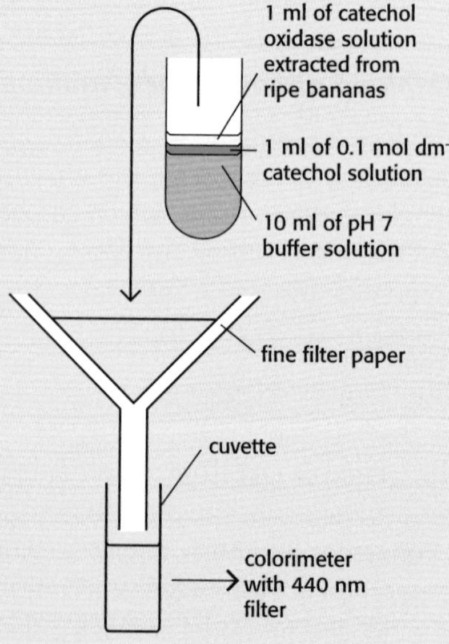

Figure 20 Apparatus for investigating catechol oxidase activity

HL

Metabolic pathways

The word "metabolism" was introduced in the 19th century by the German cytologist and physiologist Theodor Schwann, to refer to the chemical changes that take place in living cells. It is now known that a huge range of chemical reactions occur in cells, catalysed by over 5000 different types of enzyme. Although metabolism is very complex, there are some common patterns.

1 Most chemical changes happen not in one large jump, but in a sequence of small steps, together forming what is called a metabolic pathway. The word game In Figure 21 is an analogy.
2 Most metabolic pathways involve a *chain* of reactions. Figure 22 shows a reaction chain that is used by cells to convert phenylalanine into fumarate and acetoacetate, which can be used as energy sources in respiration. Phenylalanine causes severe health problems if there is an excess of it in the blood.
3 Some metabolic pathways form a *cycle* rather than a chain. The Krebs cycle (see page 89) and the Calvin cycle (see page 107) are examples.

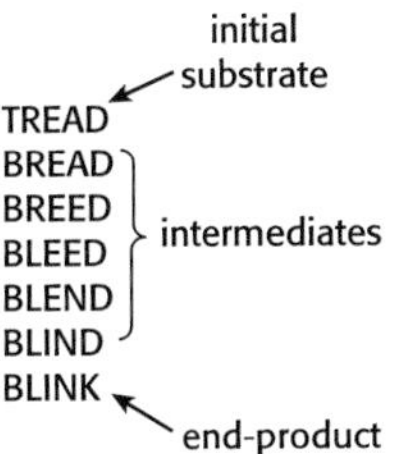

Figure 21 Word game analogy for metabolic pathways

End-product inhibition of metabolic pathways

Many enzymes are regulated by chemical substances that bind to special sites on the enzyme away from the active site. These are called **allosteric interactions** and the binding site is called an **allosteric site**. In many cases, the enzyme that is regulated catalyzes one of the first reactions in a metabolic pathway and the substance that binds to the allosteric site is the end-product of the pathway. The end-product acts as an inhibitor. The pathway works rapidly in cells with a shortage of end-product but can be switched off completely in cells where there is an excess.

To see why this is such an economical way to control metabolic pathways, we need to understand how the concentration of the product of a reaction can influence the rate of reaction. Reactions do not normally go to completion–instead an equilibrium position is reached with a characteristic ratio of substrates and products. So, if the concentration of products increases, a reaction will eventually slow down and stop. This effect reverberates back through a metabolic pathway when the end product accumulates, with all the intermediates accumulating. End-product inhibition prevents this build-up of intermediate products.

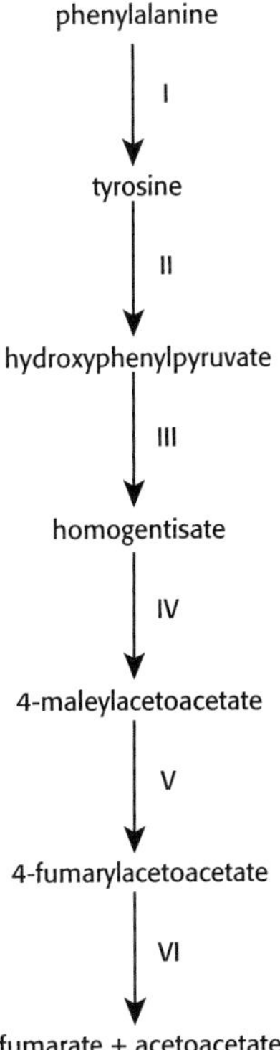

Figure 22 Example of a metabolic pathway

1 State the minimum number of enzymes that are needed to convert phenylalanine into fumarate and acetoacetate. [1]
2 Some people lack the enzyme that converts phenylalanine into tyrosine (Enzyme I in Figure 22). This causes a disease called phenylketonuria. Predict whether levels of phenylalanine and tyrosine in the blood will be high or low in phenylketonuria. [2]
3 Black urine, caused by high levels of homogentisate, is a symptom of the genetic disease alkaptonuria. Deduce the cause. [2]
4 Suggest why the pathway shown in Figure 22 is not controlled by end-product inhibition. [1]
5 Deduce, with a reason, whether end-product inhibition is:
(a) reversible or irreversible [2]
(b) competitive or non-competitive. [2]

Chapter 7 questions

1 Papain is a protease that can be extracted from pineapple fruits. Figure 23 shows the effect of temperature on the activity of papain. The experiment was performed using papain dissolved in water and then repeated with the same quantity of papain that had been immobilised by attaching it to a solid surface. The results show the percentage of the protein in the reaction mixture that was digested in a fixed time.

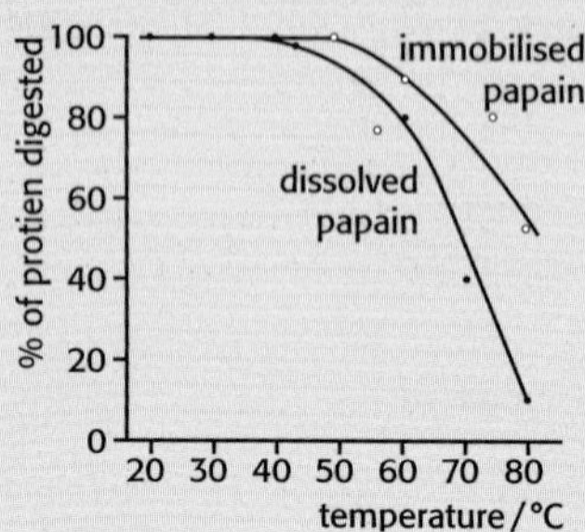

Figure 23

(a) (i) Outline the effects of temperature on the activity of dissolved papain. [2]

(ii) Explain the effects of temperature on the activity of dissolved papain. [2]

(b) (i) Compare the effect of temperature on the activity of immobilized papain with the effect on dissolved papain. [2]

(ii) Suggest a reason for the difference that you have described. [2]

(iii) In some parts of the human body, enzymes are immobilised in membranes. Suggest one enzyme and a part of the body where it would be useful for it to be immobilized in a membrane. [2]

2 Figure 24 shows how the concentration of substrates and products of a reaction changed over time after the enzyme and the substrate were mixed together.

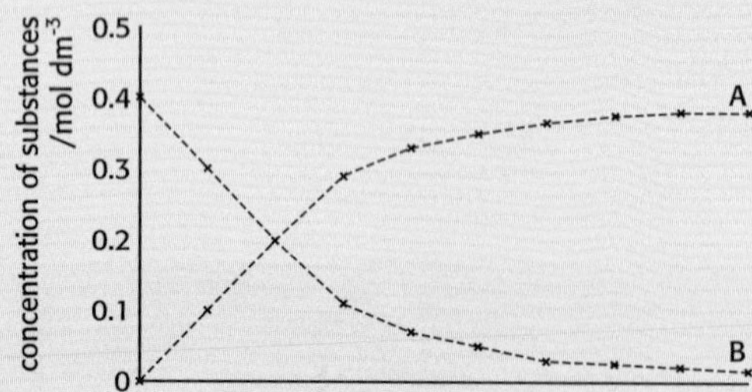

Figure 24

(a) Deduce which curve shows the concentration of substrate. [1]

(b) (i) Using the data in the graph, deduce how the activity of the enzyme changed during the experiment. [2]

(ii) Explain the change in enzyme activity during the experiment. [2]

Higher-level questions

3 (a) State the meaning of the term "metabolic pathway". [2]

Glucose phosphate (G6P) is converted to pyruvate in one of the metabolic pathways of cell respiration. This process happens whether oxygen is available or not. Figure 25 shows the concentrations of the intermediates of this pathway in rat heart tissue. The concentrations are shown as a percentage of the concentrations in the heart when it has been starved of oxygen.

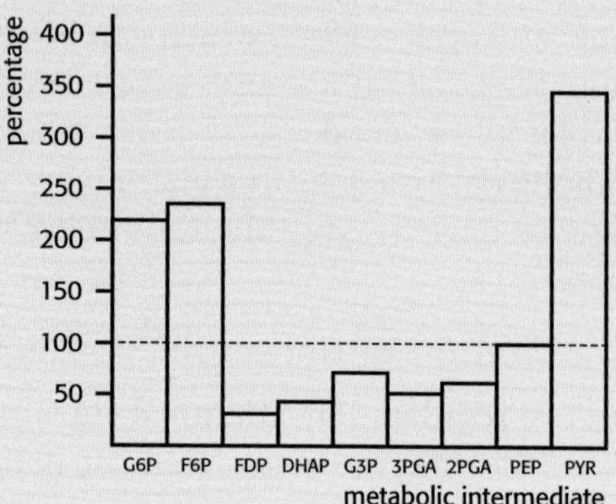

Figure 25

(b) Compared with concentrations during oxygen starvation, state which metabolic intermediate:

(i) increased in concentration most [1]

(ii) decreased in concentration most [1]

(iii) did not change in concentration. [1]

(c) (i) The concentrations shown in Figure 25 suggest that the rate of this metabolic pathway has been greater than is needed by the heart cells. Explain how the data in the bar chart shows this. [2]

(ii) Because rate of the pathway has been greater than necessary, the enzyme catalyzing one of the reactions in the pathway has been inhibited. Deduce which reaction this enzyme catalyzes, giving reasons for your answer. [3]

4 Thousands of different enzymes have been discovered.

(a) Explain the reasons for cells needing to have many different enzymes in their cytoplasm. [5]

(b) Explain how different chemical reactions are catalyzed by different enzymes. [5]

(c) Distinguish, with examples, between competitive and non-competitive inhibition. [8]

5 Some enzymes have commercial uses and a source of them is needed.

(a) Explain the reasons for using living cells to produce enzymes rather than synthesizing them in artificial cell-free systems. [6]

(b) Suggest reasons for almost all enzymes being proteins rather than other biochemicals. [4]

(c) Compare the lock and key and induced-fit models of enzyme activity. [8]

8 Cell respiration

Every living cell must carry out cell respiration in order to convert energy into a form that can be used within the cell. Cells require energy for three main types of activity.

- Synthesizing large molecules like DNA, RNA and proteins.
- Pumping molecules or ions across membranes by active transport.
- Moving things around inside the cell, such as chromosomes, vesicles, or in muscle cells the protein fibres that cause muscle contraction.

The energy for these processes is supplied by ATP (adenosine triphosphate). When ATP is split into ADP (adenosine diphosphate) and phosphate, energy is released:

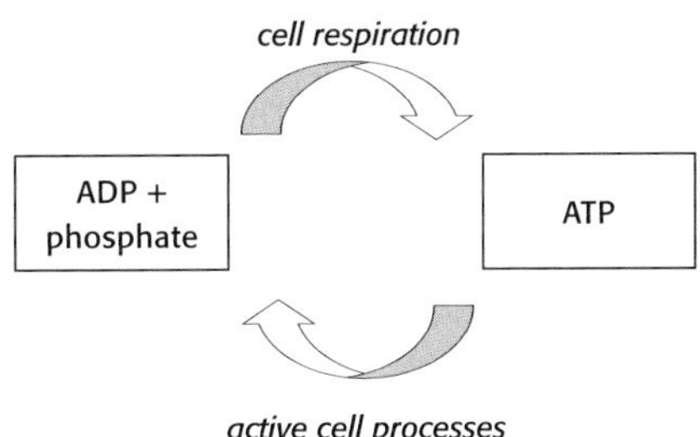

ATP cannot usually be absorbed through plasma membranes, so every cell must produce its own supplies. Organic compounds containing energy are broken down by enzymes, in a careful and controlled way, so that as much as possible of the energy released can be used to form ATP from ADP and phosphate. Cell respiration is therefore defined as *the controlled release of energy from organic compounds to form ATP.*

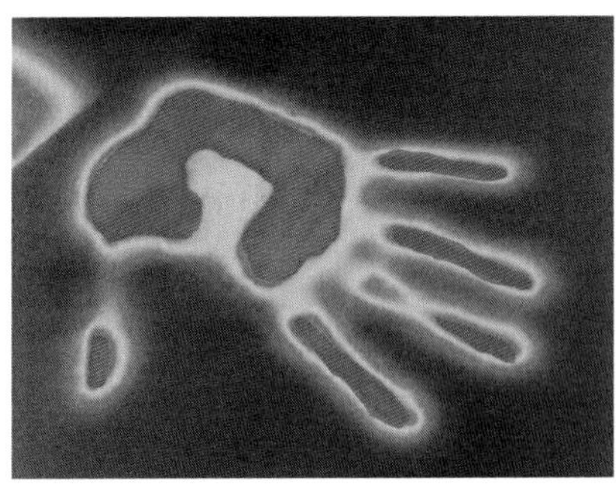

Figure 1 This image was obtained using an infrared camera. It shows the heat left by a person's hand. All organisms generate heat, with birds and mammals generating extra heat to maintain a constant body temperature. This heat is a by-product of the process of cell respiration.

Data-based questions: cell respiration in yellow-billed magpies

Figure 2 shows the results of an experiment in which yellow-billed magpies (*Pica nuttalli*) were put in a cage in which the temperature could be controlled. The birds' rate of respiration was measured at seven different temperatures, from −10 °C to +40 °C. Between −10 °C and 30 °C the magpies maintained constant body temperature, but above 30 °C body temperature increased.

1. Describe the relationship between external temperature and respiration rate in yellow-billed magpies. [3]
2. Explain the change in respiration rate as temperature drops from +10 °C to −10 °C. [3]
3. Suggest a reason for the change in respiration rate as temperature increased from 30 °C to 40 °C. [2]
4. Suggest two reasons for the variation in respiration rate between the birds at each temperature. [2]

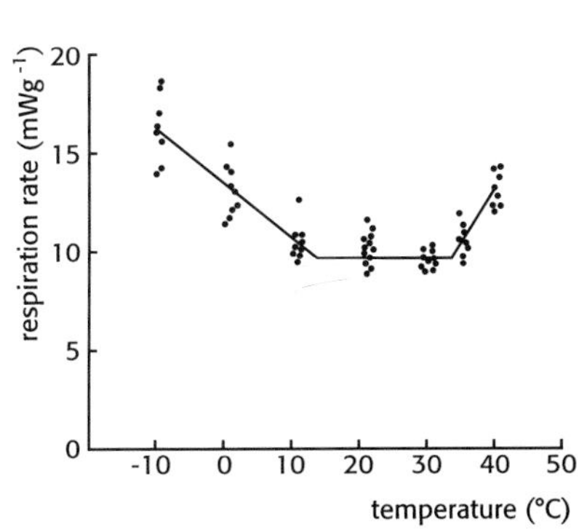

Figure 2 Cell respiration rates at different temperatures in yellow-billed magpies

Glycolysis and anaerobic cell respiration

Cell respiration can release energy from a variety of organic compounds, but carbohydrates and lipids are the usual substrates. If carbohydrates such as glucose are being used, almost all organisms begin the process of cell respiration in the same way. A chain of reactions takes place in the cytoplasm of cells, which converts the glucose into a substance called pyruvate. This chain of reactions is called **glycolysis**.

No oxygen is used in glycolysis and a small amount of ATP is produced. If no oxygen is available then this is the only ATP that can be produced in cell respiration. In these anaerobic conditions, glycolysis can only continue if the pyruvate is converted into other substances. In humans, pyruvate is converted into lactate (lactic acid). In contrast, yeast cells convert pyruvate into ethanol (alcohol) and carbon dioxide. These substances are toxic in excess, so must be removed from the cells that produce them, or be produced in strictly limited quantities.

Summary equations

glucose → pyruvate (ADP → ATP)

Occurs in almost all organisms, in aerobic or anaerobic cell respiration.

pyruvate → lactate

Occurs in humans and some bacteria in anaerobic cell respiration.

pyruvate → ethanol + CO_2

Occurs in yeast in anaerobic cell respiration.

Investigating biology: anaerobic cell respiration in milk bacteria

Some bacteria are able to live and grow in milk, using the lactose in the milk as a substrate for cell respiration. The oxygen in the milk is rapidly used up and the respiration of the bacteria is then anaerobic, with lactate (lactic acid) as a waste product. The pH of the milk therefore drops and eventually falls low enough to denature the proteins in the milk. If this process is stopped at an early stage, yoghurt is produced. If the process is allowed to continue, the proteins separate from the rest of the milk. This is called curdling and is the first stage in the production of most cheeses. Figure 3 shows apparatus that could be used to show that bacteria are causing these changes in the milk. Typically anaerobic cell respiration causes the a drop from about pH 7 to pH 4 over 24 hours.

Similar apparatus could be used to investigate a factor that might affect the rate of anaerobic cell respiration. Discuss these questions with the other students in your team:

1 What factor will you investigate? This is your independent variable.
2 How will you control the level of the independent variable during the experiment?
3 What other factors will you need to keep constant during the experiment and how will you do this?
4 How will you measure the rate of anaerobic respiration? Data logging using pH probes is not the only way of doing this.

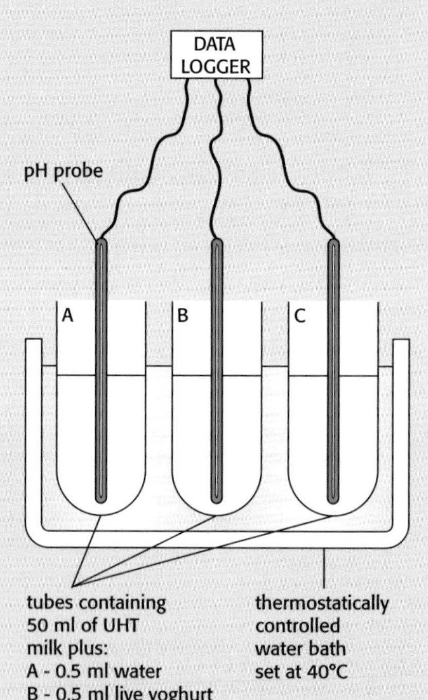

Figure 3 Apparatus to monitor anaerobic respiration in milk

1 Explain the reasons for
   (a) using UHT (Ultra Heat Treated) milk [2]
   (b) setting up a tube with milk only in it [2]
   (c) setting up a tube with milk and unboiled yoghurt [2]
   (d) setting up a tube with milk and boiled yoghurt [2]
2 Discuss whether one of these tubes is the control tube. [2]

Critical considerations: using yeast in industry

Although yeast (*Saccharomyces* spp) is a simple fungus composed of small oval cells, it is of huge and growing economic importance. In many parts of the world yeast is used in bread-making. When incorporated into the bread dough, the carbon dioxide that it produces by anaerobic cell respiration becomes trapped and causes the dough to expand or rise. This gives the bread a lighter texture (leavened bread).

In some human societies, yeast is also used to brew alcoholic drinks such as beer and wine. In this case, it is the ethanol rather than the carbon dioxide that is more significant product, though in champagne and other sparkling wines carbon dioxide is also present in high concentrations in the final product.

Another use of the ethanol produced by anaerobic cell respiration in yeast is as a fuel. Brazil has been a leader in the widespread use of ethanol in the internal combustion engines of vehicles, but with oil reserves starting to run out, other countries are also using this biotechnology. The US, for example, produces large amounts of ethanol from corn, and may soon overtake Brazil in ethanol production.

?

By what criteria do we judge whether something is environmentally friendly? What research would we need to do to decide?

Data-based question: monitoring anaerobic cell respiration in yeast

The apparatus in Figure 4 was used to monitor mass changes during the brewing of wine. The flask was placed on an electronic balance, which was connected to a computer for data-logging. The results are shown in Figure 5.

1 Calculate the total loss of mass during the experiment and the mean daily loss. [3]
2 Explain the loss of mass. [3]
3 Suggest two reasons for the increasing rate of mass loss from the start of the experiment until Day 6. [2]
4 Suggest two reasons for the mass remaining constant from Day 11 onwards. [2]

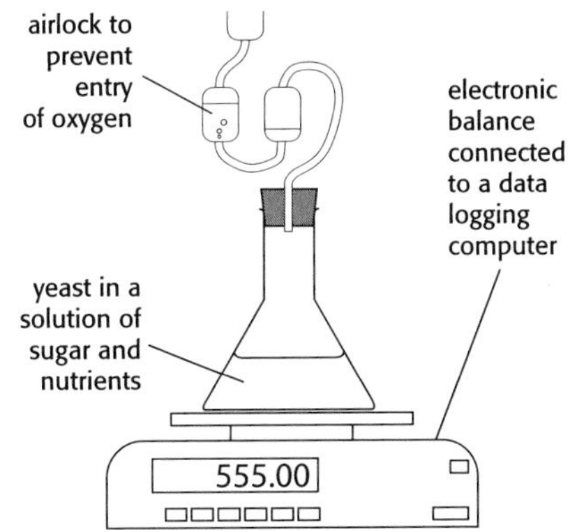

Figure 4 Yeast data-logging apparatus

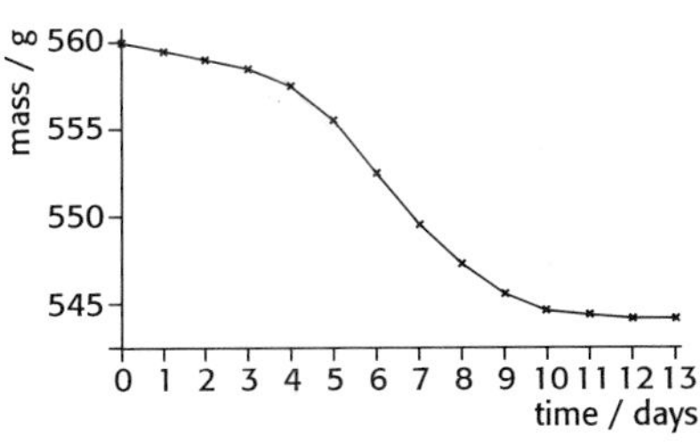

Figure 5 Monitoring anaerobic cell respiration in yeast

Investigating biology: what makes a good research question?

To plan an investigation in biology, you need a focussed research question. Here are some examples of research questions involving cell respiration in yeast. Consider how good each question is:

1 "Which foods can yeast use in anaerobic cell respiration?"
2 "Which monosaccharides is yeast able to use in anaerobic cell respiration?"
3 "Does yeast carry out anaerobic cell respiration faster at 0°C or 100°C?"
4 "If oxygen is available, does yeast use aerobic or anaerobic cell respiration?"

Try to choose a question or hypothesis that genuinely interests you – this makes an investigation both more worthwhile and more interesting. If you are investigating yeast for example, are you interested in alternative fuel sources, enzymes, baking bread or brewing beer?

Once you have decided on your research question, you must design your experiment, including answers to the following discussion points:

(a) How you will vary your independent variable (the factor that you are investigating), and what range of it you will have in your experiment?
(b) How you will measure the amount of anaerobic cell respiration in yeast? These measurements may or may not be affected by your ndependent variable, and so are called the dependent variable.
(c) What factors you will keep constant, so that they do not affect the rate of respiration?

Aerobic cell respiration

If oxygen is available to a cell, pyruvate produced by glycolysis can be oxidised to release more energy. In eukaryotic cells this happens inside the mitochondrion. Energy released from pyruvate oxidation is used to produce ATP. The yield of ATP per glucose is much higher than from glycolysis – more than ten times as much. Oxidation of pyruvate also involves the production of carbon dioxide and water. In most organisms carbon dioxide is a waste product and has to be excreted, but the water is often useful. In humans about half a litre is produced per day. Some desert animals never need to drink and eat only dry foods, because cell respiration supplies almost all their water needs. Figure 6 summarizes aerobic respiration.

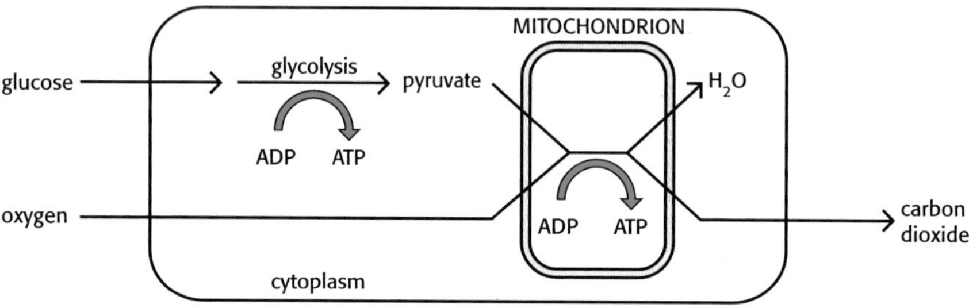

Figure 6 Summary of aerobic cell respiration

Rediscovering biology: respirometers and respiration rates

A respirometer is any device that is used to measure respiration rate. There are many possible designs. Most involve these parts:

- A sealed glass or plastic container in which the organism or tissue is placed.
- A base, such as potassium hydroxide, to absorb carbon dioxide.
- A capillary tube containing fluid, connected to the container.

One possible design of respirometer is shown in Figure 7.

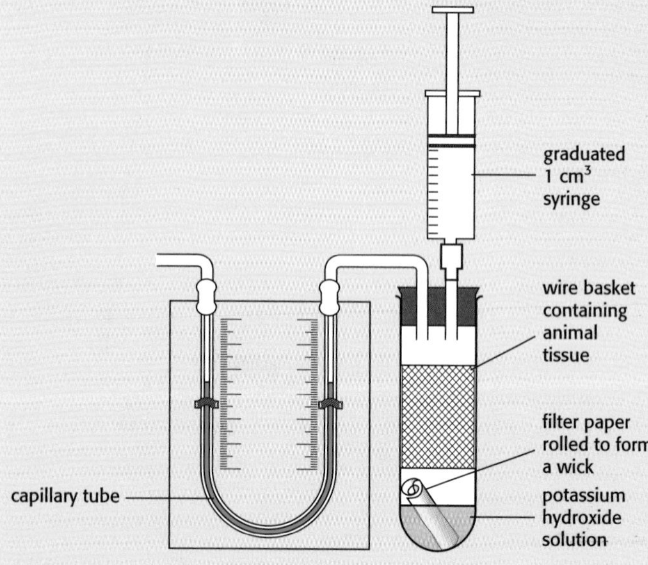

Figure 7 Diagram of a respirometer

?

1 Explain the need for a base inside the respirometer. [2]

2 Deduce, with a reason, which direction the fluid will move in on the right hand side of the capillary tube. [2]

3 Predict, with a reason, the change in nitrogen concentration inside the test tube during the experiment. [2]

4 Explain how the following changes would improve the reliability of results from the experiment:

(a) Placing the test tube in a thermostatically controlled water bath. [2]

(b) Attaching the left-hand side of the capillary tube to another test tube, identical to the right-hand tube except that it has no respiring tissue inside. [2]

HL

Oxidation and reduction

Oxidation and reduction are chemical processes that always occur together. This is because they involve transfer of electrons from one substance to another. Oxidation is the loss of electrons from a substance and reduction is the gain of electrons.

A useful example for biologists is in Benedict's test, a test for certain types of sugar. The test involves the use of copper sulfate solution, containing copper ions with a charge of two positive (Cu^{2+}). These copper ions are reduced and become atoms of copper by being given electrons. Copper atoms are insoluble and so form a red or orange precipitate. The electrons come from sugar molecules, which are therefore oxidized.

Electron carriers in cells

Electron carriers are substances that can accept and give up electrons as required. They often link oxidations and reductions in cells. The main electron carrier in respiration is NAD (nicotinamide adenine dinucleotide). In photosynthesis a phosphorylated version of NAD is used, NADP (nicotinamide adenine dinucleotide phosphate). The structure of the NAD molecule is shown in Figure 8.8. The equation below shows the basic reaction.

$$\text{NAD} + 2\text{ electrons} \longrightarrow \text{reduced NAD}$$

The chemical details are a little more complicated. NAD initially has one positive charge and exists as NAD^+. It accepts two electrons in the following way: two hydrogen atoms are removed from the substance that is being reduced. One of the hydrogen atoms is split into a proton and an electron. The NAD^+ accepts the electron, and the proton (H^+) is released. The NAD accepts both the electron and proton of the other hydrogen atom. The reaction can be shown in two ways:

$$NAD^+ + 2H^+ + 2\text{ electrons } (2e^-) \longrightarrow NADH + H^+$$

$$NAD^+ + 2H \longrightarrow NADH + H^+$$

This reaction demonstrates that reduction can be achieved by accepting atoms of hydrogen, because they have an electron. Oxidation can therefore be achieved by losing hydrogen atoms.

Oxidation and reduction can also occur through loss or gain of atoms of oxygen. There are fewer examples of this in biochemical processes, perhaps because in the early evolution of life, oxygen was absent from the atmosphere. A few types of bacteria can oxidize hydrocarbons using oxygen:

$$\underset{n\text{-octane}}{C_7H_{15}\text{—}CH_3} + \tfrac{1}{2}O_2 \longrightarrow \underset{n\text{-octanol}}{C_7H_{15}\text{—}CH_2OH}$$

Nitrifying bacteria oxidise nitrite ions to nitrate.

$$NO_2^- + \tfrac{1}{2}O_2 \longrightarrow NO_3^-$$

Adding oxygen atoms to a molecule or ion is oxidation, because the oxygen atoms have a high affinity for electrons and so tend to draw them away from other parts of the molecule or ion. In a similar way, losing oxygen atoms is reduction.

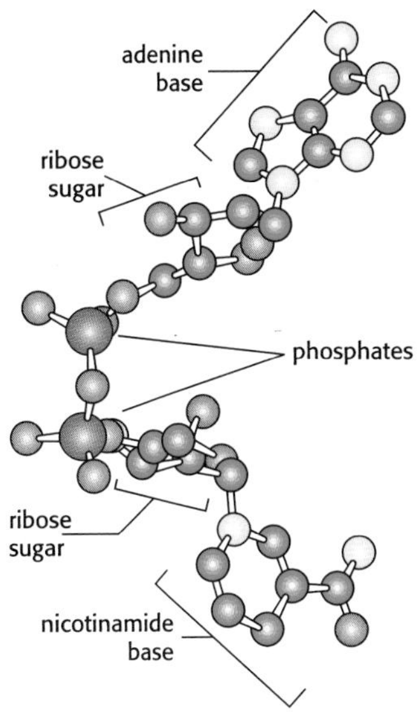

Figure 8 Structure of NAD

In both mitochondria and chloroplasts there are chains of electron carriers located in membranes. Pairs of electrons are passed along these chains, with each carrier first being oxidised and then reduced.

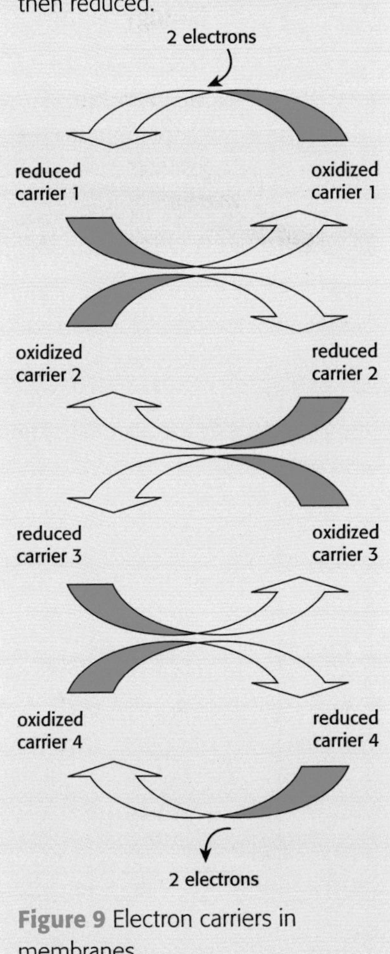

Figure 9 Electron carriers in membranes

HL

Glycolysis

The most significant consequence of glycolysis is the production of a small yield of ATP without the use of any oxygen, by converting sugar into pyruvate. This cannot be done as a single-step process and instead is an example of a metabolic pathway, composed of many small steps. The first of these may seem rather perverse: ATP is used up in phosphorylating sugar.

glucose → fructose → fructose phosphate → fructose biphosphate

ATP ADP   ATP ADP

These phosphorylation reactions reduce the activation energy required for reactions that follow and so make them much easier to perform. In the next step, the fructose bisphosphate is split to form two molecules of triose phosphate. Each of these triose phosphates is then oxidized to glycerate-3-phosphate in a reaction that yields enough energy to make ATP. This oxidation is carried out by removing hydrogen. Note that it is hydrogen atoms that are removed. If only hydrogen ions were removed (H^+), no electrons would be removed and it would not be an oxidation. The hydrogen is accepted by NAD^+, which becomes $NADH + H^+$. In the final stages of glycolysis, the phosphate group is transferred to ADP to produce more ATP and also pyruvate. These stages are summarized in the equation below, which occurs twice per glucose.

NAD^+ $NADH + H^+$

triose phosphate → glycerate-3-phosphate → pyruvate

ATP ADP   ATP ADP

1 ATP is used up in two of the reactions of glycolysis and is produced in two of the reactions. Explain how there is a net yield of 2ATP per glucose from glycolysis. [2]

2 Explain why a supply of NAD^+ is required for glycolysis. [2]

3 In anaerobic conditions, the NAD^+ that is needed in glycolysis is produced by recycling the $NADH + H^+$ produced in glycolysis back to NAD^+. This is done by removing hydrogen. The hydrogen is used to reduce another substance. Deduce the reaction in:
   (a) yeast cells [2]
   (b) humans. [2]

4 ATP inhibits the enzyme that catalyses the conversion of fructose phosphate to fructose bisphosphate. Explain the advantage of this inhibition to cells where it occurs. [2]

Working with data: aerobic cell respiration in yeast

Louis Pasteur discovered that if yeast cells are supplied with oxygen (aerobic conditions), they use glucose six to eight times more slowly than in anaerobic conditions. Figure 10 shows the results of an experiment in which a batch of yeast was cultured in aerobic conditions. The growth of the yeast was monitored, using a colorimeter. Glucose and ethanol concentrations were also monitored.

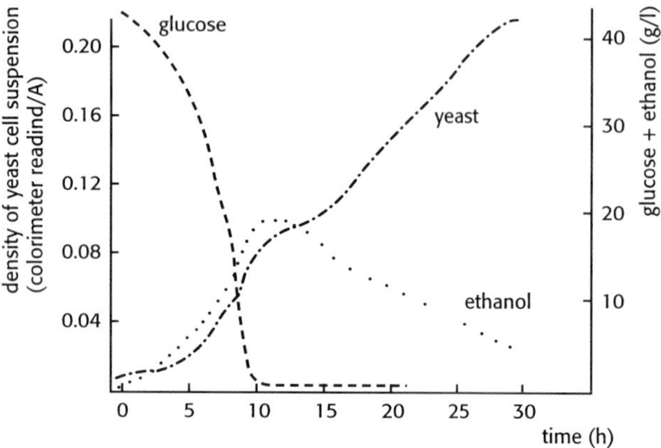

Figure 10 Results of culturing yeast in aerobic conditions

1 (a) Using only the data in the graph, describe the changes in glucose concentration during the first 10 hours of the experiment. [2]
   (b) Explain the changes in glucose concentration, by reference to changes in ethanol concentration and the growth of the yeast. [3]

2 Calculate the increase in yeast between:
   (a) 5 and 10 hours [1]
   (b) 10 and 15 hours. [1]

3 Suggest a reason for the difference in increase in yeast between 5 and 10 hours and 10 and 15 hours. [1]

4 Discuss whether the yeast used aerobic or anaerobic cell respiration during the different phases of the 30-hour period shown on the graph. [2]

HL

The Krebs cycle

Two molecules of pyruvate are produced in glycolysis per molecule of glucose. If oxygen is available, this pyruvate is absorbed into the mitochondrion, where it is fully oxidized.

$$2CH_3\text{—}CO\text{–}COOH + 5O_2 \longrightarrow 6CO_2 + 4H_2O$$
pyruvate

As with glycolysis, this is not a single-step process. The oxidation of pyruvate is achieved by the removal of pairs of hydrogen atoms. The hydrogen carrier NAD^+, and a related compound called FAD, accept these hydrogen atoms. Carbon and oxygen are removed in the form of carbon dioxide, in reactions called decarboxylations. Figure 11 summarizes the process. The first step, in which pyruvate is decarboxylated and oxidized to form an acetyl group, is called the link reaction, because it links glycolysis with the cycle of reactions that follows. This cycle has several names but is often called the Krebs cycle, in honour of the biochemist who was awarded the Nobel Prize for its discovery. The link reaction involves one decarboxylation and one oxidation. There are two more decarboxylations and four more oxidations in the Krebs cycle.

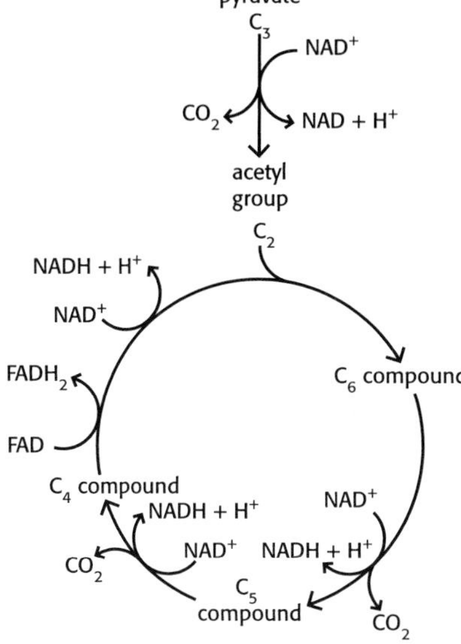

Figure 11 Summary of the Krebs cycle

If glucose or oxygen were oxidized by burning them in air, energy would be released as heat. Most of the energy released in the oxidations of the link reaction and the Krebs cycle is used to reduce hydrogen carriers (NAD^+ and FAD). The energy therefore remains in chemical form and can be passed on to the final part of aerobic cell respiration: oxidative phosphorylation.

Thinking about science: the Krebs cycle and evolution

Hans Krebs was awarded the Nobel Prize in 1953. The two final paragraphs of the lecture that he gave on this occasion are reproduced here.

Before I conclude I would like to make an excursion into general biology, prompted by the remarkable fact that the reactions of the cycle have been found to occur in representatives of all forms of life, from unicellular bacteria and protozoa to the highest mammals. We have long been familiar with the fact that the basic constituents of living matter, such as amino acids and sugars, are essentially the same in all types of life. The study of intermediary metabolism shows that the basic metabolic processes, in particular those providing energy and those leading to the synthesis of cell constituents, are also shared by all forms of life.

The existence of common features in different forms of life indicates some relationship between the different organisms, and according to the concept of evolution these relations stem from the circumstance that the higher organisms, in the course of millions of years, have gradually evolved from simpler ones. The concept of evolution postulates that living organisms have common roots, and in turn the existence of common features is powerful support for the concept of evolution. The presence of the same mechanism of energy production in all forms of life suggests two other inferences: firstly that the mechanism of energy production has arisen very early in the evolutionary process; and secondly that life, in its present forms, has arisen only once.

1 Outline the argument for similarities of metabolism as evidence for evolution.
2 Are there any alternative explanations for the similarities?
3 Do you agree that the concept of evolution "postulates that living organisms have common roots"?

HL

Mitochondrion structure

Mitochondria are found in the cytoplasm of almost all eukaryotic cells. They were discovered at the end of the 19th century and were named by German biologist Carl Benda, using a light microscope. Their detailed structure was revealed by the electron microscope: they have an outer smooth membrane within which there is a narrow space and then an inner membrane. The space inside the inner membrane is larger and is called the matrix. It contains many enzymes, ribosomes and one or more small loops of naked DNA. The inner mitochondrial membrane usually has folds or invaginations. These, together with the double membrane, usually make mitochondria easy to recognize. Four electron micrographs are shown below. The magnifications of the micrographs are as follows: (a) ×45000 (b) ×65000 (c) ×100000 (d) ×10000000

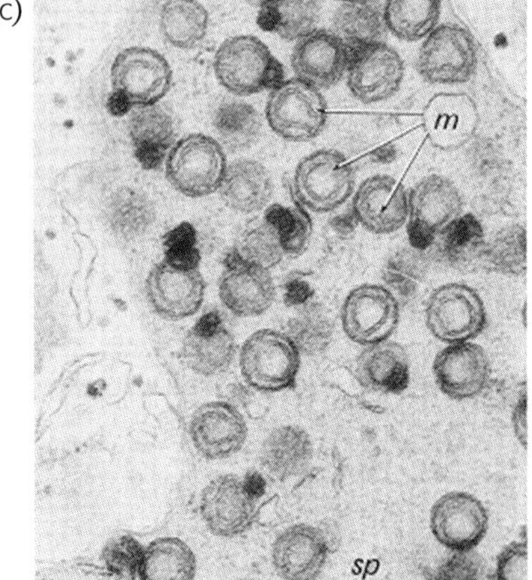

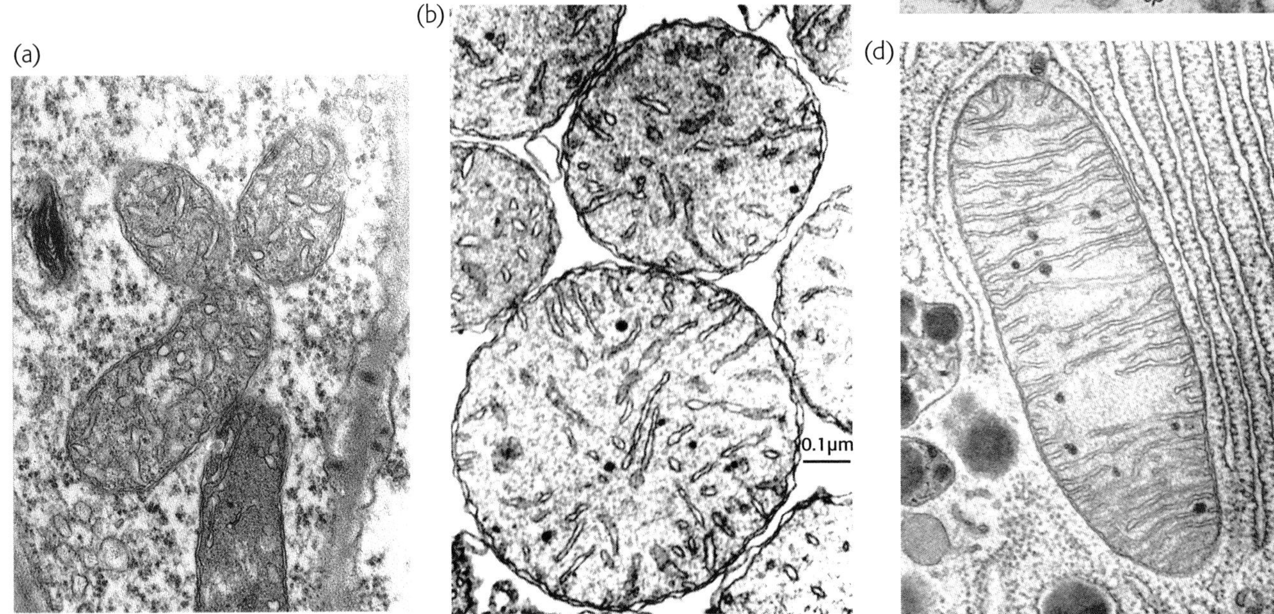

Figure 12 Electron micrographs of mitochondria: (a) from a bean plant (b) from mouse liver (c) from axolotl sperm (d) from bat pancreas

?

Study the electron micrographs in Figure 12 and then answer the multiple-choice questions.

1 The fluid-filled centre of the mitochondrion is called the matrix. What separates the matrix from the cytoplasm around the mitochondrion?
- A One wall.
- B One membrane.
- C Two membranes.
- D One wall and one membrane.

2 What is the range of actual lengths of the mitochondria shown in the electron micrographs?
- A 100nm to 650nm
- B 0.65μm to 1.4μm
- C 0.1μm to 1.4μm
- D 0.65μm to 14μm

3 The mitochondrion matrix contains 70S ribosomes, whereas the cytoplasm of eukaryotic cells contains 80S ribosomes. Which of these hypotheses is consistent with this observation?
- (i) Protein is synthesized in the mitochondrion.
- (ii) Ribosomes in mitochondria have evolved from ribosomes in bacteria.
- (iii) Ribosomes are produced by aerobic cell respiration.
- A (i) only
- B (ii) only
- C (i) and (ii)
- D (i), (ii) and (iii)

HL

Oxidative phosphorylation

The final part of aerobic respiration is called oxidative phosphorylation, because ADP is phosphorylated to produce ATP, using energy released by oxidation. The main substance oxidised is NADH + H^+. The energy is not released in a single large step, but in a series of small steps, carried out by a chain of electron carriers. In this way, more of the energy can be trapped in ATP. The mechanism used to couple the release of energy by oxidation to ATP production remained a mystery for many years, but is now known to be chemiosmosis. This happens in the inner mitochondrion membrane. It is called chemiosmosis, because a chemical substance (H^+) moves across a membrane, down the concentration gradient. This releases the energy needed for the enzyme ATP synthase to make ATP. The main steps in the process are as follows (see also Figure 13).

- NADH + H^+ supplies pairs of hydrogen atoms to the first carrier in the chain, with the NAD^+ returning to the matrix.
- The hydrogen atoms are split, to release two electrons, which pass from carrier to carrier in the chain.
- Energy is released as the electrons pass from carrier to carrier, and three of these use this energy is used to transfer protons (H+) across the inner mitochondrial membrane, from the matrix to the intermembrane space.
- As electrons continue to flow along the chain and more and more protons are pumped across the inner mitochondrial membrane, a concentration gradient of protons builds up. This proton gradient is a store of potential energy.
- To allow electrons to continue to flow, they must be transferred to a terminal electron acceptor at the end of the chain. In aerobic respiration this is oxygen, which briefly becomes O^{2-}, but then combines with two H^+ ions from the matrix to become water.
- Protons pass back from the intermembrane space to the matrix through ATP synthase. As they are moving down the concentration gradient, energy is released and this is used by ATP synthase to phosphorylate ADP.

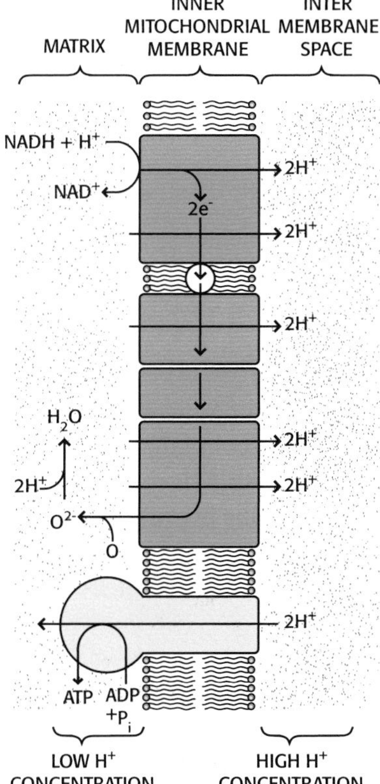

Figure 13 Summary of oxidative phosphorylation

Data-based question: oxygen consumption by mitochondria

Figure 14 shows the results of an experiment in which mitochondria were extracted from liver cells and were kept in a fluid medium, in which oxygen levels were monitored. Pyruvate was added at point I on the graph, and ADP was added at points II, III and IV.

1 Explain why oxygen was not consumed:
   (a) before pyruvate was added [2]
   (b) after pyruvate was added. [2]
2 Predict, with reasons, what would have happened if ADP had not been added at stage (iii). [2]
3 Suggest two reasons for the mitochondria not resuming consumption of oxygen after ADP was added at stage (iv). [4]

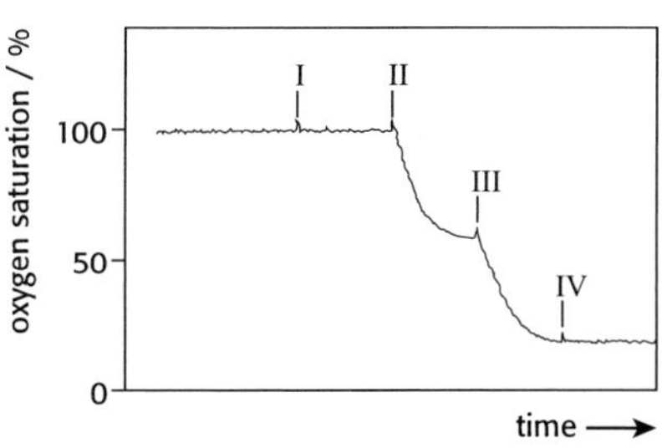

Figure 14 Results of oxygen consumption experiment

HL

Structure–function relationships in mitochondria

There is often a clear relationship between the structures of the parts of living organisms and the functions they perform. This can be explained in terms of natural selection and evolution. The mitochondrion can be used as an example (see Figure 15). If mitochondrial structure varied, those organisms with the mitochondria that produced ATP most efficiently would have an advantage. They would have an increased chance of survival and would tend to produce more offspring. These offspring would inherit the type of mitochondria that produce ATP more efficiently. If this trend continued, the structure of mitochondria would gradually evolve to become more and more efficient. This is called **adaptation** – a change in structure so that something carries out its function more efficiently.

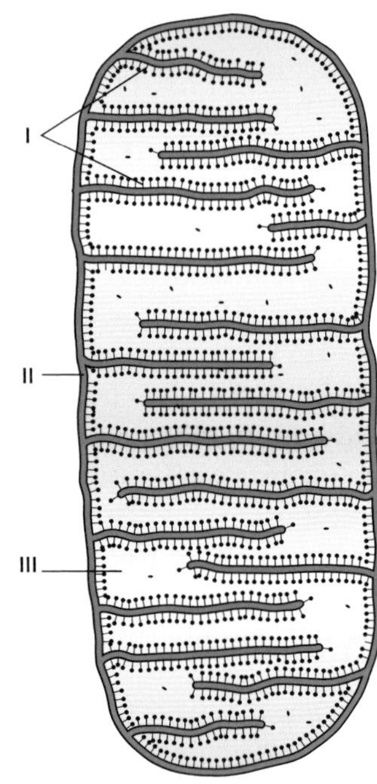

Figure 15 Mitochondrial structure

Study Figure 15 and deduce how each of the labelled structures is adapted to help in the efficient production of ATP: I many cristae, II small space between inner and outer membrane, III matrix with decarboxylases and dehydrogenases.

Thinking about science: Peter Mitchell and chemiosmosis

In 1961 Peter Mitchell proposed the chemiosmotic hypothesis to explain the coupling of electron transport in the inner mitochondrial membrane to ATP synthesis. His hypothesis was a radical departure from previous hypotheses and only after many years was it generally accepted. He was awarded the Nobel Prize for Chemistry in 1978 and part of the Banquet Speech that he gave is reproduced here.

Emile Zola described a work of art as a corner of nature seen through a temperament. The philosopher Karl Popper, the economist F.A. Hayek, and the art historian K.H. Gombrich have shown that the creative process in science and art consists of two main activities: an imaginative jumping forward to a new abstraction or simplified representation, followed by a critical looking back to see how nature appears in the light of the new vision. The imaginative leap forward is a hazardous, unreasonable activity. Reason can be used only when looking critically back. Moreover, in the experimental sciences, the scientific fraternity must test a new theory to destruction, if possible. Meanwhile, the creator of a theory may have a very lonely time, especially if his colleagues find his views of nature unfamiliar and difficult to appreciate.

The final outcome cannot be known, either to the originator of a new theory, or to his colleagues and critics, who are bent on falsifying it. Thus, the scientific innovator may feel all the more lonely and uncertain.

On the other hand, faced with a new theory, the members of the scientific establishment are often more vulnerable than the lonely innovator. For, if the innovator should happen to be right, the ensuing upheaval of the established order may be very painful and uncongenial to those who have long committed themselves to develop and serve it. Such, I believe, has been the case in the field of knowledge with which my work has been involved. Naturally I have been deeply moved, and not a little astonished, by the accidents of fortune that have brought me to this point.

1. What could Zola have meant when he said "a work of art is a corner of nature seen through a temperament?
2. The "upheaval of the established order" has been referred to by Thomas Kuhn as a "scientific revolution" as it results in a shift in the framework of how observations are interpreted. What other examples are there in the history of the biological sciences?
3. Max Planck remarked that a new scientific idea does not triumph by convincing its opponents, but rather because its opponents eventually die. Is this true, or is Planck being unduly cynical about scientists?
4. With whom does the burden of proof lie – the innovator or the scientific establishment?

Thinking about science: the role of technology

Ideas in science sometimes change gradually. But sometimes they remain stable for years or even decades and then undergo a sudden change This can be due to the insight or enthusiasm of a particular scientist, or team. The development of new techniques can sometimes be the stimulus. The technique of electron tomography has recently allowed 3-D images of the interior of mitochondria to be made. One of the leaders in this field is Dr. Carmen Mannella, Director, Division of Molecular Medicine, Wadsworth Center Albany NY: Resource for Visualization of Biological Complexity. He recently gave this brief comment on developments in our understanding of mitochondrial structure and function.

The new take-home message about the mitochondrial inner membrane is that the cristae are not simple infoldings but are invaginations, defining micro-compartments in the organelle. The cristae originate at narrow openings (crista junctions) that likely restrict diffusion of proteins and metabolites between the compartments. The membranes are not only very flexible but also dynamic, undergoing fusion and fission in response to changes in metabolism and physiological stimuli. The working hypothesis is that the observed changes in membrane shape (topology) are not random and passive but rather a specific mechanism by which mitochondrial function is regulated by changes in internal diffusion pathways, e.g., allowing more efficient utilization of ADP. It appears that there are specific proteins and lipids that actively regulate the topology of the inner membrane. This is a bit speculative at the time but it gives a sense of where things are headed in the field.

Figure 16 shows three images of the inner mitochondrial membrane of mitochondria from liver cells.

1 What does electron tomography allow us to do, which previous methods of electron microscopy could not easily do?
2 What has been revealed about mitochondria that was not clear before?
3 How should the diagram on page 92 be modified?

Figure 16 Three images of the inner mitochondrial membrane of mitochondria from liver cells

Chapter 8 questions

1 What is cell respiration?
A Taking air into the lungs.
B Removing carbon dioxide from the lungs.
C Use of ATP in cells.
D Controlled release of energy in cells.

2 What are the products of glycolysis?
A Glucose.
B Glucose and ATP.
C ATP and pyruvate.
D Glucose, ATP and pyruvate.

3 Which substances are produced by yeast in anaerobic cell respiration?
A Carbon dioxide and lactate.
B Carbon dioxide and ethanol.
C Lactate and ethanol.
D Carbon dioxide, lactate and ethanol.

4 If a blood sample from a person contains a high concentration of lactate, what is the conclusion?
A The person has been drinking too much milk and ingested large amounts of lactose.
B The person is lactose intolerant and should change to drinking lactose-free milk.
C The person has eaten live yoghurt, containing bacteria that have carried out anaerobic cell respiration.
D The person has exercised vigorously and carried out anaerobic cell respiration.

5 Where is pyruvate produced and used in a cell respiring aerobically?

	Where is pyruvate produced?	Where is pyruvate used?
A	In the cytoplasm	In the cytoplasm
B	In the cytoplasm	In the mitochondrion
C	In the mitochondrion	In the cytoplasm
D	In the mitochondrion	In the mitochondrion

6 What is the main source of energy in aerobic cell respiration?
A Oxygen.
B ATP.
C Heat.
D Organic compounds.

Higher-level questions

7 (a) State one example in aerobic cell respiration of:
(i) a six-carbon compound [1]
(ii) a three-carbon compound [1]
(iii) a two-carbon compound [1]
(iv) a one-carbon compound [1]
(v) a no-carbon compound. [1]
(b) State the stage of aerobic cell respiration that involves:
(i) a five-carbon compound [1]
(ii) a four-carbon compound. [1]
(c) State the name of a compound in cell respiration that can:
(i) accept hydrogen atoms [1]
(ii) accept phosphate. [1]

8 (a) Distinguish between the processes of oxidation and reduction. [4]
(b) Outline the process of glycolysis. [6]
(c) Explain the reasons for a much lower ATP yield per glucose from anaerobic cell respiration than from aerobic cell respiration. [8]

9 A new technique, called electron tomography, has been used recently to obtain three-dimensional images of mitochondria. The questions below refer to the three images of the cristae of a mitochondrion on page 93. The inner and outer membranes of the mitochondrion are not shown. They would have been outside the cristae.
(a) The diameter of the mitochondrion was 700 nm. Calculate the magnification of the image. [3]
(b) Compare the cristae in Figure 16 with the cristae in Figure 15 on page 93. [3]
(c) Electon tomography has shown that cristae are dynamic structures and that the volume of the intracristal compartment increases when the mitochondrion is active in electron transport. Suggest how electron transport could cause an increase in the volume of fluid inside the cristae. [2]
(d) Junctions between the cristae and boundary region of the inner mitochondrial membrane can be have the shape of slots or tubes and can be narrow or wide. Suggest how narrow tubular connections could help in ATP synthesis by one of the cristae in a mitochondrion. [2]

9 Photosynthesis

Photosynthesis is production of organic compounds, including carbohydrates, proteins and lipids, using light energy and simple inorganic substances such as carbon dioxide and water. It is an example of energy conversion, as light energy is converted into chemical energy.

Prokaryotes were the first organisms to perform photosynthesis, starting about 3500 million years ago. They were joined billions of years later by algae and plants, which have been carrying out photosynthesis ever since. One consequence of photosynthesis is the rise in the oxygen concentration of the atmosphere, which began about 2300 million years ago (see Figure 1).

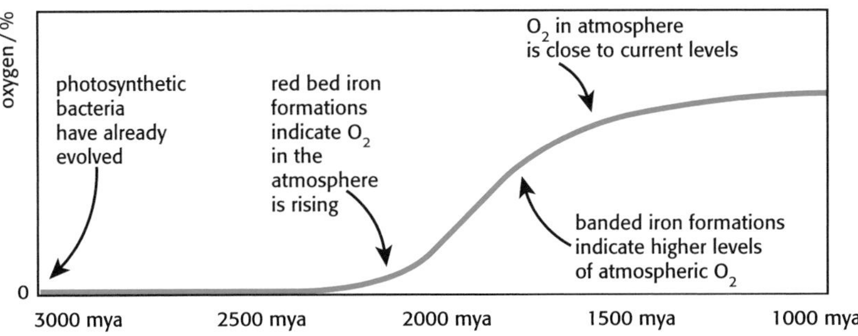

Figure 1 Possible oxygen concentrations in the Earth's atmosphere between 3000 and 1000 million years ago (mya), as indicated by reduced or oxidised iron in rocks

Table 1

Planet	Composition of atmosphere (%)				
	CO_2	N_2	Ar	O_2	H_2O
Venus	98	1	1	0	0
Earth	0.04	78	1	21	0.1
Mars	96	2.5	1.5	2.5	0.1

?

Compare the composition of the Earth's atmosphere with that of Venus and Mars.

Thinking about science: Helmont's willow experiment

One of the most famous experiments in the history of biology is that of the Flemish scientist Johannes Baptista van Helmont, published in 1648. It is regarded as the first quantitative biology experiment and also changed our understanding of the growth of plants. At this time, plants were thought to be "soil-eaters". To test this idea, van Helmont put 200 pounds (90 kg) of dry soil in large pot and in it planted a willow tree, which had a mass of 5 pounds (2.2 kg). He attempted to keep dust out of the pot by covering it with a perforated metal plate. He watered the tree with rainwater or distilled water over a period of five years. When the willow was reweighed at the end of this time it had increased to 169 pounds (76 kg). After drying the soil from the pot he found that it had remained almost unchanged in mass, having lost only one eighth of a pound (about 50g).

1 What evidence against the hypothesis that plants are soil eaters does van Helmont's experiment provide?
2 Removal of willow roots from soil is very difficult as soil particles inevitably get stuck to the roots. van Helmont's masses for the soil before and after the five-year period are therefore remarkably close. What are the likely consequences, if scientists make their data fit pre-decided conclusions?
3 van Helmont concluded from his results that, "164 pounds of Wood, Barks, and Roots, arose out of water only." (164 pounds is 73 kg.) This was not a new idea – 2000 years earlier the Greek philosopher Thales had stated that all matter arose from water. To what extent was van Helmont's conclusion correct?

Figure 2 Johannes van Helmont

Figure 3 Helmont's willow plant before (left) and after (right)

Pigments and light absorption

The first stage in photosynthesis is the absorption of sunlight. Visible light is a form of electromagnetic radiation. Sunlight is a mixture of different wavelengths of visible light, which we see as different colours, including blue, green and red. When droplets of water in the sky split sunlight up and a rainbow is formed, the different colours are made visible. Light absorption in photosynthesis involves chemical substances called pigments. A white or transparent substance does not absorb visible light. Pigments are substances that do absorb light and therefore appear coloured to us. Pigments that absorb all of the colours appear black, because they emit no light. Pigments that, for example, absorb all colours except blue appear blue to us, because this part of the sunlight is not absorbed. Instead the blue light is reflected and can pass into our eye, to be perceived by cells in the retina. Photosynthesizing organisms use a range of pigments, but the main photosynthetic pigment is chlorophyll. Figure 4 shows the amount of energy in solar radiation at wavelengths up to 3000 nm.

1 Chlorophyll appears green. Which colours of light does chlorophyll absorb and which does it reflect? [3]
2 Some flowers contain a yellow pigment. Deduce which colours it absorbs? [3]
3 The part of the spectrum that is absorbed by the retina and is therefore visible to humans is shown as the grey band in Figure 4. Plants absorb a similar range of wavelengths for use in photosynthesis. Suggest reasons for this similarity. [4]

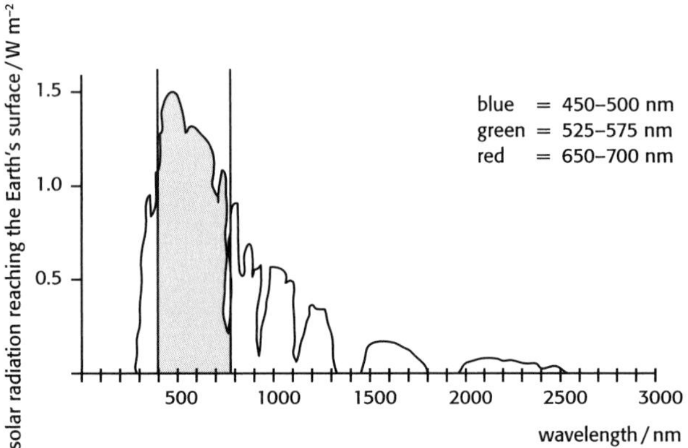

Figure 4 The spectrum of electromagnetic radiation reaching the Earth's surface

Rediscovering biology: photosynthesis rates in red, green and blue light

The apparatus shown in Figure 5 can be used to find out how much photosynthesis occurs in red, green and blue light. The intensity of light emitted by the red, green and blue LEDs (light-emitting diodes) needs to be checked and adjusted until it is equal. A powerful lamp with colour filters of the sort used for theatre lighting can also be used, but this gives a wider range of wavelengths and so is not as suitable. The pondweed must be healthy, with plenty of leaves and is placed upside down in the tube, with the upper end of the stem cut. With each colour, the number of bubbles of oxygen that emerge per unit time can be counted, to give a measure of the rate of photosynthesis.

1 Predict, with a reason, which colour of light will give the lowest rate of photosynthesis. [2]
2 Explain why the pondweed must be healthy, with plenty of leaves. [2]
3 Explain why the pondweed must be placed upside down, with the end of its stem cut. [2]
4 List the variables that must be kept constant in this experiment. [3]
5 Suggest one reason for the pondweed stopping bubbling after a few minutes in this experiment, even if it is in bright light. [1]

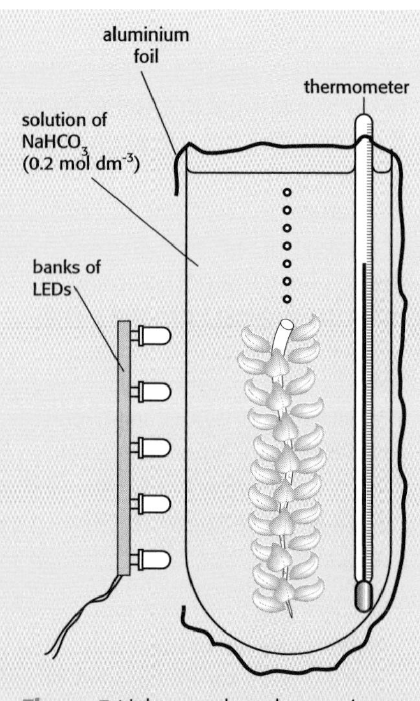

Figure 5 Light wavelength experiment

Photosynthesis–the effect of light intensity

The intensity of light can affect the rate of photosynthesis. Figure 6 shows a typical relationship. At low light intensities the rate of photosynthesis increases when the light intensity is increased, but at high light intensities, further increases in light intensity have no effect.

Several essential steps in photosynthesis are achieved using the light energy absorbed by chlorophyll:

- ATP is produced, from ADP and phosphate.
- Water molecules are split to release hydrogen (photolysis).

Hydrogen is needed in a later stage of photosynthesis to make carbohydrate. Splitting water also releases oxygen, which is a toxic waste product and is allowed to diffuse out into the surrounding water or air.

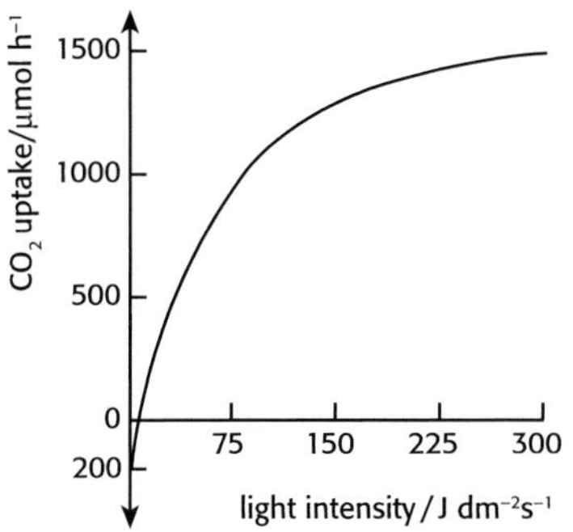

Figure 6 The graph shows the results of an experiment in which the rate of photosynthesis was found by measuring the uptake of carbon dioxide.

Rediscovering biology: the effect of light intensity on photosynthesis

Photosynthesis experiments can be done using leaf discs, cut out of young healthy leaves using a cork borer or the end of a drinking straw. If the discs are prevented from drying out, and are placed in a suitable environment, they will continue to photosynthesize for at least a few hours. The method below could be done with each student or group of students using a different light intensity.

1. Remove the barrel of a 10 ml plastic syringe, cover the end of the nozzle with a finger and then pour 10 ml of 0.2 mol dm^{-3} sodium hydrogen carbonate solution into the barrel of the syringe, to provide a supply of carbon dioxide.
2. Put ten leaf discs into the sodium hydrogen carbonate solution and replace the plunger of the syringe.
3. Hold the syringe vertically with the nozzle pointing upwards. Squeeze all the air out of the syringe, then put a finger over the nozzle and pull on the barrel of the syringe. This will create suction, which will draw air out of the air spaces inside the leaf discs. Stop applying suction and take your finger off the end of the nozzle. Tap the syringe to encourage the bubbles of gas to rise into the nozzle.
4. Repeat the previous stage several times, until the leaf discs become denser than the sodium hydrogen carbonate solution and sink to the bottom.
5. Place the syringe in a vertical position near a bright light source, ensuring that the discs are all fully illuminated. The discs should eventually rise to the top of the syringe, when enough oxygen has been produced inside them to make them less dense than the solution. Time how long it takes for each disc to rise to the surface.
6. Measure the light intensity using a lux meter.
7. Plot a graph of the results for all the light intensities that have been tested.
8. The leaf discs could be tested for starch at the end of the experiment.

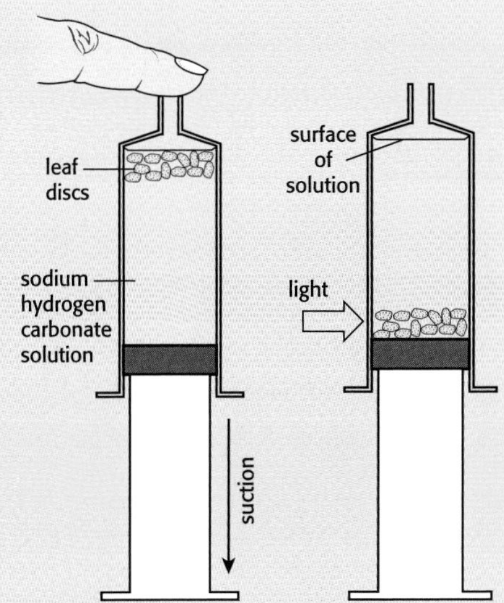

Figure 7 Method of measuring the rate of photosynthesis in leaf discs

Photosynthesis–the effect of carbon dioxide concentration

Although the concentration of carbon dioxide in the atmosphere has been rising over the past 100 years, it is still low enough to affect the rate of photosynthesis at times. Figure 8 shows a typical relationship. At low carbon dioxide concentrations, increases in concentration cause the rate of photosynthesis to rise. At high carbon dioxide concentrations, further increases in concentration do not affect the rate of photosynthesis.

Carbon dioxide is essential for photosynthesis as it is the source of the carbon that is needed to make organic molecules. The conversion of carbon dioxide into solid or liquid carbon compounds is called carbon fixation. ATP is needed to provide energy and hydrogen from the photolysis of water is also needed. In darkness carbon fixation therefore stops and in low light intensities it slows down, due to a shortage of ATP and hydrogen.

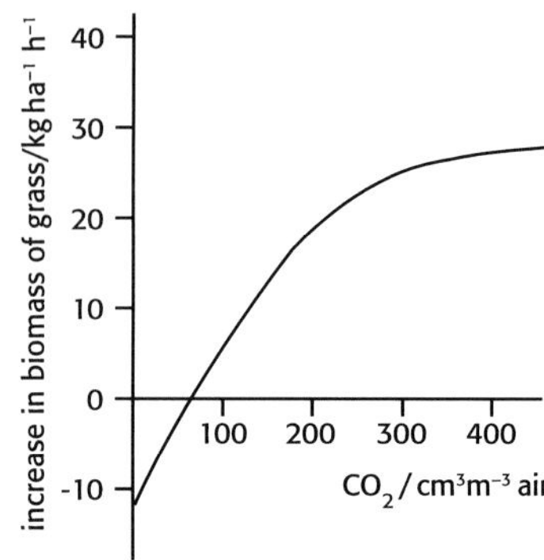

Figure 8 In this graph the rate of photosynthesis was measured indirectly by measuring the change in plant biomass.

Rediscovering biology: factors affecting photosynthesis in pondweed

If a stem of pondweed such as *Elodea, Cabomba, or Myriophyllum* is placed upside-down in water and the end of the stem is cut, bubbles of gas may be seen to escape. If these are collected and tested, they are found to be mostly oxygen, produced by photosynthesis. The rate of oxygen production can be measured by counting the bubbles. Factors that might affect the rate of photosynthesis can be varied to find out what effect this has. In the method below carbon dioxide concentration is varied.

1 Enough water to fill a large beaker is boiled and allowed to cool. This removes carbon dioxide and other dissolved gases.

2 The water is poured repeatedly from one beaker to another, to oxygenate the water. Very little carbon dioxide will dissolve.

3 A stem of pondweed is placed upside-down in the water and the end of its stem is cut. No bubbles are expected to emerge, as the water contains almost no carbon dioxide. The temperature of the water should be about 25 °C and the water should be very brightly illuminated. Suitable apparatus is shown in Figure 9.

4 Enough sodium hydrogen carbonate is added to the beaker to raise the carbon dioxide concentration by 0.01 mol dm^{-3}. If bubbles emerge, they are counted for 30 seconds, repeating the counts until two or three consistent results are obtained.

5 Enough sodium hydrogen carbonate is added to raise the concentration by another 0.01 mol dm^{-3}. Bubble counts are done in the same way.

6 The procedure above is repeated again and again until further increases in carbon dioxide do not affect the rate of bubble production.

1 Why are the following procedures necessary?
   (a) Boiling and then cooling the water before the experiment.
   (b) Keeping the water at 25°C and brightly illuminating it.
   (c) Repeating bubble counts until several consistent counts have been obtained.

2 What other factor could be investigated using bubble counts with pondweed and how would you design the experiment?

3 How could you make the measurement of the rate of oxygen production more accurate?

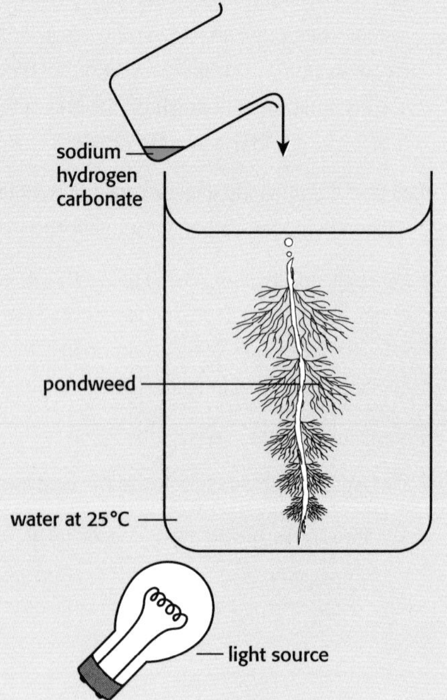

Figure 9 Apparatus for measuring photosynthesis rates in different concentrations of carbon dioxide

Photosynthesis–the effect of temperature

Like light intensity and carbon dioxide concentration, temperature can affect the rate of photosynthesis, but the relationship is rather different. At low temperatures the rate of photosynthesis is very low or zero. As temperature increases the rate of photosynthesis increases until it reaches a maximum at a temperature that is called the optimum for photosynthesis (see Figure 10). Above this temperature, the rate of photosynthesis decreases and eventually becomes zero. This is similar to the effect of temperature on enzymes. This is not coincidental. The fixation of carbon dioxide is catalysed by enzymes, which work faster as temperature increases. In most plants the optimum temperature is between 25 and 35 °C. This is lower than the temperature at which enzymes are denatured, but one of the enzymes involved in photosynthesis works inefficiently above about 30 °C.

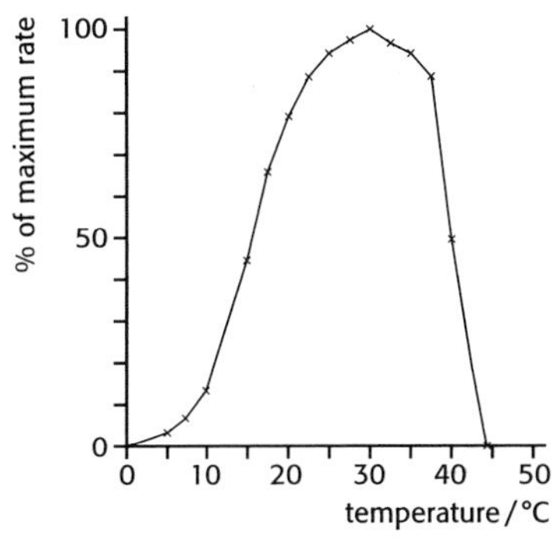

Figure 10 The graph shows the results of an experiment in which photosynthesis was measured using oxygen production rates. The results are shown as a percentage of the rate at the optimum temperature.

Data-based question: measuring the effect of temperature by data logging

Photosynthesis can be monitored by data logging.

Oxygen concentrations in water around pondweed can be monitored using an oxygen electrode, or pressure changes due to oxygen could be monitored. The apparatus in Figure 11 can be used to measure carbon dioxide uptake. Table 2 shows results obtained at five different temperatures.

Table 2

Temperature (°C)	pH (±0.05) at 1-hour time intervals				
	0 hours	1 hour	2 hours	3 hours	4 hours
22.5	9.39	9.45	9.53	9.59	9.64
25.0	9.41	9.47	9.53	9.62	9.69
27.5	9.40	9.46	9.54	9.59	9.66
30.0	9.39	9.42	9.48	9.54	9.58
32.5	9.42	9.44	9.48	9.53	9.55

1 Why was pH monitored during this experiment? [2]
2 Identify the independent and the dependent variable. [2]
3 Deduce which variable should go on the *x* axis and which variable on the *y* axis if a graph of the results is plotted. [2]
4 Discuss what conclusions can be drawn about the optimum temperature for photosynthesis in the green alga *Chlorella*. [2]
5 If this experiment is repeated using a different alga, the results may not be the same. Suggest reasons for differences in the results. [2]
6 Evaluate the methods used in the experiment and suggest improvements. [5]

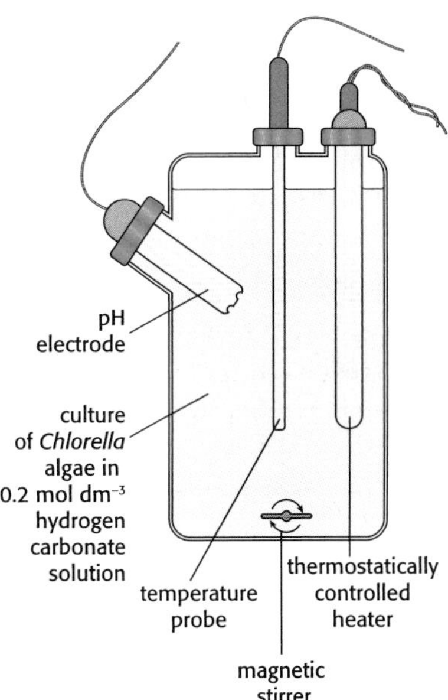

Figure 11 Fermenter used for measuring the rate of photosynthesis at different temperatures by data logging

Working with data: error bars and regression lines

In biology experiments the results are often variable, even when attempts have been made to keep conditions constant. Because of this it is often useful to repeat readings and compare the results. If any of the results are much lower or higher than the others, they are probably anomalous and a mistake may have been made in the experiment. These readings should therefore be repeated. However, it is not acceptable to discard results simply because they do not fit a hypothesis.

If a consistent set of results is obtained, the mean results can be calculated and plotted on a graph. Sometimes it is useful to show the range of results as well as the means. The easiest way to do this is to plot the mean result and then also the highest and the lowest results, joined with a line. This line is a simple example of an error bar. In scientific papers error bars are often shown, but they usually indicate the size of a statistic that gives a measure of the range of results, such as standard deviation. It is important to indicate which type of error bar is being used.

Another way of showing how much variation there is in set of results is to plot every result individually. A line can then be drawn, as close as possible to all of the points. This is called a regression line. A computer is usually used to calculate its position.

The graphs in Figure 12 show the effect of planting density of Apeldoorn tulip bulbs on the yield of bulbs large enough to sell.

The upper graph shows the mean values, and error bars showing the highest and lowest values. The lower graph shows all the results, with a regression line.

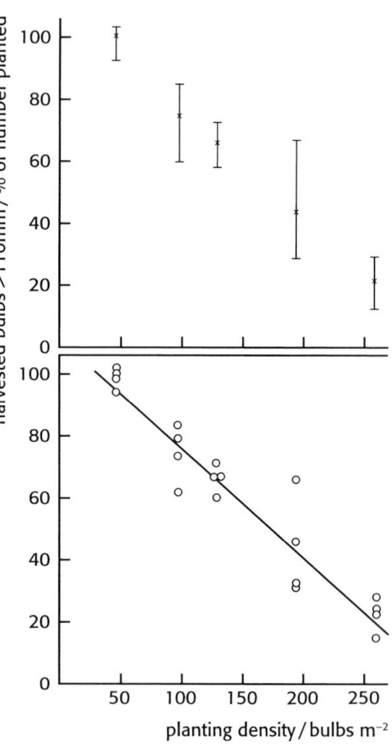

Figure 12 Effect of planting density on the yield of tulip bulbs

Data-based question: photosynthesis rates in red light

Figure 13 shows the results of an experiment in which *Chlorella* cells were given light of wavelengths from 660 nm (red) up to 700 nm (far red). The rate of oxygen production by photosynthesis was measured and the yield of oxygen per photon of light was calculated. This gives a measure of the efficiency of photosynthesis at each wavelength. The experiment was then repeated with supplementary light with a wavelength of 650 nm at the same time as each of the wavelengths from 660 to 700 nm, but with the same overall intensity of light as in the first experiment.

1 Describe the relationship between wavelength of light and oxygen yield, when there was no supplementary light. [2]
2 Describe the effect of the supplementary light. [2]
3 Explain how the error bars help in drawing conclusions from this experiment. [2]
4 The probable maximum yield of oxygen was 0.125 molecules per photon of light. Calculate how many photons are needed to produce one oxygen molecule in photosynthesis. [2]
5 Oxygen production by photolysis involves this reaction:

$$4H_2O \rightarrow O_2 + 2H_2O + 4H^+ + 4e^-$$

Each photon of light is used to excite an electron (raise it to a higher energy level). Calculate how many times each electron produced by photolysis must be excited during the reactions of photosynthesis. [2]

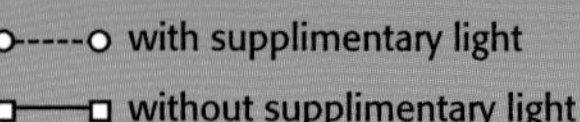

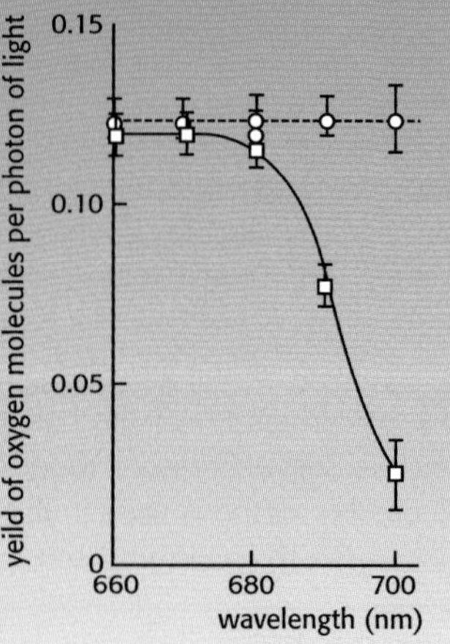

Figure 13 Photon yield of photosynthesis in different light intensities

HL

Action spectra and absorption spectra

An action spectrum is a graph showing the rate of photosynthesis at each wavelength of light. An absorption spectrum is a graph showing the percentage of light absorbed at each wavelength by a pigment or a group of pigments. It is not difficult to explain why action and absorption spectra are very similar: photosynthesis can only occur in wavelengths of light that chlorophyll or the other photosynthetic pigments can absorb.

Absorption of light by chlorophyll has been explained in terms of discrete units of light energy, each of which is called a quantum of light, or photon. The energy carried by photons depends on their wavelength: the shorter the wavelength the greater the energy. An electron in a pigment molecule can absorb the photon if the energy carried by it allows the electron to be raised up from one energy level to a higher energy level. If the photon carries either too much or too little energy, it cannot be absorbed. The electron that has been raised to a higher energy level is called an excited electron. There are different types of chlorophyll in chloroplasts, with slightly different light-absorbing properties, but all of them absorb in the red and blue parts of the spectrum. The other pigments present, called accessory pigments, absorb other wavelengths and then transfer the light energy to chlorophyll. The absorption spectra in Figure 14 show that these photosynthetic pigments together absorb a wide range of wavelengths, but green light is nevertheless absorbed less effectively than other wavelengths. The rate of photosynthesis, shown in the action spectrum (Figure 15) is therefore lower for green light.

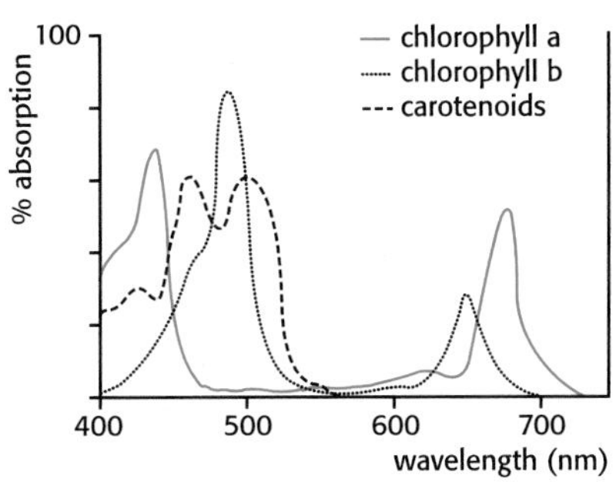

Figure 14 Absorption spectra of chlorophylls

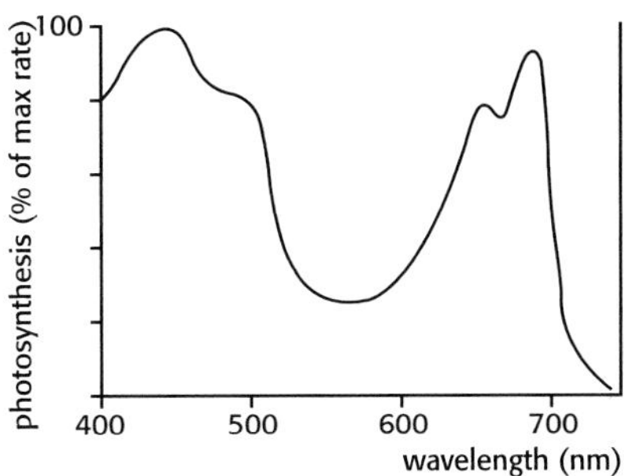

Figure 15 Action spectrum of photosynthesis

Rediscovering biology: measuring light absorption by photosynthetic pigments

To test the light absorbing properties of chlorophylls and the accessory pigments, grind leaves in an organic solvent such as propanone, with a little sand to provide abrasion. The solution obtained contains chlorophyll together with smaller amounts of the other pigments in the leaf. The amount of light absorbed at different wavelengths can then be measured using a colorimeter. Figure 16 shows the results obtained using a green leaf of *Fagus sylvatica* (beech) and a leaf of *Acer palmatum* (Japanese maple), which contained a red pigment in addition to chlorophyll.

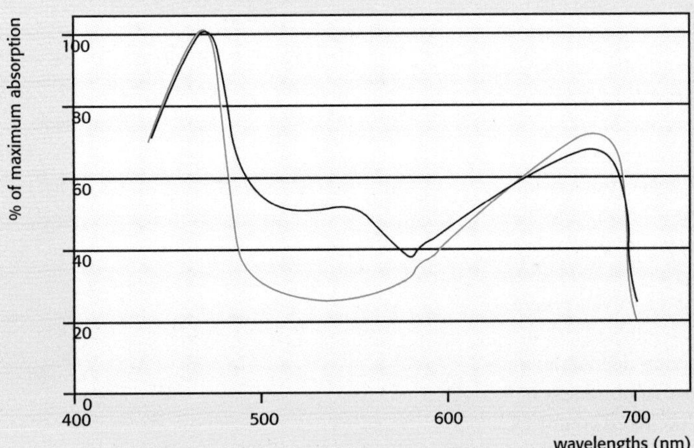

Figure 16 Absorption spectrum of pigments from two types of leaf

1 Compare the absorption of the two samples of pigments shown in the graph, including both the similarities and the differences. [3]
2 Deduce, with reasons, which curve shows the absorption of the pigments from the *Fagus* leaf and which shows the absorption of the pigments from the *Acer* leaf. [2]
3 Suggest reasons for plants using pigments that absorb light in the range 400–700nm and not higher or lower wavelengths. [3]
4 Some algae growing on rocky beaches have a brown colour. Predict, with reasons, the curve that would be obtained if light absorption of their pigments were investigated. [2]

HL

The concept of limiting factors

The effects of light intensity, temperature and carbon dioxide concentration on the rate of photosynthesis were discussed earlier in this chapter. Each of these factors can limit the rate if they are below the optimal level. These three factors are therefore called limiting factors. According to the concept of limiting factors, under any combination of light intensity, temperature and carbon dioxide concentration, only one of the factors is actually limiting the rate of photosynthesis. This is the factor that is furthest from its optimum. If the factor is changed to make it closer to the optimum, the rate of photosynthesis increases, but changing the other factors will have no effect, as they are not the limiting factor. Of course, as the limiting factor is moved closer to its optimum, while keeping the other factors constant, a point will be reached where this factor is no longer the one that is furthest from its optimum and another factor becomes the limiting factor. For example, at night, light intensity is presumably the limiting factor for photosynthesis. When the sun rises and light intensity increases, temperature will usually take over as the limiting factor. As the temperature increases during the morning, carbon dioxide concentration might well become the limiting factor.

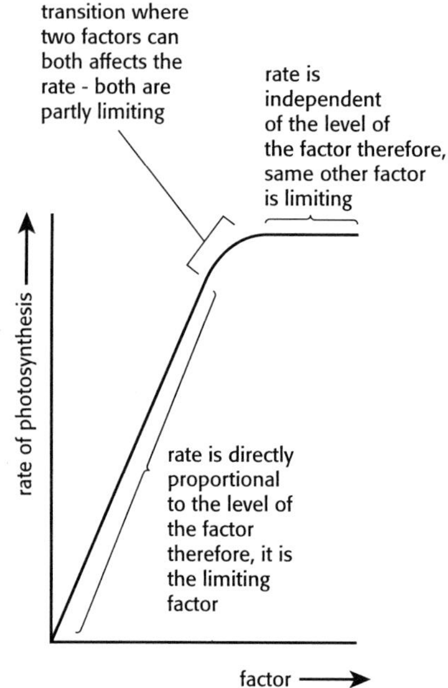

Figure 17 Graph showing the general relationship between a limiting factor and the rate of photosynthesis

The concept of limiting factors fits most experimental data, but there are some combinations of light intensity, temperature and carbon dioxide concentration where increasing either of two factors increases the rate of photosynthesis. This is where both are equally close to their optimum. It explains the curved region in Figure 17 showing the effect of a limiting factor.

The three limiting factors affect the photosynthesis at different stages in the process. To understand this we must look more carefully at the stages and these are described in the following pages.

Thinking about science: the need to control variables

It is clear that in photosynthesis experiments, light intensity, temperature and carbon dioxide concentration must all be controlled. To find the effect of one factor, the other two must be kept constant. Usually it is best to keep them at their optimum level, to ensure that they do not limit the rate.

Careful control of variables is an essential part of experimental design. It is quite easy in photosynthesis experiments, but is much more difficult in other areas of biology. The following is from a newspaper article summarizing a scientific investigation.

Sight worries? Try looking beyond carrots

According to the *Journal of Ophthalmology, sweetcorn*, broccoli, peas and squash contain the compounds lutein and zeaxanthin, which offer protection against damage to eyesight later in life. The compounds are carotenoids, the substances that give fruit and vegetables their colour – carrots contain beta-carotene. Scientists from the University of Wisconsin studied women aged 50–79 who either had high or low dietary intakes of lutein or zeaxanthin. They concluded that higher intakes over 15 years were associated with a reduced risk of age-related macular degeneration, the leading cause of blindness in old people in developed countries.

1 If we were able to compare the two groups of women with high and low intakes of lutein and zeaxanthin, would we expect them to be similar in other factors affecting health, or would be expect to find differences?

2 (a) From the information given in the article, what do you know that the researchers did?
  (b) What do you suspect that they did?
  (c) What *must* they have done, for their conclusions to be justified?

HL

Photoactivation and photolysis

Chlorophyll and the accessory pigments are grouped together in large light-harvesting arrays called photosystems. These photosystems are located in the thylakoids, an arrangement of membranes inside the chloroplast. There are two types of light harvesting arrays, called photosystems I and II. Both types of photosystem contain many chlorophyll molecules, which absorb light energy and pass it to two special chlorophyll molecules in the reaction centre of the photosystem. Like other chlorophylls, when these special chlorophyll molecules absorb the energy from a photon of light an electron within the molecule becomes excited. The chlorophyll is then **photoactivated**. However, the chlorophylls at the reaction centre have the special property of being able to donate excited electrons to an electron acceptor.

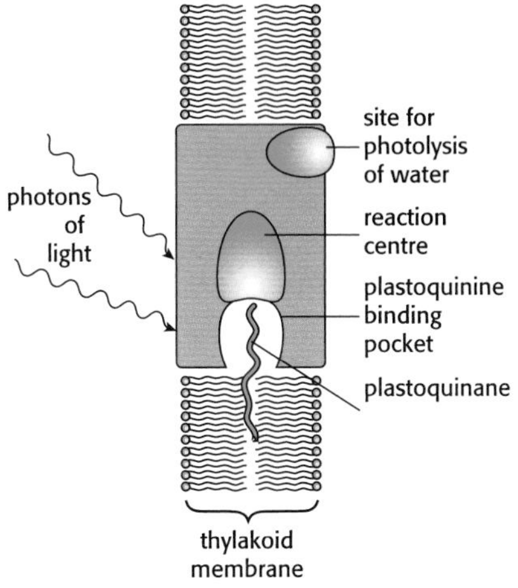

Figure 18 Photosystem II

Rather confusingly, Photosystem II, rather than Photosystem I, is where the light-dependent reactions of photosynthesis begin. The electron acceptor for this photosystem is called plastoquinone. It collects two excited electrons from Photosystem II and then moves away to another position in the membrane. Plastoquinone is hydrophobic, so although it is not in a fixed position, it remains within the membrane.

Absorption of two photons of light causes the production of one reduced plastoquinone, with one of the chlorophylls at the reaction centre having lost two electrons. Photosystem II can repeat this process, to produce a second reduced plastoquinone, so the chlorophyll at the reaction centre has lost four electrons. It is then a powerful oxidising agent and causes the water molecules nearest to it to split and give up electrons, to replace those that it has lost:

$$2H_2O \rightarrow O_2 + 4H^+ + 4e^-$$

The splitting of water, called photolysis, is how oxygen is generated in photosynthesis. Oxygen is a waste product and diffuses away. The useful product of Photosystem II is the reduced plastoquinone, which not only carries a pair of electrons, but also much of the energy absorbed from light. This energy drives all the subsequent reactions of photosynthesis.

Critical considerations: light harvesting and biological fuel

Table 3 gives estimates for energy transfers in the year 2000 (units are Joules per year).

1 According to these figures, is human food production sustainable?
2 According to these figures, could enough fuel be produced by photosynthesis to sustain current lifestyles?
3 What are the advantages of biomass fuels?

Table 3
56×10^{23} J of solar energy strikes the atmosphere. 50% of this radiation is reflected back by clouds and gases in the atmosphere. 50% of solar radiation reaching Earth's surface shines on photosynthesis organisms.
Energy fixed in global photosynthesis: 3×10^{21} J. Energy in food consumed by humans: 2×10^{19} J. Energy in fuel used by humans: 3.95×10^{19} J.
43% of solar radiation is wavelengths absorbed by photosynthetic pigments. 80% of suitable wavelengths are absorbed by an optimally arranged canopy of leaves. 28% of absorbed light energy is fixed to chemical energy in photosynthesis. 67% of energy fixed in photosynthesis is not released by plant respiration.
% of sunlight of wavelengths suitable for photosynthesis that is converted to harvestable biomass in crop plants: • wheat in the Netherlands 1.7% • cassava in Malaysia 2.0% • rice in the Philippines 2.9% • sugar cane in Hawaii 3.8% • *Pennisetum* grass in El Salvador 4.2%.

HL

Photophosphorylation and chemiosmosis

The production of ATP, using energy derived from light is called **photophosphorylation**. It is carried out by the **thylakoids**. These are regular "stacks" of membranes, with very small fluid-filled spaces inside (see Figure 19). The thylakoid membranes contain the following structures:

- Photosystem II
- a chain of electron carriers
- ATP synthase
- Photosystem I.

The main type of photophosphorylation is non-cyclic. A slightly different process, which is cyclic, will be described on page 105. To carry out non-cyclic photophosphorylation, reduced plastoquinone is needed, carrying the pair of excited electrons from the reaction centre of Photosystem II. Plastoquinone carries the electrons to the start of the chain of electron carriers and they are then passed from carrier to carrier in this chain. As the electrons pass, energy is released, which is used to pump protons across the thylakoid membrane, into the space inside the thylakoids.

A concentration gradient of protons develops across the thylakoid membrane, which is a store of potential energy. Photolysis, which takes place in the fluid inside the thylakoids, also contributes to the proton gradient. The protons can travel back across the membrane, down the concentration gradient, by passing through the enzyme ATP synthase. The energy released by the passage of protons down their concentration gradient is used to make ATP from ADP and inorganic phosphate. This method of producing ATP is strikingly similar to the process that occurs inside the mitochondrion and is given the same name: **chemiosmosis.**

When the electrons reach the end of the chain of carriers they are passed to plastocyanin, a water-soluble electron acceptor in the fluid inside the thylakoids. Reduced plastocyanin is needed in the next stage of photosynthesis.

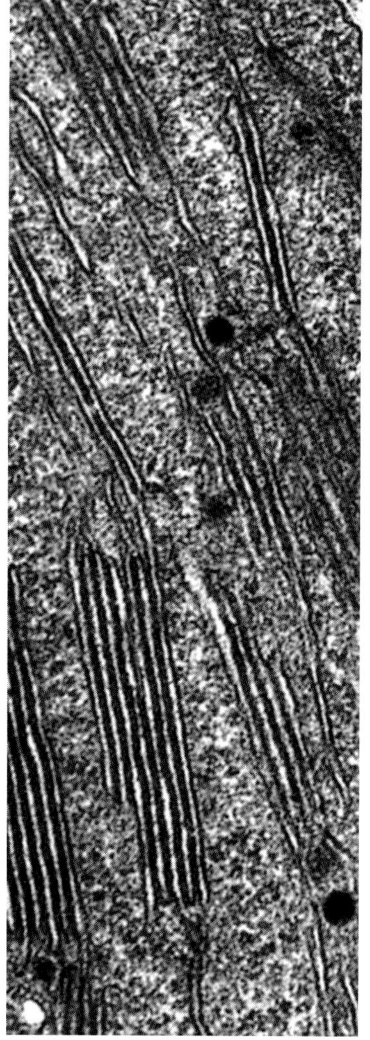

Figure 19 Electron micrograph of thylakoids × 75000

Data-based question: evidence for chemiosmosis

One of the first experiments to give evidence for ATP production by chemiosmosis was performed in the summer of 1966 by André Jagendorf. Thylakoids were incubated for several hours in darkness, in acids with a pH ranging from 3.8 to 5.2. The lower the pH of an acid, the higher its concentration of protons. During the incubation, protons diffused into the space inside the thylakoids, until the concentrations inside and outside were equal. The thylakoids were then transferred, still in darkness, into a solution of ADP and phosphate that was more alkaline. There was a brief burst of ATP production by the thylakoids. The graph shows the yield of ATP at three acid incubation pHs and a range of pHs of the ADP solution.

1 (a) Describe the relationship between pH of ADP solution and ATP yield, when acid incubation was at pH 3.8. [2]

(b) Explain why the pH of the ADP solution affects the ATP yield. [2]

2 Explain the effect of changing the pH of acid incubation on the yield of ATP. [2]

3 Explain why there was only a short burst of ATP production. [2]

4 Explain the reason for performing the experiment in darkness. [2]

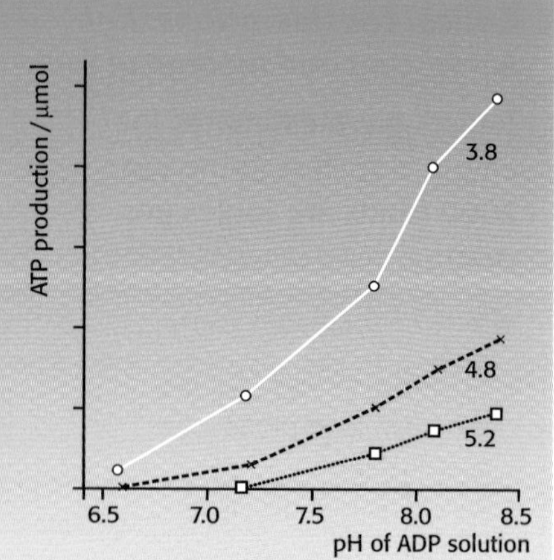

Figure 20 Results of Jagendorf experiment

HL

Completing the light-dependent reactions

The remaining parts of the light-dependent reactions involve Photosystem I. The useful product of these reactions is NADPH, which is needed in the light-independent reactions of photosynthesis. NADPH has a similar role to NADH in cell respiration: it carries a pair of electrons that can be used to carry out reduction reactions.

Chlorophylls within Photosystem I absorb light energy and pass it to the special two chlorophyll molecules in the reaction centre. This raises an electron in one of the chlorophylls to a high energy level. As with Photosystem II, this is called photoactivation. The excited electron passes along a chain of carriers in Photosystem I, at the end of which it is passed to ferredoxin, a protein in the fluid outside the thylakoid. Two molecules of reduced ferredoxin are then used to reduce $NADP^+$, to form NADPH + H^+.

The electron that Photosystem I donated to the chain of electron carriers is replaced by an electron carried by plastocyanin. Photosystems I and II are therefore linked: electrons excited in Photosystem II are passed along the chain of carriers to plastocyanin, which gives them to Photosystem I. The electrons are re-excited with light energy and are eventually used to reduce $NADP^+$. For reasons that will be described on later pages, the supply of $NADP^+$ sometimes runs out. When this happens the electrons return to the electron transport chain that links the two photosystems, rather than being passed to $NADP^+$. As the electrons flow back along the electron transport chain to Photosystem I, they cause pumping of protons, which allows ATP production. This process is **cyclic photophosphorylation.**

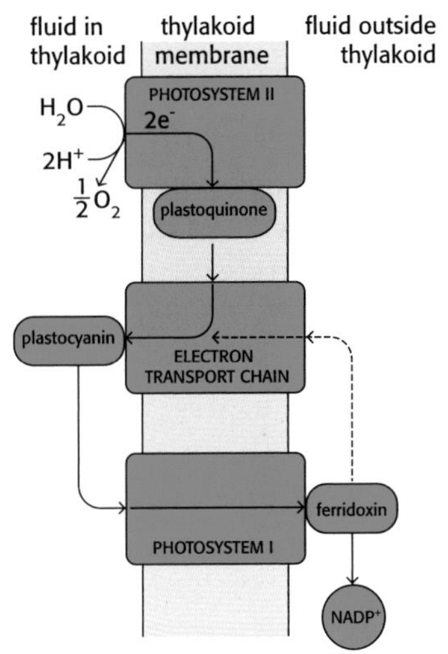

Figure 21 Summary of the light dependent reactions of photosynthesis

Investigating biology: the Hill reaction

Robert Hill did one of the key experiments that showed that the light-dependent reactions of photosynthesis can occur separately from the light-independent reactions in which carbon dioxide is fixed. His success depended on separating photosynthetically active chloroplasts from leaf cells. (A suitable method for doing this is explained right.) Hill found that chloroplasts in the light would reduce artificial electron acceptors and release oxygen, without any carbon dioxide being used. DCPIP is often used in Hill reaction experiments as an artificial electron acceptor. It changes from blue to colourless when it is reduced. The following protocol can be used to demonstrate the basic Hill reaction.

Make up three tubes with these contents:

Tube number	1	2	3
Volume of buffer solution (ml)	4.0	5.0	4.5
Volume of DCPIP solution (ml)	1.0	0.0	1.0

Place a lamp to point at the tubes. Add 0.5 ml of chloroplast suspension to tubes 1 and 2, switch on the lamp and watch the tubes for three minutes. The blue colour should disappear from Tube 1, as Photosystem I reduces all the DCPIP.

You can now design an experiment to find the effect of light intensity on the rate of reduction of DCPIP. To obtain quantitative results, a colorimeter could be used with a 600 nm filter, to assess the intensity of blue coloration remaining at the end of your experiment.

Method for obtaining chloroplasts

1 Make up a phosphate buffer by dissolving 2.8 g of anhydrous Na_2HPO_4 and 6.4 g of anhydrous KH_2PO_4 in 1 dm^3 of deionized water. This will have a pH of 6.5.
2 Make a sucrose/phosphate buffer by dissolving 51.4 g of sucrose in 0.5 dm^3 of the phosphate buffer and chill it in a refrigerator for several hours.
3 Tear up 25 g of fresh spinach leaves, not including the midrib.
4 Put the leaf pieces in a food blender, add 100 ml of sucrose/phosphate buffer and blend for one minute.
5 Place a small beaker into a larger beaker of crushed ice. Place a double layer of muslin cloth over the small beaker and strain the leaf extract though the cloth into the small beaker. Keep the extract chilled until it is used.

To make the DCPIP solution, dissolve 60 mg of DCPIP in 100 ml of phosphate buffer. To make the buffer solution for adding to the test tubes, dissolve 380 mg of KCl in 500 ml of phosphate buffer.

HL

Carbon fixation and carbohydrate synthesis

Carbon dioxide is the carbon source for all organisms that carry our photosynthesis. The carbon fixation reaction in which it is converted into another carbon compound is arguably the most important in all living organisms. In plants and algae it occurs in the stroma – the fluid that surrounds the thylakoids in the chloroplast. The product of this carbon fixation reaction is a three-carbon compound: glycerate 3-phosphate. As so often in biological research, the details of the reaction were a surprise when they were discovered. Carbon dioxide does not react with a two-carbon compound to produce glycerate 3-phosphate. Instead, it reacts with a five-carbon compound called ribulose bisphosphate (RuBP), to produce two molecules of glycerate 3-phosphate. The enzyme that catalyses this reaction is called ribulose bisphosphate carboxylase, usually abbreviated to rubisco. The stroma contains large amounts of rubisco to maximize carbon fixation.

RuBP is a 5-carbon sugar derivative, but when it is converted to glycerate 3-phosphate by adding carbon and oxygen, the amount of hydrogen in relation to oxygen is reduced. In sugars and other carbohydrates, the ratio of hydrogen to oxygen is 2:1. Hydrogen has to be added to glycerate-3-phosphate by a reduction reaction to produce carbohydrate. This involves both ATP and NADPH, produced by the light dependent reactions of photosynthesis. ATP provides the energy needed to perform the reduction and NADPH + H^+ provides the hydrogen atoms. The product is a three-carbon sugar derivative, triose phosphate.

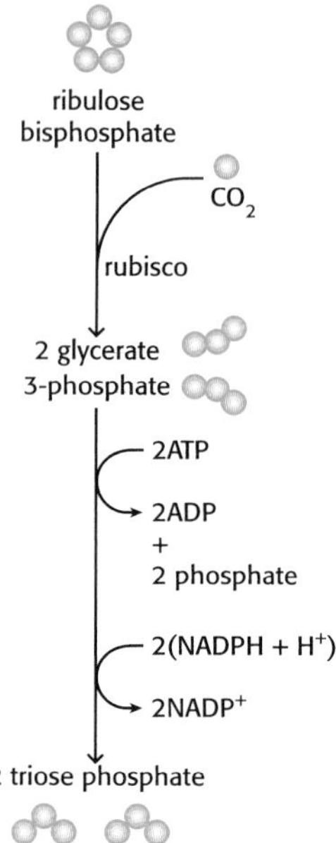

Figure 22 Summary of carbon fixation reactions

Data-based question: identifying the first products of carbon fixation

Sometimes progress in biological research suddenly becomes possible because of other discoveries. Martin Kamen and Samuel Ruben discovered ^{14}C in 1945. The half-life of this radioactive isotope of carbon makes it ideal for use in tracing the pathways of photosynthesis. Figure 23 shows apparatus used by Melvin Calvin and his team. At the start of their experiment, they replaced the $^{12}CO_2$ supplied to algae with $^{14}CO_2$. They took samples of the algae at very short time intervals and found what carbon compounds in the algae contained radioactive ^{14}C. The results are shown in Figure 24. The amount of radioactivity of each carbon compound is shown as a percentage of the total amount of radioactivity.

1 Explain the evidence from the graph that convinced Calvin that glycerate 3-phosphate is the first product of carbon dioxide fixation. [4]
2 Explain the evidence from the graph for the conversion of glycerate-3-phosphate to triose phosphate and other sugar phosphates. [4]
3 Using the data in the graph, estimate how rapidly carbon dioxide can diffuse into cells and be converted with RuBP to glycerate-3-phosphate. [2]

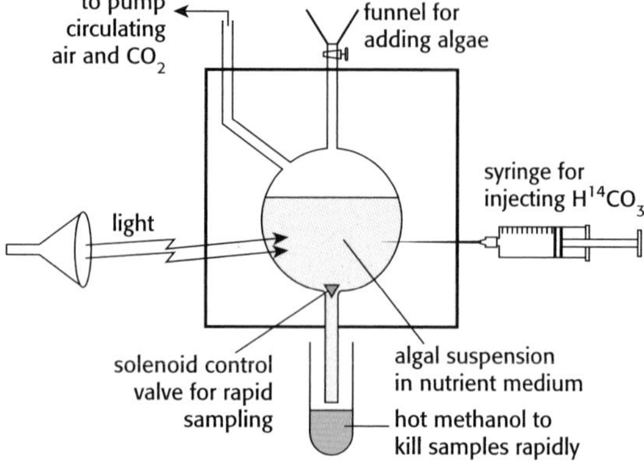

Figure 23 Calvin's lollipop apparatus

Figure 24 Graph showing Calvin's results

HL

Regeneration of RuBP

The first carbohydrate produced by the light-independent reactions of photosynthesis is triose phosphate. Two triose phosphate molecules can be combined to form hexose phosphate and hexose phosphate can be combined by condensation reactions to form starch. However, if all of the triose phosphate produced by photosynthesis was converted to hexose or starch, the supplies of RuBP in the chloroplast would soon be used up. Some triose phosphate in the chloroplast therefore has to be used to regenerate RuBP. This process is a conversion of 3-carbon sugars into 5-carbon sugars and it cannot be done in a single step. Instead a series of reactions takes place, which are summarized in Figure 25.

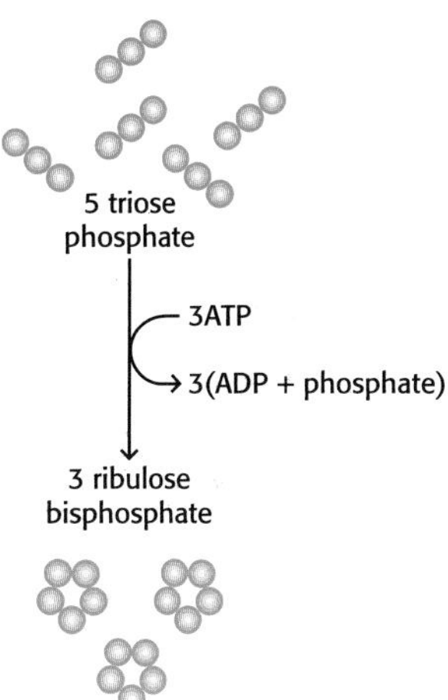

Figure 25 Summary of RuBP regeneration

As RuBP is both consumed and produced in the light-independent reactions of photosynthesis, these reactions form a cycle. It is called the **Calvin cycle** to honour Melvin Calvin, who was given the Nobel Prize for chemistry in 1961 for his work in elucidating this process. For the Calvin cycle to continue indefinitely, as much RuBP must be produced as consumed. If three RUBP molecules are used, six triose phosphates are produced. Five of these are needed to regenerate the three RuBP molecules. This leaves just one triose phosphate for conversion to hexose, starch or other products of photosynthesis. To produce one molecule of glucose for example, six turns of the Calvin cycle are needed, each of which contributes one of the fixed carbon atoms in the glucose.

Data-based question: the effect of light and dark on carbon dioxide fixation

One of the pioneers of photosynthesis research was James Bassham. The results of one of his experiments are shown in the graph below. Concentrations of ribulose bisphosphate and glycerate-3-phosphate were monitored in a culture of cells of the alga, *Scenedesmus*. The algae were kept in bright light and then in the dark.

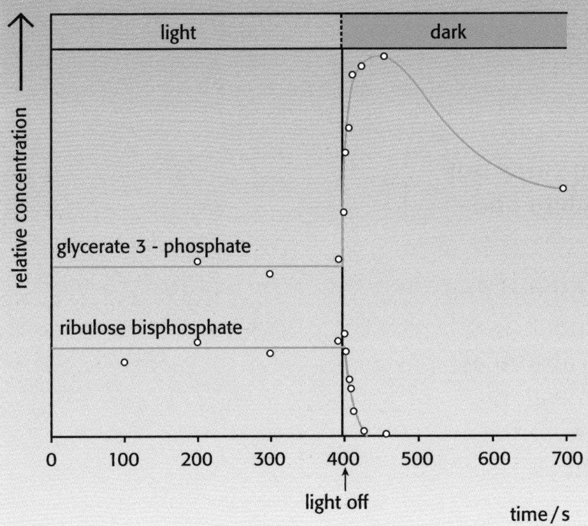

Figure 26 Results of Bassham experiment

1 Compare the effects of the dark period on the concentrations of ribulose bisphosphate and glycerate-3-phosphate. [2]

2 Explain the change that took place in the 25 seconds after the start of darkness, to the concentration of:
   (a) glycerate-3-phosphate [3]
   (b) ribulose bisphosphate. [1]

3 Predict what the effect would be of turning the light back on after the period of darkness. [2]

4 Predict the effect of reducing the carbon dioxide concentration from 1.0% to 0.003%, instead of changing from light to darkness:
   (a) on glycerate-3-phosphate concentration [2]
   (b) on ribulose bisphosphate concentration. [2]

HL

Chloroplast structure

Chloroplasts are quite variable in structure but share certain features:

- a double membrane forming the outer **chloroplast envelope**
- an extensive system of internal membranes called **thylakoids**, which are an intense green colour
- small fluid filled spaces inside the thylakoids
- a colourless fluid around the thylakoids called **stroma** that contains many different enzymes.

In most chloroplasts there are stacks of thylakoids, called **grana**.
If a chloroplast has been photosynthesizing rapidly then there may be **starch grains** or **lipid droplets** in the stroma.

?

State the location within the chloroplast of each of the following structures or substances:

(a) Photosystems I and II
(b) ATP synthase
(c) the electron transport chain
(d) enzymes of the Calvin cycle
(e) NADPH and $NADP^+$
(f) ATP and ADP.

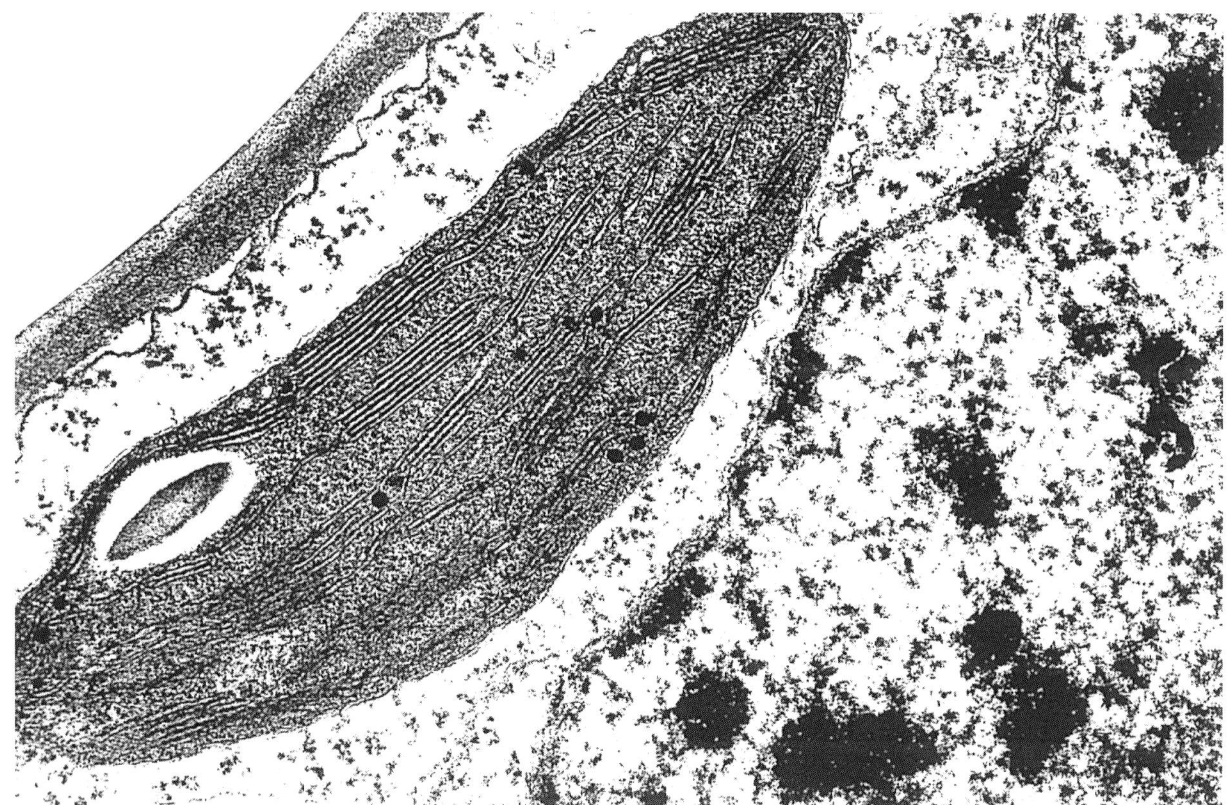

Figure 27 Electron micrograph of chloroplast with nucleus and cell wall also visible. × 30000

Structure–function relationships in chloroplasts

If chloroplasts are carefully extracted from leaves they can carry out the whole process of photosynthesis–both the light-dependent and light-independent reactions.

Consider the following requirements for efficient photosynthesis and how the structure of the chloroplast meets the requirements.

1. There must be large amounts of pigment spread out to absorb as much light as possible. The pigments are not soluble in water.
2. There must be a means of generating a steep enough proton gradient for ATP synthesis, with as few photons of light as possible.
3. There must be a site where enzymes of the Calvin cycle can be concentrated, catalyzing reactions that use ATP and NADPH + H^+ supplied by the light dependent reactions of photosynthesis.

Chapter 9 questions

1 Water with mineral nutrients dissolved in it was sterilized and then placed in a 2 dm^3 fermenter. The temperature was kept at 25 °C. The fermenter was kept in natural sunlight, but a lamp was also used to increase the light intensity. The lamp was controlled by an electronic timer, which switched it off at night. A light meter was placed against the side of the fermenter, near the base, to measure the intensity of light passing through the liquid in the fermenter. The maximum reading it could give was 1200 lux. At the start of the experiment, a small quantity of *Chlorella,* a type of algae, was added to the fluid in the fermenter. Figure 28 shows the light intensity measured over the 45 days of the experiment.

(a) The light intensity followed a similar pattern, every day from Day 12 onwards.
  (i) Outline the daily changes in light intensity over a typical day after Day 12. [2]
  (ii) Explain these daily changes in light intensity. [2]

(b) Each day there is a maximum light intensity. Outline the trends in maximum light intensity.
  (i) from Day 1 to Day 12 [1]
  (ii) from Day 13 to Day 38 [2]
  (iii) from Day 39 to Day 45. [2]

(c) Explain why the light intensity when the light was switched on was lower at the end of the experiment than at the start. [3]

(d) Suggest reasons for the trend in maximum daily light intensity between Day 39 and Day 45. [3]

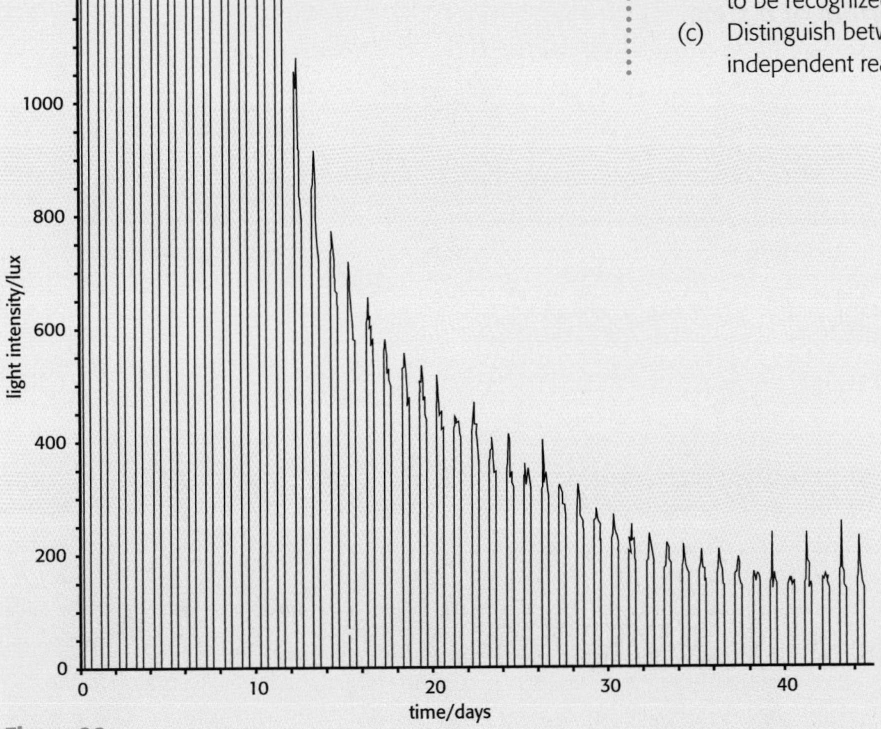

Figure 28

Higher-level questions

2 Figure 29 shows the effects of varying light intensity on the carbon dioxide absorption by leaves, at different, fixed carbon dioxide concentrations and temperatures.

(a) Deduce the limiting factor for photosynthesis at:
  (i) W (ii) X (iii) Y (iv) Z. [4]

(b) Explain why curves I and II are the same between 1 and 7 units of light intensity. [3]

(c) Explain the negative values for carbon dioxide absorption when the leaves were in low light intensities. [3]

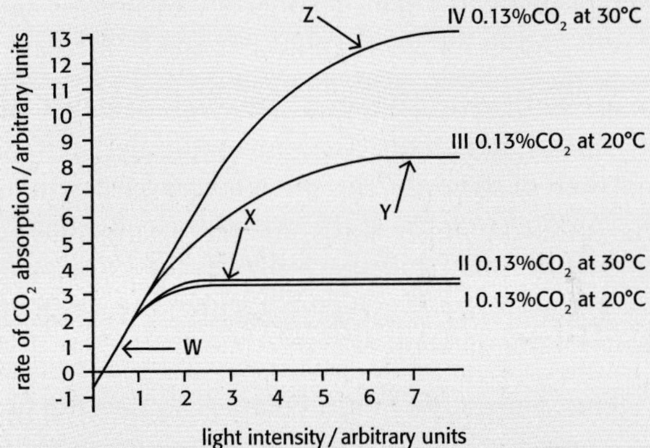

Figure 29

3 (a) Draw the structure of a leaf cell that can carry out photosynthesis at a rapid rate. [5]

(b) Outline the features of chloroplasts that allow them to be recognized in electron micrographs. [5]

(c) Distinguish between the light-dependent and light-independent reactions of photosynthesis. [8]

10 Plant science

HL The growth of a plant is an everyday phenomenon, but it is nonetheless remarkable. Without visibly ingesting anything, a plant can grow larger and more complex. Using only simple inorganic substances and sunlight, plants synthesize a huge range of substances – they are amazing chemical factories, that fuel both themselves and most food chains in terrestrial ecosystems. Plants retain groups of stem cells throughout their lives, allowing them to continue to grow indefinitely. These groups of cells are called **meristems**. Cells in meristems are small and go through the cell cycle repeatedly to produce more cells, by mitosis and cytokinesis. These new cells absorb nutrients and water and so increase in volume and mass.

Primary meristems are found at the tips of stems and roots. They are called **apical meristems**. The root apical meristem is responsible for the growth of the root. The shoot apical meristem, at the tip of the stem, is more complex. It throws off the cells that are needed for the growth of the stem and also produces the groups of cells that grow and develop into leaves. Figure 1 shows the shoot apical meristem of a **dicotyledonous** plant.

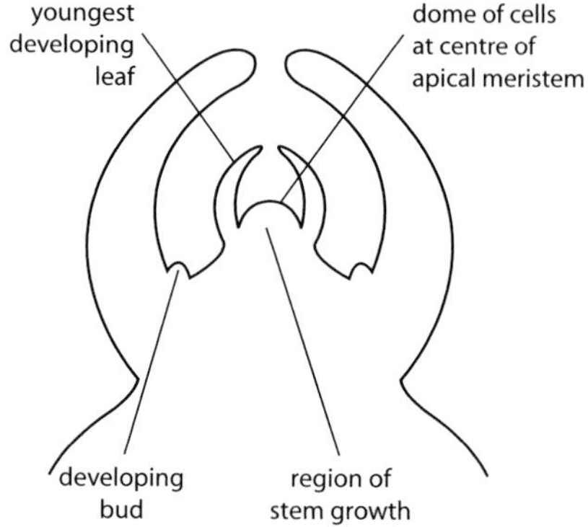

Figure 1 Structure of a shoot apical meristem

The stems, leaves and roots of plants can develop in many different ways. There is a huge diversity of plant form, which can be related to the habitats and adaptations of plants. Hydrophytes, which are plants that grow in water, are very different in structure, for example, from xerophytes (plants that grow in deserts).

Working with data: plan diagram of a stem

A plan diagram is used to show the distribution of tissues in a leaf or other organ. The lines show the edges of each area of tissue. Individual cells are not shown.

Figure 2 shows part of a young *Helianthus* (sunflower) stem in transverse section.

Look carefully at the cells and count how many different types there are. The size, shape and wall thickness of cells can be used to distinguish the different cell types.

Draw a plan diagram to show these areas of tissue, including epidermis, cortex, fibres, phloem, xylem and pith.

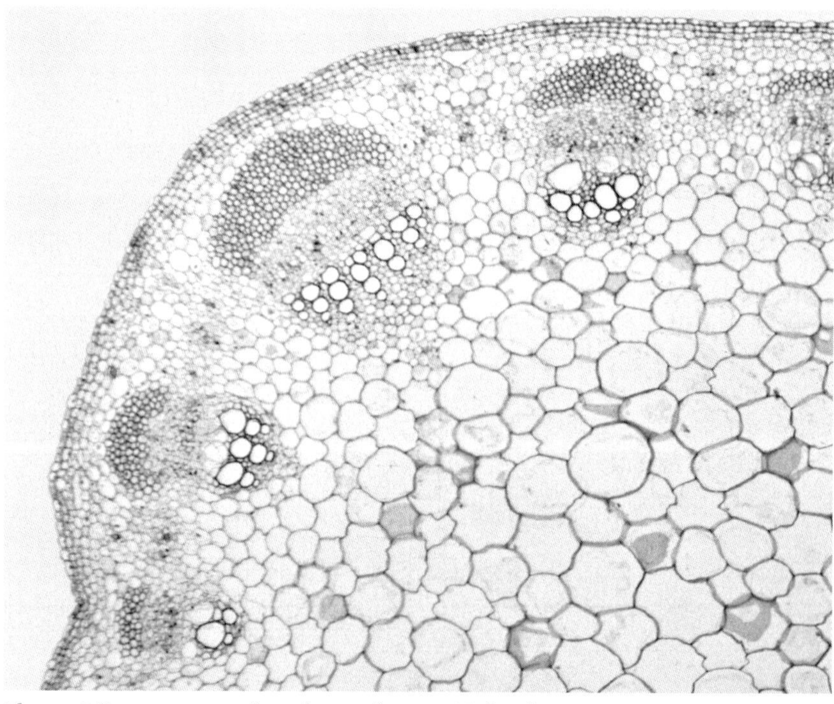

Figure 2 Transverse section of part of young *Helianthus* stem

HL

Stems

As stems grow upwards, they produce new leaves and also sometimes form branches. The lower parts of the stem therefore have to support a larger and larger mass. Terrestrial plants support themselves in three ways:

- with turgid cells, which are almost rigid because of their high pressure
- with cells that have thickened cellulose cell walls
- with xylem tissue, which has cell walls impregnated with lignin, making it woody and hard.

Flowering plants can be divided into two broad groups–monocotyledons and dicotyledons (see page 116). The stems of dicotyledons can grow wider as a plant grows taller, to support the increase in mass. To do this, a meristem develops in a complete circle around the stem. This is called a **lateral meristem**. The lateral meristem develops between the xylem and the phloem and the cells it produces develop into more of both these tissues. Xylem develops on the inner side of the lateral meristem, gradually adding to the thickness of the wood of the stem. The plants that produce the largest amount of xylem and can therefore grow tallest and compete for light most effectively are called trees. Figure 3 shows a transverse section of a stem that has produced extra xylem tissue using its lateral meristem.

Monocotyledons have stems with an apical meristem at the tip of the stem, but the tissues of the stem differentiate in a different arrangement from those of dicotyledons (see Figure 5).

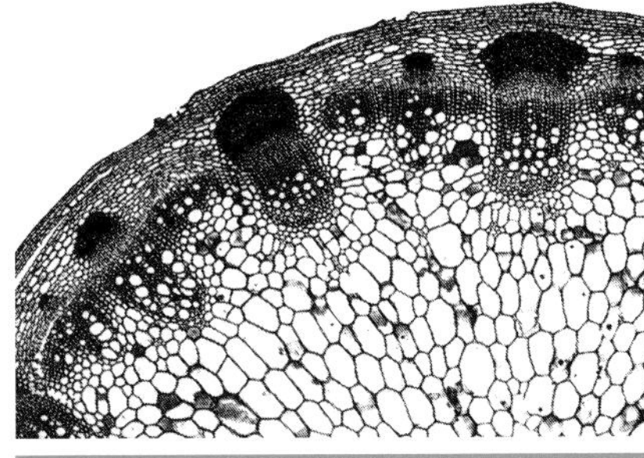

Figure 3 Transverse section of part of a thickened *Helianthus* stem

Data-based question: comparing stem structure

1. Figure 4 is a diagram of a dicotyledon stem based on the transverse section shown in Figure 3. Deduce whether the areas on the diagram represent single cells or areas of tissue. [2]
2. Outline two similarities in structure between the monocotyledon stem (Figure 5) and the dicotyledon stem. [2]
3. Compare the position of xylem and phloem in the two types of stem. [4]
4. Deduce why monocotyledon stems cannot thicken in the same way as dicotyledon stems. [2]

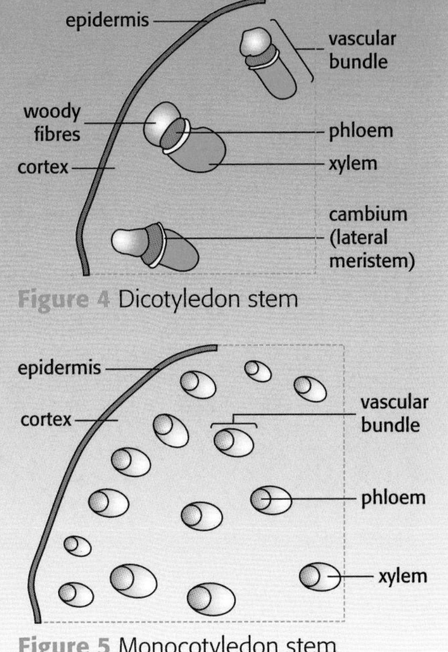

Figure 5 Monocotyledon stem

HL

Leaf structure and adaptations

In areas that are favourable for plant growth, the vast numbers of leaves make the Earth appear green from space. Leaves are produced by the apical meristem at the tip of stems. They vary in size and shape considerably, but most have a large upper surface facing the sun and a very narrow distance between the upper and lower surfaces. The commonest shape is oval (see Figure 6), with a stalk at one end linking the leaf to the stem and a downwardly curving point at the other end to shed drops of rainwater.

Leaves are organs, because they are composed of groups of different tissues (see Table 1).

Figure 6 Oval leaves of *Mirabilis jalapa* from Gerard's Herbal of 1633

Table 1 Tissues found in plant leaves.

epidermis – a single outer layer of cells	*upper epidermis* with a thick layer of wax on the outside, called the cuticle
	lower epidermis with a thinner waxy cuticle and guard cells with stomata (pores) in between
mesophyll – the tissue filling most of the leaf inside the epidermis	*palisade mesophyll* – tightly packed cells in the upper half of the leaf containing many chloroplasts
	spongy mesophyll – loosely packed cells in the lower half of the leaf, with large air spaces between them and fewer chloroplasts inside them
vascular tissue – found in the veins of the leaf	*xylem* – mostly xylem vessels, which are long, dead, tubular structures containing water and dissolved mineral ions
	phloem – lines of living cells, containing water with sucrose and other biochemicals dissolved, and pores between the cells

Explain which tissue or tissues of the leaf performs each of these functions:

1 absorption of light for photosynthesis [3]
2 gas exchange, including CO_2 uptake and O_2 release [3]
3 support, to ensure that the leaf faces the sun [3]
4 water conservation [3]
5 transport of water from the stem to leaf cells [3]
6 transport out of the leaf of sugars and other products of photosynthesis. [3]

Working with data: plan diagrams of leaves

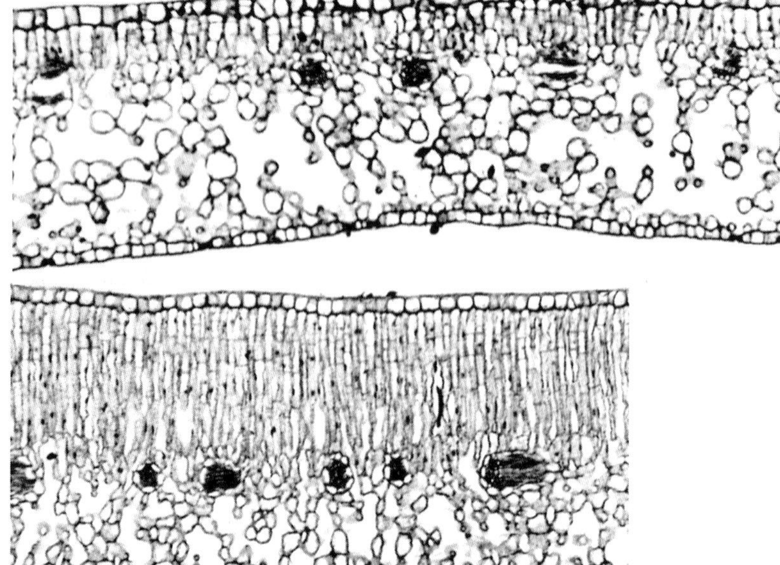

Figure 7 Micrographs of sun and shade leaves

Figure 7 shows sections of two leaves of *Prunus caroliniana*, one that grew in the sun and one that grew in the shade. Draw plan diagrams of the tissues in each.

1 Compare the features of the two leaves, including overall thickness, structure of palisade mesophyll and spongy mesophyll and the thickness of cuticle. [4]
2 (a) Deduce which leaf was grown in the sun, and which was grown in the shade. [1]
  (b) Discuss why the leaves that grew in sun and those that grew in shade have differences and similarities in their structure. [5]

HL

Translocation in phloem

Phloem tissue is found throughout plants, including the stems, roots and leaves. It links parts of the plant that need a supply of sugars and amino acids to other parts that have a surplus. Table 2 classifies parts of the plant into sources (areas where sugars and amino acids are loaded into the phloem) and sinks (where the sugars and amino acids are unloaded and used).

Table 2

Sources	Sinks
Photosynthetic tissues: • mature green leaves • green stems.	Roots that are growing or absorbing mineral ions using energy from cell respiration.
Storage organs that are unloading their stores: • storage tissues in germinating seeds • tap roots or tubers at the start of the growth season.	Parts of the plant that are growing or developing food stores: • developing fruits • developing seeds • growing leaves.

Figure 8 shows the results of a simple experiment in which two rings of bark were removed from an apple tree. The bark contains the phloem tissue. The effects on apple growth are clearly visible.

Sometimes sinks turn into sources, or vice versa. For this reason the tubes in phloem must be able to transport biochemicals in either direction and unlike the blood system of animals, there are no valves or central pump in phloem. However there are similarities between transport in phloem and blood vessels: in both systems a fluid flows inside tubes because of pressure gradients. Energy is needed to generate the pressures, so the flow of blood and the movement of phloem sap are both active processes. The movement of substances in phloem is called **active translocation**.

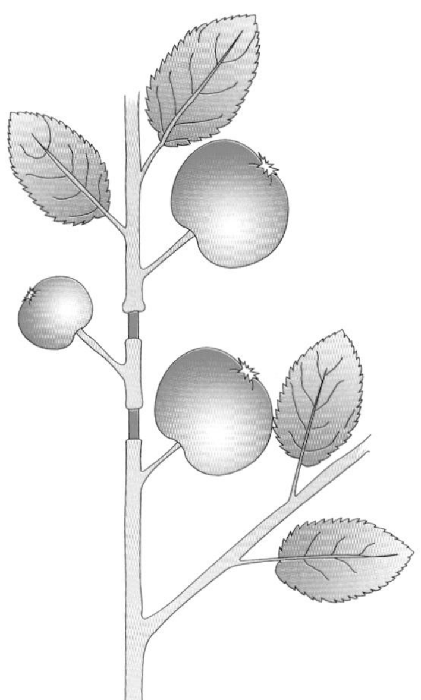

Figure 8 Results of apple tree ringing experiment

1 State which the sources and which the sinks are in this part of the apple tree. [2]

2 (a) Compare the sizes of the apples. [2]

(b) Explain the conclusions that can be drawn from the sizes of the apples. [4]

3 Suggest reasons for the swelling above the upper ring of removed bark. [2]

Working with data: carbohydrates in Cyclamen

Cyclamen persicum plants were dissected and the carbohydrate content was measured in parts of the plants. Mean results are shown in Table 3. The ± values are standard error figures. There is a 5 per cent probability that the true mean falls outside the range given by standard error values.

Table 3

Plant part	Mean carbohydrate content ($\mu g\ g^{-1}$ fresh mass ± standard error of mean)			
	sucrose	glucose	fructose	starch
Leaf blade	1312 ± 212	210 ±88	494 ±653	62 ±25
Vascular bundle in the leaf stalk, consisting of xylem and phloem	5757 ±1190	479 ±280	1303 ±879	<18
Tissue surrounding the vascular bundle in the leaf stalk	417 ±96	624 ±714	1236 ±1015	<18
Buds, roots and tubers (underground storage organs)	2260 ±926	120 ±41	370 ±242	152 ±242

Data presentation

Choose a suitable presentation format to display the data in the table, including the standard error values. You can use graphing software or you can draw graphs, tables, charts or diagrams by hand.

Conclusions

Describe the trends in the data and suggest reasons for them based on your knowledge of photosynthesis, the structure of disaccharides and polysaccharides and the transport and storage of carbohydrates in plants.

HL

Phototropism

Plants use hormones to control the growth of stems and roots. Both the rate and the direction of growth are controlled. The direction in which stems grow is influenced by two external stimuli: light and gravity. Stems grow towards the source of the brightest light or in the absence of light they grow upwards, in the opposite direction to gravity. These directional growth responses to directional external stimuli are called **tropisms**. Growth towards the bright light is called **phototropism** and will be described here as an example of the control of plant growth. Charles Darwin was the pioneer of research into phototropism, which he called heliotropism. Some of his drawings are shown in Figure 9.

Figure 9 *Phalaris canariensis*: cotyledons after exposure for 8 h in a box open on one side in front of a south-west window. Curvature towards the light accurately traced. The short horizontal lines show the level of the ground. From *The Movements of Plants by Charles Darwin*, 1882.

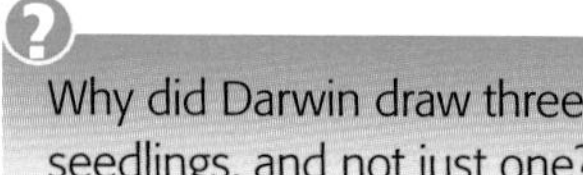

Why did Darwin draw three seedlings, and not just one?

The first stage in phototropism is the absorption of light by photoreceptors. Proteins called phototropins have this role. When they absorb light of an appropriate wavelength, their conformation changes. They can then bind to receptors within the cell, which control the transcription of specific genes. Although much research is still needed in this field, it seems likely that the genes involved are those coding for a group of glycoproteins located in the plasma membrane of cells in the stem that transport the plant hormone **auxin** from cell to cell. The position and type of these proteins can be varied, to transport auxin to where growth is needed. Auxin promotes the elongation of cells in stems, by causing loosening of the connections between cellulose microfibrils in cell walls. Auxin is synthesized by the tips of growing stems and is transported down the stem to stimulate growth. If phototropins in the tip detect a greater intensity of light on one side of the stem than the other, auxin is transported laterally from the side with brighter light to the more shaded side. Higher concentrations of auxin on the shadier side of the stem cause greater growth on this side, so the stem grows in a curve towards the source of the brighter light. The leaves attached to the stem will therefore receive more light and be able to photosynthesize at a greater rate. Figure 10 shows the rate at which the phototropic response occurs.

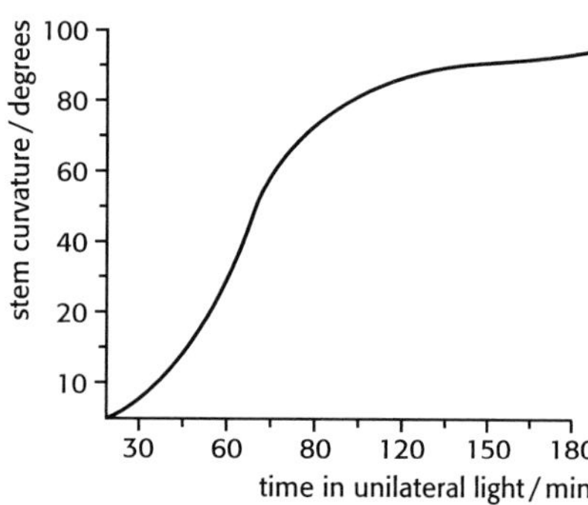

Figure 10 Graph showing the rate of a phototropic response

Rediscovering biology: the action spectrum for phototropism

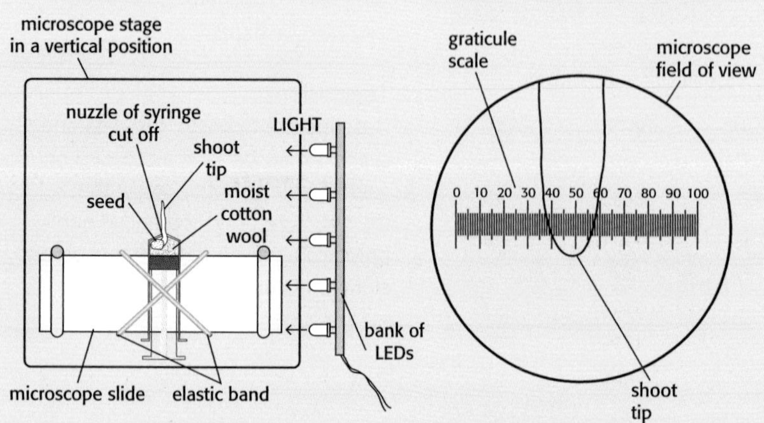

Figure 11 Apparatus for measuring rate of phototropic response

Figure 11 shows apparatus used to measure phototropism in the shoots of seedlings. The syringe is fixed securely onto a microscope stage and graticule scale in the eyepiece is used to measure movements of the shoot tip. Light from one side can be tested, using red, blue and green light emitting diodes.

1 Which colour gives the strongest phototropic response?
2 To draw the action spectrum for phototropism, which variables should be on the x and y axes?

HL

Root systems and mineral ion absorption

One of the earliest stages in the development of an embryo plant is the formation of a root. When seeds germinate, the embryonic root bursts out of the seed coat and grows downwards into the soil, anchoring the plant and absorbing water and mineral ions. This seedling root repeatedly develops **branches** in some plants, to form an extensive fibrous network. Figure 12 shows patterns of root branching. Many roots also develop **hairs**, especially in soil that is drying out. Branching and hair growth both increase the surface area of the root system and therefore increase the capacity for absorption.

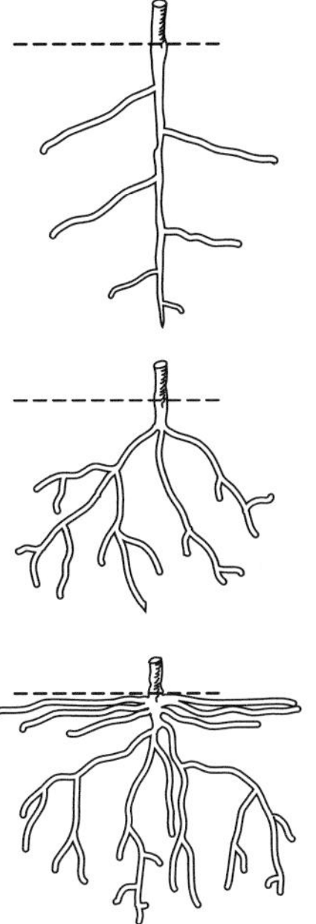

Figure 12 Patterns of root branching

Water is absorbed into root cells by osmosis. This happens because the solute concentration inside the root cells is greater than that in the water in the soil. Most of the solutes in both the root cells and the soil are mineral ions. The concentrations of mineral ions in the root can be 100 or more times higher than those in the soil. These concentration gradients are established by **active transport**, using protein pumps in the plasma membranes of root cells. There are separate pumps for each type of ion that the plant requires.

Mineral ions can only be absorbed by active transport if they make contact with an appropriate pump protein. This can occur by **diffusion**, or by **mass flow** when water carrying the ions drains through the soil. Some ions move through the soil very slowly by these methods, because the ions bind to the surface of soil particles. To overcome this problem, some plants have developed a relationship with a fungus. The fungus grows on the surface of the roots and sometimes even into the cells of the root. The thread-like **hyphae** of the fungus grow out into the soil and absorb mineral ions such as phosphate from the surface of soil particles. These ions are supplied to the roots, allowing the plant to grow successfully in mineral-deficient soils. This relationship is found in many trees, in members of the heather family and in orchids. Most, but not all of these plants supply sugars and other nutrients to the fungus, so both the fungus and the plant benefit. This is an example of a **mutualistic relationship**.

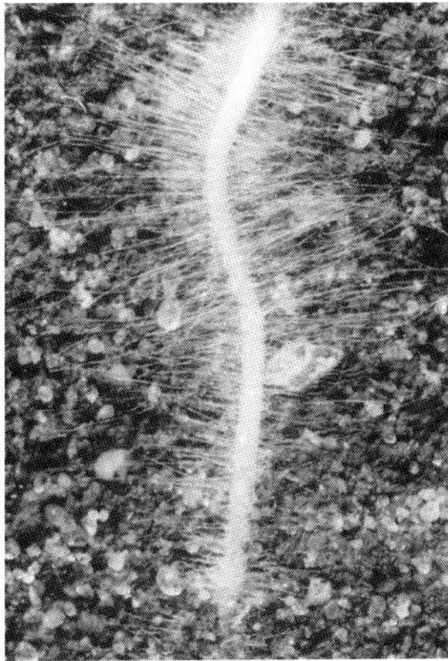

Figure 13 Micrograph of radish root hairs

Data-based question: fungal hyphae and mineral ion absorption

Figure 14 shows the results of an experiment in which seedlings of Sitka spruce, *Picea sitchensis*, were grown for 6 months in sterilized soil either with or without fungi added: C was the control with no fungi added. The species of fungi added were:
I = *Laccaria laccata;* II = *Laccaria amythestea;* III = *Thelophora terrestris* from a tree nursery; IV = *Thelophora terrestris* from a forest; V = *Paxillus involutus;* VI = *Pisolithus tinctorius.*

1 (a) Discuss the effects of the five species of fungi on the growth of the roots and shoots of the tree seedlings. [4]

(b) Explain the effects of the fungi on the growth of tree seedlings. [2]

2 (a) State the relationship between root growth and shoot growth in the tree seedlings. [1]

(b) Suggest a reason for the relationship. [1]

(c) Using the data in Figure 10.14, deduce whether the effects of closely related fungi on tree growth are the same. [2]

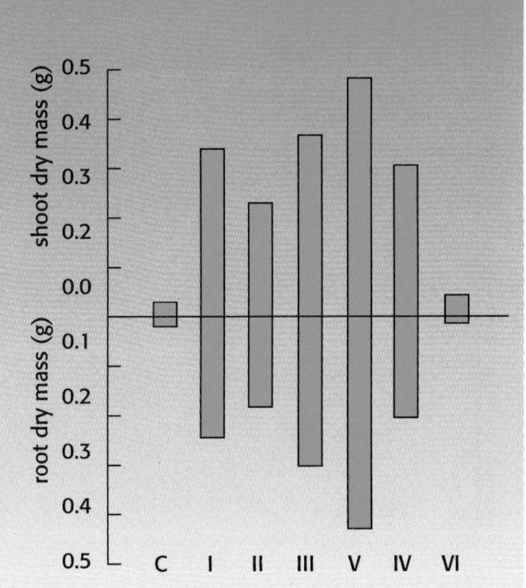

Figure 14 Results of Sitka spruce experiment

HL

Monocotyledons and dicotyledons

Plants that produce seeds enclosed inside fruits are called angiospermophytes. As with many biological terms, this word is derived from ancient Greek:

angeion – a container; *sperma* – a seed; *phyton* – a plant.

Angiospermophytes are divided into two large and important groups: **monocotyledons** and **dicotyledons**. These names refer to the number of leaves contained in the embryo, called cotyledons. The monocotyledons include the palms, gingers, lilies, irises, grasses, sedges, rushes, orchids, bananas, bromeliads and aroids. The dicotyledons include most trees and shrubs and many non-woody plants. These two groups are normally called monocots and dicots. Although the differences listed in Table 4 are not found in all monocots and dicots, the well-informed biologist should be able to distinguish between them!

Table 4

	Monocots	Dicots
Number of cotyledons	1	2
Numbers of organs in the flower	3 or 6	4 or 5
Meristems	Apical only – stems cannot grow wider	Apical and lateral so stems can widen
Leaf attachment	Around the whole stem circumference	To one side of the stem by a leaf stalk
Method of formation of new roots	By formation of roots from the stem	By branching of other roots
Leaf veins	Parallel	Branching to form a network
Vascular tissue in the stem	Vascular bundles spread throughout	Vascular bundles arranged in a ring

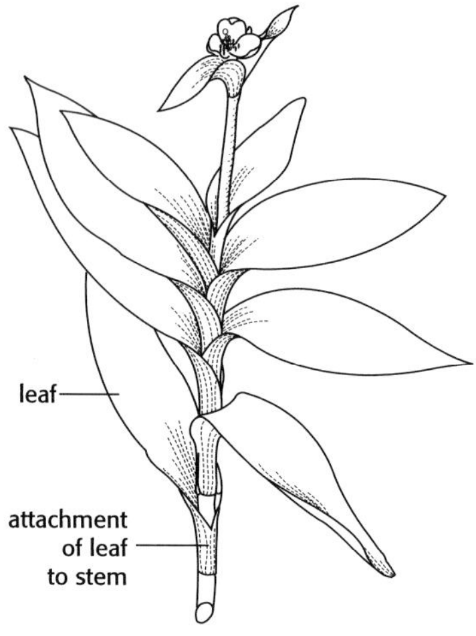

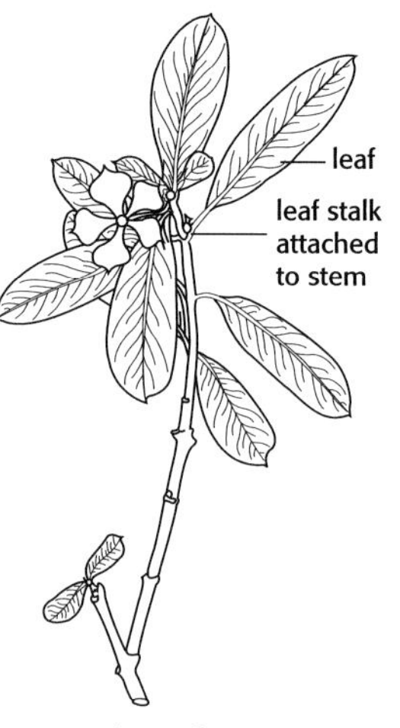

Figure 15 Drawing of *Tradescantia pallida* (a monocotyledon) and *Catharanthus roseus* (dicotyledon).

Thinking about science: classification

The text (below) is from the second edition of *Methodus Plantarum*, by John Ray, published in 1703. It is the first classification to name monocotyledons and dicotyledons as separate groups.

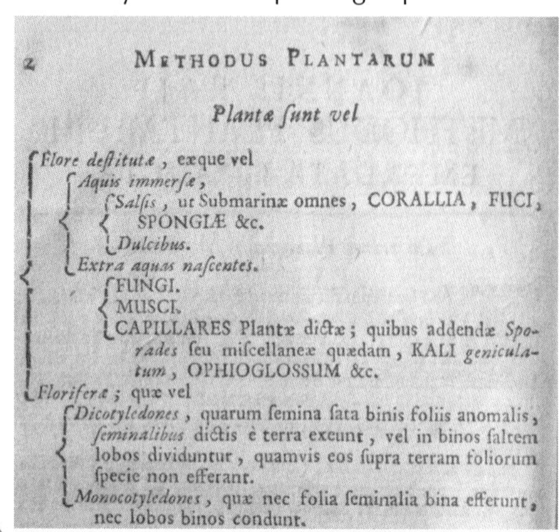

2 METHODUS PLANTARUM

Plantæ ſunt vel

Flore deſtitutæ, eæque vel
Aquis immerſæ,
Salſis, ut Submarinæ omnes, CORALLIA, FUCI, SPONGIÆ &c.
Dulcibus.
Extra aquas naſcentes.
FUNGI.
MUSCI.
CAPILLARES Plantæ dictæ; quibus addendæ *Sporades* ſeu miſcellaneæ quædam, KALI *geniculatum*, OPHIOGLOSSUM &c.
Floriferæ; quæ vel
Dicotyledones, quarum ſemina ſata binis foliis anomalis, *ſeminalibus* dictis è terra exeunt, vel in binos ſaltem lobos dividuntur, quamvis eos ſupra terram foliorum ſpecie non efferant.
Monocotyledones, quæ nec folia ſeminalia bina efferunt, nec lobos binos condunt.

1 Why did Ray write his classification in Latin and why is this language still used to publish descriptions of new species?
2 Ray classified corals and sponges with plants. To what extent is this classification justified?
3 Ray developed the plant classifications known as monocotyledon and dicotyledon that are still in use today. Suggest why this part of his system persists and his classification of corals and sponges does not persist.

HL

Adaptations of roots, stems and leaves

Roots, stems and leaves are the principal plant organs. Plants use them to build the structures that they need for water and mineral absorption, support and photosynthesis. Other functions can be carried out if roots, stems or leaves are modified. For example, roots usually absorb water and mineral ions, but in some trees they are modified into props, growing out from the trunk above soil level to help support the tree. In trees that grow in mangrove swamps some roots grow upwards above the surface of the anaerobic mud in the swamp, acting as breathing roots, or pneumatophores.

Figures 16 and 17 show modified leaves. Tendrils are narrow outgrowths from leaves that rotate through the air until they touch a solid support, to which they attach, allowing the plant to climb upwards. Bulbs consist of swollen leaf bases attached to a short stem. They are used for food storage. Figures 18 and 19 show swollen underground stems and roots that are also used for food storage, called stem tubers and root tubers.

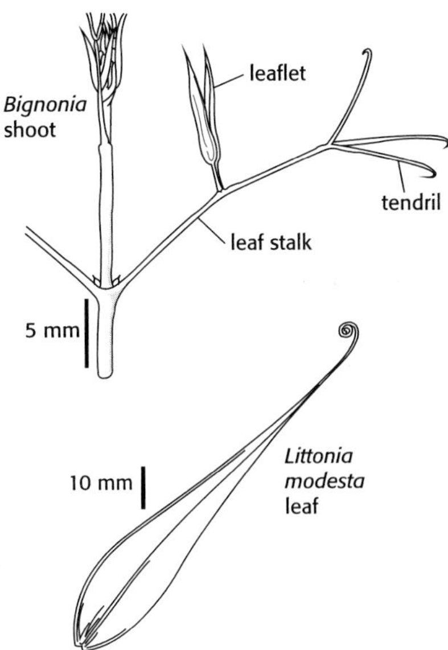

Figure 16 Leaves with tendrils

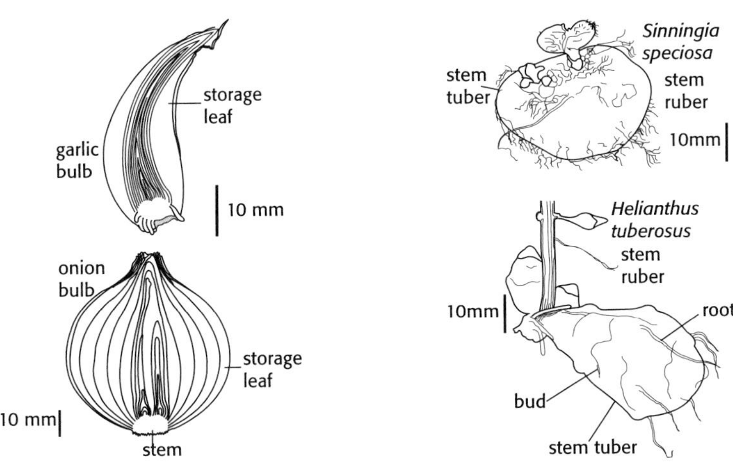

Figure 17 Bulbs

Figure 18 Stem tubers

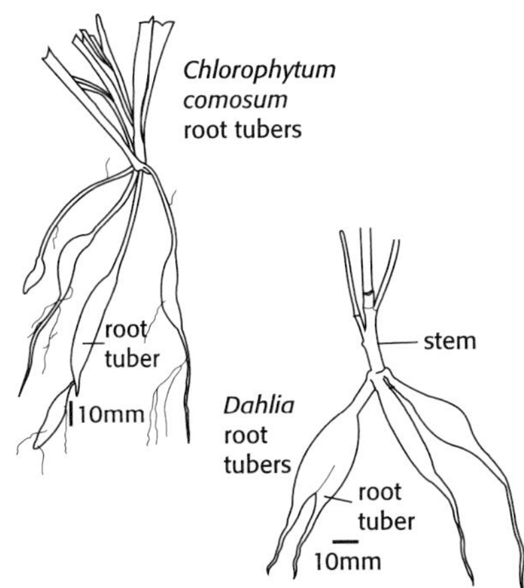

Figure 19 Root tubers

Critical considerations: stem and root tubers as food

Table 5 The content of B vitamins and vitamin C per 100 g in four types of tuber, used as a main food item in parts of Africa.
ICT can be used to present this data in charts, to make analysis easier.

Vitamin	Cassava	Yam	Potato	Sweet potato	Adult daily requirement
Thiamine (μg)	87	112	71	78	4800
Riboflavin (μg)	48	32	34	61	1200
Niacin (mg)	0.85	0.55	1.07	0.56	80
Pantothenic acid (μg)	107	314	281	800	7000
Vitamin B_6 (μg)	88	293	203	209	800
Folate (μg)	27	23	18	11	200
Vitamin B_{12} (μg)	0	0	0	0	1.5
Vitamin C (mg)	20.6	17.1	19.7	2.4	40

?

1 Identify which tuber:
   (a) has the highest content of each vitamin [3]
   (b) has the lowest content of each vitamin. [3]
2 Suggest reasons why large amounts of one food are often eaten in an area. [2]
3 Explain, using the data, the advantages of eating a variety of different foods. [2]

HL

Stomata

Plants that live on land, with stems and leaves in contact with the air, must avoid excessive water loss and death by dehydration. Water loss from the leaves and stems of plants is called **transpiration**. The waxy cuticle that covers the epidermis of leaves reduces transpiration, but it also prevents other small molecules from passing through, including carbon dioxide and oxygen. Absorption of carbon dioxide is essential for photosynthesis, so pores through the epidermis, with its waxy coating, are needed. These are called **stomata**. The physiological problem for plants is that if stomata allow carbon dioxide to be absorbed, they will usually also let water vapour escape from the leaf, with the consequent risk of dehydration and death. This is an intractable problem for plants and other organisms: gas exchange without water loss is impossible.

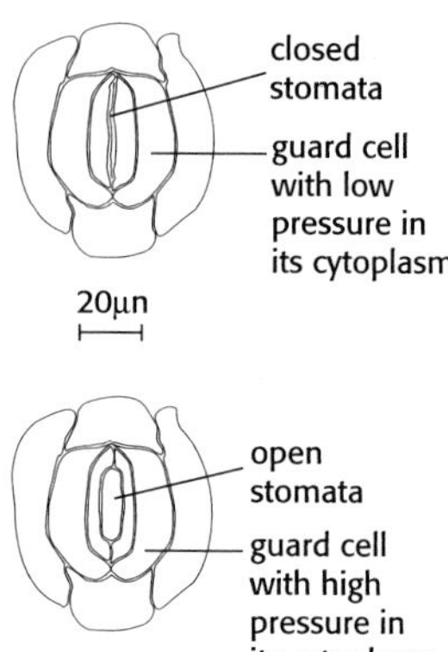

Figure 20 Open and closed stomata

Plants minimise water losses through stomata using **guard cells**. These are the cells that are found in pairs, one on either side of a stoma. Guard cells control the aperture of the stoma and can adjust from fully open to fully closed. To increase the aperture, the guard cells absorb water, which raises the pressure and pushes the guard cells apart. Conversely, movement of water out of guard cells causes them to lose pressure and surrounding epidermis cells push them together, closing the stoma. Figure 20 shows open and closed stomata.

The opening and closing of stomata is affected by various external stimuli:

- light causes stomata in most plants to open
- low carbon dioxide concentrations in the air spaces inside the leaf cause stomata to open
- shortage of water, which results in leaf cells becoming deficient in water, causes stomata to close.

When leaf cells become deficient in water, they synthesize a plant hormone called abscisic acid. This hormone causes the closure of stomata and it over-rides other external stimuli – the stomata close even if it is light and carbon dioxide levels inside the leaf are low. In this situation it is less important for the plant to continue to photosynthesize, than to avoid dehydration and death.

Investigating biology: distribution and movements of stomata

Guard cells are large enough to be visible under a light microscope and the width of the stomatal aperture can be measured. A whole leaf is too thick to view, but it is possible to peel the epidermis off from the rest of the leaf in some species (Figure 21). *Commelina* and *Tradescantia* are both suitable for this, but other species are worth trying. Pieces of epidermis can be floated on solutions to find their effect. 50 mmol dm^{-3} KCl solution is suitable as a basic solution. The effect of light intensity or of varying the carbon dioxide concentration or the abscisic acid concentration can then be investigated.

An alternative is to paint colourless nail varnish onto the leaf to form a replica of the leaf surface. This can be peeled off when it has dried and allows the shape and size of epidermis cells to be deduced, including guard cells. Unless the stomata are sunk in pits, it is possible to measure the stomatal aperture. The distribution of stomata can be investigated in the lower and upper epidermis of dicotyledon leaves or in monocotyledon leaves such as grasses. An investigation could be done to find out whether the density of stomata per unit area of leaf affects the rate of photosynthesis.

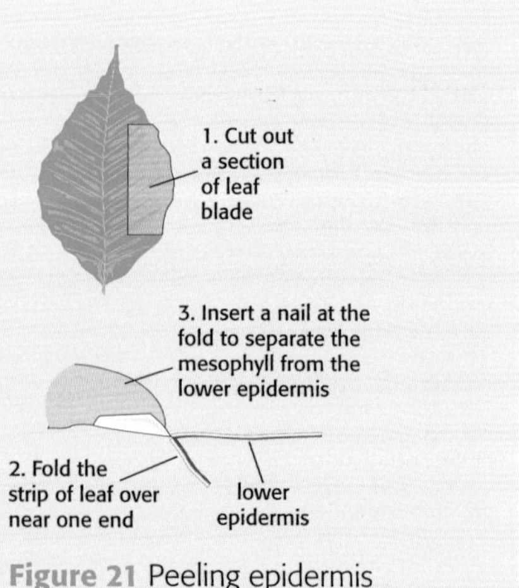

Figure 21 Peeling epidermis

HL

Abiotic factors affecting transpiration

The rate of transpiration is the amount of water vapour that a plant loses from its leaves and stems per unit time. The rate depends on several variables:

- the size of the plant
- the thickness of its waxy cuticle
- how widely spaced its stomata are
- whether the stomata are open or closed.

These are all factors that the plant can control and as the plant is a living organism, they are **biotic factors**. Other external factors are not controlled by the plant, or any other organism. They are therefore **abiotic factors**. Four abiotic factors have a major effect on the rate of transpiration.

- **Light** causes stomata to open, increasing the rate of transpiration. Stomata close in darkness in most species of plant, as there is no need to absorb carbon dioxide, and water can be conserved without reducing the rate of photosynthesis.
- **Humidity** is the water vapour content of the air. It is usually measured as a percentage of the maximum amount of water vapour that air can hold. Because of evaporation of water from moist cell walls, the humidity of air in the spaces inside the leaf is always around 100%. The lower the humidity outside the leaf, the faster the rate of diffusion of water out through the stomata and the higher the transpiration rate.
- **Wind** increases the transpiration rate. In still air or light winds, transpiration causes the humidity around the leaf to increase, reducing the rate of transpiration. In windy conditions, the rate of transpiration becomes maximal for the humidity of the air.
- **Temperature** affects the rate at which water evaporates from the moist surfaces inside the leaf. At higher temperatures, evaporation rates rise. Higher temperatures also increase the rate of diffusion between the air spaces inside the leaf and the air outside. Increases in temperature also allow the air to hold more water vapour and so reduce the relative humidity of air outside the leaf. The concentration gradient therefore increases and water is lost more rapidly.

Investigating biology: the effect of biotic and abiotic factors on transpiration

The rate of transpiration is difficult to measure directly. Instead, the rate of water uptake is usually measured, using a potometer. Figure 22 shows one type of potometer.

To design an investigation you will need to discuss the following questions.

1. How will you measure the rate of transpiration in your investigation?
2. What biotic or abiotic factor will you investigate?
3. How will you vary the level of this factor?
4. How many results do you need, at each level of the factor that you are varying?
5. How will you keep other factors constant, so that they do not affect the rate of transpiration?

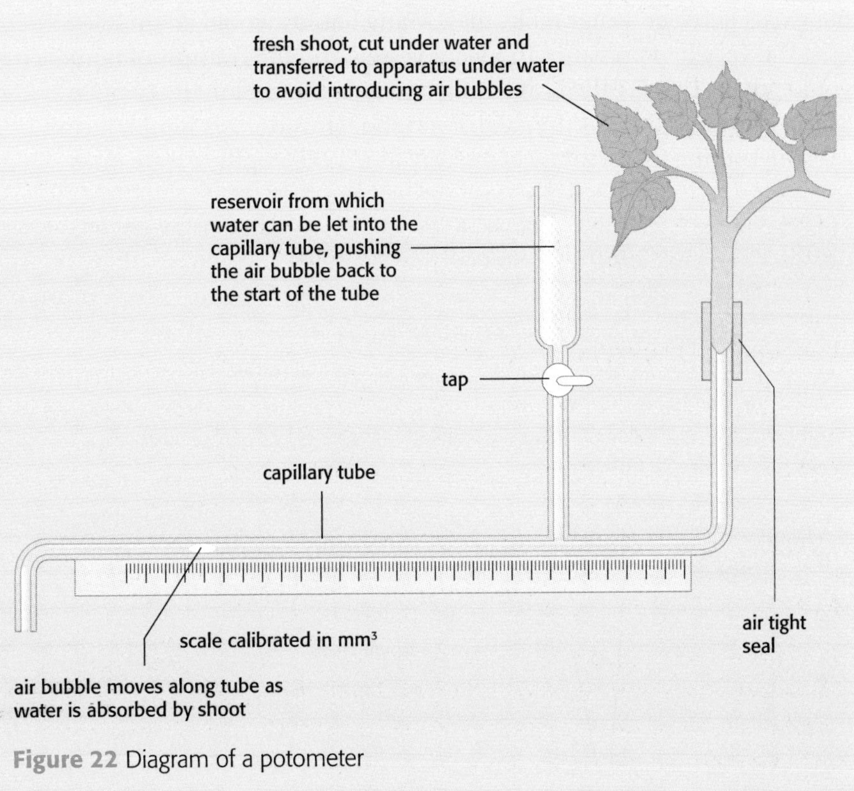

Figure 22 Diagram of a potometer

HL

Water transport in xylem

The **structure of xylem vessels** allows them to transport water inside plants very efficiently. Xylem vessels are long continuous tubes. Their walls are impregnated with lignin, strengthening them, so that they can withstand very low pressures without collapsing. Xylem vessels are formed from files of cells, arranged end-to-end. The walls between adjacent cells in the file are removed and the plasma membranes and contents of the cells beak down. When mature they are therefore non-living and the flow of water along them must be a passive process.

The pressure inside xylem vessels is usually much lower than atmospheric pressure. The walls of cells inside the leaf generate this low pressure. Water is attracted to the cellulose and other substances in the cell walls, which causes **adhesion**. When water evaporates from the surface of the wall, adhesion causes water to be drawn through the cell wall from the nearest available supply, to replace the water lost by evaporation. The nearest available supply is the xylem vessels in the veins of the leaf. Even if the pressure in the xylem is already low, the force of adhesion is strong enough to suck water out of the xylem, further reducing its pressure.

The low pressure generates a pulling force that is transmitted though the water in the xylem vessels down the stem and to the ends of the xylem in the roots. It is called the **transpiration pull** and is strong enough to move water upwards, against the force of gravity to the top of the tallest tree. For the plant it is a passive process, with all the energy needed for it coming from the thermal energy that causes transpiration.

The pulling of water upwards in xylem vessels depends on the **cohesion** that exists between water molecules. Many liquids would be unable to resist the very low pressures in xylem vessels and the column of liquid would break. This is called cavitation and it does occasionally happen even with water, but it is unusual. Even though water is a liquid, it can transmit pulling forces in the same way as a solid length of rope does.

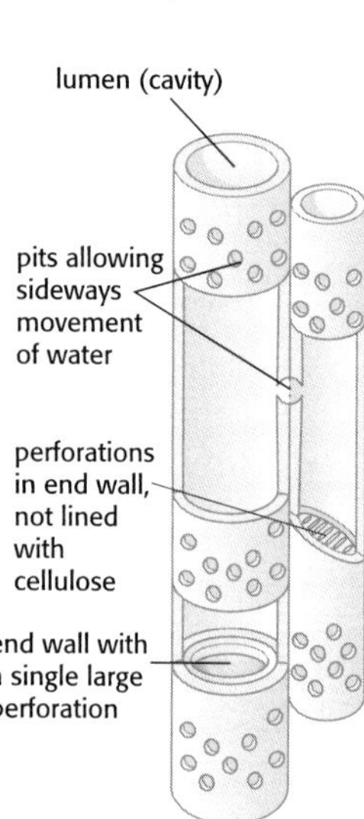

Figure 23 Structure of xylem vessels

Data-based question: the Renner experiment

Figure 24 shows the results of an experiment by the German plant physiologist Otto Renner in 1912. A transpiring woody shoot was placed in a potometer and the rate of water uptake was measured. A clamp was attached to the stem to restrict the flow of water up to the leaves. Later on, the top of the shoot, with all of its leaves, was removed. A vacuum pump was then attached to the top of the shoot.

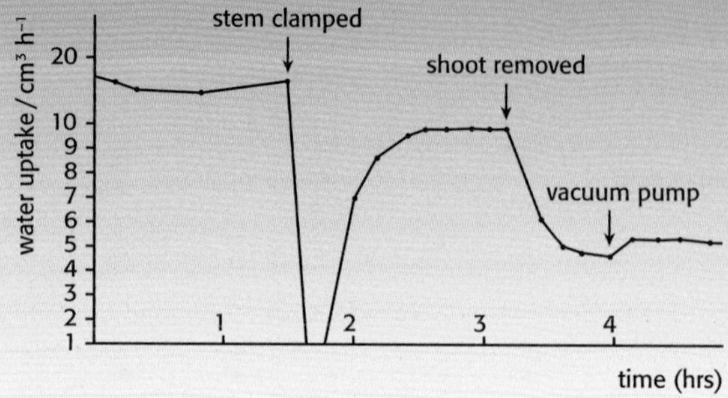

Figure 24 Results of the Renner experiment

1. Describe the effect of clamping the stem on the rate of water uptake. [3]
2. Explain the effect of cutting off the top of the shoot on the rate of water uptake. [3]
3. Calculate the difference between the rate of water uptake caused by the vacuum pump and the rate caused by the leaves immediately before the shoot top was cut off. [2]
4. The water in the potometer was at atmospheric pressure. The vacuum pump generated a pressure of zero. Discuss what the results of the experiment showed about the pressures generated in the xylem by the leaves of the shoot. [2]

HL

Xerophytes

Xerophytes are plants adapted to growing in deserts and other dry habitats. There are various strategies that plants can use to survive in these habitats, including increasing the rate of water uptake from the soil and reducing the rate of water loss by transpiration. Some xerophytes are ephemeral, with a very short life cycle that is completed in the brief period when water is available after rainfall. They then remain dormant as embryos inside seeds until the next rains, sometimes years later. Other plants are perennial and rely on storage of water in specialised leaves, stems or roots.

Most cacti are xerophytes, with leaves that are so reduced in size that they usually only consist of spines. The stems contain water storage tissue and become swollen after rainfall. Pleats allow the stem to expand and contract in volume rapidly. The epidermis of cactus stems has a thick waxy cuticle and, unlike most plant stems there are stomata, though they are spaced more widely than in leaves. The stomata usually open at night rather than in the day, when it is much cooler and transpiration occurs more slowly. Carbon dioxide is absorbed at night and stored in the form of a four-carbon compound, malic acid. Carbon dioxide is released from the malic acid during the day, allowing photosynthesis even with the stomata closed. This is called Crassulacean acid metabolism. Plants such as cacti that use this system are called CAM plants.

Cacti are native plants of North and South America. Xerophytes in other parts of the world belong to different plant families. The adaptations in these xerophytes are often very similar to those of cacti. Some African species of *Euphorbia* for example, are difficult to distinguish from cacti until they produce flowers. There are some distinctive xerophytes in South Africa that have just two swollen leaves, which only protrude a few millimetres above the surface of the stony soil where they grow, sheltering them from wind. The shape and coloration of the leaves is very similar to small stones, hence their common name of Living Stones.

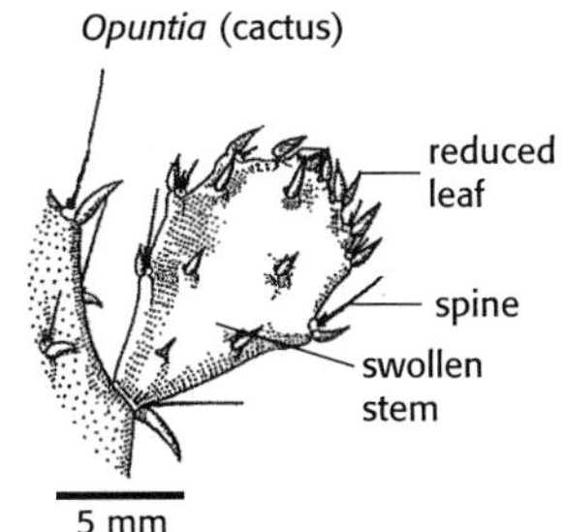

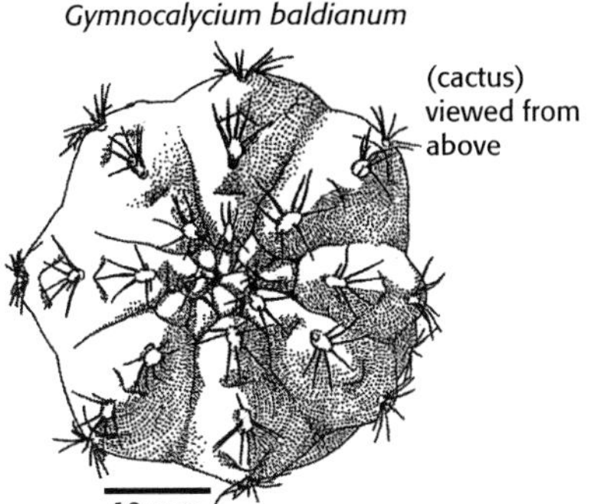

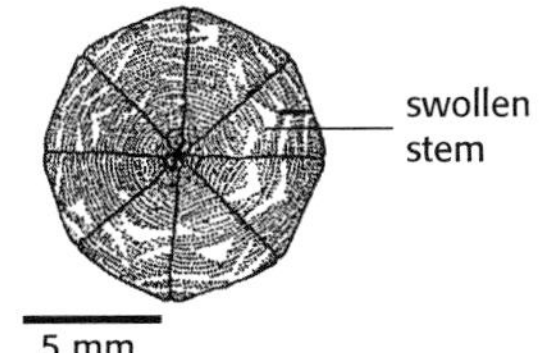

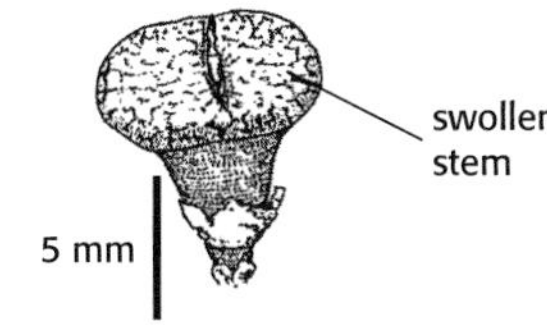

Figure 25 Xerophytes

Data-based question: water permeance of waxy cuticle

The graphs in Figure 26 show the results of investigations into the rate at which water is able to diffuse though the waxy cuticle of plants, which is called the water **permeance** of the cuticle. Figure 26a shows the relationship between temperature and water permeance of four species of plant. Figure 26b shows the relationship between the thickness of cuticular wax and water permeance. The results of the experiment show how important it is to test hypotheses, even when it may seem that this is not necessary.

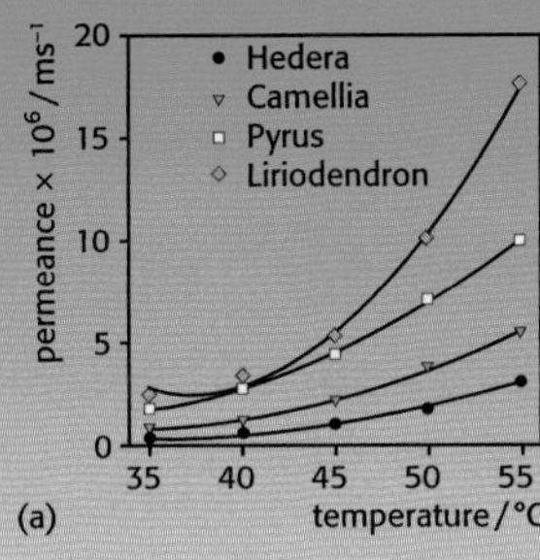

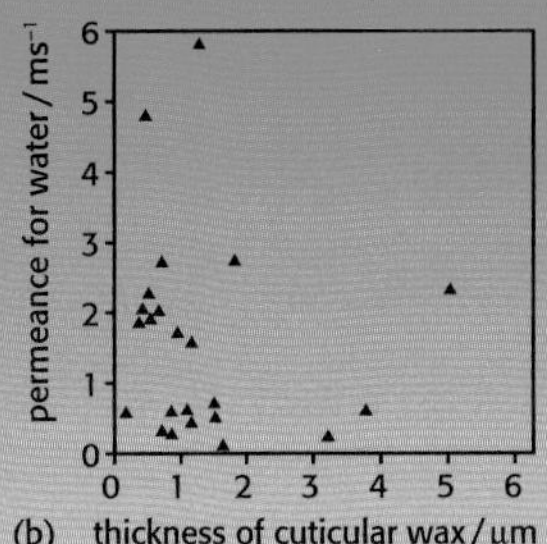

Figure 26 Factors affecting water permeance of waxy cuticle

1 Using the data in Figure 26a, describe the relationship between temperature and water permeance. [2]

2 Discuss the consequences for plants of the effect of temperature on cuticular water permeance. [3]

3 Using the data in Figure 26b, state the thickness of cuticular wax with:
(a) the highest water permeance
(b) the lowest water permeance. [2]

4 Evaluate the hypothesis that the water permeance of the cuticle is positively correlated with its thickness, using the data in Figure 26b. [3]

HL

Seed structure and dispersal

A seed is a package containing an embryo plant and food reserves, all inside a protective seed coat. The embryo plant consists of an **embryo root, embryo shoot** and one or two **cotyledons**, depending on whether the plant is monocotyledonous or dicotyledonous. The cotyledons are the embryo leaves and in many plants they contain the food reserves of the seed. In other seeds there is a special food storage tissue called endosperm. The scientific name for the seed coat is the **testa**. There is a small hole through the testa, called the **micropyle**. It is located next to a scar where the seed was attached to the parent plant. Figure 27 shows the external and internal structure of a bean seed (*Phaseolus vulgaris*).

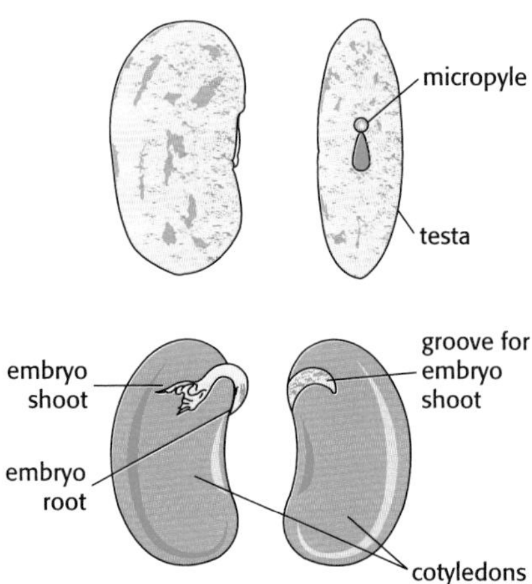

Figure 27 Structure of bean seed (*Phaseolus vulgaris*); external structure (above); internal structure (below)

Seeds cannot move themselves, but nonetheless they often travel long distances from the parent plant. This is called **seed dispersal** and it reduces competition between offspring and parent and helps to spread the species. The type of seed dispersal depends on the structure of the fruit – dry and explosive, fleshy and attractive for animals to eat, feathery or winged to catch the wind, or covered in hooks that catch onto the coats of animals.

Some seeds do not immediately germinate, even if given the conditions normally required. This is called dormancy and it allows time for seeds to be dispersed. It may also help to avoid germination at an unfavourable time.

Data-based question: fire and seed dormancy in a plant of the chaparral

Emmenanthe penduliflora grows in chaparral (shrubland) in California. It is rarely seen in unburnt chaparral, but appears after fires, growing to about 250 mm, flowering, forming seed and dying in a few months. The electron micrographs below show the results of an experiment in which seeds of the plant were treated with smoke for 3 minutes and then soaked in a solution of lanthanum nitrate hexahydrate.

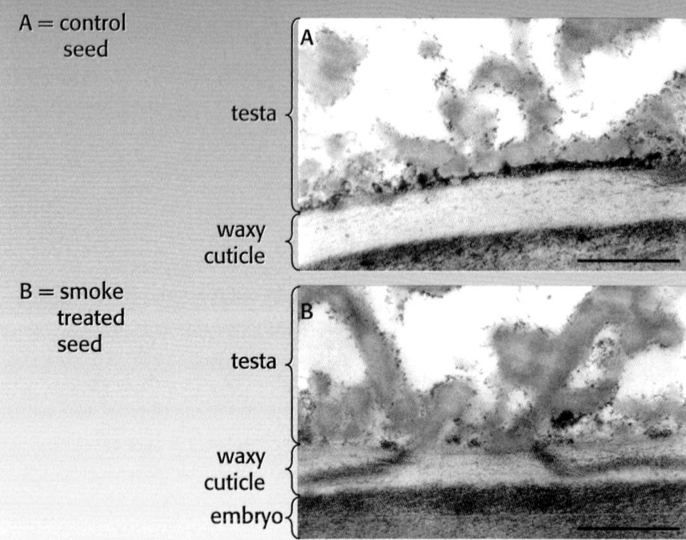

Figure 28 Two electron micrographs of *Emmenanthe penduliflora*. (a) Control seed (above); (b) smoke-treated seed (below)

1. The scale bars in the electron micrographs represent 1 μm. Calculate the thickness of waxy cuticle between the testa and the embryo and food stores inside the control seed. [2]
2. The lanthanum solution appears as dark staining in the electron micrographs and shows how far water was able to penetrate. Deduce how far water could penetrate into the control seeds. [2]
3. (a) Compare the staining of the waxy cuticle in the smoke-treated seeds with the staining of the cuticle in the control seeds. [2]
   (b) Suggest a hypothesis for the germination of plants of *Emmenanthe penduliflora* after fires, based on the differences in staining that you have described. [2]
4. Suggest **two** advantages to *Emmenanthe penduliflora* of dormancy ending after fires in the chaparral. [2]

HL

Germination

All seeds need **water** for germination. Many seeds are dry and need to rehydrate their cells. Some seeds contain a hormone that inhibits germination and water is needed to wash it out of the seed. Germination involves growth of the embryo root and shoot and this also requires water.

The metabolic rate of a dry and dormant seed is close to zero, but after absorption of water, metabolic processes begin again, including energy release by aerobic cell respiration. Another requirement for germination is therefore a supply of **oxygen**. Because germination involves enzyme-catalysed metabolic reactions, **warmth** is required and germination often fails at low temperatures.

Another metabolic process occurring at the start of germination is synthesis of **gibberellin**, a plant hormone. Several genes have to be expressed to produce the various enzymes of the metabolic pathway leading to gibberellin. This hormone stimulates mitosis and cell division in the embryo. In starchy seeds it also stimulates the production of amylase. This enzyme is needed to break down starch in the food reserves into maltose. Other enzymes convert the maltose into sucrose or glucose. Whereas starch is insoluble and immobile, sucrose and glucose can be transported from the food reserves to where they are needed in the germinating seed. The embryo root and shoot need sugars for growth, together with amino acids and other substances released from the food stores. All parts of the embryo need glucose for aerobic cell respiration.

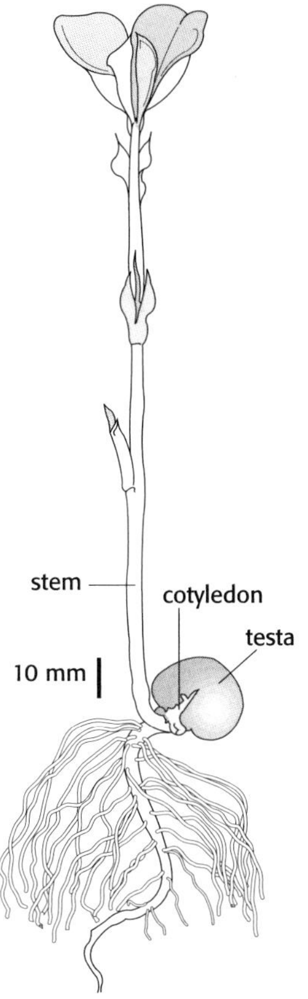

Figure 29 Structure of a germinating pea seedling

Investigating biology: factors required for germination

Most vegetable crop varieties have been bred to germinate quickly – they do not usually have long periods of seed dormancy. Nevertheless, growers of vegetable crops sometimes have difficulty in getting crops to germinate after sowing.

Choose one of the possible causes of crop failure shown in the mind-map, to investigate.

Design an experiment and see whether you obtain evidence for or against your cause. You will need to decide:

- which seed type to use
- how to vary the factor that you are investigating
- how to keep other factors constant
- how to collect your results, including how to assess whether germination has occurred.

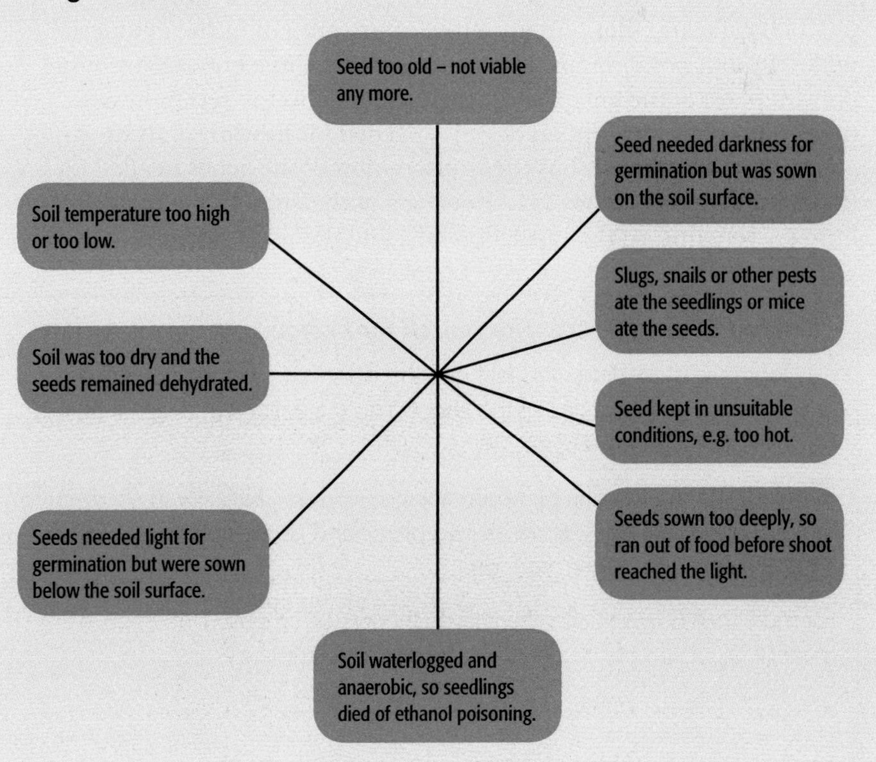

HL

Control of flowering

When a seed germinates, a young plant is formed that grows roots, stems and leaves. These are called vegetative structures, so the plant is in the vegetative phase. This can last for weeks, months or years, until a trigger causes the plant to change into the reproductive phase and produce flowers. The change from the vegetative to the reproductive phase happens when meristems in the shoot start to produce parts of flowers instead of leaves. This involves a change to the pattern of gene expression in cells produced by the meristem. Observations of flowering suggested that the trigger for this in some plants might be a particular day length. Experiments to test this hypothesis were done in growth chambers, where the length of light and dark periods can be controlled. These showed that it is the length of darkness that matters, not the length of daylight.

- Long-day plants flower in summer when the nights have become short enough.
- Short-day plants flower in the autumn, when the nights have become long enough.

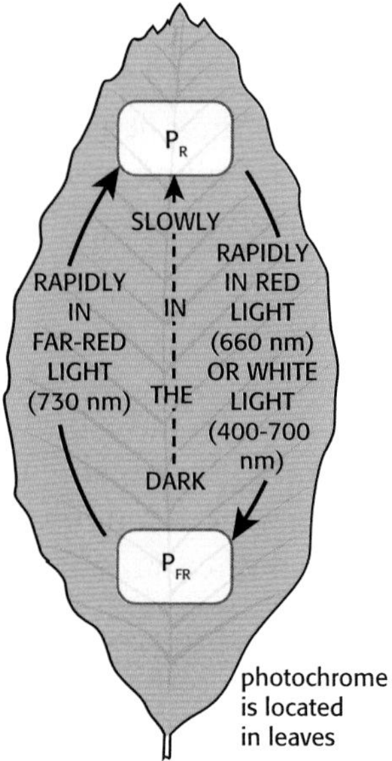

Figure 30 Interconversions of phytochrome

A pigment was discovered in leaves that plants use to measure the length of dark periods. It is called phytochrome and is unusual as it can switch between two forms, P_R and P_{FR}. When P_R absorbs red light of wavelength 660 nm it is converted into P_{FR}. When P_{FR} absorbs far-red light, of wavelength 730 nm, it is converted back to P_R. This latter conversion is not of great importance as sunlight contains more light of wavelength 660 nm than 730 nm, so in normal sunlight phytochrome is rapidly converted to P_{FR}. However, P_R is more stable than P_{FR}, so in darkness P_{FR} very gradually changes into P_R.

Further experiments have shown that P_{FR} is the active form of phytochrome and that receptor proteins are present in the cytoplasm to which P_{FR} but not P_R binds. In long-day plants, large enough amounts of P_{FR} remain at the end of short nights to bind to the receptor, which then promotes transcription of genes needed for flowering. In short-day plants, the receptor inhibits the transcription of the genes needed for flowering when P_{FR} binds to it. However, at the end of long nights, very little P_{FR} remains, so the inhibition fails and the plant flowers.

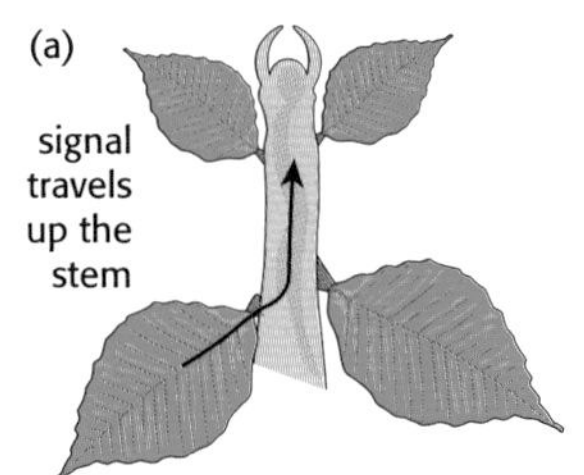

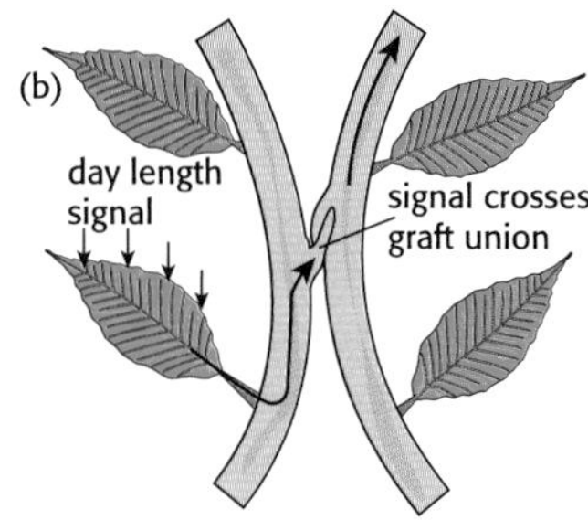

Figure 31 Evidence for florigen

Thinking about science: the search for florigen

Day length is detected in the leaf, but the cells that produce flowers are in the meristem in the tips of the shoot. This suggests that a signal travels from leaf to shoot tip (Figure 31a).

In the 1930s the Russian plant physiologist Mikhail Chailakhyan showed that a signal could pass between two plants and cause flowering (Figure 31b). He called the signal *florigen*.

For more than 60 years after the florigen theory was proposed, attempts were made, without success, to identify it as a chemical substance.

1 Explain why the florigen theory persists, even though scientists have failed to find evidence for it in 60 years.

2 (a) According to Karl Popper, a theory is scientific if it is falsifiable. Explain what this means.

(b) What would a scientist have to do to show that florigen *doesn't* exist?

3 Predict the future of the florigen theory.

HL

FLower structure, pollination and fertilization

Figure 32 shows a flower of *Prunus domestica*. In the base of the flower are nectar-secreting glands, which attract insects, especially bees. The petals are large and white, helping insects to find the flower. The anthers produce pollen, containing the male gametes. The filaments hold the anthers in a position where they are likely to brush pollen onto visiting insects. If an insect is already carrying pollen from another *Prunus domestica* flower and brushes it onto the stigma, then the flower has been **pollinated**.

The next process after pollination is **fertilization**. From each pollen grain on the stigma a tube grows down the style to the ovary. The pollen tube carries male gametes to fertilize the ovary. The ovary is located inside a small rounded structure called an ovule. The fertilized ovule develops into a seed and the ovary develops into a fruit. In this case the fruit is called a plum.

Figure 32 Structure of a plum flower

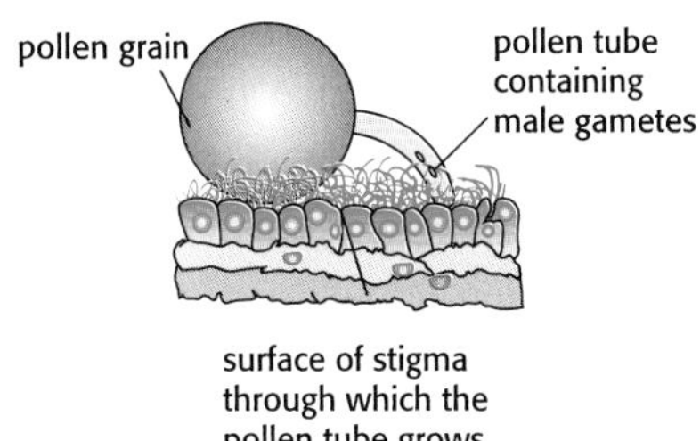

Figure 33 Pollen grain germinating on a stigma at the start of the fertilization process

Working with data: factors affecting pollen development

Pollen grains sometimes develop when they are placed in drop of fluid on a microscope slide. The composition of the fluid and its temperature affect whether this happens or not. Table 6 show the results of studies of pollen development in plant species in Hong Kong.

Table 6

Sucrose concentration (mmol dm^{-3})	Percentage of *Camellia japonica* pollen grains that developed	Copper ion concentration (ppm)	Mean growth of pollen tubes of *Bougainvillea glabra* (μm h^{-1})
0.30	22.5	0.0	33.6
0.46	23.0	1.0	25.1
0.60	13.0	2.5	15.5
0.75	0.0	5.0	10.8
0.90	0.0	25.0	0.0

Plant species	Diameter of pollen grain (μm)	Mean growth of pollen tube (μm h^{-1})	Optimal sucrose conc. (mmol dm^{-3})
Bougainvillea glabra	44.00	41.8	0.75
Delonix regia	70.30	4.9	0.45
Leucaena leucocephala	64.60	111.0	0.75
Bauhinia purpurea	71.50	69.9	0.45
Lilium bulbiferum	91.60	11.1	0.30
Gladiolus gandavensis	86.82	50.6	0.45

1 The data in the Table 6 is difficult to analyse in its current form. Choose suitable presentation formats to display the data clearly and allow you to identify any significant trends. You can use ICT or you can draw graphs, tables, charts or diagrams by hand.
2 Describe clearly any trends that you have found in the data. Try to explain each trend that you describe, using your biological knowledge.
3 Identify any weaknesses in the data obtained. Suggest how the investigation could have been improved.

HL

Chapter 10 questions

1 (a) Phloem sap is nutrient-rich compared with many other plant products, and the nutrients in it are small soluble molecules that do not need to be digested. Despite this, the only animals to consume it as the main part of their diet are insects belonging to a group called the Hemiptera, including aphids, whitefly, mealybugs and psyllids. The data in this question comes from research into aphids.

The sugar content of phloem sap is very high – often greater than 1 mol dm^{-3}.

(i) Explain how plants increase the sugar concentration of phloem sap to such high levels. [1]

(ii) Explain how high sugar concentrations cause a high pressure to develop in the phloem. [2]

(b) Aphids only ingest a small proportion of the sugar in phloem sap. The remainder passes out in the faeces, which is a liquid called honeydew. Because of the high sugar concentrations, phloem sap has a much higher solute concentration than aphid cells. Enzymes secreted into the aphid gut reduce the solute concentration of phloem sap by converting sugars into oligosaccharides. Figure 34 shows the relationship between the sucrose concentrations of phloem sap ingested by aphids and the oligosaccharide content of the honeydew.

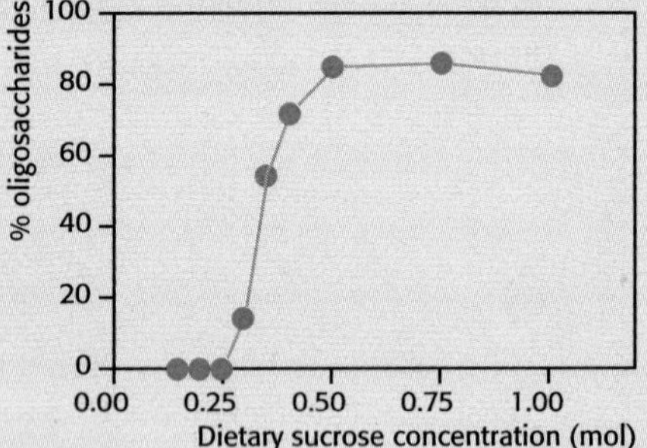

Figure 34

(i) Describe the relationship between the sucrose concentration of phloem sap ingested by aphids and the percentage of oligosaccharides in the honeydew. [3]

(ii) Suggest reasons for aphids secreting enzymes to reduce the solute concentration of the fluid in the gut. [2]

(c) Aphids ingest larger volumes of phloem sap than they need, to obtain sufficient sugar for cell respiration. This is because they also need to obtain amino acids and the concentration of amino acids in phloem sap is low. Figure 35 shows the percentages of individual amino acids in phloem sap and the percentages in aphid protein. Nine of the amino acids cannot be synthesized in aphid cells and so are called essential amino acids. The other amino acids can be synthesized from other amino acids and so are non-essential.

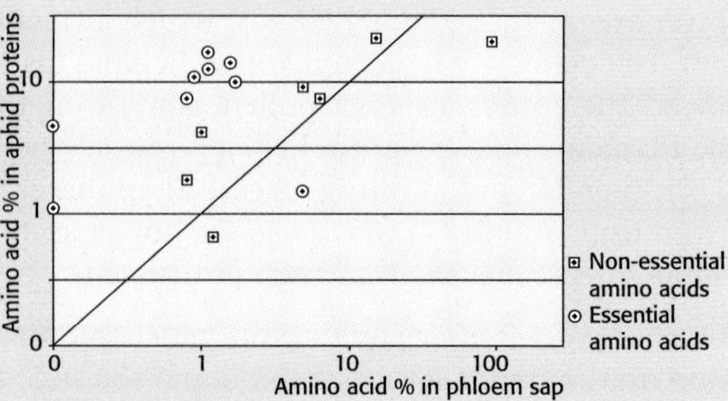

Figure 35

(i) Evaluate phloem sap as a source of amino acids for aphids. [3]

(ii) Suggest reasons for the differences in amino acid content between phloem sap and aphid protein. [2]

(d) Specialized cells have been discovered in aphids called bacteriocytes. These cells contain bacteria called *Buchnera*, which synthesize essential amino acids from aspartic acid and sucrose. Aspartic acid is a non-essential amino acid that is found in much higher concentrations in phloem sap than any other amino acid. When aphids reproduce, they pass on *Buchnera* bacteria to their offspring.

(i) Explain how antibiotics could be used to obtain evidence for the role of *Buchnera* in aphids. [2]

(ii) Using the data in this question, discuss the reasons for few animals using phloem sap as the main part of their diet. [3]

2 State one the role of each the following in plants:

(a) apical meristems	(f) P_{FR}
(b) bulbs	(g) spongy mesophyll
(c) cotyledons	(h) tendrils
(d) guard cells	(i) waxy cuticle
(e) palisade mesophyll	(j) xylem.

3 Distinguish between pollination, fertilization and seed dispersal in the reproduction of flowering plants. [6]

4 (a) Outline the structure of roots and how it allows water to be absorbed. [5]

(b) Explain how water is transported from the roots to the top of the tallest trees [8]

(c) Suggest reasons why monocotyledonous plants do not grow into large trees, whereas some dicotyledonous plants grow into very large trees. [4]

Meiosis

Sexual reproduction is a method of producing offspring that also generates genetic diversity in a species. In eukaryotic organisms, it involves the process of fertilization. Fertilization is the union of sex cells, or gametes, usually from two different parents. Fertilization would double the hereditary information each generation, if the process of creating gametes did not involve the halving of hereditary information before fertilization.

As a consequence of fertilization, humans have pairs of chromosomes, with one chromosome in a pair from each parent. A nucleus like this with two chromosomes of each type is **diploid**. A nucleus with only one chromosome of each type is **haploid**.

Meiosis is the process by which hereditary information is halved during the production of gametes. It is achieved by halving the number of chromosomes. The process can be summarized as follows (see Figure 1).

1 During **interphase**, the chromosomes replicate. Each chromosome consist of two identical **chromatids**.
2 At the start of meiosis I, homologous chromosomes pair up. The homologous chromosomes exchange genetic material with each other in a process termed **crossing over**.
3 During meiosis I, the homologous pairs of chromosomes separate. One of each pair goes to each of the two daughter cells. The result is two haploid daughter cells.
4 In meiosis II, the two daughter nuclei divide again. This time the chromatids of each chromosome separate. Meiosis II is similar to mitosis. The end result is four haploid cells.

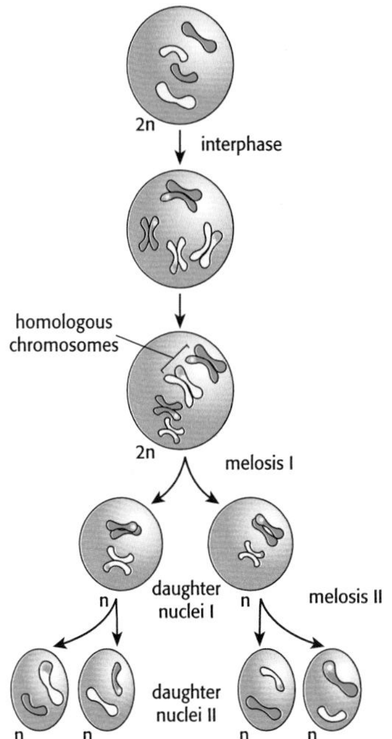

Figure 1 Outline of meiosis

Table 1 gives more detailed information about the sequence of events.

Table 1 Diagrams in the central column represent animal meiosis while most of the micrographs in the final column are cells from the anther of a Lily.

Stage	Diagram	Micrograph
Prophase I • Cell has 2n chromosomes (double chromatid): n is haploid number of chromosomes. • Homologous chromosomes pair (synapsis). • Crossing over occurs.	Prophase I Nuclear envelope disintegrates Chromosomes each consist of two sister chromatids Spindle microtubules Chiasma (point of cross over)	
Metaphase I • Spindle microtubules move homologous pairs to equator of cell. • Orientation of paternal and maternal chromosomes on either side of equator is random and independent of other homologous pairs.	Metaphase I Equator	

Anaphase I

- Homologous pairs are separated. One chromosome of each pair moves to each pole.

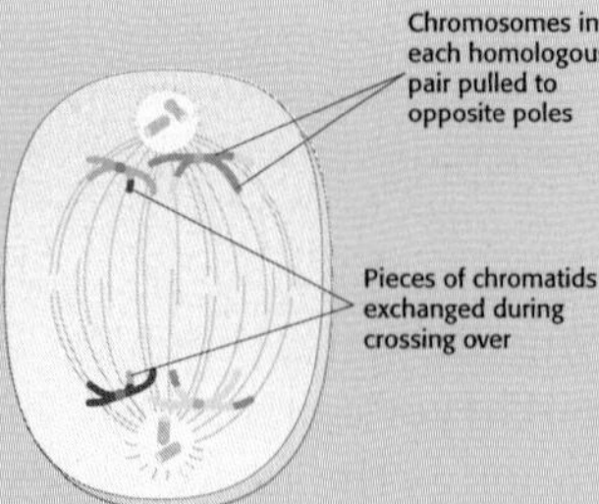

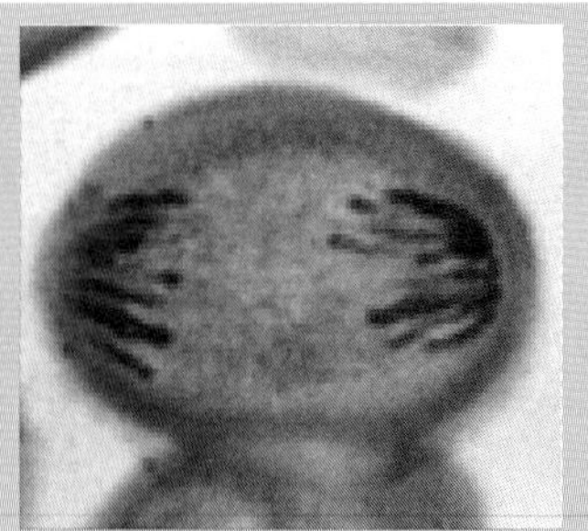

Telophase I

- Chromosomes uncoil. During interphase that follows, no replication occurs.
- Reduction of chromosome number from diploid to haploid completed.
- Cytokinesis occurs.

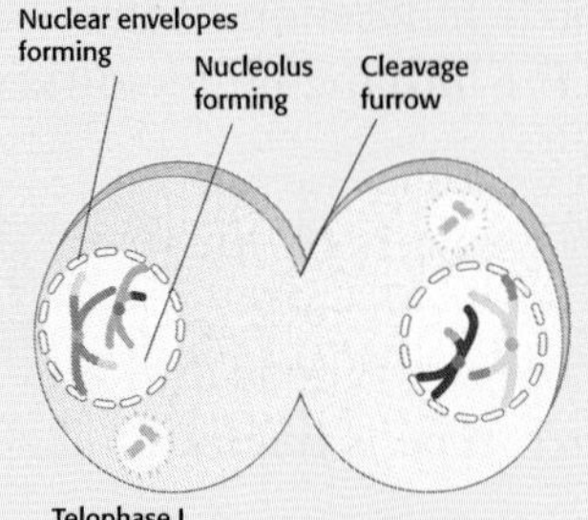

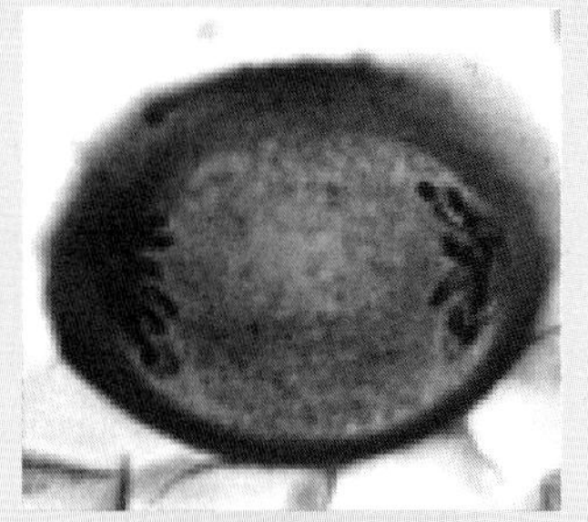

Prophase II

- Chromosomes, which still consist of two chromatids, condense and become visible.

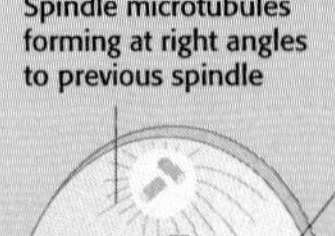

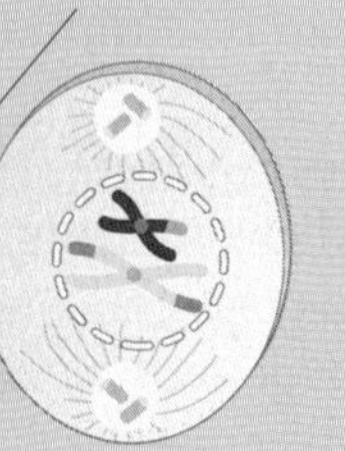

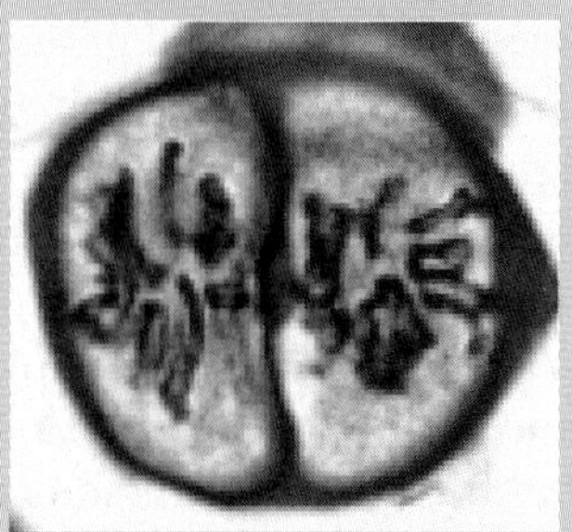

Metaphase II

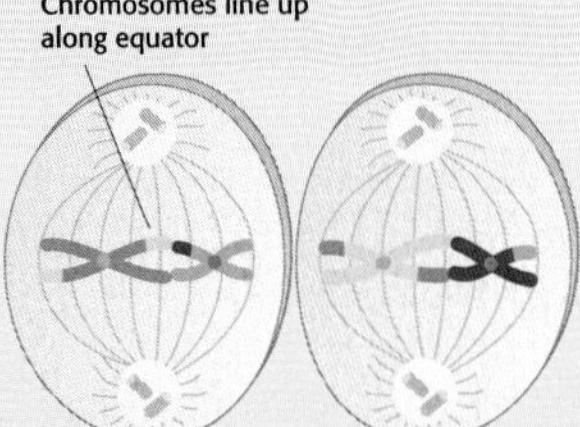

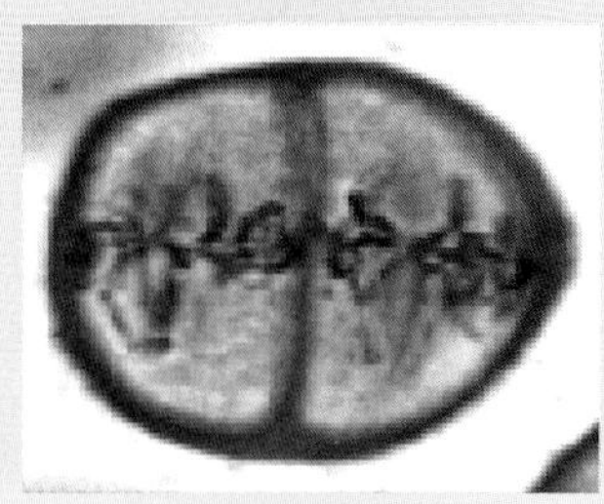

Anaphase II

- Centromeres separate and chromatids are moved to opposite poles.

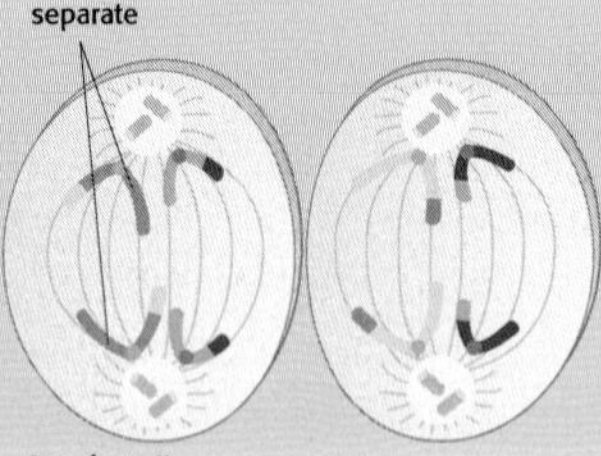

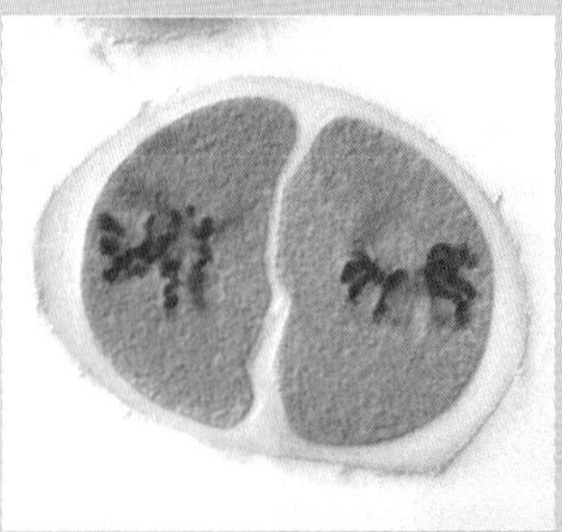

Telophase II

- Chromatids reach opposite poles
- Nuclear envelope forms
  Cytokinesis occurs

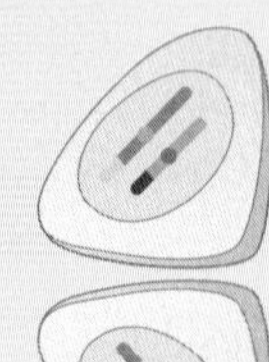

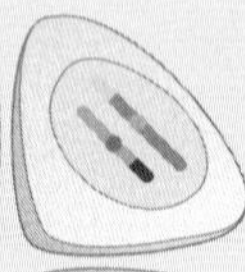

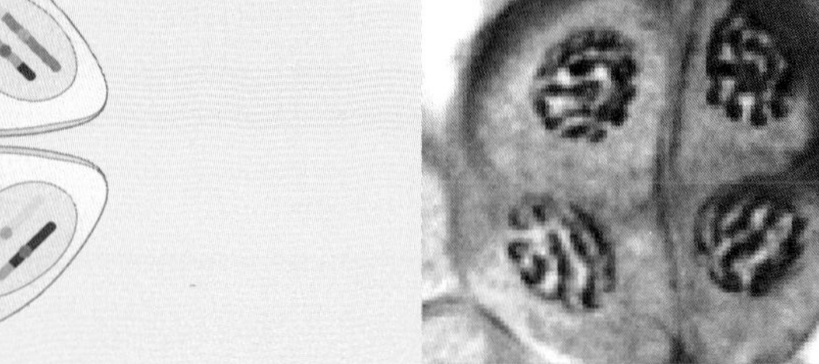

Non-disjunction

Meiosis is sometimes subject to errors. One example of this is when homologous chromosomes fail to separate at anaphase. This is termed non-disjunction. In other words, segregation does not occur for a certain pair of homologous chromosomes. The result will be a gamete that either has an extra chromosome or is deficient in a chromosome. If the gamete is involved in human fertilization, the result will be an individual with either 45 or 47 chromosomes.

An abnormal number of chromosomes will often lead to a person possessing a syndrome, i.e. a collection of physical signs or symptoms. For example trisomy 21, also known as Down syndrome, is due to a non-disjunction event that leaves the individual with three of chromosome number 21 instead of two. While individuals vary, some of the component features of the syndrome include hearing loss, heart and vision disorders. Mental and growth retardation are also common.

For each of the following syndromes, research the chromosomes involved in the non-disjunction event and some of the component features of the resulting syndrome.

a) Turner's syndrome
b) Klinefelter's syndrome
c) Patau syndrome.

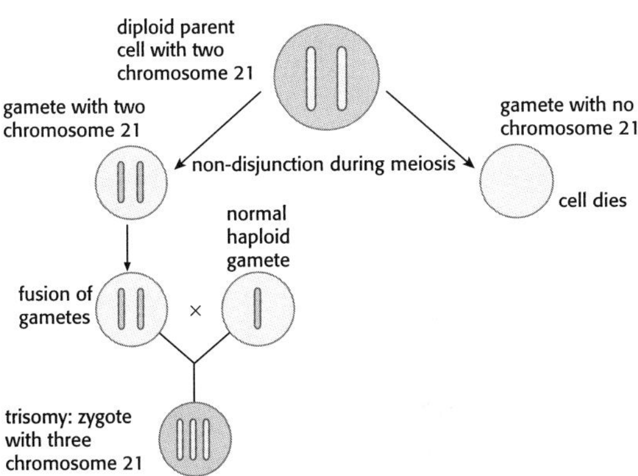

Figure 2 How non-disjunction can give rise to Down syndrome

Data-based question: risk of chromosomal abnormalities with advancing age of the parent

The data presented in Figure 3 shows the relationship between maternal age and the incidence of trisomy 21 and of other chromosomal abnormalities.

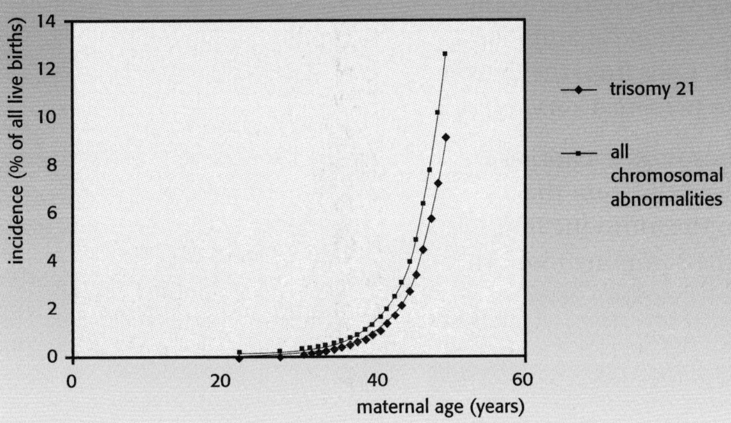

Figure 3 The incidence of trisomy 21 and other chromosomal abnormalities as a function of maternal age

1 Outline the relationship between maternal age and the incidence of chromosomal abnormalities in live births. [2]
2 (a) For mothers 40 years of age, determine the probability that they will give birth to a child with trisomy 21. [1]
(b) Using the data in Figure 3, calculate the probability that a mother of 40 years of age will give birth to a child with a chromosomal abnormality other than trisomy 21. [2]
3 Only a small number of possible chromosomal abnormalities are ever found among live births, and trisomy 21 is much the commonest. Suggest reasons for these trends. [3]
4 Discuss the risks parents face when choosing to postpone having children. [2]

Testing for Down syndrome

For older parents, a standard clinical practice is to administer a serum blood test such as the triple test or the quad test. These are blood tests performed on expectant mothers that look for unusual levels of such chemicals as alpha-fetoprotein (AFP) and human chorionic gonadotropin (HCG). AFP is produced in the yolk sac and in the liver of the developing fetus. HCG is produced by the placenta. The levels of each of these chemicals in the mother's blood varies with the gestational age of the pregnancy. With trisomy 21, the mother's blood will show levels of AFP that are about 25 per cent lower than normal levels and HCG levels that are approximately two times higher than the normal HCG level.

If the serum test raises concern, than parents will be advised of the option to have a karyotype produced which can give a more definitive diagnosis. A karyotype (see Figure 4) is an organized image of metaphase fetal chromosomes. Technicians stain the chromosomes, which results in a banding pattern. The technician can then organize the chromosomes by their length, the position of their centromere and by the banding pattern.

(a)

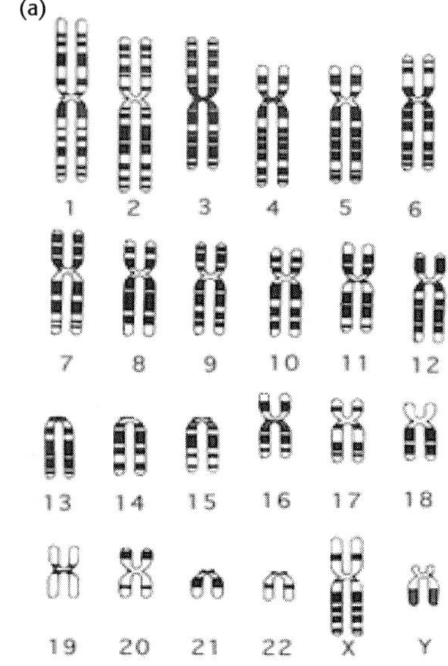

(b)

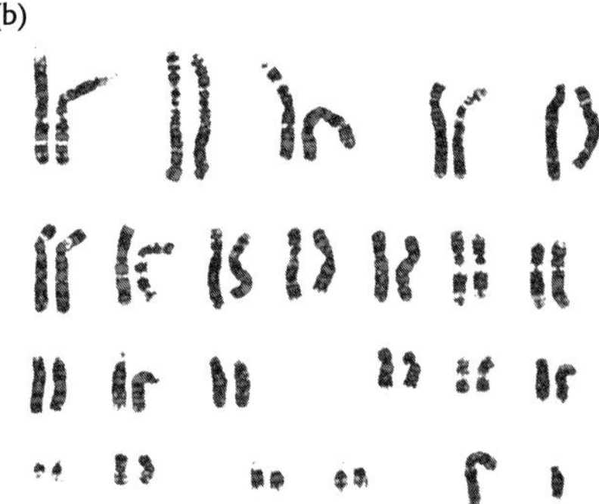

Figure 4 A diagram of the 24 types of chromosomes in humans (Figure 4a) and photograph of a human karyotype (Figure 4b)

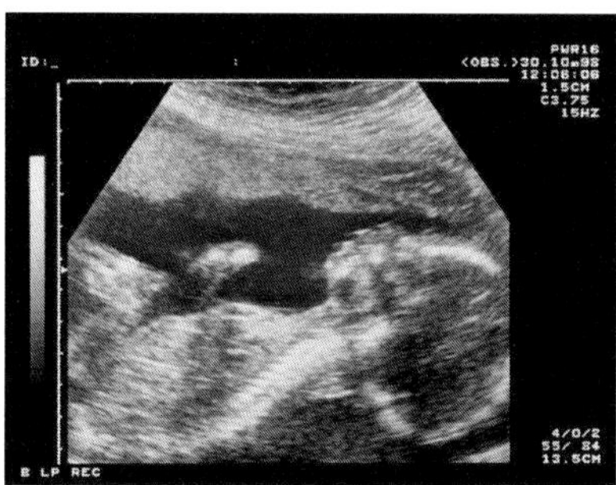

Figure 5 Ultrasound scan image of 12-week old fetus

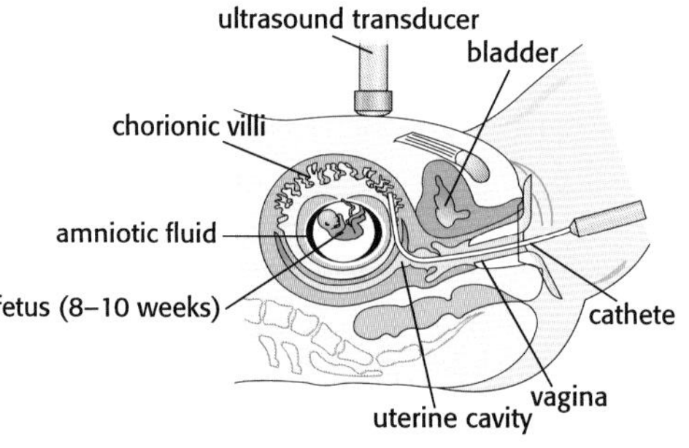

Figure 6 Chorionic villus sampling

There are two procedures for obtaining the fetal chromosomes to produce the karyotype. One procedure, called amniocentesis, involves passing a needle through the mother's abdominal wall, using ultrasound to guide the needle. Figure 5 shows a woman undergoing amniocentesis. The needle is used to withdraw a sample of amniotic fluid from the amniotic sac of a developing fetus. Cells from the fetus in the amniotic fluid are cultured and then used to prepare a karyotype.

A second procedure for obtaining fetal chromosomes is chorionic villus sampling, or cvs. This procedure samples cells from the placenta, specifically the chorion, rather than the amniotic fluid. It can be done earlier than amniocentesis and the sampling tool can enter through the vagina (see Figure 6).

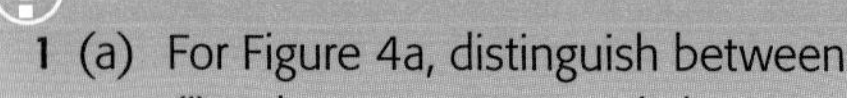

1 (a) For Figure 4a, distinguish between:
   (i) chromosome 5 and chromosome 6
   (ii) chromosome 17 and chromosome 18
   (iii) the X and Y chromosome. [3]

2 (a) State the gender of the subject of the human karyotype in Figure 4b. [1]
   (b) Identify the syndrome that this person will possess. [1]

HL

Chiasmata

During prophase I of meiosis, all of the chromatids of two homologous chromosomes become tightly associated in a process called synapsis. The resulting combined pair of homologous chromosomes is called a bivalent (referring to the two homologous chromosomes) or a tetrad (referring to the four chromatids within the structure).

The maternal and paternal chromosomes exchange corresponding sections of DNA and once crossing over is complete, new combinations of alleles will have been created. The process by which offspring possess a combination of alleles different from that of either parent is called recombination (see Figure 7).

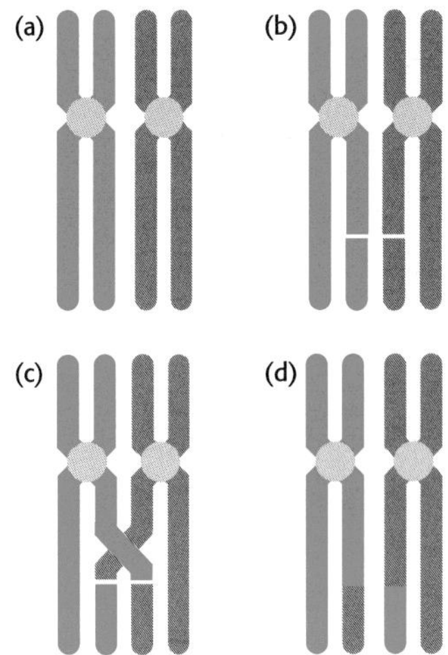

Figure 7 The process of crossing over

A chiasma is an X-shaped structure formed between non-sister chromatids during prophase I of meiosis. The chiasma is a physical manifestation of crossing over. Usually between one and three chiasmata form per homologous pair (Figure 8). The chiasmata persist through metaphase I and play a role in the prevention of non-disjunction.

Meiosis and genetic variety

The random orientation of chromosomes at metaphase I leads to variation within offspring. For every chromosome pair, the number of possible chromosome combinations doubles. For a haploid number of n, the number of possible combinations is 2^n. For humans with a haploid number of 23 this amounts to 2^{23} or over 8 million combinations. Crossing over increases this number still further – so much so that meiosis can produce an effectively limitless number of genetically different haploid cell types from one diploid cell type.

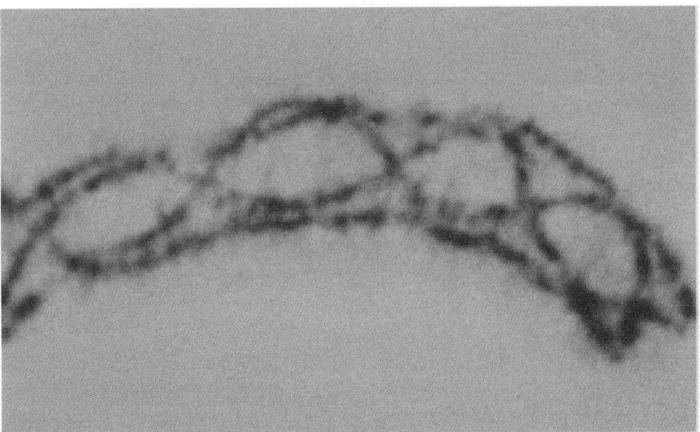

Figure 8 This image shows that multiple chiasmata can form within one tetrad.

Working with data: variation, standard deviation and the use of the graphic calculator

In any sexually reproducing population, there is variety. This can be shown using beans.

1 Measure the length of every bean in a sample of 100 beans. Design a data table to record your raw data.

2 To ensure accuracy, a set of calipers is recommended. A ruler can be used, but is less accurate. If calipers are used, obtain a bean and measure its length with a standard ruler to obtain a prediction of what the measurement should be once you use the calipers. This will ensure you are using the calipers correctly.

3 Use the calipers to measure the length of 100 beans.

4 To enter the data into your Ti-83:
   (a) push STAT
   (b) push 4 to Clear what is in list 1
   (c) push 2nd 1 (L_1) to specify which list you want cleared
   (d) push enter
   (e) push STAT
   (f) push enter
   (g) enter all data into L_1.

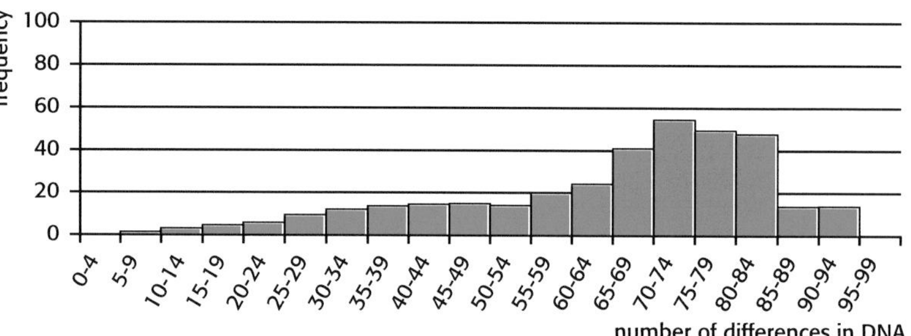

Figure 9 An example of a histogram

5 To sort the data using your calculator:
   (a) push STAT when finished
   (b) push Sort A
   (c) push 2nd 1 (L_1)
   (d) Push STAT followed by Edit to view the sorted data.

6 To determine the mean:
   (a) push STAT
   (b) move the cursor over to CALC
   (c) push enter for single variable statistics
   (d) push 2nd 1 (L_1).

7 Record:
   (a) the mean
   (b) the minimum value and the maximum value
   (c) the value of "s" (a quantity referred to as standard deviation).

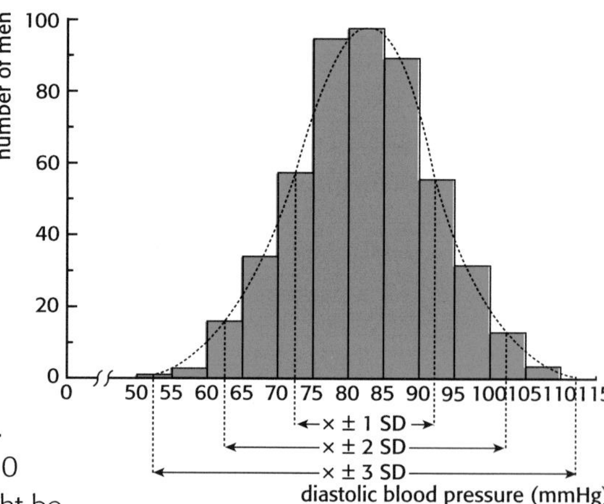

Figure 10 Normal curve calculated from blood pressures of 500 men, with a mean 82 mmHg and a standard deviation of 10 mmHg. 68% of the data falls within 72 mmHg and 92 mmHg.

8 Identify the frequency classes that you will use (between 7 and 10 classes are recommended). An example of a frequency class might be between 1.9 and 2.1 cm.

9 Create a tally chart and enter the data into their frequency classes.

10 Construct a histogram. This is a graphical representation of a frequency distribution. The frequency of occurrence is represented by a rectangular column. The height of the column is proportional to the frequency of observations within that frequency class. Figure 9 shows an example of a histogram.

11 Mark the mean on the graph.

12 What is a normal distribution? Is your data normally distributed?

13 Mark the point that is one standard deviation higher than the mean. Mark the point that is one standard deviation below the mean.

14 Colour in the area under the curve that is bounded by +/– one standard deviation.

15 Make an estimate of the area under the curve that is shaded in.

16 If your data is normally distributed, you should find that approximately 68% of your data is coloured.

Chapter 11 questions

1 For each set of four words, identify the term that does not belong and explain why:
- non-disjunction, crossing over, synapsis, tetrad
- independent assortment, metaphase I, random orientation, segregation
- length, banding pattern, centromere position, chemical composition
- sickle-cell anemia, Turner's syndrome, trisomy 21, Klinefelter's syndrome.

2 Figure 11 shows the uterus of a pregnant woman.

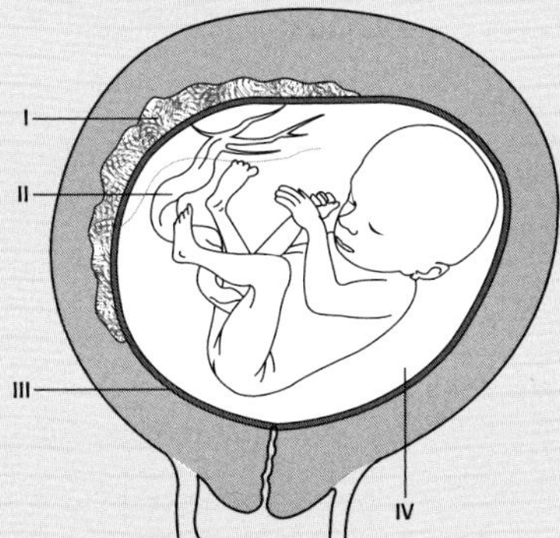

Figure 11 Fetus in uterus

Where are samples drawn from for:
(a) amniocentesis [1]
(b) chorionic villus sampling? [1]

3 Human somatic cells have 46 chromosomes, while our closest primate relatives, the chimpanzee , the gorilla and the orangutan all have 48 chromosomes. One hypothesis is that the human chromosome number 2 was formed from the fusion of two chromosomes in a primate ancestor. The image below shows human chromosome 2 compared to chromosome 12 and 13 from the chimpanzee.
(a) Compare the human chromosome 2 with the two chimpanzee chromosomes (Figure 12). [3]
(b) The ends of chromosomes, called telomeres, have many repeats of the same short DNA sequence. If the fusion hypothesis were true, predict what would be found in the region of the chromosome where the fusion is hypothesized to have occurred. [2]

C
H

Figure 12

4 Construct a Venn diagram which compares mitosis and meiosis.

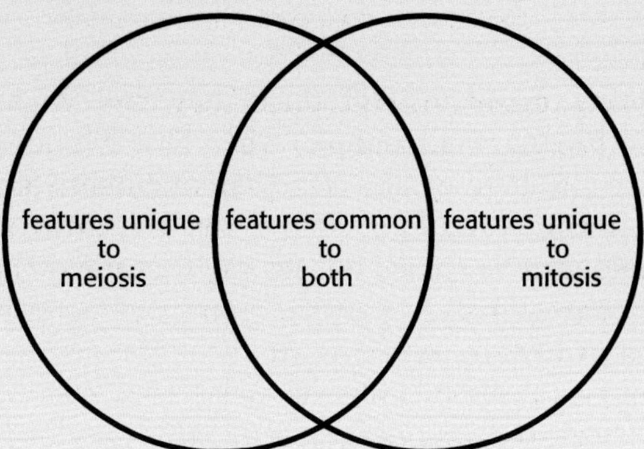

5 Which response describes the behaviour of chromosomes in metaphase I and anaphase II of meiosis?

	Metaphase I	Anaphase II
A.	Chromosomes line up at the equator	Separation of homologous chromosomes
B.	Tetrads (bivalents) line up at the equator	Separation of homologous chromosomes
C.	Chromosomes line up at the equator	Separation of sister chromatids
D.	Tetrads (bivalents) line up at the equator	Separation of sister chromatids

6 A cell with a diploid number of 12 chromosomes undergoes meiosis. What will be the product at the end of meiosis?
A 2 cells each with 12 chromosomes.
B 4 cells each with 6 chromosomes.
C 2 cells each with 6 chromosomes.
D 4 cells each with 12 chromosomes.

12 Genetics – the study of inheritance

When living organisms reproduce, they pass on characteristics to their offspring. For example, when blue whales reproduce, the young are also blue whales – they are members of the same species. More than this, variations, such as the markings on the skin of a blue whale, can be passed on. We say that the offspring inherit the parents' characteristics. However, some characteristics cannot be inherited. An example of this are the scars seen on the tails of some blue whales, caused by killer whale attacks. According to current theories, acquired characteristics like this cannot be inherited.

Inheritance has been discussed since the time of Hippocrates and earlier. For example, Aristotle observed that children sometimes resemble their grandparents more that their parents. Many of the early theories involved blending inheritance, in which offspring inherit characters from both parents and so have characters intermediate between those of their parents. Some of the observations that biologists made in the first half of the 19th century could not be explained by blending inheritance, but it was not until Mendel published his paper "Experiments in Plant Hybridization" that an alternative theory was available.

Mendel's experiments were done using varieties of pea plant, each of which reliably had the same characters when grown on its own. Mendel carefully crossed varieties of pea together by transferring the male pollen from one variety to the female parts in flowers of another variety. He collected the pea seeds that were formed as a result and grew them to find out what their characters were. Mendel did this experiment with seven different pairs of characters.

(a) Prediction based on blending inheritance

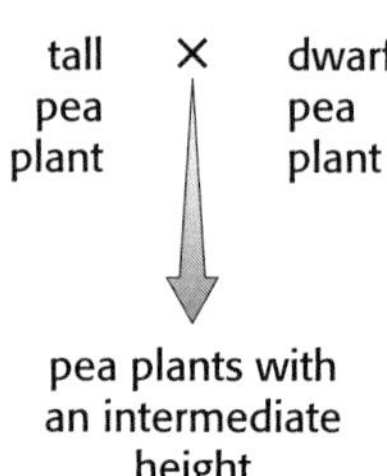

(b) Mendel's actual results

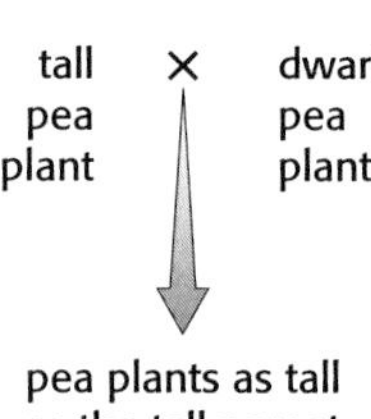

Figure 1 Example of one of Mendel's monohybrid cross experiments. All the hybrid plants produced by crossing two varieties together had the same character as one of the parents and the character of the other parent was not seen. This is a clear falsification of the theory of blending inheritance.

Thinking about science: the case of the midwife toad

Dr Paul Kammerer (1880–1926) was a Viennese biologist who became expert at breeding amphibians and reptiles in captivity.

He reported that spotted salamanders stopped producing tadpoles when kept in dry conditions with no water and started giving birth to live young.

The colour of the salamanders appeared to change depending on whether they were reared on black soil (left) or yellow soil (right).

Most toads breed in water. Males develop spiny black regions called nuptual pads on the palms of their front feet to help grip onto females during mating.

Kammerer kept the land-breeding mid-wife toad in conditions that encouraged it to breed in water. He reported that after three generations males started to develop nuptual pads, which they do not normally have.

The nuptual pads were apparently an example of inheritance of an acquired characteristic. In 1926 it was found that Indian ink had been injected into the palm of the only preserved male to make it look darker. Kammerer's experiments were all discredited.

Figure 2 Does the Indian ink prove that acquired characteristics cannot be inherited?

Mendel and particulate inheritance

Mendel continued his experiments with peas, described on the previous page, by allowing the hybrid plants to self-pollinate (see Figure 3). They therefore acted as both the male and the female parent. According to blending inheritance, the offspring would have the same phenotype as this male and female parent. Again Mendel's results were surprising: the character from the original parent that had apparently disappeared reappeared, so for example dwarf plants were seen as well as tall plants. Again, no intermediates were seen, confirming that inheritance is particulate, not blending. Mendel called the particles of inheritance factors, but we now call them **genes**.

Mendel was unusual for his time in that he quantified his results. He observed that the ratio of pea plants was always close to 3:1, with three having the character that dominated after the initial cross and one having the character that reappeared after being absent for a generation.

gene *a heritable factor that controls a specific characteristic.*

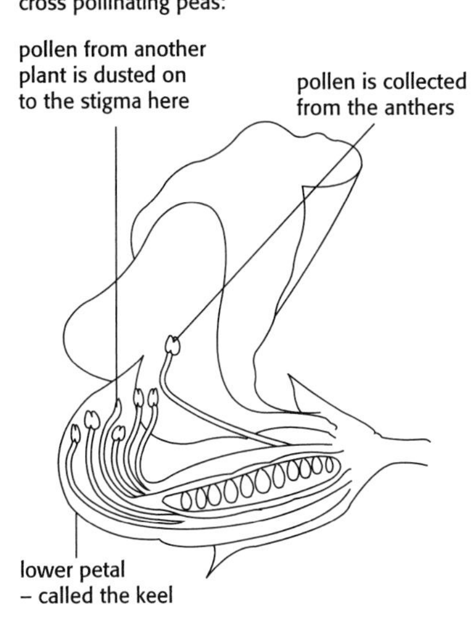

Figure 3 Cross and self pollination

Critical considerations: were Mendel's results too perfect?

Mendel's results for the self-pollinating hybrids are summarized in Table 1.

Table 1

Parental plants	Hybrid plants	Offspring from self-pollinating the hybrids	Ratio
Tall stem × dwarf stem	All tall	787 tall : 277 dwarf	2.84 : 1
Round seed × wrinkled seed	All round	5474 round : 1850 wrinkled	2.96 : 1
Yellow cotyledons × green cotyledons	All yellow	6022 yellow : 2001 green	3.01 : 1
Purple flowers × white flowers	All purple	705 purple : 224 white	3.15 : 1
Full pods × constricted pods	All full	882 full : 299 constricted	2.95 : 1
Green unripe pods × yellow unripe pods	All green	428 green : 152 yellow	2.82 : 1
Flowers along stem × flowers at stem tip	All along stem	651 along stem : 207 at tip	3.14 : 1

In 1936 the English statistician R.A. Fisher published an analysis of Mendel's data. His conclusion was that "the data of most, if not all, of the experiments have been falsified so as to agree closely with Mendel's expectations". Doubts still persist about Mendel's data – a recent estimate put the chance of getting seven ratios as close to 3:1 as Mendel's at 1 in 33,000.

1 To get ratios as close to 3:1 as Mendel's would have required "a miracle of chance". What are the possible explanations, apart from a miracle of chance?
2 If a scientist is untruthful about experimental results, what are the implications:
   (a) for this scientist?
   (b) for the advancement of knowledge?
   (c) for non-scientists?
3 Many distinguished scientists, including Louis Pasteur, are known to have discarded results when they did not fit a theory. Is it acceptable to do this? How can we distinguish between results that are due to an error and results that falsify a theory? What standard do you use as a student in rejecting anomalous data?
4 Is it ethical for a scientist to alter data slightly, so that it is convincing enough to be accepted for publication?

Thinking about science: laboratory notebooks

Almost all research scientists use a laboratory notebook to record the methods they have used in experiments and the raw data obtained, together with any other thoughts and observations that they wish to remember. Historians of science sometimes compare the data in scientists' notebooks with the data in papers published in scientific journals. The extracts below come from *The Private Science of Louis Pasteur,* by Gerald Geison. This book focuses on three episodes in Pasteur's research career: the discovery of optical isomers in tartaric acid, an experiment with an anthrax vaccine and a trial of a vaccine against rabies. In each case, Geison found discrepancies between the account in Pasteur's private notebooks and the public account that he gave of his work.

Most of the discrepancies between Pasteur's public and private science are of a sort that will come as no great surprise to working scientists. It will be obvious that such discrepancies are part and parcel of the process by which "raw data" are transformed into published "results". In the interests of brevity, clarity, logical coherence, and rhetorical power, the published record always projects a more or less distorted image of what the scientist "really" did.

For some reason, laboratory notebooks were long overlooked by historians of science, but their virtues as a strategic site of inquiry have become evident in recent years. Much remains to be done in this line of research. But in every case thus far in which records of "private science" have been closely investigated, one can detect discrepancies of one sort or another between these records and published accounts. Even the best scientists routinely dismiss uncongenial data as aberrations, arising from "bad runs," and therefore omit or "suppress" them from the published record. Equivocal experiments are sometimes transformed into decisive results. The order in which experiments were performed is sometimes reversed. And the actual nature or direction of research is otherwise simplified, telescoped, and generally "tidied up." There is rarely anything sinister about such practices, rarely any intention to deceive, and their existence has long been recognized. As long ago as the seventeenth century, Francis Bacon noted that "never any knowledge was delivered in the same order it was invented."

There are, of course, those who insist that "genuine" scientific knowledge is independent of the process by which any particular scientist arrives at his or her conclusions. In very different ways, philosophers and sociologists of science tend to be suspicious of historical studies of individual "scientific creativity." For philosophers in the tradition of Karl Popper, such studies seem to be pursuing a will-o'-the-wisp, an elusive "psychology of discovery," at the expense of a clear-cut "logic of justification." For them, the object of study is the published text, and the "scientificity" of a given text is to be assessed in terms of logical and methodological criteria that transcend particular individuals, particular social groups, or any contingent historical circumstances.

For sociologists of knowledge, by contrast, studies of individual scientific activity run the risk of ignoring the extent to which scientific knowledge is a community affair — the outcome of a complex process of social negotiations. On this view, scientific knowledge is constructed within a culturally limited space. For some, the boundaries of that space are set by the broadly cultural "interests" of participants.

Often lost from sight in such theoretical discussions is the real individual scientist who tries to navigate a safe passage between the constraints of empirical evidence on the one hand and personal or social interests on the other.

1 Why do most research scientists use laboratory notebooks, often with a hard cover, for recording their work?

2 In 1878, when Louis Pasteur was 55 years old, he told his family never to show anyone his private laboratory notebooks. What reasons might he have had for this?

3 Do you agree that papers in scientific journals should be assessed solely "in terms of logical or methodological criteria", or do we need to take into account cultural factors and the personal and social interests of the scientist?

4 If you do some research in biology, for example for an extended essay, will you publish all your results, some of your results, or your best results only?

Explaining Mendel's 3:1 ratios

Mendel explained the results of his experiments using the terms *dominant character* and *recessive character*. A hybrid plant, made by crossing one parent having the dominant character with another having the recessive character, always showed the dominant character. He realized that pea plants pass on factors in their male and female gametes and he used letters to designate these factors – an upper case for the dominant character and lower case for the recessive character. The hybrid plants inherit one factor from each parent and so have two different factors. When they form male and female gametes they pass on either the dominant or the recessive factor, not both. The factors therefore separate when male and female gametes are made. This is called segregation and it is the key to understanding why inheritance is not blending. Random fertilization of eggs and pollen with dominant or recessive factors allows different combinations of factors to come together in the next generation. The chance of a plant inheriting only recessive factors is ½ × ½ = ¼. The chance of inheriting two dominant factors is also ¼ and the chance of inheriting one dominant and one recessive factor is ½. As the plants with one or two dominant factors show the dominant character, the overall chance of this is ¾, hence the 3:1 ratio. This is shown clearly by using a 2 by 2 table, called a **Punnett grid** after the geneticist who first used this type of table. Figure 4 involves a Punnett grid in an explanation of a cross between tall and dwarf pea plants.

Mendel's *factors* are now called **alleles** and the character of an organism is carefully distinguished from its alleles using the terms **phenotype** and **genotype**.

allele *one of a number of different forms of gene.*
genotype *the alleles of an organism.*
phenotype *the characteristics of an organism.*

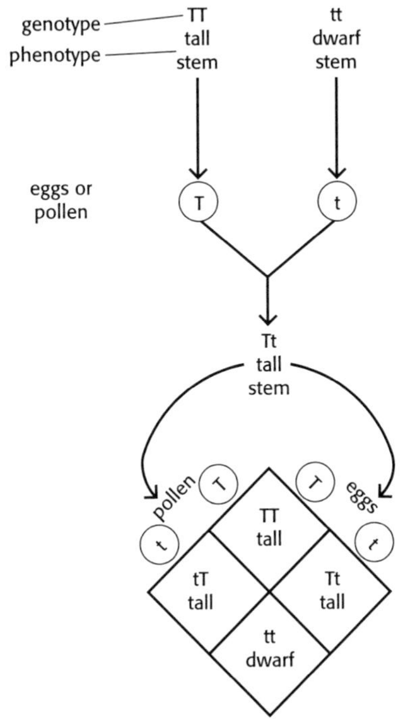

Figure 4 Explanation of Mendel's 3:1 ratio

Data-based question: coat colour in the house mouse

In the early years of the 20th century, many crossing experiments were done in a similar way to those of Mendel. The French geneticist Lucien Cuénot used the house mouse, *Mus musculus,* to see whether the principles that Mendel had discovered also operated in animals. He crossed normal grey-coloured mice with albino mice. The hybrid mice that were produced were all grey. These grey hybrids were crossed together and produced 198 grey and 72 albino offspring.

?

1 Calculate the ratio between grey and albino offspring, showing your working. [2]
2 Deduce the colour of coat that is due to a recessive allele, with two reasons for your answer. [3]
3 Choose suitable symbols for the alleles for grey and albino coat and list the possible genotypes of mice using your symbols, together with the phenotype for each genotype. [3]
4 Using the headings shown right, explain how the observed ratio of grey and albino mice was produced. [5]
5 Suggest how one gene can determine whether the mice had grey fur and black eyes or white fur and red eyes. [2]

Key to alleles

Parental phenotypes:
Parental genotypes:

Alleles in gametes:

Hybrid phenotype:
Hybrid genotype:

Alleles in gametes:

Genotypes and phenotypes of offspring of hybrid mice, shown using a Punnett grid:

Test crosses

If there are two alleles of a gene, there are three possible genotypes. For example, the gene that influences the stem height in pea plants has two possible alleles, T and t. The three possible genotypes are TT, tt and Tt. If an organism has two of the same allele, it can only produce gametes containing that allele, so it is **homozygous**. If it has two different alleles, then half the gametes will contain one of the alleles and half will contain the other. The organism is **heterozygous**.

If an organism is heterozygous, and one of its alleles is dominant and the other is recessive, the phenotype is determined by the dominant allele.

If an organism has a character caused by a dominant allele, it is not possible to know if it is homozygous simply by looking at it. One way to find out is to do a **test cross**. The individual with the unknown genotype is crossed with one or more individuals that are homozygous recessive. Figure 5 shows the two possible outcomes of a test cross in sheep where the allele for black wool is recessive to the allele for white wool.

If a small number of offspring are obtained from a test cross, all of which show the phenotype caused by the dominant allele, the cross must be repeated until there is a large enough number of offspring to be confident that no recessive allele is present.

dominant allele *an allele that has the same effect on the phenotype whether it is present in the homozygous or the heterozygous state.*

heterozygous *having two different alleles of a gene.*

homozygous *having two identical alleles of a gene.*

recessive allele *an allele that only has an effect on the phenotype when it is present in the homozygous state.*

test cross *testing a suspected heterozygote by crossing it with a known homozygous recessive.*

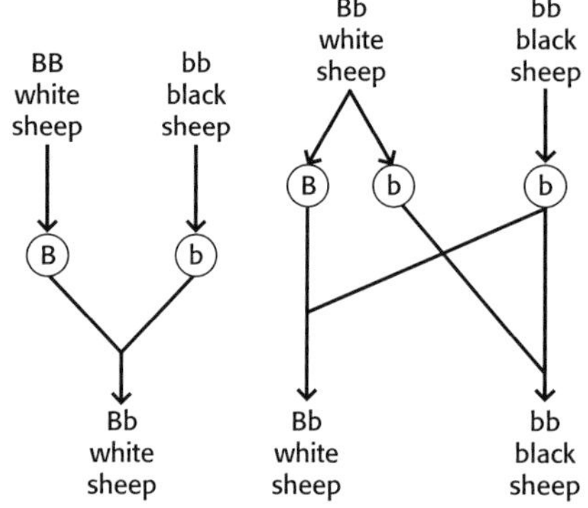

Figure 5 Possible outcomes of a test cross

Data-based question: deducing genotypes from pedigree charts

It isn't possible to investigate human genotypes with test crosses! Sometimes a person's genotype can be deduced from a pedigree chart. Figure 6 is a pedigree chart for a rare condition in humans. Individuals with the condition are shaded. Roman numerals indicate the generations and Arabic numbers are used for individuals in each generation.

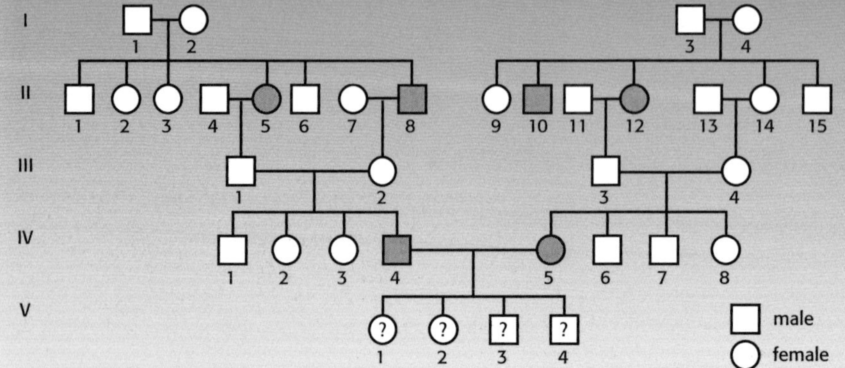

Figure 6 Example of a pedigree chart

1 Explain, using evidence from the pedigree, whether the condition is due to a recessive or a dominant allele. [3]
2 Explain what the probability is of the individuals in generation V being:
   (a) homozygous recessive
   (b) heterozygous
   (c) homozygous dominant. [3]
3 Deduce, with reasons, the possible genotypes of:
   (a) 1 in generation III
   (b) 13 in generation II. [2]
4 Explain which sets of parents in the pedigree are genetically similar to a test cross. [2]

Co-dominance

In the example of the black and white sheep on the previous page, heterozygous lambs are as black as their black parent when they are born, but their wool gradually becomes greyer as they grow older. The allele for black wool therefore does not quite fit the definition of a recessive allele and the white allele is not truly dominant, as it affects the phenotype of a heterozygous sheep. As early as 1905 William Bateson, Edith Saunders and Reginald Punnett published details of a cross between a black and a white variety of chicken, *Gallus gallus,* which produced the so-called blue Andalusian hen, which was actually a grey colour, intermediate between the parental phenotypes. There are many other cases of two alleles both affecting the phenotype in a heterozygote. This pattern of inheritance is now called co-dominance.

co-dominant alleles *pairs of alleles that both affect the phenotype when in a heterozygote.*

A well-known example is the flower colour of *Mirabilis jalapa*. If a red-flowered plant is crossed with a white-flowered plant, the offspring have pink flowers. It is important to note that this does not show that co-dominance is due to blending inheritance. When pink-flowered plants are crossed together there are red-, pink- and white-flowered offspring (see Figure 7). This is the result of particulate inheritance and segregation:

- there is an allele for red flowers, C^R
- there is an allele for white flowers, C^W
- these alleles are co-dominant so C^RC^W gives pink flowers.

When pink-flowered plants produce gametes, the alleles C^R and C^W segregate, and mating results in red, pink and white offspring in a 1:2:1 ratio.

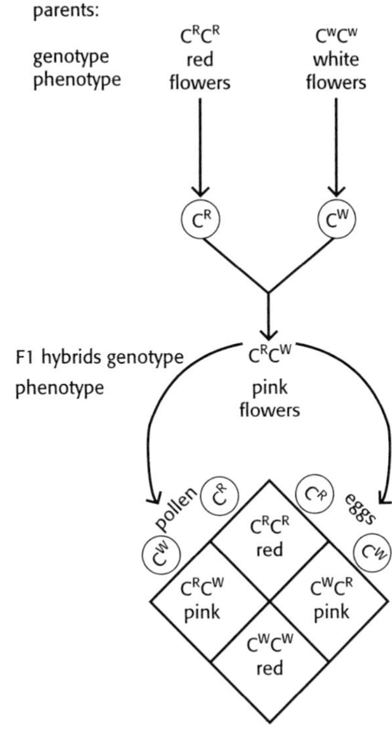

Figure 7 A cross involving co-dominance

Critical considerations: F1 hybrid seeds

Seed catalogues contain pure-breeding varieties of crop plant and also F1 hybrid varieties. F1 hybrids are produced by crossing two pure-breeding varieties together. Pure-breeding varieties are homozygous for all their genes and F1 hybrids are heterozygous for many genes.

F1 hybrids are often more vigorous than either of their parents. Part of the explanation of this involves co-dominance – the phenotype of heterozygotes is different from either homozygote and for some genes there is an advantage in having two different alleles, rather than two of the same allele. Heterozygotes therefore grow more vigorously.

Sometimes seed is collected from a crop of F1 hybrid plants. This is F2 hybrid seed. When it is sown it almost always produces a less vigorous and more variable crop. Farmers and gardeners are therefore advised to buy new F1 hybrid seed every year and not sow seed that they collected from the previous crop.

1 Why is F1 hybrid seed more expensive than seed of pure-breeding varieties?
2 Why is the vigour of F2 plants lower than their F1 parents?
3 What reasons might seed companies have for encouraging clients to buy F1 hybrid seed rather than pure-breeding varieties?

Figure 8 An example of F1 hybrid seed

Inheritance of ABO blood groups

The ABO blood group system in humans is an example of co-dominance. It is of great medical importance: before blood is transfused, it is vital to find out the blood group of a patient and ensure that it is matched. Unless this is done, there may be complications due to coagulation of red blood cells.

One gene determines the ABO blood group of a person. The genotype I^AI^A gives blood group A and the genotype I^BI^B gives group B. Neither I^A nor I^B is dominant over the other allele and a person with the genotype I^AI^B has a different blood group, called AB. There is a third allele of the ABO blood group gene, usually called *i*. A person with the genotype ii is in blood group O. The genotypes I^Ai and I^Bi give blood groups A and B respectively, showing that *i* is recessive to both I^A and I^B. The reasons for two alleles being co-dominant and the other allele being recessive are as follows:

- I^A causes the production of a glycoprotein in the membrane of red blood cells. If this glycoprotein is not present then antibodies are produced against it, so people who do not have the allele I^A have anti-A antibodies.
- I^B causes the production of a different glycoprotein in the red blood cell membrane. In a similar way, people who do not have the allele I^B produce anti-B antibodies.
- The genotype I^AI^B therefore causes the production of both glycoproteins, so neither anti-A nor anti-B antibodies are produced This genotype therefore gives a different phenotype to I^AI^A and I^BI^B so the alleles I^A and I^B are co-dominant.
- The allele *i* is recessive because it does not cause the production of a glycoprotein. I^AI^A and I^Ai therefore give the same phenotype and so do I^BI^B and I^Bi.

The ABO gene is not unusual in having three alleles and other examples are described on the next page.

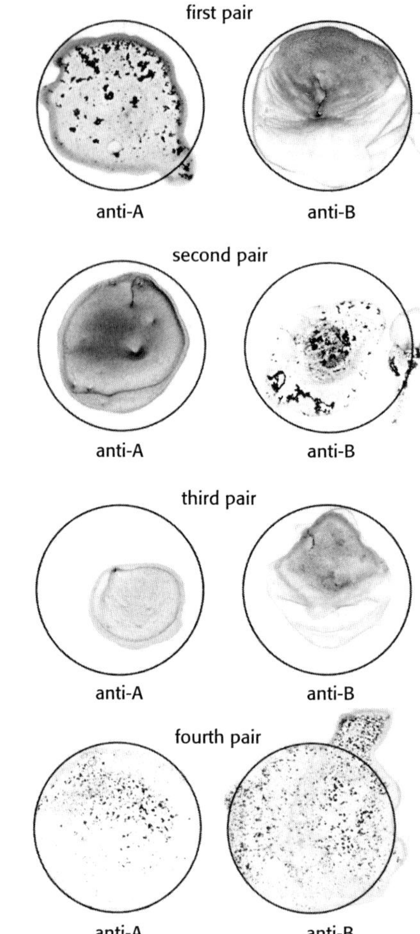

Figure 9 Blood group can easily be determined using test cards. This drawing shows typical results.

Data-based question: distribution of ABO blood groups

The map in Figure 10 shows the frequency of the I^B allele in indigenous populations around the world.

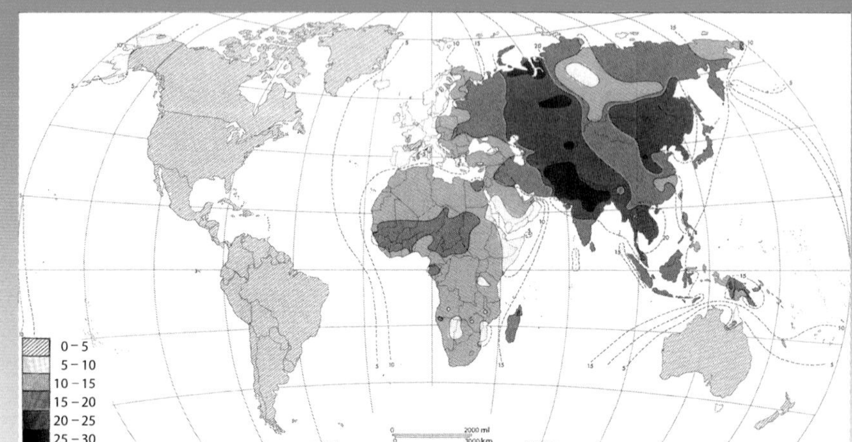

Figure 10 Map of I^B allele frequencies

1. State the frequency of I^B in:
   (a) northern India
   (b) Japan
   (c) North America. [3]
2. The frequency of the allele I^A in South America is less than 5%. Calculate the frequency of the allele *i* in South America. [2]
3. Suggest an explanation for gypsies in Europe having a much higher frequency of I^B than other indigenous groups in Europe. [2]
4. Use the data in the map to evaluate Thor Heyerdahl's theory that Pacific islands were colonized by groups from South America sailing west. The frequency of I^B on these islands is between 10% and 25%.

Multiple alleles and mutation

There are many examples of inheritance like the one described on the previous page, in which a gene has three alleles. Where a gene has more than two alleles, the pattern of inheritance is called *multiple alleles.*

Genes with different alleles are an important source of variation in every species. Without variation, natural selection could not occur and a species could not evolve in response to environmental change and might become extinct. Ironically, although natural selection depends on a species having variation in its alleles, one of the effects of natural selection is to eliminate some alleles from the species. New alleles must therefore be formed, if variation is to be maintained and the species is to retain the capacity to evolve.

A gene consists of a length of DNA, with a base sequence that can be hundreds or thousands of bases long. The different alleles of a gene have slight variations in the base sequence. Usually only one or a very small number of bases are different.

New alleles are formed from other alleles by **gene mutation**.

Mutations are random changes – there is no mechanism for a particular mutation being carried out. A random change to an allele that has developed by evolution over perhaps millions of years is unlikely to be beneficial. Almost all mutations are therefore either neutral or harmful. Some mutations are lethal – they cause the death of the cell in which the mutation occurs. Mutations in body cells are eliminated when the individual dies, but mutations in cells that develop into gametes can be passed on to offspring and cause genetic disease.

gene mutation *a change to the base sequence of a gene.*

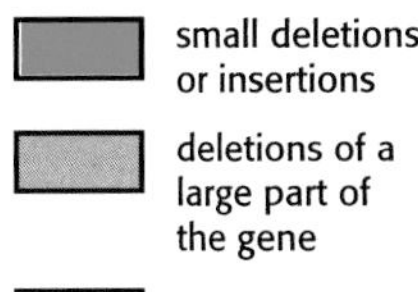

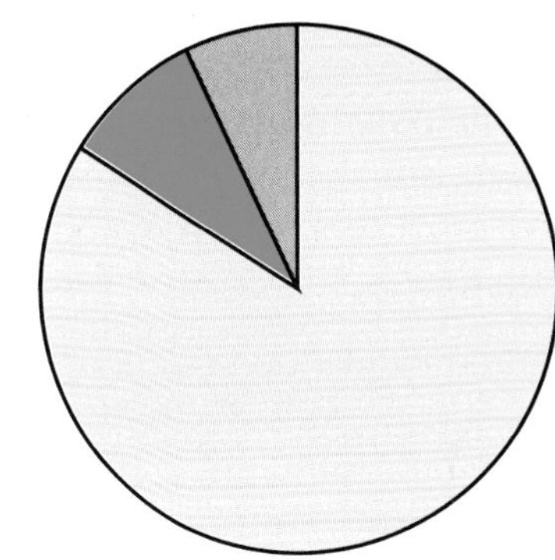

Figure 11 The types of mutation that have been found in one gene in humans. The gene codes for the enzyme ornithine transcarbamylase, which is needed for urea production. Over 140 different mutations of the gene have been found, hence there are more than 140 alleles of this gene.

Critical considerations: the aftermath of Chernobyl

Mutations can cause a cell to become a tumour cell. The release of 6.7 tonnes of radioactive material from the nuclear power station at Chernobyl in 1986 was therefore the cause of large numbers of deaths due to cancer. The UN Chernobyl Forum stated that "up to 4000 people" may ultimately die as a result of the disaster, but Green Party members of the European Parliament commissioned a report from a radiation scientist, which gave an estimate of 30,000 to 60,000 extra deaths. One way of obtaining an estimate is to use data from previous radiation exposures, such as the detonation of nuclear warheads at Hiroshima and Nagasaki in 1945. The data below is an analysis of deaths due to leukemia and cancer between 1950 and 1990 among those exposed to radiation from these warheads. It was published by the Radiation Effects Research Foundation.

Radiation dose range (Sv)	Number of deaths in people exposed to radiation	Estimate of excess deaths over control groups	Percentage of deaths attributable to radiation exposure
Leukemia			
0.005–0.2	70	10	14
0.2–0.5	27	13	48
0.5–1	23	17	74
>1	56	47	84
Cancer			
0.005–0.2	3391	63	2
0.2–0.5	646	76	12
0.5–1	342	79	23
>1	308	121	39

1 What reasons do governments, doctors and the nuclear industry have for trying to estimate the effects of release of radioactive material?
2 Why do the estimates published by different groups differ so widely?
3 Why might the effects of mutations only show up in the children of those exposed to radiation, or even in later generations?
4 What criteria can be used to decide whether a risk, such as generation of electricity by nuclear fission, is acceptable?

Genetic diseases and carriers

A genetic disease is an illness that is caused by a gene. There are over 4000 genetic diseases in humans. Most of these are caused by a recessive allele of a gene. The disease therefore only develops in individuals that are homozygous for the recessive allele. If a person is heterozygous, with one allele for the genetic disease and one dominant allele, they will not show symptoms of the disease, but they can pass on the recessive allele to their offspring. These individuals are called **carriers**.

carrier *an individual that has one copy of a recessive allele that causes a genetic disease in individuals that are homozygous for the allele.*

Genetic diseases caused by a recessive allele usually appear unexpectedly. Both parents of a child with the disease must be carriers, but as they do not show symptoms of the disease, they are unaware of this. The probability of these parents having a child with the disease is 25 per cent (see Figure 12a).

A small proportion of genetic diseases are caused by a dominant allele. It is not possible to be a carrier of these diseases – if a person has one dominant allele then they themselves will develop the disease. If one parent has the allele for the disease, the chance of a child inheriting it is 50 per cent (see Figure 12b).

Most genetic diseases reduce the chances of survival and reproduction, so the alleles causing them are not usually passed on to offspring and remain very rare. There are a small number of genetic diseases where the frequency of the allele that causes them is much higher. In these cases the allele must confer an advantage, causing its frequency to increase by natural selection. Sickle-cell anemia is an example of this and is described on the next page.

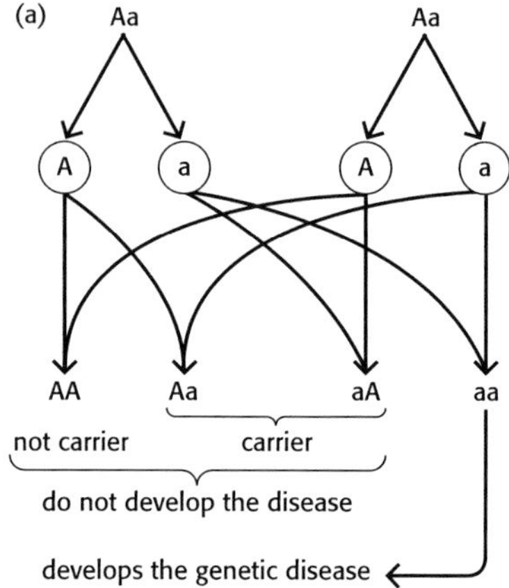

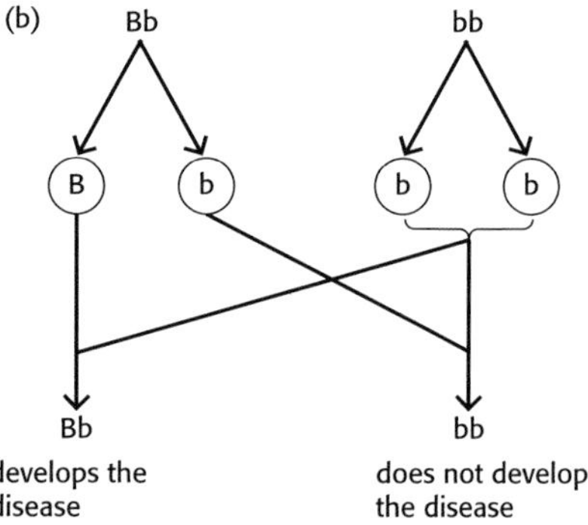

Figure 12 (a) Genetic diseases caused by a recessive allele. (b) Genetic diseases caused by a dominant allele.

Thinking about science: correlation and cause

There is a correlation between high frequencies of the sickle-cell allele in human populations and high rates of infection with Falciparum malaria. Where a correlation exists, it may or may not be due to a causal link. Consider the information below and decide whether there is a causal link between malaria and sickle-cell anemia.

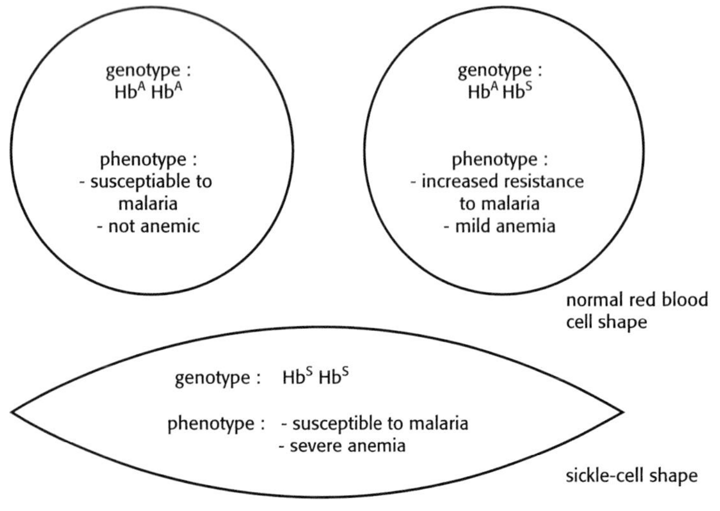

Figure 13 Effects of Hb^A and Hb^S alleles

Sickle-cell anemia

Sickle-cell anemia is the commonest genetic disease in the world. It is due to a mutation of the gene that codes for the β-globin polypeptide in hemoglobin. The symbol for this gene is *Hb*. Most humans have the allele Hb^A. At some time in the past, a base substitution mutation converted the sixth codon of the gene from GAG to GTG. This happened in an ovary or testis cell and may have occurred one or in more than one person. The allele with this mutation is called Hb^S. When the Hb^S allele is transcribed, the mRNA produced has GUG as its sixth codon instead of GAG, and when this mRNA is transcribed, the sixth amino acid in the polypeptide is valine instead of glutamic acid. This change causes hemoglobin molecules to stick together, in tissues with low oxygen concentrations. The chains of hemoglobin join together into bundles that are rigid enough to distort the red blood cells into a sickle shape. These sickle cells cause damage to tissues by becoming trapped in blood capillaries, blocking them and reducing blood flow.

When sickle cells return to high oxygen conditions in the lung, the hemoglobin chains break up and the cells return to their normal shape. These changes occur time after time, as the red blood cells circulate. Both the hemoglobin and the plasma membrane are damaged and the life of a red blood cell can be shortened to as little as 4 days. The body cannot replace red blood cells at a rapid enough rate and anemia therefore develops.

So, a small change to a gene can have very harmful consequences for individuals that inherit the gene. Despite this, there are parts of the world where up to 5 to 6 per cent of newborn babies are homozygous for the sickle-cell allele.

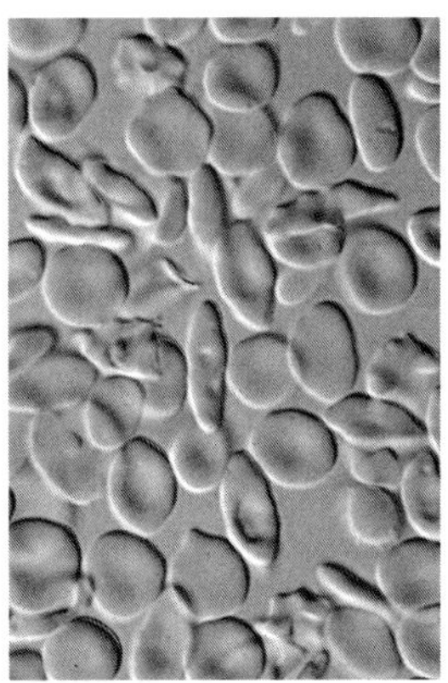

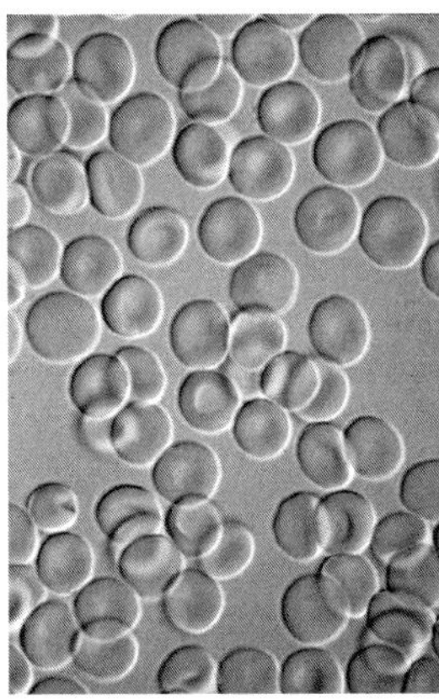

Figure 14 Micrographs of sickle cells and normal red blood cells

Critical considerations: consequences of genetic disease

There are many issues for families in which there is genetic disease. Consider the scenarios below and discuss what advice should be given.

1 A man and woman are planning to get married. Both have had genetic screening to find whether they have the allele for sickle-cell anemia. Both are carriers of the allele.

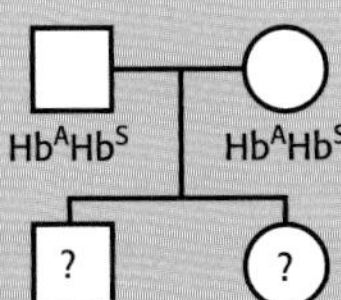

2 A gene called BRCA2 is linked to a high risk of breast cancer. A screening program is being planned, to find out which women have the gene. An ethics committee has to decide who should be able to find out the results of the screening:

- the women who have been tested
- the women's doctors
- medical researchers investigating breast cancer
- life insurance companies
- companies who are hiring workers.

3 A teenage boy's mother has just died as a result of Huntington's disease. This disease is due to a dominant allele. The onset of the disease is not usually until the age of 35. The boy isn't sure whether to agree to have genetic screening to find out whether he has the allele for Huntington's disease.

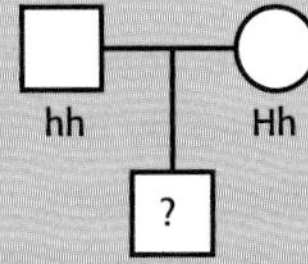

4 A 25-year-old woman would like to have a baby. Her husband, who is 30 years old, has had genetic screening and has found out that he has the dominant allele for Huntington's disease.

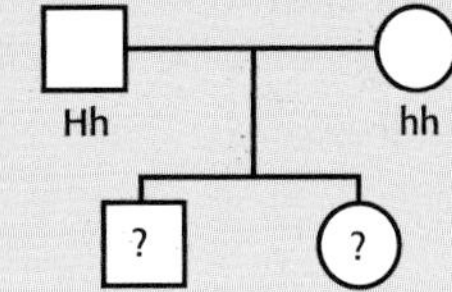

Sex linkage

Thomas Morgan was one of a group of geneticists who realized the importance of Mendel's experiments and investigated the pattern of inheritance in other species to find out if it is the same as in peas. Morgan chose to study the fruit fly, *Drosophila*, which is about 4 mm long and completes its life cycle in 2 weeks, allowing crossing experiments to be done quickly with large numbers of flies. Many of Morgan's experiments showed the same pattern of inheritance as pea plants:

- two alleles of each gene are present
- these two alleles can be the same (homozygous) or different (heterozygous)
- just one allele is passed on to offspring in gametes
- one allele is usually dominant over another allele.

Unlike pea plants, *Drosophila* has separate males and females and this allowed Morgan to perform reciprocal crosses. Examples of reciprocal crosses are:

normal-winged males × small-winged females
small-winged males × normal-winged females.

In most cases, the ratios were the same whichever way the cross was done, but in some cases there were differences. The first example of this in *Drosophila* involved eye colour and is shown in Figure 15. This pattern of inheritance, where there are differences in genotypes and phenotypic ratios between males and females, is called **sex linkage**. Geneticists had observed that the inheritance of genes and of chromosomes showed clear parallels and so genes were likely to be located on chromosomes. It was also known that female *Drosophila* have two copies of a chromosome called X and males only have one copy. Morgan deduced that sex linkage of eye colour could therefore be due to the eye colour gene being located on the X chromosome. Male *Drosophila* also have a Y chromosome, but this does not carry the eye-colour gene.

Similar examples of sex linkage due to a gene on the X chromosome have since been discovered in humans. Hemophilia is a famous example. Red–green colour blindness is a very common sex-linked condition.

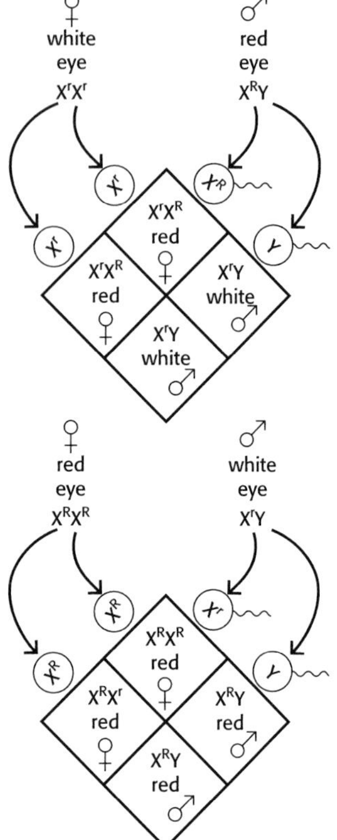

Figure 15 Reciprocal sex-linkage crosses

Thinking about science: Mendel and paradigm shifts

Mendel published his "Experiments in Plant Hybridization" in 1865, but it was almost completely ignored by the scientific community for 34 years. Mendel's work was "rediscovered" independently by Hugo De Vries, Carl Correns and Erich von Tschermak, who realized its significance. The reasons for this long gap between publication and recognition are intriguing. Consider and evaluate these arguments:

- Mendel published his paper in the *Proceedings of the Brno Society for the Study of Natural Science*, an obscure journal that few scientists read.
- Mendel's paper is long and very complicated. It includes quantitative results and algebra to explain them, which was very unusual in 1865.
- In the 19th century the universality of genes and inheritance was not understood, so Mendel's results with peas would not have been expected to apply to other organisms, including humans.
- Other distinguished scientists had proposed different theories about the nature of inheritance and as Mendel was largely unknown, his views were dismissed without proper consideration.
- Mendel's work represented a fundamental change of ideas about inheritance, and so required a paradigm shift. As so often with revolutionary ideas, Mendel's contemporaries viewed inheritance through a different framework. They failed to understand his paper and so ignored it.

Determination of gender in humans

The gender of a human is determined at the moment of fertilization, by one chromosome carried in the sperm. This can either be an X or a Y chromosome. Because X and Y determine gender, they are called the sex chromosomes.

The X chromosome is large, with many genes on it that are essential in both male and female development. The Y chromosome is much smaller, with far fewer genes. Part of the Y chromosome has the same sequence of genes as a small part of the X chromosome, but the genes on the remainder of the Y chromosome are not found on the X chromosome and are not needed for female development. One gene in particular, called TDF, is only found on the Y chromosome. It initiates the development of male features, including testes and testosterone production. Males therefore have one Y chromosome to cause male development and one X chromosome with its vital genes. Females have two X chromosomes. Because they do not have the *TDF* gene on a Y chromosome, ovaries develop instead of testes and female sex hormones are produced, not testosterone. Females pass on one of their two X chromosomes in each egg cell, so all offspring inherit an X chromosome from their mother. When sperm are formed, half contain the X chromosome and half the Y chromosome. Daughters inherit their father's X chromosome and sons inherit his Y chromosome.

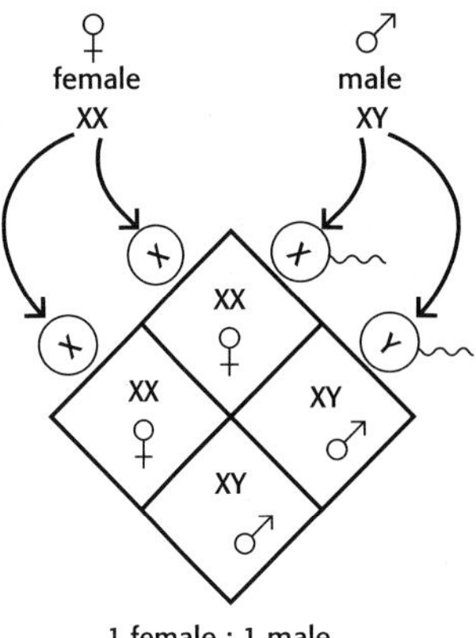

Figure 16 Determination of gender

Data-based question: hemophilia in Queen Victoria's family

The diagram is probably the most famous of all pedigree charts. It shows the incidence of hemophilia in the descendents of Queen Victoria. The blood of a person with hemophilia does not clot properly, due to the absence of a protein that is involved in the clotting process. Hemophilia is a life-threatening condition, but it can now be treated, by injecting the missing protein into the blood.

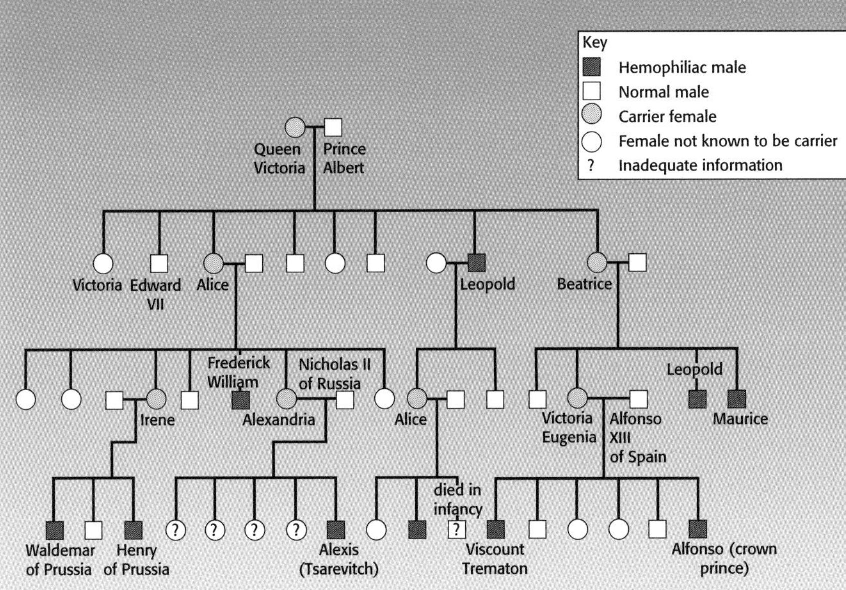

Figure 17 Pedigree chart for hemophilia

1. Explain why there are no male carriers of the allele for hemophilia in this pedigree chart. [3]
2. Explain why there are no females with hemophilia in this pedigree chart. [3]
3. Explain the evidence from the pedigree chart for hemophilia being a sex-linked condition. [3]
4. Deduce the possible genotypes and phenotypes of the daughters of Nicholas II of Russia and Alexandria. Show your working and indicate the probabilities. [3]
5. Deduce the possible genotypes and phenotypes of the second and third sons of Alfonso XIII of Spain and Victoria Eugenia. Show your working and indicate the probabilities. [3]

Chromosomes, genes and genomes

Morgan, and other geneticists working in the early part of the 20th century, discovered a group of genes that were all located on the X chromosome of *Drosophila*. By careful crossing experiments they were able to show that these genes were arranged in a linear sequence along the X chromosome. Groups of genes were then assigned to the other chromosomes in *Drosophila*, again arranged in a specific sequence. The same pattern has been found in other species – each particular gene is found in a specific position on one chromosome type. This is called the **locus** of a gene. If two chromosomes have the same sequence of genes they are **homologous.** Homologous chromosomes are not usually identical to each other because, for at least some of the genes on them, the alleles will be different.

Since Morgan's time it has been discovered that all the genes on a chromosome are part of one DNA molecule. Sophisticated techniques now allow the sequence of bases along a chromosome to be discovered. From this the sequence of genes can be deduced. This has been done for every chromosome in some species. This is called sequencing of the **genome**.

homologous chromosomes *chromosomes that have the same genes as each other, arranged in the same sequence, but not necessarily the same alleles of those genes.*

locus *the particular position on homologous chromosomes of a gene.*

genome *the whole of the genetic information of an organism.*

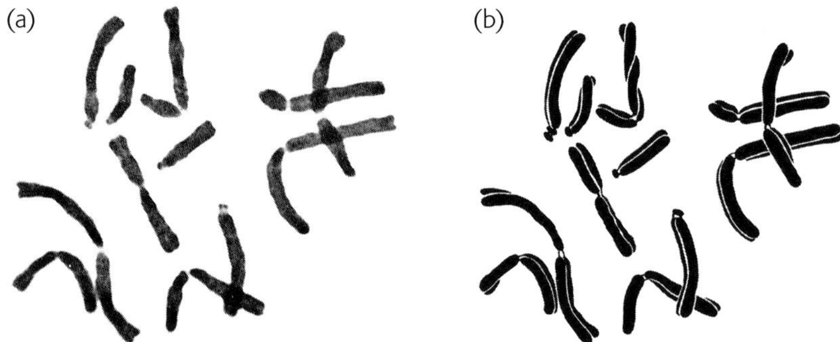

Figure 18 (a) *Trillium luteum* cell with 12 chromosomes (b) drawing to interpret figure 18a

Rediscovering biology: garlic chromosomes

1 Garlic has large chromosomes so is an ideal choice for looking at chromosomes. Cells in mitosis are needed. Garlic bulbs grow roots if they are kept for 3 or 4 days with their bases in water, at about 25 °C. Root tips with cells in mitosis are yellow in colour, not white.

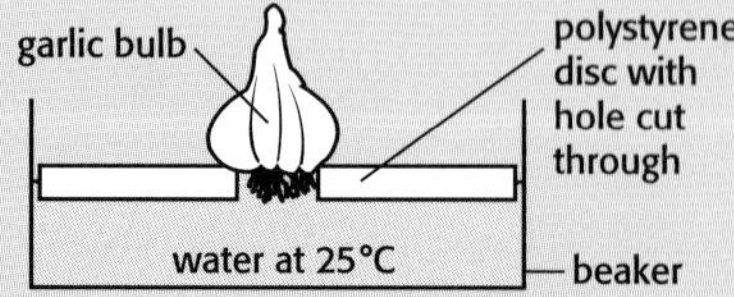

2 5-mm-long root tips are put in a mixture of a stain that binds to the chromosomes and acid, which loosens the connections between the cell walls. A length of about 5 mm long is suitable. Ten parts of aceto-orcein to one part of 1.0 mol dm^{-3} hydrochloric acid gives good results.

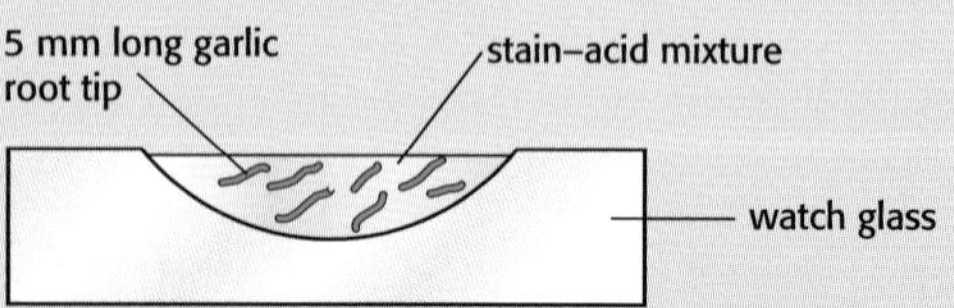

3 The roots are heated in the stain–acid mixture on a hot plate, to 80 °C for 5 minutes. One of the root tips is put on a microscope slide, cut in half and the 2.5 mm length furthest from the end of the root is discarded.

4 A drop of stain and a cover slip is added and the root tip is squashed to spread out the cells to form a layer one cell thick. The chromosomes can then be examined and counted and the various phases of mitosis should also be visible.

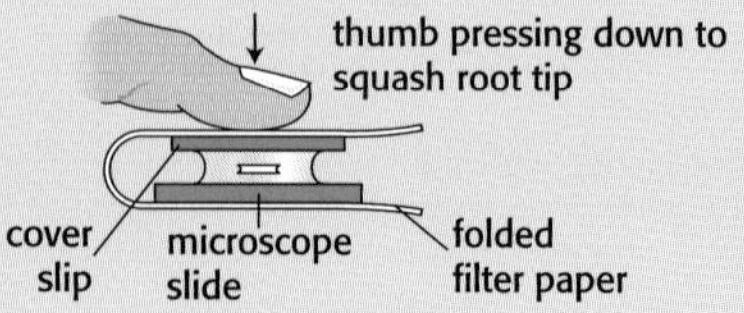

Chromosome numbers

The chromosomes of eukaryotes are large enough to be visible with a microscope during mitosis or meiosis. This is because the chromosomes are condensed – they have become shorter and fatter by a coiling process. The coiling involves histone proteins, which are associated with DNA in eukaryotes. These proteins are also involved in the control of gene transcription.

eukaryote chromosomes *DNA molecules associated with protein.*

Data-based question: differences in chromosome number

When chromosomes are visible during mitosis they can be counted. Every eukaryote species has a characteristic number of chromosomes. Table 2 shows the chromosome numbers of some species.

Table 2

Plants	Chromosome number	Animals
Haplopappus gracilis	4	*Parascaris equorum var bivalens* (horse threadworm)
Luzula purpurea (woodrush)	6	*Aedes aegypti* (yellow fever mosquito)
Crepis capillaris	8	*Drosophila melanogaster* (fruitfly)
Vicia faba (field bean)	12	*Musca domestica* (house fly)
Brassica oleracea (cabbage)	18	*Chorthippus parallelus* (grasshopper)
Citrullus vulgaris (water melon)	22	*Cricetulus griseus* (Chinese hamster)
Lilium regale (royal lily)	24	*Schistocerca gregaria* (desert locust)
Bromus texensis	28	*Desmodus rotundus murinus* (vampire bat)
Camellia sinesis (Chinese tea)	30	*Mustela vison* (mink)
Magnolia virginiana (American sweet bay)	38	*Felis catus* (domestic cat)
Arachis hypogaea (peanut)	40	*Mus musculus* (mouse)
Coffea arabica (coffee)	44	*Mesocricetus auratus* (Syrian or golden hamster)
Stipa spartea (porcupine grass)	46	*Homo sapiens* (modern humans)
Chrysoplenum alternifolium (golden saxifrage)	48	*Pan troglodytes* (chimpanzee)
Aster laevis (Michaelmas daisy)	54	*Ovis aries* (domestic sheep)
Glyceria canadensis (rattlesnake manna grass)	60	*Capra hircus* (goat)
Carya tomentosa (hickory)	64	*Dasypus novemcinctus* (nine-banded armadillo)
Magnolia cordata	76	*Ursus americanus* (American black bear)
Rhododendron keysii	78	*Canis familiaris* (dog)

?

1. There are many different chromosome numbers in the table, but some numbers are missing, for example, 5, 7, 11, 13. Explain why none of the species has 13 chromosomes. [3]
2. Discuss, using the data in the table, the hypothesis that the more complex an organism is, the more chromosomes it has. [4]
3. Explain why the size of the genome of a species cannot be deduced from the number of chromosomes. [1]
4. Suggest, using the data in Table 2, a change in chromosome structure that may have occurred during human evolution. [2]

Haploid and diploid

In the body cells of most eukaryotes there are two chromosomes of each type – two homologous chromosomes. This condition is called diploid.

The classic experiments in genetics showed that a parent only passes on one copy of each gene to offspring. This is because only one chromosome of each type is passed on in a gamete. In the gametes of eukaryotes, there is one chromosome of each type. Gametes are therefore haploid.

When a male gamete fuses with a female gamete during fertilization, the zygote that is formed is diploid. Division of the zygote results in the production of more cells by mitosis. All the body cells are therefore genetically identical and diploid. This is the usual situation in eukaryotes, including plants and animals. To produce haploid gametes, a special type of division of the nucleus is needed, called meiosis (see Chapter 11).

diploid nucleus *a nucleus that has two sets of chromosomes, and so has two chromosomes of each type.*

haploid nucleus *a nucleus that has one set of chromosomes, and so has one chromosome of each type.*

Chapter 12 questions

1 Distinguish between:
(a) genotype and phenotype [2]
(b) dominant and recessive alleles [2]
(c) homozygous and heterozygous [2]
(d) gene and genome [2]
(e) haploid and diploid. [2]

2 A three-leaf clover plant was crossed with a four-leaf clover plant (see Figure 19). All the offspring were four-leaf. When these F1 hybrid plants were crossed together, there were both three-leaf and four-leaf plants in the F2 generation.

four-leaf clover

three-leaf clover

Figure 19

(a) State the genotype of the F1 plants, using suitable symbols for the alleles. Include a key to the symbols in your answer. [2]
(b) (i) Deduce the expected ratio of genotypes and phenotypes in the F2 generation, showing your working. [3]
(ii) Discuss whether this is the expected ratio of three-leaf and four-leaf clover plants in wild populations of clover. [2]
(c) Explain how homozygous varieties of three-leaf clover and four-leaf clover could be developed, starting with wild plants. [3]

3 The pedigree in Figure 20 shows the ABO groups of three generations of a family.

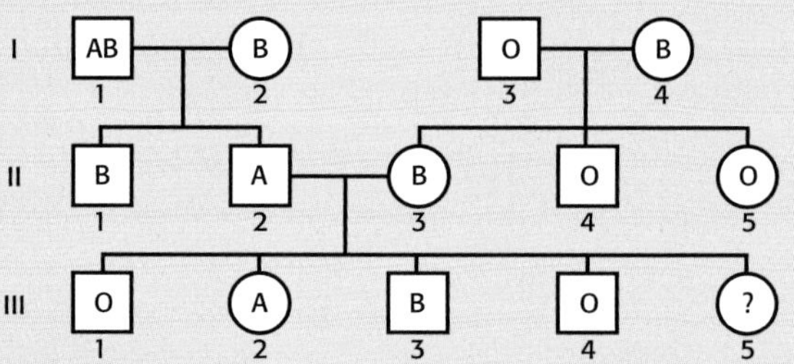

Figure 20

(a) Deduce the genotype of each person in the family. [4]
(b) Deduce the possible blood groups of individual **III** 5, with the percentage chance of each. [2]
(c) Deduce the possible blood groups and the percentage chance of each blood group:
(i) of children of individual **III** 1 and his partner who is also in blood group O [2]
(ii) of children of **III** 2 and her partner who is in blood group AB. [2]

4 Figure 21 shows two chromosomes and eight genes on them.

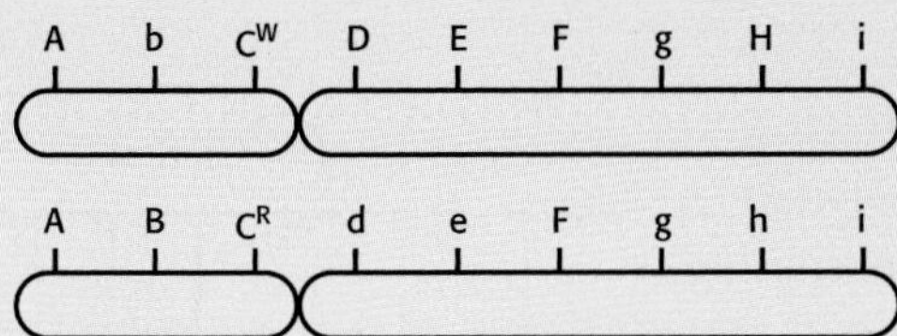

Figure 21

(a) State one allele that is recessive, one that is dominant and one that is co-dominant. [3]
(b) State one gene for which the chromosomes are homozygous and one for which they are heterozygous. [2]
(c) Explain whether the chromosomes are identical, homologous, or different. [3]
(d) Explain how the diagram shows that B and b are different alleles of the same gene. [2]

5 A mother and father have two sons, one of whom has hemophilia and the other does not. They also have a daughter. The son who has hemophilia was treated with factor VIII, the protein that is missing from his blood, but the factor VIII was contaminated with HIV and the son has developed AIDS. Hemophilia is sex-linked.
(a) Explain how the son inherited hemophilia. [4]
(b) Explain the advice for the daughter and non-hemophiliac son about the chance of them passing on hemophilia to their children. [7]
(c) Suggest possible social effects for the family of the hemophilia and AIDS. [4]

6 (a) A female magpie moth, *Abraxas grossulariata* (Figure 22) was discovered that had paler wings than normal. It was female and was crossed with a normal dark-winged male. All the offspring had dark wings. Explain what this shows about the inheritance of wing colour in this species. [2]

Figure 22

(b) The dark-coloured offspring were crossed with each other, to produce the F2 generation. Explain, using a Punnett grid, what the expected ratio of dark and light offspring is in the F2 generation. [2]
(c) The actual F2 results were:
males: all dark winged;
females: 1:1 ratio of dark and pale winged.
Explain what these results show about the inheritance of wing colour in this species. [2]
(d) Sex determination in the magpie moth is not the same as in humans. Deduce the sex chromosomes of magpie moths from the results of the cross experiment. [2]
(e) Predict the result of crossing a dark female moth with a pale male moth. Use a genetic diagram to show how you reached your answer. [2]

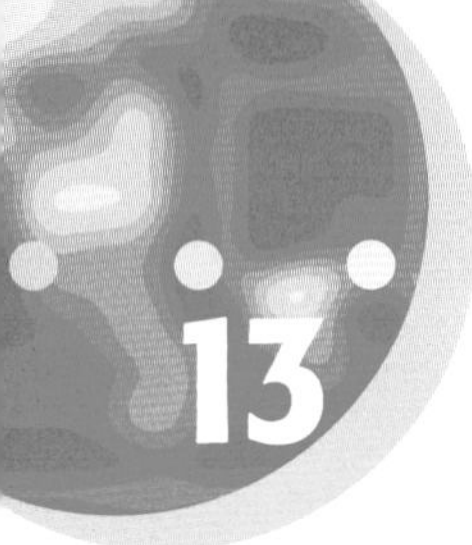

13 Advanced genetics

HL In a dihybrid cross, the inheritance of two genes is investigated together. Mendel performed dihybrid crosses, for example, he crossed pure-breeding peas that had round yellow seeds with pure-breeding peas that had wrinkled green seeds. All the F1 (first-generation) hybrids had round yellow seeds. This is not surprising, as these characters are due to dominant alleles. When Mendel allowed the F1 plants to self-pollinate, he found that four different phenotypes appeared in the F2 generation:

smooth yellow – one of the original parental phenotypes
smooth green – a new phenotype
wrinkled yellow – another new phenotype
wrinkled green – the other original parental phenotype.

If the genotype of the F1 hybrids is SsYy, the gametes produced by these hybrids could contain either S or s with either Y or y. The four possible gametes are SY, Sy, sY and sy. If the inheritance of these two genes is independent, then the chance of a gamete containing S or s will not affect its chance of containing either Y or y. The chance of a gamete containing each allele is ½, so the combined chance of containing two specific alleles is ½ × ½ or ¼. This theory that the alleles of two genes pass into gametes without influencing each other is called **independent assortment**. The Punnett grid (Figure 1) shows how a ratio of F2 phenotypes is predicted, on the basis of independent assortment. There were 556 F2 plants altogether.

	SY	Sy	sY	sy
SY	SSYY	SSYy	SsYY	SsYy
Sy	SSYy	SSyy	SsYy	Ssyy
sY	SsYY	SsYy	ssYY	ssYy
sy	SsYy	Ssyy	ssYy	ssyy

Figure 1 Punnett grid for a dihybrid cross

Mendel's actual results for this cross can be compared with the predictions from a 9:3:3:1 ratio (see Table 1). The predicted and actual results are very similar and this is also seen in the other dihybrid cross results that Mendel published. The conclusion is that these genes are inherited independently of each other, allowing new combinations of characters to be formed freely.

Table 1

Ratio	Predicted number	Actual number
9 yellow round	312.75	315
3 green round	104.25	108
3 yellow wrinkled	104.25	101
1 green wrinkled	34.75	32

Rediscovering biology: dihybrid crosses with *Drosophila* (fruit flies)

1 To make a cross, one virgin female and two males are put in a tube with artificial growth medium. After mating, the female lays eggs which hatch into F1 larvae. After 7 days the parents must be removed so they are not mistaken for F1 offspring.

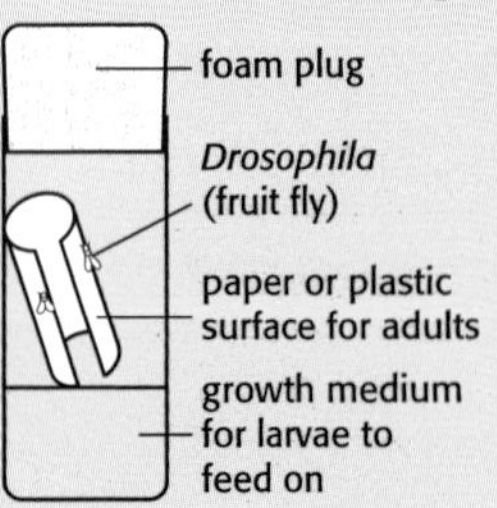

2 After 7 days the F1 larvae are fully grown. They form a pupa and after 4–7 more days the F1 adults emerge. They can be examined with a microscope, after using an anesthetic to stop them flying. Over the next few days, females must be removed each day, to ensure that they are virgins.

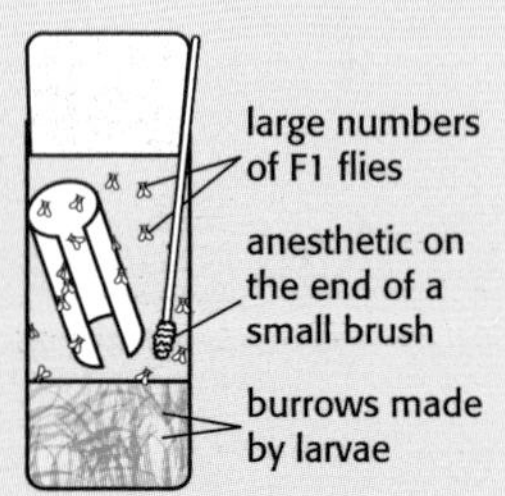

3 F1 virgin females are placed in new tubes with artificial growth medium and two males are added. These could be F1 hybrid males or double recessive males for a test cross. The stages used to produce the F1 generation are repeated to produce the F2 generation.

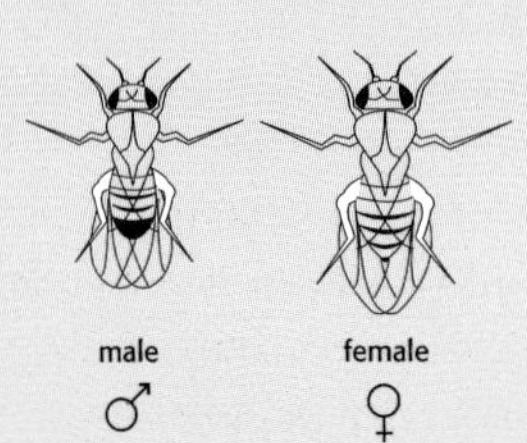

4 When the F2 generation has emerged, they can be scored. Males and females can be scored separately to check for sex linkage. What F2 ratio would you expect if the two genes assort independently:

- if two hybrids were crossed?
- if a test cross was done between a hybrid and a homozygous recessive?

phenotype	♂	♀
red eye wt body	𝍸 II	𝍸 II
red eye ebony	III	𝍸
brown eye wt body	III	II
brown eye ebony	I	

tally chart

Independent assortment and meiosis

When Mendel's work was rediscovered at the start of the 20th century, the mechanism that causes independent assortment of unlinked genes was soon identified. Observations of meiosis in a grasshopper, *Brachystola magna*, had shown that homologous chromosomes pair up during meiosis and then separate, moving to opposite poles. The pole to which each chromosome in a pair moves depends on which way the pair is facing. This is random. Also, the direction in which one pair is facing does not affect the direction in which any of the other pairs are facing. This is called independent orientation.

We can now make the link between genes and chromosomes: if a cell is heterozygous for a gene, one chromosome in a pair will carry one allele of the gene and the other chromosome will carry the other allele. The orientation of the pair of chromosomes will determine which allele moves to which pole. Each allele has a 50 per cent chance of moving to a particular pole. Similarly for another gene, located on another chromosome, for which the cell is heterozygous, there is a 50 per cent chance of an allele moving to a particular pole. Because there is random orientation of chromosome pairs, the chance of two alleles coming together to the same pole is 25 per cent (see Figure 2).

After Mendel's death, his theory about independent assortment was elevated to the status of a law – Mendel's law of independent assortment – but exceptions had already been found. These exceptions are called gene linkage and can also be explained in terms of the behaviour of chromosomes in meiosis.

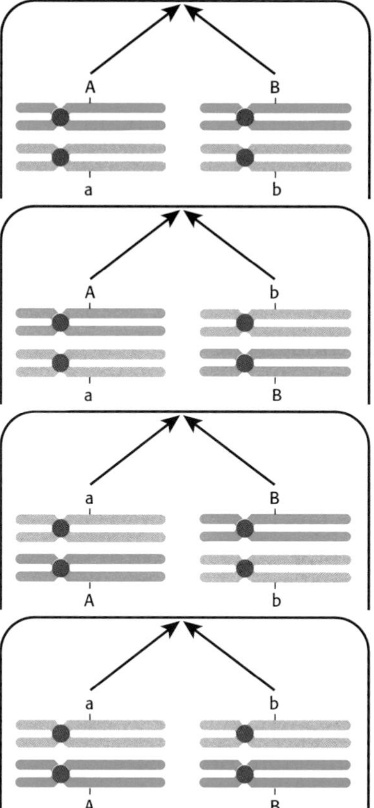

Figure 2 Random orientation

Thinking about science: hypotheses, theories and laws

At the age of 100, the distinguished American biologist Ernst Mayr published his 25th book, entitled *What makes Biology Unique? Autonomy of a Scientific Discipline*. Extracts from an interview he gave about the book are given here.

I show that biology is as serious, honest, legitimate a science as the physical sciences. All the occult stuff that used to be mixed in with the philosophy of biology, like vitalism and teleology – Kant after all, when he wanted to describe biology, he put it all on teleology, just to give an example – all this sort of funny business is out. Biology has exactly the same hard-nosed basis as the physical sciences, consisting of natural laws. The natural laws apply to biology just as much to the physical sciences. But the people who compare the two, or who like some philosophers, put biology with the physical sciences, they leave out a lot of things. And the minute you include those, you can see clearly that biology is not the same sort of thing as the physical sciences. And I cannot give a long lecture now on that subject, that's what the book is for.

I'll give you an example. In principle, biology differs from the physical sciences in that in the physical sciences, all theories, I don't know exceptions so I think it's probably a safe statement, all theories are based somehow or other on natural laws. In biology as several people have shown, and I totally agree with them, there are no natural laws corresponding to the natural laws of the physical sciences.

Now you can say, how can you have theories in biology if you don't have laws on which to base them? Well, in biology our theories are based on something else. They're based on concepts. Like the concept of natural selection forms the basis of evolutionary biology. You go to ecology and you get concepts like competition and resources, ecology is just full of concepts. And those concepts are the basis of all theories in ecology. Not the physical laws, they're not the basis. They are of course ultimately the basis, but not directly, of ecology. So that's what I do in this book. I show that the theoretical basis, you might call it, or I prefer to call it the philosophy of biology, has a totally different basis than in the theories of physics.

1. If a scientific law must allow us to predict, with certainty, what will happen under specific conditions, why are there fewer laws in biology than in physics?
2. If a theory must give us a workable explanation for an observation, why are theories usually based on concepts in biology and on scientific laws in physics?
3. How are theories tested in both biology and physics?
4. Teleology is the view that things have a final purpose, and develop so that they can carry it out. Why does Mayr reject this view?

Gene linkage

William Bateson, Edith Saunders and Reginald Punnett discovered the first exception to the law of independent assortment in 1903. When they crossed sweet pea plants with purple flowers and long pollen grains with plants with red flowers and round pollen grains, all the F1 hybrids had purple flowers and long pollen grains. When these F1 plants were self-pollinated, four phenotypes were observed in the F2 generation, but not in the familiar 9:3:3:1 ratio. The cross was repeated with larger numbers and the F2 contained the numbers of plants shown in Table 2.

Table 2

Phenotype	Observed frequency	Observed %	9:3:3:1 %
purple long	4831	69.5	56.25
purple round	390	5.6	18.75
red long	393	5.6	18.75
red round	1338	19.3	6.25

Although the observed percentages do not fit the 9:3:3:1 ratio, results like this were not unexpected. Genes were known to be part of chromosomes, and as there are far more genes than chromosomes, some genes must be found together on the same chromosome. Alleles of these genes would therefore not follow the law of independent assortment and would pass together into a gamete. This is seen in the results for the sweet pea cross – there were more of the purple long and red round plants than expected. This is because these were the original parental combinations of alleles. This pattern of inheritance is called gene linkage. Since 1903 many more examples have been found, always with a higher frequency of the parental combinations than predicted from Mendelian ratios.

A genetic diagram explaining the cross in sweet peas is shown in Figure 3, using lines to symbolize the chromosomes on which the linked genes are located. The way in which new combinations of alleles are formed, giving the purple round and red long plants, is described below (see page 153 and also page 134).

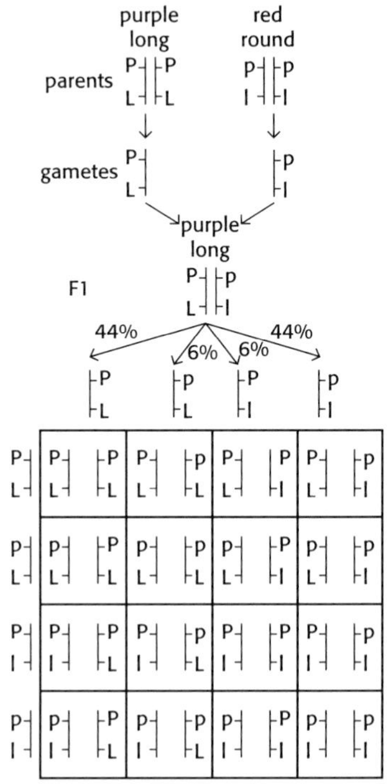

Figure 3 Cross involving gene linkage

Data-based question: gene linkage in Zea mays

Corncobs are often used for showing inheritance patterns. All the grains on a cob have the same female parent, and with careful pollination they can also have the same male parent. A variety with coloured and starchy grains was crossed with a variety with white and waxy grains. The F1 grains were all coloured and starchy. The F1 plants grown from these grains were crossed (F1 × F1).

1 Calculate the expected ratio of F2 plants, assuming that the genes for coloured/white and starchy/waxy grains are unlinked. Use a genetic diagram to show how you reached your answer. [3]

2 The actual frequencies were:

coloured starchy	1774
coloured waxy	263
white starchy	279
white waxy	420.

Using this data, deduce whether the genes for coloured/white and starchy/waxy are linked. [2]

A variety with coloured and shrunken grains was crossed with a variety with white and non-shrunken grains. The F1 grains were all coloured and non-shrunken. The F1 plants grown from these grains were test crossed using pollen from a homozygous recessive variety with white shrunken grains.

3 Calculate the expected ratio of F2 plants, assuming that the genes are unlinked, using a diagram to show how you reached your answer. [2]

4 The actual frequencies were:

coloured non-shrunken	638
coloured shrunken	21,379
white non-shrunken	21,096
white shrunken	672.

Using this data, deduce whether the genes for coloured/white and non-shrunken/shrunken are linked. [2]

5 Deduce whether the genes for starchy/waxy and non-shrunken/shrunken are linked. [1]

Crossing over and recombination of linked genes

In *Drosophila* there are eight chromosomes in diploid nuclei. In males one of these is an X and another is a Y chromosome. In females two of them are X chromosomes. The other six chromosomes are common to males and females – they are called **autosomes**.

Diploid nuclei have two of each type of autosome, so in *Drosophila* there are three types of autosome and in humans there are 22. Geneticists working early in the 20th century found four groups of linked genes in *Drosophila*, corresponding to the three types of autosome and to the X chromosome.

There are two types of linkage – **autosomal gene linkage**, when the genes are on the same autosome, and **sex linkage**, when the genes are located on the X chromosome.

The linkage between pairs of genes in a linkage group is not usually complete and new combinations of alleles are sometimes formed. This happens as a result of crossing over, which was described as a part of meiosis earlier in this Course Companion. Figure 4 shows how crossing over gives new allele combinations. The formation of a chromosome or DNA with a new combination of alleles is **recombination**. An individual that has this recombinant chromosome and therefore has a different combination of characters from either of the original parents is called a **recombinant**.

autosomes *chromosomes other than the X and Y chromosomes.*

linkage group *genes that are located on the same chromosome type and do not assort independently.*

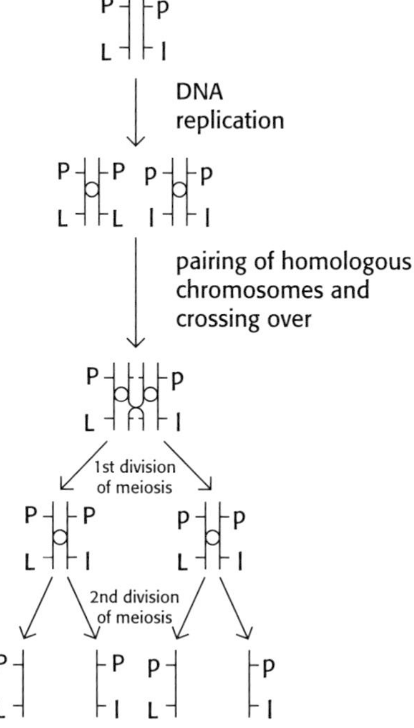

Figure 4 Formation of recombinants

Data-based question: gene linkage in mutant tomatoes

Tomato plants usually have flowers in small groups and produce fruits that are round and smooth. Three mutations were found:

peach – hairy fruits
oblate – flattened fruits
clustered – with many flowers in a group.

Cross 1

Pure-breeding varieties with peach oblate fruits and smooth round fruits were crossed together. The F1 hybrids, which all had smooth round fruits, were test-crossed with plants that were homozygous for peach oblate fruits. The numbers in the F2 generation were:

smooth round	421	smooth oblate	112
peach oblate	369	peach round	98.

Cross 2

A pure-breeding variety with smooth fruits and few flowers was crossed with a pure-breeding variety with peach fruits and clustered flowers. All the F1 hybrids had smooth fruits and few flowers. These hybrids were test-crossed with plants that were homozygous for peach fruit and few flowers. The numbers in the F2 generation were:

smooth clustered	350	smooth few	183
peach few	308	peach clustered	159.

1 Identify, with reasons, which tomato plants are recombinants in the F2 generations in:
(a) cross 1 [2]
(b) cross 2. [2]

2 Draw diagrams to show how recombinants were produced by crossing over in:
(a) cross 1 [2]
(b) cross 2. [2]

3 Calculate the percentage of recombinants in:
(a) cross 1 [2]
(b) cross 2. [2]

4 Discuss the reasons for the difference in recombination percentage between the two crosses. [3]

Polygenic inheritance and continuous variation

In the examples of inheritance considered so far in this chapter, every gene has affected a different character. There are also examples of inheritance in which two or more genes affect the same character. The genes have an additive effect. Mendel discovered an example of this in beans, where a cross between a purple-flowered plant and a white-flowered plant gave purple-flowered plants in the F1 generation, but when these were self-pollinated, the expected 3:1 ratio did not occur and instead a range of flower colours was seen. This can be explained if there are two unlinked genes, with co-dominant alleles (see Figure 5). Self-pollination of the F1 should give five different shades of flower colour in a ratio of 1:4:6:4:1. If the number of unlinked genes, with co-dominant alleles, were larger then there would be more phenotypic variants. The number and frequency of variants can be predicted using alternate rows of Pascal's triangle (Figure 6). A frequency distribution is shown (Figure 7) for a character affected by five genes with co-dominant alleles. As the number of genes increases, the distribution becomes increasingly close to the **normal distribution.** Many characters in humans and other organisms are close to the normal distribution, for example the mass of bean seeds, height in humans and intelligence in humans. The closeness to a normal distribution suggests that more than one gene is involved. This situation is called polygenic inheritance.

When the variation due to polygenic inheritance is examined carefully, it is usually found to be continuous – there is a complete range of variation, rather the distinct classes that were predicted above. This is because the differences in phenotype between the classes are subtle and the effects of the environment blur these differences so much that they are undetectable.

polygenic inheritance: *two or more genes affecting the same character.*

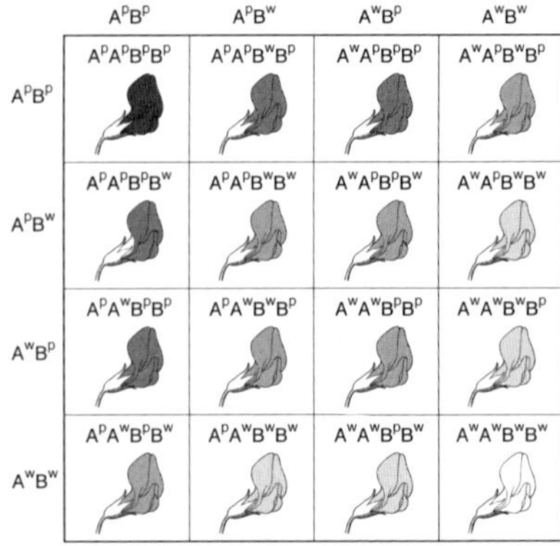

Figure 5 Results of a cross involving polygenic inheritance

1:2:1
1:4:6:4:1
1:6:15:20:15:6:1
1:8:28:56:70:56:28:8:1

Figure 6 Pascal's triangle

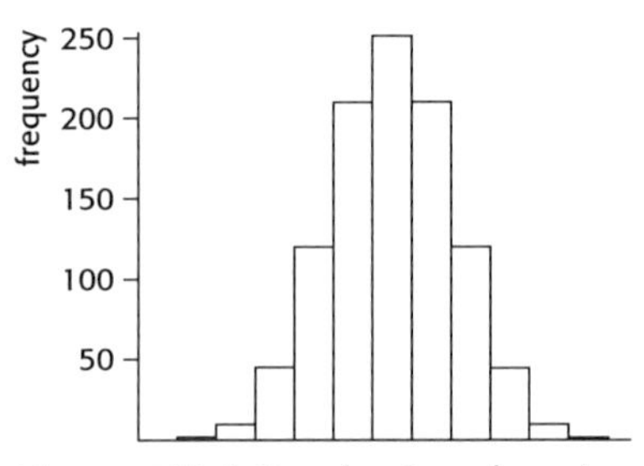

Figure 7 Variation due to polygenic inheritance

Working with data: displaying discontinuous and continuous variation

It is easy to display frequencies with discontinuous data, for example the relative numbers of people in different blood groups. Every person falls neatly into one group and pie charts or bar charts (Figure 8) can be used to display the numbers. It is much more difficult to display frequencies with continuous variation. Every individual is slightly different, so the methods used for discontinuous variation are not suitable. Classes are artificially defined, with a histogram to show the frequency of each class (Figure 9).

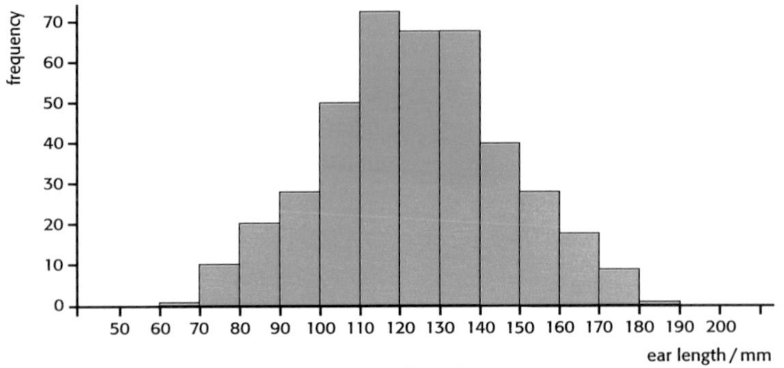

Figure 9 Frequency of classes of cob length in maize

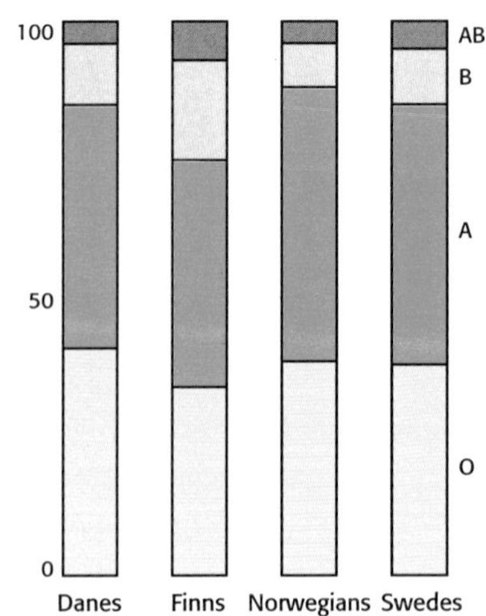

Figure 8 Frequency of blood groups in four populations in Northern Europe

Inheritance of skin colour

The theory that characteristics showing continuous variation could be due to the combined effects of many genes was suggested in the early 20th century, but little progress was made during rest of the century in identifying the genes involved. Skin colour in humans is an interesting example. A study published in 1964 of families in Liverpool that had both African and European ancestors suggested a minimum of three or four genes affecting skin colour, but much larger numbers of genes could be involved. Mutations of several genes are known that can prevent melanin production completely, leading to albinism, but genes varying the amount of melanin production are required to explain the observed continuous variation in humans. A mutation of a gene called SLC24A5 has been discovered that causes paler skin and accounts for more than 25 per cent of the variation between European and African populations, but the remaining variation must be due to other genes. When these have been discovered, it will be possible to say exactly how many genes are involved in this example of polygenic inheritance.

Figure 10 Variation in human skin colour

Investigating biology: variation in skin colour

Two forms of melanin can exist in human skin and in human hair: pheomelanin, which is red-yellow and eumelanin, which is black-brown. These two pigments together determine the light-absorbing properties of skin and hair. Computer software is available to analyse the density and proportions of red, blue and green pixels in images obtained using digital cameras. Using this software, quantitative measures of pigmentation can be obtained.

There are many possibilities for investigations of skin colour:

- Is the skin colour of a child always intermediate between that of the parents, or can it be closer to one parent?
- How quickly does skin pigmentation darken after exposure to sunlight or to artificial sources of ultraviolet radiation?
- Can human skin colour be assessed on a single linear scale, or is the range of pigmentation more complex?

Figure 11 shows the percentages of red or green and of blue pixels in images of the irises in the eyes of 10 individuals. Red and green pixels dominate in images of brown irises.

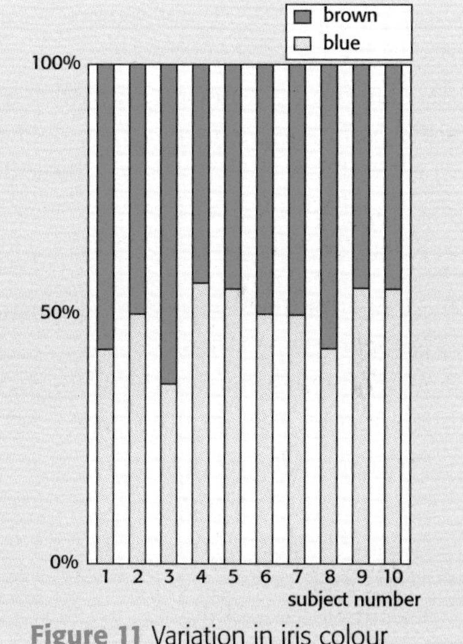

Figure 11 Variation in iris colour

Critical considerations: skin colour and racial bias

In recent years, biologists have been reluctant to investigate skin colour, perhaps for fear of accusations of racial bias. There is a long history of suspicion and hostility between human races and, as differences in skin are one of the clearest markers of racial difference, this has often been a focus. The following text is from an address to the United Nations in 1963, by Haile Selassie, Emperor of Ethiopia.

On the question of racial discrimination, the Addis Ababa conference taught, to those who will learn, this further lesson: that until the philosophy which holds one race superior and another inferior is finally and permanently discredited and abandoned; that until there are no longer first and second class citizens of any nation; that until the colour of a man's skin is of no more significance than the colour of his eyes; that until the basic human rights are equally guaranteed to all without regard to race; that until that day, the dream of lasting peace and world citizenship and the rule of international morality will remain but a fleeting illusion, to be pursued but never attained.

1. Does anyone disagree with Haile Selassie's sentiments?
2. To what extent has there been progress since 1963 in the issues that he identifies?
3. Depending on the research question, skin colour research might reinforce racial stereotypes and prejudices or it might lead to greater understanding of cellular mechanisms such as in gene expression. Suggest criteria that might be used to evaluate the legitimacy of research that looks at characteristics associated with racial differences.

Thinking about science: the evolution of skin colour

The density of skin pigmentation varies considerably between human populations. The reasons for this are intriguing. In a recent study, Jablonski and Chaplin (2004) measured the reflectance of skin in indigenous populations around the world and attempted to correlate it with other variables. The closest correlation was with the intensity of ultraviolet radiation in the area inhabited by the population. The correlation coefficient was 0.77 in the northern hemisphere and 0.70 in the southern hemisphere – these are strong correlations. The intensity of ultraviolet radiation (UVR) has some significant effects of human physiology and therefore on health, survival and reproduction.

- UVR breaks down many compounds in the skin, including folate, a B vitamin.
- UVR is mutagenic (causes mutations) and is especially associated with skin cancer.
- UVR is needed for the synthesis of vitamin D in the skin. Vitamin D is required for absorption of calcium from food in the gut.

A theory to explain the variation in human skin pigmentation can be based on these observations. In every area of the world, humans will develop the skin pigmentation that gives the greatest overall chance of survival and reproduction. This involves allowing the minimum amount of UVR penetration to allow sufficient vitamin D production and therefore maximizing the protection against destruction of folate and protection against skin cancer.

Evidence for this theory can be drawn from observations about specific populations:

- Albino schoolchildren in South Africa need less dietary vitamin D than normal children, to reach target calcium levels.
- The Greenland population of Inuit has darker skin than expected – darker than the skin colour–UVR correlation predicts. The Inuit traditionally obtain large amounts of dietary vitamin D from eating marine mammals.
- Many Inuit now eat supermarket food rather than their traditional diet and they have one of the world's highest rates of vitamin D deficiency.
- Of the 12 populations with unexpectedly dark skins, four are Bantu-speaking populations who, only about 200 years ago, migrated from equatorial regions to southern Africa, where UVR intensity is lower.
- Of the nine populations with unexpectedly pale skins, three are the indigenous populations of the Philippines, Vietnam and Cambodia, who migrated closer to the equator within the last 2000 years.
- Our nearest non-human relatives, the chimpanzees, have pale skins, but gain protection from UVR with a thick layer of darkly pigmented body hair. Humans have much less body hair.

1. What should a dark-skinned person do in areas of the world with low UVR intensity, to avoid health problems?
2. What should a light-skinned person do in areas of the world with high UVR intensity, to avoid health problems?
3. What risk might there be for a light-skinned child whose parents insisted on applying a thick layer of sun-block cream whenever the child went outside?
4. Which of these views is closest to your own?
   - We must do everything that we can to minimize risk.
   - The only acceptable risk is zero risk.
   - Risk is healthy and we all need risk in our lives.
   - We cannot eliminate risk completely, so it is best to ignore it.
5. How strong is the theory to account for human skin colour? Is it strong enough to convince us that this is what causes the variation in skin colour to develop? Can the theory be tested using scientific methods? Does it matter that there are a few populations with skin colours not explained by the theory? For example, the aboriginal population of Tasmania has unexpectedly dark skin, despite having migrated south to Tasmania 10,000 years ago.

Chapter 13 questions

Higher-level questions

1 (a) Distinguish between autosomes and sex chromosomes. [2]
(b) Compare autosomal linkage and sex linkage. [2]
(c) Distinguish between continuous and discontinuous variation. [2]
(d) Distinguish between polygenic inheritance and multiple alleles. [2]
(e) State two examples of polygenic inheritance. [2]

2 The diploid chromosome number of the pea plant, which Mendel used in his experiments, is 14.
(a) How many linkage groups are there in peas? [1]
(b) Table 3 shows the chromosome location of the genes that affect each of Mendel's character differences.
(i) State one pair of character differences that you expect to be unlinked. [1]
(ii) State one pair of character differences that you expect to show autosomal linkage. [1]

Table 3

Character difference investigated in Mendel's experiments	Chromosome on which gene located
Tall stem/dwarf stem	4
Round/wrinkled seed	7
Yellow/green cotyledons	1
Purple/white flowers	1
Full pods/constricted pods	4
Green/yellow unripe pods	5
Flowers on stem/at stem tip	4

(c) If pure-breeding plants with full green pods were crossed with pure-breeding plants with constricted yellow pods, and the F1 plants were allowed to self-pollinate, what ratio of phenotypes would you expect in the F2 generation? Use a genetic diagram to work out your answer. [5]
(d) When dihybrid crosses are performed with plants differing in cotyledon colour and flower colour, they give results suggesting that the two genes involved are unlinked. The two genes are located far apart from each other on chromosome number 1. Explain how this can cause them to behave as though they are in different linkage groups. [2]

3 (a) Explain the relationship between Mendel's law of independent assortment and the events of meiosis. [6]
(b) Explain how recombination of linked genes occurs. [6]
(c) Explain the variation and inheritance of skin colour in humans. [6]

4 Chickens have a red fleshy crest on their head, called the comb. There are four shapes of comb: single, rose, pea and walnut (see Figure 12).

Figure 12

(a) If pure-breeding chickens with rose comb and single comb shapes are crossed, all the F1 hybrids have rose combs. Using A and a for the dominant and recessive alleles, state the genotypes of the parents and the F1 hybrids with rose comb. [2]
(b) If pure-breeding chickens with pea comb and single comb shapes are crossed, all the F1 hybrids have pea combs. Using B and b for the dominant and recessive alleles, state the genotypes of the parents and the F1 hybrids with pea comb. [2]
(c) If pure-breeding chickens with rose comb and pea comb shapes are crossed together, all the F1 have walnut combs. Deduce the genotype of the F1 hybrids with walnut comb. [1]
(d) Explain, using a genetic diagram, the ratio of phenotypes you would expect in the F2 generation if F1 hybrids with walnut combs were crossed together. [3]
(e) Discuss whether comb shape in chickens is an example of polygenic inheritance. [2]

5 The scattergraph (Figure 13) shows the results of a study by Newman, Freeman and Holzinger of identical twins, who were reared apart. The intelligence of the twins was assessed by an IQ test and the differences in their education were also assessed using a numerical scale.

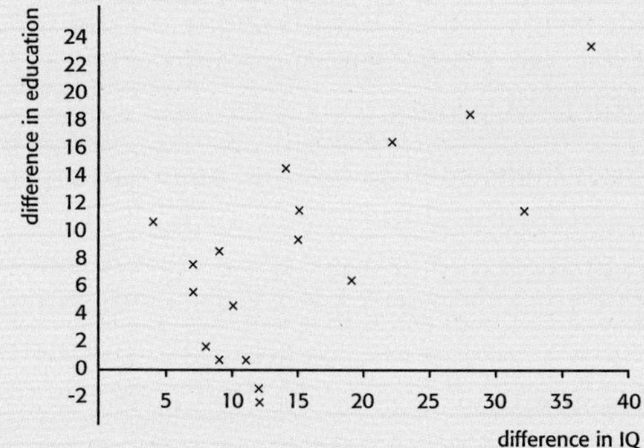

Figure 13

(a) Explain why identical twins reared apart are very useful in studies of inheritance. [2]
(b) Explain why identical twins reared together show a close correlation in intelligence. [2]
(c) Using the data in the scattergraph, state the smallest and the largest difference in intelligence of the twins reared apart. [2]
(d) The scattergraph shows a positive correlation. Suggest a cause of this correlation. [2]
(e) The points on the graph are widely scattered. Suggest reasons for this. [2]

14 Biotechnology

The Human Genome Project

The Human Genome Project (HGP) was a co-operative venture between many different independent research groups as well as Celera Genomics, a private research company. The HGP achieved the goal of obtaining a sequence of the human **genome** – all the DNA in a human cell. Far fewer protein-coding regions have been found within the genome than was expected. The estimate of the total number of human genes has been reduced from as high as 100,000 down to about 25,000 genes.

The HGP stimulated huge improvements in sequencing technology. Improved techniques have made it possible to sequence the genomes of other organisms that will help with understanding the human genome. The HGP has also provided new insights into the role of "junk DNA" (sections of DNA that are not transcribed).

Knowledge has been gained about the significance of certain sequences. The presence or absence of the sequences can be detected using microarray technology. DNA microarrays are a fast and efficient way to look for thousands of short sequences of DNA at once. A microarray is a flat plate that holds many different DNA sequences, called probes. These probes are short pieces of DNA that have a known sequence. When an unknown sample of DNA is tested using a microarray, it is broken up into small pieces, and the pieces are "tagged" with a fluorescent dye. The sample DNA is then washed over the microarray. If a section of the sample DNA is complementary to one of the probes, it binds to the probe and is held on the microarray. When light is shone on the microarray, a pattern of fluorescent dots shows where sequences from the sample have bound to the DNA probes.

New research fields such as bioinformatics and pharmacogenomics have emerged from the HGP. Bioinformatics involves the use of computers to store and analyse the huge amounts of data being generated by the sequencing of genomes. Pharmacogenomics is a field which links differences in genomic information in different populations to any differences in their response to drug treatment.

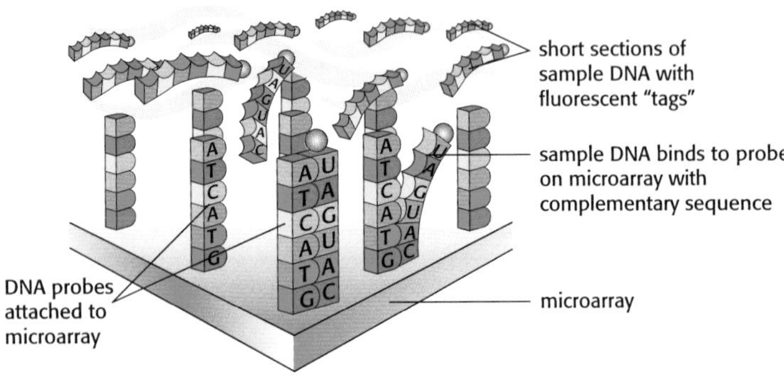

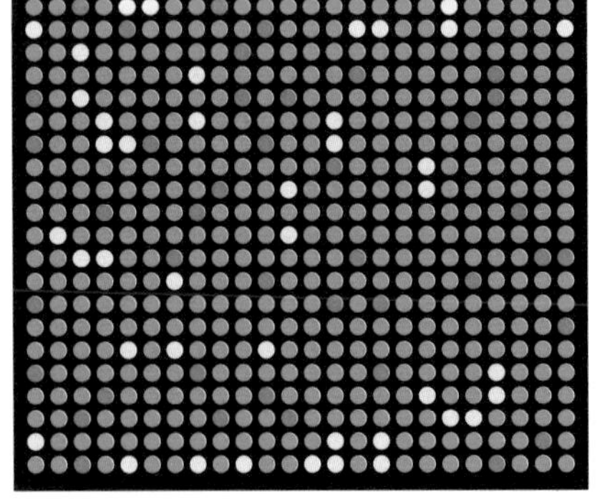

Figure 1 A microarray is a technique used to test for the presence of certain DNA sequences.

Data-based question: determining an open reading frame

Once the sequence of bases in a piece of DNA has been determined, a researcher may want to locate a gene. To do this, computers search through the sequences looking for open reading frames. An open reading frame is one that is uninterrupted by stop sequences and could therefore code for the production of a protein. The stop codons are UGA, UAA and UAG.

Table 1

DNA 3′	A T T A A C T A T A A A G A C T A C A G A G A G G G C T A G T A C
mRNA 5′	
RF1	U A A U U G A U A U U U C U G A U G U C U C U C C C G A U C A U G
RF2	A A U U G A U A U U U C U G A U G U C U C U C C C G A U C A U G
RF3	A U U G A U A U U U C U G A U G U C U C U C C C G A U C A U G

?

1 State the number of codons in the genetic code. [1]
2 Determine the fraction of codons that are stop codons in the genetic code. [2]
3 In Table 1, the codons could start with the first, second or third base. These correspond to three different reading frames (RF1, RF2 or RF3). Determine which of the reading frames, 1, 2 or 3 above, might be an open reading frame. [2]

Critical considerations: race and medicine

The excerpt below is from an article in *Nature* called "Race: A Genetic Melting-pot" (2003).

The issue of whether race is a biologically useful or even meaningful concept when applied to humans in a medical context is controversial – holders of opposing views each claim to have evidence to support them. But there is no contradiction between these two well-substantiated bodies of data, as they actually deal with two different questions that have become confused with one another. The first question is: "Is it possible to find DNA sequences that differ sufficiently between populations to allow correct assignment of major geographical origin with high probability?" The answer to this question is yes, as shown by studies of genetic polymorphisms…

The second question is: "What fraction of human genetic variation, whether based on protein-coding genes or other sequences, falls within geographically separated populations, and what fraction occurs between these populations?" The answer to this question is that most genetic diversity occurs within groups, and that very little is found between them. Why this apparent paradox? The answer is that genes that are geographically distinctive in their frequencies are not typical of the human genome in general. It has been suggested that racial categorization has a valid role in good medical practice because many medically important genes vary between populations from different regions. But although knowing a patient's ancestry is often extremely useful in diagnosis and treatment, race is both too broad and too narrow a definition of ancestry to be biologically useful.

For any species, definitions of race can lose their discriminating power when individuals migrate to different regions and mate with their counterparts there. Among humans, large-scale migrations between continents – particularly through European colonial expansion and the commercial slave trade – has resulted in matings of individuals from different continents and the creation of new populations, especially in the Western Hemisphere and Oceania. To use genotype effectively in making diagnostic and therapeutic decisions, it is not race that is relevant, but both intra- and trans-continental contributions to a person's ancestry.

Race and ancestry are confounded both by genetic heterogeneity within groups and by the widespread mixing of previously isolated populations. The assignment of a racial classification to an individual hides the biological information that is needed for intelligent therapeutic and diagnostic decisions. A person classified as 'black' or 'Hispanic' by social convention could have any mixture of ancestries, as defined by continent of origin. Confusing race and ancestry could be potentially devastating for medical practice.

This second quote is excerpts from an article in the *British Medical Journal* called "Pharmacogenetics and ethnically targeted therapies" (2005).

In the middle of this resurgence of the role of biology in concepts of race and ethnicity comes BiDil, a new drug treatment for heart failure tested solely in one particular racial group. In 2001 NitroMed began the African-American heart failure trial, the first heart failure trial conducted exclusively in African-American patients, claiming that "observed racial disparities in mortality and therapeutic response rates in Black heart failure patients may be due in part to ethnic differences in the underlying pathophysiology of heart failure". The study found that BiDil combined with standard therapy for heart failure reduced mortality by 43% among black patients. Hailed by the media as the first ethnic drug, BiDil is reported to be on the way to becoming the first drug approved by the US Food and Drug Administration to treat heart failure in African-American patients only.

The major implication of BiDil is that differential responses to treatment between racial groups, defined by using ostensibly social categories (here, patients self reported to be African-Americans), are attributed primarily to genetic differences. Many researchers and policy makers argue against the use of racial or ethnic categories in medicine, saying that classifying people according to race and ethnicity reinforces existing social divisions in society or leads to discriminatory practices. Others cite research showing that genetic differences are greater within socially defined racial groups than between groups. The relation between features that traditionally define race and contribute to ethnicity, such as skin colour, and genetic differences has been found to be inconsistent. In addition, by focusing on biological factors as the explanation for differences in response to drugs, researchers risk ignoring other possible environmental, psychosocial, and economic factors, and lifestyle factors such as diet, that are important in producing illness.
The new genetics has reopened the debate on the biological basis of race and ethnicity. Pharmacogenetics is growing fast, and it will lead to a more refined understanding of ethnic and racial differences in drug response. Many claims and counterclaims will be made. Doctors need to take an open minded but critical stance.

Imagine that you are a researcher conducting clinical trials on a new drug. Discuss the advantages and disadvantages of collecting any of the following :

1 Information about the subject's visible race on a questionnaire.
2 Information about the subject's self-reported ancestry; i.e. knowledge about which populations they might belong to.
3 A tissue sample to determine genomic information using microarray technology.

Genetic engineering

Agriculture depends upon the use of plants or animals that have desirable characteristics, such as livestock that produce higher quantities of muscle or milk. Selective breeding is a traditional method where deliberate crosses or matings of organisms were made so that offspring would have a desired characteristic derived from one or both of the parents. For example, modern varieties of wheat yield up to three times as much as the 19th century varieties.

Gene transfer technology takes advantage of the universality of the genetic code so that desirable qualities found in entirely different donor species can be transferred to a different host species. Figure 2 shows the steps involved in one method of preparing recombinant DNA. Donor DNA from another species is incorporated into a bacterial **plasmid**. An example of the use of this procedure is the introduction of the gene for the human hormone insulin into a plasmid, which is then taken up by the bacterium *E. coli*. The bacterium is then able to manufacture human insulin. The DNA for insulin has to be produced from mRNA rather than inserting the DNA gene for insulin directly. This is because the human insulin gene contains **introns**, which bacteria cannot remove.

Before the advent of gene transfer technology, patients with diabetes had to inject themselves with insulin derived from the pancreas of an animal. Now they can use human insulin, manufactured by genetic engineering.

Enzymes called restriction endonucleases are important tools in gene transfer procedures. These enzymes cut DNA at specific sequences. Restriction enzymes were first observed in bacteria. Bacteria use restriction enzymes to defend themselves against viral infection. The restriction enzymes cut up the viral DNA to prevent the virus from being replicated.

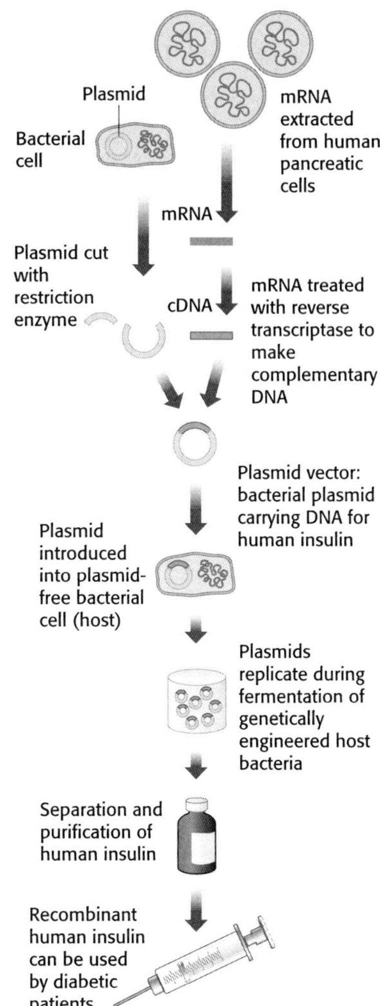

Figure 2 The procedure for producing human insulin using recombinant DNA technology

Golden rice – an example of genetic modification

An example of the use of genetic engineering in agriculture is "golden rice". This involves the introduction of three genes – two genes from daffodil plants and one from a bacterium – into rice plants, so that the orange pigment β-carotene is produced in the rice grains. Beta-carotene is a precursor to vitamin A. The development of golden rice was intended as a solution to the problem of vitamin A deficiency, which is a significant cause of blindness among children globally. Rice is the staple diet in many regions where vitamin A deficiency is common.

Critics of golden rice have argued that absorption of β-carotene from golden rice will not meet the recommended daily allowance of vitamin A. The first version of golden rice would have provided people with 15–20 per cent of the recommended daily allowance (RDA) of vitamin A. The second version of golden rice (golden rice 2) is estimated to be able to provide 50 per cent of the RDA of β-carotene. Critics suggest that the key to overcoming micronutrient deficiencies lies in improved and varied diets – including a return to traditional diets. Unpolished rice, for example, possesses large quantities of β-carotene.

Alternative recombinant methods

Figure 3 illustrates an alternative method for introducing the recombinant DNA. One of the first medical by-products of a genetically engineered animal to gain regulatory approval was an anti-clotting agent called anti-thrombin, extracted from the milk of genetically engineered goats.

The mammary glands of mammals are specialized for protein production, and some of the post-translational modifications required to make the proteins functional can only be carried out in a mammalian system. By combining the gene with regulatory regions from the gene of a milk-specific protein, it can be ensured that a gene will be expressed in the mammary gland. A single goat can produce about 3.0 kg of therapeutic protein per year.

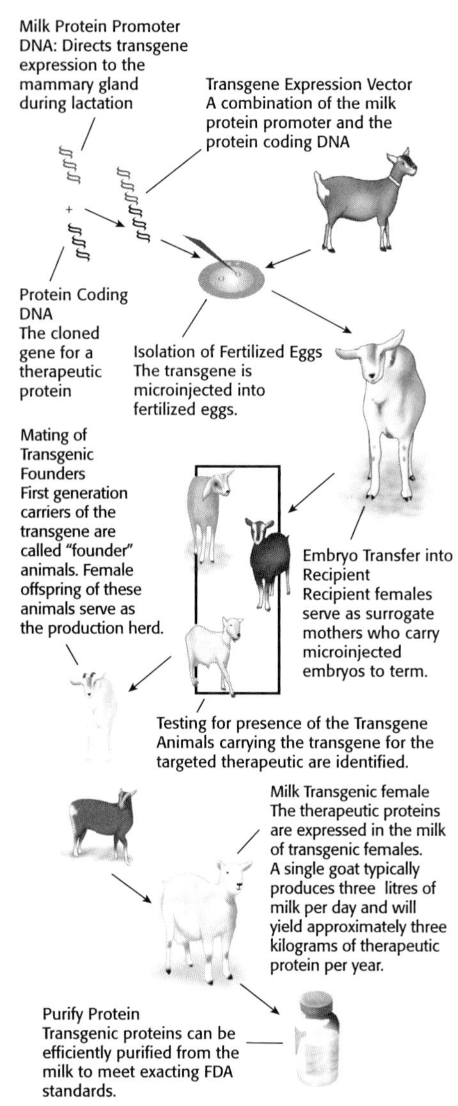

Figure 3 Procedure for genetically modifying goats so that their milk yields human protein

Thinking about science: the precautionary principle

There have been many fears expressed about the possible dangers of genetic modification. Some campaigners have argued that in the absence of certainty about risk, the precautionary principle should be followed.

The quote below is from "In Defense of the Precautionary Principle", by Carolyn Raffensperger and Katherine Barrett (Letter to the Editor, *Nature Biotechnology*, 2001).

Although there are differences in wording, three core elements are present in all statements of the precautionary principle: if there is reason to believe that a technology or activity may result in harm and there is scientific uncertainty regarding the nature and extent of that harm, then measures to anticipate and prevent harm are necessary and justifiable.

The precautionary principle is necessary and justifiable because, simply stated, our ability to predict, calculate, and control the impacts of technologies such as GM organisms is limited. The novelty and complexity associated with inserting isolated gene constructs into organisms, and releasing those organisms on a global scale demand that we acknowledge uncertainties, accept responsibility, and exercise due caution.

...the precautionary principle is neither unscientific nor anti-technology. It requires robust scientific analysis with close attention to uncertainty and to the probability of both false positive and false negative conclusions. The precautionary principle can also stimulate alternative directions for regulatory policies and technology development. Its power lies not in halting all new activities, but in heightening our attention to the potential consequences of our actions, shifting the scope of questions we ask about technologies, and finding innovative solutions to complex problems.

1 Science is an inductive process – it can never prove anything with certainty. To what extent do you accept the author's statement that the precautionary principle is consistent with the process of science?
2 The internal assessment criteria encourage us to be aware of uncertainties in the data we collect. Why?

The polymerase chain reaction

A clone is a group of genetically identical organisms, or a group of cells derived from a single parent cell. The word "cloning" can also refer to the production of a number of genetically identical DNA molecules.

The polymerase chain reaction (PCR) is a technique that can be used to "amplify" small quantities of DNA. This makes it possible to study the DNA further without the risk of using up a limited sample. For example, DNA extracted from fossils can be amplified using PCR.

PCR involves repeated cycles (see Figure 4). The first stage of the cycle involves denaturing the DNA sample (separating the two strands) using heat. The second stage involves annealing with a primer selected to match a particular target within the DNA. This is done at slightly cooler temperatures to facilitate the annealing. The final stage involves the extension of the primer using a DNA polymerase from a bacterium adapted to life at high temperatures such as *Thermophilus aquaticus*. This final stage is carried out at an intermediate high temperature. These processes takes place in a thermal cycler, which alternates between the three required temperatures.

PCR primers will not bind unless there is a complementary sequence to which the primer can bind. One test for the presence of genetically modified ingredients in foods involves the use of a primer that binds to the genetically modified DNA. Any such DNA present is amplified by the PCR, but if there is none present the PCR has no effect.

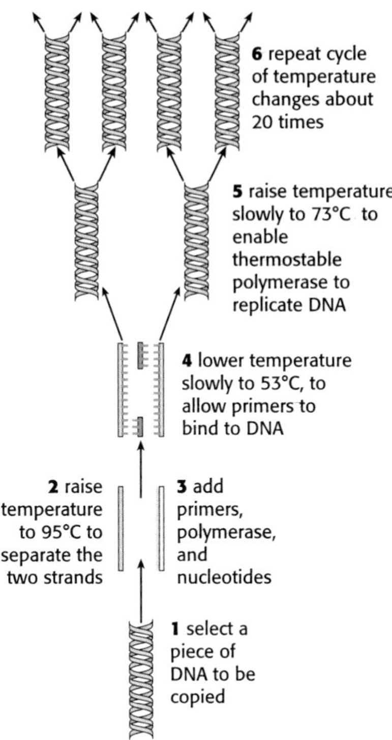

Figure 4 Stages in the polymerase chain reaction

Working with data: PCR and Neanderthals

The evolution of groups of living organisms can be studied by comparing the base sequences of their DNA. If a species separates into two groups, differences in base sequence between the two species accumulate gradually over long periods of time. The number of differences can be used as an "evolutionary clock".

Samples of DNA were recently obtained from fossil bones of a Neanderthal (*Homo neanderthalensis*). They were amplified using PCR. A section of the Neanderthal mitochondrial DNA was sequenced and compared with sequences from 994 humans and 16 chimpanzees. The bar chart in Figure 5 shows how many base-sequence differences were found within the sample of humans, between the humans and the Neanderthal and between the humans and the chimpanzees.

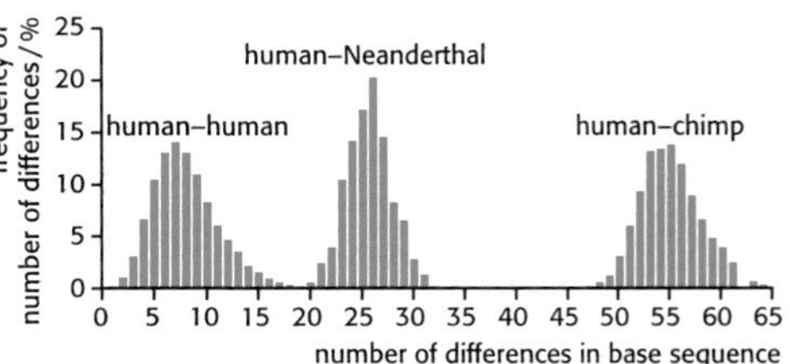

Figure 5 Number of differences in base sequences between humans, chimps and Neanderthals

1 State the most common number of differences in base sequence between pairs of humans. [1]
2 Humans and Neanderthals are both classified in the genus *Homo* and chimpanzees are classified in the genus *Pan*. Discuss whether this classification is supported by the data in the bar chart. [3]
3 Suggest a limitation to drawing any conclusion from the human–Neanderthal comparison. [1]

DNA profiling

Variations in the genetic information between individuals are found throughout the DNA, in non-coding as well as in coding areas. For example, on the short arm of chromosome 1 there is a sequence of 16 bases that is repeated a number of times. The number of repeats differs between individuals. Some people have as few as 14 repeats, while others have as many as 40. Further, the number of repeats can be different on the paternal and maternal chromosomes. Such variations exist throughout the genome and can be used to define a unique "genetic fingerprint" for each individual. Variations between the genetic information in individuals mean that the pattern of fragments produced by restriction enzymes will be unique to an individual.

Genetic fingerprints have broad applications. They are used for instance in forensic investigations, in paternity cases and in evolutionary studies. Gel electrophoresis followed by staining is a technique that allows restriction fragments of amplified DNA to be separated, visualized and analysed.

Gel electrophoresis involves separating charged molecules in an electric field. Samples are placed in wells cast in a gel. The gel is immersed in a conducting fluid and an electric field is applied. The sample will move a certain distance through the gel, according to the charge on the molecule and its size. The gel can be thought of as resisting the movement of the sample, so that small molecules will move further than molecules that are large.

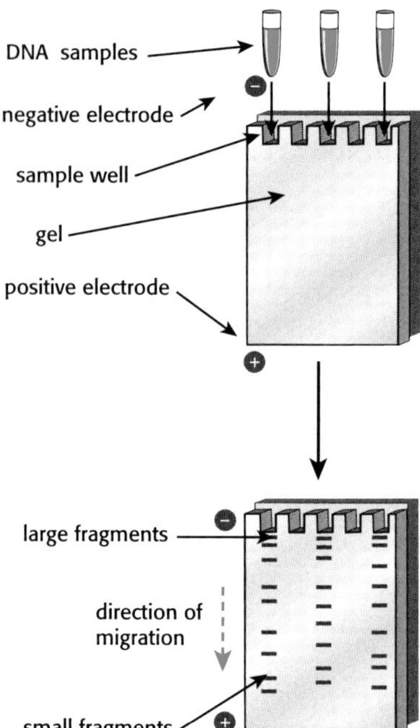

Figure 6 Procedure for creating a DNA profile

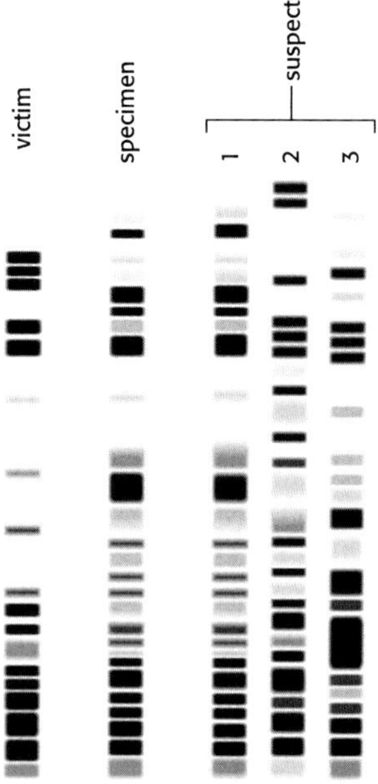

Figure 7 A DNA profie can be used as forensic evidence from a crime scene

In Figure 7, which of the three suspects' DNA fingerprints matches the specimen recovered from the crime scene?

Rediscovering biology: gel electrophoresis of dyes

Materials and equipment

- stains/dyes such as Bromophenol Blue, Safranin O, Xylene Cyanol, Orange G, Janus Green and a tube containing a dye mixture
- pipettes and pipette tips
- electrophoresis chambers
- power supplies
- 1% agarose in 1 × TAE buffer; 1 × TAE for immersing the gels.

1 Set the comb in the centre of the casting frame. Pour melted 1% agarose into the casting frame and allow the gel to harden. Remove the comb.
2 Immerse the gel in 1 × TAE buffer in an electrophoresis chamber.
3 Load each dye into a separate well in the gel. Use a separate pipette tip for each sample. One of the samples could be a mixture of dyes.
4 Attach the electrodes of the chamber to the power supply and run the gel until separation is clearly evident.

1 Identify the type of charge on each dye.
2 Which dye moved the furthest?
3 State two reasons for the different rates of movement of the dye.

Cloning

A group of genetically identical individuals is a clone. Cloning can happen naturally – asexual reproduction produces genetically identical progeny, and identical twins are genetically identical.

It is also possible to produce clones artificially. In livestock, an egg can be fertilized *in vitro* and allowed to develop into a multicellular embryo. Individual cells can be separated from the embryo while they are still **pluripotent** (capable of developing into all types of tissue) and transplanted into surrogate mothers. Only a limited number of clones can be obtained this way, because after a certain number of divisions the embryo cells are no longer pluripotent.

Another method of cloning involves the use of non-reproductive cells. The first mammal cloned from adult cells was Dolly the sheep. The cloning method used is outlined in Figure 8.

Reproductive cloning of humans using adult differentiated cells is a controversial idea. Therapeutic cloning is less controversial. It could involve production of embryonic stem cells for disease treatment. Therapeutic cloning refers to the production of embryonic stem cells from a denucleated unfertilized egg cell that has had a nucleus transplanted into it from a patient's cell or a cell from a close genetic match such as a relative. Stem cell therapies are promising for degenerative diseases like Parkinson's disease and multiple sclerosis. However if the stem cells are not genetically compatible with a person, their immune system will reject them.

Human cloning for therapeutic purposes is currently banned in many countries, but is potentially admissible in at least 13 countries.

1 Should therapies developed in societies where scientific activity is less subject to regulation be available outside of those societies?
2 Which ethical standards do scientific journals apply when choosing to publish results to an international community where variations in regulations exist?

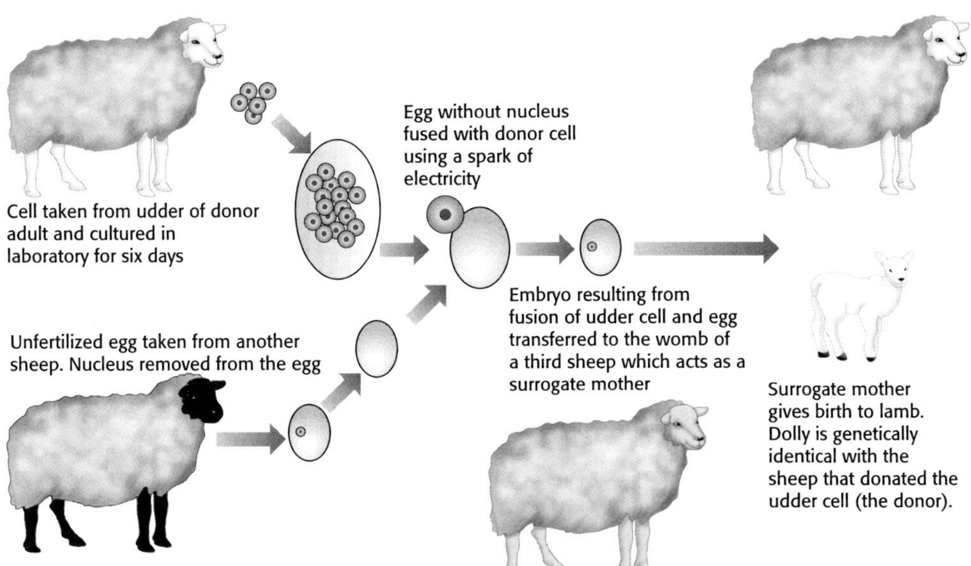

Figure 8 A method for cloning an adult sheep using undifferentiated cells

Working with data: logarithmic graphs

A logarithm is an alternative way to express an exponent. For example,

$\log 1000 = \log 10^3 = 3;$ $\quad \log 100 = \log 10^2 = 2.$

In biology, very large changes in a variable are easier to represent graphically if logarithms are used.

In the example below (Figure 10), DNA fragments are being separated using gel electrophoresis. The fragments vary in size from 100 bp (base pairs) up to 5000 bp.

Table 2

Known ladder fragment size (bp)	Distance moved (mm)
5000	58
2000	96
850	150
400	200
100	250

Figure 10 shows an electrophoresis gel. The two outside columns of the gel represent ladders, i.e. mixtures of DNA fragments of known size. These were used to obtain the data in Table 2 and create the plot shown in Figure 9. The other inner columns shown in Figure 10 are unknowns.

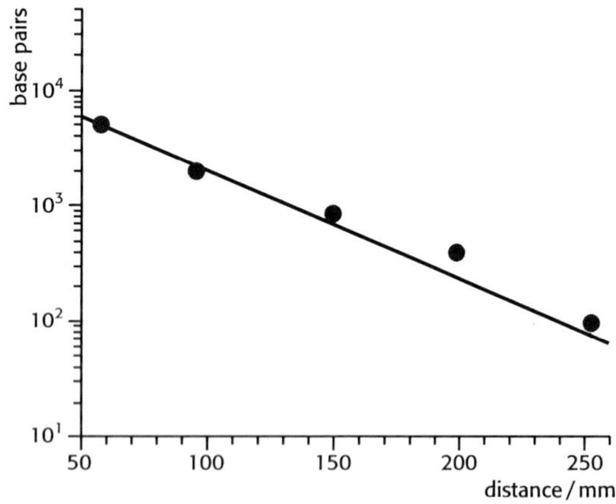

Figure 9 Distance moved as a function of fragment size in gel electrophoresis. Notice that the y-axis scale on this graph goes up in powers of 10. This is a logarithmic scale.

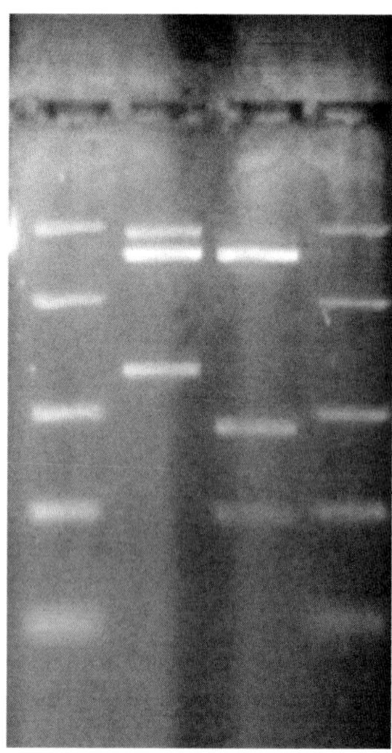

Figure 10 Gel electrophoresis. The outside columns represent ladders of known length. The two inside columns represent samples of unknown length.

1 Using Figures 9 and 10, determine the size of DNA fragments in the two centre digests:

Fragment size (bp) (column 2)	Distance moved (mm) (column 2)	Fragment size (bp) (column 3)	Distance moved (mm) (col. 3)
	60		70
	70		160
	130		200

Chapter 14 questions

1 The cheetah (*Acinonyx jubatus*) is an endangered species of large cat found in South and East Africa. A study of the level of variation of the cheetah gene pool was carried out. In one part of this study, blood samples were taken from 19 cheetahs and analysed for the protein transferrin using gel electrophoresis. The results were compared with the electrophoresis patterns for blood samples from 19 domestic cats (*Felis sylvestris*). Gel electrophoresis can be used to separate proteins using the same principles as in DNA profiling. The bands on the gel which represent forms of the protein transferrin are indicated.

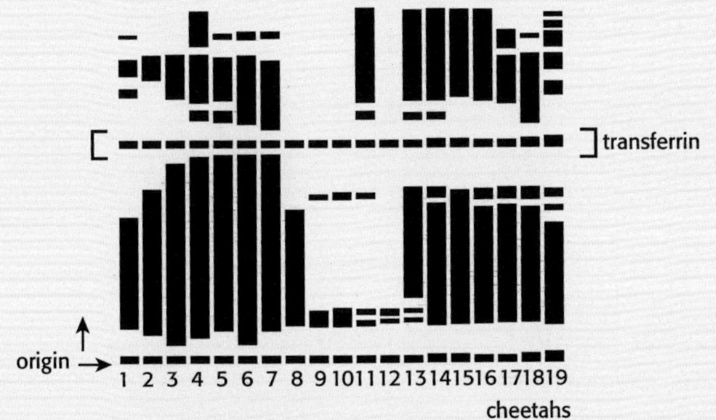

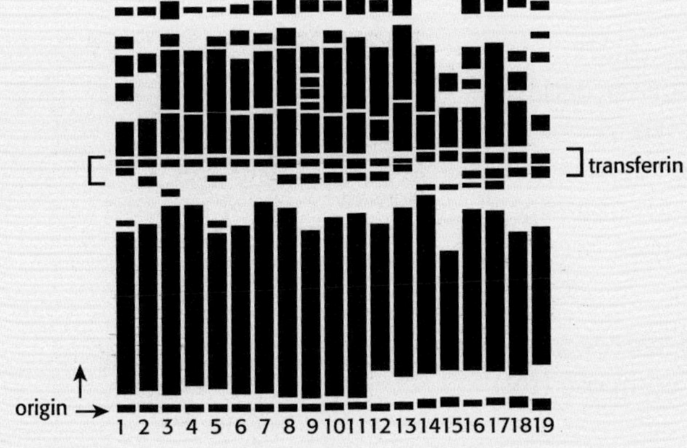

Figure 11

Using Figure 11, deduce with reasons:

(a) the number of domestic cats and the number of cheetahs that were heterozygous for the transferrin gene; [2]

(b) the number of alleles of the transferrin gene in the gene pool of domestic cats; [2]

(c) the number of alleles of the transferrin gene in the gene pool of cheetahs. [1]

2 Outline two uses of restriction enzymes in biotechnology. [4]

3 Suggest why therapeutic cloning might be opposed by advocates of women's rights. [2]

4 Compare the polymerase chain reaction with DNA replication as it occurs in cells. [3]

5 List three examples of where cloning occurs naturally. [3]

6 Figure 12 shows the results of DNA profiling of a family consisting of a man, a woman, and their four children.

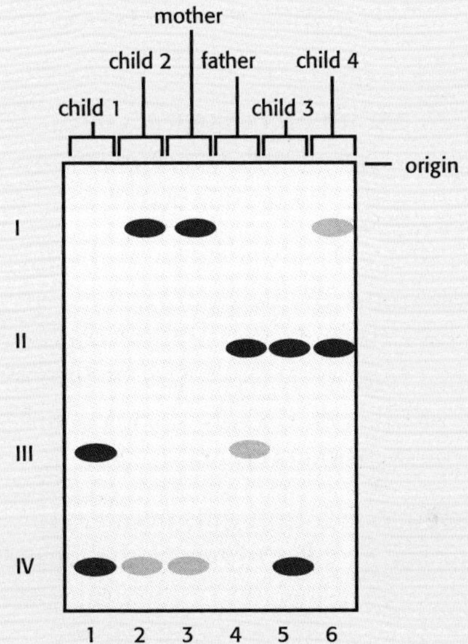

Figure 12

(a) Explain which fragment is the smallest. [2]

(b) Deduce which child is not the biological offspring of the father. [3]

7 What are functions of the polymerase chain reaction?

(i) To copy fragments of DNA.

(ii) To amplify fragments of DNA.

(iii) To translate fragments of DNA.

A (i) and (ii) only

B (i) and (iii) only

C (ii) and (iii) only

D (i), (ii) and (iii).

15 Ecology

Trophic relationships

A population is a group of individual organisms of the same species living within a certain area. A community is what emerges when different populations within a specific area interact. Individual organisms within a community interact in a number of ways. One possibility is a trophic relationship, that is, a feeding relationship. The simplest sequence along which food is transferred is called a food chain. Each organism in the sequence feeds on the organism before it. Arrows point toward the consumer, indicating the direction of nutrient and energy flow.

Trophic relationships within an ecological community tend to be complex and web-like. A food web is a representation of the complexity of feeding relationships, showing interacting food chains (see Figure 1).

An organism's trophic level is its position in the food chain. This is determined by the number of energy-transfer steps prior to that level. The terms "producer", "primary consumer" and "secondary consumer" are making reference to trophic level. The trophic level that supports all of the others is the producers. Usually these consist of photosynthetic organisms such as terrestrial green plants and phytoplankton.

Producers are referred to as autotrophs because they make their own food from inorganic sources of carbon. In the case of photosynthetic organisms, the carbon source is CO_2. Consumers are known as heterotrophs. They are organisms that obtain their carbon from organic sources. Heterotrophs occupy higher trophic levels. Animals that ingest prey, and fungi that absorb externally digested organic matter, are two examples of heterotrophs.

Detritus feeders (detritivores) such as earthworms, and decomposers such as bracket fungi, are heterotrophs that feed on or decompose (rot) dead organic matter.

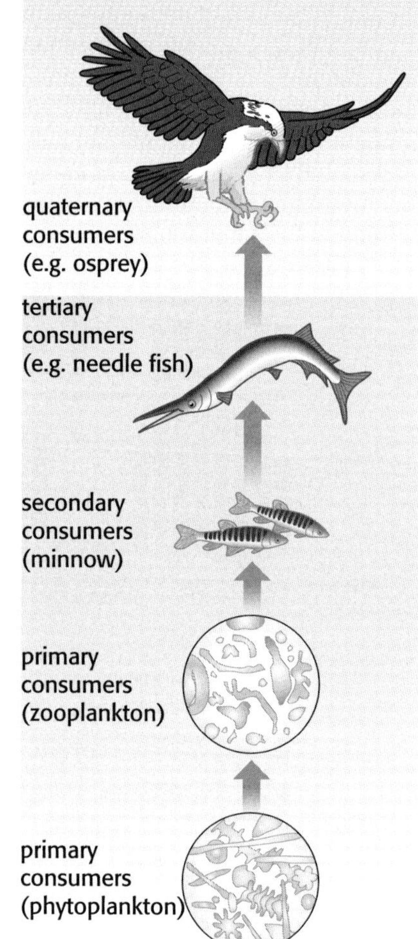

Figure 2 A marine food chain

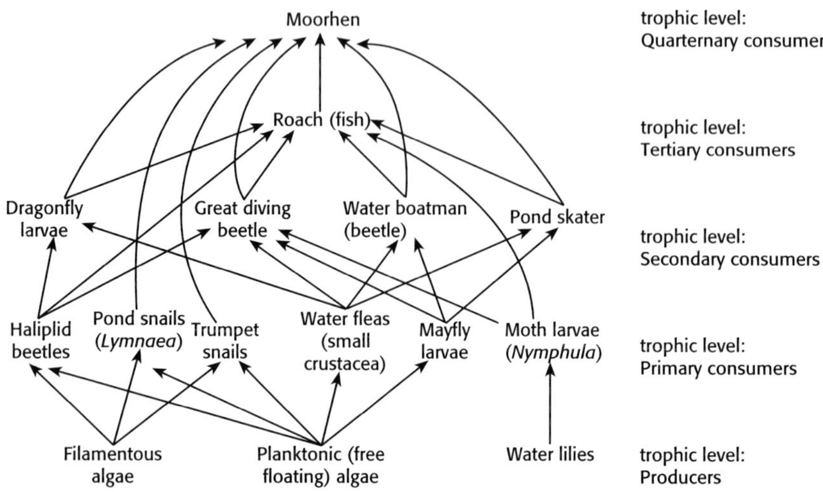

Figure 1 A pond food web

Data-based question: fishing down marine food webs

Trophic levels can be represented by a number indicating the position of a species within an ecosystem. By definition, producers occupy the first trophic level (TL) and so have a TL of 1. For primary consumers TL = 2, and so on. The larger the number, the more energy-transfer steps between the organism and the initial fixing of the sun's energy. Trophic levels are not always stated as whole numbers. Fish and other animals that feed at more than one level often have estimated mean trophic levels.

One effect of commercial over-fishing is the reduction in the number of fish that feed at higher trophic levels, such as long-lived fish. The phrase "fishing down marine food webs" refers to the increased tendency for marine landings to consist of animals that feed at lower trophic levels (see Figure 3).

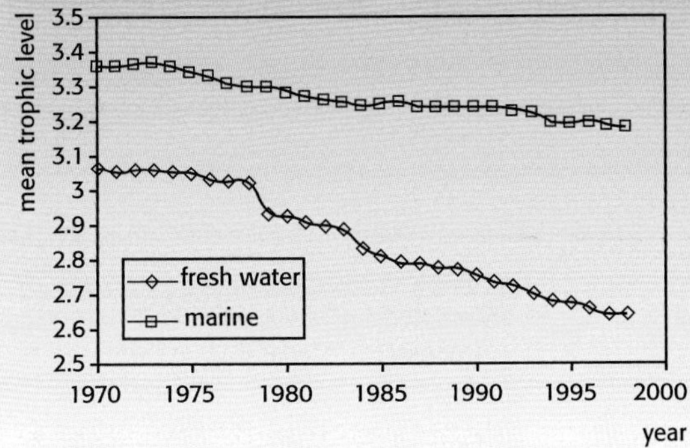

Figure 3 How the mean trophic level of landed fish has changed over a 30-year period

1 Suggest a method that might be used to deduce the trophic level of a fish once it is captured. [2]

2 (a) Compare the changes in mean trophic level of landed fish from marine and freshwater fisheries since 1970. [3]

(b) Suggest why there is a difference in the two trends. [2]

3 Explain why the mean trophic level might increase with the age of an individual fish. [2]

4 Deduce the change in age of captured fish over the period shown. [2]

5 Explain two advantages of humans catching and consuming fish at a lower mean trophic level. [4]

Ecological energetics

For most biological communities, the initial source of energy is light captured by plants undergoing photosynthesis. Light captured by plants is converted to chemical energy. A portion of this energy is used by the plant in cellular respiration and is ultimately released as waste heat to the environment. Energy stored in plant tissues can be passed to the next trophic level where plant matter is eaten by primary consumers. Alternatively, a portion of the plant material may become **detritus**, in which case stored energy can be passed on to decomposers or detritus-feeders.

The energy stored in plant matter ingested by primary consumers can be used directly as a source of energy for cellular respiration. This will ultimately result in waste heat being released. Alternatively, organic matter containing stored energy in the primary consumer can be eaten by a secondary consumer. In addition, undigested plant matter released as feces by the primary consumer contains available energy for decomposers and detritus feeders to utilize.

Energy transfer between trophic levels usually has an efficiency of between 5 and 20 per cent. Energy is passed from consumer to consumer in a food chain, but with every transformation energy is lost from the community as heat generated by respiration. Further, when an animal eats, a portion of its food is never absorbed and is egested as feces. Some material, such as bones or hair, may not be eaten. Energy contained in these kinds of material can either be utilized by decomposers and detritus feeders or it can become trapped as fossil fuels or peat.

As a result of energy losses from respiration and undigested or uneaten material, the length of food chains is limited by the loss of energy. The amount of energy converted to new biomass during a given time period by each trophic level in an ecological community can be represented by a pyramid of energy. In this case it would be most accurately called a pyramid of net production. The width of the bars is proportional to the energy in that trophic level. Figure 4 shows an example of a pyramid of energy for an aquatic ecosystem, showing the loss of energy through the food chain. To be more accurate, the boxes should be drawn to have relative widths that match the relative energy content at each trophic level.

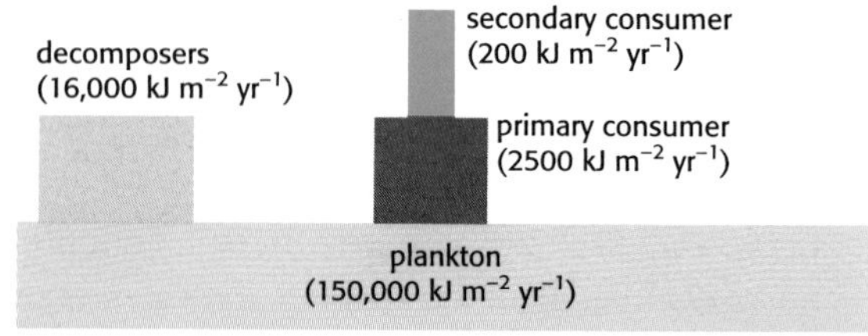

Figure 4 An energy pyramid for an aquatic ecosystem

Working with data: constructing a pyramid of energy

Table 1 shows the annual energy fixed in biomass in joules per square centimetre in each trophic level of two separate ecosystems.

1. Use the data to construct two separate pyramids of energy. They should both be drawn to the same scale.
2. Compare the two pyramids.
3. Explain the low biomass and low numbers of organisms in higher trophic levels.

Table 1

Trophic level	Cedar Bog Lake	Lake Mendota
Tertiary consumers	–	0.2
Secondary consumers	0.8	1.4
Primary consumers	3.6	35.1
Producers	27.1	104.4

Investigating biology: food web investigations

Activity A

Place a cut piece of fruit in an open container in a warm, shaded place outside. By observing the fruit over a period of time, determine which organisms obtain their energy from the fruit. Outline the fate of all of the fruit biomass over the course of the study.

Activity B

Obtain a whole, uncleaned fish from the marketplace. Dissect the fish and examine its stomach contents to determine its feeding behaviour, i.e. whether it feeds continuously or sporadically. Its trophic level may be deduced if any undigested material is present.

Activity C

Studies of leaves that have been eaten can be used to investigate trophic relationships. For example, the larvae of a diversity of insects act as leaf miners (they feed on the tissue between leaf epidermal layers, leaving characteristic tracks). Questions to be addressed could include:

1. Is there a difference in the average area of a leaf eaten by a miner versus a leaf that is not eaten on the same plant?
2. What is the proportion of leaves on a plant that have been eaten?
3. What proportion of leaves contain a larva when first examined?
4. Within a plant, are there certain areas that are preferentially mined?
5. Within a population of plants, are trees near the edge of a forest more likely to be attacked than those further from the edge?

Activity D

An owl pellet is a mass of undigested parts of the owl's diet that it regurgitates. The contents of the pellet can be used to gather information about the bird and its community without disturbing the bird. The contents might include such things as exoskeletons of insects, bones including skulls, fur and claws. Dichotomous keys might be used to identify species found within the pellet. Skeletons might be reconstructed and pasted onto a filing card.

Working with data: creating a dichotomous key

The four skulls in Figure 5 are different species of rodent that might be found in an owl pellet. Page 199 describes the procedure for creating a dichotomous key. Create a dichotomous key for the skulls, that might be used by a student to identify skulls found in an owl pellet.

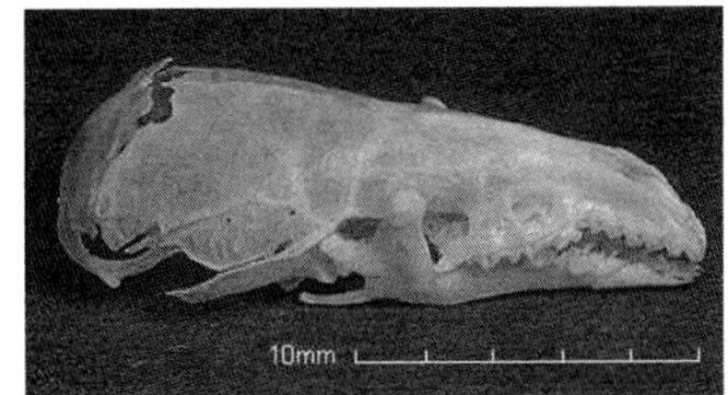

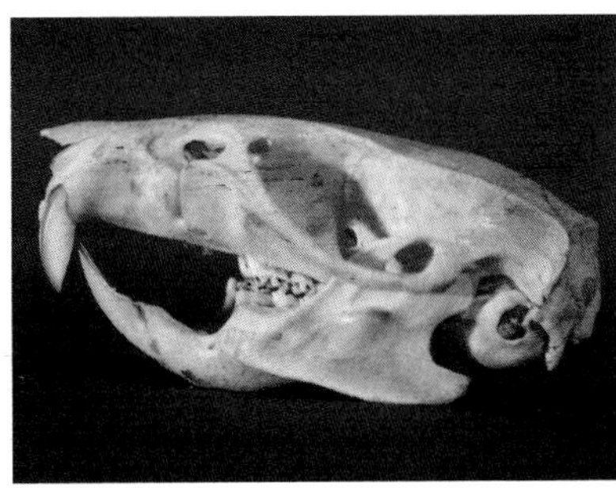

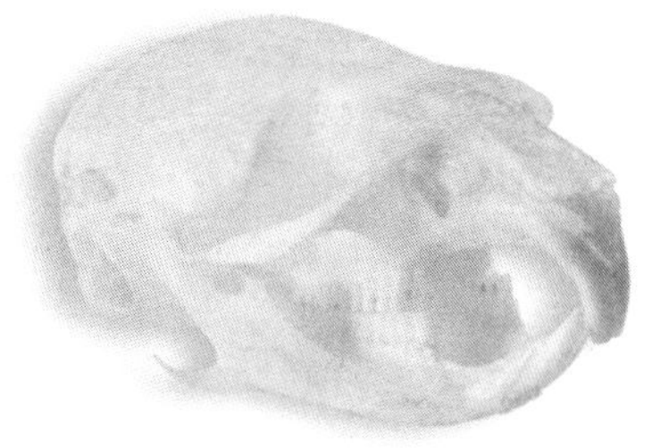

Figure 5 Rodent skulls

Data-based question: a simple food web

A sinkhole is a surface feature which forms when an underground cavern collapses. Montezuma Well in the Sonoran desert in Arizona is a sinkhole filled with water. It is an aquatic ecosystem that lacks fish, due in part to the extremely high concentrations of dissolved CO_2. The dominant top predator is *Belostoma bakeri,* a giant water insect that can grow to 70 mm in length.

Figure 6 shows a food web for Montezuma Well.

1 Compare the roles of *Belostoma bakeri* and *Ranatra montezuma* within the food web. [2]

2 Deduce, with a reason, which organism occupies more than one trophic level. [2]

3 Deduce using P values:

(a) what would be the most common food chain in this web [2]

(b) what is the preferred prey of *B. bakeri?* [1]

4 Construct a pyramid of energy for the first and second trophic levels. [3]

5 Calculate the percentage of energy lost between the first and second trophic levels. [2]

6 Discuss the difficulties of classifying organisms into trophic levels. [2]

7 Outline the additional information that would be required to complete the pyramid of energy for the third and fourth trophic level. [1]

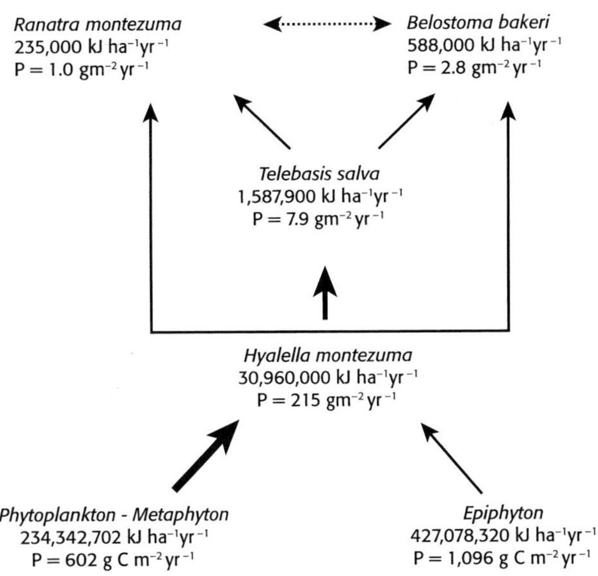

Figure 6 A food web for Montezuma Well. P values represent the biomass stored in the population of that organism each year. Energy values represent the energy equivalent of that biomass. Arrows indicate trophic linkages and arrow thickness indicates the relative amount of energy transferred between trophic levels.

Ecosystems

The biological definition of a species is a group of organisms that can interbreed and produce fertile offspring. All of the members of a species that live in a certain area comprise a population. All of the populations living and interacting in an area comprise a community. Ecological communities are influenced by non-living environmental variables such as CO_2 concentration, wind speed and light intensity. These non-living variables are called **abiotic factors**. Members of the ecological community, in turn, impact on the environment and are called **biotic factors**. An ecosystem consists of a community, its physical environment and the interactions between them. Being a system, it is more than the sum of all of its individual biotic and abiotic factors. The interactions between these elements must also be taken into account.

An organism's habitat is the place where it normally lives. Habitats are often characterized by a dominant plant form or physical characteristic. Ecology is the study of the relationships between living organisms and their environment.

Investigating biology: determining the impact of light intensity on plant variables

Light intensity is an **abiotic** factor that can influence plant phenotypes.

Some possible protocols to explore this interaction are listed below. Note that some of these require careful measurements using appropriately precise measuring procedures.

1. Compare shape ratios of leaves from a lobe-leaved tree such as an oak tree (*Quercus sp.*). Shape ratio is calculated using the formula shape ratio $= \frac{\text{sinus area}}{\text{leaf area}}$.

   To calculate leaf area, trace the leaves on centimetre-lined graph paper. Count the squares to calculate the leaf area. Calculate the sinus area by connecting the leaf lobes with straight lines and adding together areas between the dotted line and the edge of the leaf, as shown in Figure 7b.

2. Compare the specific leaf areas of outer and inner canopy leaves from a tree. Specific leaf area is leaf area per unit dry mass. Determine the surface area of a leaf. Weigh each leaf to within 0.1 g and divide the weight by leaf area.

3. Determine if mean leaf thickness is affected by light intensity. Leaf thickness can be determined using a free-standing digital microscope with a ×200 lens. Alternatively leaf thickness can be approximated by determining mass per unit surface area (Figure 2).

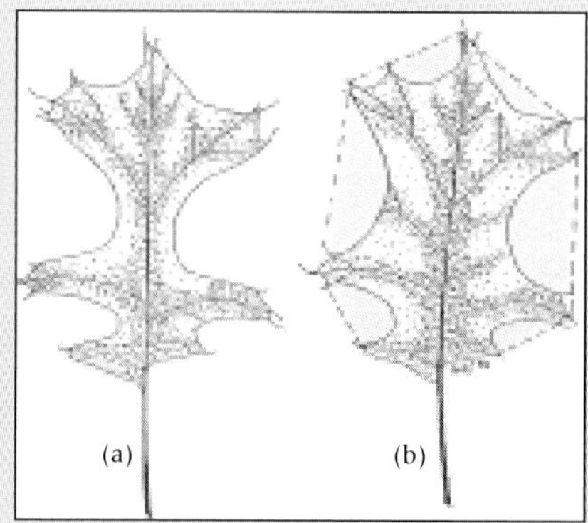

Figure 7 Light intensity can lead to changes in leaf shape.

Data-based question: temperature and rainfall

The two primary determinants of climate are temperature and rainfall. These abiotic factors are also the primary determinants of the type of terrestrial ecosystem that forms, as shown in Figure 8. The numbers in each box represent the quantity of solar energy that is fixed in the biomass in kg m^{-1} yr^{-1}.

1. Determine the range of temperatures over which tundra will exist. [2]
2. Outline the conditions necessary for the formation of a boreal conifer forest. [2]
3. Predict the possible effects if the mean annual temperature of a Northern ecosystem were to change from 0 to 3 °C. [2]

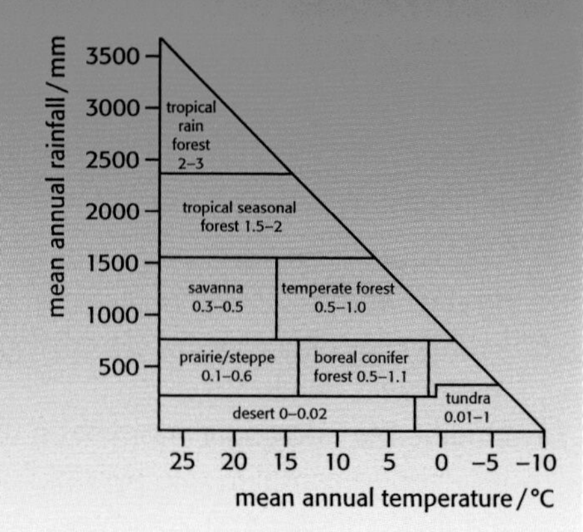

Figure 8 A climograph

Population growth

Population studies often focus on variables such as population size, density, growth and the interaction of the population with the biotic and abiotic factors of the habitat it occupies.

Figure 9 shows the population growth of a culture of the unicellular organism *Paramecium aurelia,* kept in controlled conditions, including a constant supply of food. The graph illustrates a pattern called the sigmoid, or S-shaped, growth curve.

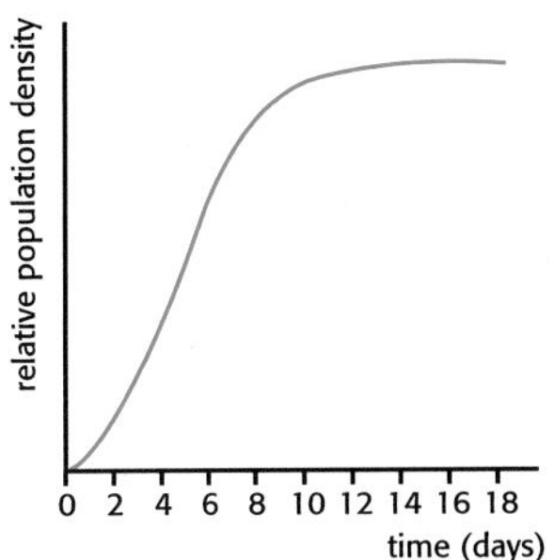

Figure 9 Population growth of a culture of *Paramecium Aurelia*

The S-curve is representative of what happens when a population colonizes a new habitat. With limited environmental resistance, a population will grow exponentially. At this stage birth rate (natality) is higher than death rate (mortality).

As population density increases, various density-dependent factors begin to limit population growth. Examples of such limiting factors include competition for resources, a build-up of the toxic by-products of metabolism, an increase in predation or an increase in the incidence of disease. The initial result is that natality slows in relation to mortality. This is the transition phase on the curve.

The maximum size of a population that an environment can support is its carrying capacity. In the sigmoid growth pattern, when a population reaches its carrying capacity, the population will stop growing and natality and mortality will be equal. This is referred to as the plateau phase of the S-curve.

Some populations can overshoot the carrying capacity of the environment. The result is a "boom-and-bust" pattern. In 1911, for example, 25 reindeer were released on St Paul, an island in the Bering Sea off the coast of Alaska. No major predators lived on the 106 km^2 island. The initial herd was composed of 21 females and 4 males. Figure 10 shows that the population grew exponentially to about 2000 animals in 1938. The herd severely overgrazed their habitat, which led to a population collapse. By 1950 the herd had been reduced to just eight reindeer.

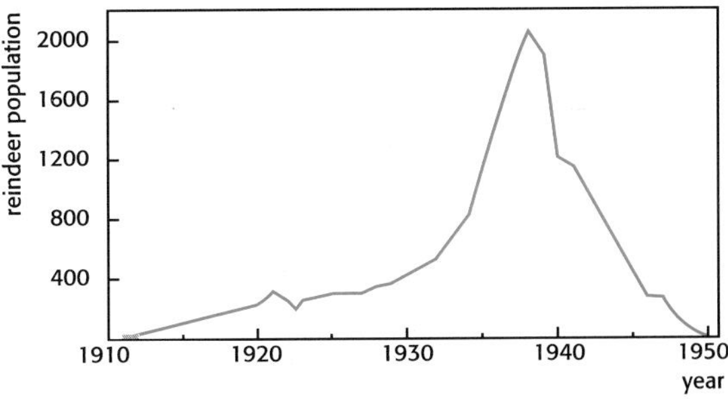

Figure 10 Reindeer population on St Paul Island

Data-based question: population growth of pheasants

Figure 11 shows the growth of a population of ring-necked pheasants (*Phasianus colchicus*) on Protection Island off the north-west coast of the USA. The original population released by the scientists consisted of two male and eight female birds. Two of the females died immediately after release.

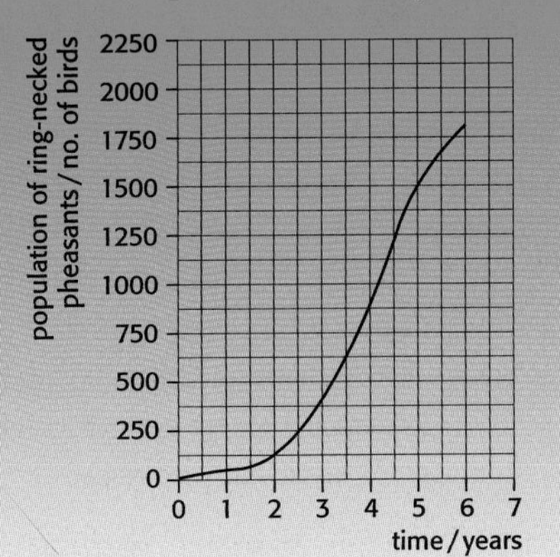

Figure 11 Ring-necked pheasant population on Protection Island

1. The scientists predicted that the population would reach its carrying capacity of 2000 by year 8. Suggest two factors that could limit the population increase between years 8 and 10. [2]
2. Predict, with a reason, how the population growth would have been different if all the female birds in the original sample had survived. [2]

Critical considerations: ten myths of human population growth

Below are excerpts from an article in *Discover*, April 1996, by Joel E. Cohen.

Population problems are entangled with economics, the environment, and culture in such complex ways that few people can resist the temptations of unwarranted simplification. The result is a loose and widely accepted collection of myths, for example:.

A single factor limits how many people Earth can support. *This myth has a long, distinguished history. In 1679, Antonie van Leeuwenhoek, the inventor of the microscope, estimated how many people the planet could support. He assumed that what limited Earth's population was population density alone – that is, the number of people per unit of land area. He further assumed that Earth could not be more densely inhabited than the Holland of his day, which had an estimated 1 million people at a density of around 300 per square mile… Therefore, he concluded, the planet could support at most 13.4 billion people. In 1989 a third of the world population lived at densities greater than 300 people per square mile…*

Another belief is that what limits global population is the availability of food. *In fact, except for people who are actually starving, humans today do not have more or fewer children according to whether they have more or less food. On the contrary, the average number of children per woman is lowest in the rich countries where food is most abundant.*

Since Leeuwenhoek, some 65 estimates of how many people Earth can support have been published, using a wide range of limiting factors – everything from food to land to freshwater, phosphorus, photosynthesis, fuel, nitrogen, waste removal, and human ingenuity. The estimates have ranged from fewer than 1 billion to more than 1 trillion, and in the past few decades they have grown increasingly divergent. But there are a number of problems with all these studies. Many of the factors are not independent of one another. Everything affects everything else. Most important, many limiting factors are subject to changing cultural values. If a peasant farmer in Kenya believes that educating her children matters greatly, and if school fees begin to rise, then she may choose to have fewer children not because land is scarce but because she values her children's future more than their labor as farmhands.

Questions for discussion

1 What are myths?
2 What are the differences between theories and myths as forms of explanation?
3 Limiting factors are those variables that define the carrying capacity of an environment for a particular species.
   (a) What are the features of a limiting factor?
   (b) What are limiting factors that might determine the carrying capacity of a sub-region of the planet?
   (c) What are limiting factors that might determine the carrying capacity of the planet?

Investigating biology: modelling population growth

1 In a series of pots, plant increasing numbers of seeds, starting first with one, then two, then three etc. Determine the total wet biomass per pot. Graph biomass versus number of seeds. Explain the pattern seen. To what extent is this a good model of population growth? If these plants are allowed to flower, what is the effect of planting density on the number of seeds produced or the timing of seed production?
2 Duckweed (*Lemna* sp.) are stemless water plants (see Figure 12). Each plant grows one to four leaf-like structures called thalli. Duckweeds reproduce asexually by growing new thalli from older thalli and when they reach a certain size, they break free from the parent plant. A number of experiments are possible:

What is the carrying capacity of a given container? What conditions of light, nutrients or container surface area are ideal for population growth?

Figure 12 Duckweed thalli

Thinking about science: using bird census data

Volunteer organizations often undertake censuses of bird populations. The Audubon Society in North America organizes an annual Christmas bird count. Some of these censuses have been occurring for years. The extracts below are from Nickens, T. Edward "The case of the disappearing warbler". *National Wildlife, 2006.*

Zigzagging up the tree, draped in its drab winter plumage, the cerulean warbler is difficult to spot. Across its range, however, from the wooded ridges of Ohio to here on the slopes of the northern Andes, Dendroica cerulea, *the "blue tree dweller", is finding itself under increasing observation. Populations of the bird have declined about 3 per cent a year, "faster than any other migratory warbler in eastern North America," reports Jason Jones, a biologist at Vassar College.*

The warbler is among a suite of birds known as Neotropical migrants, species that breed in Canada or the United States yet spend much, if not most, of their lives in Mexico, Central or South America or the Caribbean. About half the continent's birds fall into this group, including 52 species of warblers, backyard birds like scarlet tanagers, ruby-throated hummingbirds and rose-breasted grosbeaks, even raptors such as broad-winged hawks and Mississippi kites.

By analyzing data from 47 Breeding Bird Census sites collected between 1937 and 1995, biologists David King of the U.S. Forest Service Northeastern Research Station and John Rappole of the Smithsonian Institution found that 60 per cent of the ten most frequently encountered mature-forest, long-distant migrants had decreased over that period with no migrants posting population increases and only 40 per cent showing stable numbers. Previous estimates, from U.S. Breeding Birds Survey data, were based almost entirely on annual counts of birds singing by roadsides conducted only since 1965. In their report, King and Rappole single out winter habitat loss as one likely culprit. "By the middle of the twentieth century," explains King, "tropical deforestation was already very advanced. It's possible that a lot of the bird population declines had already happened by the time the Breeding Bird Survey was initiated."

Working with data: survivorship curves

Find headstones in a cemetery where people were born in the 19th century. Research the date that the cemetery was first used. From the headstones, determine birth dates, death dates and age at death. Often only two of the three data points are available. Reject headstones for people who were born before the cemetery was first used.

You are interested in a quantity called survivorship, and your aim is to construct a survivorship curve, that is, survivorship vs age at death. The formula for survivorship is total no. surviving/original no. × 1000.

Tables 2 and 3 give survivorship data for a sample of headstones in St James cemetery, Toronto, from 1850 to 1879, and provide a model for how to organize your data. Remember to include the point (0, 1000) in the curve.

The sample size should be as large as possible to get a smoother curve. Students could pool data into a class database.

Students could compare local survivorship patterns with data saved in

?

1 Data mining has been defined as "the non-trivial extraction of implicit, previously unknown, and potentially useful information from data". Is data that has been directly generated by an experimenter more trustworthy than data that has been extracted from a database? What are the advantages and dangers of data mining?

2 To what extent is the investigation carried out by King and Rappole similar to an investigation in history?

Table 2

Birth date (year)	Death date (year)	Age at death (years)
1850	1901	51
1851	1942	91
1852	1881	29
1852	1884	32
1852	1900	48

Table 3

Age at death (years)	No. of original group surviving	Survivorship
5	21	954
10	20	909
27	18	818
27	18	818
28	17	772
29	16	727
31	15	681
32	14	636
33	13	590
43	12	545
48	11	500
50	10	454
51	6	272
51	6	272
51	6	272
51	6	272
53	4	181
63	4	181
81	3	136
86	2	90
91	1	45
95	0	0

Chapter 15 questions

1 Distinguish between the words within each of the following pairs of terms:
(a) natality and mortality [2]
(b) ecosystem and community [3]
(c) heterotroph and autotroph. [3]

2 The total solar energy received by a grassland is 5×10^5 kJ m^{-2} yr^{-1}. The net production of the grassland is 5×10^2 kJ m^{-2} yr^{-1} and its gross production is 6×10^2 kJ m^{-2} yr^{-1}. The total energy passed on to primary consumers is 60 kJ m^{-2} yr^{-1}. Only 10 per cent of this energy is passed on to the secondary consumers.
(a) Calculate the energy lost by plant respiration. [2]
(b) Construct a pyramid of energy for this grassland. [3]

3 Figure 13 shows the energy flow through a temperate forest. The energy flow is shown per square metre per year (kJ m^{-2} yr^{-1}).

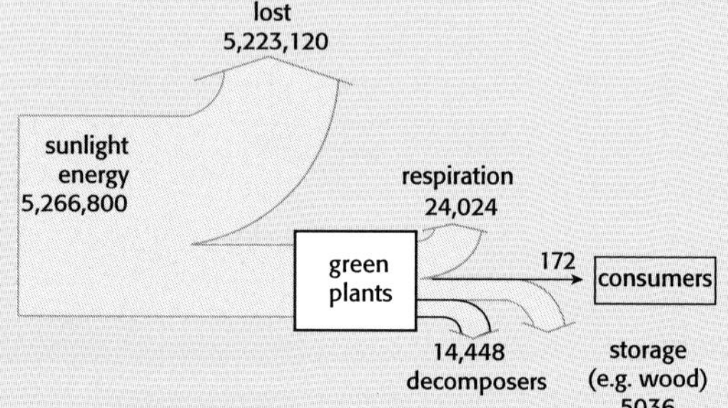

Figure 13

(a) The chart shows that 99.17 per cent of the sunlight energy in the temperate forest is lost. Predict with a reason whether a greater or lesser percentage of sunlight energy would be lost in desert. [2]

(b) Only a small part of the net production of plants in the temperate forest passes to herbivores. Explain the reasons for this. [2]

4 Cormorants (*Phalocrocorax*) are large birds which eat fish, including trout (*Salmo trutta*), from lakes and rivers. Between 1970 and 1991 the population of cormorants in Switzerland increased very greatly. During the same period there was a decrease in the number of trout caught by fishermen. The cormorants were blamed for the decreased catch. Figure 14 shows the biomass of trout caught by the fishermen and the estimated biomass caught by the cormorants in a 17 km river between Lake Walenstadt and Lake Zurich.

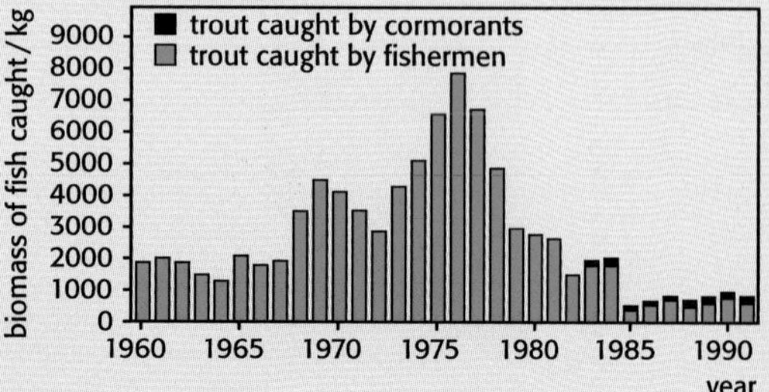

Figure 14

(a) State the year in which there was the greatest increase in biomass of trout caught by fishermen, compared with the previous year. [1]

(b) Using the data from the bar chart, discuss whether cormorants caused the decrease in the number of trout caught by the fishermen. [3]

5 Figure 15 shows an ocean food web. The arrows indicate the energy flow.

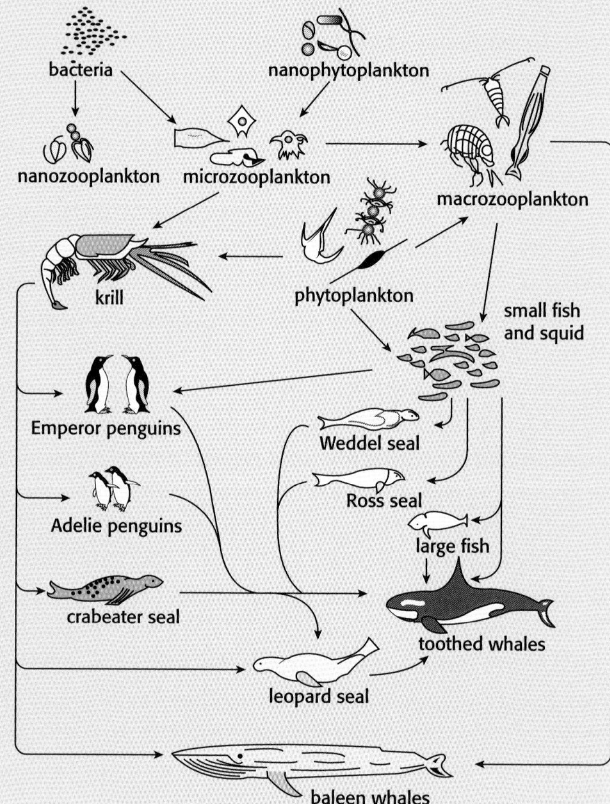

Figure 15

(a) State the trophic level of nanophytoplankton. [1]

(b) Identify in the food web:
(i) the shortest food chain [1]
(ii) the longest food chain [1]
(iii) the organism with the most predators [1]
(iv) the organism with greatest number of different types of prey. [1]

The carbon cycle, the greenhouse effect and climate change

Unlike energy, which flows through an ecosystem and must be constantly replenished, nutrients cycle within an ecosystem. This suggests a crucial role for organisms such as detritus feeders and decomposers which release nutrients trapped in detritus through feeding and rotting. In some ecosystems, trapped nutrients are released by periodic fires.

Data-based questions: nutrient cycling

In an experiment to investigate the cycling of phosphorus in an ecosystem, radioactive phosphorus (^{32}P) was introduced into an aquarium (figure 1).

1 (a) State which group of organisms was the first to absorb the ^{32}P. [1]

(b) Outline the changes in ^{32}P in the attached algae over the course of the experiment. [2]

(c) Explain why increasing quantities are found in the sediment. [3]

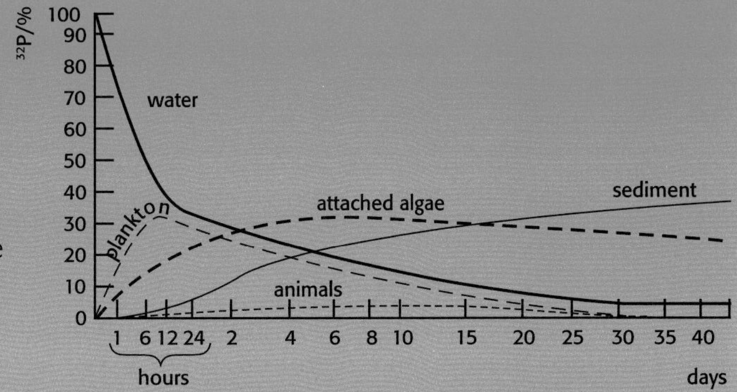

Figure 1 Movement of radioactive ^{32}P through an ecosystem

2 Within an ecosystem, nitrogen can be stored in one of three organic matter compartments: above ground, in roots and in the soil. Figure 2 shows the distribution of nitrogen in the three organic matter compartments for each of six major biomes.

(a) Deduce what the "above ground" compartment consists of in an ecosystem. [1]

(b) State which biome has the largest "above ground" compartment. [1]

(c) Explain why it is difficult to grow crops in an area where equatorial forest has been cleared of its vegetation. [2]

(d) State the name of the process carried out by decomposers and detritus feeders that releases CO_2 into the atmosphere. [1]

(e) Suggest why most of the nitrogen in a tundra ecosystem is in the soil. [1]

(f) Explain why warming due to climate change might cause a release of CO_2 from tundra soil. [2]

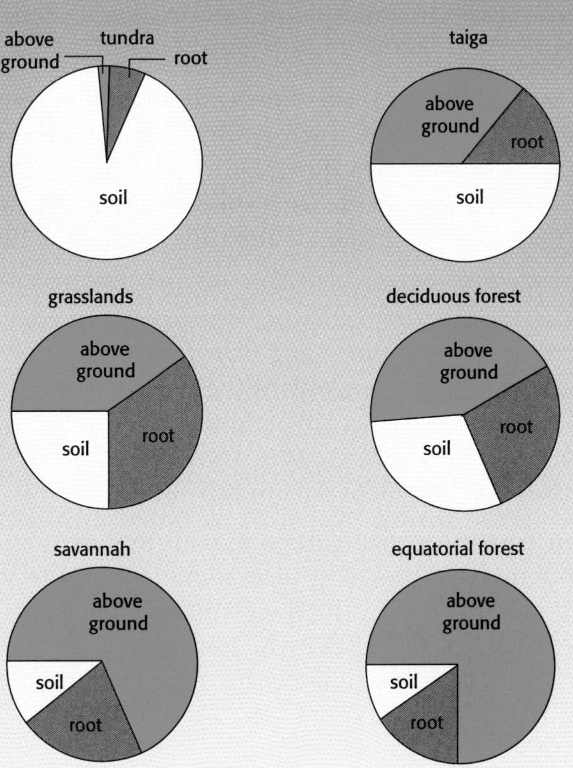

Figure 2 The distribution of nitrogen in the three organic matters compartments for each of six major biomes

The carbon cycle

Figure 3 shows the global carbon cycle.

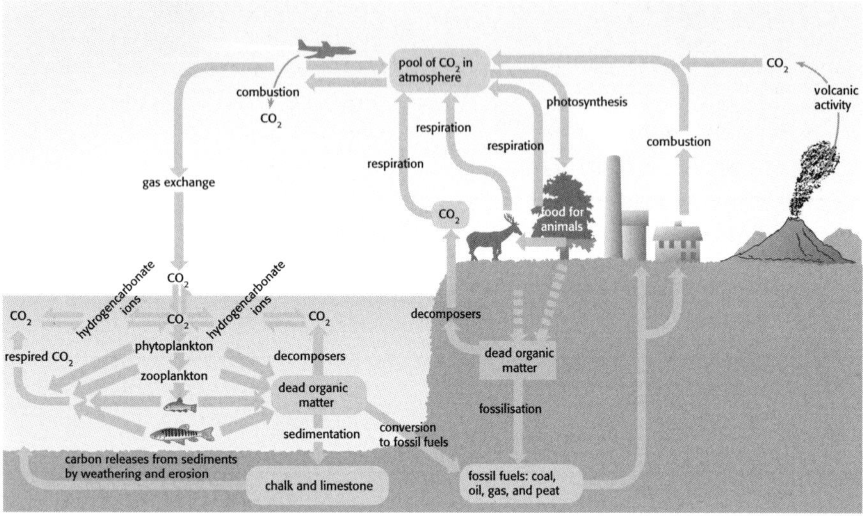

Figure 3 The global carbon cycle

1 A carbon pool is a system that has the capacity to release or store carbon. List three examples of carbon pools. [2]
2 A carbon sink is a carbon pool that has more carbon flowing into it than out of it. Suggest two carbon sinks. [2]
3 Deduce, giving a reason, whether all of the primary consumers in an ecosystem represent a carbon sink. [2]
4 Carbon flux is the flow of carbon from one carbon pool to another. List three processes in Figure 3 which contribute to carbon flux. [3]
5 Starting with the pool of CO_2 in the atmosphere, draw three possible cycles a single carbon atom could follow. One of the drawings should illustrate a cycle that includes a large number of pools and processes. [3]

CO_2 in the atmosphere

Figure 4 shows data gathered from analysing atmospheric gas bubbles trapped in ice from Greenland. Columns of ice were sampled. The ice has built up over thousands of years, so ice from deeper down is older than the ice near the surface. The graphs show long-term changes in atmospheric carbon dioxide concentration and temperature.

Figure 5 (see page 179) shows how carbon dioxide levels in the atmosphere have changed since 1958, when measurements began at Mauna Loa observatory in Hawaii. Comparison of current CO_2 levels with the levels shown in Figure 4 shows that current atmospheric levels of CO_2 are higher than at any time in the last 220,000 years.

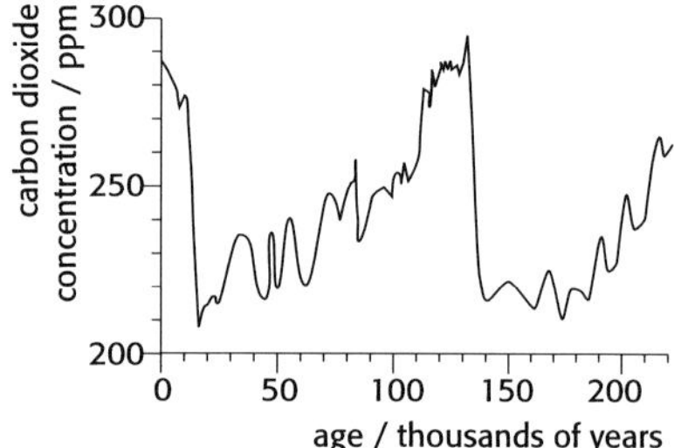

Figure 4 Atmospheric temperature and carbon dioxide concentration over the last 220,000 years

The greenhouse effect

In a greenhouse, light enters and warms up the surfaces within the greenhouse, which in turn warms the air. The greenhouse glass prevents the heat from escaping and the greenhouse warms up. A similar series of events happens in an automobile with its windows closed, when it has been parked in full sunlight.

Greenhouse gases, such as carbon dioxide and methane, are those that absorb infrared radiation (heat radiation) and re-emit it, some of it towards the Earth. The resulting global warming, called the "greenhouse effect", makes the Earth habitable. Without this greenhouse effect, it is estimated that the mean temperature at the Earth's surface would be about −18 °C!

1 To what extent is there a correlation between carbon dioxide levels and temperature change in the past 220,000 years? [2]
2 State the current concentration of CO_2 in the atmosphere (use Figure 5). [1]

Human influence on greenhouse gases

Of concern is the "enhanced greenhouse effect" caused by human additions of greenhouse gases to the atmosphere, mostly through fossil fuel burning.

The data in Figure 5 shows how the atmospheric levels of CO_2 have grown since 1958. Levels of CO_2 in the atmosphere were on an upward trend even before this date, due to the increase in industrialization that began in the 19th century. The annual fluctuation in carbon dioxide shown in the Mauna Loa data is caused by seasonal variations in carbon dioxide uptake by land plants, particularly in northern hemisphere forests. This annual cycle is shown in the inset of Figure 5.

Figure 6 shows variations in average near-surface temperatures since 1850. The historical correlation between temperature and CO_2 levels that seems evident in Figure 4 is also evident when Figures 5 and 6 are considered together. Increases in greenhouse gases in the recent past are correlated with recent increases in global average temperatures.

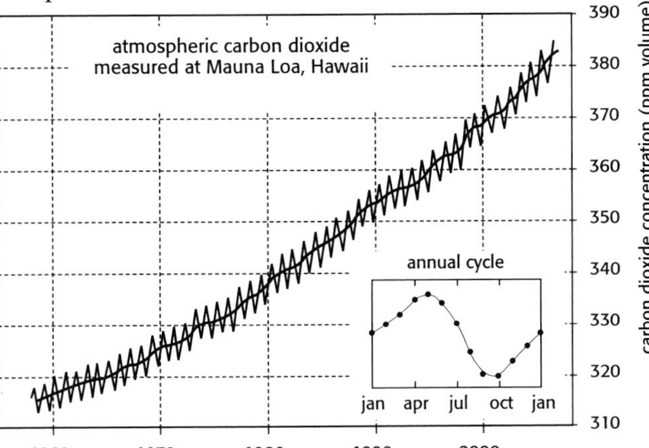

Figure 5 Growth in the level of atmospheric CO_2 since 1958

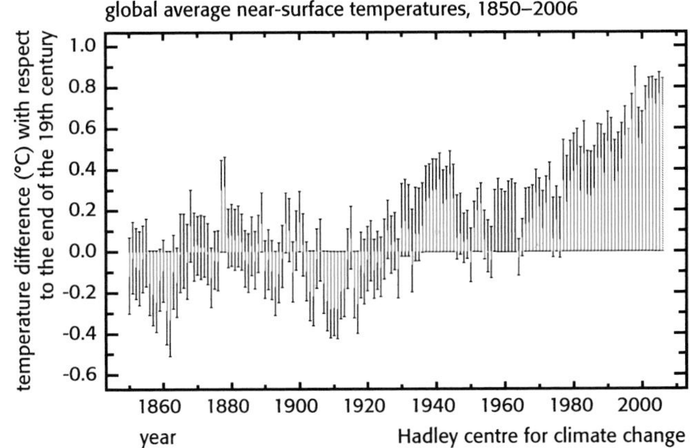

Figure 6 Global average near-surface temperatures since 1850. The values are shown as differences from average global temperatures at the beginning of the 20th century.

1 Deduce from the data in the inset of Figure 5 whether the contribution to the seasonal variation is more significant from Northern or Southern hemisphere forests.

2 Predict the effect on the seasonal variation of significant deforestation in temperate forests.

Rediscovering biology: modelling the greenhouse effect

Fill a flask with water. Displace the water in the flask with carbon dioxide, either using bottled gas or by collecting the gas generated from a chemical reaction such as sodium bicarbonate and vinegar. The water can be displaced by inverting the flask in a pan and then displacing the water with the CO_2 gas.

Stopper the flask with a one-holed stopper. Insert either a data-logging thermometer or a standard laboratory grade thermometer through the hole. As a control, set up the same apparatus but with a flask that contains just air.

Position both flasks an equal distance from a 100-W light bulb and then record the changes in temperature over a 10-minute period. Compare how the temperature changes in the two flasks over time.

This experiment could be extended over one or more 24-hour cycles by using a timer to switch a light on and off to simulate night and day.

The impact of global warming on Arctic ecosystems

It is reasonable to predict that due to global warming, climate change should occur. The inevitable result will be changes in the distribution of ecosystem types. Figure 7 shows the same location in Alaska, showing the transition from tundra (a) to wetlands (b) that has occurred over the last 20 years, presumably due to global warming.

Figure 7 This Alaskan ecosystem has undergone a transition from tundra to wetland.

Data-based question: phenology

Phenologists are biologists who study the timing of seasonal activities in animals and plants, such as the opening of tree leaves and the laying of eggs by birds. Data such as these can provide evidence of climate changes, including global warming.

The date in the spring when new leaves open on horse chestnut trees (*Aesculus hippocastaneum*) has been recorded in Germany every year since 1951. Figure 8 shows the difference between each year's date of leaf opening and the mean date of leaf opening between 1970 and 2000. Negative values indicate that the date of leaf opening was earlier than the mean. The graph also shows the difference between each year's mean temperature during March and April and the overall mean temperature for these two months. The data for temperature was obtained from the records of 35 German climate stations.

1 Identify the year in which:
   (a) the leaves opened earliest [1]
   (b) mean temperatures in March and April were at their lowest. [1]
2 Use the data in the graph to deduce the following:
   (a) the relationship between temperatures in March and April and the date of opening of leaves on horse chestnut trees. [1]
   (b) whether there is evidence of global warming towards the end of the 20th century. [2]

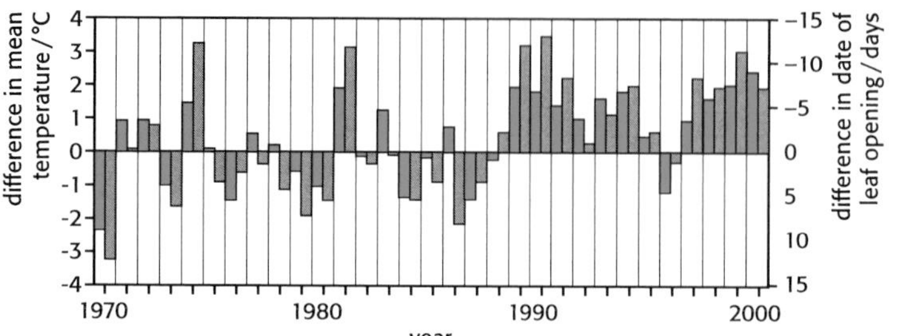

Key:
- temperature
- leaf opening

Figure 8 The relationship between temperature and horse chestnut leaf opening in Germany since 1951

Critical considerations: evaluating the evidence

Claims about the effects of global warming have sometimes been disputed and distrusted. One consequence of global warming, increasingly reported in scientific journals, is a shift in the range occupied by a species; i.e. a change in latitude and altitude of a species' habitat. Table 1 summarizes some of these reported range shifts.

Choose one of the journal articles cited in Table 1, or any journal article that discusses ecological impacts due to climate change, and obtain it from the local university library or from the Internet.

1 Summarize the article.

2 Evaluate the article against the following criteria:

(a) Consistency with other published reports. One method of doing this is to read the article's introduction, which frames the article in the context of research in the area.

(b) Frequency of citation by other scientists. This is a measure of the credibility of the article, though some of the citations could represent critiques. One method of doing this is to carefully enter the citation for the article (as it would appear in a bibliography) into a web search engine. Such engines will usually provide a count of the number of other journal articles that cite the article.

(c) Consistency between data and conclusions drawn. Read the article carefully and determine whether there are other possible interpretations of the results than those given by the article.

3 Identify any other criteria that could be used to critique a journal article.

Table 1.

Recent latitudinal and altitudinal range shifts			
Species	**Location**	**Observed changes**	**Citation**
Treeline	New Zealand	Advancement towards higher altitudes.	Wardle, P. *et al., NZ J. Bot.* **30**, 303 (1992).
Arctic shrub vegetation	Alaska	Expansion of shrubs in previously shrub-free areas.	Sturm, M. *et al., Nature* **411**, 546 (2001).
Alpine plants	European Alps	Elevational shift of 1–4 m per decade.	Grabherr, G. *et al., Nature* **369**, 448 (1994).
Antarctic plants and invertebrates	Antarctica	Distributional changes.	Kennedy, A. D. *et al., Ecol. Syst.* **26**, 683 (1995).
Zooplankton, intertidal invertebrate and fish communities	North Atlantic	Increasing abundance of warm-water species.	Southward, A. J. *et al., J. Therm. Biol.* **20**, 127 (1995).
39 butterfly species	North America and Europe	Northward range shifts.	Parmesan, C. *et al., Nature* **399**, 579 (1999).
Red fox (*Vulpes vulpes*), Arctic fox (*Alopex lagopus*)	Canada	Northward expansion of red fox range and simultaneous retreat of Arctic fox range.	Hersteinsson, P. *et al., Oikos* **64**, 505–515 (1992).
12 bird species	Britain	18.9 km average range movement northwards over a 20-year period.	Thomas, C. D. & Lennon, J. J., *Nature* **399**, 213 (1999).

Critical considerations: the new uncertainty principle

The following excerpts are from an article by David Appell that appeared in *Scientific American* in January 2001.

Observe before you project yourself on a parabolic trajectory. The weight of 28.35 grams of prevention is worth 454 grams of cure. Science certainly has much to say on taking precautions. But for the enormously complex and serious problems that now face the world – global warming, loss of biodiversity, toxins in the environment – science doesn't have all the answers, and traditional risk assessment and management may not be up to the job. Indeed, given the scope of such problems, they may never be.

Given the uncertainty, some politicians and activists are insisting on caution first, science second. Although there is no consensus definition of what is termed the precautionary principle, one oft-mentioned statement, from the so-called Wingspread conference in Racine, Wis., in 1998 sums it up: "When an activity raises threats of harm to human health or the environment, precautionary measures should be taken even if some cause and effect relationships are not fully established scientifically."

In other words, actions taken to protect the environment and human health take precedence. Therefore, some advocates say, governments should immediately ban the planting of genetically modified crops, even though science can't yet say definitively whether they are a danger to the environment or to consumers.

Critics assert that the principle's definition and goals are vague, leaving its application dependent on the regulators in charge at the moment. All it does, they allege, is stifle trade and limit innovation. "If someone had evaluated the risk of fire right after it was invented, remarked Julian Morris of the Institute of Economic Affairs in London, they may well have decided to eat their food raw."

In an October [2000] speech at the National Academy of Sciences in Washington, D.C., New Jersey governor Christine Todd Whitman averred that "policymakers need to take a precautionary approach to environmental protection.... We must acknowledge that uncertainty is inherent in managing natural resources, recognize it is usually easier to prevent environmental damage than to repair it later, and ***shift the burden of proof away from those advocating protection toward those proposing an action that may be harmful."***

1 In some jurisdictions, a relatively conservative protocol has emerged for approval of drugs so that they become available far later than in other jurisdictions. Patient advocacy groups often exert pressure for the process to be expedited. Tests and trials make drugs less risky, but the risk is never removed entirely. Urging that drugs be made available earlier is equivalent to urging that greater levels of risk be accepted. Can it be argued that there is a scientific standard for acceptable levels of risk? Can science ever prove that something poses no risk?

2 The potential consequences of climate change are so significant, some would argue that government policies should involve strong action. Ironically, this strong action itself has associated risk. Some argue that it might threaten economic development in less developed countries. Is it possible to balance environmental risks with socio-economic and livelihood risks, or do priorities need to be established?

Chapter 16 questions

1 Figure 9 shows monthly average carbon dioxide concentrations for Baring Head, New Zealand and Alert, Canada.

(a) Suggest why scientists have chosen such areas as Mauna Loa, Baring Head and Alert as the locations for monitoring stations. [1]

(b) Compare the trends illustrated in both graphs. [2]

(c) Explain why the graphs show different patterns. [3]

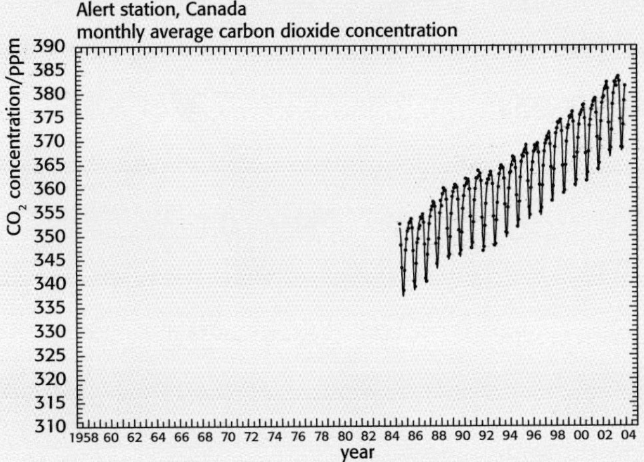

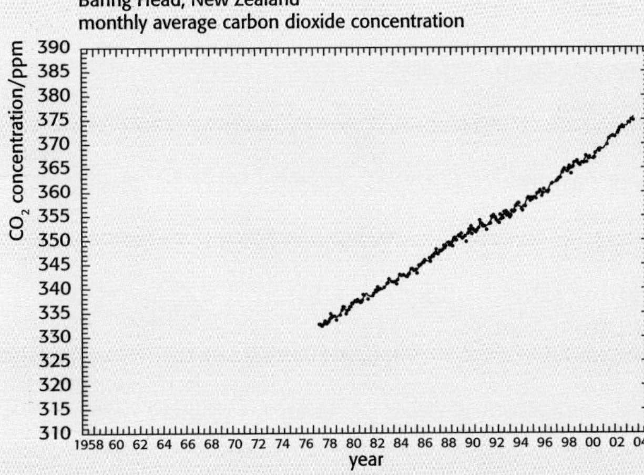

Figure 9

2 The mosquito *Wyeomyia smithii* uses day-length as a guide either to continue development of its larvae or to begin hibernation. This response to day-length is genetically controlled. Longer day-lengths maintain development whereas shorter day-lengths induce hibernation. In the northern regions of the Northern hemisphere, even though day-lengths are longer, winter arrives earlier than in regions closer to the equator. The following data is from an experiment to determine if *W. smithii* has adapted to later onsets of winter as a consequence of global warming. In 1972 and 1996, larvae were collected at various locations in the United States at latitudes 30–50° N. The larvae were examined to determine what day-length induced hibernation. Each circle on the graph in Figure 10 represents one larval population.

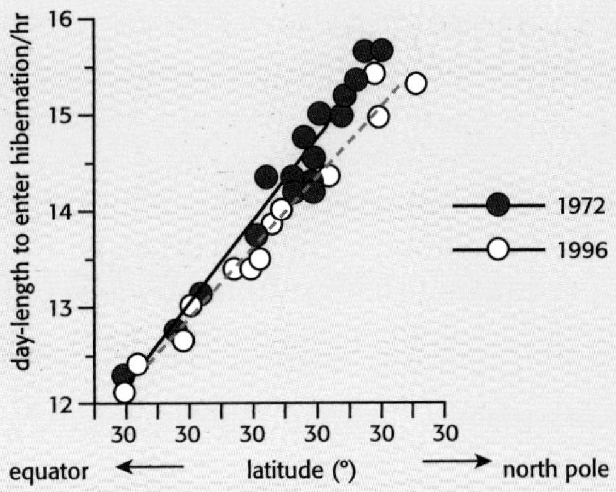

Figure 10

(a) Outline the relationship between day-length and latitude for the larval populations in 1972. [1]

(b) Compare the data of 1972 with 1996. [2]

(c) Explain how the data illustrates an evolutionary response to longer growing seasons due to a later onset of winter. [2]

3 Figure 11 shows the concentration of CO_2 in the atmosphere, measured in parts per million (ppm). In a forest, concentrations of CO_2 change over the course of the day and change with height. The top of the forest is referred to as the canopy.

(a) (i) State the highest concentration of CO_2 reached in the canopy. [1]

(ii) Determine the range of concentration found in the canopy. [2]

(b) (i) State the time of day (or night) when the highest levels of CO_2 are detected. [1]

(ii) The highest levels of CO_2 are detected just above the ground. Deduce two reasons why this is the case. [2]

(c) Give an example of an hour when CO_2 concentrations are reasonably uniform over the full range of heights. [1]

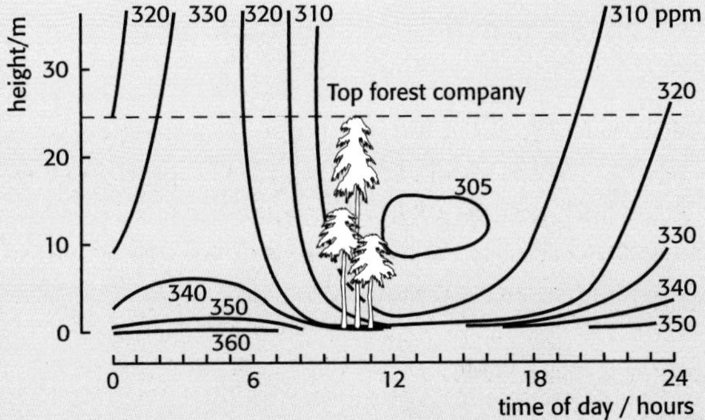

Figure 11

17 Evolution

The word evolution has several different meanings. The biological meaning is that evolution is the process by which living organisms are formed, by gradual change, from previous organisms. As currently understood, the process takes many generations and works at the level of a population. Individual organisms cannot evolve because the characteristics that they acquire during their lifetime cannot be inherited by the next generation. In *The Origin of Species*, published in 1859, Charles Darwin gives evidence for evolution and proposes a mechanism for it. The first evidence he gives concerns variation in domesticated animals.

evolution *the cumulative change in the heritable characteristics of a population.*

Figure 1 Charles Darwin

Evidence for evolution: domesticated animals

Humans have deliberately bred and used particular animal species for thousands of years. If modern breeds of livestock are compared with the wild species that they most resemble, the differences are often huge. Consider the differences between modern egg-laying hens and the jungle fowl of Southern Asia, or between Belgian Blue cattle and the aurochs of Western Asia. There are also many different breeds of sheep, cattle and other domesticated livestock, with much variation between breeds.

It is clear that domesticated breeds have not always existed in their current form. The only credible explanation is that the change has been achieved simply by repeatedly selecting for breeding the individuals most suited to human uses. This process is called **artificial selection.** Its effectiveness is shown by the considerable changes that have occurred in domesticated animals over periods of time that are very short, in comparison to geological time. It shows that selection can cause evolution, but it does not prove that evolution of species has actually occurred naturally, or that the mechanism for evolution is natural selection.

Figure 2 Pig and wild boar from Darwin's *Animals and Plants Under Domestication*

Data-based question: domestication of corn

A wild grass called teosinte that grows in Central America was probably the ancestor of cultivated corn, *Zea mays*. When teosinte is grown as a crop, it gives yields of about 150 kg per hectare. This compares with a world average yield of corn of 4100 kg per hectare at the start of the 21st century. Table 1 gives the lengths of some cobs. Corn was domesticated at least 7000 years ago.

1 Calculate the percentage difference in length between teosinte and Silver Queen. [2]
2 Calculate the percentage difference in yield between teosinte and world average yields of corn. [2]
3 Suggest factors apart from cob length, selected for by farmers. [3]
4 Explain why improvement slows down over generations of selection. [3]

Table 1

Corn variety and origin	Length of cob (mm)
Teosinte – wild relative of corn	14
Early primitive corn from Colombia	45
Peruvian ancient corn from 500 BC	65
Imbricado – primitive corn from Columbia	90
Silver Queen – modern sweetcorn	170

Evidence for evolution: fossils

In the first half of the 19th century, the sequence in which layers or strata of rock were deposited was worked out and the geological eras were named. It became obvious that the fossils found in the various layers were different – there was a sequence of fossils. In the 20th century, reliable methods of radioisotope dating revealed the ages of the rock strata and of the fossils in them. There has been a huge amount of research into fossils, which is the branch of science called palaeontology. It has given us strong evidence that evolution has occurred.

The types of evidence are summarized to the right. They offer strong evidence that evolution has occurred, but this does not amount to proof. Neither do fossils reveal a mechanism for evolution.

- The sequence in which fossils appear matches the sequence in which they would be expected to evolve, with bacteria and simple algae appearing first, fungi and worms later and land vertebrates later still. Among the vertebrates, bony fish appeared about 420 million years ago (mya), amphibians 340 mya, reptiles 320 mya, birds 250 mya and placental mammals 110 mya.
- The sequence also fits in with the ecology of the groups, with plant fossils appearing before animal, plants on land before animals on land, and plants suitable for insect pollination before insect pollinators.
- Many sequences of fossils are known, which link together existing organisms with their likely ancestors. For example, horses, asses and zebras, members of the genus *Equus,* are most closely related to rhinoceroses and tapirs. An extensive sequence of fossils, extending back over 60 million years, links them to *Hyracotherium,* an animal very similar to a rhinoceros.

Thinking about science

The text below is from *The Map that Changed the World*, by Simon Winchester.

The Reverend George Young, from the Scottish village of Coxiedean, was a theologian attracted to the mysteries of fossils. He came to prominence in 1819 with his discovery, in Yorkshire, of a gigantic reptile ichthyosaur, since identified as Leptopterygius acutirostris. *Though the find brought the enthusiastic Presbyterian minister some national fame, it equally confronted him with an interesting challenge, an acute mental and moral dilemma. It forced him to wonder about two possibilities that his religious beliefs sternly discountenanced: animal extinction on the one hand (there were no living ichthyosaurs – and so this particular species must have vanished), and animal evolution on the other. Consideration of either of these possibilities was a heresy and anathema to contemporary followers of the Bible, who regarded it as nothing less than a documentary history of the planet.*

The Reverend Young was forced to engage in some interesting spiritual gymnastics to come to terms with his problems. He eventually committed his conclusions to paper in a book with what might be considered the somewhat contradictory title Scriptural Geology. *The science he advanced in it was not overendowed with logic: the ichthyosaur he had found was not extinct, he declared, because a living specimen would probably be found sooner or later.*

1 Coelacanths were found first as fossils and were assumed to be extinct. In 1938, a living specimen was caught by a trawler. Can we be sure that living specimens of all fossils will not be eventually be found?

2 The Reverend Young was clearly influenced by a rigid set of beliefs. Is it possible for scientists to be unbiased in the way that they interpret their findings, or is it inevitable that we are all influenced by our beliefs?

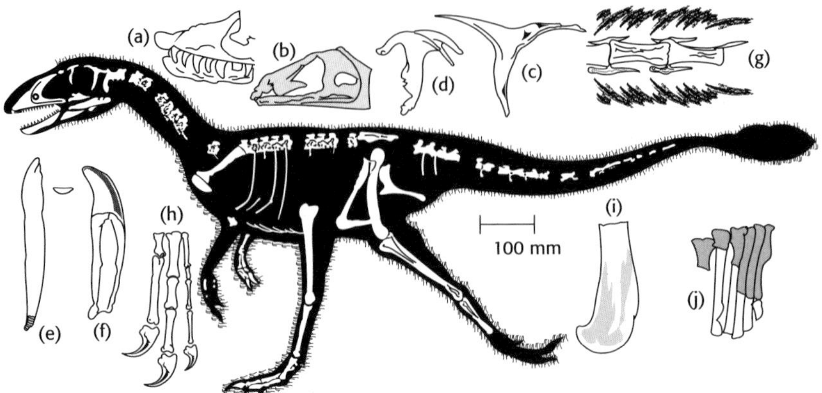

Figure 3 Drawings of fossils recently found in Western China. They show *Dilong paradoxus*, a 130-million-year-old tyrannosauroid dinosaur with protofeathers. a–d: bones of skull; e–f: teeth; g: tail vertebrae with protofeathers; h–j: limb bones

What two groups could *Dilong paradoxus* link?

Evidence for evolution: homologous structures

Darwin pointed out in *The Origin of Species* that some similarities in structure between organisms are superficial, for example between a dugong and a whale, or between a whale and a fish. Similarities like those between the tail fins of whales and fishes are known as analogous structures. When we study them closely we find that these structures are very different. An evolutionary interpretation is that they have had different origins and have become similar because they perform the same or a similar function. This is called convergent evolution.

Homologous structures are the converse of this. They are structures that may look superficially different and perform a different function, but which have what Darwin called a "unity of type". He gave the example of the forelimbs of a human, mole, horse, porpoise and bat and asked what could be more curious than to find that they "include the same bones, in the same relative positions", despite on the surface appearing completely different. The evolutionary explanation is that they have had the same origin, from an ancestor that had a pentadactyl or five-digit limb, and that they have become different because they perform different functions. This is called adaptive radiation.

There are many examples of homologous structures. They do not prove that organisms have evolved or had common ancestry and do not reveal anything about the mechanism of evolution, but they are difficult to explain without evolution. Particularly interesting are the structures that Darwin called "rudimentary organs" – reduced structures that serve no function. They are now called vestigial organs and examples of them are the beginnings of teeth found in embryo baleen whales, despite adults being toothless, the small pelvis and thigh bone found in the body wall of whales and some snakes, and of course the appendix in humans. These structures are easily explained by evolution as structures that have lost their function and so are being gradually lost.

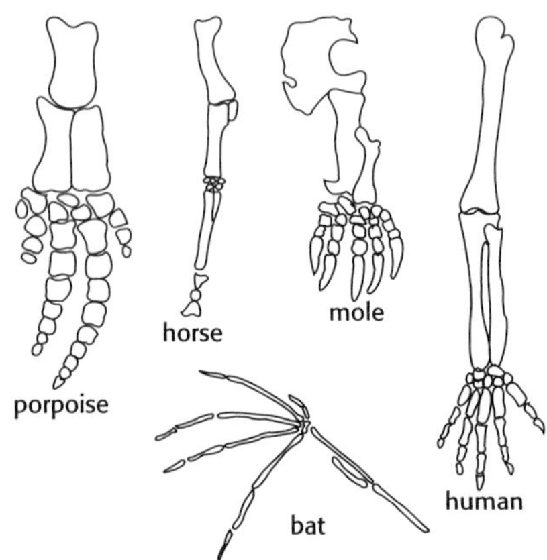

Figure 4 Pentadactyl limbs (not to scale)

Natural selection

Charles Darwin developed his understanding of evolution over many years, after returning to England from his voyage around the world on *HMS Beagle*. He probably developed the theory of natural selection in the late 1830s, but then worked to accumulate evidence for it. Darwin published his great work, *The Origin of Species*, in 1859. In this book of nearly 500 pages, he explains his theory and presents the evidence for it that he had found over the previous 20 to 30 years. The theory can be summarized in a series of observations and deductions.

Table 2

Observation	Deduction
Populations tend to reproduce rapidly and if every individual survived, there would be a geometrical or exponential increase in the population. On the other hand, when natural populations are studied, they tend to remain stable. There are natural checks to increases in population, for example, food supplies for animals. There is a limit to the size of population of a species that the environment can support.	There is a struggle for existence, in which some individuals survive and some die.
Organisms vary – there are differences between individual organisms even if they are members of the same species. These differences affect how well suited or fitted an organism is to its environment and mode of existence. This is called adaptation. Some individuals are better adapted than others because they have the favourable variations.	In the struggle for existence, the less well-adapted individuals will tend to die and the better adapted will tend to survive. This is natural selection.
Much of the variation between individuals can be passed on to offspring – it is heritable.	Because the better-adapted individuals survive, they can reproduce and pass on their characteristics to their offspring. The greater survival and reproductive success of these individuals leads to an increase in the proportion of individuals in the population that have the favourable variations. Over the generations, the characteristics of the population gradually change – this is evolution by natural selection.

Thinking about science: "Man is but a worm"

1 The existence of homologous structures is evidence that living organisms share ancestors. What is the deduction from the cartoon (figure 5)?
2 This deduction was very shocking to people living in the second half of the 19th century. Do you think that the cartoon is making fun of people who were shocked, or of Darwin for implying something so absurd?
3 Darwin was well aware that there would be hostile reactions to his theory and probably delayed publishing it for as long as he could. Which do you think Darwin feared more, hostility from other scientists or from religious groups?

Figure 5 Cartoon published in 1881 by *Punch*, an English satirical magazine

Responses to Darwin's theory

Darwin's theory of evolution by natural selection is made up of a group of linked concepts (see box right). None of these concepts has been credibly falsified, despite much research. Instead research has strengthened them. To take one example, there is much better understanding than in Darwin's time of the mechanisms that promote variation and allow it to be inherited. Sexual reproduction promotes variation in species, by random fertilization and by crossing over and random orientation of chromosomes in meiosis. Details of these processes are given in Chapters 11 and 26. The theory of evolution by natural selection is now an extremely strong one and it is highly unlikely that it will ever be falsified in a way that scientists could accept. However, this does not amount to proof that evolution is "true".

- population growth
- resource limitation
- a struggle for existence
- variation
- adaptation
- differential reproduction
- natural selection
- descent with modification
- origin of species
- extinction of species

Rediscovering biology: computer modelling of evolution

Computer modelling is widely used in some fields of science. Some programs have been designed to simulate or model evolution by natural selection. Examples of these can be found by doing an internet search.

1 What features would a program have to include for it to simulate evolution by natural selection realistically?
2 What conclusions, if any, can be drawn from these computer simulations?
3 Why are computer models used more widely in other branches of science than in biology?

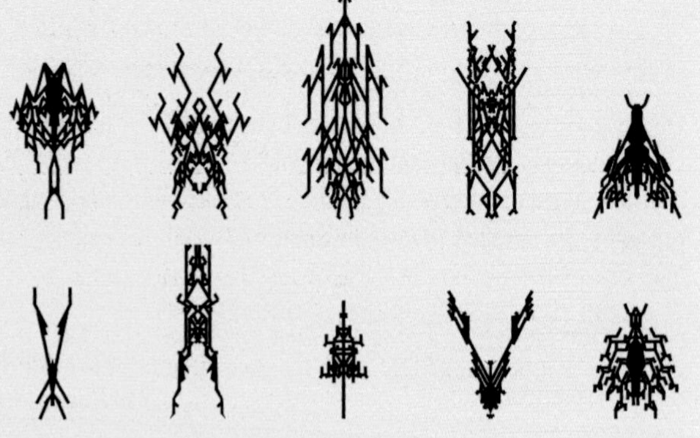

Figure 6 Sample results of computer modelling of evolution

Rediscovering biology: simulations of natural selection

1 Camouflage

- Distribute a large number of objects over a lawn, or other surface. The objects should be identical apart from their colour, for example short lengths of wool, small paper discs or plastic paper clips.
- Students act as predators and attempt to catch the objects. Allow long enough for most of them to be caught.
- The remaining objects are collected and are allowed to "breed", so that more of them are included when the objects are redistributed over the same area as before.
- Does one colour eventually predominate, because it is more camouflaged?

2 Beak size and shape

- Two or more seed types, large and small, are spread over an area, such as a laboratory bench.
- Students are given a wide variety of gripping or cutting tools, such as scissors, pliers, forceps, wire-strippers, garden shears. Each student or group of students uses one tool, for a short time, at the same time as the other students.
- The students that have caught fewest seeds, either by their number or mass, are eliminated and the seeds are re-spread, for more rounds of feeding and elimination of the students with the least adapted tool for picking up seeds. Is one tool best for all of the seeds?

3 Streamlining

- Make ten or more artificial fish using modelling clay, or some other malleable material. Drop each of them into a measuring cylinder of water and time how long each takes to reach the bottom.
- Discard the half of the models that were slowest. Pair up the fastest models and make intermediate shapes, to represent their offspring. Random new shapes can also be introduced to simulate mutation.
- Test the new generation and repeat the elimination of the slowest and the breeding of the fastest. Does one shape gradually emerge? Describe its features.

Thinking about science: belief systems and evolution

In some parts of the world, fewer than half of people accept the theory of evolution by natural selection. This is usually because of belief systems derived from religion. The questionnaire below allows you to test your understanding of the theory of evolution by natural selection and which of the parts of the theory you accept.

Section A

For each of the following statements, answer "a" if you agree, "b" if you disagree and "c" if you are unsure.

1 If every dog in a city were allowed to breed without control, the number of dogs would soon become a serious nuisance.
2 There are different breeds of dog and even dogs of the same breed look slightly different from each other.
3 Some dogs, for example German Shepherds, can breed with wolves, which they resemble and from which dogs were probably derived.
4 Because of their characteristics, some dogs are well suited for sheep-herding, some for pulling sleds, some for hunting, some as watchdogs and some for fighting.
5 People who want to buy a dog choose a particular breed and also particular characteristics in that breed. Dog breeders use males and females with these characteristics for breeding. Because characteristics are inherited, they can produce more dogs for sale, with the desirable characteristics.
6 If a group of dogs was placed together in a large pen, food would become scarce and the dogs would fight over the food.
7 If humans do not intervene, some of the dogs in the pen may be hurt or killed. The stronger and more agile male dogs will be more likely to survive and breed with any female dogs present.
8 People have developed new breeds of dog from other breeds by selecting individuals with particular characteristics, by breeding them together over many generations and by always selecting individuals with the desired characteristics for breeding the next generation.
9 Breeds of dog died out in the past, when a breed of dog was no longer needed for its particular job, or if another breed was found that did the job better.
10 There is a limit to how much food can be produced for feeding to dogs.

Section B

Consider each of the following parts of the theory of evolution by natural selection. Match each of them with one of the statements in Section A.

11 Population growth.
12 Resource limitation.
13 A struggle for existence.
14 Variation.
15 Adaptation.
16 Differential reproduction.
17 Natural selection.
18 Descent with modification.
19 The origin of species.
20 Extinction of species.

1 Assess your acceptance of the theory of evolution by natural selection and your understanding of it.
2 Count how many questions in Section A you answered with "a". The higher the number, to a maximum of ten, the more parts of the theory you seem to accept.
3 Check how many of these matches between sections A and B you made: 11–1, 12–10, 13–6, 14–2, 15–4, 16–7, 17–5, 18–3, 19–8, 20–9. The more matches you made, the better your understanding of evolution.
4 Discuss with other students whether there is a link between understanding the theory and accepting it.

Thinking about science: evolution versus creationism

The text below is an extract from a radio broadcast from the BBC on 28 April 2006 by Professor Lisa Jardine, a historian of science.

Sometimes, if you're lucky as a historian, you find a bit of evidence which illuminates a big idea. That happened to me this week in the Pepys Library at Magdalene College, Cambridge.

The thought uppermost in my mind was how odd it is that non-scientists think of science as being about certainties and absolute truth. Whereas scientists are actually quite tentative –they simply try to arrive at the best fit between the experimental findings so far and a general principle. Science, as I say, is not doctrinaire. Strongly held religious views, however, are. This week John Mackay from Queensland Australia, a passionate advocate of Creationism, has been touring halls and chapels in the UK attacking Darwin's theory.

I was looking for documents relating to attempts by the Dutch Scientist Christian Huygens to develop a pendulum clock which would enable mariners to find their longitude at sea. In 1664, shortly after the first proper scientific research institute, the Royal Society, had been established in London, its president offered to organise a series of sea-trials to be conducted by the English navy, using two of his pioneering clocks. Captain Robert Holmes agreed to take the clocks with him on a nine-month voyage down the west coast of Africa. When he got back to London in 1665 Holmes presented his results to an expectant Royal Society. The clocks had performed spectacularly well. But the inventor himself was not so sure. And his reason for being more cautious than his London colleagues was precisely the fact that the clocks had proved so spectacularly accurate. So the Royal Society asked an official from the Navy Board to check the evidence Holmes had provided. Lo and behold, it turns out that Holmes had falsified his evidence. The pendulum clocks had proved no more accurate for calculating longitude than conventional methods. Holmes had thought that by tampering with his evidence he would please the scientists at the Royal Society. Instead, the too-precise nature of the match between his results and the results they wanted alerted them to the fact that his testimony was unreliable.

And Huygens was right to be sceptical. His pendulum clocks never did prove accurate enough at sea to solve the problem of finding longitude. A scrupulous scientist like Huygens would rather be disappointed, than accept dubious evidence to provide pat confirmation of a pet theory.

That continues to be true in all areas of scientific investigation today. Which is why no scientist will take up Mackay's challenge to "prove" the truth of Darwin's theory of evolution in a public debate. They know that they cannot present a strongly held view based on a body of supporting evidence with the certainty of a revealed truth.

The most today's Royal Society is prepared to say is that a belief that all species on earth have always existed in their present form is "not consistent with the evidence". Because most of us want more certainty, we're on the side of the 17th century ship's captain, believing the experiments ought to prove the scientific theory once and for all.

Unfortunately, where arguments about ecology are concerned, time is not on our side. We cannot afford the luxury of waiting for evidence which clinches the theory. We are going to have to learn to participate in debates which are not about certainties. We have to decide now whether we should sacrifice our right to cut-price air travel in order to cut down on carbon emissions. A public understanding of science has never been more important.

1 Are there certainties in science or absolute truths? For example, can a biologist say with certainty that life has evolved by natural selection?
2 Does it weaken or strengthen scientists' views if they are unwilling to say that they are certain about them?
3 The theory that species on Earth have always existed in their present form is part of what is sometimes called creation science. Does the existence of intermediate forms in the fossil record, vestigial structures, homologous structures or selectively bred animals contradict creation theory? If so, what is the justification for creation science?
4 Do you agree that religion offers more certainty than science and if so, what are the reasons for this? What distinguishes a religious claim from a scientific claim?
5 To what extent is Lisa Jardine right that over issues such as global warming we have to act before scientists are certain of the causes?

Galápagos finches – evolution in action

Darwin visited the Galápagos Islands in 1835 and collected specimens of small birds, which were subsequently identified as finches. There are 14 species in all. Darwin observed that the sizes and shapes of the beaks of the finches varied, as did their diet. From the overall similarities between the birds and their distribution over the Galápagos islands (see Figure 7), Darwin hypothesized that "one might really fancy that from an original paucity of birds in this archipelago, one species had been taken and modified for different ends".

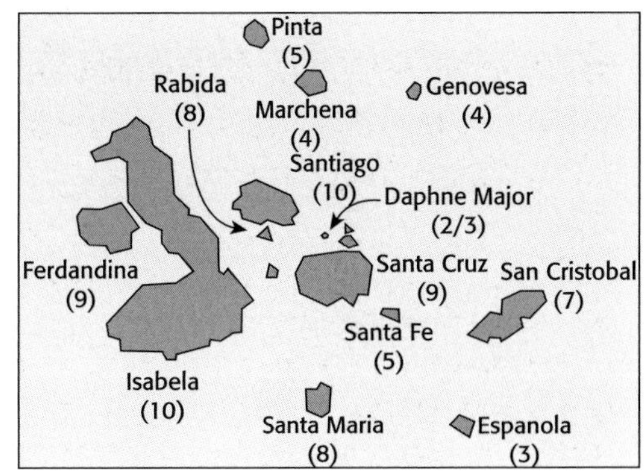

Figure 7 The Galápagos archipelago with the number of species of finch found on each island

There has since been intense research into what have become known as Darwin's finches. In particular, Peter and Rosemary Grant have shown that beak characters and diet are closely related and when one changes, the other does also.

A particular focus of Peter and Rosemary Grant's research has been a population of the medium ground finch, *Geospiza fortis,* on a small island called Daphne Major. On this island, the small ground finch, *Geospiza fuliginosa,* is almost absent. Both species feed on small seeds, though *G. fortis* can also eat larger seeds. In the absence of competition from *G. fuliginosa* for small seeds, *G. fortis* is smaller in body size and beak size on Daphne Major than on other islands.

In 1977, a drought on Daphne Major caused a shortage of small seeds, so *G. fortis* fed instead on larger, harder seeds, which the larger-beaked individuals are able to crack open. Most of the population died in that year, with highest mortality among individuals with shorter beaks.

In 1982–83 there was a severe El Niño event, causing eight months of heavy rain and as a result an increased supply of small, soft seeds and fewer large hard seeds. *G. fortis* bred rapidly, in response to the increase in food availability. With a return to dry weather conditions and greatly reduced supplies of small seeds, breeding stopped until 1987. In that year, only 37 per cent of those alive in 1983 bred and they were not a random sample of the 1983 population. In 1987, *G. fortis* had longer and narrower beaks than the 1983 averages, correlating with the reduction in supply of small seeds.

Variation in the shape and size of the beaks (see Figure 8) is mostly due to genes, though the environment has some effect. The proportion of the variation due to genes is called the heritability. Using the heritability of beak length and width and data about the birds that had survived to breed, the changes in mean beak length and width between 1983 and 1987 were predicted. The observed results are very close to the predictions. Average beak length was predicted to increase by 10 μm and actually increased by 6 μm. Average beak width was predicted to decrease by 130 μm and actually decreased by 120 μm.

One of the objections to the theory of evolution by natural selection is that significant changes caused by natural selection have not been observed actually occurring. It is unreasonable to expect huge changes to have occurred in a species, even if it had been followed since Darwin's theory was published in 1859, but in the case of *G. fortis,* significant changes have occurred that are clearly linked to natural selection.

(a) *G. fortis (large beak)*

(b) *G. fortis (small beak)*

(c) *G. magnirostris*

Figure 8 Variation in beak shape in Galapagos finches. (a) *G. fortis* (large beak). (b) *G. fortis* (small beak). (c) *G. magnirostris.*

Data-based question: Galápagos finches

When Peter and Rosemary Grant began to study finches on the island of Daphne Major in 1973, there were breeding populations of two species, *Geospiza fortis* and *Geospiza scandens*. *Geospiza magnirostris* established a breeding population on the island in 1982, initially with just two females and three males. Figure 9 shows the numbers of *G. magnirostris* and *G. fortis* on Daphne Major between 1997 and 2006.

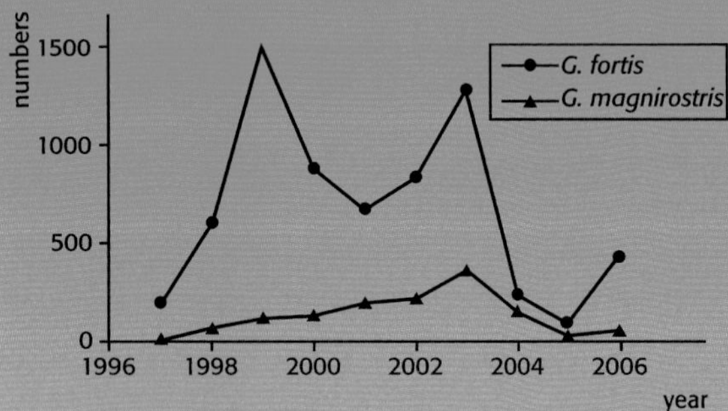

Figure 9 Changes in numbers of *G. fortis* and *G. magnirostris* between 1996 and 2006

1 (a) Describe the changes in the population of *G. magnirostris* between 1997 and 2006. [2]

(b) Compare the changes in population of *G. fortis* between 1997 and 2006 with the changes in the population of *G. magnirostris*. [3]

2 Daphne Major has an area of 0.34 km^2. 1 km^2 is 100 hectares and 1 hectare is 100×100 m. Calculate the maximum and minimum population densities of *G. fortis* during 1997–2006. [4]

Table 3 shows the percentages of three types of seed in the diets of the three finch species on Daphne Major. Small seeds are produced by 22 plant species, medium seeds by the cactus *Opuntia echios*, and large seeds, which are very hard, by *Tribulus cistoides*.

Table 3

Year	Small	Medium	Large
G. fortis			
1977	73.1	10.2	16.7
1985	80.5	0.0	19.5
1989	77.1	5.1	16.2
2004	80.4	11.3	8.2
G. magnirostris			
1985	18.5	0.0	81.5
1989	5.9	11.8	82.3
2004	4.5	26.4	69.1
G. scandens			
1977	85.2	14.8	0.0
1985	77.1	21.9	0.0
1989	23.4	69.7	0.0
2004	17.4	82.6	0.0

3 (a) Outline the diet of each of the species of finch on Daphne Major. [3]
(b) There was a very severe drought on Daphne Major in 2003 and 2004. Deduce how the diet of the finches changed during the drought, using the data in the table. [3]

4 Figure 10 shows an index of beak size of adult *G. fortis* from 1973 to 2006, with the size in 1973 assigned the value zero and the sizes in other years shown in comparison to this.

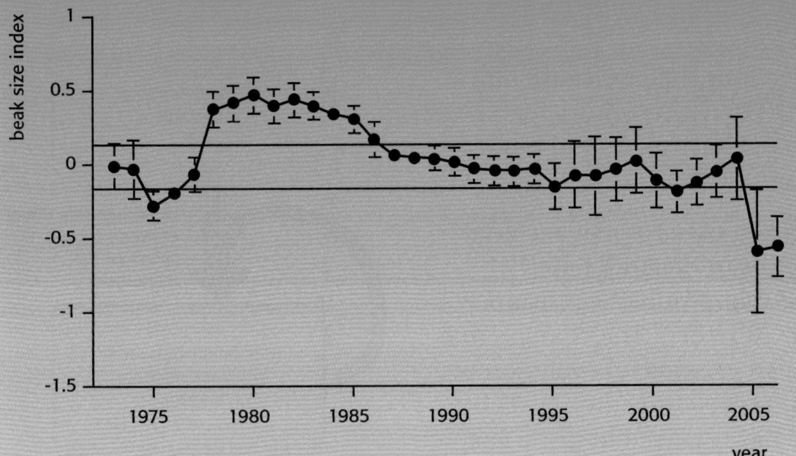

Figure 10 Relative beak size in *G. fortis* between 1973 and 2006

The graph shows two periods of very rapid change in mean beak size, both of which correspond with droughts on Daphne Major.

(a) State two periods of most rapid change in mean beak size of *G. fortis*. [2]
(b) Suggest two reasons for mean beak size changing most rapidly when there is a drought [2]
(c) In the first severe drought, the mean beak size of *G. fortis* increased, but in the second drought, it decreased. Using the data in this question, explain how natural selection could cause these changes in beak size in the two droughts. [3]

5 The intensity of natural selection on Daphne Major was calculated during the two droughts. The calculated values are called selection differentials. They range from –1.08 for beak length during the second drought, to +0.88 for beak length in the first drought, with similar selection differentials for beak width and depth and overall beak size. These are very large selection differentials, compared to values calculated in other investigations of evolution.

Suggest reasons for natural selection on the beak size of *G. fortis* being unusually intense on the island of Daphne Major. [2]

6 Discuss the advantages of investigations of evolution over long periods and the reasons for few long term investigations being done. [3]

Antibiotic resistance in bacteria – evolution in action

Antibiotics were one of the great triumphs of medicine in the 20th century. Their action is explained in Chapter 21. When they were first introduced, it was expected that they would offer a permanent method of controlling bacterial diseases, but there have been increasing problems of antibiotic resistance in pathogenic bacteria. The following trends have become established:

- After an antibiotic is introduced and used on patients, bacteria showing resistance appear within a few years.
- Resistance to the antibiotic spreads to more and more species of pathogenic bacteria.
- In each species the proportion of infections that are caused by a resistant strain increases.
- Strains of bacteria appear that are resistant to more and more different antibiotics – this is called multiple resistance.

So, during the time that antibiotics have been used to treat bacterial diseases there have been cumulative changes in the antibiotic resistance properties of populations of bacteria. Antibiotic resistance is due to genes in bacteria and so it can be inherited. The changes that have taken place in bacteria are therefore an example of evolution by natural selection. The mechanism is summarized in Figure 11. The evolution of multiple antibiotic resistance has occurred in just a few decades. This rapid evolution is due to the following causes:

- There has been very widespread use of antibiotics, both for treating diseases and in animal feeds used on farms.
- Bacteria can reproduce very rapidly, with a generation time of less than an hour.
- Populations of bacteria are often huge, increasing the chance of a gene for antibiotic resistance being formed by mutation.
- Bacteria can pass genes on to other bacteria in several ways, including using plasmids, which allow one species of bacteria to gain antibiotic resistance genes from another species.

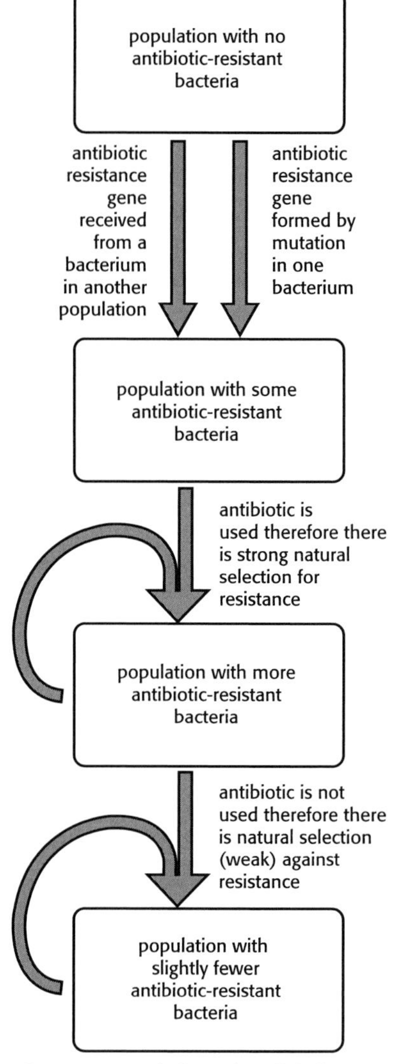

Figure 11 Evolution of antibiotic resistance

Critical considerations: the future of antibiotics

Some strains of *E. coli* can cause life-threatening infections of the blood and cerebrospinal fluid. Figure 12 shows the growth in resistance to ciprofloxacin, a quinolone antibiotic that was a very good treatment for these infections.

1 Approximately how rapidly did rates of resistance to ciprofloxacin double, in the years from 1990 to 2004?
2 If this trend continued, when would all the population of *E. coli* be resistant to ciprofloxacin?
3 Discuss the merits of each of the following methods of combating the problem of antibiotic resistance in bacteria:
   - Discovering or developing new antibiotics faster than bacteria develop resistance.
   - Withdrawing the use of certain antibiotics in the hope that bacteria will evolve to become susceptible to them, allowing their subsequent reintroduction.
   - Making inappropriate use of antibiotics illegal, including self-prescription and use in animal feeds.
4 How would the loss of all effective antibiotics affect us?

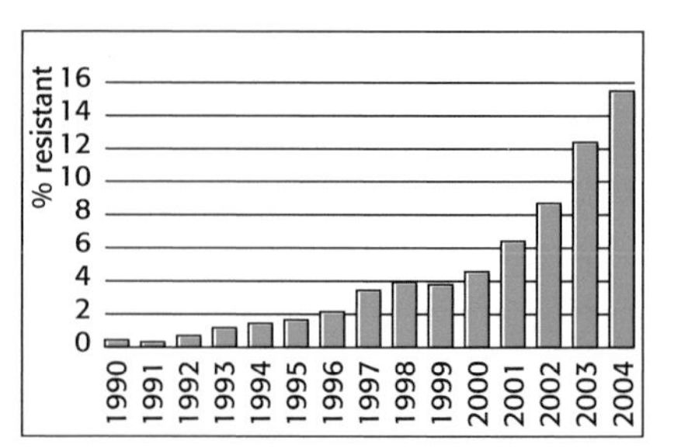

Figure 12 Percentage resistance to ciprofloxacin between 1990 and 2004

Multiple-choice questions

The bar charts in Figure 13 show the growth of three populations of an alga, *Ectocarpus siliculosus,* at different copper concentrations. One population came from an unpolluted environment at Rhosneigr in Wales. The other two came from the undersides of ships that had been painted with a copper-containing anti-fouling paint.

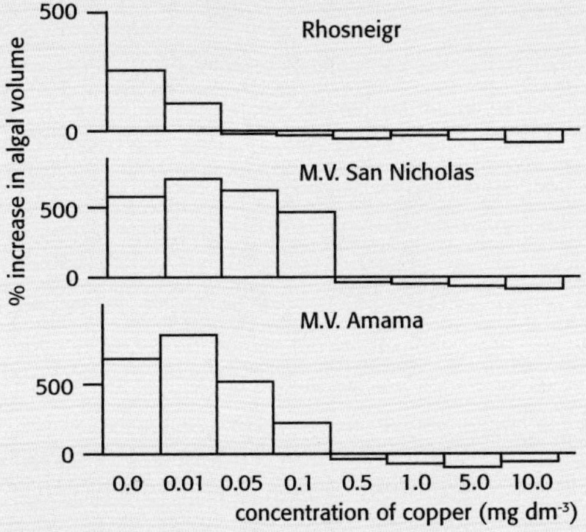

Figure 13

1 How much higher was the maximum copper concentration tolerated by the algae from ships than the algae from an unpolluted environment?

A 0.09 times higher
B 0.11 times higher
C 1.0 times higher
D 10 times higher.

2 What is the reason for results lower than zero on the bar charts?

A The volume of algae decreased.
B The algae all died.
C Increases in volume were less than 100%.
D Results were too small to measure accurately.

3 What was the reason for the difference in copper tolerance between the algae?

A The algae on the ships absorbed copper.
B The algae can develop copper tolerance and pass it on to their offspring.
C The copper in the paint caused mutations.
D The copper in the paint caused natural selection for higher levels of copper tolerance.

4 Which of the following processes are required for copper tolerance to develop in a population?

(i) variation in copper tolerance
(ii) inheritance of copper tolerance
(iii) failure of algae with lower copper tolerance to survive or reproduce.

A (i) only
B (i) and (ii) only
C (i) and (iii) only
D (i), (ii) and (iii).

Data-based question

The bar charts in Figure 14 show the results of an investigation of evolution in rice plants. F1 hybrid plants were bred by crossing together two rice varieties. These hybrids were then grown at five different sites in Japan. Each year the date of flowering was recorded and seed was collected from the plants, for re-sowing at that site in the following year.

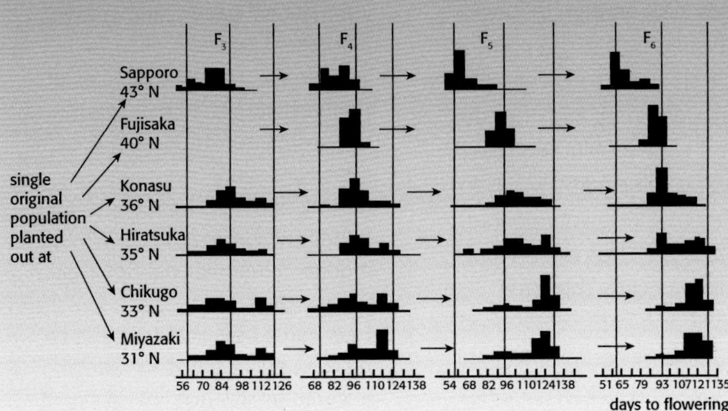

Figure 14

1 Why was the investigation done using hybrids rather than a single pure-bred variety? [2]

2 Describe the changes, shown in the chart, between the F3 and F6 generations of rice plants grown at Miyazuki. [2]

3 (a) State the relationship between flowering time and latitude in the F6 generation. [1]

(b) Suggest a reason for this relationship. [1]

4 (a) Predict the results if the investigation had been carried on until the F10 generation. [1]

(b) Predict the results of collecting seeds from F10 plants grown at Sapporo and from F10 plants grown at Miyazuki and sowing them together at Hiratsuka. [3].

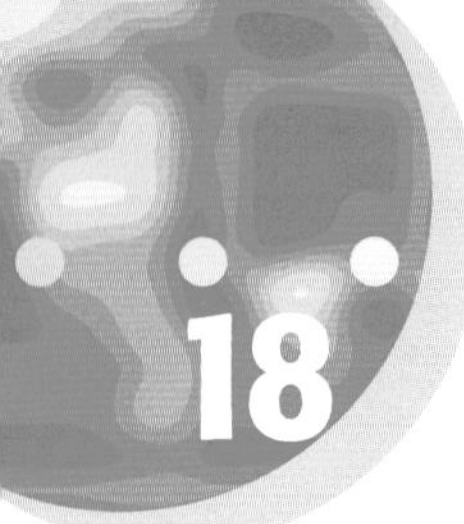

18 Classification

It is natural for humans to recognize the features of living organisms and to use these features to put organisms into groups. At a basic level, simple observation shows that there are often many organisms of the same type. If we agree on a name for a group of organisms, we can then talk or write about them. Naming organisms is called nomenclature. The idea of a group of organisms of the same type has developed into the biological concept of the species. In every language, names have been chosen for species, but science is an international venture and so names are needed that are understood throughout the world. The system that biologists use is called **binomial nomenclature**, because the international name of a species consists of two words. An example is *Linnaea borealis*. The first name is the genus name. A genus is a group of species that share certain characteristics. The second name is the species or specific name. There are various rules about binomial nomenclature:

- The genus name begins with an upper-case (capital) letter and the species name with a lower-case (small) letter.
- In typed or printed text, a binomial is shown in italics.
- After a binomial has been used once in a piece of text, it can be abbreviated to the initial letter of the genus name with the full species name, for example: *L. borealis*.
- The earliest published name for a species, from 1753 onwards, is the correct one.

Figure 1 *Linnaea borealis:* Binomials are often chosen to honour a biologist, or to describe a feature of the organism. *Linnaea borealis* is named in honour of Carl Linnaeus, the Swedish biologist who introduced the binomial system of nomenclature and named many plants and animals using it.

Critical considerations: internationalism in nomenclature

Carl Linnaeus's 1753 book *Species Plantarum* introduced consistent two-part names (binomials) for all species of the vegetable kingdom then known. Thus the binomial *Physalis angulata* replaced the obsolete phrase-name, *Physalis annua ramosissima, ramis angulosis glabris, foliis dentato-serratis*. Linnaeus brought the scientific nomenclature of plants back to the simplicity and brevity of the vernacular nomenclature out of which it had grown. Folk-names for species rarely exceed three words. In groups of species alike enough to have a vernacular group-name, the species are often distinguished by a single name attached to the group-name, as in the Ancient Greek αδιαυτου το λενκον and αδιαυτου το μεαυ (used by Threophrastus), Latin *anagallis mas* and *anagallis femina* (used by Pliny), German weiss Seeblumen and geel Seeblumen (used by Fuchs), English wild mynte and water mynte (used by Turner) and Malayan jambu bol and jambu chilli (applied by Malays to different species of *Eugenia)*.

The International Botanical Congress held in Genoa in 1892 proposed that 1753 be taken as the starting point for both genera and species. This was incorporated in the American "Rochester Code" of 1883 and in the code used at the Berlin Botaniches Museum and supported by British Museum of Natural History, Harvard University botanists and a group of Swiss and Belgian botanists. The International Botanical Congress of Vienna in 1905 accepted by 150 votes to 19 the rule that "La nomenclature botanique commence avec Linné, *Species Plantarum* (ann. 1753) pour les groupes de plantes vasculaires."

1 Why is it important to have an international system for naming species of living organism?
2 Why was Linnaeus's system for naming plants adopted as the international system, rather than any other system?
3 Why do the international rules of nomenclature state that genus and species names must be in Ancient Greek or Latin?
4 Making decisions by voting is rather unusual in science. Why is it done at International Botanical Congresses?
5 How can a botanical congress or other scientific group or meeting ensure that it is truly international?

The hierarchy of taxa

The word taxon is Greek and means things that are arranged into a group. The plural is taxa. In biology, species are arranged or classified into taxa. Every species is classified into a genus. Genera are grouped into families. An example of the genera and species in a family is shown in Figure 2. Families are grouped into orders, orders into classes, classes into phyla and phyla into kingdoms. These seven taxa form a hierarchy, as each taxon includes taxa from the level below. Going up the hierarchy, the taxa include larger and larger numbers of species, which share fewer and fewer features.

Taxonomists often disagree about how the higher taxa in the hierarchy should be organized. The best classification is the one that follows most closely the way in which species evolved. Following this principle, members of a taxon should have a common ancestor. In the past, this type of classification was attempted by looking at as many visible characteristics as possible, but new molecular methods have been introduced and these have caused significant changes to the classification of some groups. More details of this are given in Option D and Option F.

Table 1

Taxon	Grey wolf	Date palm
Kingdom	Animalia	Plantae
Phylum	Chordata	Angiospermophyta
Class	Mammalia	Monocotyledoneae
Order	Carnivora	Palmales
Family	Canidae	Arecaceae
Genus	*Canis*	*Phoenix*
Species	*lupus*	*dactylifera*

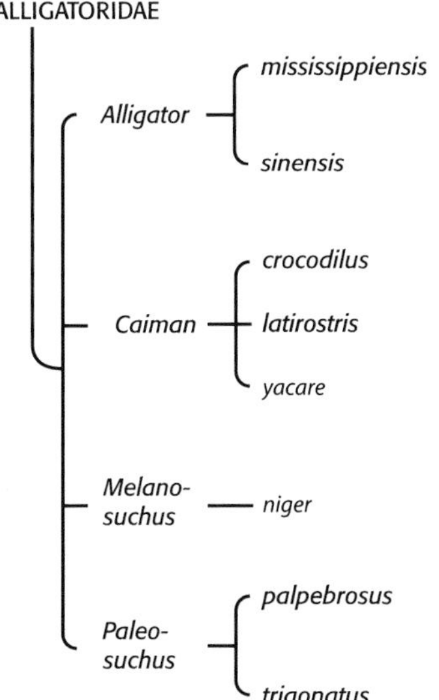

Figure 2 Classification of the alligator family

Data-based question: classifying cartilaginous fish

All the fish shown in Figure 3 are in the class Chondrichthyes. They are the most frequently found fish in this class in north-west Europe.

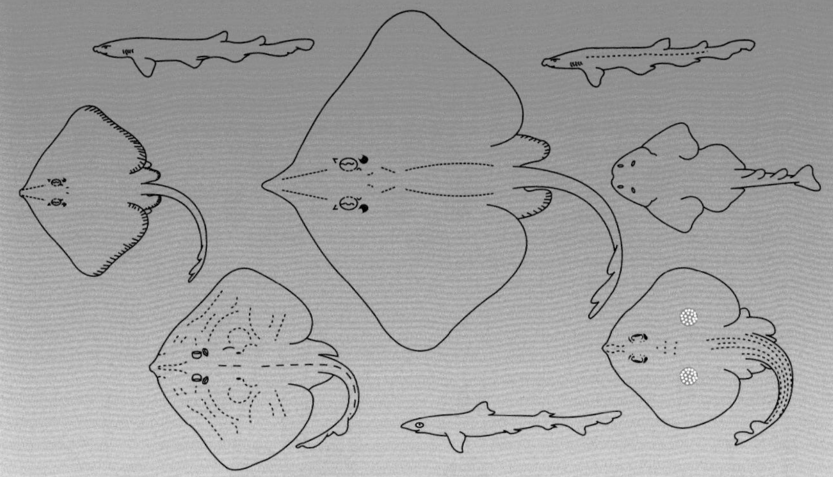

Figure 3 Cartilaginous fish in seas in North-west Europe

1 (a) Discuss how many different orders are needed to classify the eight species of fish. [2]
  (b) Discuss how many families are needed in each of these orders. [2]

2 One species is in a different order from all the others. Deduce, with reasons, which it is. [2]

3 Four of the species are classified in the same genus.
  (a) Deduce which these are. [2]
  (b) Describe the features of the genus. [2]

Dichotomous keys

The hierarchy of taxa can be used to help identify an organism. Ecologists often need to do this when they are investigating a community. Usually it is obvious what kingdom a species belongs to, and often the phylum and class are easily recognized as well. Going down through the lower taxa, through the order, family, genus and species, a point is often reached where the ecologist needs help to assign a species to the correct group. Keys are often then used.

A dichotomy is a division into two; a dichotomous key consists of a numbered series of pairs of descriptions. One of these should clearly match the species and the other should clearly be wrong. The features that the designer of the key chooses to use in the descriptions should therefore be reliable and easily visible. Each of the pair of descriptions leads either to another of the numbered pairs of descriptions in the key, or to an identification. An example of a key is shown in Table 2.

Table 2 Key to groups of marine mammals

1	Fore and hind limbs visible, can emerge on land	2
	Only fore limbs visible, cannot live on land	6
2	Fore and hind limbs have paws	3
	Fore and hind limbs have flippers	4
3	Fur is dark	sea otters
	Fur is white	polar bears
4	External ear flap visible	sea lions and fur seals
	No external ear flap	5
5	Two long tusks	walruses
	No tusks	true seals
6	Mouth breathing, no blowhole	dugongs and manatees
	Breathing through blowholes	7
7	Two blowholes, no teeth	baleen whales
	One blowhole, teeth	dolphins, porpoises and whales

Rediscovering biology: designing dichotomous keys

Keys are usually designed for use in a particular area. All the groups or species that are found in that area can be identified using the key. There may be a group of organisms in your area for which a key has never been designed.

- You could design a key to the trees in the local forest or on your school campus, using leaf descriptions or bark descriptions.
- You could design a key to birds that visit bird-feeding stations in your area.
- You could design a key to the invertebrates that are associated with one particular plant species.
- You could design a key to the footprints of mammals and birds (Figure 4). They are all right front footprints and are not shown to scale.

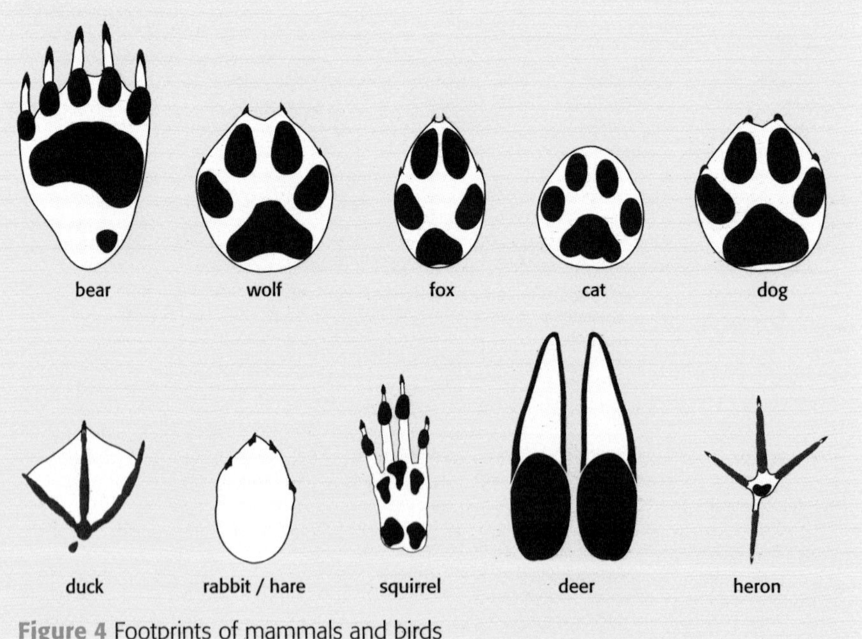

Figure 4 Footprints of mammals and birds

Plant classification

All plants are classified together in one kingdom. In the life cycle of every plant, male and female gametes are formed and fuse together. The zygote formed develops into an embryo. The way in which this embryo develops depends on the type of plant it is. The different types of plants are put into phyla. Most plants are in one of four phyla, but there are other smaller phyla. The *Ginkgo biloba* tree for example, is in one of the smaller phyla. The four main plant phyla are listed (right) and differences between them are shown in Table 3. D indicates the dispersal stage.

The four major plant phyla

- Bryophyta – mosses, liverworts and hornworts
- Coniferophyta – conifers
- Filicinophyta – ferns
- Angiospermophyta – flowering plants

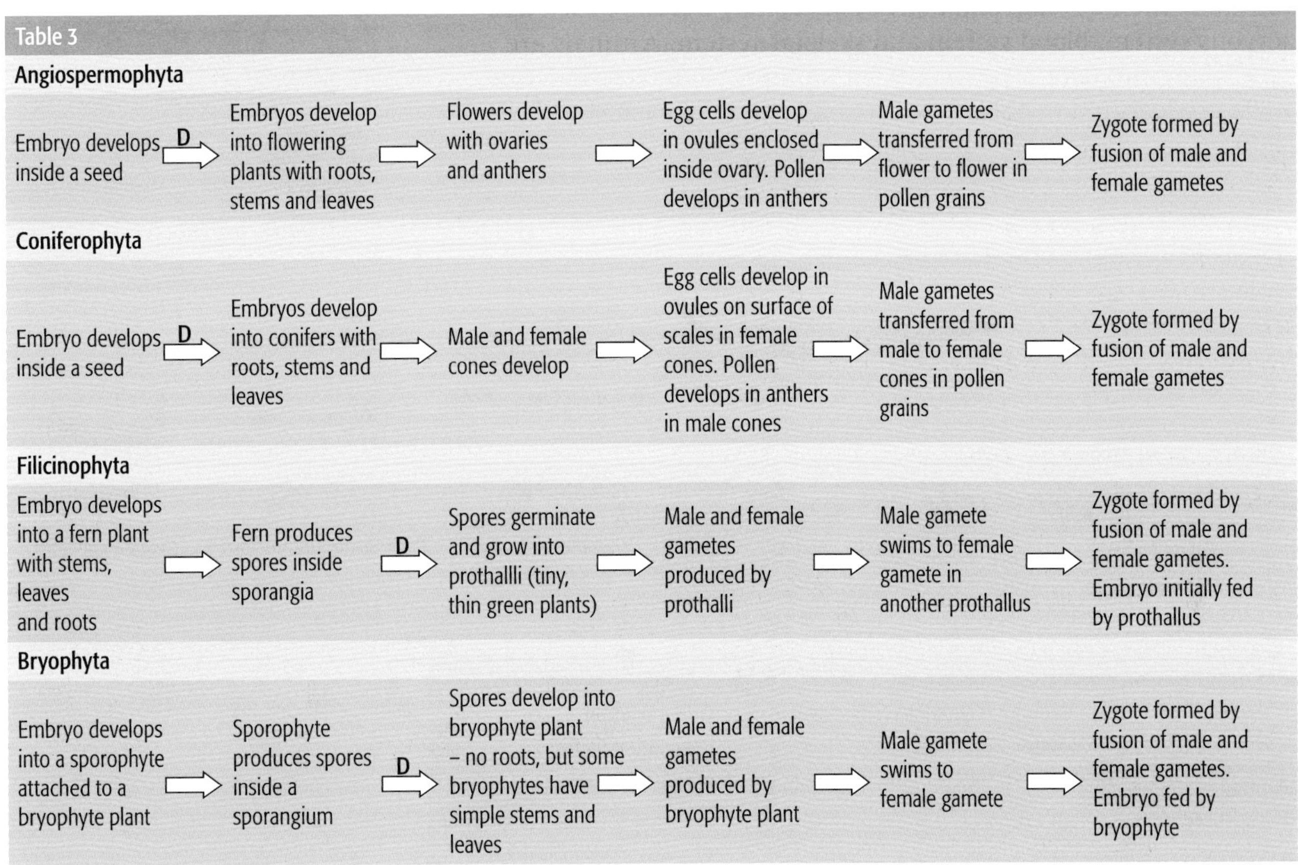

Table 3

Phylum	Stage 1		Stage 2		Stage 3		Stage 4		Stage 5		Stage 6
Angiospermophyta	Embryo develops inside a seed	D →	Embryos develop into flowering plants with roots, stems and leaves	→	Flowers develop with ovaries and anthers	→	Egg cells develop in ovules enclosed inside ovary. Pollen develops in anthers	→	Male gametes transferred from flower to flower in pollen grains	→	Zygote formed by fusion of male and female gametes
Coniferophyta	Embryo develops inside a seed	D →	Embryos develop into conifers with roots, stems and leaves	→	Male and female cones develop	→	Egg cells develop in ovules on surface of scales in female cones. Pollen develops in anthers in male cones	→	Male gametes transferred from male to female cones in pollen grains	→	Zygote formed by fusion of male and female gametes
Filicinophyta	Embryo develops into a fern plant with stems, leaves and roots	→	Fern produces spores inside sporangia	D →	Spores germinate and grow into prothallli (tiny, thin green plants)	→	Male and female gametes produced by prothalli	→	Male gamete swims to female gamete in another prothallus	→	Zygote formed by fusion of male and female gametes. Embryo initially fed by prothallus
Bryophyta	Embryo develops into a sporophyte attached to a bryophyte plant	→	Sporophyte produces spores inside a sporangium	D →	Spores develop into bryophyte plant – no roots, but some bryophytes have simple stems and leaves	→	Male and female gametes produced by bryophyte plant	→	Male gamete swims to female gamete	→	Zygote formed by fusion of male and female gametes. Embryo fed by bryophyte

Rediscovering biology: recognition features of plant phyla

- Examine a specimen of each of the four main phyla of plants. These could be live specimens in the lab or outside, or images on the Internet. For each specimen find out whether the following external features are present: root, stem, leaf, flower, cone, sporangium, seed, and fruit. True fruits have seeds enclosed inside them.
- Design a key to the four main phyla, using reliable and easily visible external differences to distinguish them.
- Grow a plant from each phylum, starting with seeds or spores.

If possible, grow a local species, for example a tree that you can plant out, to help conserve the species.

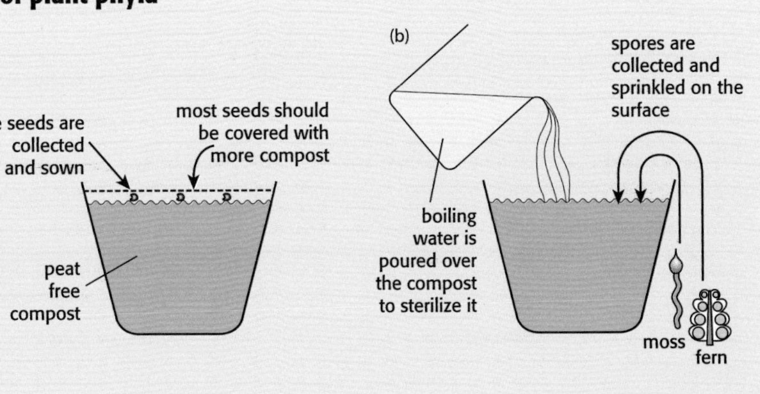

Figure 5 (a) sowing seeds (b) sowing spores

Animal classification

There are similarities between plants and animals in the way that their lives begin. In animals, as in plants, male and female gametes fuse together and a zygote is formed, which develops into an embryo. However, embryonic development is very different in animals and plants, and of course, there are significant differences between the structure of animal and plant cells. Animals are therefore classified into a separate kingdom.

In all animals, the embryo develops into a solid ball of cells and then into a hollow ball of cells, called a blastula. A mouth, gut and anus may then develop and body systems may develop, such as a nervous system, blood system and skeletal system. Animals are divided up into over 30 phyla, based on their characteristics. Six phyla are featured in Table 4, with an example of each shown in Figure 6.

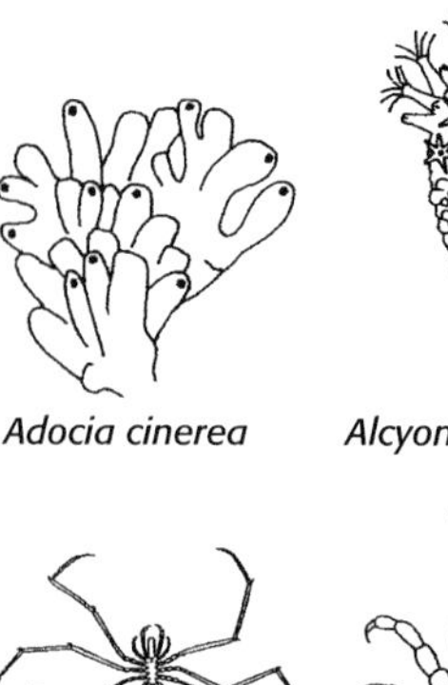

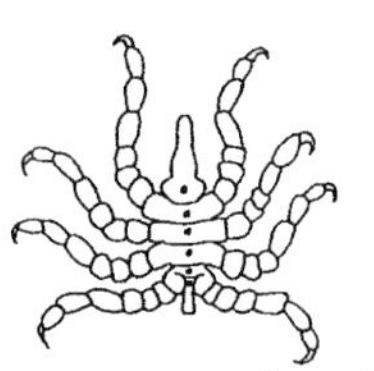
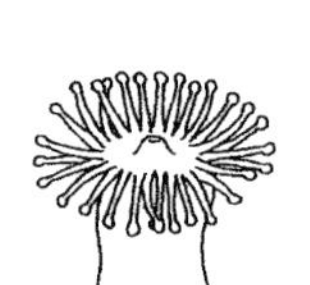
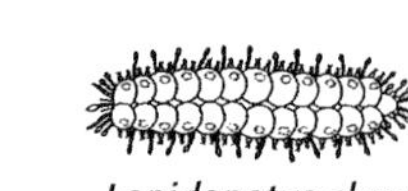
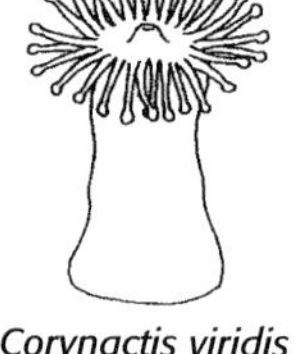
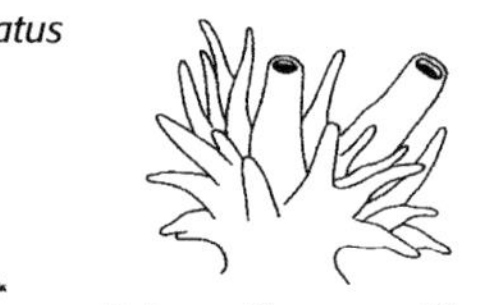
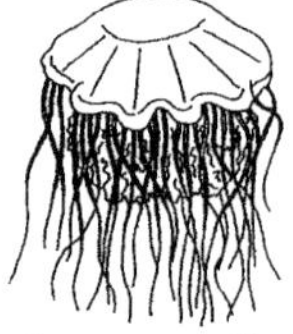

Adocia cinerea · *Alcyonium glomeratum* · *Nymphon gracilis* · *Pycnogonum littorale* · *Corynactis viridis* · *Lepidonotus clara* · *Procerodes littoralis* · *Arenicola marina* · *Prostheceraeus vittatus* · *Loligo forbesii* · *Caprella linearis* · *Gammarus locusta* · *Polymastia mammiliaris* · *Cyanea capillata*

Figure 6 Invertebrate diversity

Table 4

Phylum	Mouth/ anus	Symmetry	Skeleton	Other external recognition features
Porifera – fan sponges, cup sponges, tube sponges, glass sponges	No mouth or anus	None	Internal spicules (sketetal needles)	Many pores over the surface through which water is drawn in for filter feeding. Very varied shapes
Cnidaria – hydras, jellyfish, corals, sea anemones	Mouth and anus	Radial	Soft, but hard corals secrete $CaCO_3$	Tentacles arranged in rings around the mouth, with stinging cells. Polyps or medusae (jellyfish)
Platyhelminthes – flatworms, flukes, tapeworms	Mouth and anus	Bilateral	Soft, with no skeleton	Flat and thin bodies in the shape of a ribbon. No blood system or system for gas exchange
Mollusca – bivalves, gastropods, snails, chitons, squid, octopus	Mouth and anus	Bilateral	Most have shell made of $CaCO_3$	A fold in the body wall called the mantle secretes the shell. A hard rasping radula is used for feeding
Annelida – marine bristleworms, oligochaetes, leeches	Mouth and anus	Bilateral	Internal cavity with fluid under pressure	Bodies made up of many ring-shaped segments, often with bristles. Blood vessels often visible
Arthropoda – insects, arachnids, crustaceans, myriapods	Mouth and anus	Bilateral	External skeleton made of plates of chitin	Segmented bodies and legs or other appendages with joints between the sections

1 Study the organisms shown in the drawings and assign each one to its phylum. [7]
2 List the organisms that are:
(a) bilaterally symmetrical [2]
(b) radially symmetrical. [2]
3 List the organisms that have
(a) jointed appendages [1]
(b) stinging tentacles [1]
(c) bristles. [1]
4 List the organisms that filter feed water pumped through tubes inside their body. [1]

Classification of humans

The classification of humans has caused more controversy than any other species. One reason for this is that a natural classification indicates the probable evolutionary origins of a species. Using standard taxonomic procedures, humans are assigned to the order Primates and the family Hominidae. There has been much debate about which, if any, of the great apes to include in this family. Originally all the great apes were placed in another family, the Pongidae, but research has shown that chimpanzees and gorillas are closer to humans than to orang-utans and so should be in the same family. This would just leave orang-utans in the Pongidae. Most evidence suggests that chimpanzees are closer than gorillas to humans, so if humans and chimpanzees are placed in different genera, gorillas should also be in a separate genus. A summary of this scheme for human classification is shown in Figure 7. More details of human evolution and of extinct species of hominid are given in Option D.

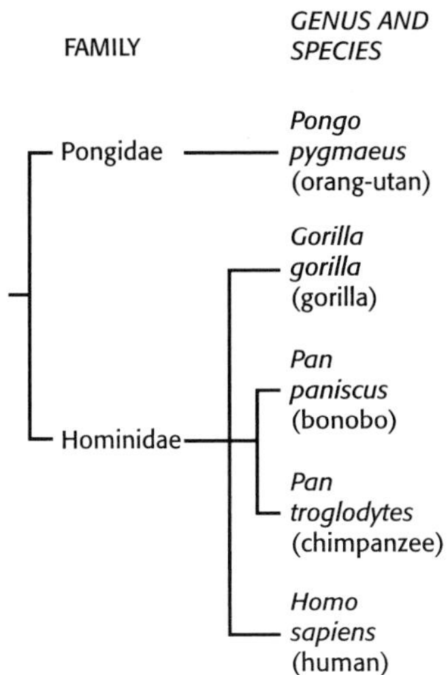

Figure 7 Classification of humans

Thinking about science: the unity of life

The words below were spoken in his garden by Marshall Nirenberg, who was awarded the Nobel Prize for Physiology or Medicine in 1968 for his work on the genetic code.

The finding that the code is universal had a terrific philosophic effect on me. I knew everything about evolution at the time, but these findings were so immediate and so profound, because I understood that most or all forms of life on this planet use the same genetic instructions and so we are all related. We're related to all living things and when I came in the garden and saw the plants, the squirrel and some of the birds, it really had a profound effect on me, which lasts to this day. I think that the feeling of being one with nature is very real and in fact is very true: we all use the same genetic language.

1 Why did the discovery of the universality of the genetic code have such a profound effect on Marshall Nirenberg and others at the time?
2 What are the implications of the recognition of the unity of life to scientists and other people?
3 Are there other examples of scientific discoveries causing a profound change in attitudes?

Critical considerations: race and racism

1 When Linnaeus did his pioneering work on the naming and classification of living organisms, he divided *Homo sapiens* into four varieties:

American: copper-coloured, choleric, regulated by custom
Asiatic: sooty, melancholic, governed by opinions
African: black, phlegmatic, governed by caprice
European: fair, sanguine, governed by laws.

2 UNESCO organised a symposium on the scientific status of race in 1952. The declaration was made that all men come from a common stock and belong to a single species. Insignificant differences between them give no support to claims of racial hierarchy.
3 Recent research has shown that humans differ from each other at only about 0.1% of the 3 million bases in the human genome.
4 There have been many investigations of human variation, including blood groups controlled by nearly 60 different genes, complex sets of skull measurements, and detailed studies of relationships between languages. These studies show a complex pattern of variation, especially in Africa. The closeness of relationships between indigenous humans in different parts of the world mirrors the history of human migration and the effects of natural selection.

1 Was Linnaeus racist, in his division of humans into four varieties or in his description of these varieties?
2 Was the UNESCO symposium correct in its statement that the differences between humans around the world are insignificant?
3 To what extent should efforts be made to preserve differences between humans?

Chapter 18 questions

1 Identify the phylum to which each of the plants shown in Figure 8 belongs, giving reasons for your answer.

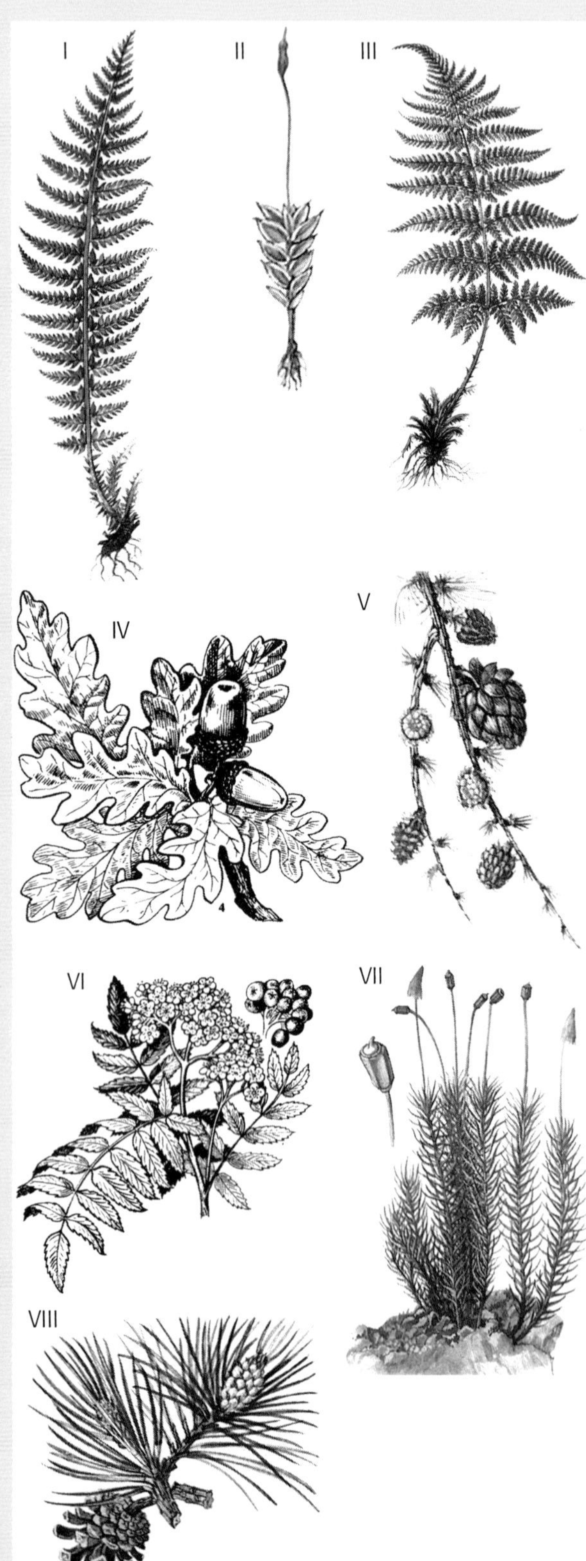

Figure 8

2 In Figure 9, each number represents a species. The closer that two numbers are on the diagram the more similar the two species. The circles represent taxonomic groups. For example, the diagram shows that 2, 3, 4 and 5 are in the same genus.

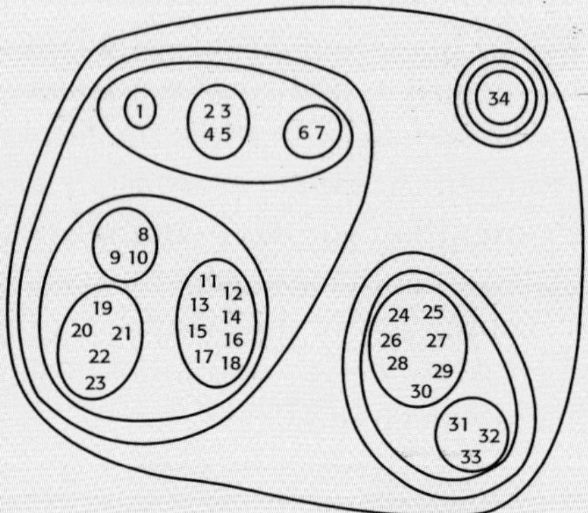

Figure 9

(a) State one species that is in a genus with no other species. [1]
(b) State the species that are in a family with two genera. [2]
(c) State the species that are in an order with two families. [2]
(d) State the species that are in a class with three orders. [2]
(e) Deduce whether species 8 is more closely related to species 16 or species 6.
(f) Explain why three concentric circles have been drawn around species 34 on the diagram. [2]

3 (a) Distinguish between Porifera and Mollusca using external recognition features. [6]
(b) Compare the external recognition features of Annelida and Arthropoda. [6]
(c) Compare the external recognition features of Cnidaria and Platyhelminthes. [6]

4 (a) State the group that humans are placed in at each of the seven levels in the hierarchy of taxa. [7]
(b) Outline the binomial system that is used for naming living organisms. [5]
(c) Explain how keys are designed to allow organisms to be identified. [6]

Digestion

Humans can eat a wide variety of foods, containing many different biochemicals. Most of these biochemicals are useful to the body but cannot be absorbed or used in the form they are in when they are eaten–they have to be digested. Proteins, polysaccharides and lipids are all digested. This is essential, because they consist of large molecules, which cannot be absorbed through the wall of the gut into the bloodstream. The membranes of cells in the wall of the gut are impermeable to large molecules. Digestion is a chemical process. The large food substances are converted by chemical reactions into smaller molecules such as monosaccharides and amino acids. These small molecules can pass easily through the gut wall.

There is a second reason for digesting food molecules. Plants, fungi or other animals have produced them, to perform functions in those organisms. They are unlikely to be ideally suited to perform functions in the human body. By breaking down proteins and other macromolecules, the human body can obtain all the subunits that it needs to build up its own macromolecules. For example, there are nine amino acids that humans must obtain by digesting proteins in food. Other amino acids can be made from these nine and all human proteins can be synthesized. Anthropologists have observed that all of the world's great civilizations have eaten and digested both legumes and cereals. Figure 1 shows how legumes and cereals supply all of the nine essential amino acids.

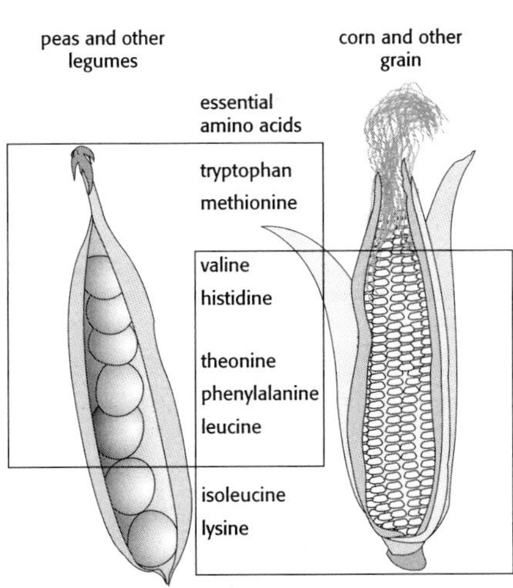

Figure 1 Essential amino acids obtained from legumes and grains

Rediscovering biology: the need for digestion

Set up the apparatus shown in Figure 2 and leave it for one hour.

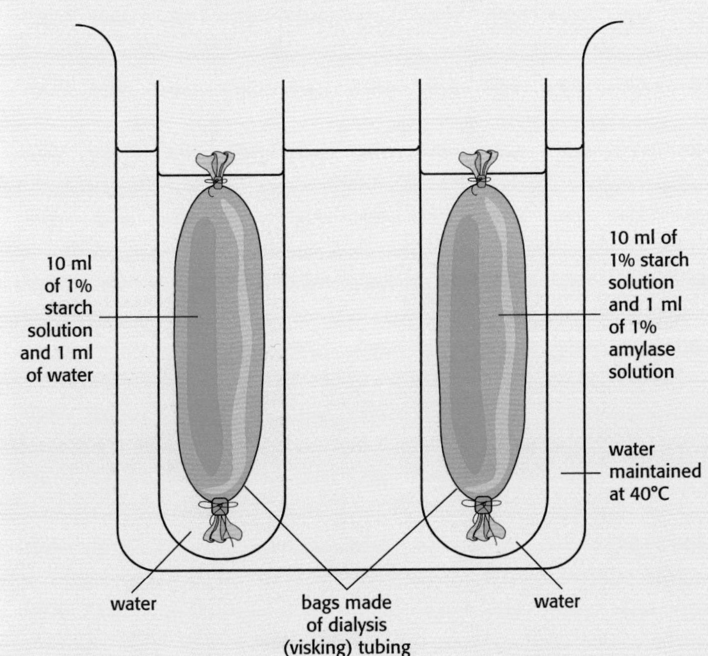

Figure 2 Apparatus for showing the need for digestion

Results

To obtain the results for the experiment, take the bags out of each tube, open them and pour the solutions from them into separate test tubes from the liquids in the tubes. You should now have four samples of fluid. Divide each of these samples into two halves and test one half for starch and the other half for sugars.

Record all the results in the way that you think is most appropriate.

Conclusions and evaluation:

State carefully all the conclusions that you can make from your results.

Discuss the strengths and weaknesses of this method of investigating the need for digestion.

Suggest improvements to the method, or suggest an entirely different method of investigating the need for digestion.

Enzymes in digestion

Enzymes are used to speed up the process of digestion. The activity of enzymes was described in Chapter 7, including the effect of temperature. By maintaining a constant body temperature of 37°C, digestive enzymes are encouraged to work relatively quickly, but it should be emphasized that 37°C is lower than the optimum for these enzymes.

Another characteristic of enzymes is specificity, though some digestive enzymes are actually rather broader in the range of substrates that they will work on than typical enzymes. For example, proteases will cope with proteins generally, rather than a single one such as hemoglobin. The main digestive reactions that are catalysed are shown in the equations (right).

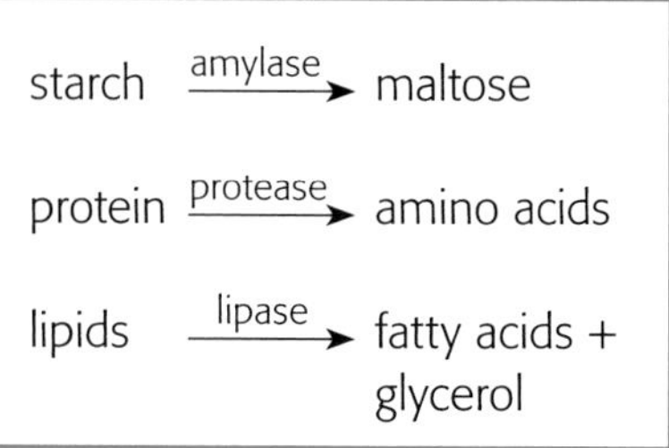

The enzymes of digestion are all made in glands and are secreted into the gut, where they mix with the food. Table 1 shows the glands involved, the optimum pH and where each enzyme works.

Table 1

Enzyme	Main secreting glands	pH optimum of enzyme	Region where enzyme works	pH of region where enzyme works
amylase	1. salivary glands	7	mouth	6.0–7.0
	2. pancreas	7	small intestine	7.0–8.3
protease	1. stomach wall	2	stomach	1.0–3.5
	2. pancreas	7	small intestine	7.0–8.3
lipase	pancreas	7	small intestine	7.0–8.3

Investigating biology: saliva and starch digestion

1 The effect of starch on iodine solution

Put 5 ml of 1% starch solution in a test tube, add five drops of iodine solution and observe the colour. Repeat using a second tube and water instead of starch solution. If a colorimeter is available, measure the absorption using a filter that you think is suitable. If either of the solutions is too opaque to give a colorimeter reading, dilute it with water.

2 The effect of saliva on starch

Collect as much saliva as possible in a small beaker. Half fill a 250 ml beaker with water at 37°C. Put 5 ml of 1 per cent starch solution into each of two test tubes. Add 1 ml of saliva to one of the tubes and mix it well. Put the tubes into the beaker of warm water. Leave them for 20 minutes and then add five drops of iodine solution to each of the tubes. Either note the colours, or measure the absorbance with a colorimeter.

3 The effect of boiled saliva on starch

Boil a small sample of the saliva for 2 minutes, then repeat the experiment using this boiled saliva.

4 Designing an investigation

Choose a research question concerning the digestion of starch by saliva (see Figure 3). An example would be: "Does smoking reduce the starch digesting activity of saliva?" Design a method for investigating your research question. Think carefully about the variables that can affect starch digestion and how you are going to vary or control them.

Please note:
local regulations for the use of body fluids in science investigations must be observed.

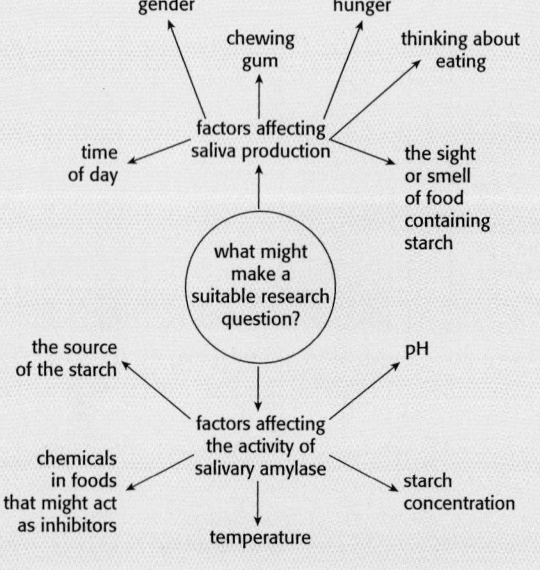

Figure 3 Mind map of ideas for research questions

Structure of the digestive system

The part of the human body used for digestion can be described in simple terms as a tube through which food passes from the mouth to the anus. The wall of the tube is made of living tissues, which include:

- muscle fibres, to push the food along
- secretory cells, which produce mucus for lubrication and enzymes for digestion
- a single layer of epithelium cells, which form the inner lining and absorb digested food.

This tubular structure is usually called the alimentary canal. It consists of a series of sections, each composed of tissues and having specific functions. The sections are therefore organs and together they form a system. There are also some associated glands that are part of the digestive system. Figure 4 shows the organs, their positions and interconnections.

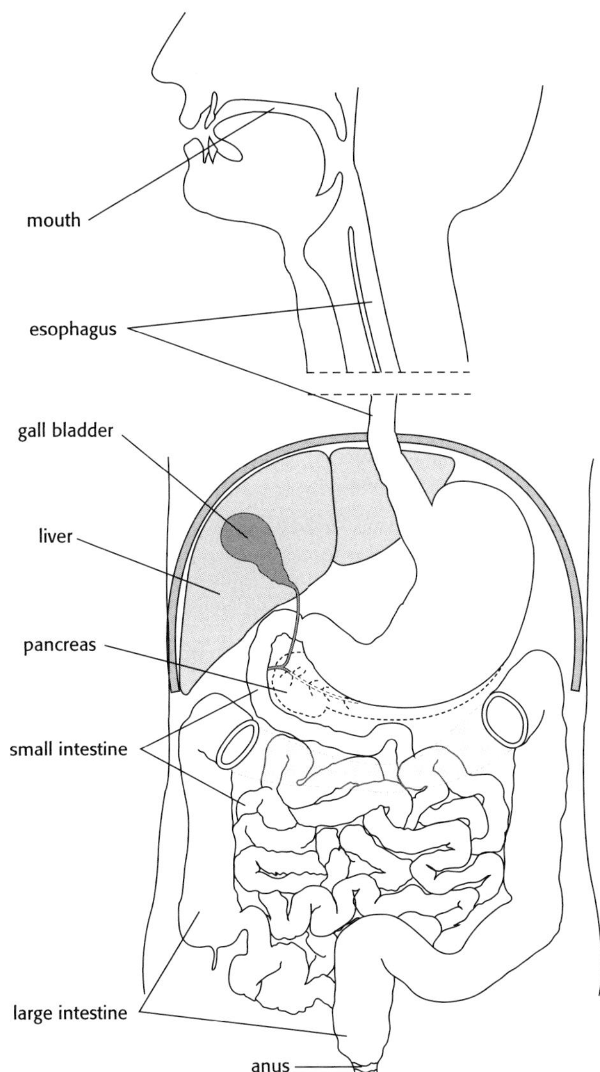

Figure 4 The human digestive system

Data-based question: the wall of the esophagus

Figure 5 shows the wall of the esophagus in transverse section.

1 Draw a plan diagram to show the layers of tissue in the wall. [5]

2 The inner tissue consists of many layers of epithelium cells, whereas in the stomach and intestines there is only a single layer of epithelium cells. Suggest reasons for the difference. [2]

3 There are two layers of muscle fibres in the wall, labelled X and Y. One of them contains circular muscle fibres and the other contains longitudinal muscle fibres. Deduce from the appearance of the muscle fibres in the micrograph which layer is which. [2]

4 Food only remains in the esophagus for a few seconds. There are small glands in the wall of the esophagus, with ducts leading from the glands to the lumen of the esophagus. Suggest a fluid that it would be useful for the small glands to secrete, and its function. [2]

5 There are small blood vessels in the wall of the esophagus. Suggest two reasons for the esophagus wall needing a supply of blood. [1]

6 In the lower part of the esophagus, the outer layer of muscle fibres is smooth muscle, but in the upper part of the esophagus, nearest to the mouth, this outer layer is striated muscle. Suggest reasons for this difference. You may need to research this question. [2]

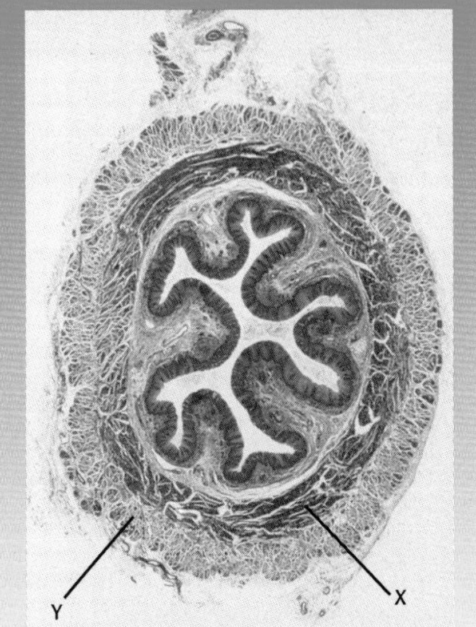

Figure 5 Transverse section of the esophagus

Functions of the stomach and small intestine

Stomach

1 The wall of the stomach is glandular, secreting mucus, hydrochloric acid and protease (see Figure 6).
2 Hydrochloric acid creates very acid conditions that kill bacteria in food that has been swallowed. Without this, we would often catch food poisoning or other diseases from the food that we eat.
3 Acid conditions denature proteins and suit protein digestion by hydrolysis. Protease secreted in the stomach, called pepsin, begins the process by breaking up polypeptides into shorter chains of amino acids.
4 The wall of the stomach is elastic, so large meals can be swallowed quickly, stored and then gradually released into the small intestine.
5 The wall of the stomach is muscular and contracts rhythmically to squeeze the food, mixing it and breaking it into smaller lumps.

Small intestine

1 Pancreatic juice is secreted into the small intestine. It contains protease, amylase and lipase, so the three main groups of macromolecule are all digested here.
2 The wall of the small intestine secretes a variety of other enzymes, which digest other substances. Nucleases, for example, are secreted to digest DNA and RNA in foods eaten. Because of the great length of the small intestine, food spends hours passing through, giving time for digestion to be completed.
3 Digested foods are absorbed by the small intestine. This process is described on the page after next.
4 Areas in the wall of the small intestine called Peyer's patches contain lymphocytes that detect pathogenic organisms in the food in the small intestine and help in the production of antibodies against them.

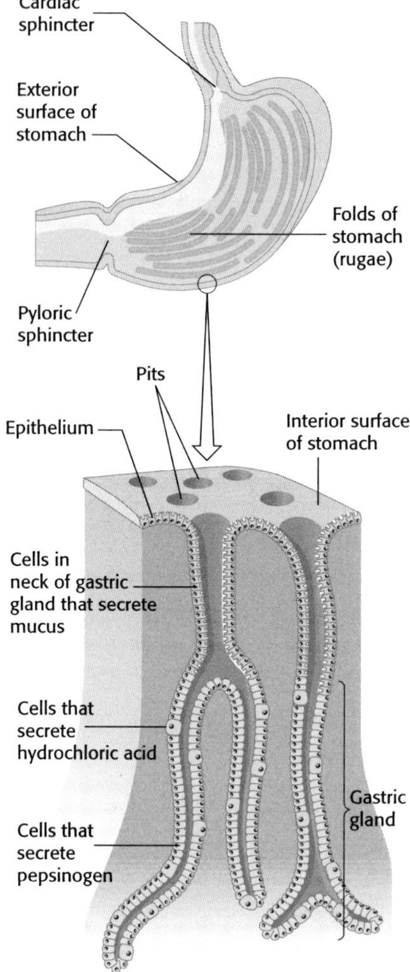

Figure 6 Structure of the stomach

Rediscovering biology: digestion of lipids

Digestion of lipids takes place in the small intestine, with lipase secreted by the pancreas. An investigation of the activity of lipase was described in Chapter 7 on page 74. A modified version of this experiment can be used to monitor lipid digestion by data logging. Double cream is used, to give a higher ratio of lipid to protein and other components of milk. The apparatus used in the experiment is shown in Figure 7.

As an alternative, samples could be taken at intervals from the reaction mixture and tested with pH indicator on a spotting tile.

The concentration of lipid can be reduced by diluting the cream with water. The results in the table below were obtained using a dilution of 1 part of cream to 9 parts of water (10%).

Table 2

Time after start/ minutes	0	10	20	30	40	50
pH	8.2	7.6	6.8	6.3	5.9	5.7

1 Why does the pH fall when lipase digests lipids?
2 If the results in the table had been obtained with undiluted cream, or with 99 per cent diluted cream, how would they have been different?
3 If a lump of butter was used in the experiment, with the same mass as the lipid in the 10% diluted cream, how would the results be different?

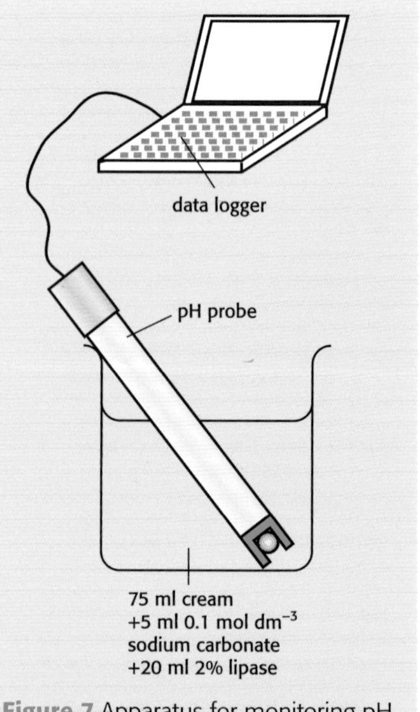

Figure 7 Apparatus for monitoring pH change

Absorption and assimilation

The lining of the digestive system must form a barrier to harmful substances, while at the same time being permeable enough to allow the food substances needed by the body to pass through. The process of taking in substances is called **absorption** and it happens principally in the small intestine. There are structures called villi, which are adapted to carry out absorption. Villi are described on the next page. Some harmful or unwanted substances and even some bacteria do pass through into the blood. However, blood from the small intestine flows directly to the liver (see Figure 8). As the blood perfuses through the liver, harmful substances are detoxified and phagocytic cells engulf bacteria.

After blood has passed through the liver, it is pumped to the lungs and then on to the rest of the body, carrying with it the useful substances. Cells absorb the substances that they need. Energy-containing lipids and glucose may be used in cell respiration. Other substances, including amino acids, fatty acids and glycerol, are built up into macromolecules, which then become part of the structure of the body. This process is called **assimilation**. By the end of the small intestine, most useful substances have been absorbed and all that remains is indigestible matter, bacteria, water and mineral ions. Most of the remaining water and mineral ions are absorbed by the large intestine, leaving just the undigested matter and bacteria, which form the feces that pass out periodically.

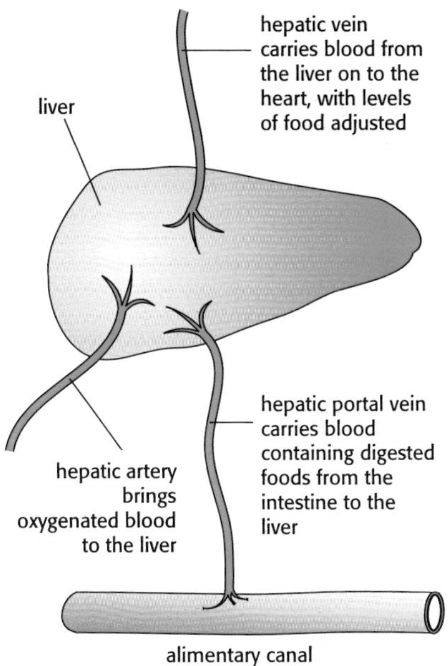

Figure 8 Blood flow to and from the liver

Thinking about science: finding the best treatment for childhood diarrhea

Acute diarrhea causes an estimated two million deaths annually in developing countries, most of them children under five. Large volumes of liquid feces are lost, causing potentially fatal dehydration. The World Health Organization has recommended giving children a solution of 90 mmol dm^{-3} of sodium with 111 mmol dm^{-3} of glucose to drink. The sodium and glucose encourage water absorption in the intestines. This treatment is called oral rehydration. Some clinicians have suggested that a lower concentration of sodium and glucose would provide a more effective treatment and have conducted trials. In 2001, medical researchers searched the literature to find the results of these trials. The effectiveness of the different treatments was assessed by how frequently the children needed emergency intravenous rehydration. A meta-analysis of 12 trials was done (Figure 9). The oral rehydration recommended by the WHO is the control and the lower concentration rehydration therapy is the treatment.

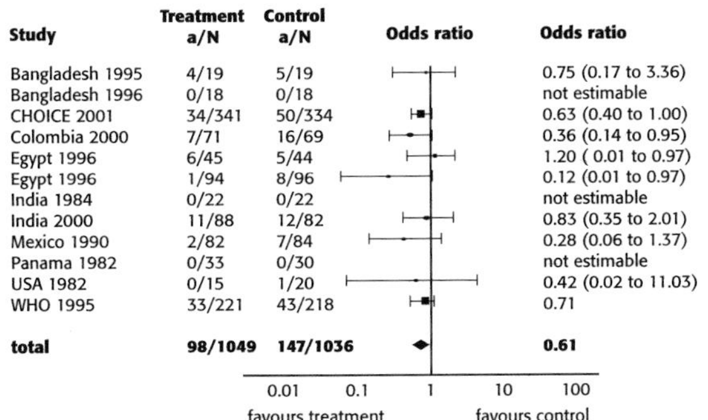

Study	Treatment a/N	Control a/N	Odds ratio
Bangladesh 1995	4/19	5/19	0.75 (0.17 to 3.36)
Bangladesh 1996	0/18	0/18	not estimable
CHOICE 2001	34/341	50/334	0.63 (0.40 to 1.00)
Colombia 2000	7/71	16/69	0.36 (0.14 to 0.95)
Egypt 1996	6/45	5/44	1.20 (0.01 to 0.97)
Egypt 1996	1/94	8/96	0.12 (0.01 to 0.97)
India 1984	0/22	0/22	not estimable
India 2000	11/88	12/82	0.83 (0.35 to 2.01)
Mexico 1990	2/82	7/84	0.28 (0.06 to 1.37)
Panama 1982	0/33	0/30	not estimable
USA 1982	0/15	1/20	0.42 (0.02 to 11.03)
WHO 1995	33/221	43/218	0.71
total	**98/1049**	**147/1036**	**0.61**

Figure 9 Meta-analysis summarizing data from studies of diarrhea treatments

1. Determine how many of the trials show the lower concentration treatment as more effective.
2. Discuss the advantage of a meta-analysis over the use of the results of only one trial.
3. Explain why only randomized, controlled trials were included in the meta-analysis.
4. Only trials where the patients did not know whether they had been given the higher or lower concentration of oral rehydration were included in the analysis. This is called concealment of allocation. Explain why other trials were not included.
5. Some of the trials involved a double blind protocol, where neither the patient nor the person giving the oral rehydration knew whether it was the higher or lower concentration. Explain why this makes the results of a trial more reliable.

Adaptations of villi for absorption

Villi are small finger-like projections that increase the area of exchange surfaces. In the small intestine they are between 0.5 and 1.0 mm long and there can be as many as 40 of them per square millimetre of small intestine wall. They increase the surface area by a factor of about 10. Apart from simply increasing the surface area for absorption, there are other ways in which the structure of villi in the small intestine is related to their role:

- Their outermost tissue is a single layer of epithelium cells, which are permeable to digested foods.
- They have a dense network of blood capillaries, close to the epithelium, so that the distance substances have to travel to reach the blood system after absorption is very small. Capillary walls always consist of a single layer of thin cells, with pores between adjacent cells, but these capillaries have larger pores than usual, allowing more rapid entry of foods.
- There is a branch of the lymphatic system in the centre of the villus called a lacteal. This absorbs lipoprotein droplets, which are not easily absorbed into blood capillaries. Lipoproteins are formed from absorbed fatty acids and glycerol, with a coating of protein to help them remain suspended in the lymph.
- There are muscle fibres around the lacteal, which contract to make the villus shorter, squeezing fluid along the lacteal.

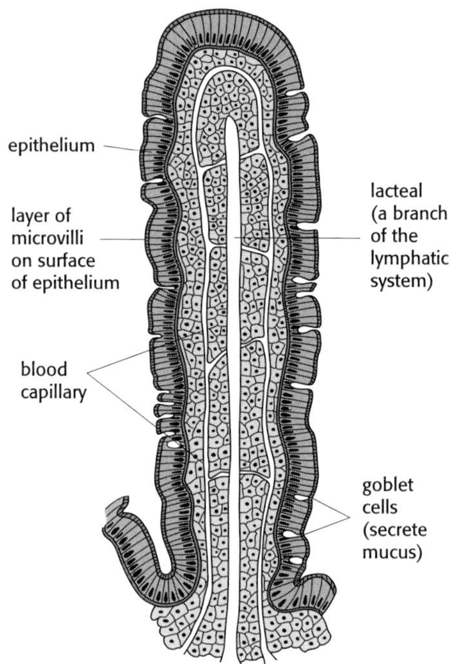

Figure 10 Structure of an intestinal villus

Thinking about science: fruit juice and senile dementia

Richard Smith, a former editor of the *British Medical Journal*, wrote the following article. It was published *The Guardian* on 5 September 2006.

Who knows how many people have started guzzling fruit juice thanks to the blizzard of publicity surrounding an article in the American Journal of Medicine *last week that suggested that drinking the stuff might help fend off Alzheimer's disease. A survey of 1836 Japanese Americans, the journal reported, found that those who drank fruit juice at least three times a week were not as likely as those who consumed it less than once a week to develop 'probable Alzheimer's disease'. It may be that fruit juice does protect against dementia, but my bet is that it doesn't. We've been here before.*

A study like this has many potential problems, but the biggest is that those who drink lots of fruit juice will be different in many ways from those who drink little. The people who drink more juice are likely to be better educated, richer, more concerned with their health, less likely to smoke, more likely to exercise and generally healthier. It's not surprising that they are less likely to develop dementia. The authors 'risk adjusted' the data, but there may be other factors that the scientists don't know about and cannot adjust for.

In the 80s, several studies showed that people who had high intakes of the antioxidant beta-carotene from eating fruit and vegetables were less likely to develop cancer. The evidence was 'convincing'. Scientists then conducted a randomized controlled trial in which some people were given beta-carotene and their chance of developing cancer was compared with those who weren't given the treatment. Such trials are the best scientific way of testing whether a treatment works. Sadly, those given beta-carotene proved more *likely to develop cancer.*

Science is cruel and hard. Many things that seem bound to do you good when properly tested prove to be harmful. Most Guardian *readers probably drink three glasses of fruit juice a week anyway, and I wouldn't stop. I drink as much myself. But nor should you bet your brain that it will prevent dementia. It may even make it more likely.*

1. What does Richard Smith mean when he writes, "Science is cruel and hard"?
2. Explain why he questions the conclusions of the study of Japanese Americans.
3. Explain the finding that those who had high intakes of beta-carotene from eating fruit and vegetables were less likely to develop cancer, when a randomized controlled trial showed that beta-carotene does not reduce the risk of cancer.
4. Describe how a randomized controlled trial could be done to test whether drinking fruit juice really does reduce the chance of dementia.

Chapter 19 questions

1 Which substance can increase in concentration while food is in the stomach?
A amylase B hydrochloric acid
C glucose D protein.

2 Where is the gall bladder located, and to which part of the digestive system is it connected by a duct? Choose from options A to D in the table below.

	Location	Connection
A	abdomen	small intestine
B	small intestine	liver
C	liver	pancreas
D	thorax	liver

3 What is a difference between absorption and assimilation?
A Food is assimilated before it is absorbed.
B All substances are absorbed but only some are assimilated.
C Assimilation involves substances entering the blood and absorption involves removing substances from the blood.
D Food is absorbed by the stomach and assimilated by the small intestine.

4 The disease cystic fibrosis often involves the pancreatic duct becoming blocked by sticky mucus. What is a consequence of this in the small intestine?
(i) Digestion of lipid is incomplete.
(ii) Digestion of protein is incomplete.
(iii) Digestion of starch is incomplete.
A (i) only
B (i) and (ii) only
C (ii) and (iii) only
D (i), (ii) and (iii).

5 During some illnesses, the body temperature rises from 37 °C to 40 °C or even higher. What is the consequence for digestion?
A Enzymes of digestion work slightly faster.
B Enzymes of digestion are denatured.
C Bacteria breed faster in the gut and cause food poisoning.
D Vitamins in food are destroyed so vitamin supplements should be eaten.

6 Why does food vomited from the stomach taste so unpleasant as it passes though the mouth?
A It contains hydrochloric acid.
B It contains partly digested proteins.
C It contains dead bacteria.
D It contains fatty acids.

7 The main organs of the digestive system are the esophagus, large intestine, mouth, small intestine and stomach.
(a) State the order in which food passes through these organs. [2]
(b) State which glands secrete fluid into each of these organs. [3]

8 (a) Draw and label a diagram of the digestive system. [6]
(b) Explain the need for enzymes in digestion. [4]
(c) Distinguish between the functions of the small intestine and the large intestine. [8]

9 An experiment was done in which genetically modified soya was fed to seven volunteers. The soya had a transgene of 2266 base pairs. An indigestible marker called PEG 4000 was also fed, to test how rapidly food passed through the small intestine. Each of the volunteers had a colostomy, so after passing through their small intestine, food passes out into a bag. This allowed food to be collected easily before it entered the large intestine. Figure 11 shows the percentage of the indigestible marker and of the transgene in the soya that had been recovered 6 hours after feeding.

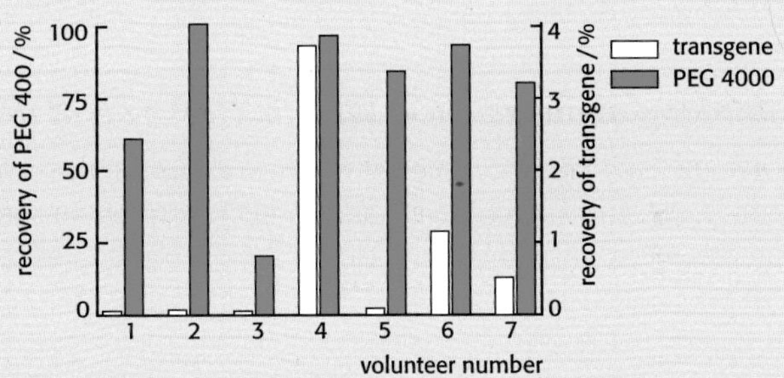

Figure 11

(a) (i) State the highest percentage of transgene that was recovered. [1]
(ii) Calculate the percentage of transgene that was not recovered from this person. [2]
(b) (i) Explain what has happened to transgenes not recovered from food. [3]
(ii) Predict what happens to the subunits of the transgenes. [2]
(c) Suggest **two** reasons for the variation in results between the seven people. [2]
(d) Explain why an indigestible marker was used in the experiment. [2]
(e) Evaluate the results of this experiment in terms of the safety of eating genetically modified food. [3]

20 The transport system

The main transport system in humans consists of blood, the heart to pump it and blood vessels through which it is pumped. It is called the cardiovascular system. Blood is unusual as it is a liquid tissue. Its structure is described on page 215. There are three types of blood vessels: arteries, capillaries and veins. They are described on page 214. Blood circulates through these blood vessels, with valves ensuring a one-way flow. There are in fact two separate circulations:

- the pulmonary circulation, to and from the lungs
- the systemic circulation, to and from all other organs, including the wall of the heart itself.

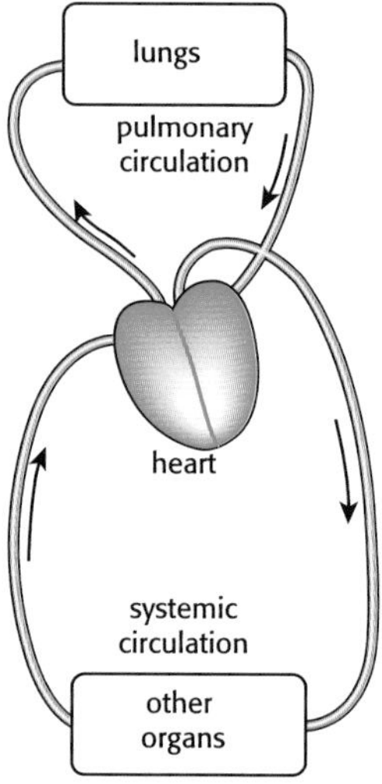

Figure 1 The double circulation

Figure 1 shows the double circulation in a simplified form. The pulmonary circulation receives deoxygenated blood that has returned from the systemic circulation, and the systemic circulation receives blood that has been oxygenated by the pulmonary circulation. It is therefore essential that blood flowing to and from these two circulations is not mixed. The heart ensures this by being a double pump, delivering blood under pressure separately to the two circulations.

Thinking about science: William Harvey and the circulation of blood

During the Renaissance, interest was reawakened in the classical writings of Greece and Rome. This stimulated literature and the arts, but in some ways it hampered progress in science. It became almost impossible to question the doctrines of such writers as Aristotle, Hippocrates, Ptolemy and Galen. According to Galen, blood is formed in the liver and is pumped to and fro between the liver and the right ventricle of the heart. A little blood passes into the left ventricle, where it meets air from the lungs and becomes "vital spirits". The vital spirits are distributed to the body by the arteries. Some of the vital spirits flow to the brain, to be converted into "animal spirits", which are then distributed by the nerves to the body. William Harvey was unwilling accept these doctrines without evidence. He made careful observations and did experiments, from which he deduced that blood circulates through the pulmonary and systemic circulations. He predicted the existence of capillaries, linking arteries and veins, even though the microscopes of the time were not powerful enough for him to see them. The following extract is from Harvey's book *On the Generation of Animals,* published in 1651 when he was 73.

And hence it is that without the due admonition of the senses, without frequent observation and reiterated experiment, our mind goes astray after phantoms and appearances. Diligent observation is therefore requisite in every science, and the senses are frequently to be appealed to. We are, I say, to strive after personal experience, not to rely of the experience of others: without which no one can properly become a student of any branch of natural science.

I would not have you therefore, gentle reader, to take anything on trust from me concerning the Generation of Animals: I appeal to your own eyes as my witness and judge.

The method of pursuing truth commonly pursued at this time therefore is to be held erroneous and almost foolish, in which so many enquire what things others have said, and omit to ask whether the things themselves be actually so or not.

1 William Harvey refused to accept doctrines without evidence. Are there any academic disciplines where doctrines can be accepted on the basis of authority, rather than evidence gathered from primary sources? Identify criteria that might be used when deciding whether to accept evidence provided by authority.
2 Harvey welcomed questions and criticism of his theories when teaching anatomy classes. Why do you think he did this?
3 Can you think of examples of the "phantoms and appearances" that Harvey refers to?
4 Why does Harvey recommend "reiteration" of experiments?
5 Harvey practised as a doctor, but after the publication in 1628 of his work on the circulation of the blood, far fewer patients consulted him. Why might this have been?
6 What advice might Harvey have given about using internet sources to obtain information?

The development and action of the heart

The human heart has developed from simpler hearts in earlier organisms. The description below and Figure 2 show how a complex organ like the heart could evolve in a series of stages.

(a) A first stage is the enlargement of a section of blood vessel with a wider lumen and a thicker, more muscular wall. This forms a **pumping chamber** called a **ventricle**.

(b) Two valves are needed, one to prevent blood flowing backwards into the vein when the ventricle is pumping, and another to prevent blood from flowing back into the ventricle from the artery when the ventricle stops pumping.

(c) The valve at the opening of the ventricle is closed when it is pumping blood, so a **collecting chamber**, called the **atrium**, is needed to hold blood arriving from the vein until the ventricle can receive it. The atrium has a thin wall as it only has to pump blood into the ventricle.

(d) The opening to the artery is positioned on the upper side of the ventricle, so that the artery and vein are attached to the same side of the heart, allowing it to contract more easily.

(e) The heart becomes a **double pump**, allowing a **double** circulation, with the pulmonary and systemic circulations supplied with blood separately.

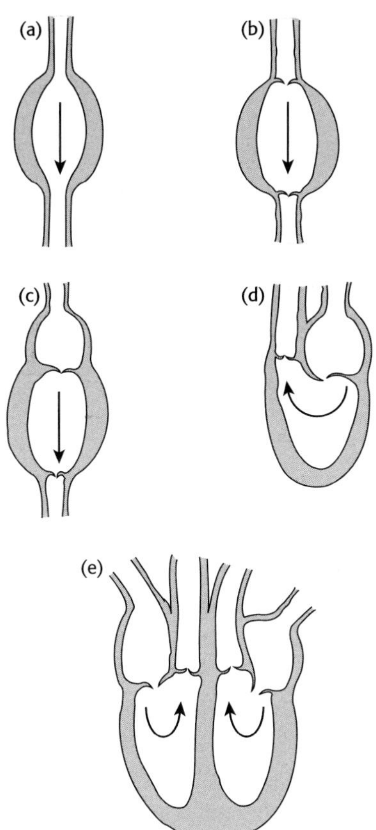

Figure 2 Stages in the development of a double heart

Data-based question: heart action and blood pressures

Figure 3 shows the pressures in the atrium, ventricle and artery on one side of the heart, during one second in the life of the heart.

1 State when blood is being pumped from the atrium to the ventricle. Give both the start and the end times. [2]
2 State when the ventricle starts to contract. [1]
3 The atrioventricular valve is the valve between the atrium and the ventricle. State when the atrioventricular valve closes. [1]
4 The semilunar valve is the valve between the ventricle and the artery. State when the semi-lunar valve opens. [1]
5 State when the semilunar valve closes. [1]
6 State when blood is being pumped from the ventricle to the artery. Give both the start and the end times. [2]
7 Deduce when the volume of blood in the ventricle is:
  (a) at a maximum [1]
  (b) at a minimum. [1]

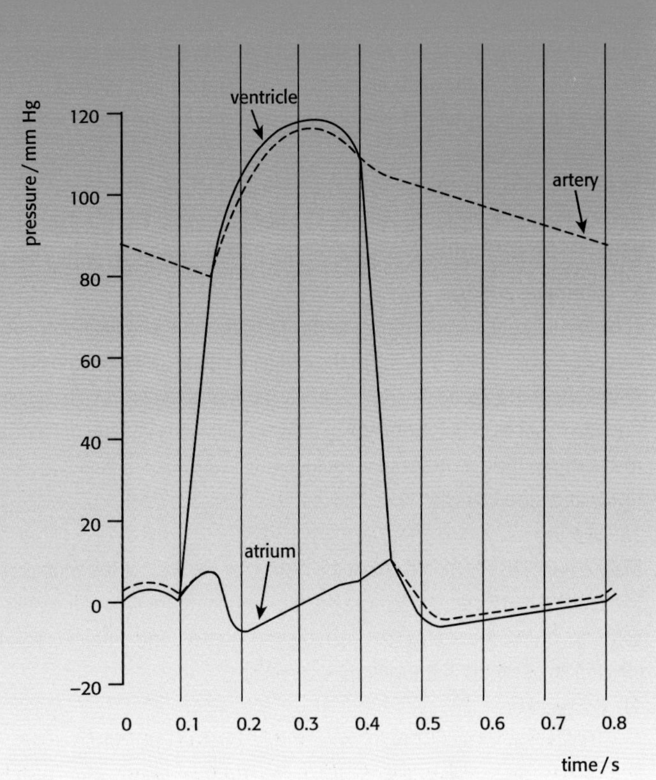

Figure 3 Pressure changes during the cardiac cycle

Rediscovering biology: dissection of the heart

A fresh specimen of a mammalian heart is needed, with blood vessels still attached, a dissecting dish or board and dissecting instruments.

1 Arteries and veins

Tidy up the blood vessels attached to the heart by removing membranes and other tissue from around them. Identify the thick-walled arteries and the thin-walled veins.

2 Pulmonary artery and aorta

Push a glass rod or other blunt-ended instrument into the heart through the arteries and feel through the wall of the heart to where the end of the rod has reached. Identify the pulmonary artery, through which you will reach the thinner-walled right ventricle, and the aorta, through which you will reach the thicker-walled left ventricle.

3 Dorsal and ventral sides

Lay the heart so that the aorta is behind the pulmonary artery, as in Figure 4. The ventral side is now uppermost and the dorsal side underneath. The dorsal side of an animal is its back.

4 Left ventricle

Identify the left ventricle. It has a smooth wall, with a tree-like pattern of blood vessels. Using a sharp scalpel, make an incision as shown by the dashed line X in Figure 4. This should open up the left ventricle. Look at the thick muscular wall that you have cut through.

5 Atrioventricular valve

Extend the incision further towards the atrium if necessary until you can see the two thin flaps of the atrioventricular valve. Tendons attached to the sides of the left ventricle prevent the valve inverting into the atrium.

6 Left atrium and pulmonary vein

Identify the left atrium. It will look surprisingly small as there is no blood inside it. The outer surface of its wall has a wrinkled appearance. Extend the incision that you have already made, either with the scalpel or with scissors, to cut through the wall of the left atrium as far as the pulmonary vein. Look at the thin wall of the atrium and the opening of the pulmonary vein or veins (there may be two).

7 Aorta

Find the aorta again and measure the diameter of its lumen, in millimetres. Using scissors, cut through the wall of the aorta, starting at its end and working towards the left ventricle. Look at the smooth inner surface of the aorta and try stretching the wall to see how tough it is.

8 Semilunar valve

Where the aorta exits the left ventricle, there will be three cup-shaped flaps in the wall. These form the semilunar valve. Try pushing a blunt instrument into the flaps to see how blood flowing backwards will get caught and push the flaps together, closing the valve.

9 Coronary artery

Look carefully at the inner surface of the aorta, near the semilunar valve. A small hole should be visible, which is the opening to the coronary arteries. Measure the diameter of the lumen of this artery. The coronary arteries supply the wall of the heart with oxygen and nutrients. Try pushing plastic filaments or tubing of different diameters into the opening of a coronary artery, to investigate the thickness of the route through which blood passes as the coronary arteries divide and form branches.

10 Septum

Make a transverse section through the heart near the base of the ventricles, along the dotted line marked Y in Figure 4. Measure the thickness in millimetres of the walls of the left and right ventricles and of the septum between them (Figure 5). The septum contains conducting fibres, which help to stimulate the ventricles to contract.

11 Pacemaker

A special region in the wall of the right atrium, marked on Figure 4, acts as the pacemaker. Dissect the right atrium to see if you can find the region where the pacemaker is located.

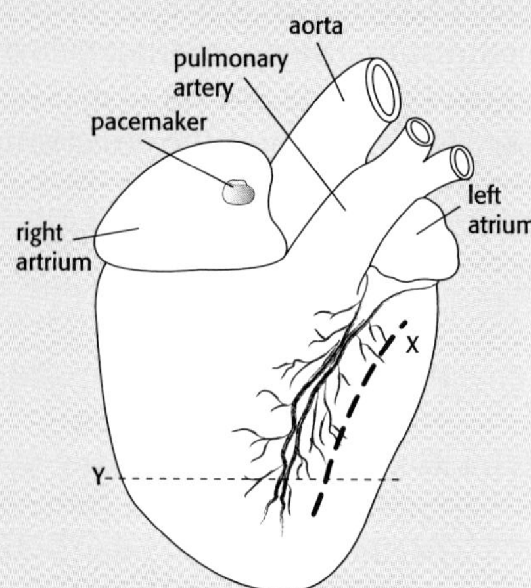

Figure 4 Ventral view of the exterior of the heart

1 Why is the wall of the atrium thinner than the wall of the ventricle?
2 What prevents the atrioventricular valve from being pushed into the atrium when the ventricle contracts?
3 Why is the left ventricle wall thicker than the right ventricle wall?
4 Does the left side of the heart pump oxygenated or deoxygenated blood?
5 Why does the wall of the heart need its own supply of blood, brought by the coronary arteries?
6 Does the right side of the heart pump a greater volume of blood per minute, a smaller volume, or the same volume?

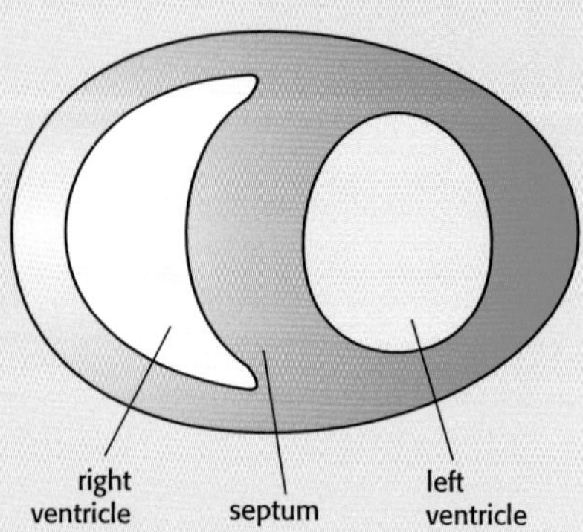

Figure 5 Transverse section through the ventricles

The control of the heart beat

Heart muscle tissue is unique in the body as it can contract without stimulation from a nerve. The contraction is called **myogenic**, meaning that the muscle itself stimulates it. If one heart muscle cell contracts, it stimulates any other cells that it is touching, so they also contract. A group of cells therefore contracts almost simultaneously at the rate of the fastest. The region of the heart that sets the overall pace of contraction is the **pacemaker** – a small group of special muscle cells in the wall of the right atrium.

The pacemaker can respond to signals from outside the heart. Branches of two nerves are connected to it, which bring signals from a region in the medulla of the brain. Impulses from one of the nerves cause the pacemaker to increase the rate, and impulses from the other nerve decrease the rate, rather like the throttle and brake of a car. The pacemaker also responds to the hormone adrenalin, by increasing the heart rate.

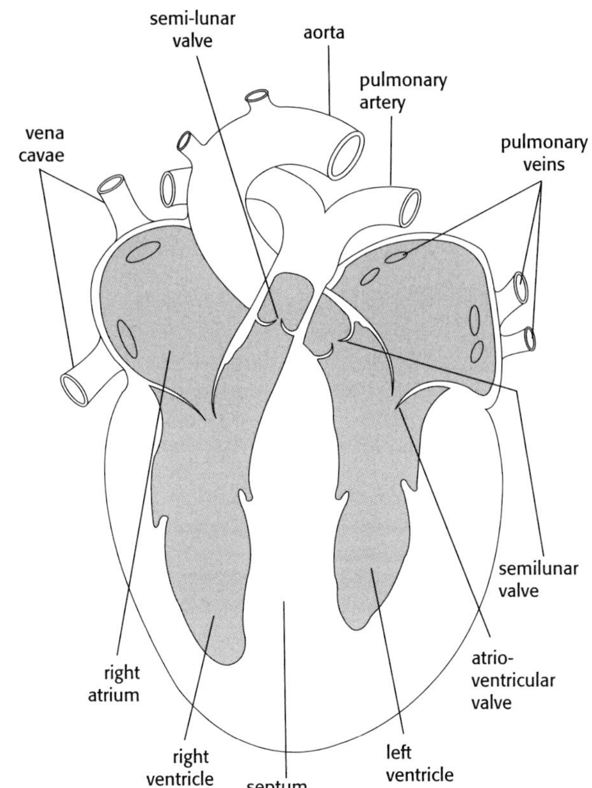

Figure 6 Structure of the heart

Investigating biology: factors affecting heart rate

Arteries have a pulse, which is equal to the heart rate and so can be used to measure heart rate.

1 Find your pulse while sitting down. You can either use the radial pulse in your wrist or the carotid pulse in your neck (see Figure 7). It may be sufficient to count for 15 or 30 seconds and multiply to get a rate per minute, rather than count for an entire minute. Repeat the count two or three times.
What is the reason for repeating the count?

2 Stand up and then take your pulse again.
What is the reason for the difference in heart rate between sitting and standing?

3 Repeat the counts using an electronic pulse-rate monitor.
Are the results the same as the counts that you took manually?
Are they more or less precise? This can be judged by how closely the repeats are clustered.
Are the results more or less accurate? This is how close the results are to a true rate that you would obtain with no errors in your counts.
Are there more or less errors and uncertainties than with manual counts?

4 Design an investigation of a factor that might affect the heart rate. Make sure that you do the following things:
   - define a clear and focused research question
   - develop methods for controlling all the variables that might influence heart rate during the experiment
   - develop methods for collecting sufficient data to allow you to answer your research question.

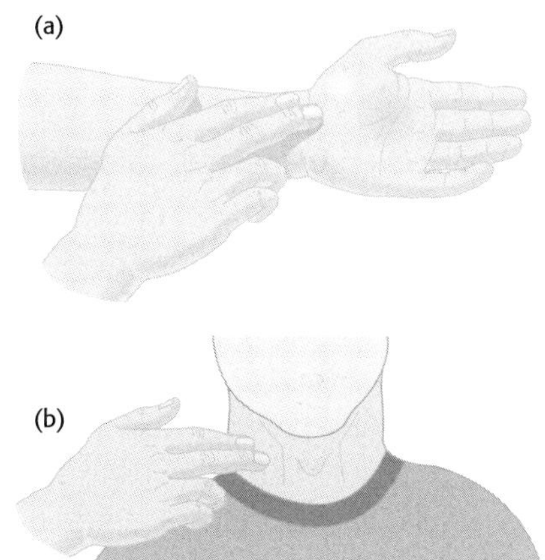

Figure 7 Taking the pulse: (a) radial pulse (b) carotid pulse

Blood vessels

The walls of blood vessels consist of a series of layers:

- tunica intima – an inner layer consisting of a single layer of epithelium cells, which forms a low-friction lining
- tunica media – a layer of tissue containing circular muscle fibres and elastin fibres, which can stretch to accommodate blood and then contract to pump the blood onwards
- tunica adventitia– an outer layer of tissue containing longitudinal collagen and elastin fibres, which increases the overall strength of the wall and links it to surrounding connective tissue.

These layers vary considerably in thickness and are not all present in all blood vessels. There is a close relationship between the structure of blood vessel walls and their functions. Use the information in Table 1 to study the relationship between the functional requirements and the structural features of arteries, capillaries and veins.

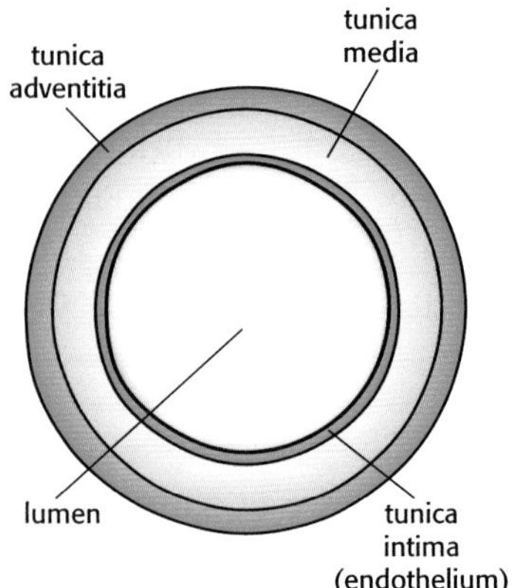

Figure 8 General structure of blood vessels

Table 1

Vessel	Functional requirements	Structural features
Arteries	1 Accommodate large volumes of blood pumped out by the ventricles of the heart and then pump it on to capillary networks. 2 Withstand high blood pressures without the artery wall bulging outwards (aneurysm) or bursting (hemorrhage).	A No valves apart from the semilunar valves where the artery and ventricle are connected. B Thick tunica media, with many muscle fibres and elastin fibres. C Moderately thick tunica adventitia, which contains many collagen and elastin fibres.
Capillaries	1 Transport blood as close as possible to all living cells of the body. 2 Allow the substances that blood is transporting to enter and leave rapidly. 3 Allow part of the blood plasma to leak out into tissues, but not proteins or blood cells.	A Wall consists only of one layer of very thin endothelium cells, coated by a filter-like protein gel and with pores between the cells. B Very narrow diameter – about 10 µm. C Much branching and rejoining to form a capillary networks, with a huge total length.
Veins	1 Transport of blood, without leaks, from capillary networks to the atria of the heart. 2 Prevention of backflow when blood pressure or flow rate is low. 3 Flexibility to be squeezed by adjacent muscles, to help pump blood.	A Pocket valves, consisting of three cup shaped flaps of tissue. B Thin tunica media containing a few muscle and elastic fibres. C Moderately thick tunica adventitia, which contains many collagen fibres.

Rediscovering biology: comparing arteries and veins

1 Structure

Examine a transverse section of an artery and vein, using a microscope. Draw labelled tissue plans to show the layers that make up the wall of the artery and vein.

2 Elasticity of the wall

If you can obtain samples of artery and vein, for example from a specimen of a lamb's heart, their elasticity can be compared. Hang a weight from each blood vessel and measure how much the wall of the blood vessel stretches. Remove the weight and re-measure to see if the wall of the blood vessel has returned to its original width.

3 Tensile strength of the wall

Using the same apparatus and samples of artery and vein, measure the strength by testing the samples to destruction. Add more and more weights until each of the blood vessels breaks.

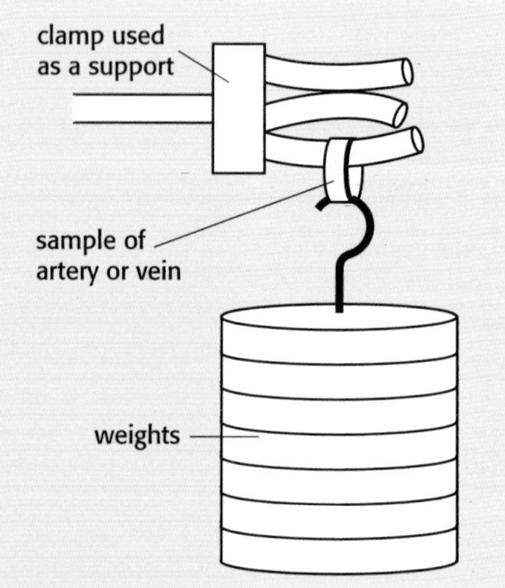

Figure 9 Testing blood vessels

The structure of blood

Most tissues consist mainly of cells, sometimes also with some solid material between the cells. Fluid is also found between cells, but the volume of this is usually small. In contrast, blood normally has a greater volume of intercellular fluid than of cells. Blood can therefore flow – it is a liquid tissue.

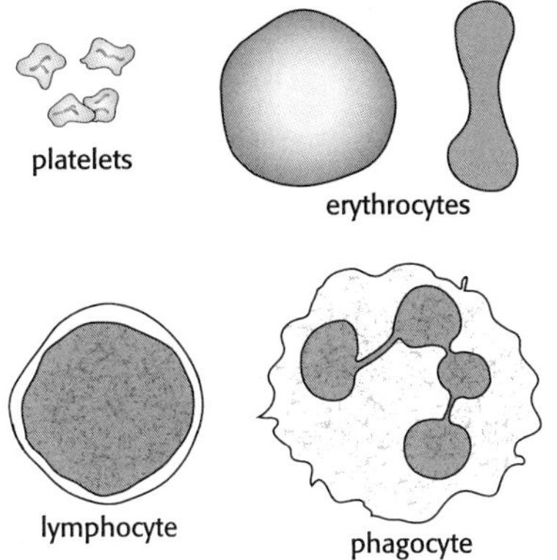

Figure 10 Blood cells

The fluid between blood cells is called **plasma**. It is mainly water, but also contains a wide range of other substances: dissolved nutrients such as glucose, some dissolved oxygen and carbon dioxide, waste products such as urea, hormones, mineral salts and plasma proteins, including antibodies. Of these, mineral salts and plasma proteins circulate continuously in the blood, but the other substances are being transported from one part of the body to another. In addition to transporting these substances, blood also takes heat from warmer parts of the body and carries it to cooler parts.

Plasma makes up 55 per cent of the blood volume. The remainder consists of blood cells and **platelets**. The majority of the cells, about 99.8 per cent of them, are red blood cells (erythrocytes). These are the cells that contain hemoglobin and greatly increase the blood's oxygen-carrying capacity. The remaining 0.2 per cent are white blood cells, or **leucocytes**, whose job it is to fight disease. There are many types of leucocyte, but a simple classification of them is into **phagocytes**, which engulf foreign material, and **lymphocytes**, which produce antibodies. The activities of these cells are described in Chapter 21.

Rediscovering biology: microscopic examination of human blood

Check local regulations on taking samples of human blood before continuing with this activity. The risk of blood-to-blood infections must be remembered at all times and the protocols used must ensure that cross-infection is impossible.

1. Sterilize an area of skin on the end of a finger.
2. Use a sterile lancet to puncture the skin and then either put the lancet into a beaker of disinfectant or a secure sharps container.
3. Blood should emerge from the skin puncture. Squeeze the finger if insufficient blood emerges. Place a small drop of the blood near the end of a sterile microscope slide. A drop of about 2 mm diameter is enough.
4. Use a second microscope slide to draw the drop of blood across the first slide, as shown in Figure 11. This must be done as quickly as possible before the blood starts to clot. Dispose of this second slide by putting it in a beaker of disinfectant
5. Wait until the blood has dried.
6. Fix the cells by covering the area of blood with drips of 100 per cent methanol. Pour off the methanol.
7. Cover the area of blood with Leishman's stain solution and leave for 3–5 minutes. The solution is made by diluting Leishman's stain 50:50 with pH 6.8 buffer solution.
8. Rinse the slide by dripping pH 6.8 buffer solution over the slide.
9. Wait until the slide is dry and then examine using a microscope.
10. After examining the slide, put it into a beaker of disinfectant.

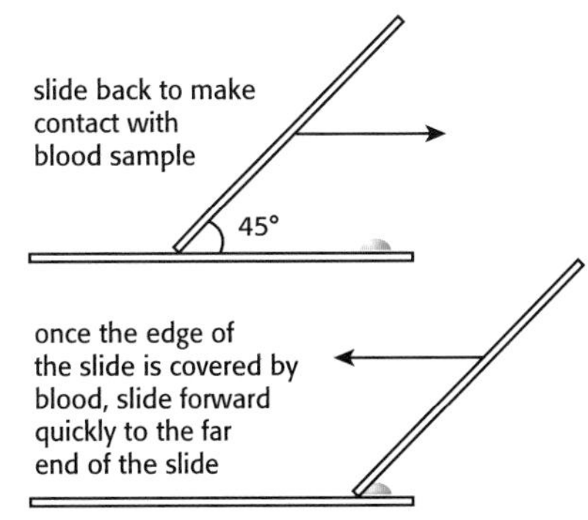

Figure 11 Making a blood smear

1. What infections can be transmitted by blood-to blood contact?
2. Why is it necessary to use sterile lancets and sterile or new microscope slides in this activity?
3. After reading the instructions (left), can you think of any ways in which cross-infection by blood-borne diseases could occur?

Chapter 20 questions

1 The micrograph below (Figure 12) shows human blood.
   (a) (i) Identify the cells labelled I and II. [2]
      (ii) Outline the characteristics of these cells that allowed you to identify them. [2]
   (b) Distinguish between the cells labelled II and III. [2]
   (c) (i) The actual diameter of the erythrocytes is 7 μm. Calculate the magnification of the micrograph. [2]
      (ii) Using this magnification, calculate the actual diameters of the largest and smallest leucocytes in the micrograph. [2]

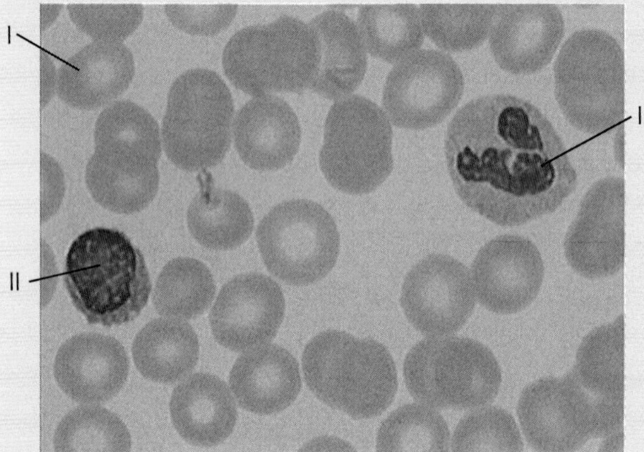

Figure 12

2 To pump blood, the heart goes through a series of actions, called the cardiac cycle. These actions are listed below in alphabetical order. Arrange them in the correct sequence, starting with the contraction of the atria. Some actions happen simultaneously:

- atria start to contract
- atria start to fill with blood
- atrioventricular valves close
- atrioventricular valves open
- blood is pumped from atria to ventricles
- blood is pumped from ventricles to artery
- semilunar valves close
- semilunar valves open
- ventricles start to contract
- ventricles start to fill with blood. [10 × 1]

3 (a) State where:
      (i) carbon dioxide is removed from the blood
      (ii) oxygen diffuses into the blood
      (iii) oxygen diffuses out of the blood
      (iv) glucose and other nutrients enter the blood
      (v) hormones enter the blood
      (vi) urea is removed from the blood. [6 × 1]
   (b) (i) Suggest two parts of the body where blood usually loses thermal energy. [2]
      (ii) Suggest one part of the body where blood usually gains thermal energy. [1]
   (c) Suggest one component of blood plasma that increases in concentration during infections. [1]

4 The electrical signals that are used by the heart to control its beating can be detected using electrodes placed on the chest. A recording of the signals can be produced, called an electrocardiogram, or ECG. On an ECG there is a repeating series of waves. The normal pattern of a waves is shown in Figure 13.

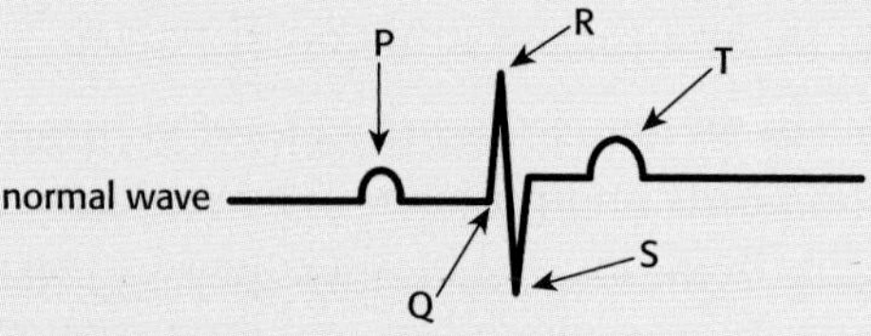

Figure 13

P is caused by the contraction of the atria.
QRS is caused by the contraction of the ventricles.

   (a) Deduce what is happening at the part of the wave called T. [1]
   (b) If the gaps between adjacent P waves on an ECG are measured, what can be deduced about the action of the heart? [1]

The two ECG waves in Figure 14 show typical waves for a person in the early stage and a later stage of an acute heart attack.

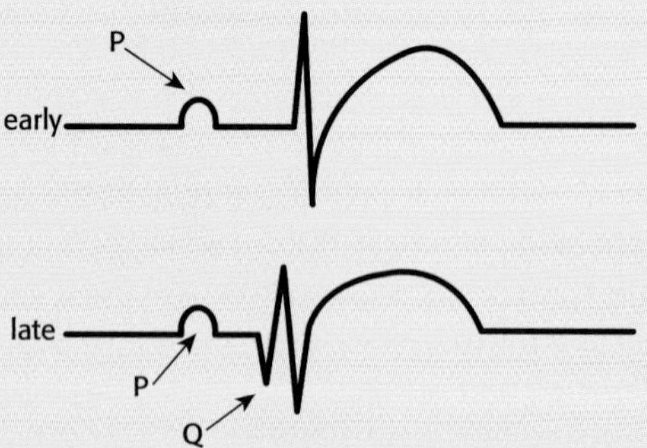

Figure 14

   (c) Compare the normal waves with the waves seen in the early stage of a heart attack. [4]
   (d) Distinguish between the waves seen in the early and later stages of a heart attack. [2]
   (e) Suggest who should be trained to recognize the signs of heart attack in ECG recordings. [2]

21 Defence against infectious disease

A pathogen is a disease-causing organism or an infectious biological agent such as a virus. A broad range of organisms can act as pathogens in humans, such as bacteria, fungi, protozoa and animals such as flatworms and roundworms. Table 1 lists the "big three" pathogens, i.e. the three infectious diseases that claim the most lives globally.

Table 1 The "big three" diseases

Disease	Cause	Pathogen
tuberculosis	bacterial infection	*Mycobacterium tuberculosis*
AIDS	viral infection	human immunodeficiency virus (HIV)
malaria	protozoan infection	*Plasmodium* sp.

Critical considerations

Table 2 lists 13 diseases and the pathogens that cause them. These diseases are classified as the "neglected tropical diseases" and share the following criteria.

- They have burdened humanity for centuries.
- They are poverty-promoting conditions.
- They are associated with social stigma.
- They are most common in low-income countries.
- Effective, low cost treatments are available.

1 Suggest factors that might lead these diseases to be neglected.
2 Suggest reasons why these diseases do not occur in high-income countries.
3 To what extent does using terms such as "big three diseases" and "neglected diseases" limit or affect our thinking?
4 Discuss how "social stigma" might be a barrier to eradication programmes and suggest how it might be overcome.

Table 2 The 13 "neglected tropical diseases" in Africa

Disease	Pathogen
Protozoan infections	
African trypanosomiasis	*Trypanosoma gambiense, T. rhodiense*
Kala-azar (visceral leishmaniasis)	*Leishmania donovani*
Helminth (worm) infections	
Soil-transmitted infections:	
Ascariasis	*Ascaris lumbricoides*
Trichuriasis	*Trichuris trichiura*
Hookworm infection	*Necator americanus*
Schistosomiasis:	
Urinary schistosomiasis	*Schistosoma haematobium*
Hematobiliary schistosomiasis	*Schistosoma mansoni*
Lymphatic filariasis	*Wuchereria bancrofti*
Onchocerciasis	*Onchocerca volvulus*
Dracunculiasis	*Dracunculus medinensis*
Bacterial Infections	
Trachoma	*Chlamydia trachomitis*
Leprosy	*Mycobacterium leprae*
Buruli ulcer	*Mycobacterium ulcerans*

Antibiotics

Both fungi and bacteria act as decomposers and so sometimes compete with one another for food. Some microorganisms produce chemicals called antibiotics to inhibit growth of their competitors. Antibiotics such as those in Table 3 interfere with bacterial metabolic processes such as DNA replication, transcription, translation, ribosome function and cell wall formation.

Viruses are not living and they can only reproduce when they are inside other living cells. They use the chemical processes inside these cells instead of having metabolism of their own. Since antibiotics interfere with the metabolic functions that viruses lack, antibiotics are not effective against viruses.

Table 3

Organism	Antibiotic produced	Mechanism of action
Fungi:		
Penicillium sp.	penicillin	inhibits cell wall synthesis
Cephalosporium sp.	cephalosporin	inhibits cell wall synthesis
Streptomyces sp.	tetracycline,	inhibits translation
	streptomycin,	inhibits translation
	erythromycin	inhibits translation
Bacteria:		
Bacillus sp.	polymyxin	disrupts cell membranes
	bacitracin	disrupts cell wall synthesis

Data-based question: antibiotic resistance

Bacterial resistance to antibiotics is a direct consequence of the overuse of these drugs. In the USA, currently more than half of the doctor visits for upper respiratory tract infections (URIs) are prescribed antibiotics, despite knowledge that most URIs are caused by viruses.

In the early 1990s, Finnish public health authorities began discouraging the use of the antibiotic erythromycin for URIs in response to rising bacterial resistance to the antibiotic, and the national erythromycin consumption per capita dropped by 43 per cent.

The data in Figure 1 shows the incidence in Finland, over a 10-year period, of *Streptococcus pyogenes* strains that are resistant to the antibiotic erythromycin. *S. pyogenes* is responsible for the condition known as "strep throat".

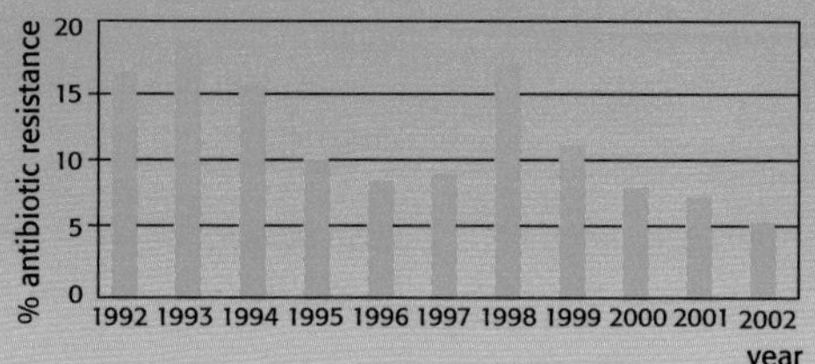

Figure 1 The incidence of *Streptococcus pyogenes* strains that are resistant to the antibiotic erythromycin over a 10-year period in Finland

1 (a) Describe the pattern of erythromycin resistance over the period from 1992 to 2002. [3]
(b) Suggest a reason for the pattern shown. [2]
2 Calculate the percentage difference in antibiotic resistance between 2002 and 1992. [2]
3 Evaluate the claim that reduction in the use of erythromycin has led to a reduction in the incidence of antibiotic resistance in *S. pyogenes*. [3]

Non-specific immunity

The ability to resist infection by a disease is termed immunity. Non-specific immunity allows the body to resist infection by a wide range of pathogens. The outermost layer of skin is tough and provides a physical barrier against the entry of pathogens and protection against physical and chemical damage. Sebaceous glands are associated with hair follicles and they secrete a chemical called sebum, which maintains skin moisture and slightly lowers skin pH. The lower pH inhibits the growth of bacteria and fungi.

Mucous membranes line internal cavities that are exposed to the external environment such as nasal passages. Mucous membranes secrete mucus. The mucus has antiseptic properties because of the presence of the anti-bacterial enzyme lysozyme. Mucus also acts as a physical barrier in the sense that pathogens and harmful particles are trapped in it and either swallowed or expelled.

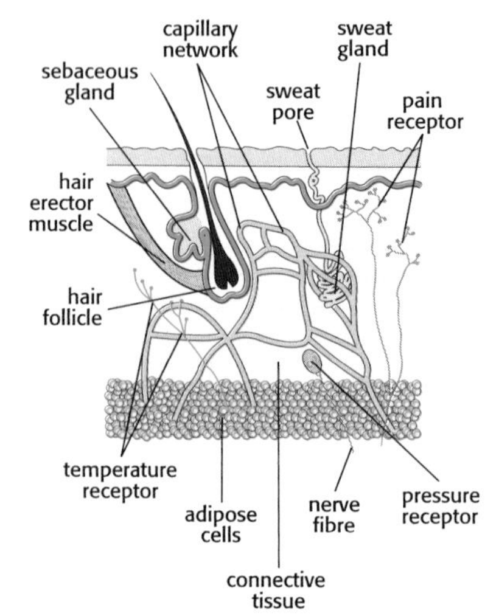

Figure 2 Anatomy of the skin

If microorganisms get past the physical barriers and enter the body, phagocytic cells engulf them by endocytosis. The microorganisms are then digested by the enzymes in the lysosomes of the phagocyte.

Rediscovering biology: imaging human skin

A digital microscope with a ×100 or ×200 lens can be used to provide a 3-D image of the skin surface.

1 Examine an area of the skin where a wound is healing. Deduce, from examining the scab, how wound-healing prevents the entry of pathogens.
2 Examine a patch of dry skin. How does dryness affect the integrity of skin? Friction can cause breaks in the skin. How do areas subjected to constant friction such as "writer's bumps" or calluses appear to be different?

Data-based question: skin pH

Figure 3 shows how the surface pH of human skin varies between different areas of the body. It also shows differences between adults and newborn infants (neonates). Skin pH protects the skin from colonization by certain microorganisms.

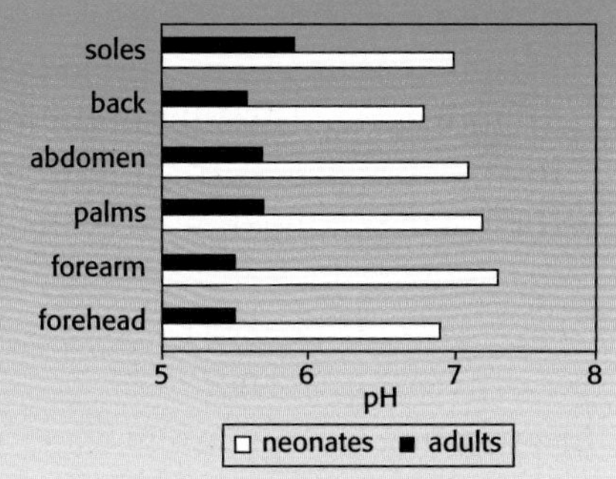

Figure 3 How the surface pH of human skin varies between different areas of the body

1 Compare the skin pH of neonates and adults. [2]
2 Suggest how the adult skin pH might be established. [1]
3 Suggest why the use of soaps (which are basic) might have a more irritating effect on the skin of a neonate. [2]
4 Deduce how basic soaps might undermine the skin's defensive function. [2]

Specific immune responses

If microorganisms get past the physical barriers of the skin, a second level of immunity exists. Proteins and other molecules on the surface of pathogens are recognized as foreign by the body and they stimulate a specific immune response. Any chemical that can stimulate an immune response is referred to as an antigen.

The body is capable of responding in a highly targeted way to pathogens. A protein called an antibody binds to antigens. Our bodies can produce a vast array of different antibodies, each one specific to a certain antigen.

Some of the outcomes of the binding of antibodies to antigens include:

- Making the pathogen more recognizable to phagocytes so that they are more readily engulfed.
- Preventing viruses from docking to host cells so that they cannot be taken up by host cells.
- Causing sticking together or "agglutination" of pathogens so that they are prevented from entering cells and are easier for phagocytes to ingest.

HL

Lymphocytes are white blood cells responsible for the production of antibodies. Two categories of lymphocyte are B-lymphocytes, which differentiate in the bone marrow, and T-lymphocytes, which differentiate in the thymus.

Clonal selection

The stages of antibody production are given in outline below (see Figure 4).

- Once a pathogen has been engulfed by a phagocyte, antigens from the destroyed pathogen are displayed on the surface of the phagocyte, bound to a membrane protein called MHC protein.
- A receptor on the surface of a specific T-lymphocyte, or T-cell, binds to the displayed antigen and becomes activated.
- The T-cell will then bind to a specific B-cell and will stimulate that B-cell to divide to form a clone of plasma B-cells or a clone of memory B-cells.
- Plasma B-cells secrete antibodies specific to the invading pathogen.
- Memory B-cells will remain long after the infection and will be involved in producing a more rapid response, in the event that the same pathogen infects the body again.

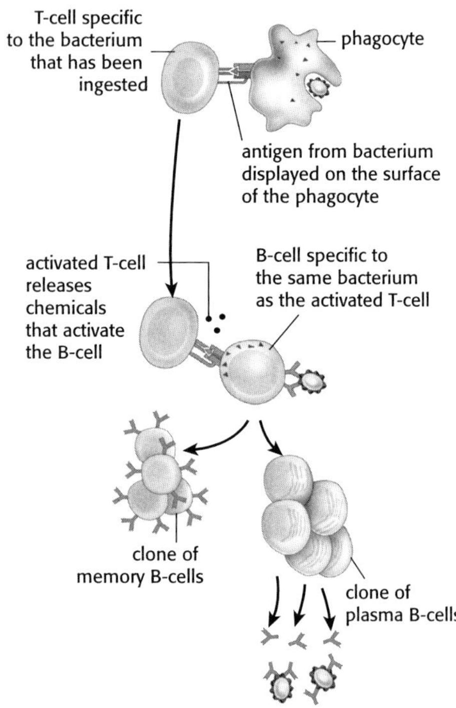

Figure 4 The stages of antibody production

Investigating biology: white blood cell count

White blood cells (WBC) possess a nucleus while mature red blood cells do not. A slide with a drop of blood spread out and stained is referred to as a blood smear.

A temporary increase in the number of WBC may indicate an infection, and an abnormally high WBC might indicate leukemia. A WBC count is usually expressed as quantity of cells per unit volume. A normal white blood cell count for an adult might be between 4×10^3 and 10×10^3 cells per litre. A WBC count of 11×10^3 per litre or above could indicate infection, stress or cancer.

Examine a commercially prepared blood smear. Since the dilution factor is not known, determine the fraction of cells that are white blood cells (leucocytes) and the fraction that are red blood cells. Observe carefully, as the white blood cells sometimes get concentrated at one edge of the slide. Figure 5 shows some of the cell types that might be observed.

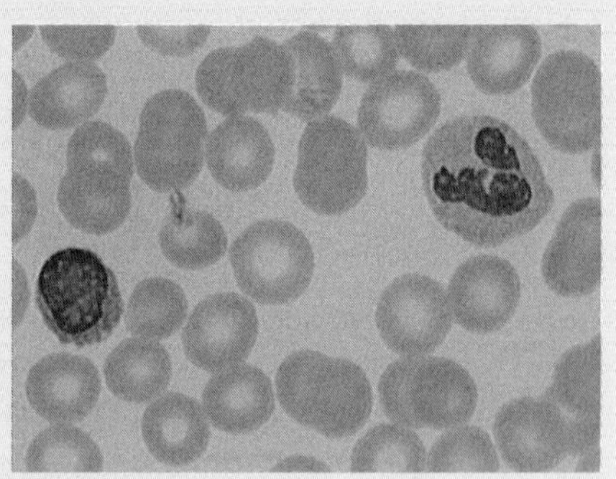

Figure 5 Types of blood cells

HL

Types of immunity

Production of antibodies during an infection is the natural means of developing active immunity.

Vaccination is an artificial means of inducing active immunity by encouraging the body to produce antibodies. Passive immunity can be achieved if an individual receives antibodies that have developed outside of the body. For example, passive immunity can be conferred on babies through antibodies crossing the placenta during pregnancy or through colostrum during breast-feeding. Colostrum is the milk produced in the first few days after childbirth. Some mothers elect not to breast feed their children, but healthcare practitioners often encourage these mothers to breast feed at least immediately after delivery, to ensure transfer of antibodies through the colostrum.

Artificial passive immunity can be achieved through the injection of antibodies that have been produced through biotechnology.

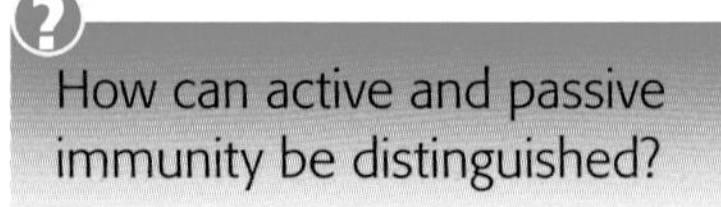

Monoclonal antibodies

Monoclonal antibodies are highly specific, purified antibodies that are produced by a clone of cells, derived from a single cell, and they recognize only one antigen. Figure 6 provides details on how monoclonal antibodies are produced.

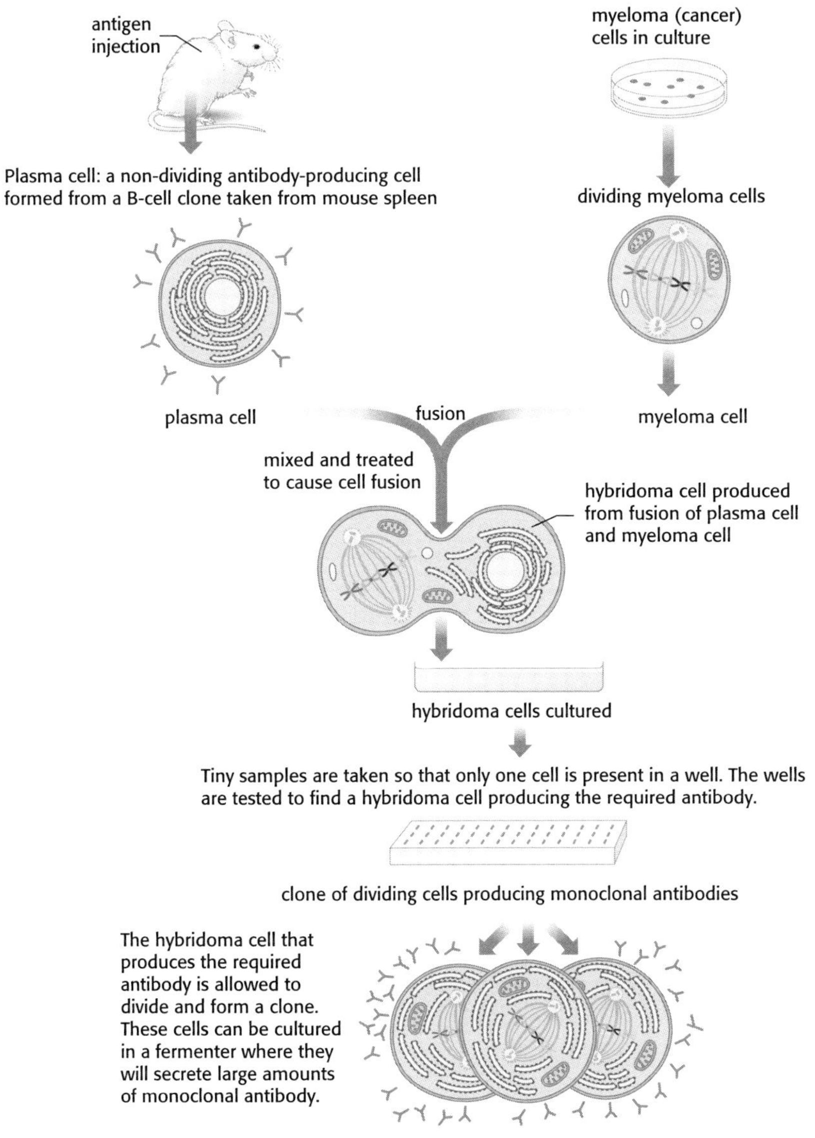

Figure 6 How monoclonal antibodies are made

Monoclonal antibodies are used for a broad range of applications, including the treatment of rabies infections and HIV diagnosis. One ingenious technique involves injecting radioactive antibodies specific to tumour cells into cancer patients, to deliver targeted chemotherapy.

Polyclonal antibodies are antibodies that are derived from different cell lines. Often a larger mammal will be injected with an antigen. Because the antigen will often have several different domains, the consequence will be different types of B-cell lines all targeting different parts of the same antigen. The antibodies can then be purified from the blood serum of the animal.

HL

Defence against infectious disease

Active immunity results from an organism being challenged by an antigen and producing antibodies in response. Memory B-cells and memory T-cells ensure that the second time an antigen is encountered, the body is ready to respond rapidly. Figure 8 can be used to compare a primary and a secondary response to antigen exposure.

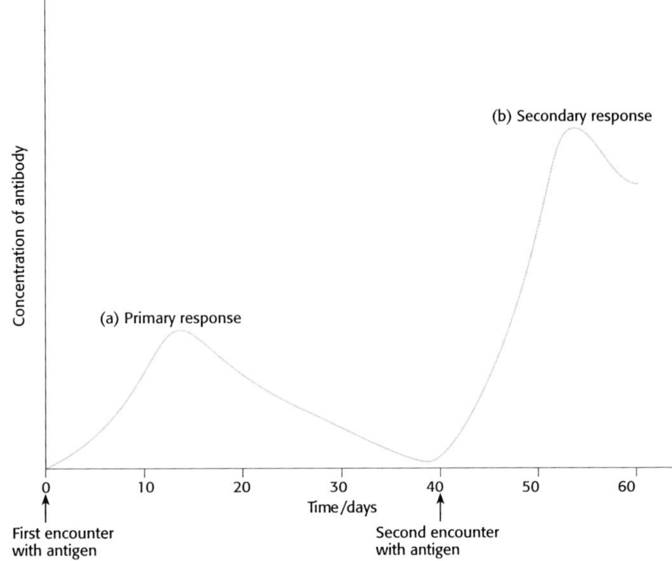

Figure 8 The secondary immune response

Active immunity can be acquired artificially through vaccination. A vaccine is introduced into the body, usually by injection. The vaccine may contain a live attenuated (weakened) version of the pathogen, or some derivative of it. This stimulates a primary immune response. If the actual microorganism enters the body as a result of infection, it will be destroyed by the antibodies in a secondary immune response. Alternatively, a second vaccination or "booster shot" may be used to cause this secondary response.

If an individual chooses to be vaccinated, then not only does the person avoid contracting the disease, but the risk of an epidemic is lowered and the cost borne by public healthcare systems is reduced. In economic terms, vaccines represent a merit good – a good that once consumed has both public and private benefits. Governments often get involved in the provision of merit goods because the market will undersupply them. Vaccination programmes to eradicate diseases require most of the population to participate to be successful. Such a successful programme led to the eradication of smallpox in the late 1970s.

Some concerns about vaccination programmes exist. Thimerosal is a preservative that contains a mercury compound. It has been used in some vaccine formulations, such as those against influenzas and the triple diphtheria–tetanus–pertussis (DTP) vaccine. While the use of thimerosal is gradually diminishing, it has not been phased out completely and trace amounts still remain in many vaccines.

Vaccines may generate side-effects such as allergic reactions or problems in the fetuses of pregnant women. However, the concerns about vaccinations are hugely outweighed by the benefits they bring.

Data-based question: antibodies in colostrum

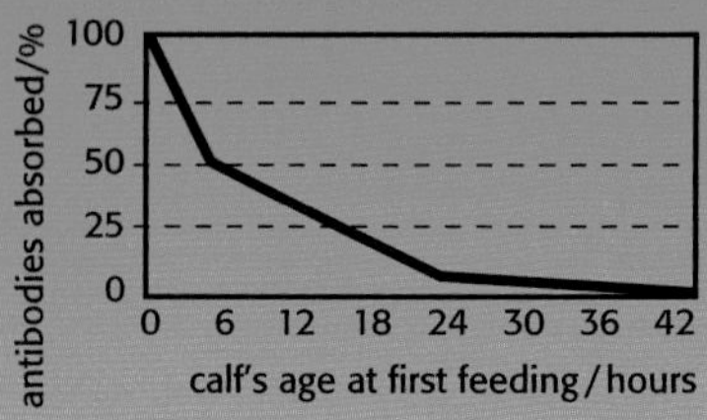

Figure 7 The ability of a calf (*Bos taurus*) to absorb antibodies decreases quickly after birth.

Figure 7 shows how the ability of a calf (*Bos taurus*) to absorb antibodies decreases quickly after birth.

1 Describe how the ability of a calf to absorb antibodies changes over the initial hours after birth. [2]
2 Suggest why calves that have endured a long difficult birth are more likely to suffer from infection. [2]
3 Predict how the composition of antibodies might vary in the cow's colostrum over the first 24 hours after birth. [2]

1 With reference to Figure 8, compare the primary and the secondary responses in terms of:
   (a) The maximum concentration of antibody produced.
   (b) The length of time between the encounter with the antigen and the maximum concentration of antibody.

Data-based question: the benefits and dangers of vaccination

A link was postulated between the use of the triple measles, mumps and rubella (MMR) vaccine and the rise of childhood autism. The MMR vaccine was introduced in Britain in 1988 after a measles outbreak killed 17 people. Though very considerable research has shown there to be no connection, some people continue to believe in such a link and they propose use of single vaccines instead.

1 Suggest reasons for doctors advocating triple rather than single vaccines. [2]

The MMR vaccination is optional in the UK. The data in Figure 9 shows the percentage of children vaccinated (MMR uptake) over a 13-year period in the UK.

2 Outline the pattern of MMR uptake over the period shown in the graph. [3]
3 Suggest, giving a reason, a year in which the UK media may have carried a story about research making a link between MMR and autism. [2]

In Japan, the MMR vaccine was introduced in 1989, but the programme was terminated in 1993 and only single vaccines used thereafter. The data in Figure 10 shows the incidence of autism in Japan for children grouped by the year that they were born in.

4 Outline the trend in the incidence of autism over the period shown in the graph. [2]
5 Evaluate using the data the claim that autism is linked to the MMR vaccine. [3]

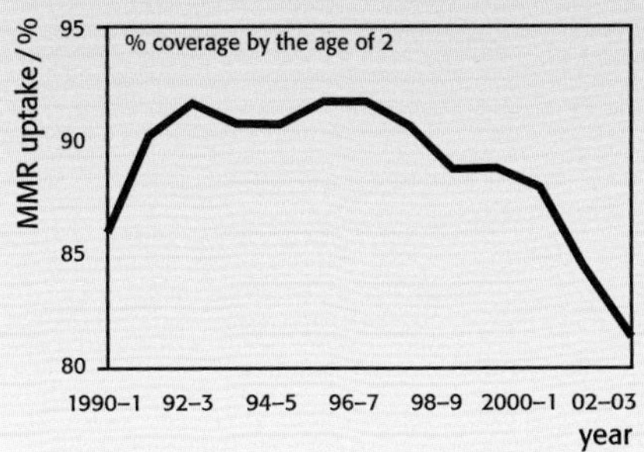

Figure 9 The percentage of children vaccinated (MMR uptake) over a 13-year period in the UK

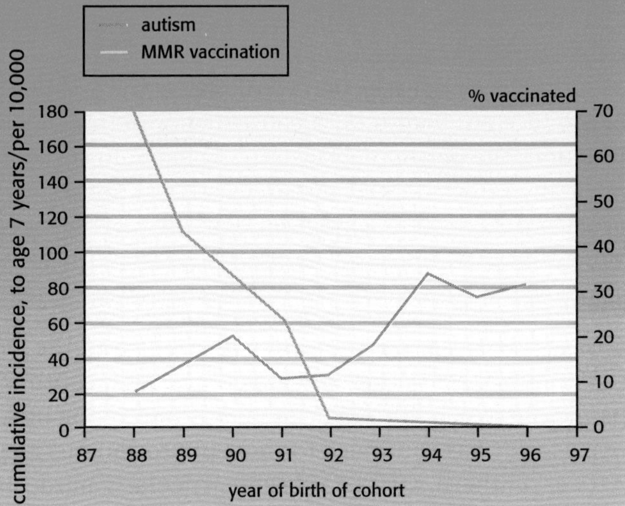

Figure 10 Changes in the uptake of the MMR vaccine and changes in the incidence of autism in Japan from 1988 to 1996

Figure 11 provides information about the history of the triple DTP vaccine (diptheria, tetanus and pertussis) in England and Wales over a 50-year period. The *y*-axis records the incidence of pertussis (whooping cough). A second curve on the same graph shows the pattern of DTP vaccine uptake from the mid-1960s to the mid-1990s. The grey bar shows a period when the British media carried many stories about the alleged side-effects of DTP vaccination.

6 Using the data in the graph, explain the changes in rates of whooping cough in the 50 years from 1940 to 1990. [3]

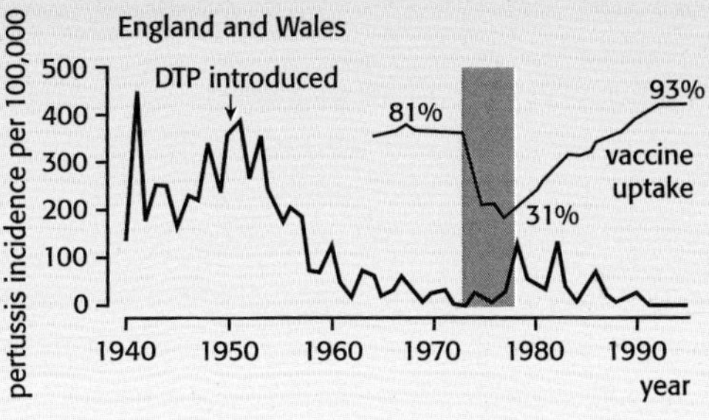

Figure 11

HIV and AIDS

The human immunodeficiency virus (HIV) targets helper T-cells. Figure 12 shows HIV particles on the surface of a T-cell. The resulting reduction in the number of circulating T-cells means a reduced capacity for clonal selection mediated by the T-cells and a loss of ability to produce antibodies. As a consequence of this impaired immune function, the person may eventually begin to suffer from several, normally rare, "marker" diseases such as Kaposi's sarcoma. A collection of several diseases or conditions existing together is called a syndrome. When the syndrome of conditions is present, the person is said to have acquired immunodeficiency syndrome (AIDS).

Cases of AIDS are not uniformly distributed in the world. The data in Figure 13 shows that for much of the 20th century, until the AIDS pandemic struck, life expectancy was improving in Africa. While life expectancy has continued to show improvements in some parts of Africa, southern Africa has been particularly hard-hit by the AIDS pandemic with a resulting catastrophic effect on the populations of these affected countries. The consequences to these societies include an undermining of social cohesion, as sufferers are stigmatized and families lose members. Increased poverty occurs because of loss of economic productivity.

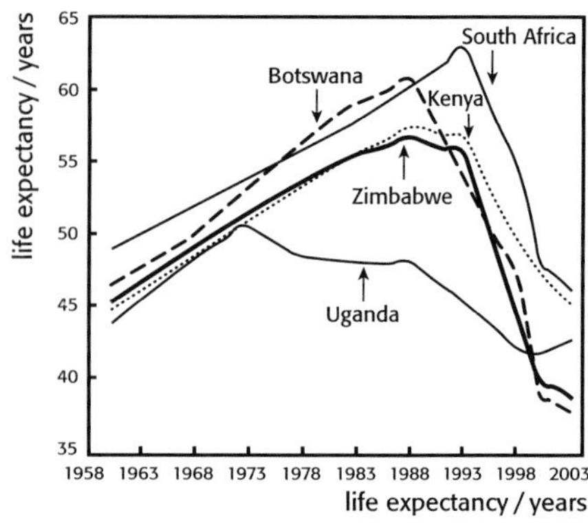

Figure 13 Changes in average life expectancy in five African nations since 1958

1. Assuming the precipitous decline in life expectancy was due to the AIDS pandemic, identify the decade when the effect began to be observed. [1]
2. Suggest why the data for Uganda is anomalous. [1]

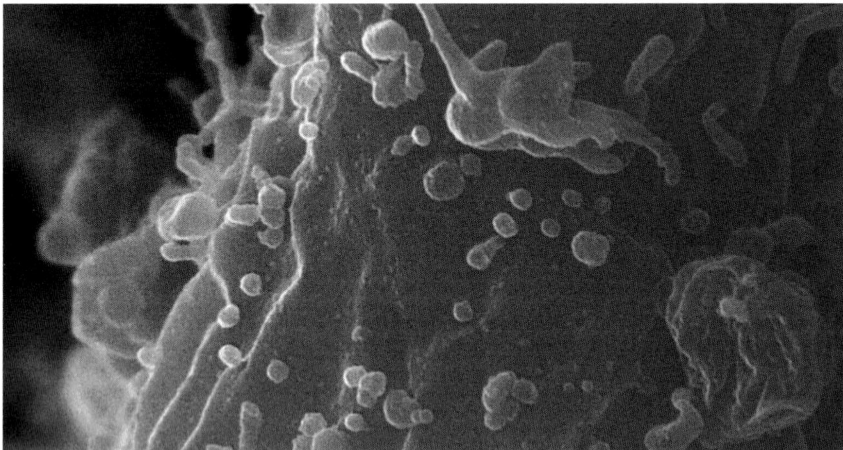

Figure 12 HIV particles on the surface of a T-cell

Rediscovering biology: epidemic spread simulation

1. Your tutor will prepare a number of test tubes, one for each student in the class. Most of the tubes are water, but one is "infected" with NaOH. This represents a pathogen.
2. Combine some of the contents of your test tube with the test tubes of any three members of your class.
3. When you have finished exchanging solutions, check if your tube has been "infected" by adding a drop of phenolphthalein to the tube.
4. Identify the non-infected individuals. Work backwards from all of the people that the non-infected people contacted to rule out more students as the source of infection. Can you discover the source?
5. To what extent is this simulation a true model of disease transmission?

Thinking about science: using inductive reasoning to solve a problem

The following quote is from an article in the *American Journal of Public Health*, by Elaine Larson, January 1989. "Innovations in Health Care: Antisepsis as a Case Study". *American Journal of Public Health* Vol. 79, number 1.

Ignaz Semmelweis, a Hungarian obstetrician, joined the staff of the Vienna Lying-in Hospital and noted a large discrepancy in death rates among women attended by midwives and midwifery students and those attended by physicians and medical students. Puerperal fever, a disease that develops after childbirth, was endemic and rampant in this hospital, and women were reputed to have their babies on the streets rather than risk the dangers of the hospital. Death rates among women attended by midwives were about 2–3 per cent and among those attended by medical staff, two to five times that rate. Semmelweis tested several hypotheses (e.g. variations in positioning of mothers during birthing and the larger hand size of male as compared to female examiners) to explain these differences, but to no avail. When one of his medical colleagues died of sepsis in 1846 after being inadvertently stuck with a sharp instrument being used for an [autopsy] of a woman who had died of PF, Semmelweis realized that whatever was causing PF was being transmitted by contact with autopsy material. He demonstrated that the discrepancy between the rates of PF among women attended by midwives and physicians was explained by the fact that only physicians performed post-mortem examinations. This discovery prompted Semmelweis to require all students and physicians to soak their hands in chlorinated lime after autopsies, before examining antepartal patients. Within a few months after enforcement of this practice, the PF deaths among women attended by medical staff had fallen to levels comparable to those of women attended by midwives.

Handwashing was certainly not a new practice in health care: physicians had been recommending general hygiene and cleanliness for decades. The unique contributions of Semmelweis were his recognition that the agent or cause of PF was being directly transmitted from one individual to another on the hands, and the concept of one necessary cause for the disease. He was incorrect in thinking that the cause was only dead and decomposing matter, as he later discovered when PF was clearly propagated from a living patient with a uterine infection to the women in surrounding beds. Despite his incomplete understanding of the etiology of PF, Semmelweis appropriately intervened to prevent transmission of the disease that we now know to be caused by Streptococcus *bacteria.*

1 (a) Distinguish inductive reasoning from deductive reasoning.
  (b) Explain the role of inductive reasoning in Semmelweis's conclusion that PF was caused by "cadaveric matter".

2 Pragmatic truth is truth which gives us predictive power, though it may not actually correspond to reality: application of the false conclusion that only dead and organic matter was the cause did lead to a reduction in the number of women who got PF.
  (a) To what extent do you think science creates pragmatic truth?
  (b) John Ziman, author of *Reliable Knowledge: An Exploration of the Grounds for Belief in Science* has argued that textbook science may be only about 90% correct. Why bother learning science when you don't know which 10% is wrong?

HL

Clotting

When human tissues are injured, blood vessels are also commonly injured. A non-specific defence results, called clotting. This serves to seal up the wound and to prevent loss of blood and blood pressure.

A clot is a semi-solid. Platelets are cellular fragments that circulate in the blood. When an injury occurs, they aggregate at the site forming a temporary plug. The platelets as well as the damaged tissue release clotting factors. These in combination with calcium and vitamin K convert prothrombin to its active form called thrombin. Thrombin in turn converts the soluble protein fibrinogen into the insoluble fibrin. The fibrin forms a mesh that traps further platelets, forming a longer-lasting plug. Figure 14 shows red blood cells trapped in this fibrous mesh.

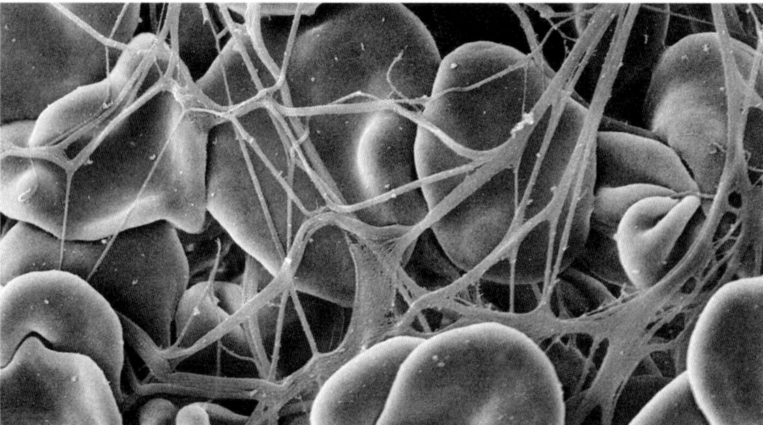

Figure 14 Micrograph of a blood clot

Working with data: using Excel to calculate the t-test

When a person is recovering from heart surgery, the formation of blood clots is a concern, so patients are often given anti-coagulants (also called blood thinner).

An experiment was conducted to determine if a drug had an impact on clotting time. One group was assigned a placebo and the other was given the drug. Are the differences in mean clotting time significant, or are they due to chance? We can conduct the *t*-test to check.

1 Open Excel and enter the following data:
Drug B: 8.7, 8.4, 7.9, 8.7, 9.1, 9.5
Placebo: 9.6, 9.6, 10.4, 8.7, 9.9, 9.1, 11.1.
2 Click on the f_x icon in the toolbar (see Figure 15). An "insert function" menu will appear. Search for *t*-test (TTEST) and choose it from the list.
3 You will be asked to specify ranges. Highlight the first column for array 1 then highlight the second column for array 2.
4 Choose a two-tailed test. Our question is "Are the means different?", whereas for a one-tailed test the question would be "Is one mean higher than the other?"
5 Choose "3" for unequal variances.
6 The value generated is the P-value.
7 A value of 0.0179 is below 0.05, so the drug appears to have a statistically significant effect on clotting time.

Microsoft Excel

File Edit View Insert Format

B9 fx

	A	B	C
1	drug	placebo	
2	8.7	9.6	
3	8.4	9.6	
4	7.9	10.4	
5	8.7	8.7	
6	9.1	9.9	
7	9.5	9.1	
8		11.1	
9			
10			
11			
12			

Figure 15 Using Excel to perform the *t*-test

Chapter 21 questions

1 Annotate each item in the following list as a component of non-specific or specific immunity:
(a) lysozyme
(b) phagocytes
(c) memory B-cells
(d) antibodies.

2 Leprosy is caused by *Mycobacterium leprae* and causes damage to the skin and peripheral nerves. The World Health Assembly set a target for the "elimination of leprosy as a public health problem" by 2000. Elimination was defined as a prevalence of less than one case per 10,000 population.
Figure 16 shows the prevalence and new-case detection rate for leprosy in India over an 18-year period.
(a) Deduce the year in which the programme was implemented. [1]
(b) Discuss the success of the programme in:
(i) reducing the number of people suffering from the disease [1]
(ii) preventing new cases. [2]

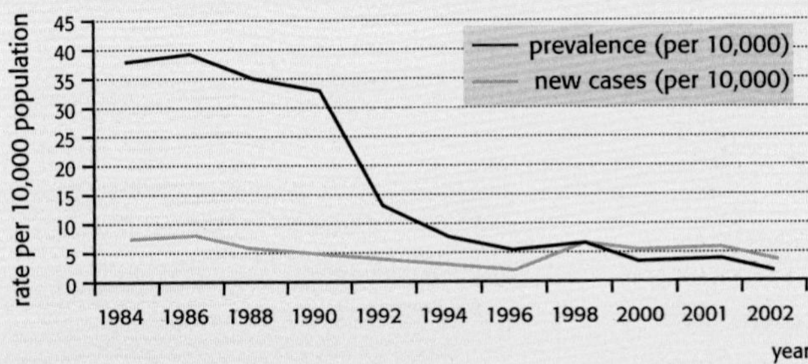

Figure 16 The prevalence and new-case detection rate for leprosy in India over an 18-year period

3 Which of the following is/are necessary to produce monoclonal antibodies:
(A) tumour cells
(B) plasma (B) cells
(C) macrophages?
(a) (B) only
(b) (A) and (B) only
(c) (B) and (C) only
(d) (A), (B) and (C).

4 Which of the following explains clonal selection?
(A) Memory cells are present at birth.
(B) Antigens activate specific immune responses.
(C) The body selects which antigens it will respond to.
(D) People with similar genes respond to antigens in a similar way.

5 (a) Explain why infection with HIV leads to a decrease in the effectiveness of the immune system. [4]
(b) Outline one specific disease in humans caused by a prokaryotic pathogen, including the name of the pathogen and its mode of transmission and possible effects. [5]
(c) Explain the production of antibodies against a pathogen. [8]

6 Figure 17 represents a simplification of the clotting process.
State what is represented by each letter. [2]

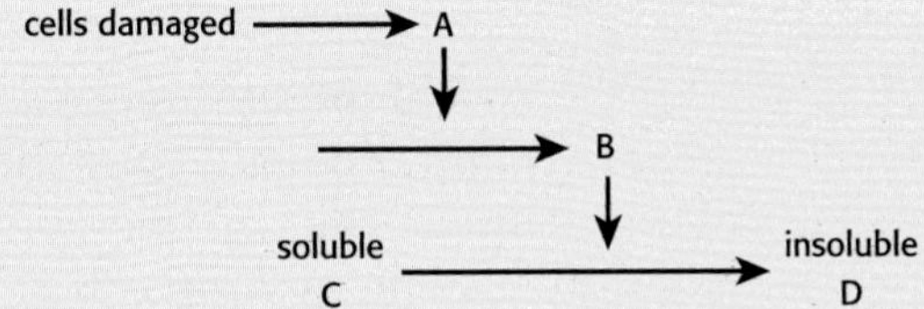

Figure 17 An outline of the clotting process

22 Gas exchange

All organisms absorb one gas from the environment and release another one. This process is called **gas exchange**. Aquatic or marine organisms exchange gases with the water around them and terrestrial organisms exchange gases with the air. Leaves absorb carbon dioxide to use in photosynthesis and release the oxygen produced by this process. Humans absorb oxygen from the air and transport it to all living cells, which use it in **cell respiration** (see Chapter 8). Carbon dioxide is produced in cell respiration and is carried by blood to the lungs, where it is released. So, plants and other photosynthesizing organisms exchange carbon dioxide for oxygen, unless their respiration rate exceeds their photosynthesis rate. Humans and other heterotrophic organisms exchange oxygen for carbon dioxide.

Gas exchange happens by diffusion, at a surface where the organism is in contact with its environment. The rate of gas exchange depends on the area of the surface. In small organisms, the outer surface is usually large enough, but because surface area to volume ratios decrease as the size of organisms increases, bigger organisms need special organs with a large surface area dedicated to gas exchange. The rate of diffusion also depends on the concentration of gases across the exchange surface. Pumping fresh air or water to the gas exchange surface, to replace the air or water already there, helps to maintain concentration gradients. This process is called **ventilation**.

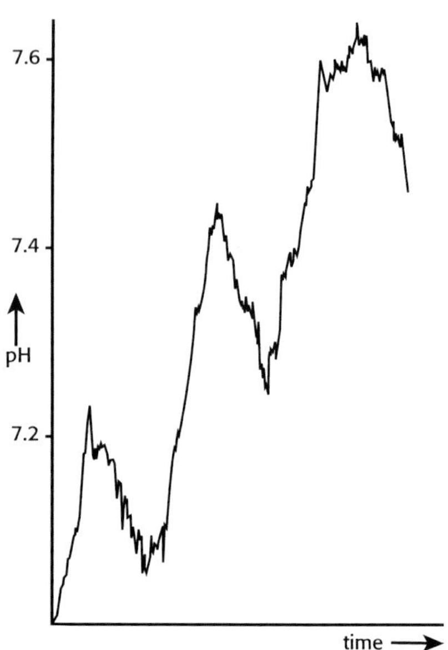

Figure 1 Graph obtained by placing a pH electrode in a pond and linking it to a data logger

1. What caused the falls in pH?
2. What caused the rises in pH?
3. Over how many days was the pH monitored?
4. Can you deduce whether days were unusually sunny or unusually cloudy?

Data-based question: emphysema and gas exchange

Figure 2 shows healthy lung tissue and tissue from a lung with emphysema, at the same magnification. Smoking usually causes emphysema. Breathing polluted air makes the disease worse.

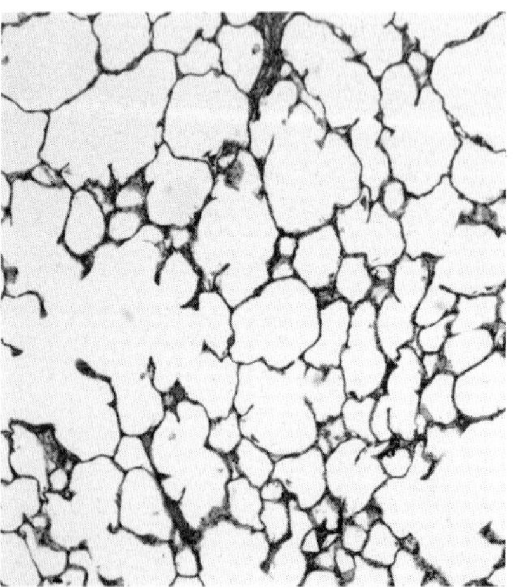

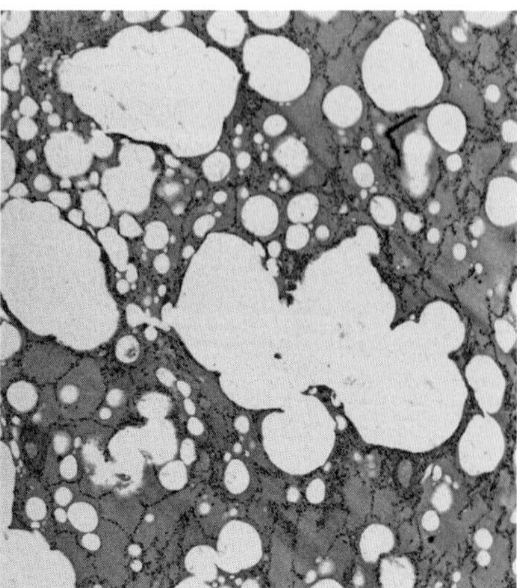

Figure 2 Healthy lung tissue (left) and lung tissue showing emphysema (right)

1. (a) Place a ruler across each micrograph and count how many times the edge of the ruler crosses a gas exchange surface. Repeat this several times for each micrograph, in such a way that the results are comparable. State your results using suitable units. [3]
   (b) Explain the conclusions that you draw from the results. [3]
2. Explain why people who have emphysema feel tired all the time. [3]
3. Suggest why people with emphysema often have an enlarged and strained right side of the heart. [1]

Alveoli and gas exchange

The lungs contain large numbers of air sacs, approximately 200 µm in diameter, called alveoli. These alveoli are arranged in groups, attached to a bronchiole, through which the alveoli are ventilated. Figure 3 shows alveoli in a small portion of lung tissue. The walls of these alveoli consist of a single layer of cells, called an epithelium. There are two types of cells in the epithelium:

- Type I cells are very flattened, so that the thickness of cytoplasm is only about 0.15 µm.
- Type II cells are rounded in shape and only occupy about 5% of the alveolar surface area. These cells secrete a surfactant, which coats the inner surface of the alveoli in a film of moisture, preventing collapse of alveoli during expiration and helping oxygen to dissolve.

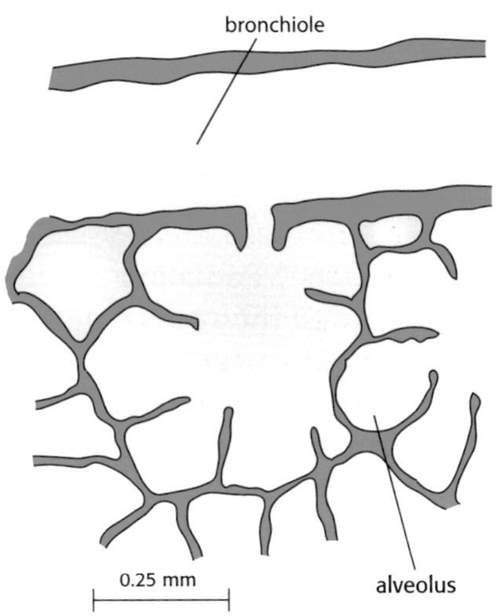

Figure 3 Air spaces in lung tissue

There is a dense network of blood capillaries between adjacent alveoli. The walls of these capillaries are a single layer of flattened cells, called an endothelium, and are also approximately 0.15 µm thick. A layer of protein gel, called the basement membrane, lies between the epithelium of the alveolus and the endothelium of the blood capillaries. It is about 0.1 µm thick, so the total distance between air and blood is only 0.4 µm. This is the gas exchange surface. Oxygen diffuses into the body across this surface, and carbon dioxide diffuses out. Figure 4 shows a small part of the wall of an alveolus and an adjacent blood capillary.

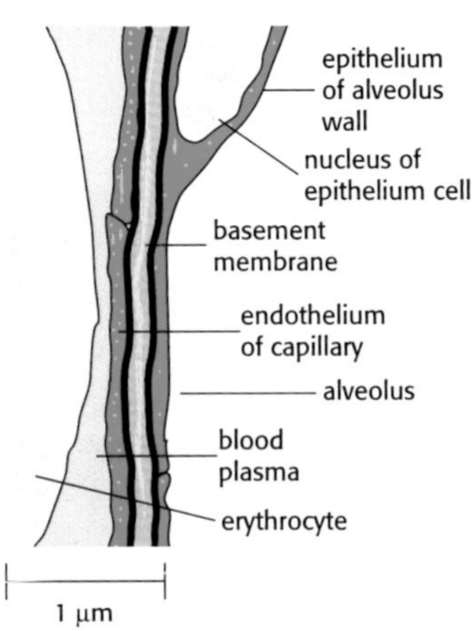

Figure 4 Part of the wall of an alveolus

Although the surface area of a single alveolus is very small, there are hundreds of millions of alveoli in the lungs in total, so the surface area for gas exchange is huge – approximately 80 m^2. As a result, gases are exchanged rapidly enough to sustain cell respiration during times of gentle to moderate exercise, and to restore oxygen and carbon dioxide concentrations inside the body after vigorous exercise.

Data-based question: concentration gradients

Figure 5 shows the typical composition of atmospheric air, air in the alveoli and gases dissolved in air returning to the lungs in the pulmonary arteries.

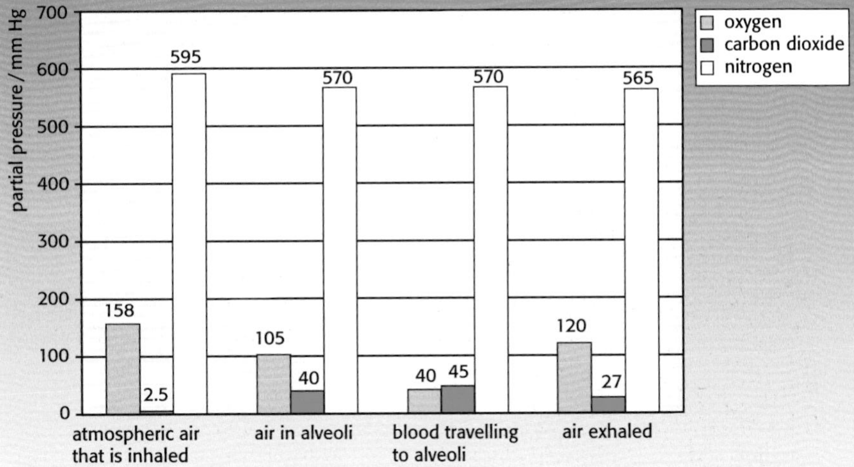

partial pressure / mm Hg
0, 100, 200, 300, 400, 500, 600, 700
oxygen
carbon dioxide
nitrogen
atmospheric air that is inhaled: 158, 2.5, 595
air in alveoli: 105, 40, 570
blood travelling to alveoli: 40, 45, 570
air exhaled: 120, 27, 565

Figure 5 Partial pressures of gases in the pulmonary system

?

1 Explain why the oxygen concentration in the alveoli is not as high as in fresh air that is inhaled. [2]

2 (a) Calculate the difference in oxygen concentration between air in the alveolus and blood arriving at the alveolus. [1]

(b) Deduce the process caused by this concentration difference. [1]

(c) (i) Calculate the difference in carbon dioxide concentration between air inhaled and air exhaled. [1]

(ii) Explain this difference. [2]

(d) Despite the high concentration of nitrogen in air in alveoli, little or none diffuses from the air to the blood. Suggest reasons for this. [2]

The ventilation system

The human ventilation system consists of the lungs, the airways that bring air in and out of the lungs from the atmosphere, and the muscles that are used to move the air in and out (see Figure 6).

Air enters the ventilation system through the nose or mouth and then passes down the **trachea**. This has rings of cartilage in its wall to keep it open even when air pressure inside is low or pressure in surrounding tissues is high. The trachea divides to form two **bronchi**, also with walls strengthened with cartilage. One bronchus leads to each lung. Inside the lungs the bronchi divide repeatedly to form a tree-like structure of narrower airways, called bronchioles. The bronchioles have smooth muscle fibres in their walls, allowing the width of these airways to vary. At the end of the narrowest bronchioles are groups of alveoli, described on page 229.

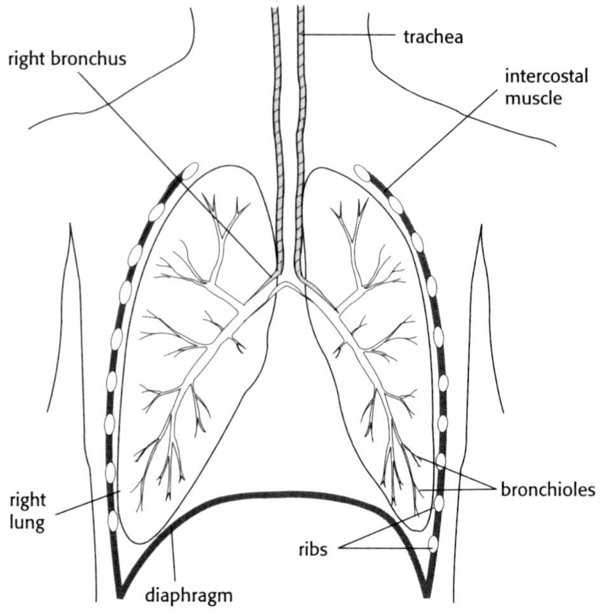

Figure 6 The ventilation system

Data-based question: treating chronic obstructive pulmonary disease

Chronic obstructive pulmonary disease (COPD) is the narrowing of the airways due to chronic bronchitis, emphysema, or both. Millions of people suffer from COPD worldwide. Most cases are caused by smoking cigarettes. The symptoms are a persistent cough with large amounts of mucus, frequent chest infections, wheezing and a feeling of breathlessness. The cause of breathlessness is low arterial oxygen saturation. This can be monitored using a pulse oximeter. Some patients can be treated using an oxygen mask to deliver air with an increased oxygen concentration. Figure 7 shows the percentage saturation of hemoglobin of a patient with COPD, while breathing air with oxygen at different concentrations.

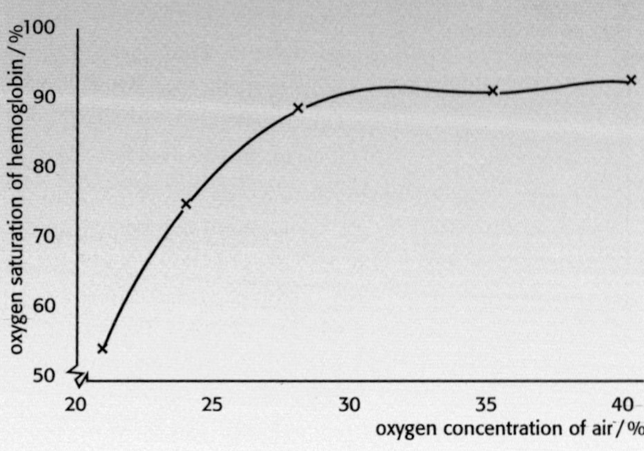

Figure 7 Relationship between oxygen concentration and saturation in COPD treatment

1 Calculate the increase in oxygen saturation:
   (a) from 21% to 24%
   (b) from 24% to 28%
   (c) from 28% to 35%
   (d) from 35% to 40%. [4]
2 Explain how breathing air with higher concentrations of oxygen than in the atmosphere causes an increase in arterial oxygen concentrations. [3]
3 Explain, using the data in the graph, why patients are not given 100% oxygen. [2]
4 The usual target when treating COPD is to raise the oxygen saturation to 90%, at the lowest possible oxygen concentration. Discuss what oxygen concentration should be used to achieve this. [3]
5 Suggest why so many people start smoking during their teenage years and do not stop even when the symptoms of COPD start to develop. [3]

The mechanism of ventilation

Ventilation of the lungs involves some basic physics. If particles of gas spread out to occupy a larger volume, the pressure of the gas becomes lower. Conversely, if a gas is compressed to occupy a smaller volume, the pressure rises. If gas is free to move, it will always flow from regions of higher pressure to regions of lower pressure.

If pressures inside the lungs are monitored during ventilation, they are found to fall below atmospheric pressure and then rise above it. These pressure changes are due to muscle action increasing and decreasing the volume inside the thorax. Figure 8 shows the thorax during inspiration and expiration. Table 1 shows how ventilation occurs at rest and during vigorous activity.

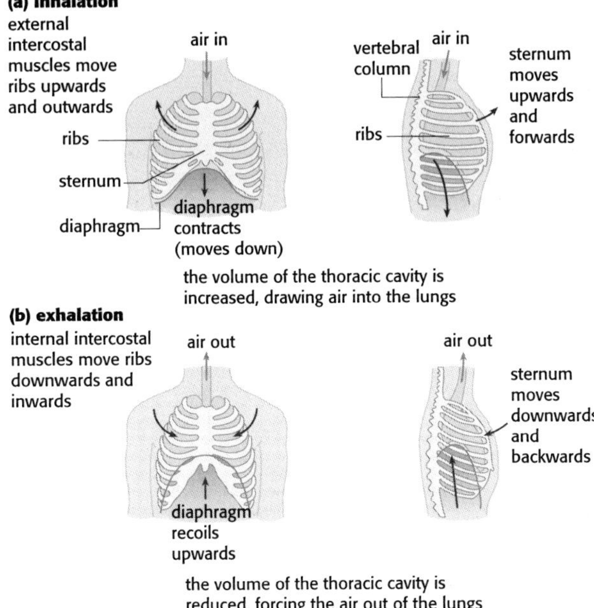

Figure 8 Ventilation of the lungs

Table 1

Quiet inspiration Contraction of the diaphragm causes it to flatten from its relaxed dome-like shape, lowering the floor of the thorax and so increasing the volume of the lungs. This reduces pressure in the lungs, causing air to flow in until the lung pressure rises to atmospheric. Elastic fibres in the lung tissue are stretched.	**Vigorous inspiration** In addition to the contraction of the diaphragm, the external intercostal muscles contract, moving the ribcage up and out. This expands the width of the thorax and gives a considerable increase in lung volume. Consequently a greater volume of air is drawn in.
Quiet expiration When the diaphragm stops contracting, the elastic fibres in the expanded lungs recoil, increasing pressure inside the lungs. This pushes out air and pulls the diaphragm back to its dome shape. No muscle contraction is needed during quiet expiration.	**Active expiration** Contraction of the internal intercostal muscles pulls the ribcage down and in. Contraction of abdominal muscles compresses the organs of the abdomen, pushing the diaphragm up into its dome shape. The reduced volume and increased thorax pressure force air out of the lungs.

Rediscovering biology: measuring ventilation rate and lung volumes

Ventilation parameters can be measured using sensors and monitored using data-logging techniques.

1 Ventilation rate

An inflatable chest belt is placed around the thorax and air is pumped in with a bladder. A differential pressure sensor is then used to measure pressure variations inside the belt due to chest expansions. The relative size and the rate of ventilations can be deduced.

2 Lung volumes

A spirometer is a device that measures lung volumes. Spirometers are available for use in data logging. They measure flow rate into and out of the lungs and from these measurements lung volumes can be deduced. Figure 9 shows the different lung volumes that can be measured.

1. Explain what each of the lung volumes labelled I–IV is.
2. Explain why volume IV is difficult to measure.
3. Which increases more during exercise, lung volume or breathing rate?
4. Do lung volumes vary between different individuals or are they a fixed human characteristic? If they vary, what factors influence lung volumes?

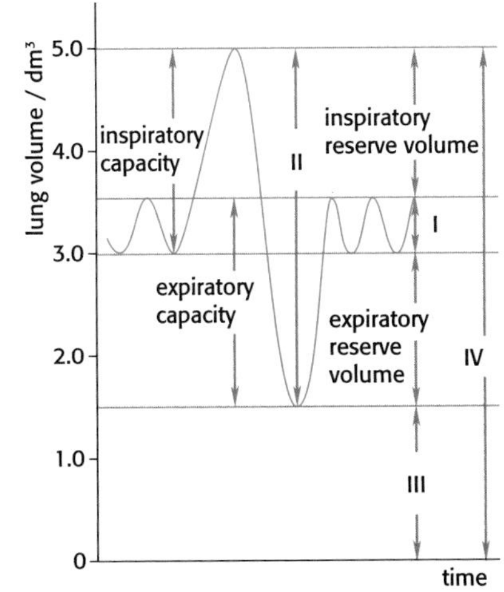

Figure 9 Typical lung volumes in a healthy adult

Chapter 22 questions

1 What is a ventilation system needed for in living organisms?
A To pump oxygen into the pulmonary artery.
B To pump oxygen into the pulmonary vein.
C To increase the rate of cell respiration in the alveoli.
D To maintain high concentration gradients in the alveoli.

2 What causes air to flow through the airways to the lungs?
A Contraction of the internal intercostal muscles.
B Contraction of muscles in the abdomen wall.
C An increase in volume in the thorax.
D An increase in pressure in the thorax.

3 What muscles must relax (release) when the external intercostal muscles contract?
A Internal intercostal muscles.
B Jaw muscles.
C Smooth muscle in bronchiole walls.
D The diaphragm.

4 Which comparison is correct?
A There are more bronchi than bronchioles in the ventilation system.
B The bronchi are wider than the trachea.
C There are more alveoli than bronchioles in the ventilation system.
D The alveoli are wider than the lungs.

5 What is the minimum number of plasma membranes that oxygen has to diffuse across to pass from air in the alveolus, to hemoglobin inside a red blood cell?
A two
B three
C four
D five.

6 Figure 10 shows male life expectancy at different ages for smokers and non-smokers.

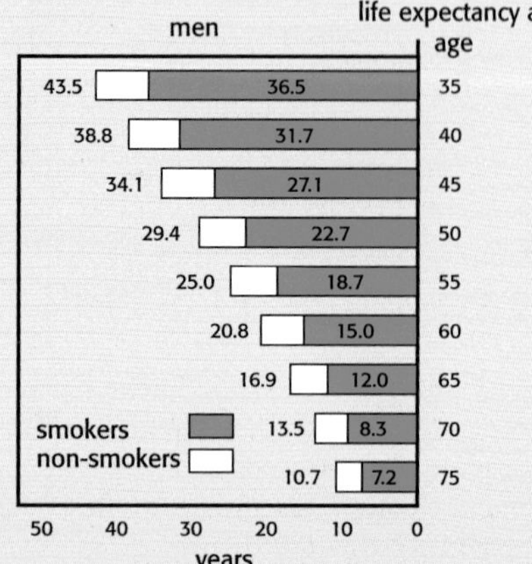

Figure 10

(a) Calculate the difference in life expectancy at age 35 and at age 70. [2]
(b) Explain the reasons for the difference in life expectancy between smokers and non-smokers. [3]

7 Sometimes the ventilation of the lungs stops. This is called apnea. One possible cause is the blockage of the airways by the soft palate during sleep. This is called obstructive sleep apnea. It has some potentially harmful consequences, including an increased risk of accidents during the daytime as a result of disrupted sleep and tiredness. Figure 11 shows the percentage oxygen saturation of arterial blood during a night of sleep in a patient with severe obstructive sleep apnea.

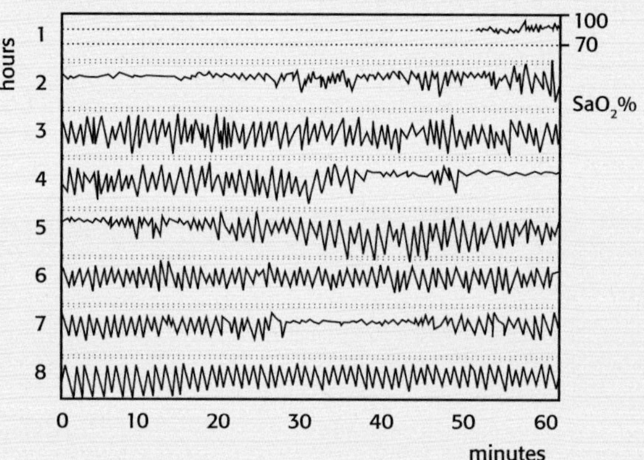

Figure 11

(a) Hour 8 shows a typical pattern due to obstructive sleep apnea.
(i) Explain the causes of falls in saturation. [2]
(ii) Explain the causes of rises in saturation. [2]
(iii) Calculate how long each cycle of falling and rising saturation takes. [2]
(b) Estimate the minimum oxygen saturation that the patient experienced during the night, and when it occurred. [2]
(c) Deduce the sleep patterns of the patient during the night when the trace was taken. [2]

23 The kidney

HL

The kidneys are responsible for removing substances from the blood that are not needed or are harmful. These substances can be divided into four groups.

1 Toxins and other substances that are ingested and absorbed but are not fully metabolized by the body, for example betain pigments in beets, and also drugs.
2 Excess water, produced by cell respiration and absorbed from food in the gut.
3 Excess salt, absorbed from food in the gut.
4 Nitrogenous wastes – mainly urea.

Substances in groups 2 and 4 are excretory products, as they are metabolic products of cells. Groups 1 and 3 are not excretory products because they are not produced by body cells.

Removal of excess water and salt is part of osmoregulation, which is described on page 238.

The kidneys filter off about one fifth of the volume of plasma from the blood flowing through them. This filtrate contains all of the substances in plasma apart from large protein molecules. The kidneys then actively reabsorb the specific substances in the filtrate that the body needs. The result of this process is that unwanted substances pass out of the body in the urine.

excretion *removal of the waste products of metabolic pathways from the body.*

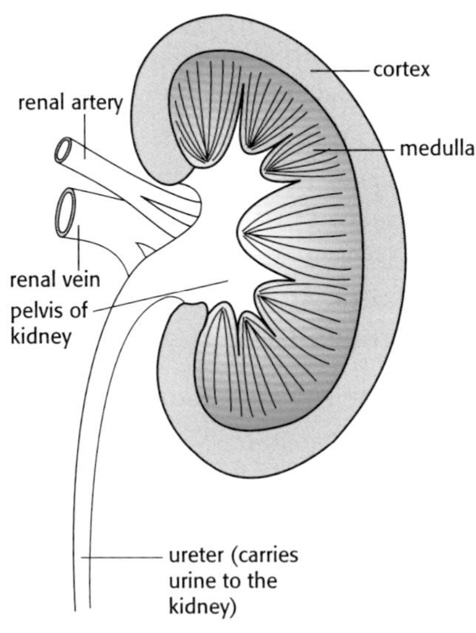

Figure 1 Structure of the kidney

Data-based question: blood supply to the kidney

Table 1 shows the flow rate of blood to the kidney and other organs, the rate of oxygen delivery and oxygen consumption. All of the values are given per 100 g of tissue or organ. The rates are for a person in a warm environment.

Table 1

	Blood flow rate (ml min^{-1} 100 g^{-1})	Oxygen delivery (ml min^{-1} 100 g^{-1})	Oxygen consumption (ml min^{-1} 100 g^{-1})
Brain	54.0	10.8	3.70
Skin	13.0	2.6	0.38
Skeletal muscle (resting)	2.7	0.5	0.18
Heart muscle	87.0	87.0	11.0
Kidney	420.0	84.0	6.80

?

1 Compare the rate of blood flow to the kidney with flow to the other organs. [2]
2 Calculate the volume of oxygen delivered to the organs per litre of blood. [2]
3 In the brain, 34 per cent of the oxygen that is delivered is consumed. Calculate the same percentage for the other organs. [4]
4 Discuss the reasons for the difference between the kidney and the other organs in the volume of blood flowing to the organ, and the percentage of oxygen in the blood that is consumed. [4]
5 Some parts of the kidney have a high percentage rate of oxygen consumption, for example the outer part of the medulla. This is because active processes requiring energy are being carried out. Suggest one process in the kidney that requires energy. [1]
6 Predict, with a reason, one change in blood flow that would occur if the person were moved to a cold environment. [2]

HL

Nephron structure

The basic functional unit of the kidney is the nephron. This is a tube with a wall consisting of one layer of cells. This wall is the last layer of cells that substances cross to leave the body – it is an epithelium. There are several different parts of the nephron, which have different functions and structures (see Figure 2):

- **Bowman's capsule** – a cup-shaped structure with a highly porous inner wall, which collects the fluid filtered from the blood.
- **Proximal convoluted tubule** – a highly twisted section of the nephron, with cells in the wall having many mitochondria and microvilli projecting into the lumen of the tube.
- **Loop of Henlé** – a tube shaped like a hairpin, consisting of a descending limb that carries the filtrate deep into the medulla of the kidney, and an ascending limb that brings it back out to the cortex.
- **Distal convoluted tubule** – another highly twisted section, but with fewer, shorter microvilli and fewer mitochondria.
- **Collecting duct** – a wider tube that carries the filtrate back through the cortex and medulla to the renal pelvis.

Associated with the nephron are blood vessels. Blood flows though them in the following sequence:

- **Afferent arteriole** – brings blood from the renal artery.
- **Glomerulus –** a tight, knot-like, high-pressure capillary bed that is the site of blood filtration.
- **Efferent arteriole** – a narrow vessel that restricts blood flow, helping to generate high pressure in the glomerulus.
- **Peritubular capillaries** – a low-pressure capillary bed that runs around the convoluted tubules, absorbing fluid from them.
- **Vasa recta** – unbranched capillaries that are similar in shape to the loops of Henlé, with a descending limb that carries blood deep into the medulla and an ascending limb bringing it back to the cortex.
- **Venules** – carry blood to the renal vein.

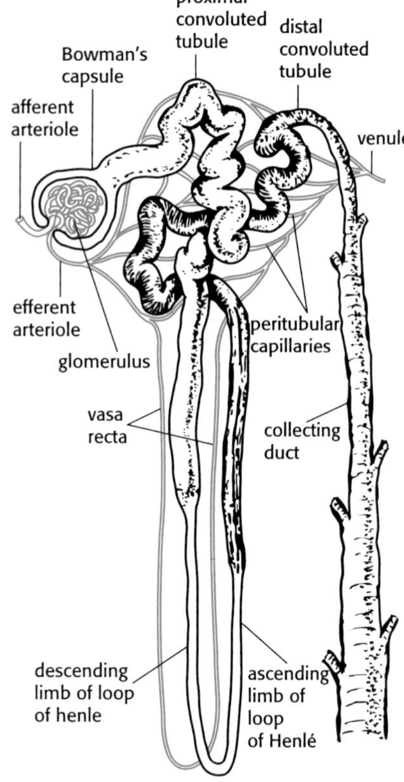

Figure 2 The nephron and associated blood vessels. The human kidney contains about a million nephrons.

Rediscovering biology: examination of kidney tissue

Focus at low power on a microscope slide with a section of kidney tissue and find a similar area to the one shown in Figure 3. Examine this area at high power. Figure 3 shows the appearance of cortex tissue at high power.

Try to find clear examples of these structures:

1. Glomerulus, with many nuclei of the capillary wall cells.
2. Bowman's capsule, with an outer wall of very thin epithelium cells.
3. Proximal convoluted tubule, with a fuzzy inner surface – the microvilli. There will be many capillaries adjacent to this section of the nephron.
4. Distal convoluted tubule, with a smoother inner surface as there are fewer microvilli than in the proximal convoluted tubule.

Return to low power and scan over the slide to find an area of medulla where the nephrons are sectioned at right angles, so they appear round. Look for these structures:

5. Loop of Henlé, with very thin walls and narrow lumen.
6. Collecting duct, which is wider, especially near the renal pelvis.
7. Vasa recta, with narrow walls and blood cells inside.

If a fresh specimen of a kidney is available, cut it in half, to reveal the areas of cortex and medulla. If hydrogen peroxide is dripped onto the cut surfaces, the nephrons inflate with oxygen and become visible. Avoid skin contact with hydrogen peroxide.

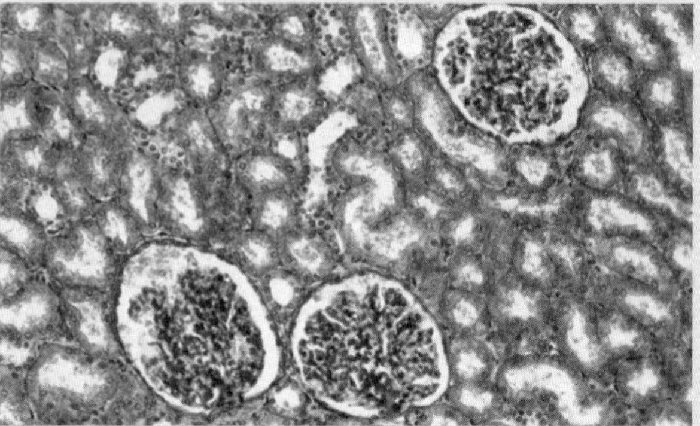

Figure 3 Micrograph of tissue in the cortex of the kidney

HL

Ultrafiltration

Blood in capillaries is at high pressure in many of the tissues of the body, and the pressure forces some of the plasma out through the capillary wall, to form tissue fluid. In the glomerulus of the kidney, the pressure in the capillaries is particularly high and the capillary wall is particularly permeable, so the volume of fluid forced out is about 100 times greater than in other tissues. The fluid forced out is called glomerular filtrate. The composition of blood plasma and filtrate is shown in Table 2. The data in the table shows that most solutes are filtered out freely from the blood plasma, but almost all proteins are retained in the capillaries of the glomerulus. This is separation of particles differing in size by a few nanometres and so is called **ultrafiltration**. All particles with a relative molecular mass below 65,000 can pass through. The permeability to larger molecules depends on their shape and charge. Almost all proteins are retained in the blood, along with all the blood cells. The structure of a section of the filter unit is shown in Figure 4. There are three parts to the ultrafiltration system.

1. **Pores** between the cells in the wall of the capillaries. These are about 100 nm in diameter. They allow fluid to escape, but not blood cells.
2. **The basement membrane** that covers and supports the wall of the capillaries. It is made of negatively charged glycoproteins, which form a mesh. It prevents plasma proteins from being filtered out, due to their size and negative charges.
3. **Podocytes** forming the inner wall of the Bowman's capsule. These cells have extensions that wrap around the capillaries of the glomerulus and many short side branches called foot processes. Very narrow gaps between the foot processes help prevent small molecules from being filtered out of blood in the glomerulus.

If particles pass through all three parts they become part of the glomerular filtrate.

Table 2

	Content (per dm^{-3} of blood plasma)	
Solutes	plasma	filtrate
Na^+ ions (mol)	151	144
Cl^- ions (mol)	110	114
glucose (mol)	5	5
urea (mol)	5	5
proteins (mg)	740	3.5

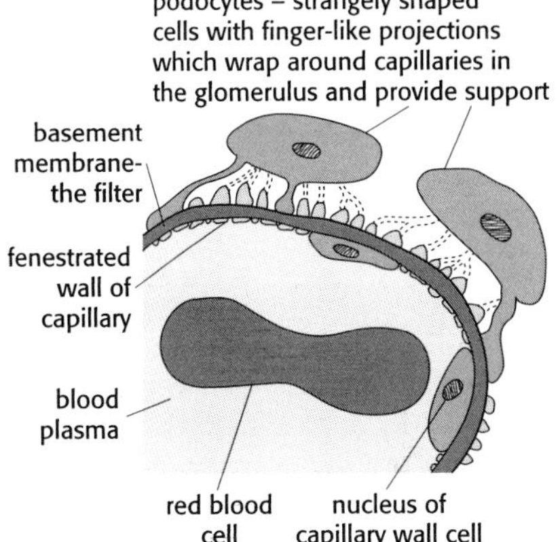

Figure 4 Structure of the filter unit of the kidney

Data-based question: ultrafiltration of charged and uncharged dextrans

Dextrans are polymers of sucrose. Different sizes of dextran polymer can be synthesized, allowing their use to investigate the effect of particle size on ultrafiltration. Neutral dextran is uncharged, dextran sulfate has many negative charges, and DEAE is dextran with many positive charges.

Figure 5 shows the relationship between particle size and the permeability of the filter unit of rat glomeruli. Animal experiments like this can help us to understand how the kidney works and can be done without causing suffering to the animals.

1. State the relationship between the size of particles and the permeability to them of the filter unit of the glomerulus. [1]
2. (a) Compare the permeability of the filter unit to the three types of dextran. [3]
   (b) Explain these differences in permeability. [3]
3. One of the main plasma proteins is albumin, which is negatively charged and has a particle size of approximately 4.4 nm. Using the data in the graph, explain the diagnosis that is made if albumin is detected in a rat's urine. [3]

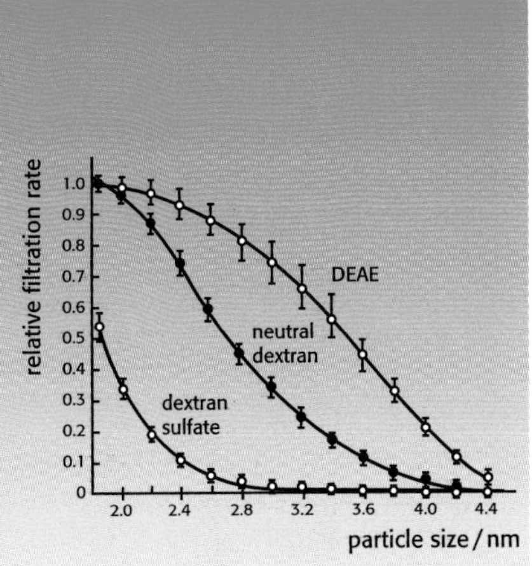

Figure 5 Relationship between particle size of dextrans and filtration rate

HL

Selective reabsorption in the proximal convoluted tubule

The volume of glomerular filtrate produced per day is huge – about 180 dm^{-3}. This is several times the total volume of fluid in the body and it contains nearly 1.5 kg of salt and 5.5 kg of glucose. As the volume of urine produced per day is only about 1.5 dm^{-3} and it contains no glucose and far less than 1.5 kg of salt, almost all of the filtrate must be reabsorbed into the blood. Most of this reabsorption happens in the first part of the nephron – the proximal convoluted tubule. Figure 6 shows this structure in transverse section. The methods used to reabsorb substances in the proximal convoluted tubule are described in Table 3. By the end of the proximal tubule all glucose and amino acids and 80 per cent of the water, sodium and other mineral ions have been absorbed.

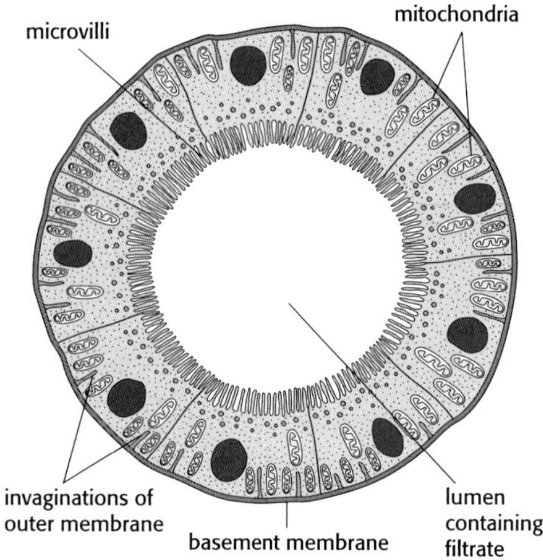

Figure 6 Transverse section of the proximal convoluted tubule

Table 3

Sodium ions: are moved by active transport from filtrate to space outside the tubule. They then pass to the peritubular capillaries. Pump proteins are located in outer membrane of tubule cells.
Chloride ions: are attracted from filtrate to space outside the tubule because of charge gradient set up by active transport of sodium ions.
Glucose: is co-transported out of filtrate and into fluid outside the tubule, by co-transporter proteins in outer membrane of tubule cells. Sodium ions move down concentration gradient from outside tubule into tubule cells. This provides energy for glucose to move at the same time to fluid outside the tubule. The same process is used to reabsorb amino acids.
Water: pumping solutes out of filtrate and into the fluid outside the tubule creates a solute concentration gradient, causing water to be reabsorbed from filtrate by osmosis.

Critical considerations: use of animals in kidney research

Table 4 shows some of the techniques that have been used to investigate kidney function. The animals used include rats, mice, cats, dogs and pigs.

1 What are the reasons for carrying out kidney research?
2 What criteria should be used to decide if a research technique is ethically acceptable or not?
3 Apply your criteria to the three techniques outlined in Table 4 to determine whether they are ethically acceptable.
4 Who should make the decisions about the ethics of scientific research?
5 If a person is opposed to animal research, what level of action is justified?

Table 4

	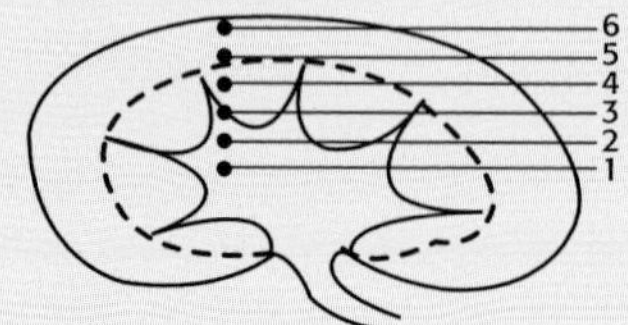	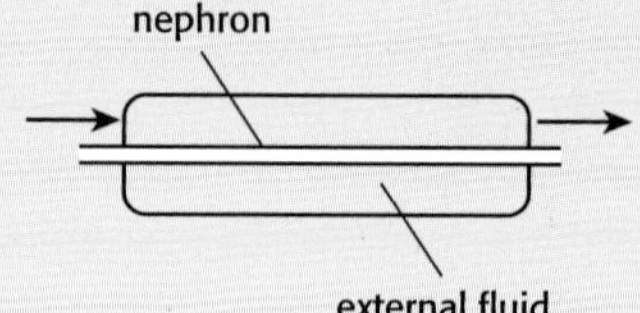
Living animal is anesthetized and its kidney is exposed by surgery. Fluid is sampled from nephrons using micropipettes. Animal is then sacrificed so that the position of the sample point in the kidney can be located.	Animal is sacrificed and kidneys are removed and frozen. Samples of tissue are cut from regions of kidney that can be identified. Temperature at which thawing occurs is found, to give a measure of solute concentration.	Animal is sacrificed and kidneys are dissected to obtain samples of nephron. Fluids are perfused through nephron tissue, using experimental external fluids to investigate the action of the wall of the nephron.

HL

The role of the loop of Henlé

The overall effect of the loop of Henlé is to create a gradient of solute concentration in the medulla. The energy to create the gradient is expended by wall cells in the ascending limb. Here sodium ions are pumped out of the filtrate to the fluid between the cells in the medulla – called the interstitial fluid. The wall of the ascending limb is unusual in that it is impermeable to water, so water is retained in the filtrate, even though the interstitial fluid now has a higher solute concentration.

Normal body fluids have a concentration of 300 mOsm. The pump proteins that transfer sodium ions out of the filtrate can create a gradient of up to 200 mOsm, so an interstitial concentration of 500 mOsm is clearly achievable. The cells in the wall of the descending limb are permeable to water, but are impermeable to sodium ions. As filtrate flows down the descending limb, the increased solute concentration of interstitial fluid in the medulla causes water to be drawn out of the filtrate until it reaches the same solute concentration as the interstitial fluid. If this was 500 mOsm, then filtrate entering the ascending limb would be at this concentration and the sodium pumps could raise the interstitial fluid to 700 mOsm. Fluid passing down the descending limb would therefore reach 700 mOsm, and the sodium pumps in the ascending limb could cause a further 200 mOsm rise. The interstitial fluid concentration can therefore rise further and further, until a maximum is reached, which in humans is 1200 mOsm.

This system for raising solute concentration is an example of a countercurrent multiplier system. It is a *countercurrent* system because of the flows of fluid in opposite directions. It is a countercurrent *multiplier* because it causes a steeper gradient of solute concentration to develop in the medulla than would be possible with a concurrent system. There is also a countercurrent system in the vasa recta. This prevents the blood flowing through this vessel from diluting the solute concentration of the medulla, while still allowing the vasa recta to carry away the water removed from filtrate in the descending limb, together with some sodium ions.

Units for solute concentrations

One *osmole* (Osm) is the solute potential of one mole of dissolved solutes in a litre.

One *milliosmole* (mOsm) is one thousandth of one osmole.

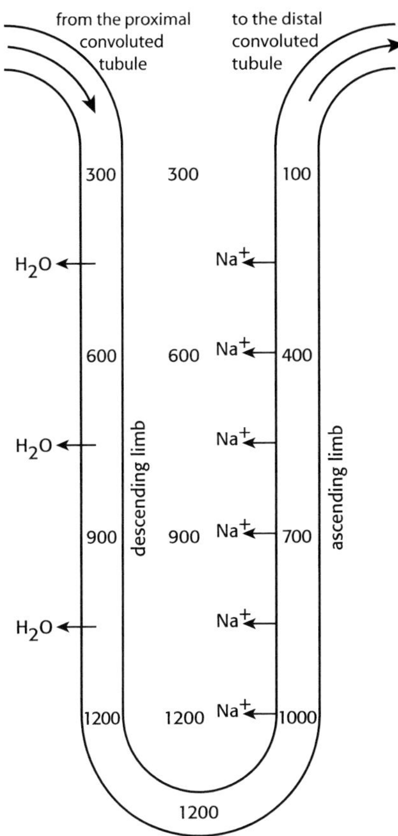

Figure 7 Solute concentrations in the loop of Henlé

Data-based question: medulla thickness and urine concentration

Table 5 shows the relative medullary thickness (RMT) and maximum solute concentration (MSC) of the urine in mOsm for 14 species of mammal. RMT is a measure of the thickness of the medulla in relation to the overall size of the kidney. All the species in the table that are shown with binomials are desert rodents.

1 Discuss the relationship between maximum solute concentration of urine and the habitat of the mammal. [3]

2 Plot a scattergraph of the data in the table, either by hand or using computer software. [7]

3 (a) Using the scattergraph that you have plotted, state the relationship between RMT and the maximum solute concentration of the urine. [1]

(b) Suggest how the thickness of the medulla could affect the maximum solute concentration of the urine. [4]

Table 5

Species	RMT	MSC (mOsm)
beaver	1.3	517
pig	1.6	1076
human	3.0	1399
dog	4.3	2465
cat	4.8	3122
rat	5.8	2465
Octomys mimax	6.1	2071
Dipodomys deserti	8.5	5597
Jaculus jaculus	9.3	6459
Tympanoctomys barrerae	9.4	7080
Psammomys obesus	10.7	4952
Eligmodontia typus	11.4	8612
Calomys mus	12.3	8773
Salinomys delicates	14.0	7440

HL

ADH and osmoregulation

osmoregulation *the control of the water balance of the blood, of tissues or of the cytoplasm of a living organism.*

When filtrate enters the distal convoluted tubule from the loop of Henlé, its solute concentration is lower than that of normal body fluids – it is hypotonic. This is because proportionately more solutes than water have passed out of the filtrate as it flows through the loop of Henlé in the medulla.

If the solute concentration of the blood is too low, relatively little water is reabsorbed as the filtrate passes on through the distal convoluted tubule and the collecting duct. The wall of these parts of the nephron can have an unusually low permeability to water. A large volume of urine is therefore produced, with a low solute concentration, and as a result the solute concentration of the blood is increased (see Figure 8a).

If the solute concentration of the blood is too high, the hypothalamus of the brain detects this and causes the pituitary gland to secrete a hormone – anti-diuretic hormone or ADH. This hormone causes the walls of the distal convoluted tubule and collecting duct to become much more permeable to water, and most of the water in the filtrate is reabsorbed. This is helped by the solute concentration gradient of the medulla. As the filtrate passes down the collecting duct, it flows deep into the medulla, where the solute concentration of the interstitial fluid is high. Water continues to be reabsorbed along the whole length of the collecting duct and the kidney produces a small volume of concentrated urine (Figure 8b). As result the solute concentration of the blood is reduced. The action of the kidney therefore helps to keep the relative amounts of water and solutes in balance at an appropriate level. This is called **osmoregulation**.

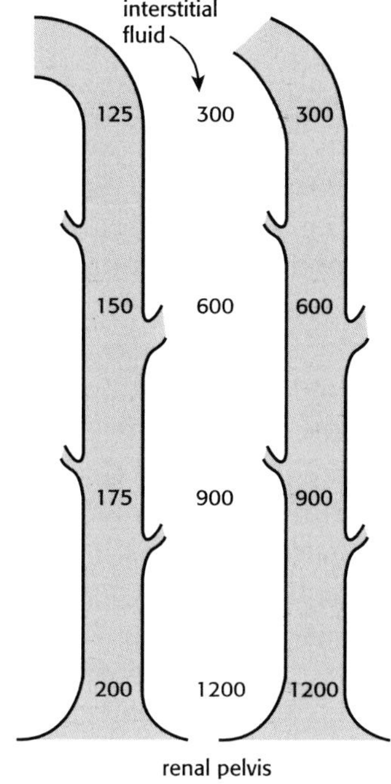

Figure 8 Solute concentrations in the collecting duct

Rediscovering biology: blood cells and osmoregulation

The importance of osmoregulation can be shown by mixing samples of animal blood with a range of salt solutions.

1 Label six centrifuge tubes **I** to **VI**. Put water and 0.25 mol dm^{-3} sodium chloride solution into the tubes as shown in Table 6.
2 Add 1 ml of defibrinated horse blood to each tube. Shake to mix and leave the tubes for five or more minutes.
3 Centrifuge each tube for two or more minutes.
4 Compare the appearance of the tubes after centrifuging.
5 Without shaking, pour off some fluid from each tube into a cuvette and measure its absorbance using a colorimeter with a blue filter.

Table 6

	I	II	III	IV	V	VI
Volume of 0.25 mol dm^{-3} NaCl solution (ml)	0	1	2	3	4	5
Volume of water (ml)	5	4	3	2	1	0
Concentration of NaCl (mol dm^{-3})	0.0	0.05	0.1	0.15	0.2	0.25

Table 7

	I	II	III	IV	V	VI
Volume of cells (ml)	0.00	0.00	0.40	0.65	0.55	0.50

1 Explain the differences in the appearance of the tubes after centrifuging.
2 Deduce from the colorimeter results the concentration of NaCl that is closest in solute concentration to blood plasma.
3 When this experiment was done using graduated centrifuge tubes, the total volume of blood cells at the bottom of the tubes was as shown in Table 7. Explain the differences in terms of movements of water by osmosis.
4 List the factors that should be kept constant in this experiment.
5 Explain how the method could be modified to find the solute concentration of blood more precisely.

HL

Chapter 23 questions

1 Table 8 shows the content of four solutes in blood plasma, glomerular filtrate and urine. Some of the figures have been rounded up or down.

Table 8

	Concentration (mg per 100 ml of blood)		
Solutes	**Plasma**	**Filtrate**	**Urine**
Na^+ ions	900	900	1200
glucose	90	90	0
urea	30	30	1200
proteins	740	0	0

(a) Explain the difference in protein concentration between blood plasma and glomerular filtrate. [2]
(b) Explain the difference in glucose concentration between glomerular filtrate and urine. [3]
(c) Explain how the urea concentration becomes 40 times more concentrated in the filtrate as it flows through the nephron to become urine. [3]
(d) Explain reasons for the sodium ions becoming concentrated by a much smaller factor than urea as filtrate flows along the nephron. [3]
(e) The concentration of sodium ions in urine can be higher or lower than 1200 mg per 100 ml of blood. Explain how the concentration of sodium ions can vary. [3]

2 The blood glucose concentration of a person with untreated diabetes often rises to 300–500 mg per 100 ml of blood. It can even rise to concentrations above 1000 mg per 100 ml. When the blood glucose level rises above 225 mg per 100 ml, glucose starts to appear in the urine. The volumes of urine produced become larger than normal, making the person dehydrated and thirsty.
(a) Explain how glucose is completely reabsorbed from the glomerular filtrate of people who do not have diabetes. [3]
(b) Explain why glucose is not all reabsorbed from the glomerular filtrate of diabetic patients. [4]
(c) Suggest why untreated diabetics tend to pass large volumes of urine and often feel thirsty. [3]

3 Discuss whether each of the following processes is a part of excretion or osmoregulation. Some may be involved in both processes, or neither.
(a) Production of urine. [1]
(b) Production of saliva. [1]
(c) Sweating in hot weather. [1]
(d) Exhaling air from the lungs containing carbon dioxide and water vapour. [2]
(e) Drinking water after feeling thirsty. [1]
(f) Removing water from unicellular organisms such as *Paramecium*, using a contractile vacuole. [1]
(g) Diffusion of oxygen and water vapour out through the stomata of plant leaves. [2]
(h) Animals licking salt from rocks or soils. [1]

4 Figure 9 shows the flow rate of filtrate in the nephron, as a percentage of the original volume of filtrate formed. It also shows the solute concentration of the filtrate in mOsm. ADH was circulating when the figures were obtained.

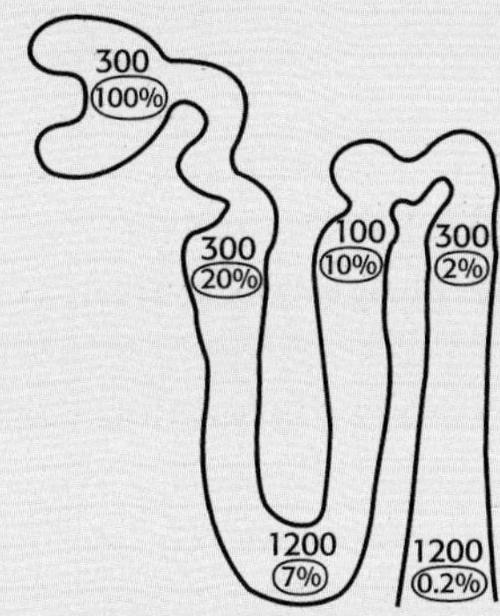

Figure 9

(a) Explain how the flow rate drops to 20% as the filtrate flows through the proximal convoluted tubule. [2]
(b) Explain how the filtrate becomes more concentrated as it flows down the descending limb of the loop of Henlé. [2]
(c) Explain how the filtrate becomes less concentrated as it flows up the ascending limb of the loop of Henlé. [2]
(d) Explain how the fluid becomes more concentrated as it flows through the distal convoluted tubule and the collecting duct. [2]
(e) When ADH is not circulating, typical figures for the start of the collecting duct are 10% and 150 and for the end of the collecting duct are 2% and 200. Explain the differences between these figures and those shown in Figure 9. [2]

5 ADH binds to a receptor protein in the membranes of nephron cells in the distal convoluted tubule and the collecting duct. There is a genetic disease in which this receptor protein is altered and ADH fails to bind to it.
(a) Predict the health problems that would result from this genetic disease, which can be serious enough to cause death if appropriate measures are not taken. [4]
(b) Children can develop this genetic disease even if their parents show no symptoms. Explain the conclusion that you draw from this observation. [2]
(c) Boys are affected by this genetic disease far more frequently than girls. Explain the conclusion that you draw from this observation. [2]
(d) Suggest how the disease could be treated. [2]

24 Nerves and hormones

The nervous system is involved in homeostasis as it regulates the body's responses to external and internal stimuli. It is involved in the coordination of information from the senses as well as responses to sensory information. It is also involved in the coordination of muscle activity and the monitoring of the functioning of the other organ systems.

The two main divisions of the nervous system are the central nervous system and the peripheral nervous system. The central nervous system (CNS) is composed of the brain and the spinal cord while the peripheral nervous system is composed of all the other nerves that are not part of the central nervous system. Afferent nerves convey sensory information to the CNS, for example from skin or the eyes, while efferent nerves conduct signals from the CNS to the muscles and the endocrine system.

The nervous system has cells called neurons. While the structure of a neuron can vary, they are typically composed of four main structural regions: a network of dendrites, a cell body, an axon and an axon terminus. Figure 1 shows an example of a neuron called a motor neuron. One challenge when drawing a picture of a motor neuron is to convey the length of the axon in relation to the length of the cell body, as the axon of a motor neuron can be many times longer than the length of the cell body. In humans, the axon connecting the spinal cord to the toes can be up to a metre in length!

Figure 2 shows the path by which nerve impulses are conducted. Signals carried from receptors such as sense organs in the skin travel to the CNS in sensory neurons, within the CNS in relay neurons, and from the CNS to effectors in motor neurons. An effector is an organ of the body such as a muscle or a gland that responds to an impulse from a motor neuron.

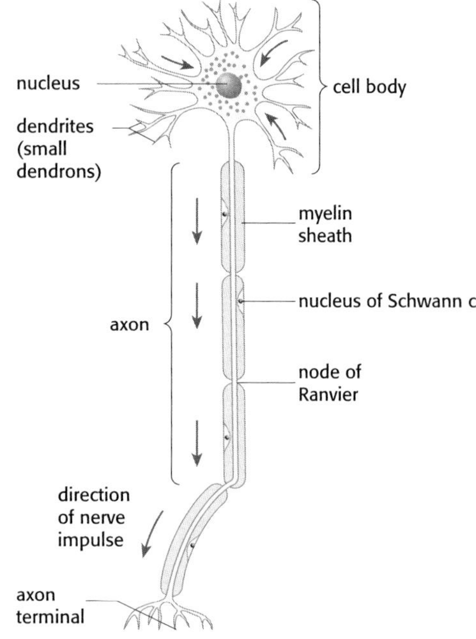

Figure 1 A motor neuron

Using Figure 2, compare the structure of the three types of neuron (motor neuron, relay neuron, and sensory neuron).

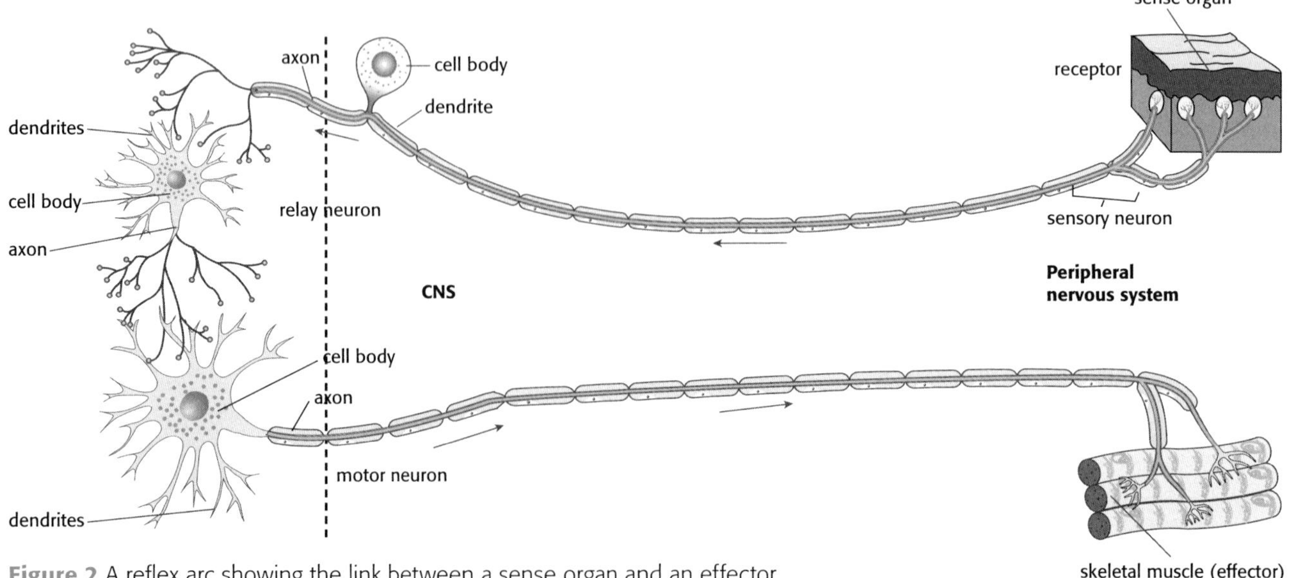

Figure 2 A reflex arc showing the link between a sense organ and an effector

Investigating biology: two-point discrimination

Bend a paper clip into a U-shape so that the two tips are separated by 2 cm. If the tips are touched to the finger tips, a person will sense it as two points, but if the tips of the paper clip are touched to the back of the shoulder, the person is likely to sense it as a single point.

Each sensory neuron connects to the central nervous system (CNS). A given CNS neuron responds to all information from its input area, for example an area of the skin, as if it were coming from one point. This input area is called the receptive field of the CNS neuron. On the shoulder, each sensory receptor gathers information from a much larger skin area than a receptor on the fingertip; i.e. the receptive field is larger on the shoulder. In order for a person to feel two points, two separate CNS neurons must be activated by stimulation of their respective receptive fields.

1 Design an experiment to investigate an aspect of two-point discrimination. Is it fixed or can it be influenced by external variables?
2 Can you hypothesize why there is superior discrimination in some areas over others? Is it possible to test this hypothesis? How many times will you repeat the test in each area of skin? How large must the difference be in the results for two areas to give reliable evidence of difference in two-point discrimination?

Thinking about science: the limits to perception

Knowledge claims in science are justified empirically through observation. Perception is the process of gathering sense data and interpreting it. The two-point discrimination activity illustrates that there are limits to the range of things that humans can sense. We can't see infrared radiation as a snake can. We can't hear sounds that a dog can.

1 If humans are sensitive only to certain ranges of stimuli, what consequences, including limitations, might this have for the acquisition of scientific knowledge?

This next activity illustrates that even if what we sense is in the range of what we can perceive, our senses can be fooled.

Obtain three containers. Place ice water into one, room-temperature water into the next and hot water into the third. Place one hand in the ice water and one hand in the hot water at the same time for 30 seconds. Then place both hands into the water at room temperature. Is the room-temperature water warm or cold?

2 How can a thermometer help in this case?
3 How does the use of technology in science modify and extend the capabilities of the senses?
4 To what extent is there a difference between data gathered with unaided senses and data gathered through the medium of technology?

Resting potentials

A neuron that is not transmitting a signal is said to be at rest. When it is at rest, there is an electrochemical gradient called the resting potential across the membrane (see Figure 3). The inside of the membrane is negative relative to the outside. This gradient exists because:

- A protein called the sodium/potassium (Na^+/K^+) pump actively transports Na^+ ions out of the cell and K^+ ions in. For every three Na^+ ions pumped out, only two K^+ ions are pumped in.
- The membrane is about 50 times more permeable to K^+ ions than Na^+ ions. K^+ ions then are more likely to diffuse down their concentration gradients.
- Na^+ ions are largely prevented from moving across the membrane.
- A higher concentration of organic anions (negatively charged ions) is found on the inside of the membrane than on the outside.

The resting membrane potential of a neuron is about –70 mV.

? Examine Figure 3 and explain four reasons why the cytoplasm side of the neuron membrane is negatively charged in relation to the outside of the membrane when the membrane is at its resting potential.

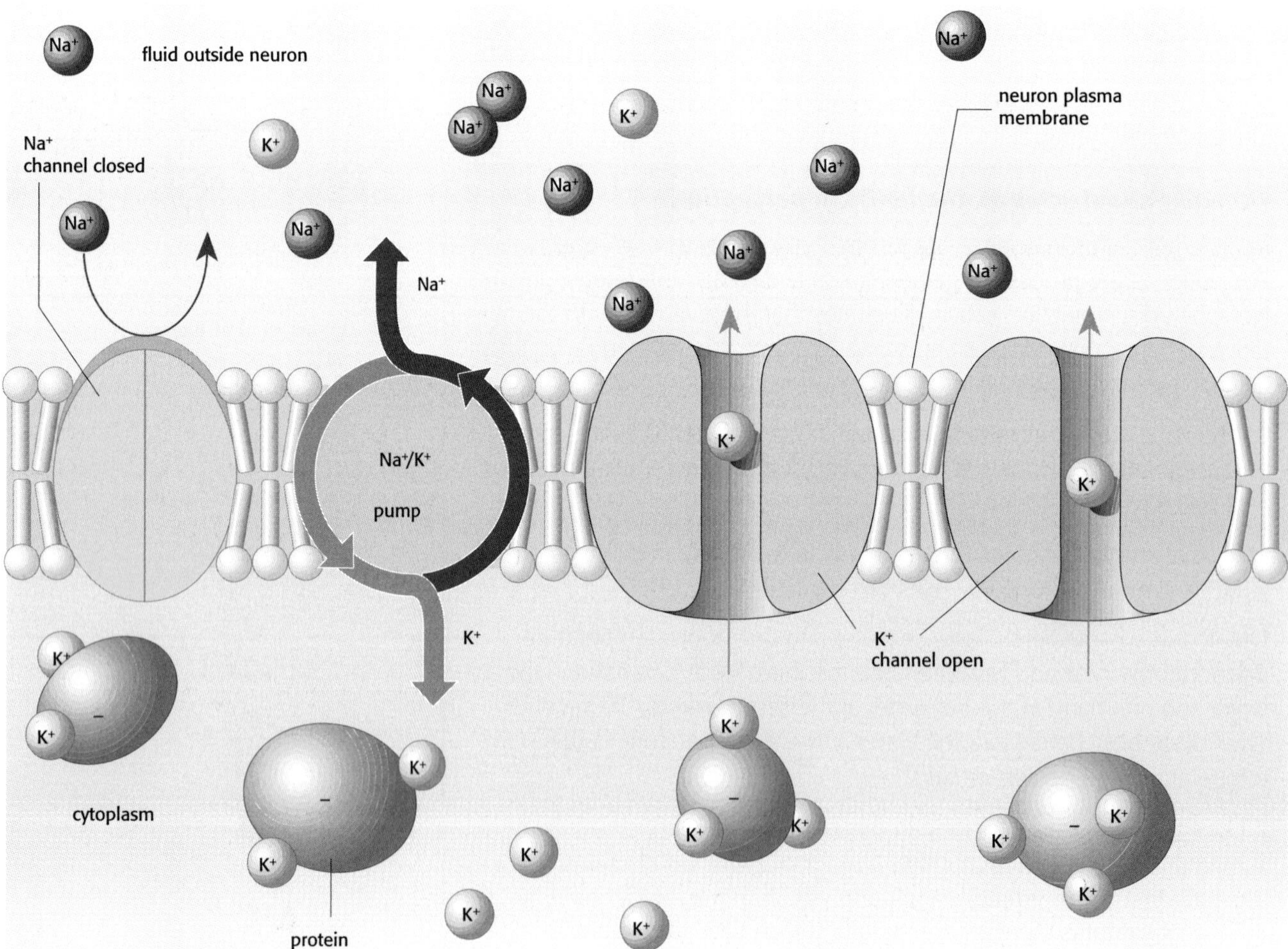

Figure 3 During the re-polarization phase, Na^+ channels are closed and K^+ channels are open.

Action potentials

The signals that travel along nerves are due to a moving action potential along the length of a neuron. An action potential is a rapid change in membrane potential from negative to positive and back to negative again. The sequence of events is as follows (see Figure 4):

1 When a stimulus is applied to a neuron, some sodium ion channels open and some sodium ions enter. This causes the membrane potential to rise; i.e. to become less negative. If the potential reaches the threshold potential, –50 mV, an action potential is triggered. Depolarization occurs when most Na^+ channels in an area of a neuron open, while the K^+ ion channels remain closed. Na^+ ions diffuse in rapidly down their concentration gradient, raising the membrane potential to a positive value.

2 At the peak potential, the Na^+ gates close and the K^+ channels open.

3 K^+ ions flow down their concentration gradient making the inside of the cell negative again.

4 The K^+ gates remain open, making the membrane more permeable to K^+ than when the membrane is at rest. The result is a membrane potential that is temporarily lower than resting potential.

5 The eventual closing of the K^+ gates and the activity of the Na^+/K^+ pump return the membrane to resting potential.

An action potential that occurs in one part of the neuron triggers an action potential in a neighbouring area of the neuron. This is due to the diffusion of Na^+ ions from the area where an action potential has just occurred into the area at rest, which then leads to the events described in step 1 above. In this way, the nerve impulse moves down the axon. The undershoot, or hyperpolarization (see Figure 5), helps to ensure the one-way movement of the impulse.

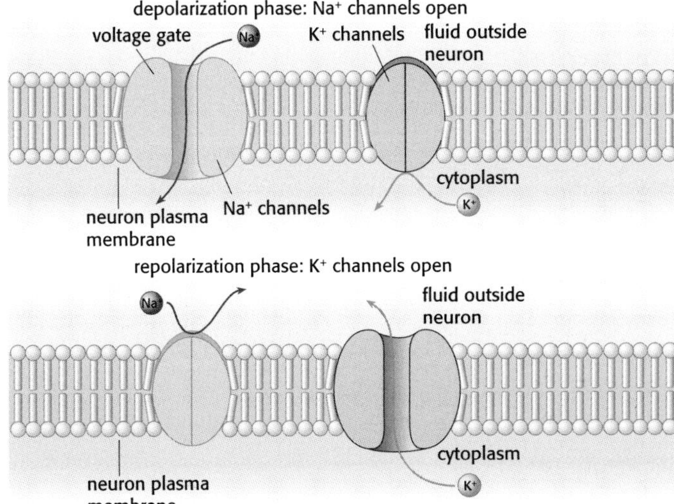

Figure 4 Na^+ channels are open during depolarization while K^+ channels are open during repolarization

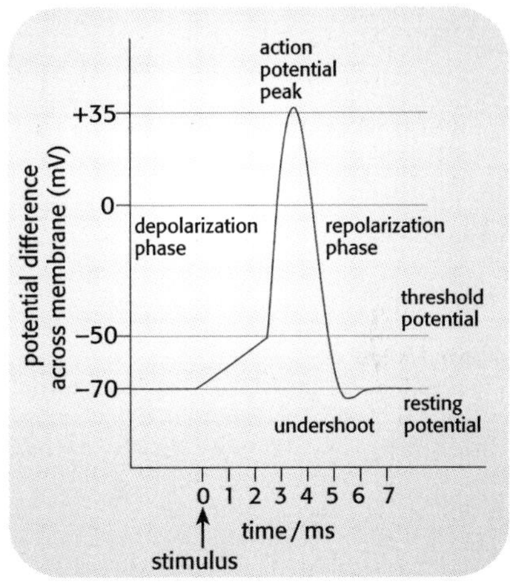

Figure 5 Changes in membrane polarity during an action potential

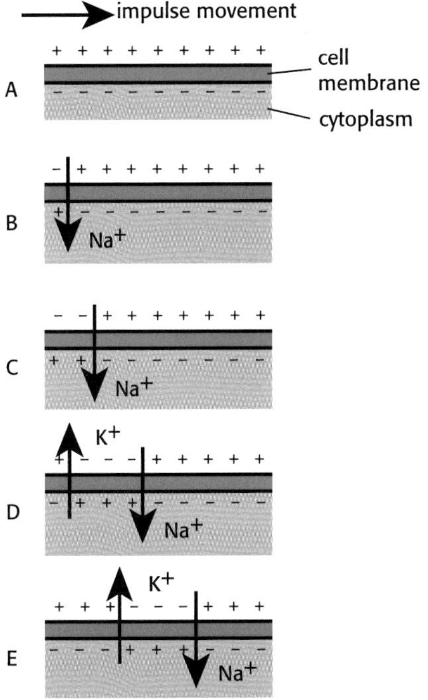

Figure 6 Nerve impulses move through an axon by temporary reversal of membrane polarity.

Data-based question: ouabain

David Livingstone was a 19th century explorer who led the British government-funded Zambesi Expedition. His team included several scientists, who gathered information about the botany, geology and ethnography of the region. The data was later analysed by scientific institutions in the UK. During his exploration they observed the use of a poison extracted from seeds of a plant from the genus *Strophanthus*. The poison was so potent that, applied to an arrow tip, it could kill a large animal. Scientists later isolated the chemical strophantine from plants from the same genus to produce the drug ouabain. In very small doses, ouabain is used to stimulate the heart in the treatment of heart failure.

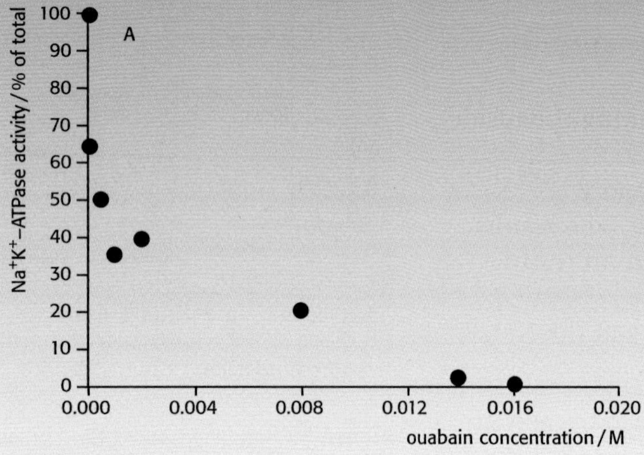

Figure 7 The effect of increasing ouabain concentration on the function of the Na^+/K^+ pump in the Antarctic sea urchin, *Sterechinus neumayeri*

1 (a) Outline the relationship between Na^+/K^+ pump function and ouabain concentration. [2]

(b) Predict the ouabain concentration at which Na^+/K^+ function would be 25%. [1]

(c) Explain why ouabain might be able to kill a large mammal. [2]

Critical considerations: bioprospecting

Bioprospecting is the collection of biological samples as well as the use of indigenous knowledge to help in the discovery and exploitation of biological resources. The 1992 Convention on Biological Diversity asserts the right of biodiversity-rich countries to share in the benefits of products such as pharmaceuticals and other biotechnology.

1 If the discovery of strophantine were a contemporary story, to what extent should the indigenous people or their country be entitled to share in the benefits of the development of a drug?

2 To what extent should the source country have a say in the exploitation of biological resources? To what extent should indigenous populations have a say in the exploitation of biological resources? What if the values of the indigenous population extend to non-exploitation of the resource?

3 To what extent does responsibility for the conservation of biodiversity extend beyond the borders of that nation where biodiversity is found? Because the whole planet benefits, should a source country be compensated by the international community for the conservation of its natural heritage?

Synapses

At the junction between two neurons there is a fluid-filled gap that separates the plasma membranes of the two neurons. This junction is called a synapse (see Figure 8). A nerve impulse that reaches the axon terminus of one neuron is propagated across the gap by chemicals called neurotransmitters.

1 The nerve impulse causes Ca^{2+} ions to diffuse through channels in the membrane.
2 The calcium influx causes vesicles containing neurotransmitters to fuse with the pre-synaptic membrane.
3 Neurotransmitter is released into the synaptic cleft. The fusion of vesicles leading to the release of materials outside the cell is called exocytosis.
4 The neurotransmitter diffuses across the synaptic cleft and binds to receptors on the post-synaptic membrane.
5 The binding of the neurotransmitter to the receptors causes adjacent ion channels to open, and ions diffuse down their concentration gradient.
6 The local depolarization leads to an impulse being generated in the post-synaptic neuron.
7 The neurotransmitter is either recycled, degraded or diffuses out of the synapse.

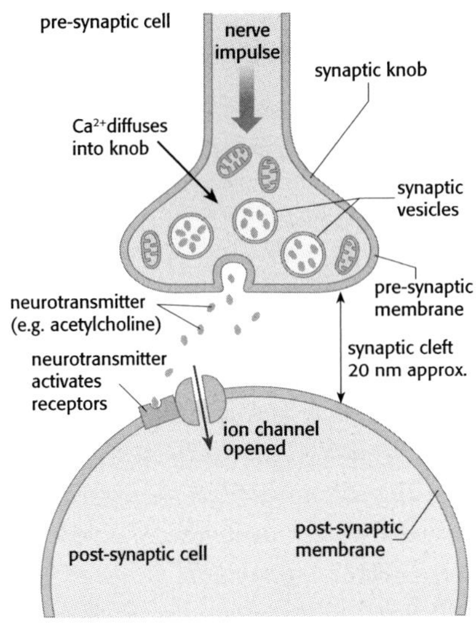

Figure 8 A nerve impulse is propagated across a synapse by the release, diffusion and post-synaptic binding of neurotransmitter

Investigating biology: nervous control of hand movements

1 Data-logging EKG sensors can be used as to produce an electromyogram (EMG), which is a plot of the electrical activity produced by muscular contraction. The lower plot in Figure 9 shows the EMG measurements that were made as a student squeezed a force sensor with different degrees of force. A similar investigation a student might undertake would be to determine the pattern of electrical activity associated with muscle fatigue.

2 Two students working together can assess reaction time. The subject rests their elbow on a table with their hand extended over the edge. The other student should hold a metre stick at the 0 cm mark between the subject's thumb and index finger. As soon as the ruler is let go, the subject should attempt to catch it. The distance the ruler falls can be used to give a measure of reaction time. Factors that might affect reaction time, such as the effect of auditory distraction or whether the subject has one or two eyes open, can be investigated. Variables will need to be carefully controlled.

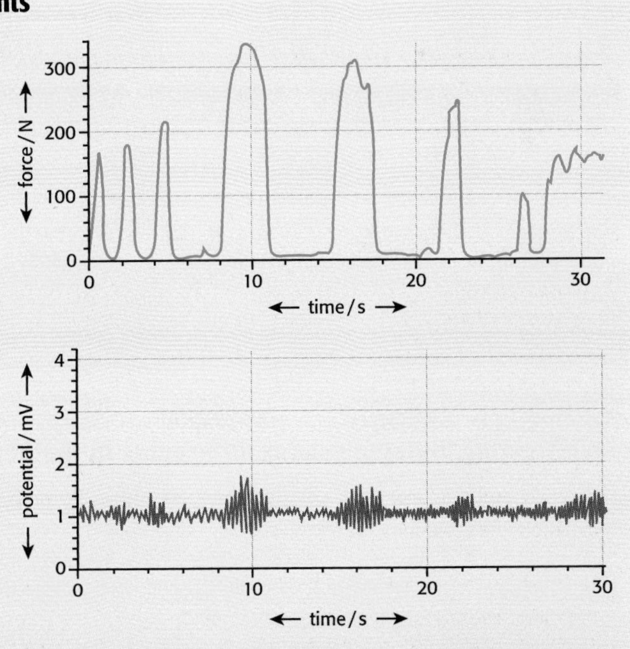

Figure 9 The relationship between hand gripping force and electrical activity in the muscle

Equation

To determine reaction time, t:

$$t = \sqrt{\frac{2d}{g}}$$

where g is the acceleration due to gravity (980 cm s^{-2}) and d is the distance measurement from the ruler.

Critical considerations: gender differences in the brain

This is an extract from an article, "His Brain, Her Brain" by Larry Cahill, in *Scientific American* (May 2005).

On a gray day in mid-January, Lawrence Summers, the president of Harvard University, suggested that innate differences in the build of the male and female brain might be one factor underlying the relative scarcity of women in science. His remarks reignited a debate that has been smoldering for a century, ever since some scientists sizing up the brains of both sexes began using their main finding – that female brains tend to be smaller – to bolster the view that women are intellectually inferior to men.

To date, no one has uncovered any evidence that anatomical disparities might render women incapable of achieving academic distinction in math, physics or engineering. And the brains of men and women have been shown to be quite clearly similar in many ways. Nevertheless, over the past decade investigators have documented an astonishing array of structural, chemical and functional variations in the brains of males and females.

Differences in the size of brain structures are generally thought to reflect their relative importance to the animal. For example, primates rely more on vision than olfaction; for rats, the opposite is true. As a result, primate brains maintain proportionately larger regions devoted to vision, and rats devote more space to olfaction. So the existence of widespread anatomical disparities between men and women suggests that sex does influence the way the brain works.

Investigations are finding anatomical sex differences at the cellular level. For example, Sandra Witelson and her colleagues at McMaster University discovered that women possess a greater density of neurons in parts of the temporal lobe cortex associated with language processing and comprehension. On counting the neurons in postmortem samples, the researchers found that of the six layers present in the cortex, two show more neurons per unit volume in females than in males. Similar findings were subsequently reported for the frontal lobe. With such information in hand, neuroscientists can now explore whether sex differences in neuron number correlate with differences in cognitive abilities – examining, for example, whether the boost in density in the female auditory cortex relates to women's enhanced performance on tests of verbal fluency.

Justifiable claims of difference in one domain are often used to support claims of difference that are not justified. This is the basis of sexism and racism.

1 Investigate the "discipline" of craniometry. What are some claims made by this "discipline"?
2 Research into brain variations between genders could lead to sex-specific treatments for disorders such as depression and schizophrenia. What criteria can be used to determine when focusing on differences is valid, and when it is not?

Investigating biology: gender differences in mathematical tests

Gender differences have been reported in two categories of mathematical exercise: calculation and reasoning.

Find subjects who are willing to participate, and understand that the purpose of the experiment is to investigate mathematical ability.

1 Give each question separately.
2 Determine the length of time required to answer the question. If the subject gets the answer wrong, continue timing until they get the answer correct.
3 Discard the trial if the subject would prefer to stop before obtaining the answer.

A suggestion is to use a statistical test such as the *t*-test to determine whether observed differences between genders are statistically significant. Ensure that your sample size is sufficiently large.

Example:

Example of a calculation test:

$\frac{24}{4} + 3(11 - 9) - 12$

Example of a reasoning test:
If only 30 per cent of hatchlings will survive, how many fish must be introduced into a pool to get 330 adult trout?

(Source of tests: "His Brain, Her Brain", http://www.exn.ca/brain/)

Homeostasis

The human body maintains many variables within defined limits, such as body temperature, blood pH, carbon dioxide concentration, and blood glucose concentration. Both the nervous system and the endocrine system play a role in homeostasis by monitoring levels of variables and correcting changes in levels by negative feedback mechanisms. Negative feedback is also known as end-product inhibition and it involves the product of a process inhibiting the process that creates the product (see Figure 10).

Thermoregulation is an example of a homeostatic mechanism. Cells in the hypothalamus, a region of the brain, detect changes in core body temperature and effect different responses depending on whether the core body temperature is above or below the set point (around 37 °C). For example, when the core temperature of the body rises above the set point, nerve impulses cause the muscles in the walls of skin surface arterioles to relax. As a consequence, their diameter increases and more blood flows near the surface. The heat of the blood warms up the skin, allowing heat to dissipate to the environment. Figure 11 summarizes some of the mechanisms of thermoregulation.

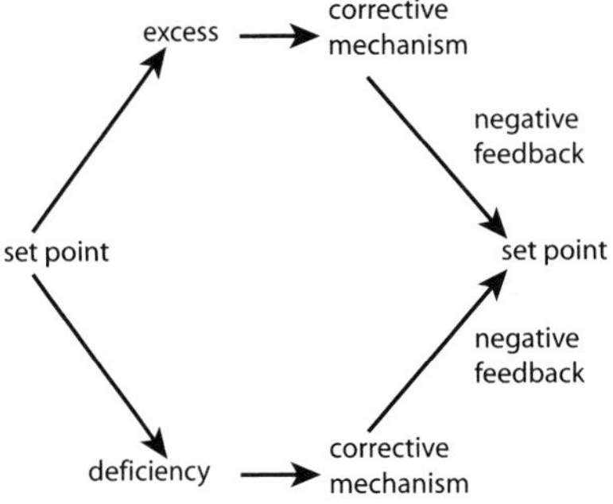

Figure 10 Negative feedback

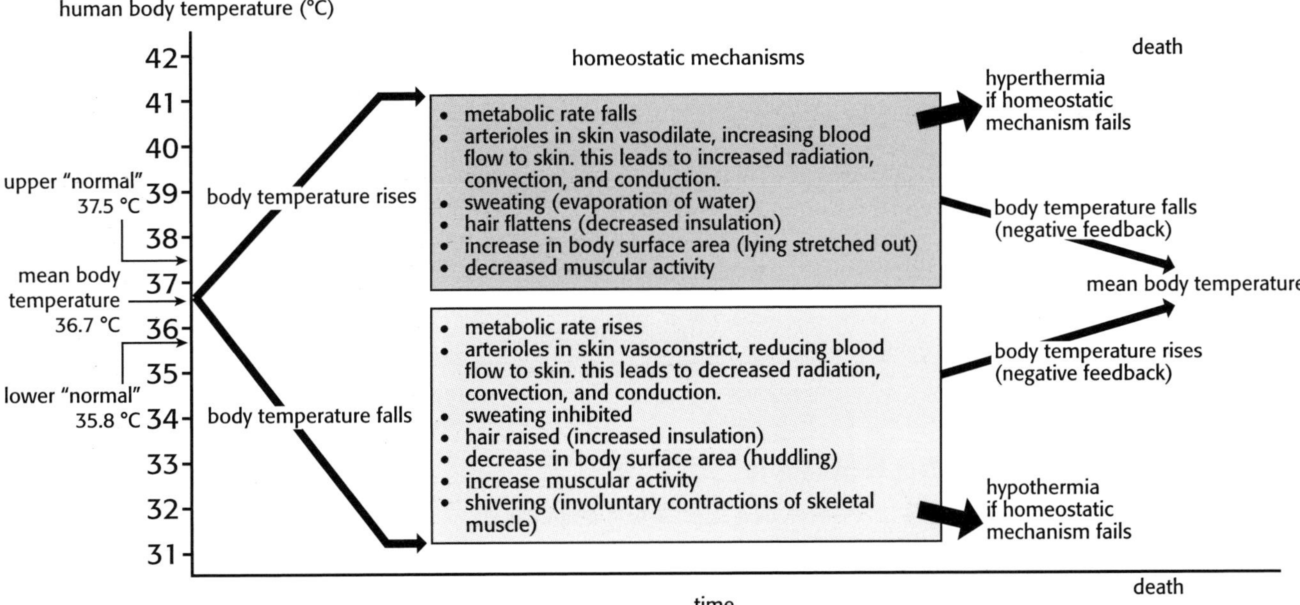

Figure 11 A summary of the mechanisms of thermoregulation

Data-based question: wind-chill nomogram

Humans must maintain a body temperature close to 37 °C. This is achieved despite wide variations of climatic conditions and amounts of heat produced by the body. Air temperature and air movement both affect heat loss from the body. Although the effect of wind has been recognized for a long time, no satisfactory means of combining air temperature and air movement was devised until Siple and Passel introduced the Wind-chill Scale.

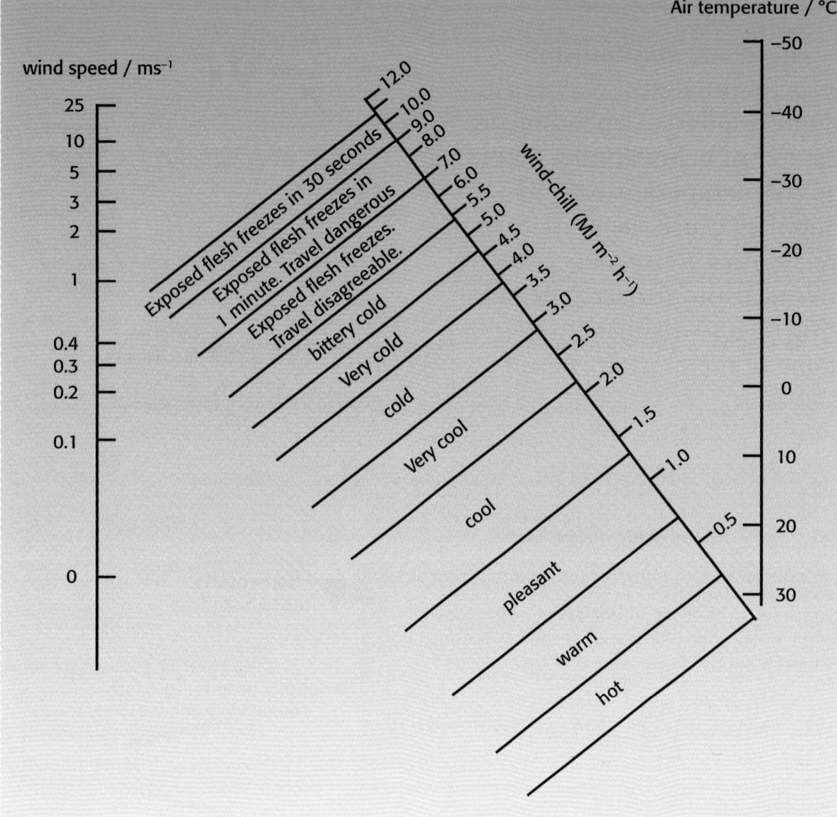

Figure 12 A nomogram used to relate temperature and air speed to wind-chill

1 Estimate, using Figure 12, the wind-chill for a wind speed of 5 m s^{-1} at a temperature of −30 °C. [1]
2 Identify how a person would feel if the wind speed is 0.2 m s^{-1} at a temperature of 10 °C. [1]
3 Calculate the energy lost in one hour for a person with a surface area of 2 m^2, when the air temperature is 20 °C and the wind speed is 1 m s^{-1}. [2]
4 Compare the effects of air temperature with wind speed on the value of the wind-chill. [2]

Investigating biology: skin surface temperature

A digital surface temperature thermometer has an extremely rapid response time to temperature changes and so can be used in a number of investigations related to thermoregulation.

1 Tape the thermometer to the surface of the bicep of a study subject and have the subject repeatedly curl a body-building weight. Monitor the temperature of the bicep. This will illustrate the production of waste heat such as occurs during shivering.
2 Expose two different areas of skin to ice wrapped in absorbant material. The two areas should differ in the degree of vascularization at the skin surface. Compare the time for the two areas of skin to recover a baseline temperature after removing the ice.
3 Design an experiment to determine if the red flush associated with embarrassment is caused by vasodilation (see Figure 13).

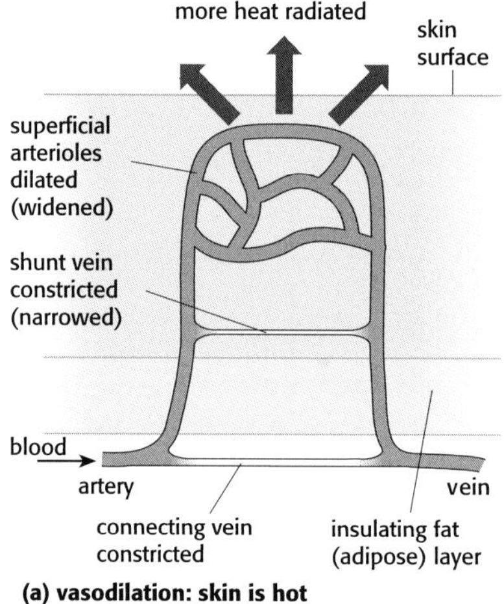

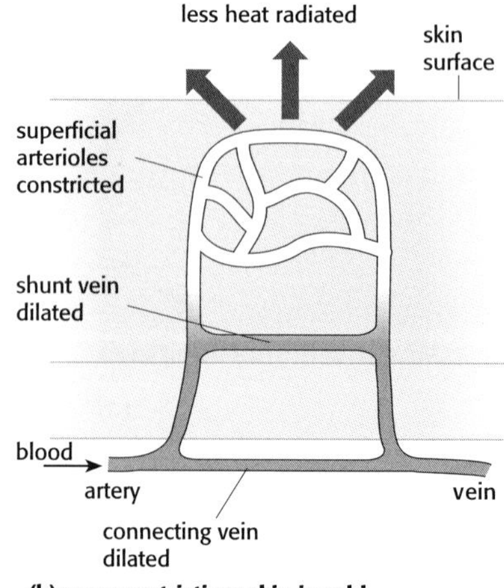

Figure 13 (a) Vasodilation of skin surface arterioles occurs when the blood needs to release heat. (b) The body can conserve heat through the vasoconstriction of skin surface arterioles.

Blood glucose

Blood glucose levels are monitored by cells in the pancreas. If the concentration deviates from the set point of about 5 mmol dm^{-3}, then homeostatic mechanisms involving the pancreatic hormones insulin and glucagon are initiated.

Diabetes is a condition whereby an individual has consistently elevated levels of sugar in the blood. Type I diabetes, or early-onset diabetes, is characterized by an inability to produce sufficient quantities of insulin. For this reason it is often referred to as insulin-dependent diabetes. Type II diabetes, or adult-onset diabetes, is most often characterized by an inability to process or respond to insulin because of a deficiency of insulin receptors on target cells.

> **?** Referring to Figure 14, outline the role of each of the following in blood glucose homeostasis:
> (a) glucagon;
> (b) insulin;
> (c) glycogen;
> (d) α and β pancreatic cells.

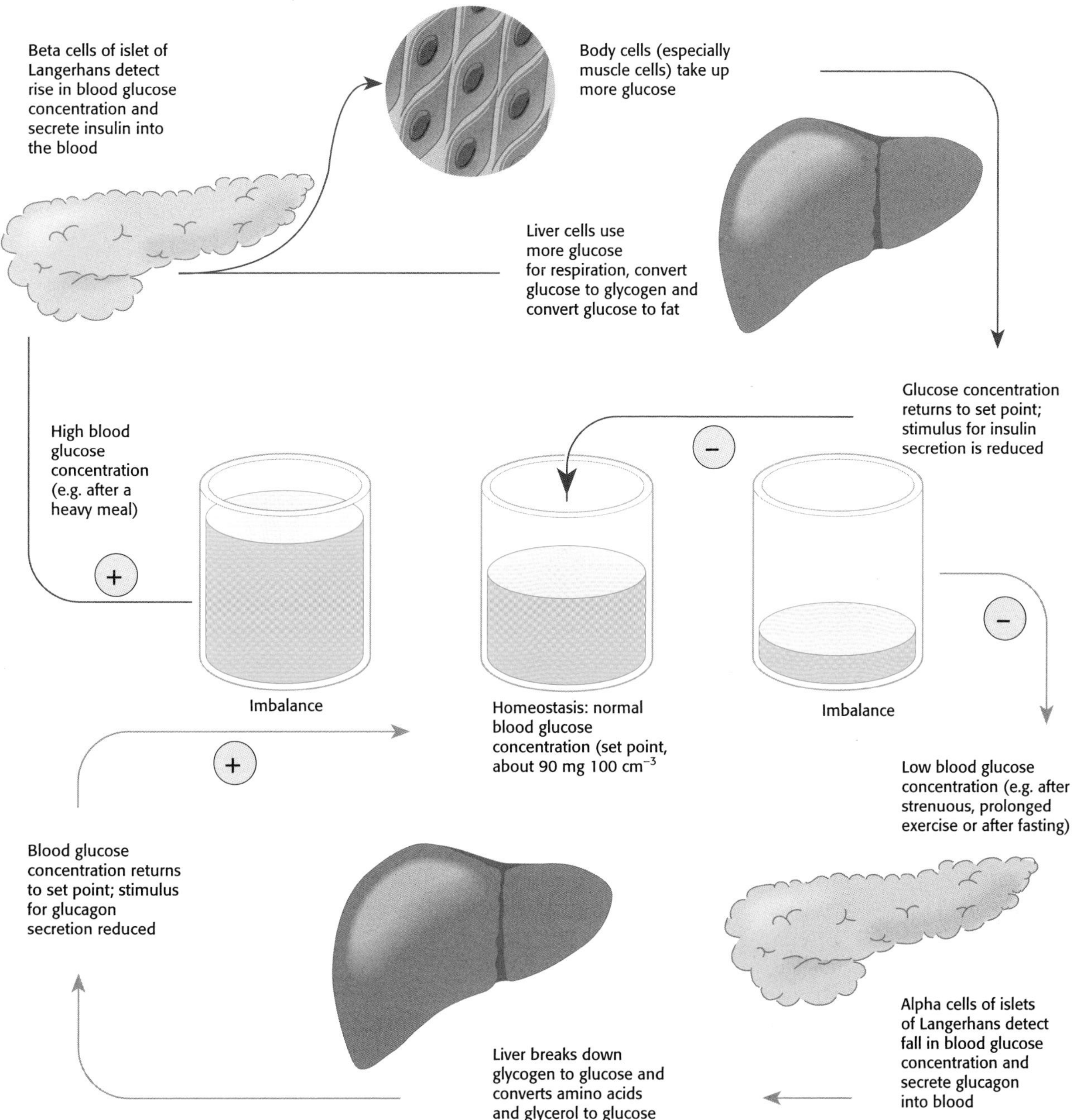

Figure 14 Blood sugar homeostasis

Data-based question: the glucose tolerance test

The glucose tolerance test is a method used to diagnose diabetes. In this test, the patient drinks a concentrated glucose solution. The blood glucose concentration is monitored to determine the length of time required for excess glucose to be cleared from the blood.

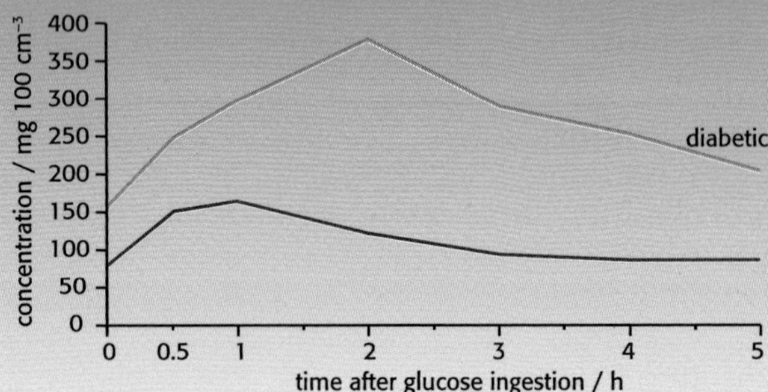

Figure 15 A person with diabetes and an unaffected person give very different responses to the glucose tolerance test.

With reference to Figure 15, compare the person with normal glucose metabolism to the person with diabetes with respect to:

(a) The concentration of glucose at time zero, i.e. before the consumption of the glucose drink.
(b) The length of time required to return to the level at time zero.
(c) The maximum glucose level reached.
(d) The time before glucose levels start to fall.

Investigating biology: the diving reflex

When marine mammals such as seals dive, their metabolism responds by undertaking a number of oxygen-saving measures collectively known as the diving reflex. The reflex is more pronounced when the water is cold. One of these responses is known as bradycardia, i.e. heart rate suppression. This reflex is observed in all mammals, including humans to some extent, though it is less pronounced. Diving can be simulated by facial immersion, though waterproof watches with heart-rate monitors such as those used by triathlon competitors might be used in a pool or lake if a lifeguard is present.

Design an experiment to investigate the diving reflex. The independent variable could be water temperature, depth, facial versus full-body immersion or many other factors. Heart rate does not necessarily have to be the variable – the length of time a person can hold their breath in cold water versus warm water could be compared, for example.

Data-based question: the diving reflex

To investigate the diving reflex in humans, heart rate changes in ten healthy subjects were monitored during facial immersions in water ranging from 3 °C to 37 °C. The data for this experiment is shown below.

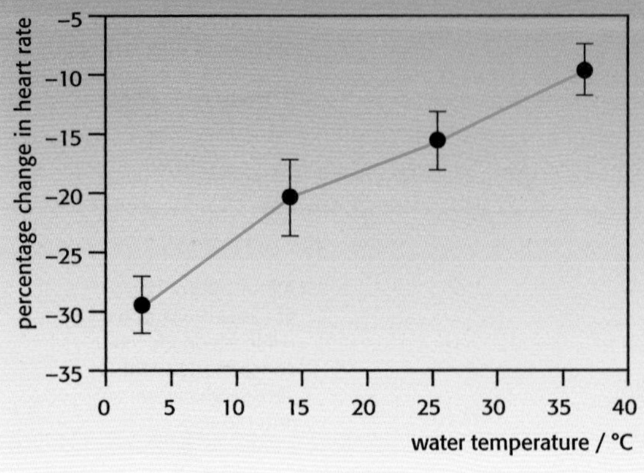

Figure 16

(a) (i) State the effect of facial immersion on heart rate over the range of temperatures shown on the graph. [1]
(ii) Suggest one reason for the relationship between facial immersion and heart rate. [1]
(b) Outline the effect of the water temperature on heart rate. [1]
(c) Calculate the heart rate of a person immersed in water at a temperature of 15 °C, if their heart rate before immersion was 70 beats per minute. [2]

Critical considerations: the epidemic of diabetes

The excerpts below are from "Unravelling the Causes of Diabetes" by Jean Marx (26 April 2002. *Science*).

"People cringe at the use of the word 'epidemic' for a chronic disease, but by all criteria, there's [a diabetes 2] epidemic" in the United States, says Allen Spiegel, who directs the National Institute of Diabetes and Digestive and Kidney Diseases (NIDDK) in Bethesda, Maryland.

The number of adults with diabetes in the United States increased by 49% between 1991 and 2000, according to data from the Centers for Disease Control and Prevention (CDC) in Atlanta, Georgia. Type II diabetes, formerly known as maturity-onset or non-insulin-dependent diabetes, accounts for practically all of that increase. Some 16 million to 17 million people now have the condition, and an equal number are thought to be "prediabetic," having early symptoms but not yet the full-fledged version. Even children are no longer immune to diabetes 2, which until recently rarely affected people before middle age.

Driving this epidemic, say Spiegel and other experts, is the continuing increase in obesity that is, in turn, fueled by a relatively new development in human history: an ample food supply coupled with a sedentary lifestyle. In the past, humans who wanted food "had to grow it, harvest it, or hunt it," says diabetes researcher Roger Unger of the University of Texas Southwestern Medical Center (UT Southwestern) in Dallas. The current overabundance of easily available food is, he adds, "a surprise to nature," one that our bodies aren't designed to handle.

Diabetics of both types develop serious complications, including kidney failure, blindness, damage to the feet and legs serious enough to require amputation, and a high risk of heart attack and stroke. Over the years, numerous studies have pointed to obesity as a major risk factor for diabetes.

Recent studies also point to some disturbing new trends. For one, diabetes is on the rise in many developing countries, as they adopt more Westernized lifestyles and diets. The World Health Organization predicts that the number of cases worldwide—now 150 million—will double by 2025. And even more alarming, obesity-driven diabetes 2 is increasingly striking younger people, including children—a situation Spiegel describes as "potentially devastating," because those who contract the disease early have longer to develop the sometimes deadly complications.

Some recent results have been encouraging, however. In a large, multicenter clinical trial, the Diabetes Prevention Program (DPP) Research Group found that it's possible to stave off diabetes 2 in people at high risk of getting the disease. The trial included 3234 people, who were divided into three roughly equal groups. The controls received a placebo plus standard recommendations for improving their diets and exercise regimens. A drug treatment group took an anti-diabetes 2 drug called metformin, and a second treatment group received intensive counseling about eating better and exercising regularly.

As reported in the 7 February 2002 NEJM, *the intensive lifestyle counseling reduced the incidence of diabetes 2 by 58%, and metformin treatment produced a 31% reduction. The treatments proved so effective, Nathan says, that the trial was halted a year early.*

1 An epidemic is a disease that appears in a population at a rate that substantially exceeds what is "expected", based on the normal incidence rate, though in a biological sense it is usually limited to infectious disease. To what extent is it helpful to describe the growing incidence of Type II diabetes as an epidemic?

2 (a) Explain why the DPP research study was halted early.
  (b) Both of the treatment groups in the DPP study reported significant reduction in the incidence of diabetes. Discuss the ethical issues associated with trials involving placebo treatments.

Chapter 24 questions

1 Figure 17 shows variations in liver glycogen over the course of one day.
  (a) Explain the variation in liver glycogen. [3]
  (b) Evaluate the contribution of glycogen to blood sugar homeostasis. [2]

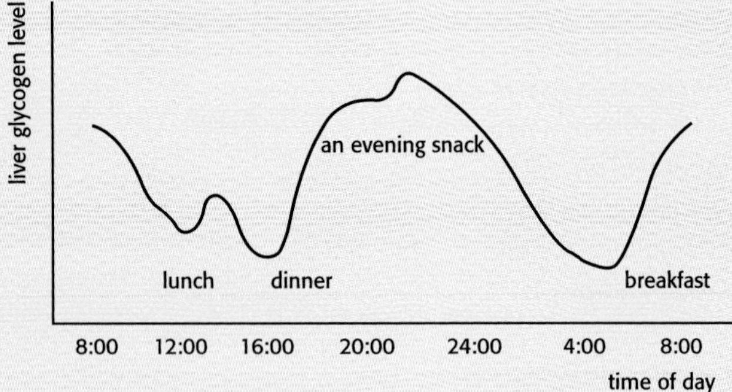

Figure 17 Changes in the level of liver glycogen over the course of a day

2 (a) Describe the events of synaptic transimission. [6]
  (b) Draw and label the structure of a motor neuron. [4]
  (c) Explain how a nerve impulse passes along a non-myelinated neuron. [8]

3 Distinguish between the following pairs of terms:
  (a) Efferent and afferent neuron. [2]
  (b) Action potential and resting potential. [2]
  (c) Sodium–potassium pump and ion channels. [2]
  (d) Axon and dendrite [2]
  (e) Glucagon and insulin. [2]
  (f) Negative feedback and positive feedback. [2]
  (g) Vasoconstriction and vasodilation. [2]
  (h) Type I diabetes and Type II diabetes. [2]

4 Figure 18 shows the variation in finger skin surface temperature when the finger is immersed in ice water.
  (a) Explain the skin temperature changes in terms of vasoconstriction and vasodilation. [3]
  (b) With reference to the data, explain homeostasis in terms of negative feedback. [3]
  (c) Note that with extended exposure to ice water, the skin surface temperature settles at a temperature well below core body temperature.
    (i) State the temperature that the skin surface appears to be settling at. [1]
    (ii) Suggest the adaptive advantages of this response. [3]

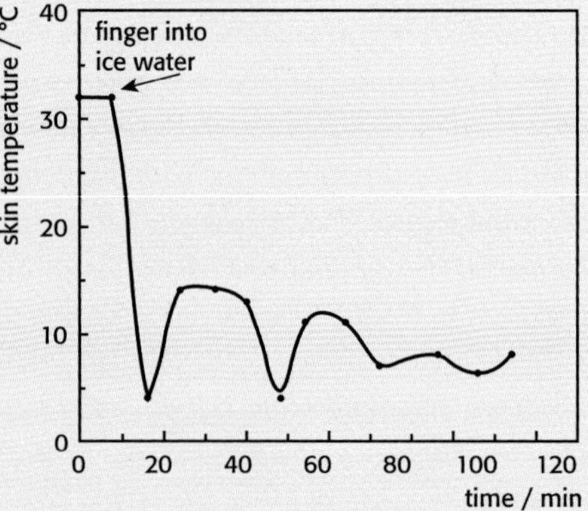

Figure 18 Variations in skin surface temperature when the skin is exposed to cold for prolonged periods

Higher-level question

5 Ouabain is used to stimulate the heart muscle. It works by binding to the Na^+/K^+ pump and inhibiting its action. The result of this is that Na^+ leaks back into the cell. When this happens, the cell tries to maintain the osmotic balance, pumping the sodium out using a protein called the Na^+/Ca^{2+} pump, which pumps Na^+ out while pumping Ca^{2+} in. Explain why this triggers heart muscle contraction. [3]

25 Muscles and movement

Bones have a number of structural functions, including supporting the body, facilitating movement and protecting internal organs. They also have a number of physiological roles including blood cell formation and metabolism of calcium and other minerals.

Muscles attach to bones and provide the forces necessary to change the position of bones. Muscles are attached to bones by tendons. Bones are attached to other bones by ligaments. Many skeletal muscles occur in antagonistic pairs, where one muscle extends the joint and the other flexes it (see Figure 1).

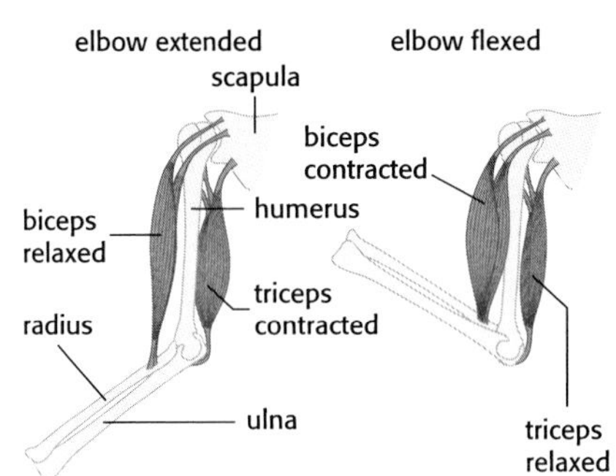

Figure 1 The biceps and triceps are antagonistic muscles.

Data-based question: flight muscles

In one research project, pigeons (*Columba livia*) were trained to take off, fly 35 metres and land on a perch. During the flight the activity of two muscles, the sternobrachialis (SB) and the thoracobrachialis (TB), was monitored using electromyography. The traces are shown in Figure 2. The spikes show electrical activity in contracting muscles. Contraction of the sternobrachialis causes a downward movement of the wing.

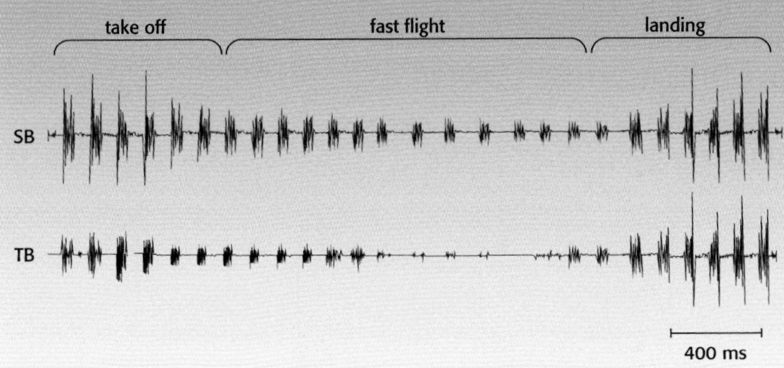

Figure 2 Electrical activity in the sternobrachialis (SB) and the thoracobrachialis (TB) muscles during flight of a pigeon

?

1. Deduce the number of downstrokes of the wing during the whole flight. [1]
2. Compare the activity of the sternobrachialis muscle during the three phases of the flight. [3]
3. Deduce from the data in the electromyograph how the thoracobrachialis is used. [1]
4. Another muscle, the supracoracoideus, is antagonistic to the sternobrachialis. State the movement produced by a contraction of the supracoracoideus. [1]
5. Predict the pattern of the electromyograph trace for the supracoracoideus muscle during the 35-metre flight. [2]

Investigating biology: the palmaris longus muscle

Face your palm upwards, pinch your thumb and your smallest finger together and then bend your wrist toward you. Wrist tendons will then become prominent. In most people, there will be three tendons present. Less commonly, one of the tendons is missing. In some cases, the third tendon is missing in one arm but present in the other. If there are only two tendons present, the person lacks the palmaris longus muscle and its associated tendon. There are a number of possible investigations.

- Visit the database OMIM (Online Mendelian Inheritance in Man) at www.ncbi.nlm.nih.gov/omim/ and do a search under palmaris longus muscle to determine the inheritance pattern and incidence of the various phenotypes. Conduct an investigation to determine the incidence of these phenotypes in your own school.
- If you find some individuals who lack the muscle, a grip-strength data logger can be used to determine if there is any effect on the grip strength or pinch strength when the muscle is absent.

Joints

The point where bones meet is called a joint. An example of a joint is the elbow. Three bones meet at the elbow – the humerus, the radius and the ulna (Figures 3 and 4).

The bones at a joint fit together in a number of different ways and so the range of movement varies. For example, in a pivot joint the movement is limited to rotation. This is the type of joint between the radius and the ulna of the forearm. In a hinge joint, movement is restricted to a single plane. This is the kind of joint that exists between the humerus and the head of the ulna.

The hip joint, between the pelvis and the femur, is a ball and socket joint. It has a greater range of motion than the elbow joint in that the leg can flex and extend, rotate and move sideways and back. This latter type of motion is known as abduction and adduction.

The joint at the knee is unique. It is a complex, compound condyloid joint. In this kind of joint, there is a greater range of motion about the joint when it is flexed than when it is extended.

As the joints move, the risk of injury due to friction is lessened by a number of adaptations. The knee, the elbow and the hip are synovial joints because they are separated by a fluid-filled cavity. The fluid is called synovial fluid and it lubricates the joint. The synovial fluid is bound by the joint capsule which contains the fluid. Lastly the ends of the bones are covered in cartilage, a type of smooth connective tissue which absorbs shocks more easily. Figure 4 shows these structures for the elbow joint.

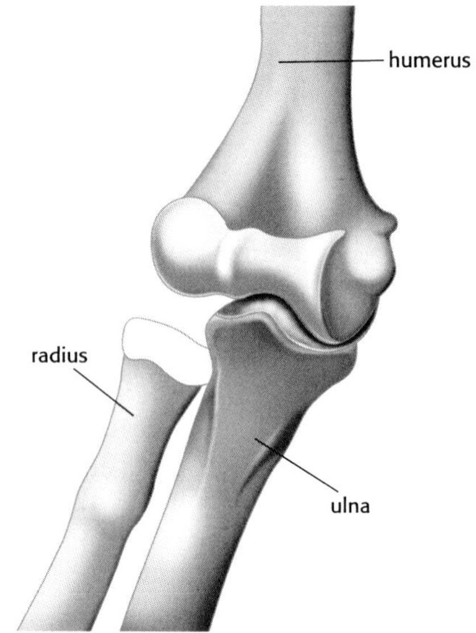

Figure 4 The pivot joint between the radius and the ulna permits rotation while the hinge joint between the humerus and the ulna permits motion in one plane only.

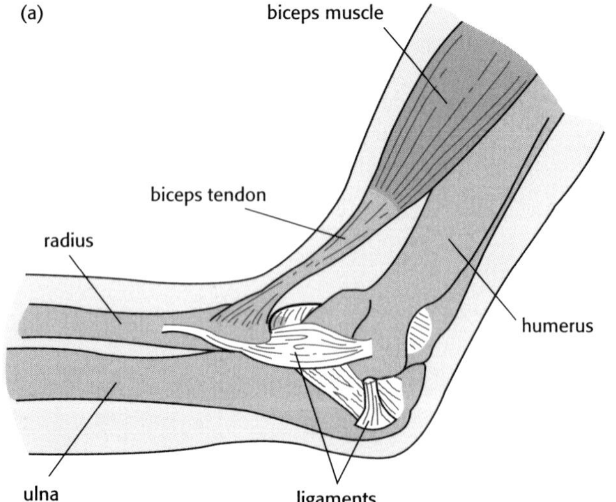

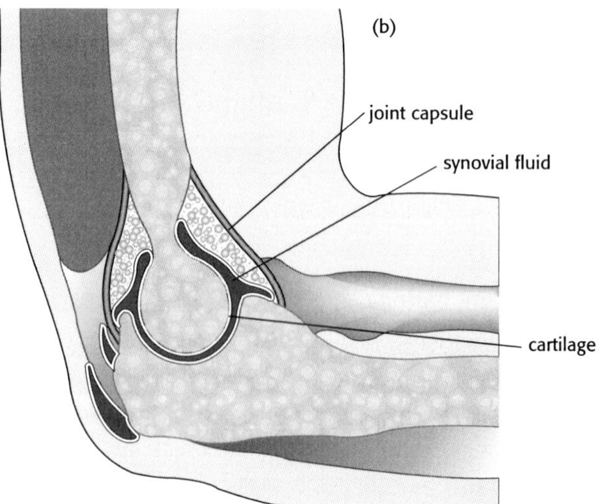

Figure 3 Structures within the human elbow joint

Thinking about science: the effects of knuckle-cracking

Popular wisdom is that knuckle-cracking promotes arthritis. To test this hypothesis, a physician named Dr Unger cracked the knuckles on one hand but not the other for 50 years. The following is a letter to the journal *Arthritis and Rheumatism* evaluating the experiment with some humour (1998).

Dr. Unger exercised amazing self-control by performing 50 years of knuckle-cracking on his left hand at least twice daily, "while those on the right cracked only rarely and spontaneously." No evidence of arthritis in either hand was found at the end of 50 years. The basic study designed by Dr. Unger is a two-arm trial without randomization. Although it is not clear, it appears that the study was not blinded. Blinding would only be possible if the investigator didn't know left from right. This is not likely since studies indicate that only 31% of primary care physicians don't know left from right. (The figure is reportedly somewhat higher for most specialists.) The lack of randomization suggests the need for a multivariate analysis to reduce bias. Controlling for knuckle-to-knuckle variation in race, sex, socio-economic status, initial severity, co-morbidities, and Ecuadorian barometric pressure at the time of measurement would be advisable. The sample size appears too small to support accurate inference. Typically, sample sizes of roughly twice the available research budget are required for valid inference. Restrictive eligibility criteria and convenience sampling limit generalization of the results to knuckle-cracking physicians with a lot of time on their hands.

1. Does Dr Unger's experiment prove that knuckle-cracking does not cause arthritis?
2. How could his experiment be altered to provide firmer evidence?

Rediscovering biology: dissection of a poultry wing

The anatomy of a poultry wing such as a chicken or turkey wing is homologous to the human arm. In this dissection, focus on the elbow joint of the poultry wing.

1. Rinse the wing under running water and thoroughly dry it with a paper towel as the surface may be contaminated with *Salmonella* bacteria.
2. Cut the skin along the entire length of the wing, pointing the scissors up so as not to cut tissues below the skin.
3. Remove the skin from the wing by placing your finger under the skin and lightly tearing at the connective tissue below it.
4. Use a blunt probe to separate the individual muscles from each other without tearing them.
5. Pull on each of the muscles and note the movement that results. Determine pairs of muscles that are antagonistic.
6. Follow a muscle to where it connects to the bones. Note the appearance of the tendons.
7. Carefully remove the muscles and tendons to expose the ligaments, which are white in appearance.
8. Identify the humerus, the radius and the ulna.
9. Separate the bones at the joint and note the appearance of the cartilage.
10. Note the oily texture of the surface of the cartilage. This is due to synovial fluid.

Muscle function

Skeletal muscles are composed of bundles of muscle fibres. Muscle cells are unusual in that they can have many nuclei and can be quite long due to the fact that embryonic muscle cells fuse together. Muscle fibres are composed of many parallel elongated fibres called myofibrils.

Figure 6 shows the ultrastructure of the muscle fibre in more detail. The sarcolemma is the muscle fibre membrane and the sarcoplasmic reticulum is a modified endoplasmic reticulum that tends to wrap around individual myofibrils. Large numbers of mitochondria are found dispersed between the myofibrils. Disc-shaped structures appearing in some myofibrils as dark lines are called Z-lines.

The functional unit of the myofibril is the sarcomere, which is the portion of the myofibril between two Z-lines. The pattern observed is due to the arrangement of two different kinds of filament, thin filaments composed of the protein actin, and thick filaments composed of the protein myosin (see Figure 7).

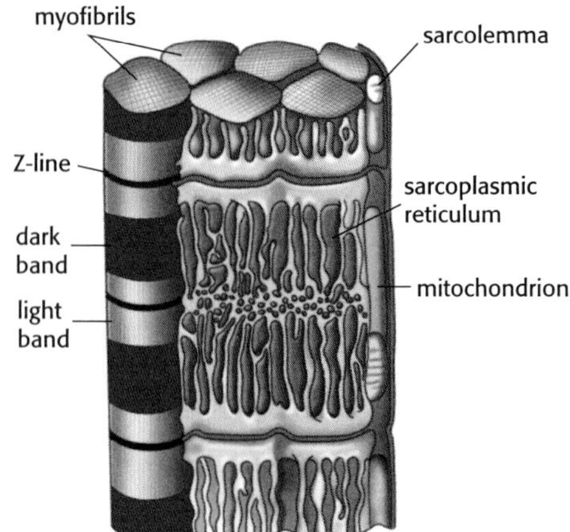

Figure 6 The ultrastructure of the muscle fibre

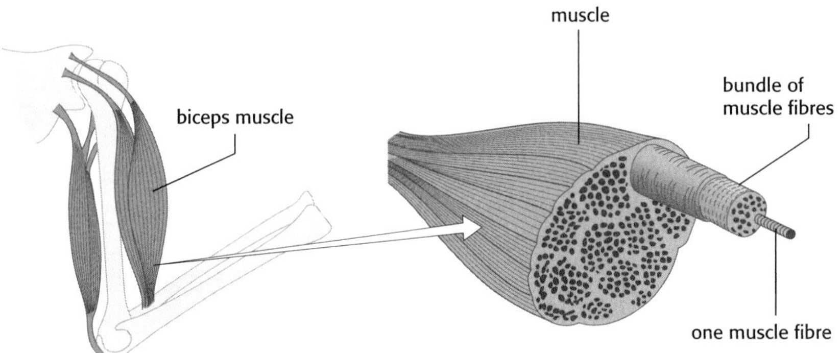

Figure 5

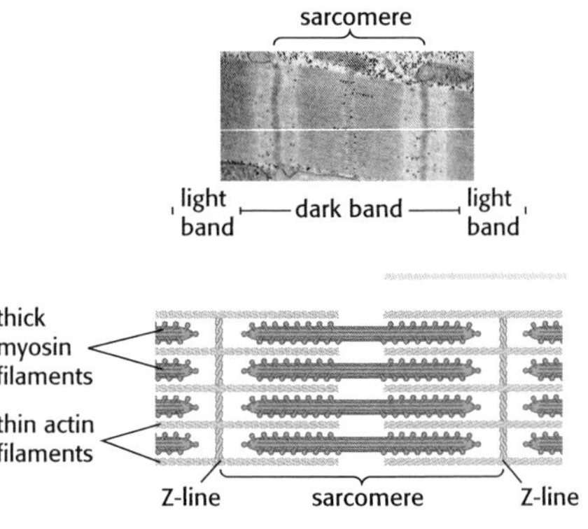

Figure 7 The structure of a sarcomere

Contraction of skeletal muscle

Figures 8 and 9 show micrographs and diagrams of relaxed and contracted sarcomeres. How is contraction achieved?

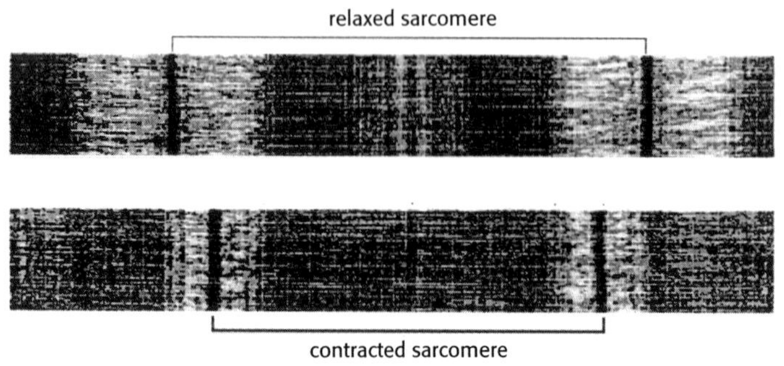

Figure 8 Photo of relaxed and contracted sarcomeres

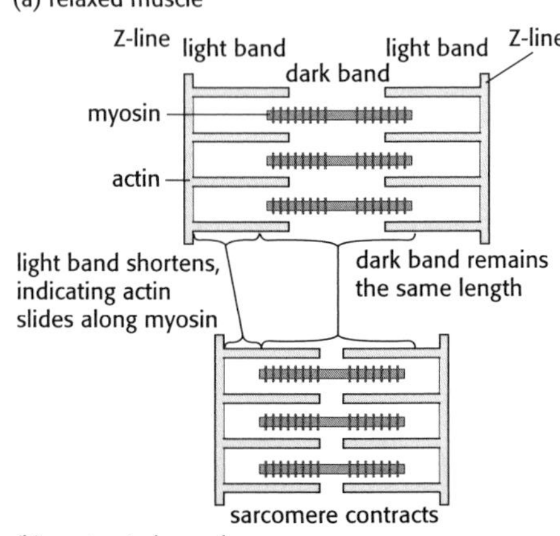

Figure 9 Diagram of relaxed and contracted sarcomeres

Role of ATP in muscle contraction

When a motor neuron sends a signal to the muscle to contract, it causes the sarcoplasmic reticulum to release Ca^{2+}. The binding sites on the actin filaments where myosin binds are normally blocked by a regulatory protein. The calcium ions lead to the binding sites being exposed. ATP plays a role in muscle contraction. The ATP attaches to the part of the myosin molecule called the myosin head. The ATP then undergoes hydrolysis to form ADP and P_i (inorganic phosphate). This causes the myosin head to cock and then form a cross-bridge with the actin filament. The release of the ADP + P_i causes the head to change its conformation, sliding the actin toward the centre of the sarcomere. Binding of a new ATP molecule causes the cross-bridge to break and the cycle is ready to repeat.

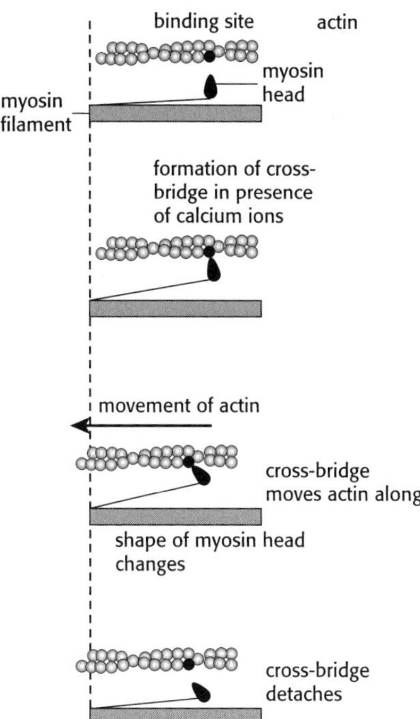

Figure 10

Data-based question: factors affecting muscle growth

Muscles often increase in mass if the amount that they are used increases. An experiment was performed to examine the effect of flight on muscle mass in European starlings (*Sturnus vulgaris*). Study birds were randomly assigned to three groups. Over 6 weeks, each group was subjected to 34 1-hour study periods. The exercise group was trained to fly for 1 hour by receiving food rewards. Control group 1 was allowed to feed freely but placed into cages that prevented flying. Control group 2 was fed the same food rewards at the same time as the exercise group, but was also placed into cages that prevented flying. Body mass was monitored before and during the experiment (see Figure 11a). At the end of the experiment, the mean mass of the birds' pectoralis muscles was compared (Figure 11b).

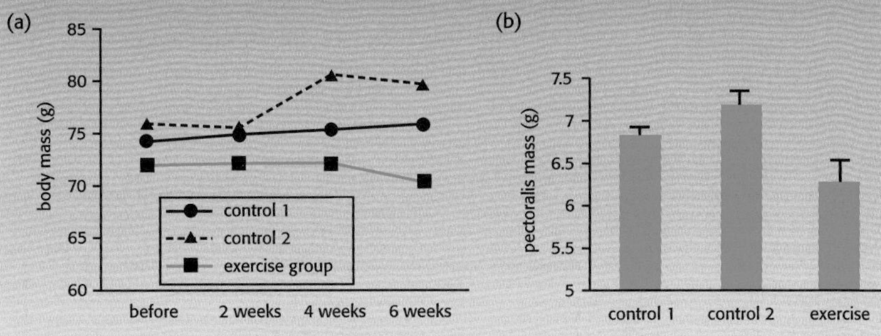

Figure 11 The effect of exercise on body mass and muscle mass in starlings

1 Compare the changes in body mass in control group 2 and the exercise group. [2]
2 Evaluate the claim that preventing exercise increases pectoralis muscle mass. [3]
3 Suggest how the mass of the birds' pectoralis muscle could be determined. [2]
4 One hypothesis that might be generated from this experiment would be that reducing motion in birds might lead to greater muscle mass per bird. Such knowledge might be used in the farming of poultry. Greater meat production per bird would result from the motion of the birds being restricted. Discuss the ethics of designing and carrying out experiments to test this hypothesis. [3]

Chapter 25 questions

1 How are muscles attached to bone?
A through ligaments
B through tendons
C through myofibrils
D through sarcomeres.

2 The unit between one Z-line and the next is termed a:
A sarcomere
B myofibril
C sarcoplasmic reticulum.
D sarcolemma

3 Which is an example of an extensor muscle?
A metacarpal
B phalange
C bicep
D tricep.

4 What type(s) of joints occur at the human elbow joint?
A pivot
B hinge
C ball and socket.

(a) A only
(b) A and B only
(c) A and C only
(d) A, B and C.

5 Distinguish between each of the following word pairs:
(a) Hinge joint and pivot joint.
(b) Radius and humerus.
(c) Hip and knee joint.
(d) Actin and myosin.
(e) ADP and ATP.
(f) Ligament and tendon.
(g) Extension and flexion.

6 (a) The structure of joints can permit or prevent movements at a joint. Describe the movements that can occur at the hip joint. [2]
(b) Explain how a skeletal muscle contracts. [8]
(c) Draw a labelled diagram of the human elbow joint. [4]

7 Members of the German women's athletic team were grouped according to the distance they normally run in competitions. Each group was asked to run at different speeds for a set time, and the concentration of lactate in the blood was measured. If the level of lactate in the blood rises above 4.0 mmol dm^{-3}, the athlete is relying on anaerobic respiration to provide the energy for running. If the lactate level is less than 4.0 mmol dm^{-3} then the athlete respires aerobically.

Figure 12 can be used to estimate the level of lactate for each group at different speeds of running.
(a) Estimate the speed at which the 400-metre runners would have a lactate level of 4.0 mmol dm^{-3}. [1]
(b) Identify which groups of runners could run at 4.2 m s^{-1} while respiring aerobically.
(c) Compare, by giving one similarity and one difference, the respiration of the 100-metre runners with the respiration of the marathon runners as the speed of running increases. [2]
(d) Explain the trend in the amount of anaerobic respiration in the muscles of runners as the distance of the race that they train for increases. [3]

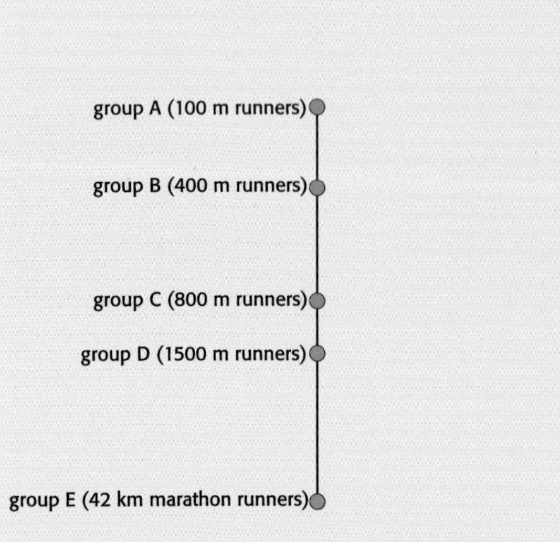

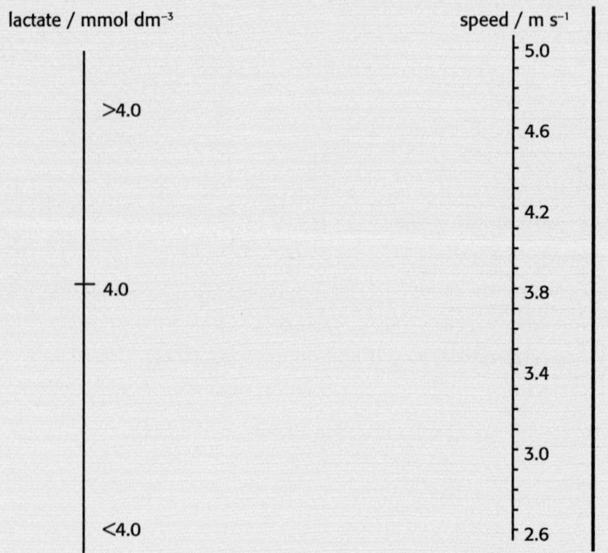

Figure 12 Nomogram showing the effect of distance of race and speed on blood lactate levels

26 Reproduction

Human reproduction involves the fusion of a sperm from a male with an egg from a female. If one particular gene called *tdf* is present in the resulting human embryo, its embryonic gonads develop into testes in about the eighth week of pregnancy, and produce testosterone until about the fifteenth week. Testosterone causes male genitalia to develop, which are shown in Figure 1. At puberty the secretion of testosterone increases. This causes the development of secondary sexual characteristics and also the sex drive in men – the desire to have sexual intercourse.

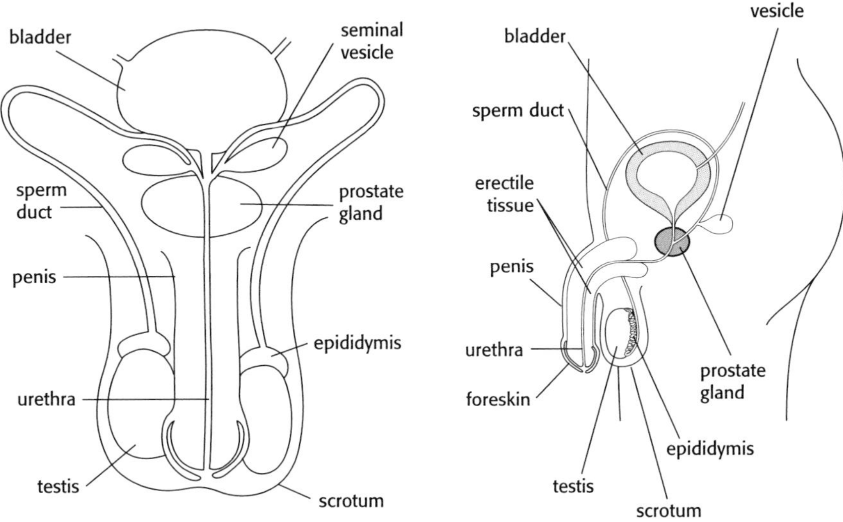

Figure 1 Male reproductive system in front and side view

Critical considerations: gender selection

It is possible to identify the gender of a fetus by ultrasound scanning. Parents who are considering terminating their pregnancy by abortion may be influenced in their decision if they know the gender of the fetus. Prabhat Jha from the University of Toronto and Rajesh Kumar at the Institute of Medical Education and Research in Chandigarh, India, and colleagues, studied data from 1.1 million households in India. They found that the likelihood of having a baby girl as a second or third child was significantly lower in households where there was no boy. "We conservatively estimate that prenatal sex determination and selective abortion account for 0.5 million missing girls yearly", Professor Jha said.

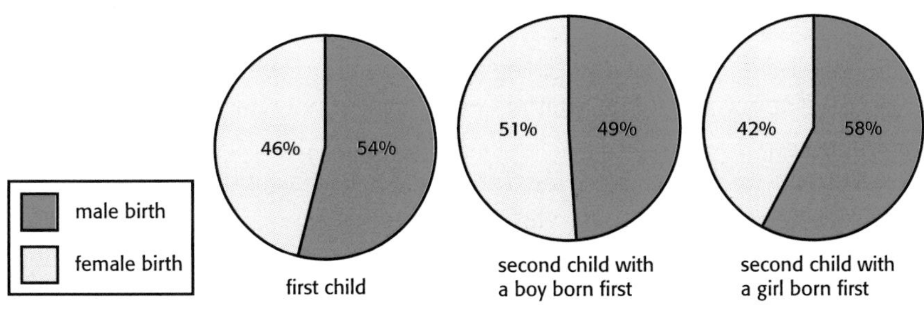

Figure 2 Gender ratios of children born in India

1 What is the expected ratio of male to female births?
2 Suggest possible social consequences of there being an excess of male births over female births in a country?
3 Discuss the ethics of prenatal gender selection.
4 Suggest a possible code of ethics for ultrasound practitioners. Should information about gender be withheld? Should private ultrasound clinics be government-regulated?
5 There are techniques for sorting sperm in semen before it is used in IVF or artificial insemination, so that the chance of a particular gender of baby is greatly increased. Is this procedure ethical?

Female reproductive system

Babies develop as females if the gene *tdf* is not present in their genome. This gene is usually located on the Y chromosome, so if an embryo has two X chromosomes, it will not have a copy of *tdf*. The embryonic gonads therefore develop as ovaries, and testosterone will not be secreted. The two female hormones, estrogen and progesterone, are always present in pregnancy, secreted first by the mother's ovaries and later by the placenta. In the absence of testosterone and the presence of estrogen and progesterone, female genitalia develop. The diagrams below show the female reproductive system in front and side view.

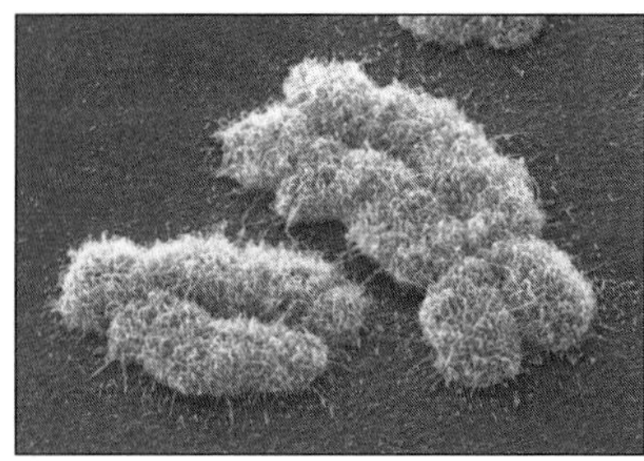

Figure 3 X and Y chromosomes

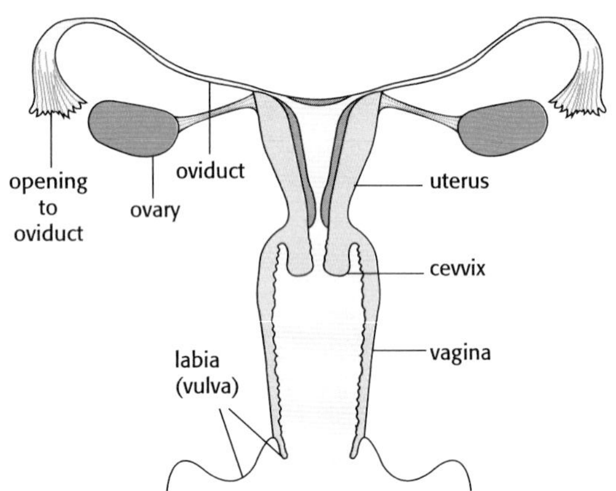

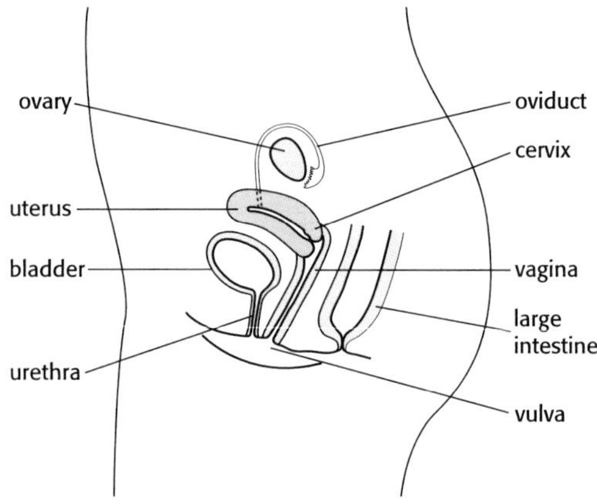

Figure 4 Female reproductive system

Thinking about science: Harvey and the mystery of reproduction

William Harvey is chiefly remembered for his discovery of the circulation of the blood, but he also had a lifelong obsession with how life is transmitted from generation to generation by reproduction.

According to the "seed and soil" doctrine of Aristotle, reproduction is accomplished by the male producing a seed, which forms an egg when it mixes with menstrual blood. William Harvey tested Aristotle's doctrine using a natural experiment. Deer are seasonal breeders – they only become sexually active during the autumn. Harvey examined the uterus of female deer during the mating season by slaughtering and dissecting them. He expected to find eggs developing in the uterus immediately after mating, but only found signs of anything developing in females two or more months after the start of the mating season. He concluded "the fetus doth neither proceed from the seed of male or female in coition, nor yet from any commixture of that seed".

Harvey was left with a mystery. "When I plainly see nothing at all doth remain in the uterus after coition, … no more than remains in the braine after sensation, …I have invented this Fable. Let the learned and ingenious flock of men consider of it; let the supercilious reject it: and for the scoffing ticklish generation, let them laugh their swinge. Because I say, there is no sensible thing in the uterus after coition; and yet there is a necessity, that something should be there, which may render the animal fruitful." Harvey was understandably reluctant to publish his findings, but he did eventually do so in his work *De Generatione Animale*, published in 1651 when he was 73 years old.

1 Why did Harvey use deer in his investigations, rather than animals that breed at any time of year?
2 Why was Harvey unable to see anything in the uterus of deer for two months after they had mated?
3 To what extent did Harvey's observations provide falsification for the Aristotelian doctrine of reproduction?

Hormonal control of the menstrual cycle

Whereas some mammals, including deer, are only able to breed during a particular season, humans can reproduce at any time of year. In the females this involves the production of an egg (oocyte) inside a fluid-filled sac called a follicle. It also involves the thickening of the endometrium – the lining of the uterus. If no embryo is formed, this thickening breaks down and is passed out of the body during menstruation. The process then begins all over again. This sequence of events is called the menstrual cycle. The role of hormones that help to control the cycle is outlined in Table 1.

During the first part of the menstrual cycle, a group of follicles, and the eggs within them, are stimulated to grow. While the follicles are developing, the endometrium is thickening, too. One follicle (the most developed one) breaks open and releases an egg into the oviduct. The other follicles now degenerate, while the follicle that released the egg becomes a body called the corpus luteum. Further development of the endometrium prepares it for the implantation of an embryo. If fertilization does not occur, the corpus luteum breaks down and then the thickening of the endometrium breaks down, to be shed during menstruation.

Table 1

Pituitary hormones: FSH and LH *bind to FSH and LH receptors in the membranes of follicle cells.*
FSH stimulates these changes: • development of oocyte in the follicle • thickening of follicle wall • secretion of follicular fluid • secretion of estrogen by follicle wall.
LH stimulates these changes: • completion of meiosis in the oocyte • partial digestion of the follicle wall allowing it to burst open (ovulation) • growth of the corpus luteum which secretes estrogen and progesterone.
Ovarian hormones: estrogen and progesterone *are absorbed by female cells, where they influence gene expression and therefore development.*
Estrogen causes these changes: • thickening of endometrium • blood vessel growth in endometrium • increase in FSH receptors in follicle • inhibition of FSH secretion and stimulation of LH secretion when estrogen levels are high.
Progesterone causes these changes: • maintenance of thickening of the endometrium with many blood vessels • inhibition of FSH and LH secretion.

Data-based question: the female athlete triad

The female athlete triad is a syndrome consisting of three interrelated disorders that can affect female athletes: osteoporosis, disordered eating and menstrual disorders. Osteoporosis is reduced bone mineral density. It can be caused by a diet low in calcium, vitamin D or energy, or by low estrogen levels. Figure 5 shows the bone mineral density in two parts of the femur for female runners who had different numbers of menstrual cycles per year. The *t*-score is the number of standard deviations above or below mean peak bone mass for young women.

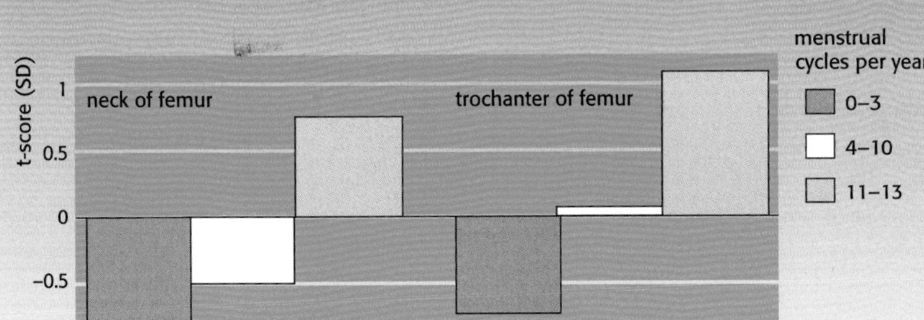

Figure 5 Bone mass in women grouped by number of menstrual cycles

1 (a) Outline the relationship between number of menstrual cycles per year and bone density. [3]
(b) Compare the results for the neck of the femur with the results for the trochanter. [3]

2 Explain the reasons for some of the runners having:
(a) higher bone density than the mean [2]
(b) lower bone density than the mean. [4]

3 (a) Suggest reasons for female athletes having few or no menstrual cycles. [2]
(b) Suggest one reason for eating disorders and low body weight in female athletes. [1]

Figure 6 shows hormone levels in a woman over a 36-day period, including one complete menstrual cycle. The pattern of changes is typical for a woman who is not pregnant. The hormone levels are measured in mass per millilitre. The actual masses are very small, so progesterone, FSH and LH are measured in nanograms (ng) and estrogen is measured in picograms (pg).

Figure 6 also shows the state of the ovary and of the endometrium. The first half of the menstrual cycle is called the follicular phase, as during this period a follicle is developing in the ovary. The second half of the cycle is called the luteal phase, as in this period a corpus luteum develops in the ovary. The endometrium breaks down and is released during menstruation. It then thickens again in preparation for the implantation of an embryo.

Hormone levels during the menstrual cycle

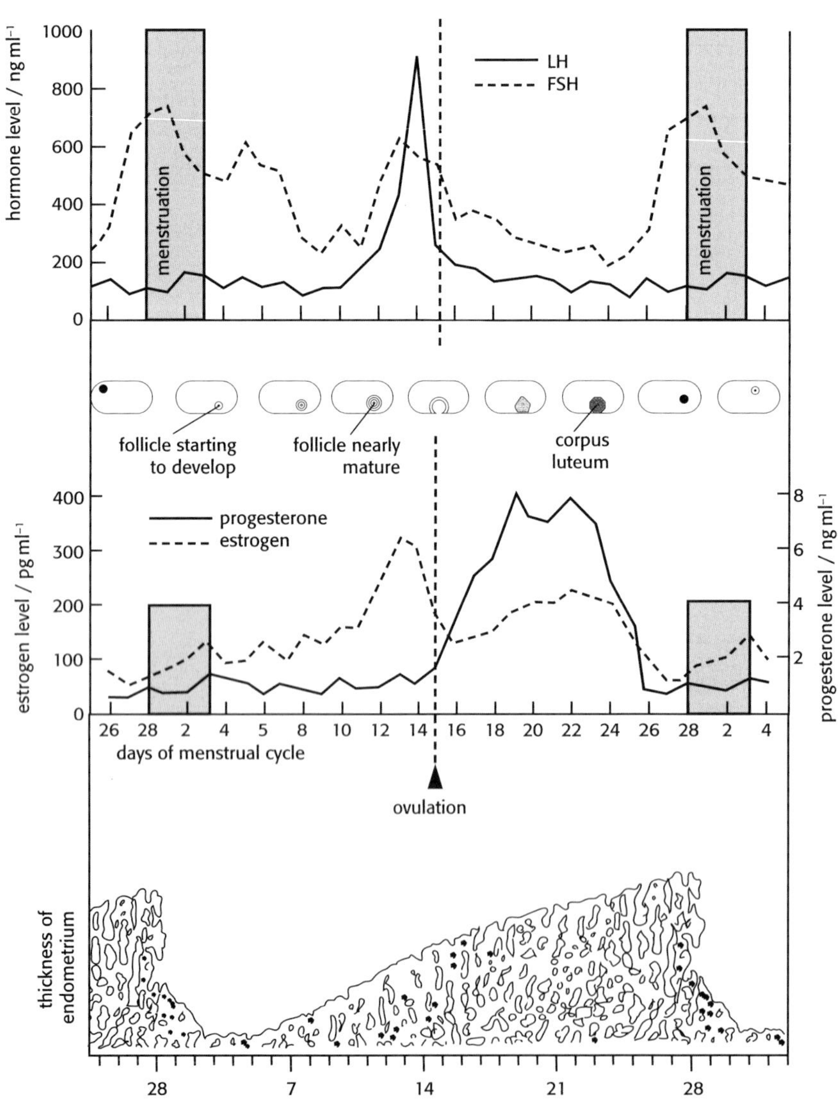

Figure 6 The menstrual cycle

1. Explain what stimulates the development of follicles in the first two weeks of the menstrual cycle.
2. Explain what causes the level of estrogen to rise to higher and higher levels between days 1 and 13.
3. Deduce from the graph the level of estrogen needed to stimulate LH secretion.
4. Outline one effect, shown on the diagram, of rising levels of estrogen, apart from LH secretion.
5. Suggest one way, involving testing the level of a hormone, in which the day of ovulation could be predicted. This is useful if a couple are trying to conceive and want to have intercourse when the woman is most fertile.
6. Outline one effect, shown on the diagram, of high levels of LH, apart from stimulating ovulation.
7. (a) Outline the trends in the level of progesterone over the 28 days of the menstrual cycle.
   (b) Explain the differences in the levels of progesterone at different stages of the cycle, in terms of the source of progesterone and the hormonal stimulation of its secretion.
8. Explain the low and falling levels of LH and FSH secretion between days 15 and 24.
9. Outline the cause of menstruation in terms of changes in hormone levels.
10. Explain the rise in FSH level from day 24 onwards.

IVF – getting the gametes

The natural method of fertilization in humans is *in vivo* –inside the living tissues of the body. Fertilization can also happen outside the body in carefully controlled laboratory conditions. This is called *in vitro* fertilization, almost always abbreviated to IVF.

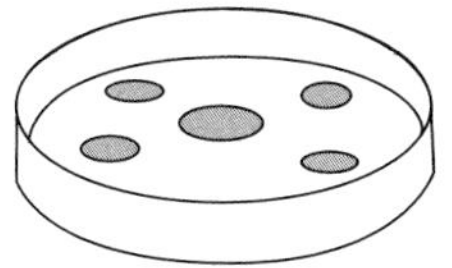

Figure 7 Plastic petri dish used for IVF

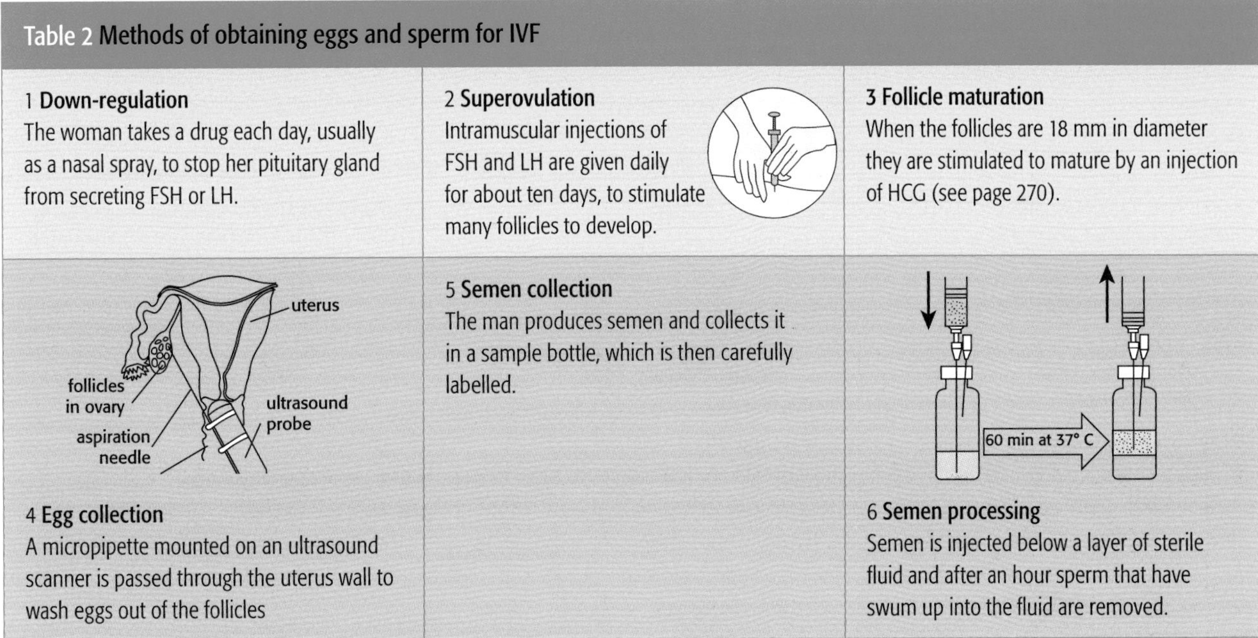

Table 2 Methods of obtaining eggs and sperm for IVF

1 **Down-regulation** The woman takes a drug each day, usually as a nasal spray, to stop her pituitary gland from secreting FSH or LH.	2 **Superovulation** Intramuscular injections of FSH and LH are given daily for about ten days, to stimulate many follicles to develop.	3 **Follicle maturation** When the follicles are 18 mm in diameter they are stimulated to mature by an injection of HCG (see page 270).
uterus; follicles in ovary; ultrasound probe; aspiration needle 4 **Egg collection** A micropipette mounted on an ultrasound scanner is passed through the uterus wall to wash eggs out of the follicles	5 **Semen collection** The man produces semen and collects it in a sample bottle, which is then carefully labelled.	60 min at 37° C 6 **Semen processing** Semen is injected below a layer of sterile fluid and after an hour sperm that have swum up into the fluid are removed.

Critical considerations: attitudes to IVF

Views on IVF vary greatly around the world and also between individuals. Cultural attitudes and religious teachings can have a great impact on views. The quotes shown here represent some of this diversity of opinion.

1 Can you deduce the religious or cultural background of the person expressing each view?
2 Which of the views do you agree with?
3 Do you think your views would change if you discovered that you and your partner were unable to conceive a child naturally?

"Children are a gift of God and should only be conceived naturally."

"At least 60% of embryos conceived naturally die in the first few days of life but, like embryos in IVF, they have no nervous system and do not suffer in any way – more human cells die when we pull a hair out!"

"Infertility can cause deep psychological suffering and we should do all that we can to help infertile couples."

"Life begins at conception and destroying an embryo, however it was conceived, is taking a human life."

"For most childless couples IVF offers only false hope – its chances of success are still low – about one in five treatments, and it cannot overcome many of the reasons for infertility."

"When a woman cannot get pregnant naturally it is God's will and it is wrong to try to overcome this artificially."

"We had twins by IVF 14 years ago and they have brought immense joy to us and to each other."

"The costs of IVF are enormous – either it is only possible for the rich or it drains money away from other more important medical treatments."

"Genetic screening of embryos conceived through IVF offers the best chance of reducing the suffering due to genetic disease."

IVF – fertilization and implantation

The mixing of egg and sperm is done in sterile conditions in a hospital or fertility centre. If *in vitro* fertilization is successful then one or more embryos are placed in the uterus. If they implant and continue to grow then the pregnancy that follows is no different from a pregnancy that began by natural conception (Table 3).

Table 3

1 Insemination Each egg is mixed with 50,000 to 100,000 sperm cells in a shallow dish, which is then incubated at 37 °C until the next day.	**2 Checking for fertilization** Fertilization can be detected 12–20 hours after mixing eggs and sperm, as two nuclei are visible inside the egg.	**3 Embryo selection** On the next day the embryos are checked and healthy four-cell embryos are selected for transfer when 48 hours old.
4 Embryo transfer Up to three embryos are drawn up into a plastic tube which is pushed into the uterus and the embryos are expelled. uterus embryo transfer catheter	**5 Luteal support** Extra progesterone is usually given as a tablet placed in the vagina, to ensure that the uterus lining is maintained.	**6 Pregnancy testing** Two weeks after embryo transfer a pregnancy test can be done to see if an embryo is developing.

Thinking about science: the difficulties of assessing risks

It is often difficult to distinguish between genuine risks and irrational fears, and to assess the level of a risk. One example of this is the risk of birth abnormalities with IVF. The Society for the Protection of Unborn Children posted this statement as an argument against IVF on its website in 2006: "An analysis of 25 studies published in the medical journal *Human Reproduction* concluded that there is a 30–40 per cent increased risk of birth defects associated with assisted reproductive technologies (ART)."

1 What percentage of babies conceived naturally do you think are born with birth defects? What percentage conceived by IVF do you think are born with birth defects?

2 The analysis referred to was done by Michelle Hansen and four other researchers. Independent reviewers were used to decide which studies to include in the analysis. Why does it matter which studies are included in an analysis of risk?

3 Using the data from all the studies, the researchers calculated the number of births from ART that would be needed for one additional child to be born with a birth defect. The answer depends on the percentage of birth defects from natural conception, which is not known with certainty. If it is 1 per cent then 250 births due to ART would be needed, but if it is 4 per cent then the number falls to 62. Do you consider this level of increased risk of birth defects to be acceptable? How do we decide what an acceptable level of risk is?

4 The cause of the increased risk is uncertain – it could be due to the underlying infertility of the parents and not to the ART methods used to conceive. If scientific evidence for this hypothesis was obtained, do you think it would alter the views of members of the Society for the Protection of Unborn Children?

HL

Spermatogenesis

Spermatogenesis is the production of sperm. It happens in the testes, which are composed of a mass of narrow tubes, called **seminiferous tubules**, with small groups of cells filling the gaps between the tubules. These gaps are called interstices, so the cells in them are **interstitial cells**. They are also sometimes called Leydig cells. The seminiferous tubules are also made of cells. The outer layer of cells is called the **germinal epithelium**. This is where the process of sperm production begins. Cells in various stages of sperm production are found inside the germinal epithelium, with the most mature stages closest to the fluid-filled centre of the seminiferous tubule. Cells that have developed tails are called **spermatozoa**, though this is almost always abbreviated to sperm. Also in the wall of the tubule are large nurse cells, called **Sertoli cells**. Figure 8 shows a small area of testis tissue, in which the structures described above can be seen.

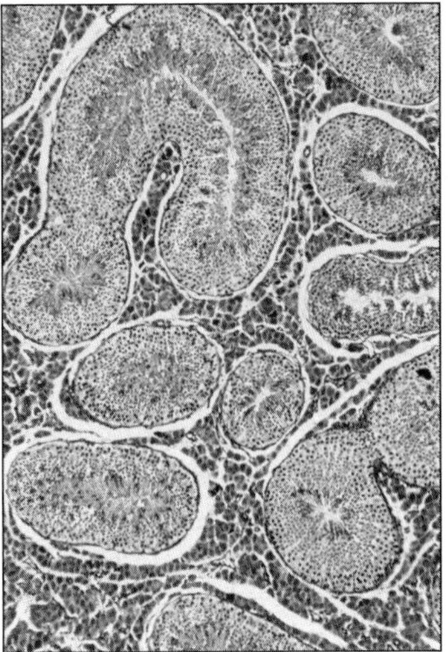

Figure 8 Micrograph of testis tissue

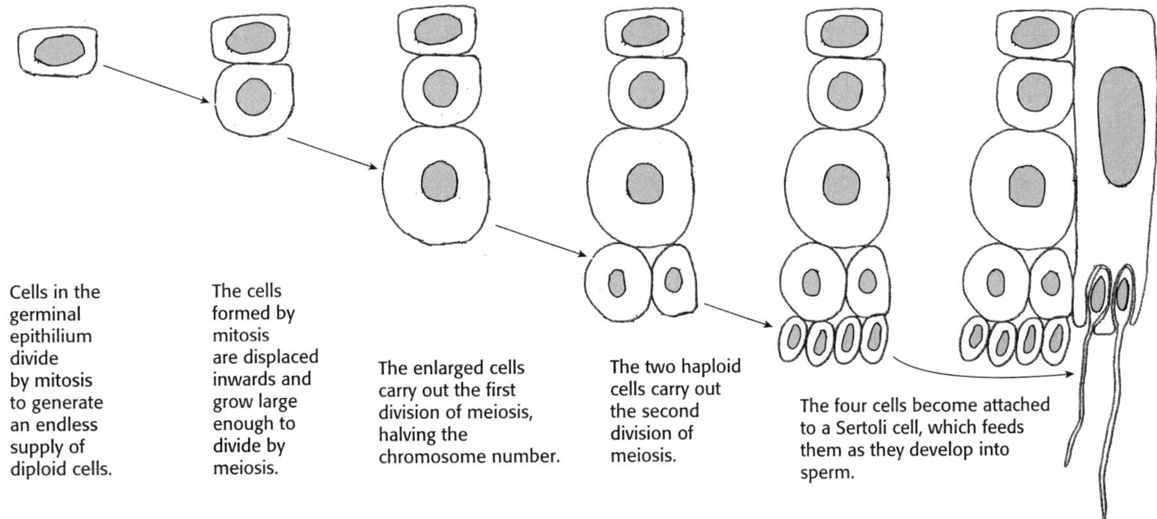

Figure 9 Stages in spermatogenesis

Data-based question: sperm counts and male fertility

A negative correlation has been found between rates of testicular cancer in a population and sperm counts. The rate of testicular cancer varies between countries, with Denmark having one of the highest rates and Finland one of the lowest rates in the world. Table 4 shows the result of two sperm count surveys: from males aged 20–45 whose partners were pregnant; and from healthy 18–20-year-old males reporting for military service. Sperm counts are in millions of sperm per millilitre of semen.

Table 4

	Men aged 20–45 with pregnant partners				Men aged 18–20	
	Summer		Winter			
Country	Sperm count	% normal sperm	Sperm count	% normal sperm	Sperm count	% normal sperm
Denmark	69	48	98	48	41	6.4
Finland	93	52	132	52	54	8.9

1 (a) Compare sperm production in summer and in winter. [3]
(b) Suggest reasons for the difference. [2]

2 (a) Compare the sperm production of men aged 20–45 with pregnant partners, with the sperm production of men aged 18–20. [3]
(b) Suggest two reasons for the difference. [2]

3 Evaluate the hypothesis that sperm production should be lower in Denmark than in Finland because the rate of testicular cancer is higher. [5]

HL

Oogenesis

Oogenesis is the production of egg cells in the ovaries. There are two main types of ovary tissue: a central tissue called the **medulla**, and surrounding tissue called the **cortex**. Oogenesis starts in the ovaries of a female fetus. It was formerly understood that the outer layer of the ovary was the **germinal epithelium**, but this is probably not so. Nevertheless, germ cells in the fetal ovary divide by mitosis and the cells formed move to distribute themselves through the cortex. When the fetus is four or five months old, these cells grow and start to divide by meiosis. By the seventh month, they are still in the first division of meiosis and a single layer of cells, called follicle cells, has formed around them. No further development takes place until after puberty. The cell that has started to divide by meiosis, together with the surrounding follicle cells, is called a **primary follicle**. There are about 400,000 in the ovaries at birth. No more primary follicles are produced, but at the start of each menstrual cycle a small batch are stimulated to develop by FSH. Usually only one goes on to become a **mature follicle**, containing a **secondary oocyte**. Stages in oogenesis and follicle development are shown in Figure 11.

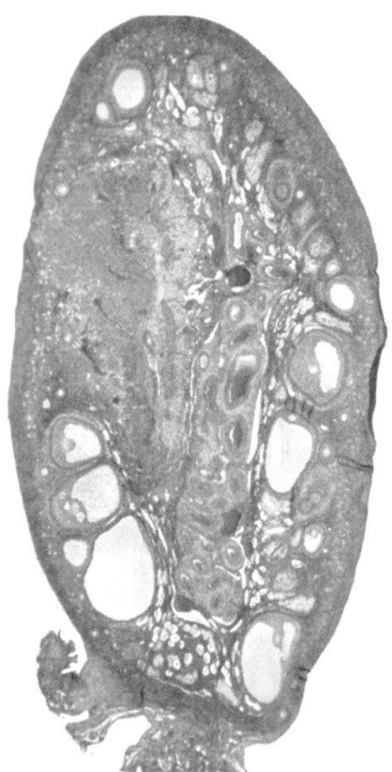

Figure 10 Micrograph of the ovary

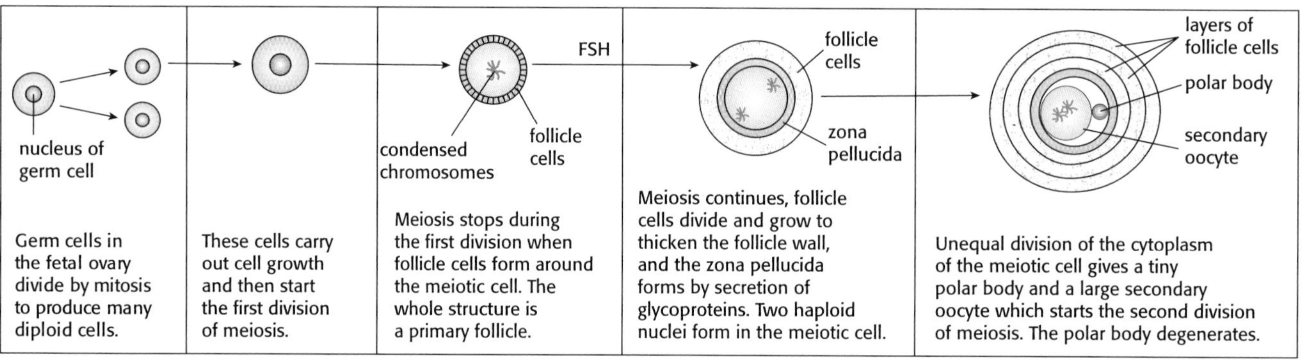

Figure 11 Stages in oogenesis

Critical considerations: ethics of egg donation

Human eggs can be obtained by using FSH to stimulate the ovaries, then collecting eggs from the ovaries using a micropipette. Women have sometimes undergone this procedure to produce eggs for donation to another woman who is unable to produce eggs herself.

Recently stem-cell researchers have used eggs in therapeutic cloning experiments. The nucleus of an egg is removed and replaced with a nucleus from an adult. If the resulting cell developed as an embryo, stem cells could be removed from it and cloned. It might then be possible to produce tissues or organs for transplanting to the adult who donated the nucleus. There would be no danger of tissue rejection because the stem cells would be genetically identical to the recipient.

There is a shortage of eggs both for donation to other women and for research. In 2006, scientists in England got permission to offer women cut-price IVF treatment, if they were willing to donate some eggs for research. In Sweden only travel and other direct expenses can be paid to egg donors, and in Japan egg donation is banned altogether.

1 Is there a distinction to be drawn between donating eggs for therapeutic cloning experiments and donating eggs to a woman who is unable to produce eggs herself, for example because her ovaries have been removed? Can the same act be judged differently depending on motives?

2 Research volunteers who donate bone marrow for use in research are usually paid, because of the time and discomfort involved. Would it be ethical to pay women who produce eggs for use in research? Would it be acceptable for a market in eggs to develop? What if economically vulnerable women were the source of the eggs?

HL

Gametogenesis – comparing male and female

Male and female gametes are very different in structure as shown in Figures 13 and 14. They are also different in the numbers produced and in the timing of formation and release.

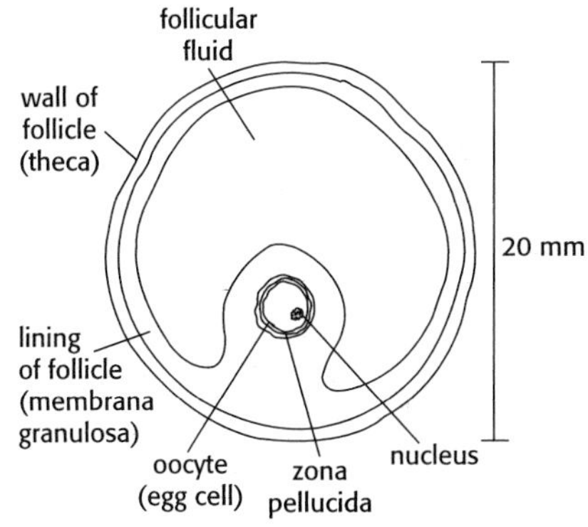

Figure 12 Structure of a mature follicle

Numbers of female gametes and timing of release

In females, the first division of meiosis produces one large cell and one very small cell. The small cell is the first polar body and it degenerates and dies. The large cell goes on to the second division of meiosis, completing it after fertilization. Again one large cell and one very small cell are produced. The small cell is the second polar body and it also degenerates and dies. Only the large cell, which is the female gamete, survives. This process happens once per menstrual cycle and usually only one egg cell is produced. During the years from puberty to the menopause only a few hundred female gametes are likely to be produced.

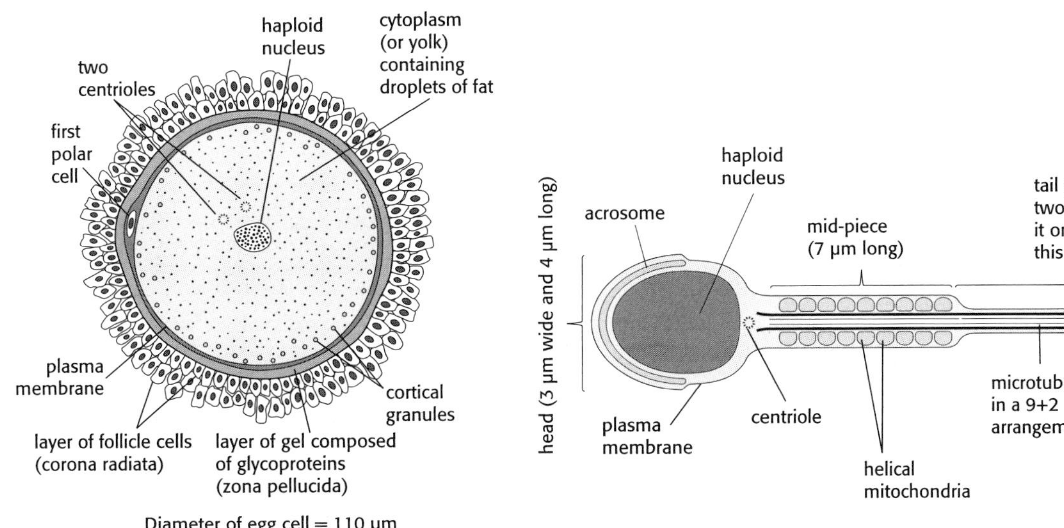

Figure 13 Structure of the female gamete

Figure 14 Structure of the male gamete

Numbers of male gametes and timing of release

From puberty onwards, the testes produce sperm continuously. At any time, there are millions of sperm at all stages of development. When sperm are fully formed, though still non-motile, they pass out of the testis, through narrow tubes, into the **epididymis**. This is a long coiled duct adjacent to the testis. Here the sperm become mature and capable of motility. They are also stored in the epididymis. The timing of their release depends on the sexual activity of the man. Whenever the male ejaculates, millions of sperm are moved by muscle action along the sperm duct from the epididymis to the urethra. Fluid is added by both the seminal vesicle and the prostate gland, to produce a total volume of about 3.5 ml of semen. The fluid secreted by the seminal vesicle includes alkali to help the sperm cope with the acidity of the vagina, fibrinogen to coagulate the semen, and fructose for sperm cell respiration. The secretion from the prostate gland contains citric acid and hydrolytic enzymes including fibrinolysin, which liquefies the coagulated sperm in the vagina.

?

1 How does the number of gametes produced in males differ from the egg production in females?

2 How does the timing of the formation and release of female gametes differ from the timing of the formation and release of male gametes?

HL

Fertilization

Sperm released into the vagina during intercourse are stimulated to swim by calcium ions diffusing from vaginal fluids into their tail. The membranes of sperm have receptors that can detect chemicals released by the egg, allowing directional swimming towards the egg. Once the egg is reached, these events take place (see Figure 15).

1 The acrosome reaction
The **zona pellucida** is a coat of glycoproteins that surrounds the egg. The **acrosome** is a large membrane-bound sac of enzymes in the head of the sperm. When the sperm makes contact with the zona pellucida, the contents of the acrosome are released and the enzymes from it loosen the zona pellucida. This allows sperm to force their way through by vigorous tail beating.

2 Penetration of the egg membrane
The acrosome reaction exposes an area of membrane on the tip of the sperm that has proteins that can bind to the egg membrane. The first sperm that gets through the zona pellucida therefore binds and the membranes of sperm and egg fuse together. The sperm nucleus enters the egg cell. This is the moment of fertilization.

3 The cortical reaction
Not only does the sperm bring parental genes, it also causes the activation of the egg. The first effect of this is on the **cortical granules** – vesicles located near the egg membrane. There are thousands of these vesicles and when activation of the egg has taken place their contents are released from the egg by exocytosis. The chemicals in the cortical granules interact with the glycoproteins in the zona pellucida, turning it into an impenetrable **fertilization membrane**. This prevents further entry of sperm and protects the developing embryo during its first days of life.

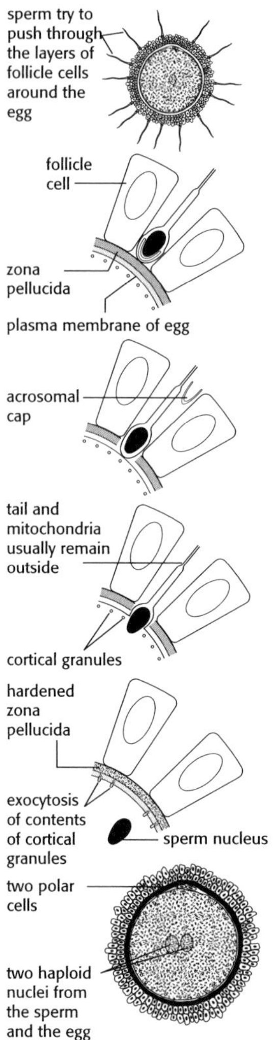

Figure 15 Stages in fertilization

Thinking about science: beliefs about conception

The traditional belief of the Trobrianders of Papua New Guinea is that conception occurs when a spirit child enters a woman's uterus and mixes with her menstrual blood. The spirit child is a Trobriander who has died. Conception does not therefore involve a man, or sexual intercourse. It does involve the return of spirits from the dead and a connection with ancestors. Even children who have been educated in Western schools, and have been taught in biology classes about intercourse and fertilization, have retained their traditional beliefs about conception.

Other tribes in Papua New Guinea do believe that sexual intercourse results in conception and pregnancy, but not always in the way that scientists understand. Dr Quentin Shaw, an English doctor working in Papua New Guinea, was approached by a magistrate and village elders of the Abalam tribe in 1987 and asked to settle a dispute over the paternity of a child. The question was how many times sexual intercourse must take place with a woman to conceive a child. Dr Shaw's answer, "Once only", was met with incredulity and derision and the question was asked again, repeatedly. Eventually Dr Shaw asked what the elders thought. "Well, maybe seven times if she is thin and about twenty times if she is fat" was the answer.

1 What are the implications of traditional Trobriand beliefs about conception for the status of men?
2 What are the implications of a belief that women need intercourse more than once to become pregnant?
3 Can and should science deny the validity of these and other alternative beliefs about conception?
4 To what extent can science be thought of as a Western belief system that is dominating and suppressing other belief systems in cross-cultural encounters?

HL

Early embryo development

After the sperm nucleus has entered the egg membrane, the female nucleus completes the second division of meiosis and one of the resulting two haploid nuclei is removed from the cell in a polar body and does not develop further. The other nucleus joins with the sperm nucleus to form a group of 46 chromosomes. These divide by mitosis to form two diploid nuclei and the cytoplasm of the fertilized egg cell divides equally to form a two-cell embryo. These two cells replicate their DNA, carry out mitosis and divide again to form a four-cell embryo (Figure 16). The embryo is about 48 hours old at this point. Further cell divisions occur, but some of the divisions are unequal and there is also migration of cells, giving the embryo the shape of a hollow ball. It is called a **blastocyst** (Figure 17). At 7 days old the blastocyst consists of about 125 cells and it has reached the uterus, wafted down the oviduct by the cilia of cells in the oviduct wall. At this age the fertilization membrane, which has surrounded and protected the embryo, breaks down. The blastocyst has used up the reserves of the egg cell and needs an external supply of food. It obtains this by sinking into the endometrium or uterus lining in a process called **implantation**. The outer layer of the blastocyst develops finger-like projections, called trophoblastic villi. These can digest uterus cells, allowing the blastocyst to penetrate the uterus lining. These villi also exchange materials with the mother's blood, including absorbing foods and oxygen. The embryo grows and develops rapidly and by eight weeks has started to form bone tissue. It is then considered to be a fetus rather than an embryo. It is recognizably human and soon visibly either male or female.

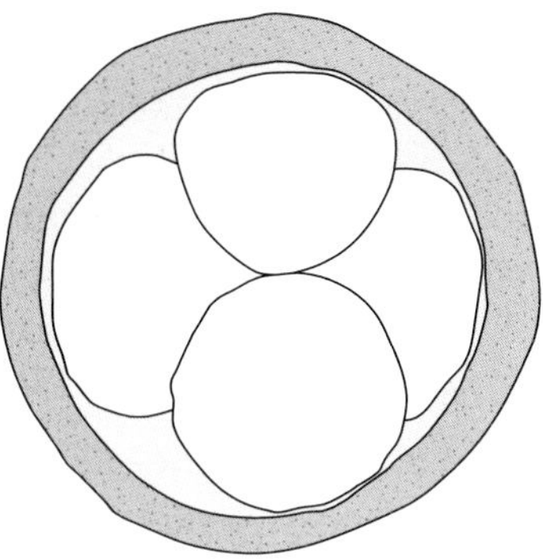

Figure 16 Four-cell embryo

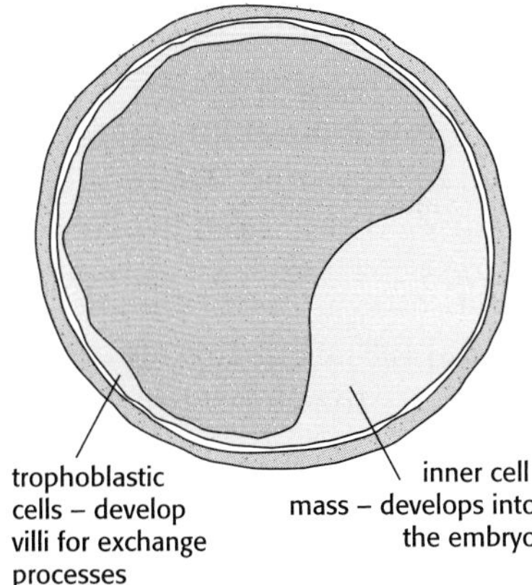

Figure 17 Blastocyst

Critical considerations: ethics of embryo research

In some countries, including Germany and the US, research involving the use of human embryos has been banned, or government funds are withheld, preventing the research. This is because some embryos are inevitably killed as a result of the research and this is considered to be unethical.

Many medical researchers are keen to use human embryos because they are a source of stem cells. Robert Lanza and colleagues from Advanced Cell Technology in Worcester, Massachusetts, claimed to have made two lines of embryonic stem cells by removing single cells from embryos, but the embryos from which the cells were taken were destroyed. The ethical objection to the destruction of potential human lives therefore remains.

Another approach to obtaining "ethical" embryonic stem cell lines is to take cells from embryos that have stopped developing. The cells of more than half of the embryos produced by IVF stop dividing naturally, leading to the death of the embryo. Even in Italy, where the law states that all embryos produced by IVF must be implanted, these embryos are discarded. In 2006, a team led by Miodrag Stojković at Newcastle University, England obtained a cell from an embryo in which division had stopped and encouraged it to start dividing again. The cells obtained developed into various tissue types, suggesting that they were stem cells. Objections have already been raised, as it is not clear that embryos in which cell division has stopped have died. There is still a remote possibility that they could revive and start to develop again.

1. Embryo research is allowed in Britain until the first signs of the backbone development are visible, or until the embryo is 14 days old, whichever is earlier. What is the rationale for this cut-off point? Do you agree with it?
2. What is the rationale for implanting all embryos with dividing cells? Do you prefer this policy or an embryologist selecting the most promising embryos for implantation?
3. Why is there likely to be less ethical objection to obtaining stem cells from embryos whose cells have stopped dividing?
4. How should death be defined in embryos? Can the same definition be used as in terminally ill patients or in critically injured patients, who could be used as donors of kidneys or other organs?

HL

Pregnancy

Pregnancy depends on the continued production of progesterone and estrogen. Early in pregnancy the embryo produces human chorionic gonadotropin – HCG. This hormone stimulates the corpus luteum in the ovary to continue to secrete progesterone and estrogen. These hormones stimulate the continued development of the uterus wall, which supplies the embryo with everything that it needs.

Humans are placental mammals – by the stage when a kangaroo or other marsupial would be born, a human fetus has developed a placenta and so can remain in the uterus for months longer. The placenta is needed because the surface area to volume ratio becomes smaller as the fetus grows larger. The structure of the placenta and its role in exchange processes are described on the next page.

By about the ninth week of pregnancy, the placenta has started to secrete estrogen and progesterone in large enough quantities to sustain the pregnancy, and the corpus luteum is no longer needed for this role. There is a danger of miscarriage at this stage of pregnancy if this switch-over fails.

The placenta is made of fetal tissues, in intimate contact with maternal tissues in the uterus wall. The fetus also develops membranes that form the amniotic sac. This contains amniotic fluid, which supports and protects the developing fetus.

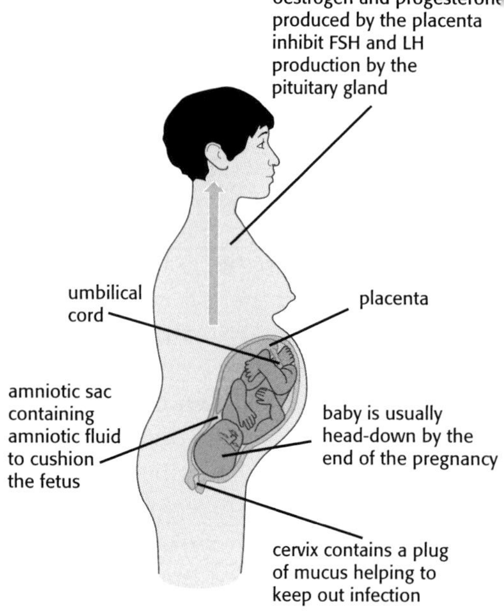

Figure 18 Mother and fetus during pregnancy

Data-based question: maternal age and pregnancy

Table 5 shows the outcomes of pregnancy, as a percentage of the total number of pregnancies, by maternal age at conception in Danish women from 1978 to 1992.

Table 5

Maternal age	Live births	Miscarriage	Ectopic pregnancy	Stillbirth
12–19	87.3	10.6	1.6	0.44
20–24	89.2	8.9	1.5	0.38
25–29	88.1	9.5	2.0	0.36
30–34	84.5	12.0	3.1	0.37
35–39	75.3	19.7	4.6	0.39
40–44	52.5	40.8	6.3	0.35
>44	17.2	74.7	7.9	0.15

Table 6 shows the relationship between social factors and maternal age, for women in their first pregnancy in Oklahoma from 1988 to 1995. The figures are percentages.

Table 6

Social factor	Age of mother (years)	
	20–29	35 or more
Mothers with a college education.	52.0	74.6
Mothers below the federal poverty line.	22.8	8.1
Married when the baby was conceived.	64.7	78.6
Mothers who were divorced or separated.	18.2	8.0
Mothers who intended to become pregnant.	59.2	83.0

?

1 Describe the trend in the percentage of live births as the age of the mother increases. [3]

2 Explain this trend, using the data in Table 5. [3]

3 A small percentage of the mothers were younger than 15. They had a miscarriage rate higher than 20%. Suggest reasons for this. [2]

4 Suggest two reasons for the difference in percentage of college-educated mothers in the two age groups (see Table 6). [2]

5 Discuss whether there is an ideal age at which to have children. [5]

HL

Structure and function of the placenta

The basic functional unit of the placenta is a finger-like piece of fetal tissue called a placental villus. These villi increase in number during pregnancy to cope with the increasing demands of the fetus for the exchange of materials with the mother. Maternal blood flows in the inter-villous spaces around the villi. This is a very unusual type of circulation as elsewhere blood is almost always confined in blood vessels. Fetal blood circulates in blood capillaries, close to the surface of each villus. The distance between fetal and maternal blood is therefore very small – as little as 5 μm. The cells that separate maternal and fetal blood form the **placental barrier**, shown in the diagram of a placental villus (Figure 19). This must be selectively permeable, allowing some substances to pass, but not others (Figure 20).

Microvilli projecting from placental cells into the maternal blood increase the surface area for these processes. So, as fetal blood flows through the placental villi it is oxygenated, recharged with nutrients and has excretory products removed from it. This blood flows back to the fetus through the umbilical vein and then through the body of the fetus, returning to the placenta via the umbilical arteries.

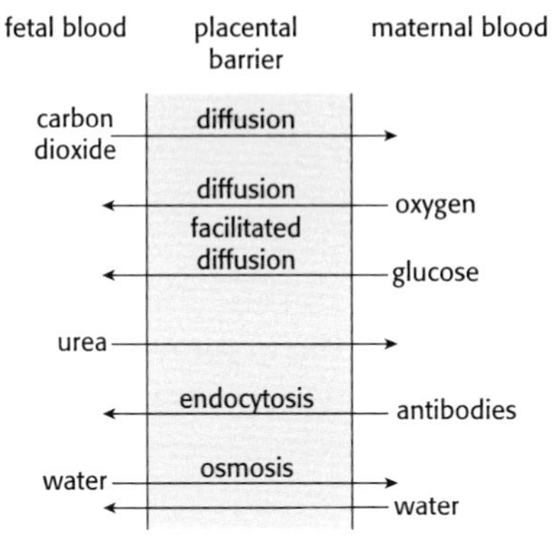

Figure 20 Exchange processes in the placenta

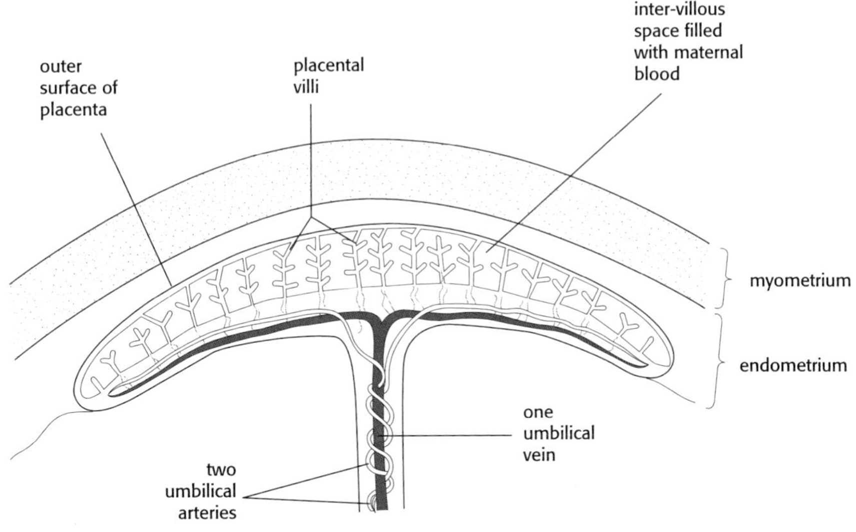

Figure 19 Structure of the placenta

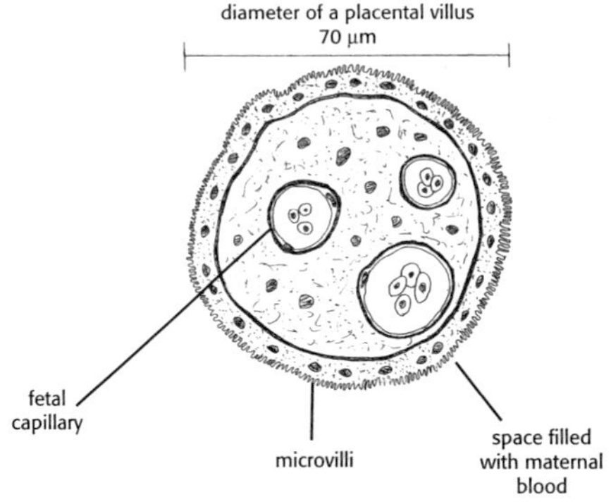

Figure 21 Transverse section of a placental villus

Data-based question: electron micrograph of placenta

Figure 22 shows a small region at the edge of a placental villus. The magnification is ×17,000.

1 (a) Identify the structures that are visible in the upper part of the micrograph. [1]
(b) Explain the functions of these structures. [3]

2 In much of the area of the electron micrograph there are rounded structures, surrounded by a single membrane. These are parts of a system of tubules called the smooth endoplasmic reticulum (sER). Its function is the synthesis of lipids, including steroids. Suggest a function for the sER in the placenta. [3]

3 Identify, with reasons, the structure in the lower left part of the micrograph. [3]

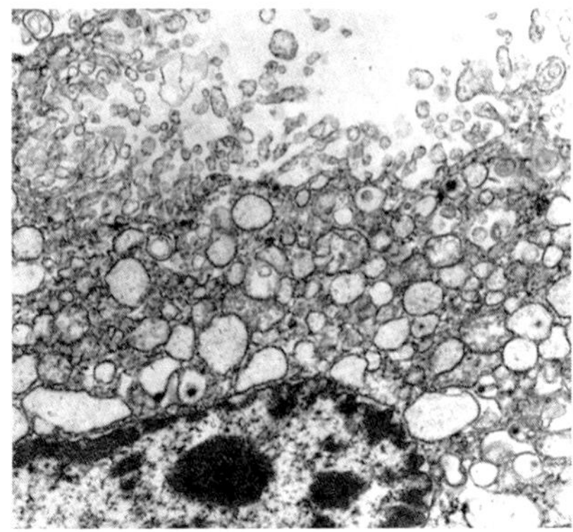

Figure 22 Small region at the edge of a placental villus

HL

Childbirth

During pregnancy, progesterone inhibits secretion of oxytocin by the pituitary gland and also inhibits contractions of the muscular outer wall of the uterus – the myometrium. At the end of pregnancy, hormones produced by the fetus signal to the placenta to stop secreting progesterone, and oxytocin is therefore secreted. This hormone stimulates contractions of the muscle fibres in the myometrium. These contractions are detected by stretch receptors, which signal to the pituitary gland to increase oxytocin secretion. Increased oxytocin makes the contractions more frequent and more vigorous, causing more oxytocin secretion. This is an example of a positive feedback system – a very unusual control system in human physiology. In this case it has the advantage of causing a gradual increase in the myometrial contractions, allowing the baby to be born with the minimum intensity of contraction. Relaxation of muscle fibres in the cervix causes it to dilate. Uterine contraction then bursts the amniotic sac and the amniotic fluid passes out. Further uterine contractions, usually over hours rather than minutes, finally push the baby out through the cervix and vagina. The umbilical cord is broken and the baby takes its first breath and achieves physiological independence from its mother.

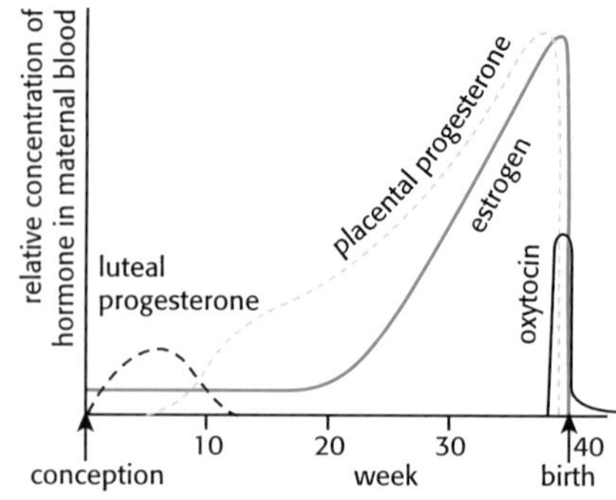

Figure 23 Hormone levels in pregnancy

Thinking about science: testing the safety of cesarean section

There is no agreement among obstetricians about whether mothers should be allowed to have their babies by cesarean section when there are no clinical reasons for this procedure. A randomized controlled trial could be used to test whether elective cesarean section is as safe as vaginal birth. This would involve women being assigned randomly to cesarean section or vaginal birth. A survey was done of 660 obstetricians to find out how much support there was for a randomized controlled trial.

Table 7

	Yes	No
1 Do you believe that women in their first pregnancy should choose their method of delivery?	49.7%	51.3%
2 Is a randomized controlled trial of elective cesarean section versus vaginal birth ethical?	40.1%	59.9%
3 Is a randomized controlled trial of elective cesarean section versus vaginal birth feasible?	26.8%	73.2%
4 Is a randomized controlled trial of elective cesarean section versus vaginal birth desirable?	48.0%	52.0%
5 Would you recruit women to a randomized controlled trial of cesarean section versus vaginal birth?	39.6%	60.4%
6 Would you consider an elective cesarean section for yourself/your partner?	40.6%	59.4%

1 What are the differences between a randomized controlled trial being feasible, desirable and ethical?

2 What do you deduce by comparing the results of these pairs of questions: (a) 3 and 4; (b) 2 and 5; (c) 1 and 6?

3 Why are women who demand the right to have their baby by cesarean section sometimes called "too posh to push"?

1 Baby positions itself before birth so that its head rests close to the cervix

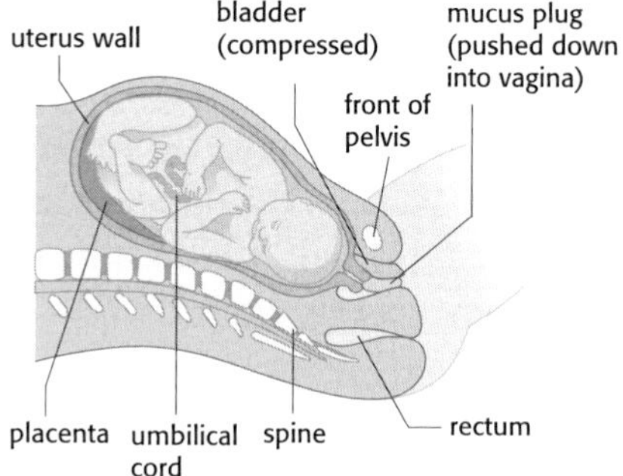

2 Baby passes into vagina and amniotic fluid is released

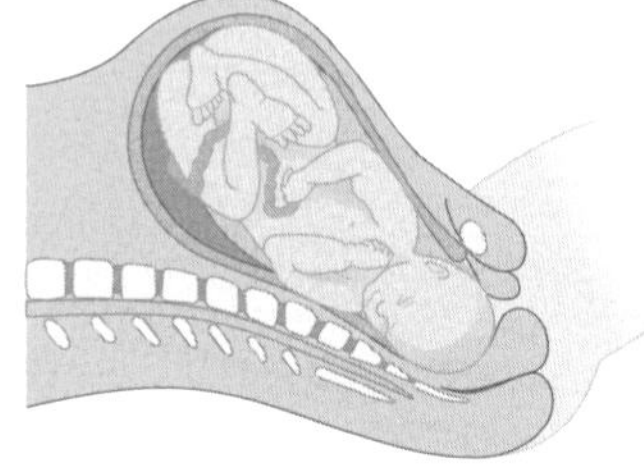

3 Baby is pushed out of mother's body

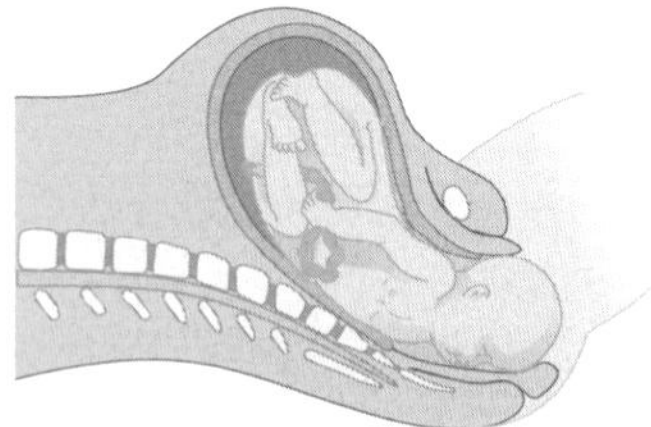

4 Placenta and umbilical cord are expelled from body

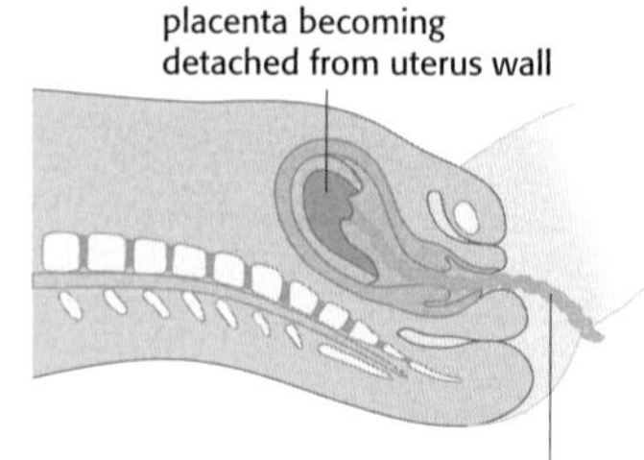

Figure 24 Stages in childbirth

Chapter 26 questions

1 (a) Draw and label a diagram of the male reproductive system. [5]
(b) Outline the roles of testosterone in males when they are fetuses and when they are adult. [5]
(c) Explain the reasons for expecting the number of births of male and female babies to be equal. [8]

2 (a) Draw and label a diagram of the female reproductive system. [5]
(b) Explain the changes in the endometrium during a menstrual cycle in terms of hormone levels and the timing of ovulation and implantation. [8]
(c) Outline the techniques used to stimulate egg production for *in vitro* fertilization. [5]

3 Using the data in Table 8:
(a) outline the relationship between the age of the mother and the success rate of IVF [3]
(b) outline the relationship between the number of embryos transferred and the chance of having a baby as a result of IVF [3]
(c) discuss how many embryos fertility centres should be allowed to transfer. [4]

Table 8

Age of mother	Percentage of pregnancies per IVF cycle according to the number of embryos transferred					
	1	2		3		
	single	single	twins	single	twins	triplets
<30	10.4	20.1	9.0	17.5	3.6	0.4
30–34	13.4	21.8	7.9	18.2	7.8	0.6
35–39	19.1	19.1	5.0	17.4	5.6	0.6
>39	4.1	12.5	3.5	12.7	1.7	0.1

4 Figure 25 shows birth rates in regions of the world from 1950 to 1990.
(a) Suggest reasons for the differences in birth rate between regions. [3]
(b) Identify the region of the world with:
(i) the smallest decline in birth rate [1]
(ii) the largest decline in birth rate. [1]
(c) Suggest reasons for declining birth rates. [2]
(d) Explain why it is impossible to deduce from the birth rate whether a population has risen or fallen. [3]

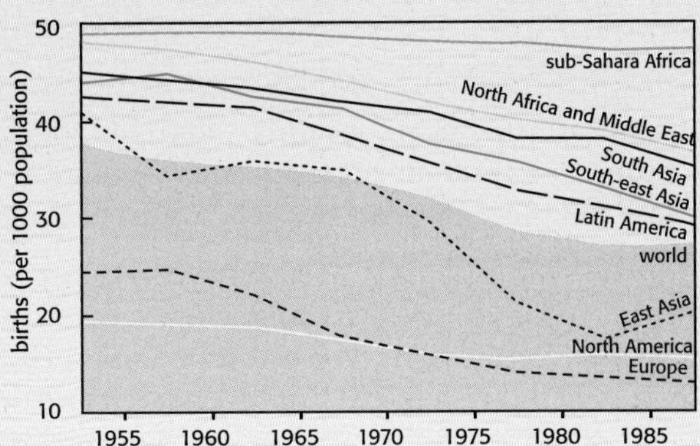

Figure 25

5 Figure 26 shows the concentrations of FSH and LH during the life of a woman. The time axis is not to scale. Only a representative sample of menstrual cycles is shown during the reproductive phase.
(a) Using the data in the graph, explain the reasons for pre-pubertal girls not menstruating. [3]
(b) (i) Compare FSH levels during pregnancy with the levels at other times in the woman's life. [3]
(ii) Explain how levels of other hormones control the FSH level during pregnancy. [3]
(c) (i) When the woman's reproductive life ends, the levels of FSH and LH in her blood rise. Calculate the percentage increase in FSH level, compared with the highest level during the reproductive phase. [2]
(ii) Explain the reasons for the levels of FSH and LH rising at the end of the reproductive phase. [2]
(iii) Suggest reasons for these high levels of FSH and LH not causing hyper-ovulation in women. [2]

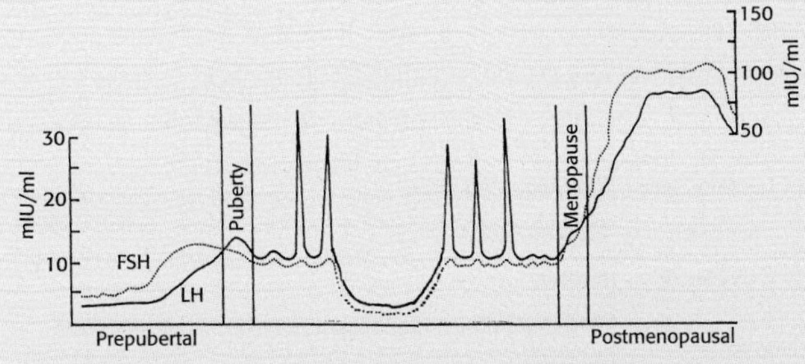

Figure 26

Option A: Human nutrition and health

Nutrients are chemical substances, found in foods, that are used in the human body. Some nutrients are **essential** in the human diet, because foods are the only possible source of the nutrient. The box (right) lists essential nutrients.

Other nutrients are **non-essential**, either because another nutrient can be used for the same purpose or because they can be made in the body from another nutrient. Glucose, starch and other carbohydrates are non-essential, because they are used in respiration to provide energy and lipids can be used instead. Of the 20 amino acids in proteins, about half are essential, because they cannot be synthesized in sufficient quantities, but the other half can be made from other simpler nitrogen compounds. For example, phenylalanine is an essential amino acid, but tyrosine is non-essential because it can be synthesized from phenylalanine, by the reaction shown below:

phenylalanine → tyrosine (tyrosine hydroxylase)

Essential nutrients

- Some amino acids
- Some unsaturated fatty acids
- Some minerals, e.g. calcium
- Vitamins
- Water

Inquiry

One of the most common genetic diseases involves a deficiency or complete absence of the enzyme tyrosine hydrolase.

1 State the name of the genetic disease.
2 Explain the consequences of the disease if it is not treated early in a child's life.
3 Explain how the disease can be diagnosed and treated, to avoid the harmful consequences.

Critical considerations: protein-deficiency malnutrition

If there is a shortage of one or more of the essential amino acids in the diet then the body cannot make enough of the proteins that it needs. Essential amino acids may be lacking due to an overall insufficiency of protein in the diet or due to an imbalance in the types of protein. Protein-deficiency malnutrition causes stunting (reduced height), wasting (serious weight loss) and general developmental disability. Figure 1 shows the incidence of stunting, wasting and developmental disability for eight regions of the world. The statistic used is YLD per 1000, which is years lost to disease per thousand members of the population. The chart shows the results for males in the years 1990 and 2000. The figures for females showed the same trends.

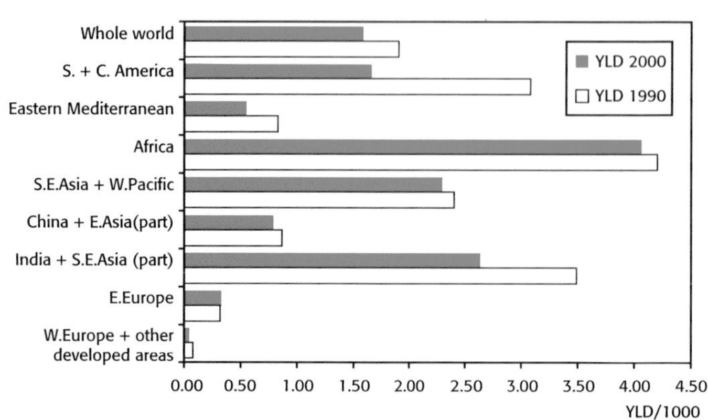

Figure 1 YLD due to protein deficiency malnutrition

1 (a) Identify the region with the greatest evidence of protein-deficiency malnutrition.
(b) Suggest reasons for this.
2 (a) Determine the percentage difference in YLD for the year 2000 between the region with the highest rate and the region with the lowest rate.
(b) What could be done to narrow the difference?
3 (a) Outline what the data reveals about the worldwide trend in protein-deficiency malnutrition.
(b) Identify the regions of the world where the trend has been most pronounced.
(c) Suggest reasons for this.
4 Predict, with reasons, how the pattern might appear in 2010.

Full details of which countries are included in each epidemiological region are given on the WHO website.

Fatty acids and health

Fatty acids vary in several ways. These include the following.

- How many carbon atoms there are in the molecule.
- Whether or not there are double bonds between carbon atoms in the hydrocarbon chain. Saturated fatty acids have no double bonds and unsaturated fatty acids have at least one.
- Whether there is one double bond or more than one. This is the difference between monounsaturated and polyunsaturated fatty acids.
- Where the first double bond is located in the molecule, in relation to the opposite end of the hydrocarbon chain from the COOH group. Omega-3 fatty acids have the first double bond between the third and fourth carbons, omega-6 have it between the sixth and seventh, and so on.

The molecular diagrams (Figure 2) show examples of the differences.

In unsaturated fatty acids in living organisms, the hydrogen atoms are nearly always on the same side of the two carbon atoms that are double bonded – these are celled called *cis* fatty acids. The alternative is for the hydrogens to be on opposite sides – called *trans* fatty acids. These two conformations are shown in Figure 3.

Trans fatty acids are produced artificially by hydrogenating vegetable or fish oils, to give solid fats for use in margarine and some other processed foods.

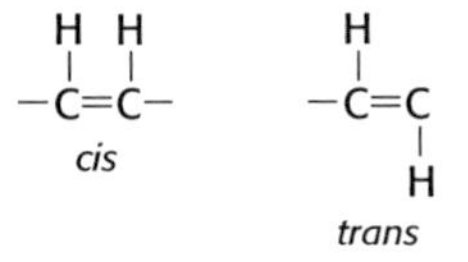

palmitic acid
• saturated
• non-essential

linolenic acid
• polyunsaturated
• all *cis*
• essential
• omega 3

palmitoleic acid
• monounsaturated
• *cis*
• non-essential
• omega 7

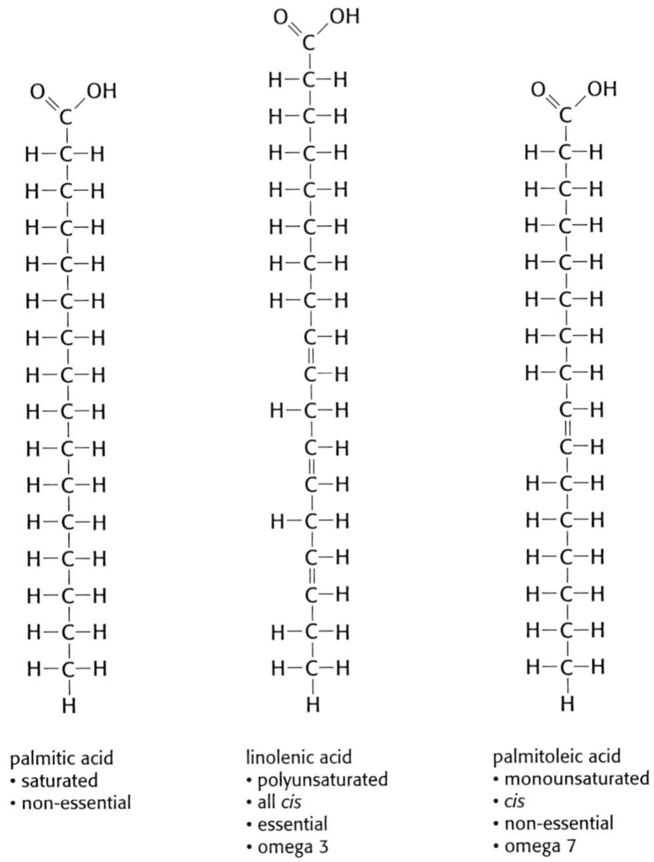

Figure 2 Examples of fatty acids

cis

trans

Figure 3 Double bonds in fatty acids

Inquiry

Many claims have been made about the health benefits and health risks of the various types of fatty acid. Dietary supplements containing particular fatty acids are available and are widely used in some parts of the world.

Choose one of the following claims and find evidence for or against the claim.

A Saturated fatty acids cause coronary heart disease.
B *Cis*-monounsaturated fatty acids prevent coronary heart disease.
C Omega-3 fatty acids promote brain development and intelligence.
D *Trans* fatty acids are a major cause of coronary heart disease.

Data-based question: saturated fats and coronary heart disease

Table 1			
Populations ranked by % calories as saturated fat			
Population	% Calories as saturated fat	Death rate/100,000 yr^{-1}	
		CHD	All
E. Finland	22	992	1727
W. Finland	19	351	1318
Zutphen	19	420	1175
USA	18	574	1088
Slavonia	14	214	1477
Belgrade	12	288	509
Crevalcor	10	248	1241
Zrenjanin	10	152	1101
Dalmatia	9	86	758
Crete	9	9	543
Montegiorgio	9	150	1080
Velika	9	80	1078
Rome	8	290	1027
Corfu	7	144	764
Ushibuka	3	66	1248

1 (a) Plot a scattergraph of the data in Table 1. [5]
(b) Outline the trend shown by the scattergraph. [2]
2 Compare the results for
(a) East and West Finland [2]
(b) Crete and Montegiorgio. [2]
3 Evaluate the evidence from this survey for saturated fats as a cause of coronary heart disease. [4]

Vitamin C

In addition to proteins, lipids and carbohydrates, which are consumed in large quantities, the human body also requires a variety of nutrients in smaller quantities. Minerals and vitamins are two groups of these. They can be distinguished by their chemical nature: minerals are chemical elements, usually in an ionic form, and vitamins are organic compounds. Although many of the vitamins are needed in minute quantities, they cannot be made in the body and are therefore essential components of the diet. For example vitamin C is a compound called ascorbic acid (Figure 4). It is needed for the synthesis of the collagen fibres that form part of many tissues in the body, including skin and blood vessel walls. A variety of symptoms develop as a result of vitamin C deficiency, which are collectively known as scurvy.

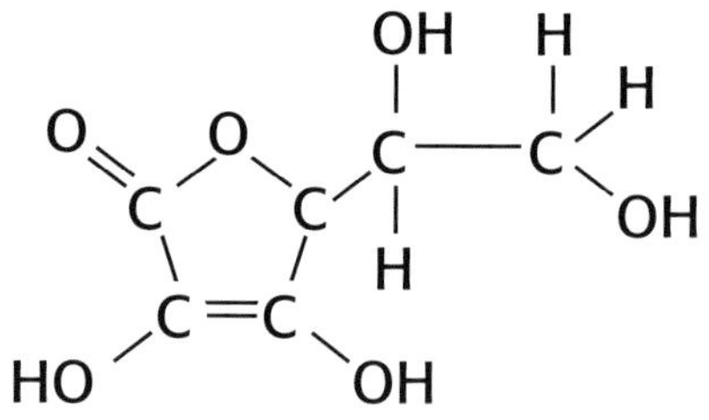

Figure 4 The structure of vitamin C

Experiments have been done to try to discover the minimum amount of vitamin C needed in the diet to prevent scurvy. These are described on page 277. Adults were found to require only about 10 mg of vitamin C per day, but the recommended daily allowances that national governments have set are much higher than this – typically 30–60 mg for adults. These levels of intake are set to give a safety margin and allow for individual variation. Intake levels above about 75 mg do not cause any increases in blood plasma concentrations of vitamin C, but despite this there have been repeated claims of health benefits from higher intakes, and recommended daily allowances may be revised upwards.

Inquiry

Linus Pauling was a distinguished American chemist and double Nobel Prize laureate. He claimed that taking massive doses of vitamin C protects the body against upper respiratory tract infections, such as the common cold. Other scientists have issued counter-claims that these doses do not give protection and could even cause problems of deficiency when they are stopped. The body becomes accustomed to excreting the excess – called systemic conditioning. If intake drops back down to normal levels, the symptoms of scurvy can develop. This is called rebound malnutrition.

1 Is there evidence to back Linus Pauling's claim?
2 Is there any evidence that rebound malnutrition is a danger?
3 Shouldn't we accept Linus Pauling's claim without question, given that he was such a distinguished scientist?

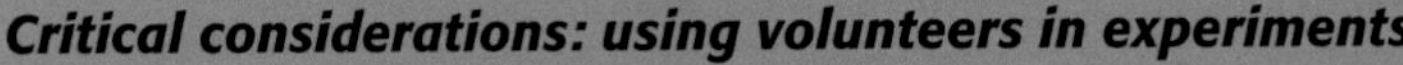

Figure 5 Linus Pauling.

Critical considerations: using volunteers in experiments

During the Second World War, experiments were conducted both in England and in the US using conscientious objectors to military service as volunteers. The volunteers were willing to sacrifice their health to help extend medical knowledge. A vitamin C trial in England involved 20 volunteers. For six weeks they were all given a diet containing 70 mg of vitamin C. Then, for the next eight months, three volunteers were kept on the diet with 70 mg, seven had their dose reduced to 10 mg and ten were given no vitamin C. All of these ten volunteers developed scurvy. Three-centimetre cuts were made in their thighs, with the wounds closed up with five stitches. These wounds failed to heal. There was also bleeding from hair follicles and from the gums. Some of the volunteers developed more serious heart problems. The groups given 10 mg or 70 mg of vitamin C fared equally well and did not develop scurvy.

Experiments on requirements for vitamin C have also been done using real guinea-pigs, which ironically are suitable because guinea-pigs, like humans, cannot synthesize ascorbic acid. During trial periods with various intakes of vitamin C, concentrations in blood plasma and urine were monitored. The guinea-pigs were then sacrificed and collagen in bone and skin was tested. The collagen in guinea-pigs with restricted vitamin C had less cross-linking between the protein fibres and therefore lower strength.

1 Is it ethically acceptable for doctors or scientists to perform experiments on volunteers, where there is a risk that the health of the volunteers will be harmed?
2 Sometimes people are paid to participate in medical experiments, such as drug trials. Is this more or less acceptable than using unpaid volunteers?
3 Is it better to use animals for experiments or are the ethical objections the same as with humans?
4 Is it acceptable to kill animals, so that an experiment can be done?

Vitamin D, iodine and fibre

Vitamin D is needed for calcium absorption from food in the intestines, so the symptoms of vitamin D deficiency are similar to those of calcium, with children developing the skeletal deformities known as rickets. Vitamin D does not fit the definition of a vitamin very well, as it can be synthesized in the skin. This only happens when sunlight, or another light source containing ultraviolet light with wavelengths in the range 290–310 nm, strikes the skin. If teenagers and adults spend enough time outside, with their skin uncovered, no vitamin D is required in the diet. Children, pregnant women and elderly people are recommended to eat 10 µg per day, to supplement the amount made in their skin. There are few dietary sources of vitamin D. Oily fishes including herring, mackerel, sardines and tuna are rich sources. Eggs and liver also contain some, and certain foods such as margarine and milk are artificially fortified with vitamin D.

Ultraviolet light has some harmful consequences, including mutations that can lead to skin cancer. Melanin in the skin intercepts and absorbs light, including the ultraviolet wavelengths. Dark skins therefore give good protection against cancer, but they also reduce vitamin D synthesis. In indigenous human populations, skin colour balances the twin risks of vitamin D deficiency and cancer or other damage due to ultraviolet light. After population migrations there can be problems. In the 1970s immigrants with dark skin from the Indian subcontinent living in the United Kingdom started to show symptoms of vitamin D deficiency. Immigrants from northern Europe with light skin living in Australia were found to have high rates of malignant melanoma. Australians with light skin were then advised to stay out of bright sunlight, cover their skin or apply sun-block creams.

Inquiry

Iodine is an essential mineral in the diet. In many parts of the world children show symptoms of iodine deficiency.

1 How is iodine used in the human body?
2 What are the symptoms of iodine deficiency?
3 What would be the benefits of dietary supplementation with iodine?
4 Would it be feasible to develop a worldwide programme of dietary iodine supplementation?
5 Are there other mineral deficiencies that are widespread causes of malnutrition?

Inquiry

Fibre is an essential component of a balanced healthy diet, but it is not a nutrient.

1 Why is fibre not considered to be a nutrient?
2 Why is it needed in a balanced diet?
3 What claims have been made about lack of dietary fibre causing health problems and what is the evidence for these claims?

Data-based question: vitamin D deficiency in multicultural Australia

Table 2 shows the results of a survey conducted at a hospital in Melbourne. Measurements were made of the concentration of vitamin D in the blood plasma of pregnant women who were dark-skinned or who for religious reasons wore a veil that covered their skin. The table shows the numbers and percentage that had less than the lowest acceptable concentration of 22.5 nmol dm^{-3} and the sample sizes.

Table 2

	Skin colour		
Skin covering*	**Very dark**	**Intermediate**	**Light**
Consistently covered	6/6 (100%)	1/2 (50%)	23/25 (92%)
Inconsistently covered	3/5 (60%)	1/3 (33%)	18/24 (75%)
Uncovered	2/2 (100%)	2/3 (67%)	0/0 –

* Consistently covered – women always covered up, including arms, hair and neck, when outdoors; inconsistently covered – women did not usually cover fully in their own garden; uncovered – women did not generally cover their arms, hair and neck when outdoors.

?

1 Explain why it is important for pregnant women to avoid vitamin deficiencies. [2]
2 Evaluate the data in the table to decide whether there should be concerns about vitamin D deficiency among any parts of the Australian population. [3]
3 Explain the dietary advice that should be given to pregnant women in Australia:
   (a) who have a dark skin [1]
   (b) who wear a veil. [1]
4 Predict, with reasons, whether light-skinned women who were uncovered would have had high rates of vitamin D deficiency. [3]

Energy, appetite and obesity

Cell respiration involves the release of energy from foods such as fat, carbohydrate or protein. The amounts of energy that these nutrients contain are shown in Table 3. Fat and carbohydrate are traditionally the main sources of energy, so diets contain large amounts of foods that can supply these nutrients. Modern diets in the developed world sometimes contain large amounts of protein, which is then used as an energy source.

Foods eaten in bulk to supply energy are called staple foods. Human societies are very diverse in their choice of staple energy sources. Some current and past examples are given here:

- Naga tribes of north-east India – rice.
- Italians – wheat in the form of pasta, pizza and bread.
- Yoruba tribe of Nigeria – cassava.
- Hopi Indian tribe of Arizona – maize.
- Blasket Islanders of Ireland – fish, shellfish, seal meat.
- Maasai of Kenya – meat, fat, blood and milk.

The great variation in staple foods is partly due to availability and partly to cultural preferences and traditions.

Table 3

Type of nutrient	Energy/kJ per 100g
carbohydrate	1760
protein	1720
fat	4000

Inquiry

The amounts of carbohydrate and protein in human diets varies considerably. What are the health consequences of diets rich in:
(a) carbohydrate?
(b) protein?
(c) fat?

Working with data: body mass index

The body mass index, usually abbreviated to BMI, was developed by a Belgian statistician, Adolphe Quelet. It is used to assess whether a person's body mass is at a healthy level, or is too high or too low. BMI is calculated using this formula:

$$\text{BMI} = \frac{\text{mass in kilograms}}{(\text{height in metres})^2}$$

Units for BMI are kg m^{-2}.

Table 4

BMI	Status
below 18.5	underweight
18.5–24.9	normal weight
25.0–29.9	overweight
30.0 or more	obese

Data-based question: the obesity epidemic

1 WHO reports that adult BMI levels are 22–23 kg m^{-2} in Africa and Asia, whereas levels of 25–27 kg m^{-2} are prevalent across North America, Europe and parts of Latin America. Suggest three reasons for the difference in BMI between these regions. [3]

2 About 300 million adults in the world are obese, but the percentage of obese adults varies from below 5 per cent in China, Japan and some African nations to over 70 per cent in urban Samoa.
Suggest two reasons for the high percentage of obese adults on Samoa. [2]

3 In the US the percentage of obese boys aged 12–17 has increased from 5 per cent in 1966–70 to 13 per cent in 1988–91. The equivalent figures for girls are 5 per cent and 9 per cent. Suggest three reasons for increasing rates of obesity among children in the US. [3]

4 Obesity increases the risk of cancer of the breast, colon, prostate, uterus, kidney and gall bladder. It contributes to osteoarthritis, ischemic heart disease and Type II diabetes. Approximately 85 per cent of diabetics have Type II diabetes and 90 per cent of these are obese or overweight.
Calculate the percentage of all people with diabetes who have Type II diabetes and are overweight or obese. [2]

Inquiry

Anorexia nervosa is a disease with psychological causes. Someone who has anorexia nervosa becomes chronically underweight because they do not eat enough food, even though plenty is available. There are many consequences.

1 What are the physiological consequences of anorexia nervosa in relation to:
(a) muscle mass?
(b) hair growth?
(c) blood pressure?
(d) fertility?
(e) menstrual cycles?
(f) dental health?

2 What are the consequences for the family and friends of someone who develops anorexia nervosa?

Breast- and bottle-feeding babies

It is natural for human babies to feed on their mother's breast milk for some months after birth. Artificial milk is now available as a powder in many parts of the world, which can be dissolved in water and fed using a bottle and teat. Table 5 shows differences in the composition of typical breast milk and artificial milk.

Advantages of breast-feeding
- Promotes bonding between mother and baby
- Acts as a natural birth-control method, reducing the chance of conception while the mother is lactating and therefore allowing more time between the birth of one child and the next
- Helps mothers to lose weight after pregnancy
- Avoids the allergies to proteins in cows' milk or soya that can develop when babies receive artificial milk

Table 5

	Human milk	Artificial milk
carbohydrate	lactose	lactose or glucose polymers
protein source	65% whey proteins, 35% casein	18% bovine whey and 82% bovine casein; or soya proteins
fatty acids	human butterfat	palm, coconut, soy or safflower oils
antibodies	antibodies present in the first milk (colostrum)	no antibodies for fighting human diseases present

It seems reasonable to assume that the composition of human breast milk is ideal for human babies. Table 5 shows that the composition of artificial or formula milk is close but not identical. For this reason most healthcare professionals recommend breast-feeding. There are other claimed benefits, apart from nutritional, for breast-feeding. These are shown in the box (top right).

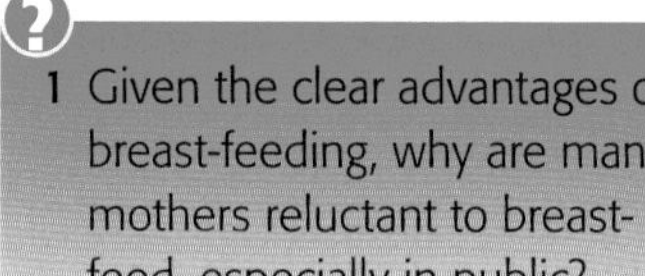

1 Given the clear advantages of breast-feeding, why are many mothers reluctant to breast-feed, especially in public?
2 How could breast-feeding be promoted more effectively?

There are also some possible disadvantages of breast-feeding:
- Some studies suggest higher rates of post-natal depression among mothers who breast-feed rather than bottle-feed.
- Mothers who are susceptible to arthritis may suffer flare-ups while they are breast-feeding.

Thinking about science: breast-feeding and intelligence

Many studies from 1929 onwards have shown higher IQ scores in children who were breast-fed, compared with children who were bottle-fed on artificial milk. Mothers have often been told that by breast-feeding they would increase the intelligence of their babies. In 2006, the results of a new study were published, together with a re-analysis of previous studies. The new study looked at factors apart from whether babies were breast-fed or bottle-fed. It was found that women who breast-fed tended to have a higher IQ than women who bottle-fed their babies. When the effects of this difference in maternal IQ were eliminated from the difference in the IQ of breast-fed and bottle-fed babies, it was found that the remaining difference was not statistically significant. Breast-fed babies were only more intelligent on average, because of inheritance of intelligence from their parents. An assumption is being made here that IQ tests are a valid measure of intelligence.

The authors of the study were careful to avoid recommending bottle-feeding: "The broader context of breast-feeding should also be considered. Evidence showing the many benefits of breast-feeding to mother and child led the World Health Organization and UNICEF to formulate the Innocenti Declaration, which includes breast-feeding for 4–6 months as a global goal. Even if it does not promote intelligence, breast-feeding remains an unequalled way of providing ideal food for the healthy growth and development of infants."

1 Was it wrong to tell mothers that they would increase the intelligence of their babies by breast-feeding them?
2 Using the example of breast-feeding and intelligence, explain the distinction between a correlation between two variables, and a causal link between the variables?
3 Why are women who breast-feed their babies more intelligent on average than mothers who bottle-feed?
4 Why were the authors of the 2006 study careful to avoid recommending bottle-feeding, despite their research findings?

Diabetes, cholesterol and coronary heart disease

There are several diseases involving excessive excretion of urine, all of which are forms of diabetes. In the commonest form, sugar is present in the urine. This is **diabetes mellitus**, and it affects hundreds of millions of people worldwide. There are two ways in which this sort of diabetes can develop:

1 Auto-immune destruction of insulin-secreting cells in the pancreas (Type I diabetes).
2 Decreased responsiveness of body cells to insulin due to "burn-out" (Type II diabetes).

Prevalence rates of Type II diabetes are rising rapidly in many countries. The study of the rates and distribution of a disease, to try to find its causes, is known as epidemiology. Epidemiological studies of Type II diabetes have implicated increased blood concentrations of fatty acids, linked to the following risk factors:

- diets rich in fat and low in fibre
- obesity due to overeating and lack of exercise
- genetic factors which affect fat metabolism.

There is huge variation between ethnic groups in rates of Type II diabetes, from less than 2 per cent in China to 50 per cent among the Pima Indians. The symptoms are not always recognized, so not all people with diabetes are diagnosed. The main symptoms are:

- elevated levels of blood glucose
- glucose in the urine – this can be detected by a simple test
- dehydration and thirst resulting from excretion of large volumes of urine.

Unless carefully managed, diabetes can cause other health problems to develop, several of which relate to the cardiovascular system:

- atherosclerosis (narrowing of arteries by fatty deposits)
- hypertension (raised blood pressure)
- coronary heart disease (narrowing of the coronary arteries with the associated risk of heart attacks).

There also seems to be a link between these cardiovascular problems and blood lipid concentrations. Links have been suggested between high concentrations of cholesterol, high concentrations of LDL and low concentrations of HDL. There has been much controversy about the role of cholesterol, in particular in the development of coronary heart disease (CHD) – see Inquiry above.

Inquiry

1 What foods are suitable for a patient with Type II diabetes, and what foods should be avoided?
2 What overall dietary advice should be given to a patient who has recently been diagnosed with Type II diabetes?

Inquiry

Some people have high concentrations of cholesterol in their blood.

1 What are the causes of high blood cholesterol?
2 In what forms does cholesterol occur in the blood?
3 What is the difference between HDL and LDL (high and low density lipoprotein)?
4 Should we try to reduce our blood cholesterol levels to a minimum?
5 Does reducing dietary intake of cholesterol reduce the concentration in the blood?
6 How strong is the evidence for dietary cholesterol as a cause of

Data-based question: diabetes and ethnic groups

Table 6 shows prevalence rates of diabetes, hypertension and CHD in six ethnic groups in Great Britain. The non-white groups are ethnic minorities. Rates are shown per 100.

Table 6

	White	African Caribbean	South-east Asian	Indian	Pakistani or Bangladeshi	Chinese
type II diabetes	1.7	5.3	6.2	4.7	8.9	3
hypertension	12	17	9	8	11	5
CHD	4.2	3.7	4.2	3.3	6	3

1 Compare the rates of Type II diabetes in the six ethnic groups. [4]
2 Compare rates of Type II diabetes and hypertension in each group, to find out whether the data shows any correlation between the rates of these health problems. [3]
3 Discuss whether data in the table prove that genetic factors cause diabetes, hypertension and CHD. [3]

Ethical issues in human diets

There are some important ethical issues concerned with the choice of food that we eat. For example, some people believe that it is wrong to eat food if its production involves animal suffering. Vegetarians do not eat meat because an animal must be slaughtered to produce it. Some vegetarians are willing to drink milk, because the cow or other animal that is milked does not have to be killed. Some vegetarians eat eggs, again because the hen does not die when it lays an egg. Stricter vegetarians, called vegans, do not drink milk or eat eggs, and they do not eat honey either.

Inquiry

1 What ethical objections do vegans have to eating:
   (a) milk
   (b) eggs
   (c) honey?
2 Are there ethical arguments to support eating animal products?

Critical considerations: food miles

If we grow our own food, and carry it into the kitchen to cook and eat it, no energy has been used in transport, apart from the energy used in our bodies in cell respiration. Much of the food that is now eaten is transported by road, rail or air, with the consequent air and noise pollution, traffic congestion and road accidents. Large amounts of fuel are used to provide the energy needed to transport food, causing the release of carbon dioxide and contributing to global warming. Food transport in the UK caused the emission of 19 million tonnes of carbon dioxide in 2002, for example, and food transport now accounts for 25 per cent of heavy trucks on UK roads.

Environmentalists have been campaigning for less transport of food and more consumption of locally produced food. The concept of **food miles** was developed to promote this. Food miles are simply a measure of how far food has been transported from its site of production to the point of consumption. Some foods travel thousands of air miles, with a greater mass of aviation fuel being burned than the mass of the food itself. Consumers are being urged to reduce food miles by careful choice of foods.

A recent statement by a member of the British Government (Sustainable Food and Farming Minister Lord Bach) gave some counter-arguments.

> *A range of factors have an effect on the overall impacts of food transport, not purely the distance travelled by individual products. For example internet buying and home delivery can cut vehicle kilometres and reduce road congestion. Buying local products has the potential to greatly reduce the distance food is transported but the benefits can be offset by the way that they are transported to the consumer's home. Transport and trade in food has the potential to lead to economic and social benefits, for example through economic gains for both developed and developing countries, reduced prices for consumers and increased consumer choice.*

The BBC published an article by Gareth Edwards-Jones recently, questioning the validity of food miles as a measure of environmental damage. He included these arguments:

- support for locally grown food is a sort of ethical "trade barrier"
- energy costs of machinery, fertilizer and pesticide use and packaging, storing and cooking food should be considered
- soils emit greenhouse gases and some farming systems cause more of these emissions than others, for example paddy fields emit more greenhouse gases than grassland.

We therefore need much more knowledge about the environmental costs of different foods before we make our choices.

1 Can consumers affect the environment by the choices that they make, for example when buying food?
2 Should we consider ethical issues when we buy food, or should we leave it up to national governments to legislate to ensure environmental protection?
3 Is it right that buying locally produced food is a form of protectionism, which can harm farmers in the developing world?
4 Evaluate the claim that internet shopping and home delivery of food reduce the use of energy in food transport.
5 Referring to the precautionary principle (Chapter 16), discuss the arguments made by Gareth Edward-Jones. Should we wait until more research has been done before changing our pattern of consumption?

28 Option B: Physiology of exercise

Exercise in humans involves the use of muscles and joints. These topics are described in Chapter 25 and must be studied for Option B. Muscles need to be supplied with oxygen for aerobic cell respiration and to get rid of the carbon dioxide that this process produces. The pulmonary system therefore also plays a vital role during exercise.

Inquiry
Find definitions for these terms:
- vital capacity
- total lung capacity.

What is the difference between these two terms?

Rediscovering biology: the effect of exercise on ventilation

During exercise, the rate of gas exchange in the alveoli needs to be increased, to provide more oxygen for cell respiration, especially in the muscles. To do this, more fresh air needs to be inhaled into the lungs from outside and more stale air exhaled. This could be done by increasing the tidal volume, increasing the ventilation rate, or by a combination of the two.

1 Choose a hypothesis, involving tidal volume and/or ventilation rate, that explains how more fresh air is inhaled and exhaled per minute during exercise than at rest.

2 A volunteer is needed who is willing to have the effect of exercise on their tidal volume and breathing rate tested. There are many ways to measure these parameters, but this is a good opportunity to use data-logging.
   - A ventilation monitoring belt is fitted around the chest and it records inhalations and exhalations by the changes in pressure of the chest on the belt.
   - A spirometer, consisting of a gas pressure or flow rate sensor can be used to measure the volume of each inhalation and exhalation.

3 Recordings are made with the volunteer at rest. The volunteer then takes some exercise and the recordings are made again. Ideally a series of levels of exercise are taken to see the effects of increasing intensity of exercise.

The data in Table 1 was obtained by monitoring the tidal volume and ventilation rate of a volunteer who was running on a treadmill.

Table 1

Treadmill speed (km h^{-1})	Tidal volume (dm^3)	Ventilation rate (min^{-1})
7	1.75	30
9	1.90	31
11	2.15	33
13	2.30	38
15	2.45	47
17	2.50	58
19	2.50	75

1 Calculate the total volume of air taken into or out of the lungs per minute at each speed. [4]
2 Plot a graph of the results in Table 1. [4]
3 Compare the effects of increases in the intensity of exercise on tidal volume with the effects on ventilation rate. [4]
4 Predict the results if the intensity of exercise had been increased to higher and higher levels. [3]

ventilation rate *the number of inhalations or exhalations per minute.*

tidal volume *the volume of air taken in or out with each inhalation or exhalation.*

Data-based question: training and the pulmonary system

Special exercise programmes, called respiratory muscle endurance training, have been devised to try to increase oxygen absorption during exercise. A recent study involved 20 fit cyclists. The volume of air breathed in or out was measured continuously, while they cycled at a speed close to their maximum, for as long as they could endure. They were then divided into three groups: RMET (a group given respiratory muscle training), C (a control group given no training) and P (a group given placebo training that did not affect respiratory muscles). After a 4-week period the lung function tests were repeated. During the 4-week period, 20 training sessions were given to the RMET and C groups. Figure 1 shows the results

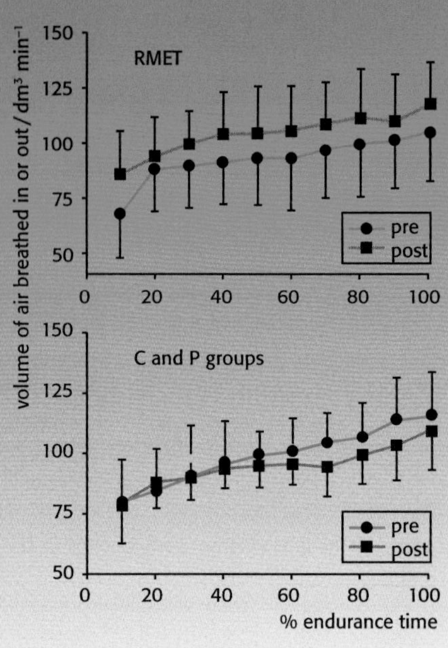

Figure 1 Results of investigation of RMET

1 There were no significant differences between the control group and the placebo group in the study. What conclusion can be drawn from this? [2]
2 Outline the trend in the amount of air breathed in or out per minute during the time of a typical endurance trial. [2]
3 The RMET group showed a significant improvement in a time trial after the training period, whereas the combined control and placebo group did not. Explain this difference using the data in the graphs. [3]
4 Training usually only increases vital capacity slightly. Discuss how training causes an increase in air breathed in or out per minute. [3]

The cardiovascular system

Oxygen absorbed in the lungs is carried to all parts of the body by the cardiovascular system. The rate at which blood is pumped varies, depending on the need for oxygen in the tissues of the body. The cardiac output is the volume of blood pumped out by the heart per minute. The greater the oxygen requirements of the body, the higher the cardiac output. Two things together determine the cardiac output: heart rate and stroke volume. Heart rate is the number of contractions per minute and stroke volume is the volume of blood pumped out with each contraction. Table 2 (on page 285) gives typical figures for different categories of person, at rest and at a maximum rate of exercise.

The mechanism for controlling cardiac output involves receptor cells in some of the main arteries, the brain and the pacemaker. The receptor cells monitor the pH of the blood. This gives a measure of the amount of dissolved carbon dioxide in the blood. When the body's overall cell respiration rate rises, for example during exercise, the carbon dioxide content of the blood rises, so the pH falls slightly. The receptor cells detect this and send impulses to the cardiovascular centre in the brain. This centre sends impulses to the pacemaker in the heart, which causes cardiac output to increase. The converse occurs when carbon dioxide levels in the blood fall below normal.

Inquiry

Venous return is the volume of blood returning to the heart via the veins per minute.

1 Why is it essential for venous return to increase during exercise?
2 Some veins are positioned between muscles that are used during exercise. How will this help with venous return?

Effect of training on heart function

The data in Table 2 shows the effect of exercise and training on the heart. Training involves sessions of rigorous physical exercise that raise the heart rate to a high level.

Table 2

Activity level	Category of person	Heart rate (min^{-1})	Stroke volume (ml)	Cardiac output ($ml\ min^{-1}$)
at rest	male before training	72	70	5040
	female before training	75	60	4500
	male after training	50	100	5000
	female after training	55	80	4400
during maximum exercise	male before training	200	110	22,000
	female before training	200	90	18,000
	male after training	190	180	34,200
	female after training	190	125	23,750

1 Using the data in Table 2, deduce the effect of exercise on:
(a) heart rate [2]
(b) stroke volume [2]
(c) cardiac output. [2]

2 Using the data in Table 2 deduce the effect of training on:
(a) heart rate at rest [1]
(b) heart rate during maximum exercise [1]
(c) stroke volume at rest [1]
(d) stroke volume during maximum exercise [1]
(e) cardiac output at rest [1]
(f) cardiac output during maximum exercise. [1]

3 Explain the effects of training on heart rate and stroke volume, in terms of changes to the structure of the heart. [4]

4 Calculate the percentage change in cardiac output caused by training, for males and for females, at rest and during maximum exercise. [4]

Data-based question: exercise and blood distribution

The distribution of blood to the systems of the body is controlled using the arterioles. These blood vessels can widen or narrow by relaxation or contraction of muscle fibres in their walls. These two movements are called vasodilation and vasoconstriction. The volume of blood in the body is less than would be needed to supply all systems simultaneously with the maximum amount, so there must always be some systems that receive less than their maximum. Figure 2 shows changes that accompany increasing levels of exercise.

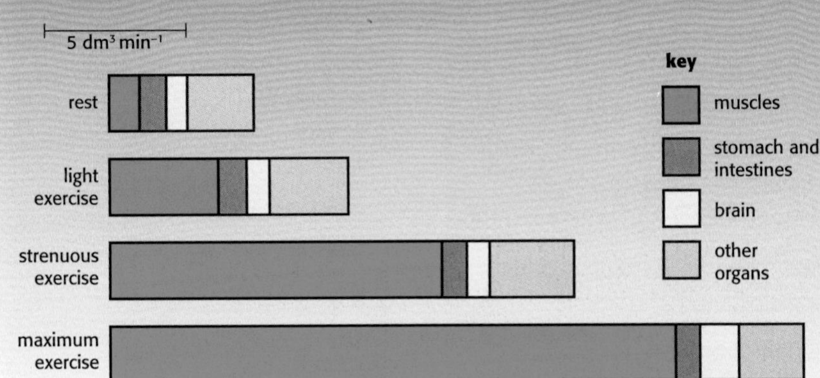

Figure 2 Effect of exercise on distribution of blood to organs

1 Explain the relationship between the total volume of blood distributed to the tissues, and intensity of exercise. [4]

2 Explain the changes in blood flow to the stomach and intestines during exercise. [3]

3 Comment on blood flow to the brain, using the data in Figure 2. [3]

4 The other organs that are grouped in the data include the skin, kidneys and heart wall. Suggest, with reasons, whether blood flow to each of these will increase, decrease or remain the same during exercise. [5]

Energy and exercise

Muscles need ATP for contraction. Aerobic cell respiration can only happen if oxygen and glucose or fat are present. Glucose can be stored in muscles in the form of glycogen, which can rapidly be broken down when it is needed. Small amounts of oxygen can be stored in some muscles, but these stores are soon depleted and aerobic cell respiration can then only happen as quickly as oxygen is supplied by the pulmonary and cardiovascular systems. The volume of oxygen that is absorbed by the body per minute and supplied to the tissues is called VO_2. As the intensity of exercise rises, VO_2 rises until it reaches a maximum. This is called $VO_{2\,max}$. With low-intensity exercise, at low VO_2 values, fat tends to be used in aerobic cell respiration. As the intensity of exercise rises and VO_2 moves towards its maximum, use of fat in cell respiration falls and use of glucose rises.

The second method of ATP production is anaerobic cell respiration. Only glucose can be used as a substrate, and lactate is produced as a toxic waste product. Lactate accumulates in muscles and blood. Only limited amounts can be tolerated, so anaerobic cell respiration can only be used for short periods of intense exercise. Afterwards the lactate must be converted into other substances. The lactate is first converted back into pyruvate, which requires oxygen. The pyruvate can then either be converted into glucose or it can be used in aerobic cell respiration. The removal of lactate therefore always involves oxygen use and so accumulation of lactate incurs an oxygen debt. To help repay it we have to continue deep ventilation after vigorous exercise.

The third method of ATP production involves this reaction:

creatine phosphate + ADP ⟶ creatine + ATP

This reaction is anaerobic. Muscle fibres only contain enough creatine phosphate for ATP to be produced for 8–10 seconds during vigorous exercise. During the period of rest afterwards ATP has to be used to remake the creatine phosphate.

Data-based question: finding the lactate threshold

Blood lactate concentrations can be measured easily using specially designed test strips and meters. This involves taking blood samples, and local regulations should be checked before it is attempted in schools.

1 A volunteer undergoes a lactate test and then does a 2-minute warm-up.

2 The blood lactate test is repeated and then a period of exercise begins at a measurable level, for example running slowly on a treadmill.

3 The blood lactate is tested again and the rate of exercise is increased. This is repeated again and again until the volunteer is exercising at their maximum rate.

Typical results are shown in Figure 3, with intensity of exercise in terms of the athlete's $VO_{2\,max}$. Discuss where the lactate threshold is, and why it is at this level of $VO_{2\,max}$.

Inquiry

Some muscles contain myoglobin, a protein that is similar in structure to hemoglobin. Another similarity is that myoglobin also binds oxygen, but a difference is that myoglobin remains inside muscle cells, rather than being carried around in the blood.

1 What is the main role of myoglobin?
2 Is there more myoglobin in muscles that specialize in aerobic or anaerobic cell respiration?
3 Why is it particularly useful to have oxygenated myoglobin in a muscle when that muscle is suddenly needed to carry out vigorous contractions?

lactate threshold *the point during increasingly intensive exercise at which blood lactate begins to accumulate above resting levels.*

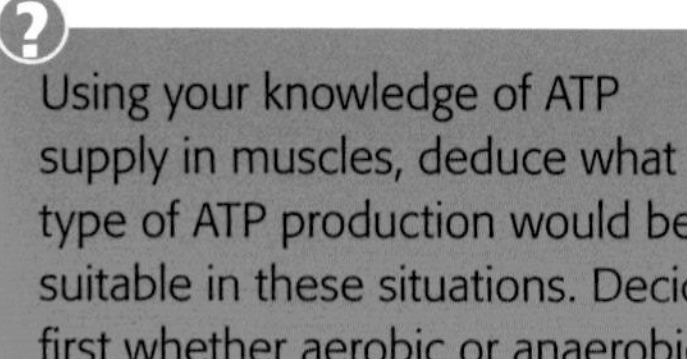

Using your knowledge of ATP supply in muscles, deduce what type of ATP production would be suitable in these situations. Decide first whether aerobic or anaerobic methods will be used and, if the method is anaerobic, whether creatine phosphate can provide some or all of the ATP required.

1 Walking for 10 km.
2 Sprinting for 10 s.
3 Running a 10 km race.
4 Running a 400 m race.
5 Weightlifting competitions.
6 Lifting weights many times.
7 Gymnastics.
8 Basketball.
9 Cross-country skiing.
10 Sprint finish to a marathon.

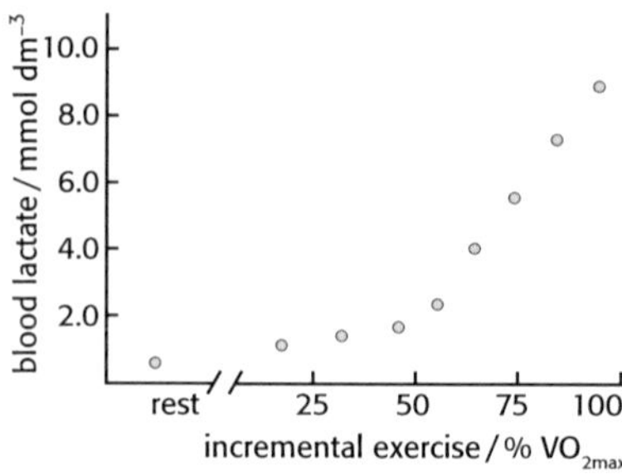

Figure 3 Lactate concentrations during incremental exercise

Muscle types and fitness

Fitness is the condition of a person's body that makes it competent to perform the physical activities that the person wants to perform. Fitness is therefore specific to particular activities – fitness for sprinting is not fitness for marathon running, for example. Training can improve fitness, especially by improving pulmonary and cardiovascular function, but there may also be some inherited elements to fitness, in particular the distribution of fast and slow muscle fibres.

Fast muscle fibres are sometimes called Type IIb fibres. They have few mitochondria, are rich in enzymes of glycolysis, have a moderate blood supply, and little myoglobin. They have a high anaerobic capacity, high strength and low stamina. Slow muscle fibres are also called Type I fibres. They have many mitochondria and so are rich in oxidative enzymes, they have an excellent blood supply with many blood capillaries and large stores of myoglobin. They have a high aerobic capacity, moderate strength and high stamina. Fast muscles fibres contract more rapidly and exert more force per unit of cross-sectional area than slow muscle fibres. Muscles vary in the proportions of the different types of fibre, both within a person's body and between people.

Inquiry

Fitness can be measured in several different ways. Two measures that are often used are strength and stamina.

1. Distinguish between strength and stamina.
2. Suggest a test of strength that could be used as a measure of fitness.
3. Suggest a test of stamina that could be used as a measure of fitness.
4. Discuss whether strength or stamina is a better measure of fitness.

Data-based question: fast and slow muscle fibres

Figure 4 shows the proportions of fast and slow fibres in the muscles of four people.

1 (a) Explain how the proportions of fast and slow muscle fibres make each of the four people fit for the activities that they perform. [4]

(b) Discuss whether the data shows each person chose activities to which they were fitted or became fit for an activity by training in it. [2]

2 Figure 5 shows the relationship between velocity of movement and power output of muscles containing mostly fast and mostly slow muscle fibres.

(a) Outline the relationship between velocity and power for muscles with 80% fast fibres. [2]

(b) Identify which muscle type generates more power at each velocity. [1]

(c) Deduce which muscle type can contract with greater velocity, when they are generating equal force. (Power = force × velocity). [1]

(d) Discuss the advantages of muscles with a high proportion of slow muscle fibres. [3]

Experiments have been done with rats to test whether the proportion of fast and slow muscle fibres in the digitus longorum muscle changes in response to training. Table 3 shows the percentage of slow (Type I) and fast (Type IIb) fibres before training and after 30, 60 or 90 minutes per day of endurance exercise training over 10 weeks.

3 Discuss the conclusions that can be drawn from the results in Table 3. [3]

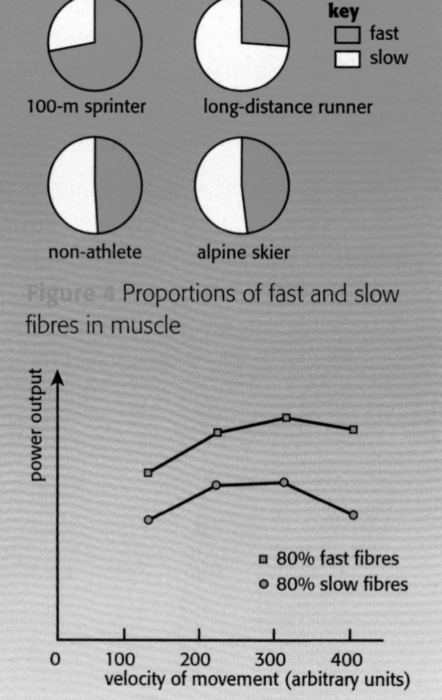

Figure 4 Proportions of fast and slow fibres in muscle

Figure 5 Velocity–power graph

Table 3

	No training (control)	Endurance training per day (minutes)		
		30	60	90
Slow fibres	3.6	4.7	5.5	8.6
Fast fibres	35.5	30.6	25.6	18.8

Injuries in sport

During training, and also in competitive events, injuries can occur, ranging from minor conditions that heal quickly to permanent and disabling damage. Muscles and joints are both vulnerable to injury. Five types of injury are described here:

- **Sprains** are injuries to joints, caused by excessive joint movement, that result in pain and disability (see Figure 6). The word comes from the French *espraindre*. Sprains involve stretching and damage to the ligaments of the joint, with pain and swelling. Ankle, finger, toe and wrist are all vulnerable to sprain injuries.
- **Torn ligaments** are caused by severe sprains. Often a fall or other accident causes excessive joint movement with sufficient force to tear ligaments. Surgery is sometimes required to brace the joint and allow the ligament to heal. A common example is tearing the ligament on the outer side of the ankle, by twisting the foot inwards.
- **Joint dislocation** is movement of the bones at a joint beyond their normal range. The ligaments that restrict movement at the joint are torn in the process, so the bones may therefore either remain dislocated or easily become dislocated again. Shoulders, fingers, knees and toes can all be dislocated, usually by a sudden accident.
- **Torn muscles** are due to excessive muscle exertion. The injury happens suddenly, with a sharp pain and sometimes also a snapping sensation. Bruising and inflammation develop later. Complete muscle tears are very rare – usually only some fibres are torn. The calf, quadriceps and hamstrings are commonly affected.
- **Intervertebral disc damage** is commonest in the lower back, caused by bending forwards while standing, at the same time as twisting sideways. The resulting pressure causes the soft pulpy core of an intervertebral disc to bulge out through the side wall. The bulge often presses on a spinal nerve, with consequent pain or numbness in the area served by the nerve.

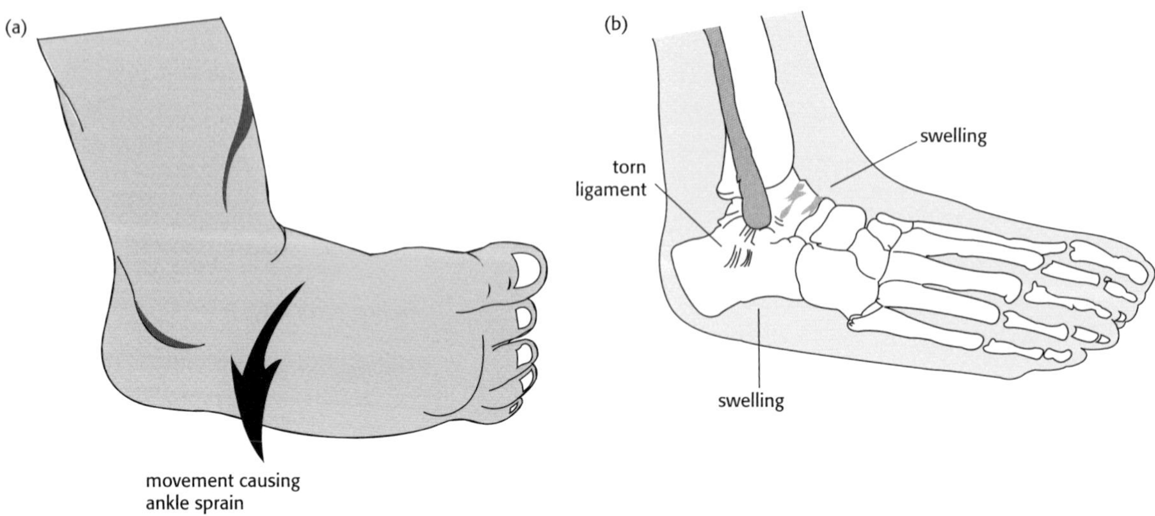

Figure 6 Sprained ankle

Thinking about science: benefits of warm-ups

Todd Jesdale is a highly successful college rowing coach. He was interviewed recently about his views on warm-ups.

What warm-up routine do you use with your rowing crews?
I don't believe in warming up with runs or other exercise. We warm up by getting in the boat and rowing. We do a series of exercises, breaking down the stroke into components and practising each in turn. We do this for about 20 minutes, aiming to get the heart rate up to about 140. Then we do four minutes at half pressure to prepare the whole body. On a race day we do some sixes – six of the crew row while the other two sit, to increase the resistance, then half pressure for four minutes and then some hard strokes to tire the body – it lets the body know what it's going to be doing.

What is the aim of warming up?
I believe that body and muscle temperature rise, but I don't know these things. If one is inadequately warmed one might get into all sorts of difficulties, including the tightening of muscle groups. Performance may be sub-par. We use this procedure not to prevent injuries, but we believe there could be increased injuries. Injuries such as tendonitis appear due to overuse rather than inadequate warm-up.

Why do you use these particular protocols for warming up?
It's a system that I've been living with for over 30 years – it has evolved and seems to work in terms of not accumulating injuries. I steal ideas! Some of the ideas were stolen from East German coaches. Those coaches were very successful.

Do you have evidence to support your methods?
The evidence is based on observation – we have raced successfully. But I easily beg the question, that is, would another system work – the answer is almost certainly yes. Sometimes because of circumstances we can't use the system without interruption, and we still do well.

?

1 Explain how Todd Jesdale's warm-up techniques allow his crews to perform as well as they can – what changes are happening in the rowers' bodies?
2 Is the evidence for the effectiveness of the warm-up routine scientific or anecdotal? Anecdotal evidence is based on observation of particular cases.
3 Todd recognizes that other techniques might also be effective. Why would Todd probably be reluctant to participate in a controlled trial to see whether his techniques really do enhance performance and prevent injury?
4 Todd uses the word "evolved" to refer to his training system. To what extent is evolution an appropriate metaphor for changes to a training system?
5 Todd's techniques have changed over the years by a process that he calls evolution. How does this happen?

Performance-enhancing substances

Many different substances have been used to try to enhance performance in sport. Some of them are effective, legal and safe. Others are ineffective or have been banned, because of the risks involved. We will investigate three examples here.

Creatine phosphate is taken as a dietary supplement, usually as creatine monohydrate.

EPO is erythropoietin, a hormone that stimulates red blood cell production in bone marrow.

Anabolic steroids are a group of chemicals related to testosterone that cause tissue growth, especially in muscles.

Inquiry

Creatine phosphate

1 Why is it reasonable to expect eating supplements of creatine phosphate to enhance performance in short-duration and high-intensity activities such as sprint cycling?
2 Is there any scientific evidence that eating creatine phosphate supplements causes:
   (a) concentrations of creatine phosphate in muscle to increase?
   (b) performance in short-duration and high-intensity activities to be enhanced?
3 Some studies have shown that creatine phosphate supplements cause weight gain by water retention. How would this affect performance?

Critical considerations: ethics of using performance-enhancing substances

How should we decide whether it is ethical to use anabolic steroids, or any other performance-enhancing substance? A utilitarian approach would be to decide on the basis of the consequences. If are they harmful, or if the harmful consequences outweigh the good ones, then a substance should be banned. In practice, it can be very hard to decide if consequences are good or not. It can also be very hard to weigh up harmful and good consequences to come to an overall decision, especially if the harmful consequences affect a minority of people and the good consequences affect the majority.

An alternative approach is to look for the inherent wrongness or rightness of an action, ignoring whether the consequences are good or bad. This approach implies that there are moral absolutes that can be defined.

These arguments have been used to support an end to all bans:

A It is impossible to eliminate use of illegal substances by testing.
B Bans encourage and reward criminal activity.
C Covert use of drugs makes their safe use less likely.
D Natural health problems pose a greater risk, e.g. asthma.
E Athletes should have health tests, not drug tests.
F The current system rewards cheats who use banned substances.
G Use of a substance should be a personal choice for the athlete.
H Variation in athletes' genes causes unfair variation in ability.
I Levels of testosterone and other endogenous substances vary.
J Legal use of drugs might equal out ability and make sport fairer.
K Drugs are cheap and do not discriminate according to wealth.
L Enhanced performance makes sport more enjoyable to watch.

1 Which of these arguments do you accept?
2 What are the counter-arguments?
3 Which approach to making a decision about the ethics of using a performance-enhancing substance do you favour:
   (a) either deciding whether using the substance is inherently wrong
   (b) or weighing up the harmful and beneficial consequences.
4 Use the example of anabolic steroids and discuss whether it is ethically acceptable for:
   (a) an athlete to take anabolic steroids to raise performance
   (b) a coach to encourage an athlete to use anabolic steroids
   (c) a drug company to manufacture and sell anabolic steroids for use in sport.

Inquiry
EPO

1 What effect does EPO have on PCV – the packed cell volume of red blood cells?
2 What effect will an increase in PCV have on rates of oxygen transport in blood, and on performance in endurance events?
3 How can blood transfusions be used to increase PCV? Whose blood is transfused into the athlete and why does it increase PCV?
4 What effect will an increase in PCV have on the risk of blood clot formation and on the risks of strokes and heart attacks?
5 In the early 1990s, several Dutch cyclists died in their sleep from unexplained cardiac arrest. What is the most likely cause?

Inquiry
Anabolic steroids

1 Find out about the side-effects of anabolic steroids. Can these health problems be caused by them?
   (a) Abnormal menstrual periods.
   (b) Shrunken testes.
   (c) Low sperm counts.
   (d) Liver disease.
   (e) Emotional problems – "road rage".
   (f) Torn muscles and tendons.
2 In which sports are the problems of anabolic steroid abuse greatest, and why is the problem worse in some sports than others?
3 The East German swimming team were outstandingly successful in the 1976 Olympic Games, winning 11 out of 13 team events. Why did they later sue the East German government?

29 Option D: Evolution

The origin of life

One of the tenets of the cell theory is that all cells come from pre-existing cells. Tracing back the ancestry of cells over billions of years, we must eventually reach the first cells to have existed. These were the first living things on Earth. For them to have arisen from non-living material, four requirements would have had to be met:

- simple organic molecules would need to be available
- the simple molecules would need to have been assembled into polymers
- these abiotically produced molecules would need to have been packaged into structures with a different internal chemistry from their surroundings
- a mechanism that makes inheritance possible would need to have developed.

A necessary precursor for life would be the **availability of simple organic molecules**. The experiments of Stanley Miller and Harold Urey illustrated that simple organic molecules such as amino acids can be produced in abiotic conditions. This was achieved experimentally by passing steam through a mixture of methane, hydrogen and ammonia. At the time of their experiment, this mixture was thought to be representative of the atmosphere of the early Earth. Electrical discharges were used to simulate lightning.

Another possible site for the origin of the first simple organic molecules is around deep-sea vents. These are cracks in the Earth's surface, characterized by gushing hot water carrying reduced inorganic chemicals such as iron sulfide. These chemicals represent readily accessible supplies of energy and raw materials for the creation of monomer organic chemicals. They are also a source of energy for the **assembly of these monomers into polymers**.

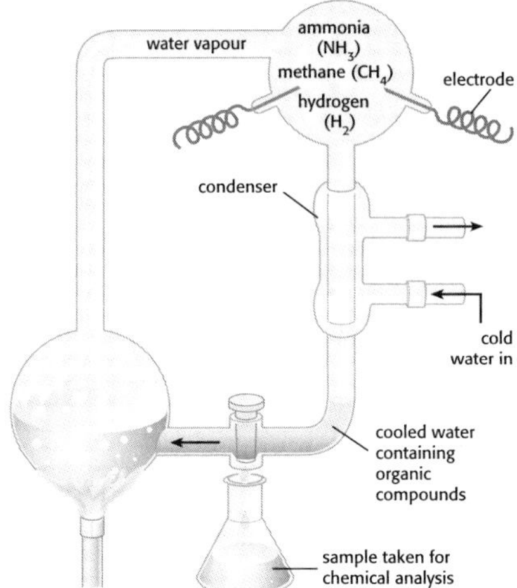

Figure 1 Miller and Urey's apparatus

Data-based question: the Murchison meteorite

The theory of panspermia states that some of the first organic molecules may have been delivered to Earth by meteorites. Certain categories of meteorites are rich in the element carbon and some of these meteorites have been found to contain many of the same molecules observed in the Miller–Urey experiment.

A meteorite that struck near Murchison, Australia, was analysed for the presence of amino acids. Similar studies have been carried out on liquid samples taken from the Miller–Urey experiment. Table 1 summarizes the results from both sources. Each dot represents a relative amount of a particular amino acid, with one dot representing a small amount and eight dots representing a large amount.

Table 1

Amino acid	Murchison meteorite	Miller–Urey experiment
Glycine	••••••••	••••••••
Alanine	••••••••	••••••••
α-amino-N-butyric acid	••••••	••••••••
α-aminoiso-butyric acid	••••••••	••••
Valine	••••••	••••
Norvaline	••••••	••••••
Isovaline	••••	••••
Proline	••••••	••
Pipecolic acid	••	•
Aspartic acid	••••••	••••••
N-ethylglycine	••••	••••••
Sarcosine	••••	••••••

Source: C. Mitchell, *Life in the Universe* (1995)

1 Compare the amino acids found in the meteorite with those produced in the Miller–Urey experiment. Refer to named examples. [3]

2 Suggest a conclusion based on your comparison. [1]

Protobionts and the role of RNA

The packaging of abiotically produced molecules into structures with a different internal chemistry from their surroundings would have been a further necessary stage in the origin of cells. Such structures are termed protobionts. They exhibit some the properties associated with life, including response to the environment, simple reproduction and basic metabolism.

A fourth requirement for the origin of cells would be a mechanism that makes inheritance possible. In modern biological systems, replication and catalysis require enzymes. However, for enzymes to be created, genetic material is necessary. RNA has two properties that make it the most likely original molecule of inheritance: it is self-replicating and it has catalytic properties. Ribozymes are enzyme-like RNA molecules. RNA also has a variable sequence, which means that it can store information.

Inquiry
Protobionts can be modelled by adding a large droplet of vegetable oil to a beaker of water. Following the addition of the oil, add a drop of food colouring. Note how the water-soluble food colouring is excluded from inside the oil droplet.

Oxygen

Early cells would have increasingly competed for energy sources and this probably provided the selection pressure necessary for the evolution of photosynthesis. Initially the oxygen produced would have dissolved in the oceans or would have been used up in chemical reactions. Eventually, it began to be released into the atmosphere. While these early prokaryotes contributed to significant rises in atmospheric oxygen, it is thought that the emergence of eukaryotes with chloroplasts was responsible for further significant increases in atmospheric oxygen. It is likely that rising quantities of oxygen led to the extinction of certain groups of prokaryotes, while other groups were able to adapt by developing mechanisms to deal with the oxygen. Cellular respiration is thought to be one of these mechanisms.

"Life on earth consists of both chemistry and information"

J.L Bada

Thinking about science: the theory of endosymbiosis

Lynn Margulis is well known for her endosymbiosis theory. This theory advances that chloroplasts and mitochondria were once free-living organisms that gave up their independence to live inside other cells. Mitochondria and chloroplasts have their own circular DNA and their own ribosomes (which resemble prokaryotic ribosomes). Their size, their shape and their double membranes are typical of some groups of bacteria. Asked "what she believes but cannot prove", Margulis had this response:

I think the common ancestor of the cilium...was a free-swimming entity... This bacterium, who still has many live relatives, entered into symbiotic partnerships with other very different kinds of bacteria. Together this two-component partnership swam and stuck together and persisted. What kind of bacterium became an attached symbiont that impelled its partner forward? None other than a squirming spirochete bacterium.

The spirochete group of bacteria includes many harmless mud-dwellers but it also contains a few scary freaks: the treponeme of syphilis and the borrelias of Lyme disease. We animals got our exquisite ability to sense our surroundings to tell light from dark, noise from silence, motion from stillness and fresh water from brackish brine – from a kind of bacterium whose relatives we despise. Even though the concept that cilia evolved from spirochetes has not been proved, I think it is true. Not only is it true but, given the powerful new techniques of molecular biology, I think the hypothesis will be conclusively proved.

1 What is the distinction between belief and knowledge?
2 To what extent is Margulis' belief falsifiable? In other words, to what extent is it a scientific claim?

Polymorphism

The gene pool is all of the genes in a population at a certain time.

Alleles are varieties of a gene. If more than one allele is common in a population, the gene is said to be **polymorphic**. If two alleles each confer some survival advantage on the organism, then it is likely that these alleles will persist in the population. This is known as balanced polymorphism.

Sickle-cell hemoglobin is a form of hemoglobin in which the conformation (shape) is altered in low-oxygen conditions so that the function is impaired. The shape of red blood cells containing the hemoglobin is also altered, interfering with circulation. People who are homozygous for sickle hemoglobin (Hb^sHb^s) suffer from sickle-cell disease. This is a serious illness that can be fatal. People who are heterozygous (Hb^AHb^s) are largely asymptomatic except when oxygen levels are low.

However, the presence of the Hb^s allele confers a degree of resistance to malaria. This means that in regions where malaria is common, a person who is heterozygous for sickle hemoglobin has an advantage over someone who is homozygous for normal hemoglobin (Hb^AHb^A). Even though some of their children may die from sickle-cell disease, they have more chance of surviving malaria. Balanced polymorphism is therefore observed in malaria-stricken areas, as shown in Figure 2.

A transient polymorphism is not stable, because one of the alleles is less advantageous. This allele is gradually removed from the population by selection pressure. The Hb^s allele would be transient in an area where malaria was not endemic.

Evolution could be defined simply as a change in a species over time. In the cross depicted in Figure 3, in the first generation, 50% of the population is red and 50% is white. In the second generation, 100% of the flowers are pink. In the third generation, there is 50% pink, 25% white and 25% red. It might appear that the population is changing, but if you look at the allele frequency in each generation, it is always 50% C^R and 50% C^W. A better definition of evolution then is a change in allele frequency over time.

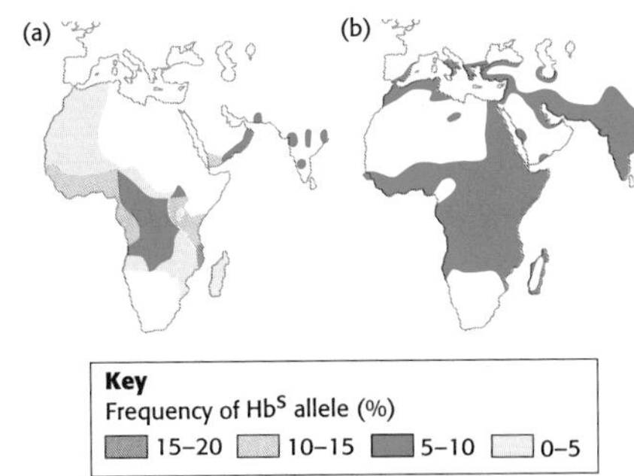

Figure 2 The frequency of the Hb^s allele (Figure 2a) is correlated with malaria-affected areas in Africa (Figure 2b).

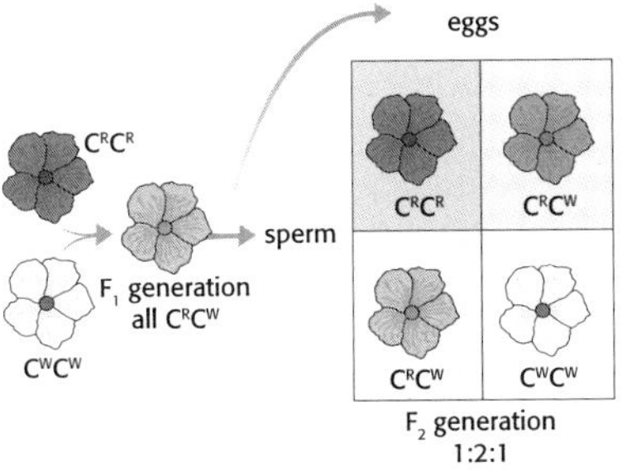

Figure 3 A change in phenotypic frequency between generations does not necessarily indicate that evolution is occurring.

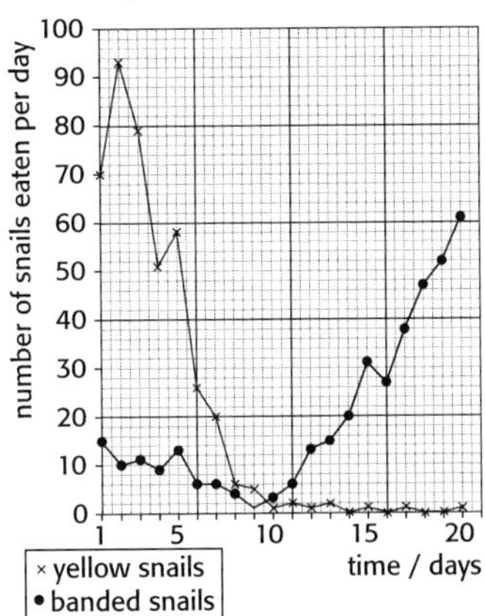

Figure 4 A comparison of the number of each type of snail eaten after release

Data-based question: selection pressures on land snails

To investigate the selection pressures acting on the land snail (*Cepaea nemoralis*), 500 of each of two forms of this snail were released in a small forest. One form had a yellow shell, although this appeared green with the animal living inside. The other form had a light brown shell with darker brown bands round the shell. The snails are predated by a bird which hits the snails against a stone to break the shell. The body of the snail inside can then be eaten. By counting the number of broken shells at the stone, the number of snails eaten each day after release was estimated (Figure 4). During the period of the experiment, the background colour of the snail's habitat changed from different shades of brown to a single continuous green colour.

1 Outline how the number of each form eaten varies during the experiment. [3]
2 Explain two reasons for the variation in the number of brown banded snails eaten during the experiment. [2]
3 Predict, with a reason, whether the two forms of snail interbreed. [1]
4 Suggest why the two forms normally show balanced polymorphism. [2]

HL

Hardy–Weinberg equilibrium

Either a population is evolving or it is in genetic equilibrium, sometimes called Hardy–Weinberg equilibrium. The conditions necessary for genetic equilibrium are: no mutation, no migration, no natural selection, random mating and maintenance of a population of sufficient size.

If a population is in genetic equilibrium, predictions can be made about genotypic, phenotypic and allele frequencies for genes that cause discrete variation.

The frequency of the dominant allele is usually represented by p. The frequency of the recessive allele is usually represented by q.

The probability of any individual having the homozygous recessive allele is $(q) \times (q)$.

The probability of having the homozygous dominant genotype is $(p) \times (p)$.

The probability of having the heterozygous genotype is $2 \times (p) \times (q)$. This is because the dominant allele can come from either parent and the recessive allele can come from either parent.

Note that $p + q = 1$ and $p^2 + 2pq + q^2 = 1$.

?

1 Red blood cells can possess a surface antigen called the Rh factor. Rh^+ (presence of the antigen) is the dominant allele and Rh^- (absence of the antigen) is the recessive allele. In a sample of 400 people drawn from the Basque region, 170 were Rh^+ and 230 were Rh^-.
   (a) Determine the allele frequencies.
   (b) Determine the probability of being homozygous dominant.

2 In a randomly breeding population of mice, 640 had black fur and 360 had brown fur. Black fur is dominant to brown fur.
   (a) Calculate the frequency of the recessive allele. [1]
   (b) Calculate the number of homozygous black mice in the sample. [2]

3 Calculate the percentage of carriers in a population where the occurrence of the condition cystic fibrosis is 1 in 2500 individuals. [2]

4 Absence of the palmaris longus muscle in both arms is recessive (see page 253). Determine the frequency of this allele in a sample of students from your school.

5 (a) Identify three morphological differences between the two species of Treecreeper shown in Figure 5. [3]
   (b) Discuss what makes these two different species. [2]

Species and species formation

The most commonly accepted definition of species is known as the biological species concept, which defines a species as a group of potentially interbreeding populations, with a common gene pool that is reproductively isolated from other populations. The biological species concept is difficult to apply in some cases because of the following problems.

- Some species only reproduce asexually.
- Some species occasionally mate to produce hybrids.
- Some species cannot interbreed but they are so close in appearance that they are difficult to tell apart. These are known as sibling species.

Speciation is the formation of a new species by the splitting of an existing species. Various barriers can isolate the gene pool of one population from that of another population. Speciation may occur when this happens. If the isolation occurs because of geographic separation of species, then the speciation is termed **allopatric speciation**.

The cichlids (fish) are one of the largest families of vertebrates. Most cichlids occur in three East African lakes, Lake Victoria, Lake Tanganyika and Lake Malawi. Fluctuations in water levels lead to isolation of populations that are then subject to different selection pressures. This results in the formation of new species (see Figure 8 on page 297).

Sometimes isolation occurs within the same geographic area. The short-toed treecreeper *Certhia brachydactyla* is closely related to the common treecreeper, *C. familiaris* (see Figure 5). The habitats of these two species can overlap, but they have different behavioural patterns and live in different niches within the habitat. The result is that hybridization between the two is rare. If isolation occurs within the same habitat then the speciation is termed sympatric speciation. For example, sexual selection occurs when closely related individuals differ in their courtship behaviour and are only successful in attracting members of their own population.

Figure 5 *C. brachydactyla* and *C. familiaris* are sibling species.

Data-based question: lacewing songs

Songs are part of the process of mate selection in members of different species within the genus *Chrysoperla* (lacewings). Males and females of the same species have precisely the same 'song' and during the pre-mating period take turns making the songs. The oscillograph for two species of lacewings are shown in Figure 6.

1 Compare the songs of the two species of lacewings. [3]
2 Explain why differences in mating songs might lead to speciation. [3]
3 The ranges of the two species currently overlap. Suggest how differences in song could have developed:
(a) by allopatric speciation
(b) by sympatric speciation. [4]

(a)

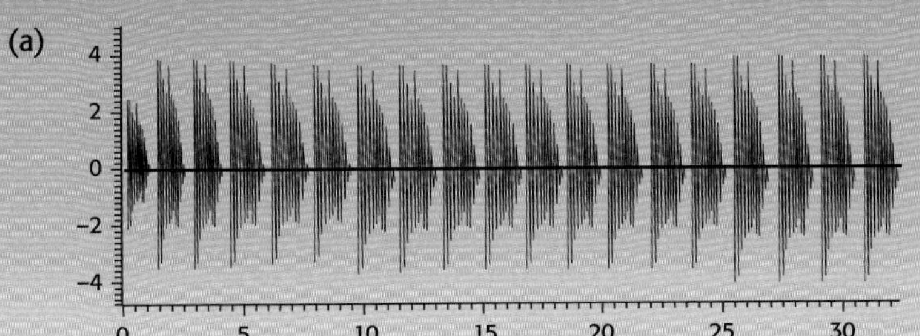

(b)

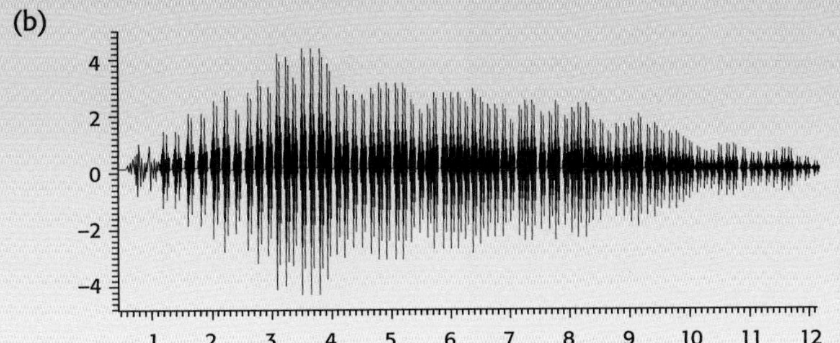

Figure 6 Pre-mating songs of lacewings: (a) *C. lucasina* and (b) *C. mediterranea*. *C. lucasina* ranges across most of Europe and eastward into western Asia, as well as across the northern quarter of Africa. *C. mediterranea* ranges across southern to central Europe and across the north African Mediterranean.

Polyploidy

Polyploidy is a method of sympatric speciation. A polyploid organism is one that has more than two sets of homologous chromosomes. This occurs most commonly in plants though it does occur in less complex animals. There is even a tetraploid mammal, the red viscacha rat (*Tympanoctomys barrerae*), a rodent from Argentina.

Investigating biology: cell size in polyploid organisms

Devise a method to determine whether mean cell size in polyploidy plants is different than mean cell size in diploid plants such as in different varieties of marigold. A graticule could be used on prepared slides of polyploid and related non-polyploid species. Alternatively a digital image can be taken of a prepared slide and a picture analysis tool in software such as logger pro can be used to measure the size of a range of cells.

Divergent evolution and adaptive radiation

Speciation involves a process of divergence where initially similar organisms become different. When a population invades a new territory, the members can spread out and as a result of being exposed to different selection pressures, diversify into several species occupying different ecological niches. This process is termed adaptive radiation. For example, an ancestral marsupial similar to the modern oppossum gave rise to a huge range of marsupials, including the koala, the wombat, the kangaroo and extinct forms such as *Diprotodon* (a marsupial resembling a modern rhinoceros) and *Thylacoleo* (the marsupial "lion", which is reputed to have had the strongest bite force of any mammal that has ever existed).

Beetles are the most numerous and diverse order of organisms. The phylogenetic tree in Figure 7 shows a partial evolutionary history of adaptive radiation events in beetles. Ancestral beetles fed on non-angiosperm plants. At several points in evolutionary history, branches of beetles made the switch to feeding on angiosperms and each time, rapid speciation and diversification resulted.

Data-based question: convergent evolution

Species with very different evolution histories can evolve similar traits when they are subject to the same selection pressures. This is termed convergent evolution. For example, species of cactus in America and *Euphorbia* in Africa are strikingly similar.

Figure 8 on page 297 shows two columns of cichlids. The column of fish on the left is from Lake Tanganyika and the fish in the right column are from Lake Malawi. The genus of the fish is listed below each picture. The phylogenetic tree below the fish shows the evolutionary relationships between the fish, based on differences in mitochondrial sequences.

1 State one example of convergent evolution and one example of divergent evolution referring to the fish. [2]
2 The *Rhamphochromis* shown in the picture hunts for fish. Deduce, with a reason, which other species hunts for fish. [1]
3 Suggest which lake the ancestor of all of the fish shown lived in. [1]
4 The nuchal hump above the eye on the *Cyphotilapia* likely serves the same function as the horizontal strips on the *Julidochromis* shown. Suggest what that function might be. [1]

Data-based question: adaptive radiation of beetles

(a) percentages of described species

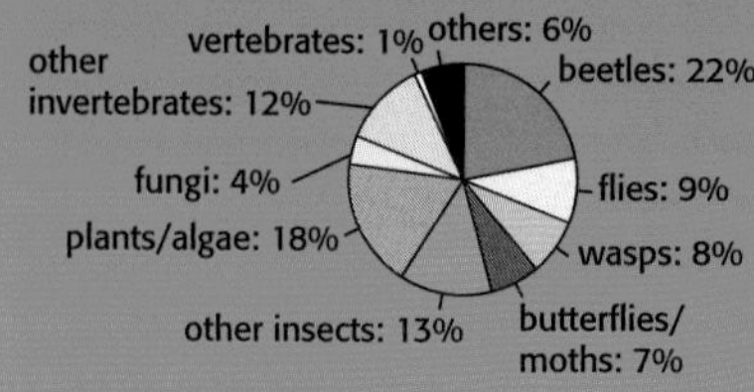

(b) number of species

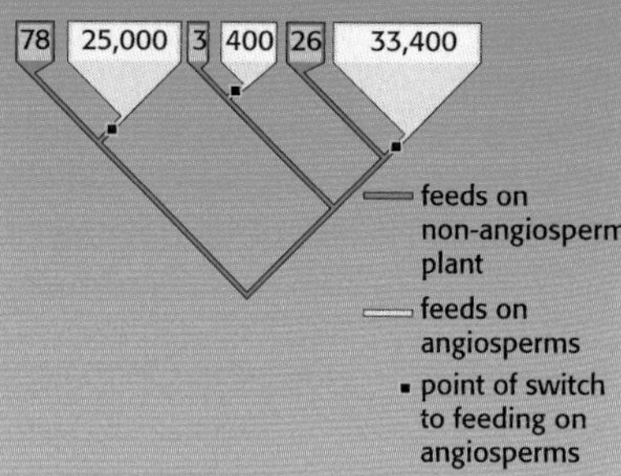

Figure 7 Figure 7a shows the percentage of described species belonging to different groups. Figure 7b at several points in evolutionary history diversification within angiosperms has lead to a diversification of beetles.

1 Using the pie chart in Figure 7a, determine the fraction of all described species that are insects.
2 What could J.B.S Haldane have meant when he said that God had "an inordinate fondness for beetles"?
3 Suggest why a swich to feedling on angiosperms could lead to such rapid speciation and diversification.

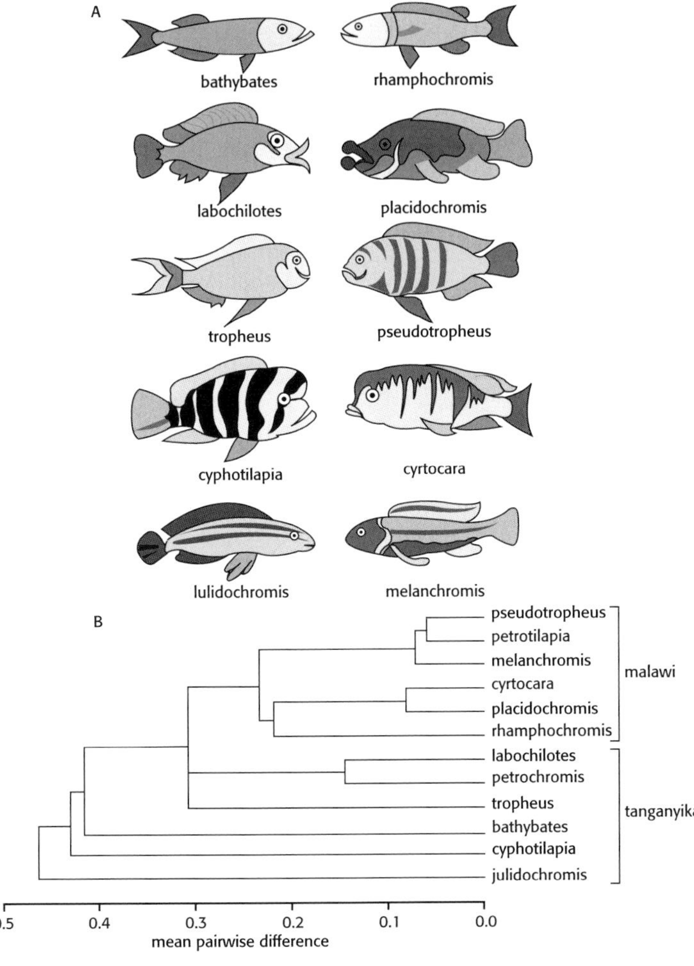

Figure 8 Convergent evolution in cichlids from Lake Tanganyika and Lake Malawi

Punctuated equilibrium and gradualism

There are two theories about the pace of evolutionary change. Gradualism, as depicted in Figure 9, is the idea that species slowly change through a series of intermediate forms. The axis label 'structure' might refer to such things as beak length in birds or cranial capacity in hominids. Punctuated equilibrium holds that long periods of relative stability in a species are 'punctuated' by periods of rapid evolution.

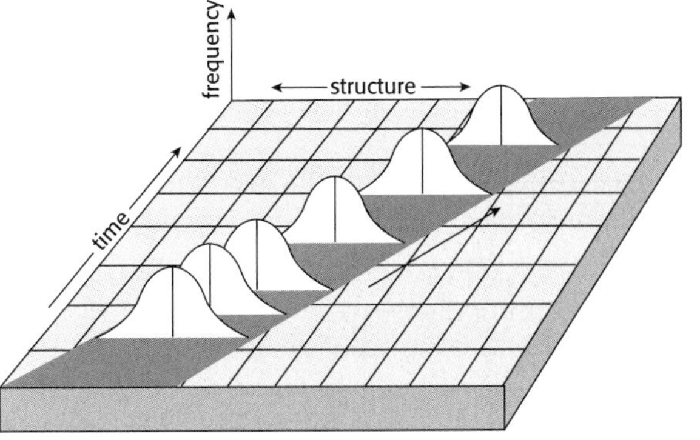

Figure 9 In the gradualist framework, new species emerge from a long sequence of intermediate forms.

Gradualism was, for a long time, the dominant framework in palaeontology. However, it was confronted by gaps in the fossil record, i.e. an absence of intermediate forms. Gradualism predicted that for the origin of a new species to occur, a long sequence of continuous intermediate forms would be needed. The absence of these intermediate forms was explained as imperfections in the fossil record. According to the theory of punctuated equilibrium, gaps in the fossil record might not be gaps at all, as there was no long sequence of intermediate forms. Events such as geographic isolation (allopatric speciation) and the opening of new niches within a shared geographic range can lead to rapid speciation.

?

How might Figure 9 be re-drawn to represent punctuated equilibrium?

HL

Phylogeny and systematics

Taxonomy is the branch of biology involved with the classification of organisms. It is useful because it allows for effective communication between scientists. Popular terms for organisms are often culturally based and can lead to confusion. The name daddy long-legs can refer to a spider (*Pholcus* sp.) in Australia, the eight-legged harvestman "spider" (not a spider at all) in Canada and the crane fly (Tipulidae) in the UK. There is (usually) much greater consensus around scientific names than folk taxonomies.

Traditionally taxonomy was done by using morphology, i.e. by examining form and structure. One problem with this approach is that convergent evolution can lead organisms to have similar morphology, suggesting a closer evolutionary relationship than might actually be the case. Similarities between organisms can either be homologous or analogous. Homologous structures are similar because of similar ancestry; for example the chicken wing and the human arm are homologous structures. Analogous structures are similar because of convergent evolution. The human eye and the octopus eye are analogous because they evolved independently (see Figure 10).

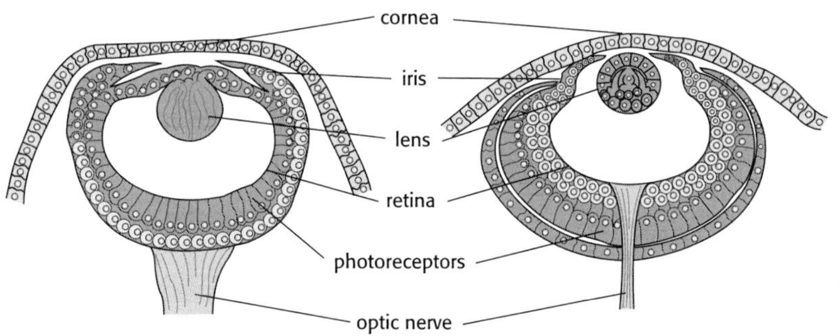

Figure 10 The human eye and the octopus eye are analogous because they are quite similar yet evolved independently.

Increasingly, classifications are being based on theories or evidence of evolutionary relationships. The advantage of this approach is that it gives greater predictive power: if several members of a group have a characteristic, it is likely to be present in a relative. The drug paclitaxel was originally derived from the bark of the Pacific yew (*Taxus brevifolia*). Over-harvesting during the development of the drug made this species increasingly rare. As a consequence, closely related species were explored and the European yew (*Taxus baccata*) yielded a chemical that became an important ingredient in the development of synthetic paclitaxel.

Figure 11 is a cladogram. This is a tree-like diagram that divides in a binary pattern. A clade is defined as a group that includes an ancestral species and all the descendents from that species. In Figure 11, birds, non-avian dinosaurs and ancestral species A form a clade called dinosauria. Birds, non-avian dinosaurs, crocodiles and ancestral species B are part of a clade called archosaurs. Lizards, snakes and ancestral species C form a clade called squamates. Members of a clade share a set of features which were not present in more distant ancestors, but which are shared by members of the

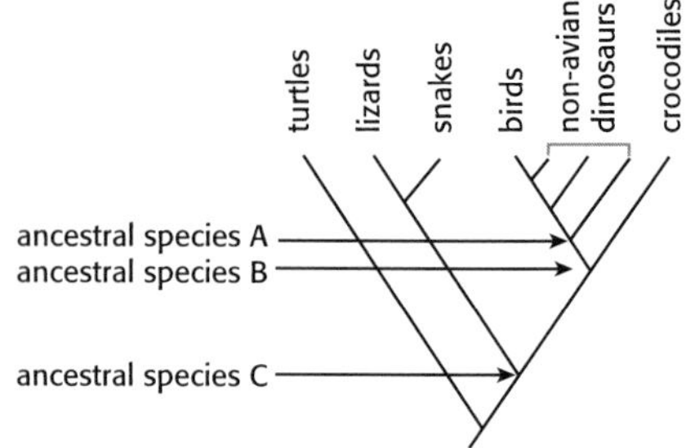

Figure 11 A cladogram showing the hypothesized relationship between birds and the traditional taxonomic group "the reptiles"

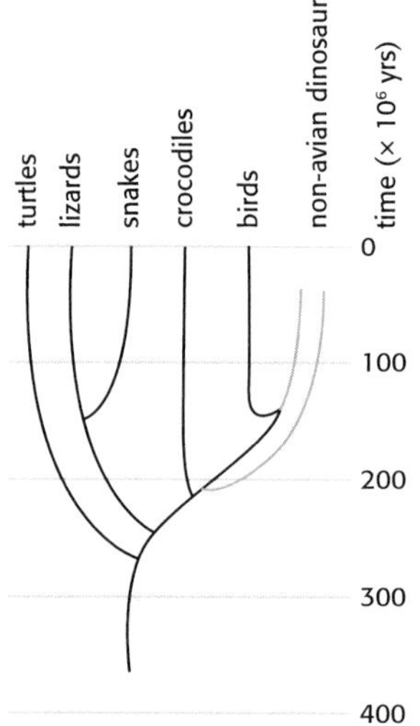

Figure 12 A phylogenetic tree shows evolutionary relationships and includes a time scale

?

1. Annotate the cladogram with the clade "aves" (the birds). [1]
2. Discuss whether "reptilia" represents a clade. [3]
3. Determine from the phylogenetic tree when the non-avian dinosaurs went extinct. [1]
4. Determine from the phylogenetic tree whether ancestral crocodiles or dinosaurs appeared first. [1]

clade. If the shared features are homolgous, then the cladogram can be the basis of a phylogenetic tree. A phylogenetic tree also includes a time scale, which is not included in a cladogram. Figure 12 shows the same groups in a phylogenetic tree.

Biochemical evidence of evolution

Evidence for evolution is found in the universality of certain biochemicals.

- DNA and RNA are universal hereditary molecules.
- The genetic code is the same in nearly every organism.
- ATP is energy 'currency' in all life.
- Optical isomers are molecules that are mirror images of one another. The two forms are D- and L- (see Figure 13). All amino acids in living things are in the L-form, while all carbohydrates in DNA and RNA are in the D-form.
- Physiological processes that are important to living things, such as cellular respiration, follow very similar metabolic pathways.

Differences between molecules can be used as part of the evidence to deduce phylogenetic relationships. Evidence suggests that mutation rates in DNA (which may result in a change in base sequence) occur with predictable frequency. Differences can then be used as a "molecular clock" to develop a phylogeny.

D-glucose L-glucose

Figure 13 Optical isomers are mirror images of each other. All carbohydrates in DNA and RNA are in the D-form.

Data-based question: inferring evolutionary relationships

A comparison was made of the base sequences of genes coding for the same four proteins found in three different mammals: the cow, sheep and pig. Figure 14 shows the differences in base sequence expressed as a percentage of base substitutions.

1 (a) Using Figure 14 identify a protein that is identical in all three mammals. [1]
  (b) Using Figure 14 deduce, giving a reason, whether the ancestors of pigs or sheep diverged more recently from those of cows. [2]
  (c) Using Figure 14 explain how the variations in these molecules can indicate the phylogeny of these groups of mammals. [2]

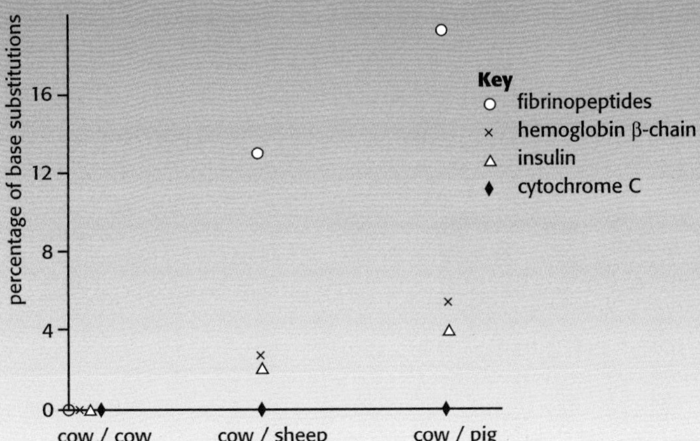

Figure 14 Percentage of base substitutions in the genes of four proteins in cows, pigs and sheep

Table 1 The number of differences between the human beta chain of hemoglobin and the beta chain of hemoglobin in other species.

Table 1

Comparison	Number of differences
human–human	0
human–chimpanzee	0
human–gorilla	1
human–rhesus monkey	8
human–dog	15
human–horse	25
human–chicken	45
human–frog	67

2 State what additional information would be helpful in constructing a phylogenetic tree for the species in Table 1. [2]

3 Discuss whether the number of differences is correlated with differences in morphology. [2]

Data-based question: dating fossils using radioisotopes

Figure 15 shows the decay curve for a radioisotope and the curve of accumulation of atoms formed by radioactive decay.

1 The half-life of ^{40}K is 1250 million years. Calculate the percentage of atoms of ^{40}K that remain in a sample
   (a) 2500 million years after it was formed. [1]
   (b) 5000 million years after it was formed. [1]
2 Estimate, using the decay curve, the percentage of ^{40}K atoms that remain 1875 million years after the formation of a sample. [1]
3 ^{40}K decays into ^{40}Ar. Explain, with reference to the decay curve, the reason why ^{40}Ar/^{40}K ratios cannot be used to obtain accurate ages for specimens that are less than 10 000 years old. [3]
4 (a) State the radioisotope that can be used to give accurate ages for specimens less than 10 000 years old. [1]
   (b) Explain why this radioisotope is useful for this purpose. [1]

Inquiry:
Outline the method for dating rocks and fossils using radioisotopes.

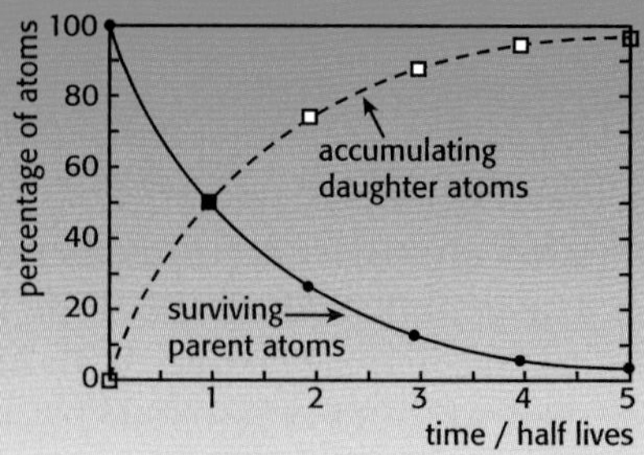

Figure 15 Generalized decay curve for a radioisotope

Hominid evolution

Humans are categorized as primates because of a number of features: opposable thumbs and toes, fingernails, forward-facing eyes, rotatable shoulder joints and colour vision. These features are regarded as adaptations for life in the trees.

While the question of gaps in the fossil record has been reviewed in the earlier punctuated equilibrium discussion, it is also true that the fossil record is sparse and that evidence of many of the species that once existed may never be uncovered. Early hominids did not practice cultural rituals such as burial, so bones would have been scavenged or scattered by geological processes. Paleoanthropology is a data-poor science. Fossil finds such as the disputed diminutive *Homo floresiensis* may lead to significant changes in theories of human evolution.

What the fossil record does suggest is that at various stages in hominid evolution, several species of hominid may have coexisted (Figure 16).

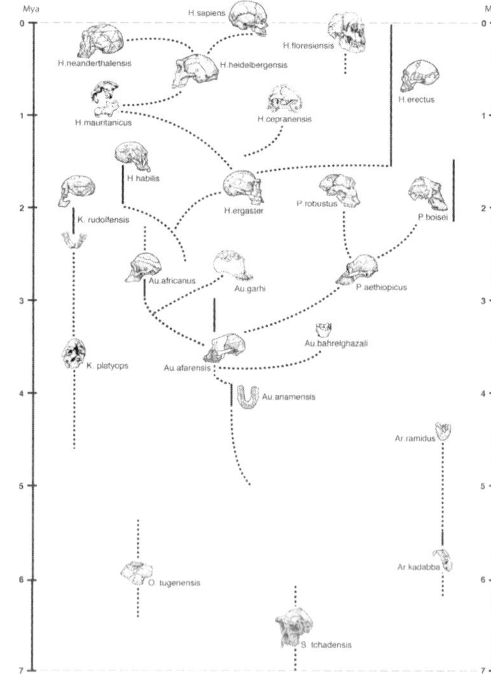

Figure 16 Hypothesized hominid phylogenetic tree

The scale to the left of Figure 16 indicates millions of years.

1 According to the figure, identify the earliest known members of the *Homo* genus. [2]
2 Identify two members of the *Homo* genus that were alive at the same time. [1]
3 State the names of four hominids that may have been alive at the same time 2 million years ago. [2]

Rediscovering biology: hominid skull comparison

Either through the examination of replicas of hominid skulls, or images such as those in Figure 17, answer the following questions. Remember to keep in mind the concept of allometry, i.e. the change in proportion of various body parts as a consequence of their growth at different rates. An increase in cranial capacity, for example, can be more significant if the overall volume of the skull has decreased.

1 Estimate the cranial capacity of the skulls by using the formula for the volume of a sphere. Measure the diameter of the back of the skull, and from this obtain the radius value for your formula.

2 Changes in cranial capacity do not necessarily imply greater intelligence but significant changes do suggest an increase in brain mass. Why would this require dietary changes?

3 Rank the skulls as to their degree of prognathism (protrusion of the lower face). Hypothesize how this might indicate changes in the relative importance of certain senses.

4 Describe the sagittal crests found on each skull. (looks like the keel of a boat) What is the relationship between the size of the sagittal crest and the size of the lower jaw. Hypothesize a function for the sagittal crest.

5 Describe the brow ridges found on each skull.

6 (a) Refer to a diagram of the teeth. Create a table to describe the nature, number and distribution of the teeth. Use the term diastema (gaps in teeth). The arrows in the diagram indicate the location of significant diastema. If only a small fragment of a hominid jaw is found and diastema is shown, what might be concluded about diet?

(b) What are molars for? How is this related to diet? How have the size of molars changed in hominid evolution?

7 Compare the "roundedness" of the forehead. Changes here indicate changes in the size of the frontal lobe. What is centred in the frontal lobe of the brain?

8 Locate the foramen ovale. Compare the changes in the position of the foramen ovale through evolution. Compare the position of the foramen in hominids to the position in a cat or dog. What is the significance of this position?

9 Swing your arms in big circles above and below your shoulder. Can your dog or cat do this? What is the significance of a rotatable shoulder joint?

10 Determine the angle made by the upper lip, the nose and the point of attachment of the jaw. What affect would changes in jaw angle have on the ability of the jaw to generate a force? How is this related to diet?

11 Hypothesize what changes you would expect to see in the pelvis, the backbone and the rib cage in the course of hominid evolution.

Thinking about science: genetic and cultural evolution

The following extract is from "The Riddle of Coexistence" by Ann Gibbons (*Science*, 2001).

Forget first contact with aliens. For real drama, consider close encounters of the human kind. Forty thousand years ago, for example, our ancestors wandered into Europe and met another type of human already living there, the brawny, big-brained Neanderthals. Such a collision between groups of humans must have happened many times. Several early human species coexisted in Africa, and when our ancestors left Africa and spread around the globe, they probably came across other kinds of humans, such as Homo erectus, who had left Africa in a previous migration. But the European encounter with Neanderthals was probably the last such meeting.

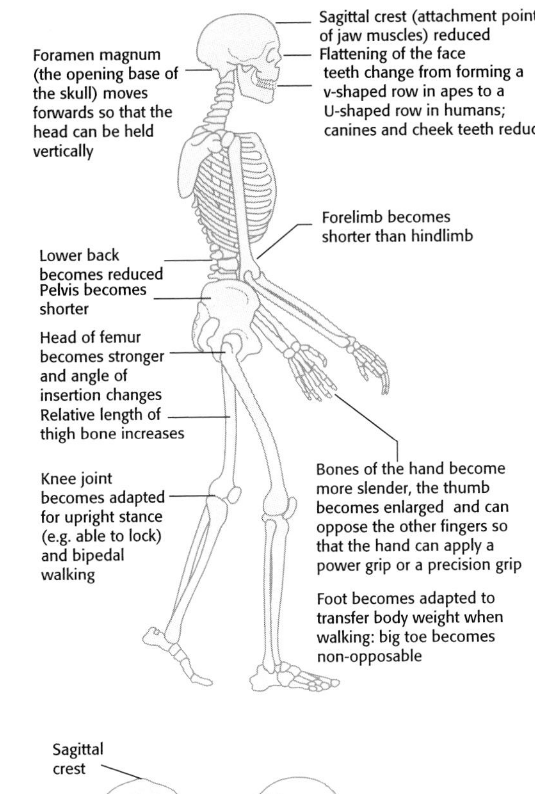

Sagittal crest

(a) (b)

Front view of skulls of (a) Homo erectus and (b) H. sapiens. Note the reduction of the sagittal crest; flattening of the facial bones; the side view shown in spread 20.12 figure 3, shows the more backward position of the lower jaw in H. sapiens

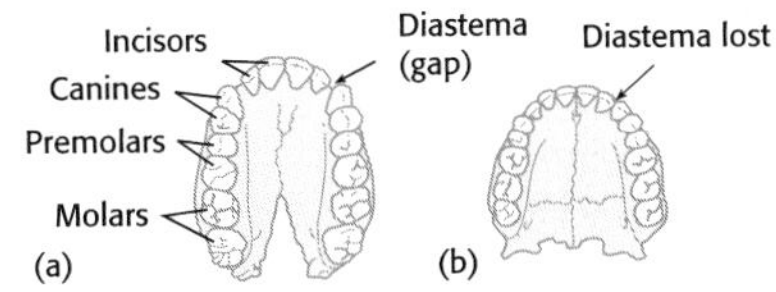

Upper jaws of (a) Australopithecus afarensis and (b) Homo sapiens. Note the loss of the diastema in H. sapiens, and the smaller canines and cheek teeth.

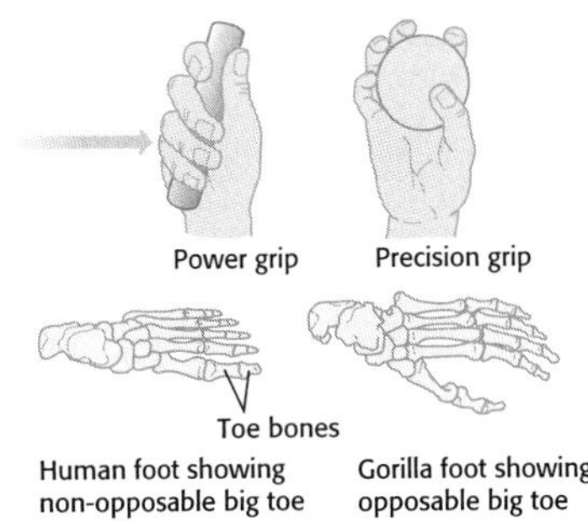

Figure 17 Some trends in the evolution of the hominid skeleton

There's no doubt, however, that the modern humans' lifestyle quickly surpassed that of the Neanderthals. Soon after they arrived in Europe, the modern newcomers made barbed projectile points and bone needles, painted vivid scenes on cave walls, carved animals out of ivory, and adorned themselves with bone pendants. Meanwhile, although some Neanderthals experimented with new technologies, they generally continued using much simpler artifacts.

Eventually, the Neanderthals disappeared, perhaps because they were unable to rebound when the climate turned frigid starting 28,000 years ago and competition for prime land became harsher. Moderns had some subtle advantage — perhaps slightly better language or abstract reasoning skills, or even a shorter interval between births of babies — that meant the difference between survival and extinction.

1 The author writes "There's no doubt, however, that the modern humans' lifestyle quickly surpassed that of the Neanderthals." To what extent is such certainty possible?

2 (a) With reference to the article, give one example of genetic evolution and one example of cultural evolution.
  (b) Which has been more important for humans recently: genetic or cultural evolution?
  (c) To what extent has language become more important than genes for humans in passing on successful adaptations?

Data-based question: cranial capacity and diet

Figure 18 shows the relationship between the brain's energy requirement and cranial capacity in several hominid species. Figure 19 shows the diet quality in a range of primate species. The higher the diet quality score, the greater its energy content.

1 (a) Outline the relationship between cranial capacity and dietary energy. [2]
  (b) Suggest a reason for this relationship. [1]

2 (a) Analysis of fossils suggests that there was an increased consumption in meat in the diet of hominids with time. State what kind of fossil evidence might show this. [2]
  (b) Explain why this change to a more carnivorous diet might have been important in the evolution of humans. [3]

3 (a) State the relationship between relative diet quality and relative brain size. [1]
  (b) Suggest a reason for humans being above the trend line for this relationship. [1]

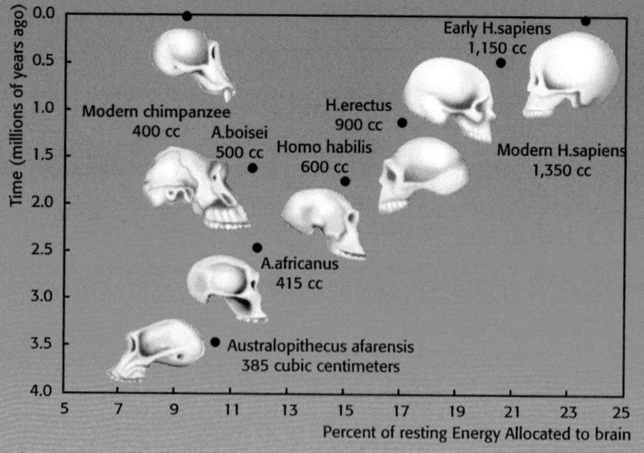

Figure 18 The evolutionary trend in hominid cranial capacity and dietary energy allocated to the brain

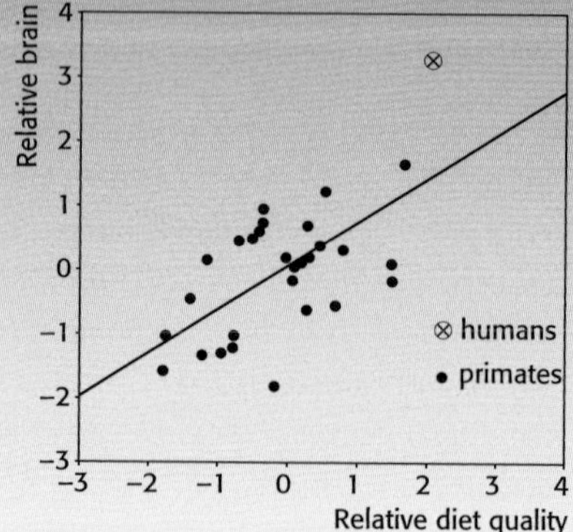

Figure 19 The relationship between brain size and diet quality

30 Option E: Neurobiology and behaviour

Neurobiology is the study of the nervous system of animals. Nervous systems are used in responses to stimuli. A stimulus is a change in the environment, either internal or external, that is detected by a receptor and elicits a response. A response is a change in an organism, produced by a stimulus. Behaviour is the pattern of responses in an animal. The simplest example of animal behaviour is the reflex. A reflex is a rapid unconscious response. An example of a reflex takes place when a young rabbit touches a stinging nettle plant for the first time with the tip of its nose. The rabbit pulls its nose away from the nettle in a fraction of a second, as a result of the following events.

1 Nerve endings in the skin of the rabbit's nose detect the pain caused by the stings. These cells are called pain receptors. The pain receptors are nerve endings of sensory neurons.
2 These sensory neurons carry impulses from the nose of the rabbit to its central nervous system.
3 The impulses travel to the ends of the sensory neurons where there are synapses with relay neurons. Messages are passed to the relay neurons by synaptic transmission.
4 The relay neurons have synapses with motor neurons, which carry impulses out of the central nervous system, to muscles in the rabbit's body.
5 Messages are passed across synapses from motor neurons to muscle fibres, which contract and pull the rabbit's nose away from the nettle. It is the connections between sensory, relay and motor neurons that ensure the response is appropriate to the stimulus – this is known as co-ordination.

Inquiry

On a large sheet of paper, build up a diagram of the whole pathway of neurons that would be used in a pain withdrawal reflex in humans, for example when someone steps on a pin. Include the spinal cord with white and grey matter, spinal nerves with their dorsal and ventral roots and ganglia, receptor cell, sensory neuron, relay neuron, motor neuron and effector—a muscle.

Inquiry

Find a local example of natural selection affecting an animal response. For natural selection to cause a response to become established in a population, the response must be genetically programmed and not due to animals copying each others' behaviour. There must be variation in behaviour, with natural selection favouring one type of behaviour rather than another.

Data-based question: garter snakes and prey selection

Some populations of the garter snake, *Thamnophis elegans* feed on slugs, while other populations do not. In coastal environments in California, slugs are abundant, but in inland areas they are scarce. Female garter snakes bearing fertilized eggs were obtained from three populations in California. Young snakes were reared from the eggs and for 10 days they were offered slugs to eat. The snakes had not seen their parents eating slugs. The number of days on which each snake ate a slug was recorded. Table 1 shows the results with the snakes divided into two groups—those that ate a slug on less than 5 days and those that ate one on 5 or more days.

Table 1

Location of population	Type	Slug-eating days	
		<5	≥ 5
Lassen County	Inland	567	114
Humboldt County	Coastal	149	314
Santa Cruz	Coastal	30	165

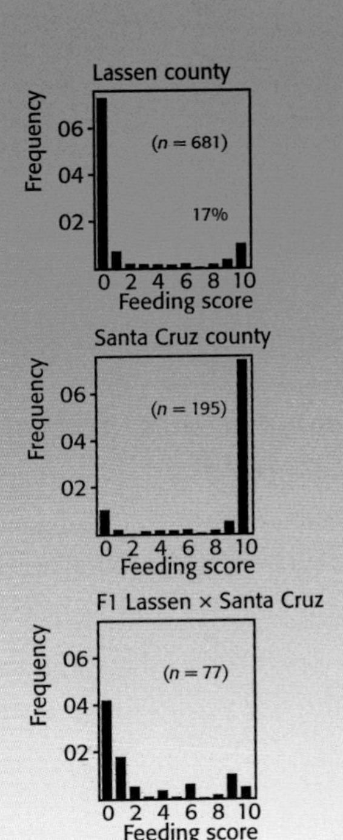

Figure 1 Slug eating in garter snakes

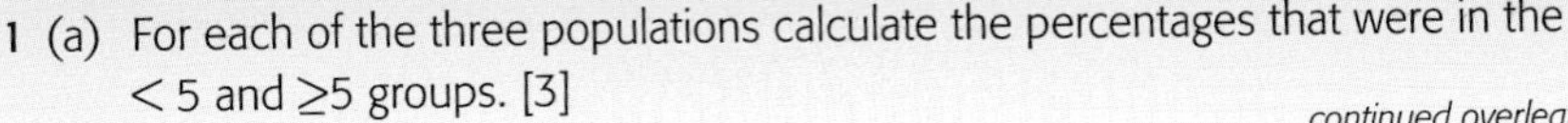

1 (a) For each of the three populations calculate the percentages that were in the <5 and ≥ 5 groups. [3]

continued overleaf

(b) Explain the differences in the behaviour of inland and coastal populations, in terms of natural selection. [4]

2 Predict, with reasons, the result if garter snakes were moved from Lassen County to Santa Cruz. [3]

Snakes from Lassen County and Santa Cruz were cross-bred. The F1 hybrids were tested in the same way as the original populations. Figure 1 shows the numbers of snakes that accepted slugs each of the possible numbers of days from 0 to 10.

3 Discuss the evidence from the bar charts for:
(a) the slug-eating response being inherited [2]
(b) the alleles for slug eating being recessive. [2]

4 Suggest a type of receptor for the slug-eating snakes to detect slugs. [1]

Inquiry

1 Find out which types of receptor are in each of these human sense organs:
(a) eye
(b) nose
(c) tongue
(d) ear
(e) skin.

2 Find out which types of receptor are in these parts of the brain:
(a) pineal gland
(b) hypothalamus.

Perception of stimuli

Stimuli are detected by receptors. Some receptors are nerve endings of sensory neurons, e.g. pain receptors. Others are special cells located in a sense organ. Animals can detect a wide variety of stimuli, using different types of receptor. Humans have the following types of receptor.

- Mechanoreceptors perceive movements, for example sound vibrations or movements due to pressure or gravity.
- Chemoreceptors perceive chemical substances, either dissolved in water or as vapours diffusing through the air.
- Thermoreceptors perceive heat and temperature.
- Photoreceptors perceive electromagnetic radiation, usually light.

Structure of the ear

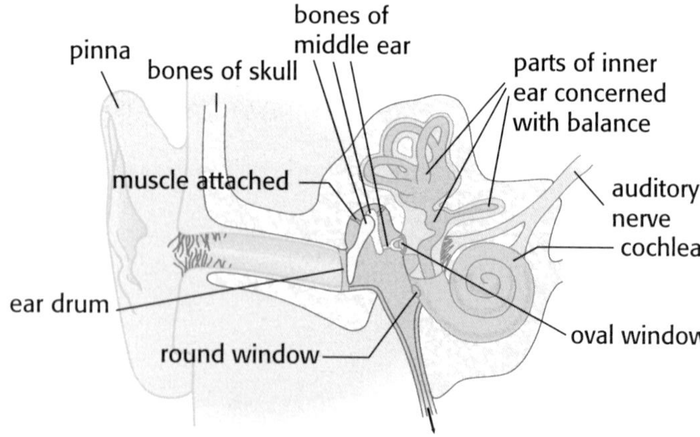

Figure 2 The structure of the ear

Inquiry

Using our ears we can hear a wide range of sounds. This involves collecting and amplifying sound vibrations and then perceiving the different frequencies of vibration.

1 How do the eardrum, the three bones in the middle ear and the oval window amplify sounds?

2 Why is the round window needed?

3 How do the hair cells in the cochlea convert sound vibrations into nerve impulses in the auditory nerve?

4 How are different frequencies of sound distinguished?

Thinking about science: perception and reality

Figure 3 shows the frequency sensitivity of six land mammals. The solid area shows where frequency sensitivity is best, while the lines indicate how much louder other frequencies need to be in order to be heard.

1 Does the world sound the same to any of the animals?

2 Which is the real world–the one we perceive or the world perceived by the bat?

3 Animals also differ considerably in their visual perception. Is what each animal sees what is really there, is it a construction of reality, or is reality a false concept?

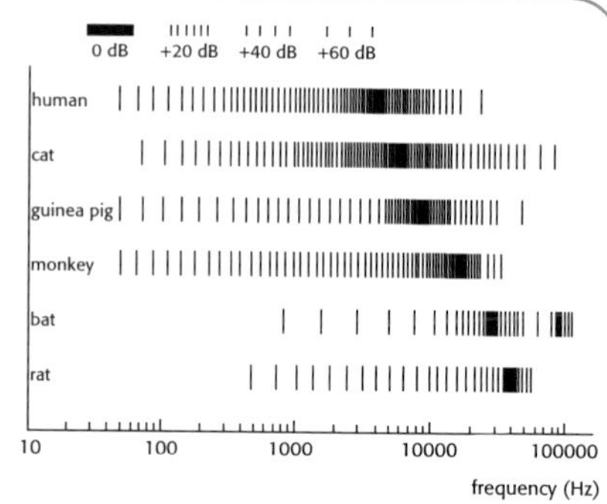

Figure 3 Sensitivity of mammals to frequencies of sound

Visual perception

Light entering the eye is focused to form an image on the retina. Figure 4 shows the cell types in the retina. Two types of light-sensitive cell detect the image and convert it into nerve impulses. These are the rod and cone cells. Rods are very sensitive to light, so work well in dim light conditions. In bright light the pigment in them is temporarily bleached, and they do not work. Rod cells absorb all the visible wavelengths of light and do not distinguish between them so they give us black and white vision.

Groups of rod cells send impulses to the brain via a single neuron, so the brain cannot distinguish which rod absorbed the light. There is therefore less detail in images transmitted to the brain by rods than by the cones, each of which sends impulses to the brain via an individual neuron. There are three types of cone, which absorb different ranges of wavelength of light. They are named according to the colour that they absorb most: red, blue or green. When any colour of light reaches the retina, the red, blue and green cones will each be stimulated by a certain amount. By analysing the amount that each of the three cone types are stimulated, the colour of light can be very precisely determined. Cones are only stimulated by bright light and therefore the colour vision that they give us does not work well in dim light.

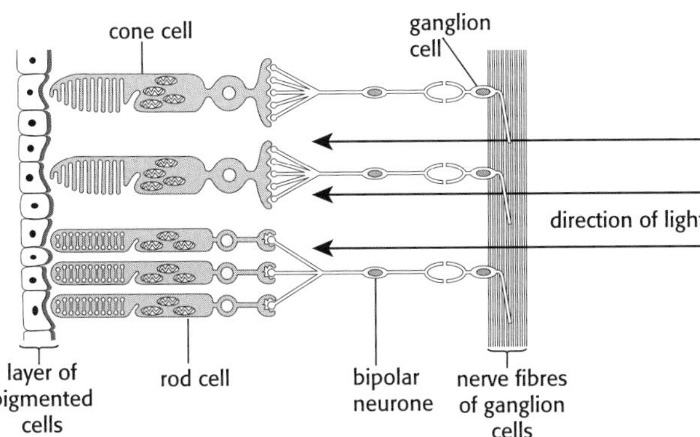

Figure 4 Rods and cones in the retina

Processing of visual stimuli

The retina begins the task of processing the visual stimuli that are collected by the rods and cones and passed via the bipolar cells to the ganglion cells. Each ganglion cell is stimulated when light falls on a small circular area of retina called the receptive field. There are two types of ganglion cell. In one type, the ganglion is stimulated if light falls on the centre of the receptive field, but this stimulation is reduced if light also falls on the periphery. In the other type, light falling on the periphery of the receptive field stimulates the ganglion cell, but this stimulation is reduced if light also falls on the centre. Both types of ganglion cell are therefore more stimulated if the edge of light/dark areas is within the receptive field. This type of processing is called edge enhancement and it is the explanation of an optical illusion called the Hermann grid (Figure 5).

Contra-lateral processing of visual stimuli means that some impulses from the left eye pass to the right side of the brain, and vice versa. This exchange happens in the optic chiasma, shown in Figure 6. The right side of the brain processes visual stimuli from the left side of the visual field, and vice versa.

Inquiry

1 Draw a cross-sectional diagram of the eye, including the following:
- sclera—the tough outer wall
- cornea—the convex transparent front wall
- conjunctiva—the transparent skin on the front of the eye
- retina—where the image forms
- choroid—the black-coloured backing to the retina
- lens—convex and transparent;
- iris—pigmented and with muscle fibres to change its size
- optic nerve—connected to the retina.

2 Add these labels: aqueous humour, pupil, vitreous humour, fovea, blind spot, eyelid.

Figure 5 Hermann grid

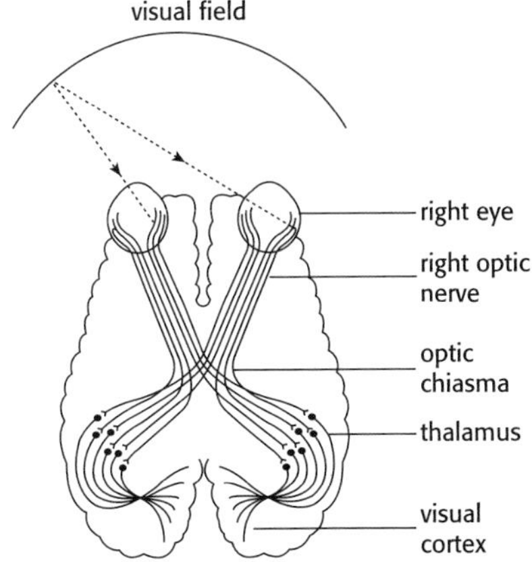

Figure 6 The optic chiasma

Inquiry

Some brain lesions cause *homonymous hemianopsia.* What is this condition, and in which parts of the brain would lesions cause it?

Innate behaviour

Some types of animal behaviour develop whatever external influences an animal experiences—they develop independently of the environmental context. They are genetically programmed and are called **innate behaviour**. An example is movement of *Planaria* flatworms towards food. Other types of behaviour develop as a result of the animal's experiences. They are examples of **learned behaviour**. An example is the use by chimpanzees of narrow twigs to extract termites from termite mounds.

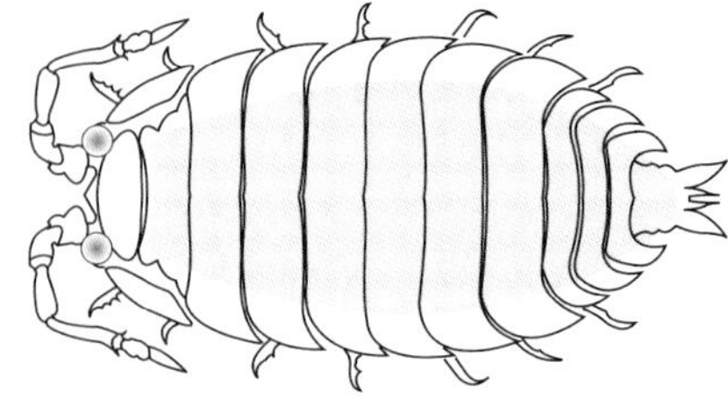

Figure 7 Woodlouse (slater)

Investigating biology: innate behaviour in invertebrates

The behaviour of invertebrate animals can be investigated by carefully designed experiments. Two types of behaviour involving movement have been defined:

- Taxis is movement towards or away from a directional stimulus. An example is movement of a woodlouse or slater (Figure 7) away from light.
- Kinesis also involves movement as a response, but the direction of movement is not influenced by the stimulus. Instead, the speed of movement or the number of times the animal turns is varied. An example is slower movement, with more frequent turning when woodlice are transferred from drier to more damp conditions.

Stages in designing an investigation:

1 Place the animals in conditions that are similar to the natural habitat.
2 Observe the behaviour and to see what stimuli affect movement.
3 Choose one stimulus that appears to cause a taxis or kinesis.
4 Devise an experiment to test responses to the stimulus.
5 Ensure that other factors do not have an effect on the movement.
6 Decide how you will measure the movement of the invertebrates.

Choice of invertebrate species

Many different invertebrates can be used in experiments. Planarian flatworms, woodlice, blowfly larvae, snails and beetles are often used. Some species can be purchased from suppliers but it is also possible to use invertebrates from local habitats. These should be kept for a short time only, protected from suffering during the experiments and then returned to their habitat. Endangered species should not be used.

Data-based question: chemotaxis in woodlice

Figure 8 shows apparatus that can be used to test the response of invertebrates to scents. The animals are placed in the syringe W, a pump is attached to tube X, to draw air gently through arms Y and Z.

The apparatus was used to test whether woodlice were attracted to the smell of other members of their own species. Three species were tested. In each case air was drawn through a container of the woodlice into one of the two arms. This air was therefore scented with the woodlice. Unscented air was drawn into the other arm. The number of woodlice that moved from syringe A into each arm was counted. The results are shown in Table 2.

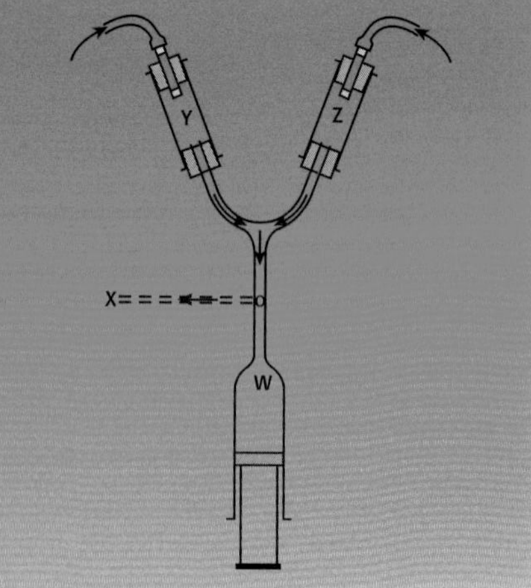

Figure 8 Testing chemotaxis

1 Suggest another method for encouraging the woodlice to move out of the syringe and into one of the arms. [2]
2 Outline the trends shown by the data in the Table 2. [2]
3 Tests show that the differences between the numbers collecting in each arm are statistically significant (chi-squared test). Deduce, with a reason, the type of receptor that the woodlice must have. [2]
4 Discuss, in terms of survival and reproduction, the possible reasons for the woodlice entering:
(a) the scented arm of the apparatus [2]
(b) the unscented arm of the apparatus. [2]

Table 2

Species	Numbers collecting in each arm	
	Scented	Unscented
Oniscus asellus	148	69
Porcellio scaber	101	62
Armadillidium vulgare	115	55

Learned behaviour

Learned behaviour involves the central nervous system and is more difficult to investigate than innate behaviour. The Russian physiologist Ivan Pavlov carried out an important series of experiments using dogs.

Pavlov inserted a tube through the cheek of his experimental animals that could be used to collect saliva from the mouth. He was then able to give stimuli and measure the volume of saliva secreted by the dogs. He found that saliva was secreted in response to the sight or smell of food. These types of stimulus, to which all dogs respond without learning, are called **unconditioned stimuli** and the secretion of saliva that results is the **unconditioned response**. Pavlov observed after a while that the dogs were starting to secrete saliva before they received the unconditioned stimulus. Something else had become a stimulus that allowed the dogs to anticipate the arrival of food. He found that the dogs could learn to use a variety of signals in this way, including the ringing of a bell, the flashing of a light, a metronome ticking or a musical box playing. These are examples of **conditioned stimuli** and the secretion of saliva that these stimuli elicit is the **conditioned response**.

Conditioned responses are used extensively in many animal's behaviour and can greatly increase survival chances. There are other more sophisticated types of learning, which also aid help animals to survive, reproduce and pass on their genes.

Inquiry

There are obvious benefits for an animal that is able to learn from its experiences.

1 Find animal examples of:
  (a) learning to repeat successful strategies for obtaining food
  (b) learning to avoid a painful experience
  (c) learning to avoid attacks by predators
  (d) learning to avoid dangerous activities.

2 Discuss why some behaviour patterns in animals are innate and some are learned. What is the advantage in having learned behaviour that is not programmed into the genes?

Data-based question: birdsong – innate or learned?

The sonograms in Figure 9 are a visual representation of birdsong, with time on the *x*-axis and frequency or pitch on the *y*-axis.

1 Compare sonograms I and II, which are from two populations of white-crowned sparrows (*Zonotrichia leucophyrys*). [2]

2 Sonogram III is from a white-crowned sparrow that was reared in a place where it could not hear any other birdsong.
  (a) Compare sonogram III with sonograms I and II. [2]
  (b) Discuss whether the song of white-crowned sparrows is innate, learned or due to both innate factors and learning. [3]

3 In 1981 Martin Morton and Luis Baptista published a very unusual discovery—a white-crowned sparrow had learned to imitate the song of another species. Sonogram IV is from a strawberry finch (*Amandava amandava*). Sonogram V is from a white-crowned sparrow that had been hand-reared by itself until it was 46 days old and then placed in an aviary with other white-crowned sparrows and a strawberry finch.
  (a) Compare sonogram V with sonogram IV. [2]
  (b) Compare sonogram V with sonograms I and II. [2]
  (c) Suggest two reasons why birds rarely imitate other species. [2]
  (d) Discuss whether Morton and Baptista's observation is evidence for innate or learned development of birdsong. [2]

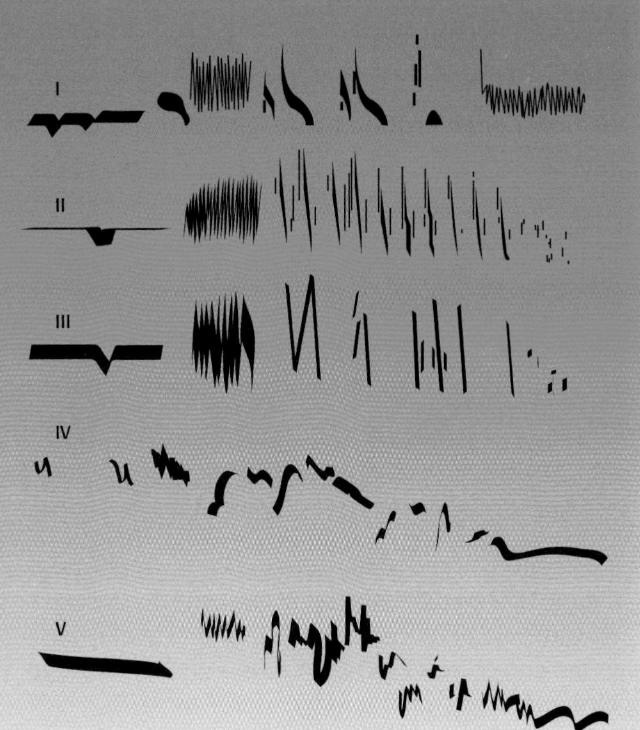

Figure 9 Sonograms of birdsong

Neurotransmitters and synapses

The basic principles of synaptic transmission were described in Chapter 24—an action potential in the pre-synaptic neuron causes release of neurotransmitter into the synaptic cleft, and an action potential is stimulated in the post-synaptic neuron when the neurotransmitter binds to receptors in its membrane. Two additional features of synapses give a fuller understanding of the functions of synapses.

- More than one pre-synaptic neuron can form a synapse with the same post-synaptic neuron (see Figure 10). Sometimes there are hundreds of pre-synaptic neurons! Usually a single release of neurotransmitter from one of the pre-synaptic neuron is insufficient to trigger an action potential. Either one pre-synaptic neuron must repeatedly release neurotransmitter, or several different pre-synaptic neurons must release neurotransmitter together. The additive effect from multiple releases of neurotransmitter is called summation.
- Neurotransmitters that stimulate action potentials in the post-synaptic neuron are called excitatory. Some neurotransmitters have a different effect—they inhibit action potentials. Neurons only release one type of neurotransmitter. Pre-synaptic neurons therefore either excite or inhibit post-synaptic transmission—they cannot do both.

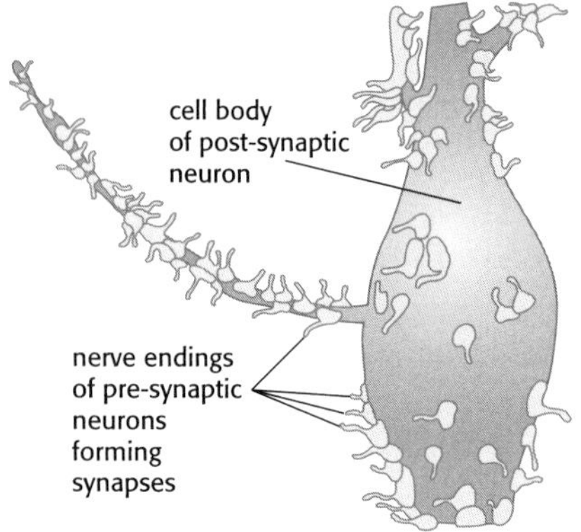

Figure 10 Multiple synapses

Where there are many synapses between pre-synaptic neurons and a post-synaptic neuron there can be interaction between the activities of the excitatory and inhibitory neurons (Figure 12). Whether a post-synaptic action potential is excited or not is a summation of the effects of all of these neurons. This is the basis of decision-making processes in the central nervous system.

Critical considerations: abuse of psychoactive drugs

Psychoactive drugs affect the brain and so can cause changes to the way we behave. They sometimes cause changes to mood and personality. The way in which they act is described on the next page. Some people enjoy experiencing the effects of psychoactive drugs and take them for this reason. Attitudes to the use of psychoactive drugs vary, depending on the drug and on individual views. However, it is widely accepted that some drugs have harmful effects and should not be used unless prescribed by doctors. The harmful effects may impact the user, his or her family and the wider society.

Figure 11 shows the engraving Gin Lane, by William Hogarth. He produced it in 1751, when London was in the grip of the Gin Craze. Gin is a drink with a high alcohol content, produced by distillation.

1 What social problems of alcohol abuse are depicted in the engraving by Hogarth?
2 To what extent have the social problems caused by drug abuse changed since 1751?

Figure 11 Gin Lane

Psychoactive drugs

The brain has many synapses, perhaps as many as 10^{16} in children. These synapses vary in their organization and use a wide variety of neurotransmitters. Over a hundred different brain neurotransmitters are known. Psychoactive drugs affect the brain and personality by altering the functioning of some of these synapses. Some drugs are excitatory, because they increase post-synaptic transmission. Others are inhibitory because they decrease it. One example of each is described here.

Cocaine

Cocaine acts at synapses that use dopamine as a neurotransmitter. It binds to dopamine reuptake transporters, which are membrane proteins that pump dopamine back into the pre-synaptic neuron. Because cocaine blocks these transporters, dopamine builds up in the synaptic cleft and the post-synaptic neuron is continuously excited. Cocaine is therefore an excitatory psychoactive drug. Synapses that use dopamine are part of what is known as the reward pathway that gives us pleasurable feelings during certain activities. Cocaine gives feelings of euphoria that are not related to any particular activity. Excitatory drugs can also work by mimicking excitatory neurotransmitters, for example, heroin mimics endorphins.

Examples of excitatory drugs

1. **Nicotine** contained in cigarettes and other forms of tobacco, derived from the plant *Nicotiana tabacum*.
2. **Cocaine** derived from the leaves of a Peruvian plant, *Erythroxylon coca*.
3. **Amphetamines** a group of compounds that are synthesized artificially.

Examples of inhibitory drugs

1. **Benzodiazepines** a group of compounds including Valium that are synthesized artificially.
2. **Alcohol** in the form of ethanol, which is obtained by using yeast to ferment sugar.
3. **Tetrahydrocannabinol (THC)** obtained from the leaves of the *Cannabis sativa* plant.

Tetrahydrocannabinol (THC)

THC binds to cannabinoid receptors in pre-synaptic membranes. Binding inhibits the release of neurotransmitters that cause excitation of post-synaptic neurons. THC is therefore an inhibitory psychoactive drug. Cannabinoid receptors are found in synapses in various parts of the brain, including the cerebellum, hippocampus and cerebral hemispheres. Several naturally occurring or endogenous cannabinoids have been discovered. They are an unusual example of retrograde signalling as they are released by the post-synaptic neuron and convey a message to the pre-synaptic neuron. People have reported a wide variety of feelings due to THC. The main effects are disruption of psychomotor behaviour, short-term memory impairment, intoxication and stimulation of appetite.

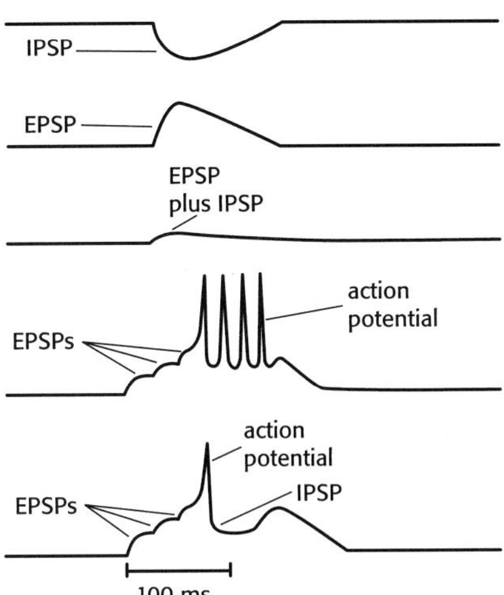

Figure 12 Excitatory post-synaptic potentials (EPSP), inhibitory post-synaptic potentials (IPSP)

Drug addiction and its causes

The American Psychiatric Association has defined addiction as:

"a chronically relapsing disorder that is characterized by three main elements:
(a) compulsion to seek and take the drug
(b) loss of control in limiting intake and
(c) emergence of a negative emotional state when access to the drug is prevented."

Only certain drugs cause addiction and usually repeated use over a prolonged period of time is needed. With a few drugs, addiction can develop more rapidly. Some people seem much more vulnerable to addiction than others. Also, addiction is more prevalent in some parts of society than others. The causes of addiction are clearly not simple and three areas need to be considered.

1. **Dopamine secretion** is associated feelings of well-being and pleasure. Many addictive drugs, including opiates, cocaine, nicotine and alcohol affect dopamine secreting synapses.

2 **Genetic predisposition** is the increased chance of developing an addiction that some people have because of their genes. One example is the gene, DRD2, that codes for the dopamine receptor protein. There are multiple alleles of this gene and a recent study showed that people with one or more copies of the A1 allele consumed less alcohol than those homozygous for the A2 allele.

3 **Social factors** affect drug use and addiction. Cultural traditions are very important and help to explain why different drugs cause problems in different parts of the world. Peer pressure, poverty and social deprivation, traumatic life experiences and mental health problems may also contribute.

Critical considerations: abuse of psychoactive drugs

Tobacco remains the leading cause of preventable death globally, with 5 million deaths every year. Besides the human loss, there are heavy social and economic costs. It is the poor and the poorest who tend to smoke the most, and the developing countries bear the highest burden…

We are faced with a unique challenge in that tobacco products remain largely unregulated. The basic rules that apply to all other consumer goods in order to ensure that their safety has been tested and that ingredients or contents are disclosed, do not apply to tobacco products. This condition complicates our efforts to advance tobacco control measures that will have a positive, life-saving impact.

We look now to the vital network of leaders, decision-makers and healthcare professionals at the global, country and community levels: we will continue to work together towards improved tobacco control measures, such as smoking bans in public places, warning labels on packages, as well as stricter regulation of the growing range of tobacco products. Our shared goal is to reduce the number of people using tobacco in all its forms, which will result in public health gains for us all.

World No Tobacco Day is an annual event sponsored by the World Health Organization. Catherine Le Galès-Camus is WHO Assistant Director-General, Non-communicable Diseases and Mental Health. She released a communiqué on World No Tobacco Day in 2006, part of which is reproduced here.

The following text is part of an article written by Jeff Collin and published by the Institute for Development Studies and Sussex University.

Recent trends in the globalization of the tobacco industry are reflected in the shifting of the burden of tobacco-related disease and deaths towards developing countries. Tobacco companies have proved sufficiently powerful to thwart comprehensive control programmes in all but a handful of countries. What new strategies are needed to control the epidemic in developing countries?

One billion people smoke worldwide and around 3.5 million die from tobacco related illnesses annually. By 2030, this figure will rise to 10 million, with 70% of deaths in lower and middle-income countries. Four companies now control 75% of global cigarette sales, as sophisticated strategies for supply, production and sales have produced increasingly popular global brands.

The onward march of Marlboro man epitomises this globalization, exploiting the opportunities presented by trade liberalisation, regional organizations and the communications revolution. Control efforts are undermined by the industry's success in developing favourable relationships with many governments, the magnitude of their foreign direct investments and the scale of advertising, marketing and sponsorship campaigns. In addition, large-scale cigarette smuggling, which comprises one third of total exports, depletes tax revenues and further jeopardizes public health.

The political sensitivities surrounding tobacco control reflect the complex array of vested interests involved.

Economists define an externality as a cost or a benefit arising from an economic activity that affects people other than the buyer and the seller.

1 Who benefits from cigarette smoking?
2 Outline the negative externalities associated with smoking.
3 Market forces tend to oversupply goods with negative externalities such as cigarettes and undersupply goods with positive externalities such as vaccinations. Discuss the potential role for governments in correcting these market failures.

HL

The human brain

There has been intense research into the structure and function of the human brain, most recently by brain scans. An example of an MRI scan is shown in Figure 13, together with a diagram identifying the following structures.

- **Medulla oblongata** controls automatic and homeostatic activities, such as swallowing, digestion, vomiting, breathing and heart activity.
- **Cerebellum** co-ordinates unconscious functions, such as movement and balance.
- **Hypothalamus** maintains homeostasis, co-ordinating the nervous and endocrine systems, synthesising the hormones secreted by the posterior pituitary, and releasing factors that regulate the secretion of hormones by the anterior pituitary.
- **Pituitary gland** the posterior lobe stores and releases hormones produced by the hypothalamus and the anterior lobe produces and secretes hormones that regulate many body functions.
- **Cerebral hemispheres** act as the integrating centre for high complex functions such as learning, memory and emotions.

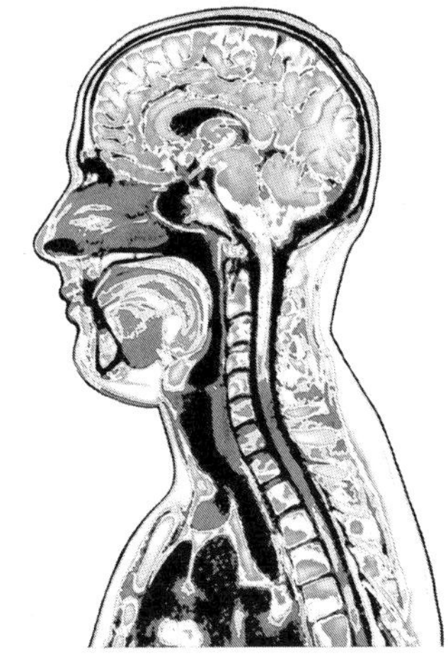

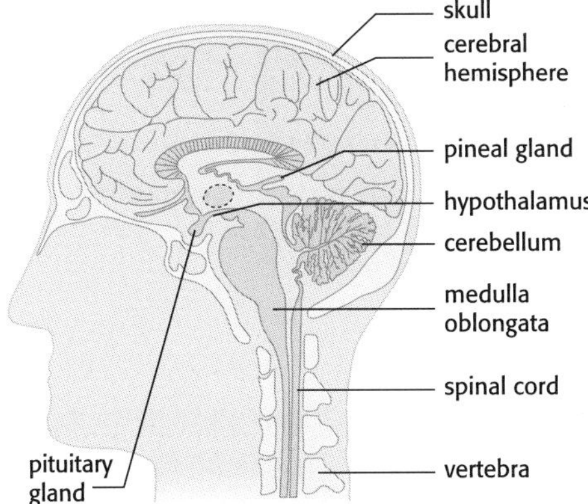

Figure 13 MRI scan and diagram of the brain

Thinking about science: techniques in brain research

Lesion studies gave the first useful information about brain functions. For example, in the 19th century, after the death of a patient who could only say the word "Tan", the French neurologist Charcot found a single large tumour damaging the lower left side of the patient's brain. He deduced that this part of the brain is involved with speech. Another famous case involved Phineas Gage, who suffered severe damage to the frontal lobes of his brain in 1848 when a large metal pin passed through his skull. The lesion radically altered his personality and capacity for social interaction. Many lesions due to tumours or accidental damage have been investigated, but rather than always wait for these fortuitous opportunities, many neuroscientists have in the past performed experiments on the brains of living animals, including experiments involving vivisection. There are widespread objections to these types of experiment today.

Increases in the flow of blood to active regions of brain can be detected using a fairly new scanning technique, functional magnetic resonance imaging (fMRI). This is a safe, non-invasive technique that can be used repeatedly on healthy living human experimental subjects. It has revolutionized the study of brain function.

1 Why is the evidence from natural lesions weaker than evidence from experiments?
2 Why are there widespread objections to experiments involving vivisection?
3 The French philosopher René Descartes believed in the duality of mind and body—the mind exists separately from the body. Is this view tenable, given the evidence that we now have from fMRI?
4 There is a curious circularity in brain research—if our thought processes are complex enough for us to understand the brain, are our thought processes too complex for the brain ever to understand them fully?

HL

fMRI experiments

Functional magnetic resonance imaging (fMRI)is a technique for determining which parts of the brain are activated by specific thought processes. Active parts of the brain receive increased blood flow, which fMRI records. The experimental subject is placed in the scanner and a high-resolution scan of the brain is taken. A series of low-resolution scans is then taken while the subject is being given a stimulus. These scans show which parts of the brain are activated during the response to the stimulus.

The three images in Figure 14 were obtained from an fMRI experiment where experimental subjects watched a moving target on a screen and moved a cursor to follow it. This task caused the cerebellum to be strongly activated because it coordinates eye and hand movements.

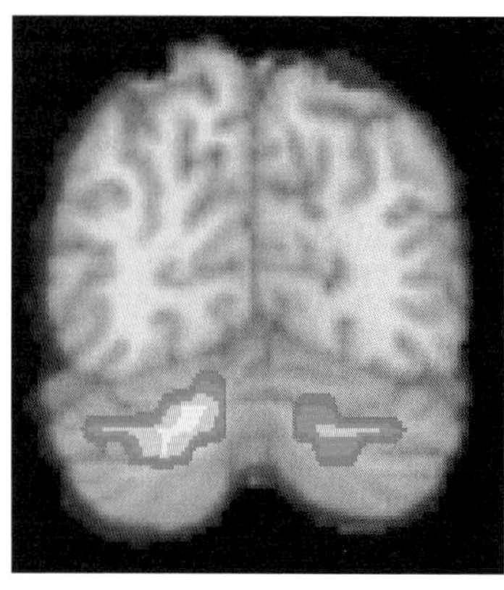

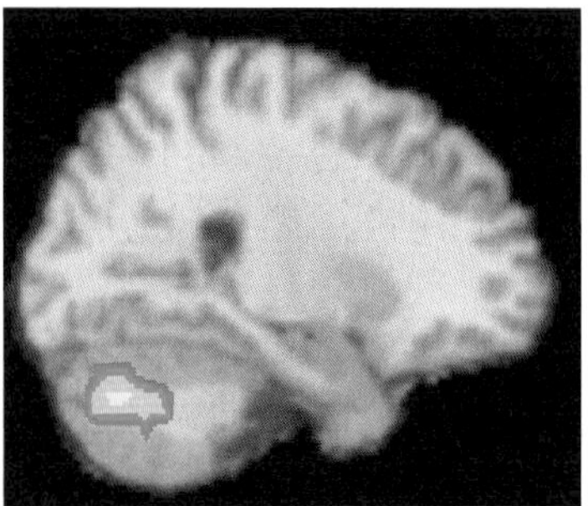

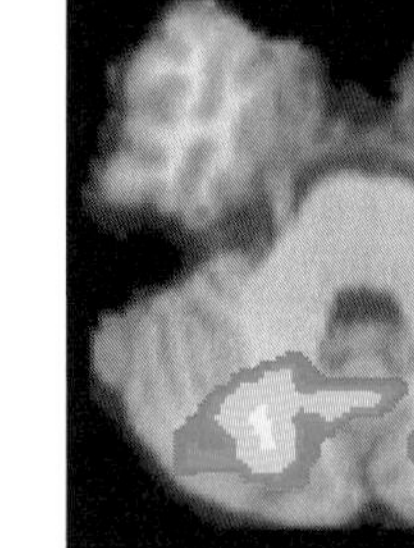

Figure 14 fMRI scans

Sympathetic and parasympathetic control

The peripheral nervous system comprises all of the nerves outside the central nervous system. It is divided into two parts: the voluntary and the autonomic nervous systems. Unconscious processes are controlled by the autonomic nervous system, which is itself divided into sympathetic and parasympathetic systems. These often have contrary effects on an unconscious process. For example, parasympathetic nerves cause an increase in blood flow to the gut wall when the gut is active. Sympathetic nerves cause a decrease in blood flow, during fasting or when blood is needed elsewhere, for example in muscles during vigorous exercise.

Inquiry

1 Find out the effect on the size of the eye's pupil of the eye of the autonomic nervous system?
2 Draw a flow diagram to show the stages in the pupil reflex that takes place when a light is shone into the eye including:
   (a) where and how the stimulus is detected
   (b) which part of the brain is co-ordinates the response
   (c) where in the eye the response occurs and how impulses are conveyed there.

Critical considerations: medical definitions of death

For legal and ethical reasons, it is often necessary to clearly define when a human life has ended. For example, a person may have expressed a wish that once dead, their healthy organs may be used for organ transplants. Human death can be established by several different criteria.

1 The biological standard: cessation of life of the whole organism.
2 The cardiopulmonary standard: when the heart and lungs cease to function. This standard is problematic today because heart and lung function can often be maintained indefinitely through artificial means.
3 The whole-brain standard: if all parts of the brain including the brain stem are non-functional. The pupil reflex is one way to check if the brain stem is non-functional. Patients have survived for years with only the brain stem functioning—known as a permanent vegetative state.
4 The higher brain standard: death is considered to have occurred if the cerebral hemispheres have ceased to function, even if the brain stem is functioning and allowing homeostasis to continue.

1 Which of these standards is:
   (a) possible in practice
   (b) most appropriate for ethical or other reasons.
2 Some would argue that what defines a person is higher brain function. If this is so, is it possible, or reasonable, to distinguish between the death of the body and the death of a person?
3 Supporters of euthanasia might argue for a quality of life standard. Why is this definition problematic? Discuss the ethical issues associated with euthanasia.

HL

Pain and painkillers

Pain receptors in the skin and other parts of the body detect stimuli such as the chemical substances in a bee's sting, excessive heat from a flame or the puncturing of skin by a hypodermic needle. These receptors are the nerve endings of sensory neurons and they convey impulses to the central nervous system. When impulses reach sensory areas of the cerebral cortex we experience the sensation of pain. At times it is necessary to block feelings of pain and the body uses natural painkillers to do this. One group, called endorphins, are oligopeptides that are secreted by the pituitary gland. They bind to receptors in synapses in the pathways used in the perception of pain. This blocks synaptic transmission and prevents the pain being felt.

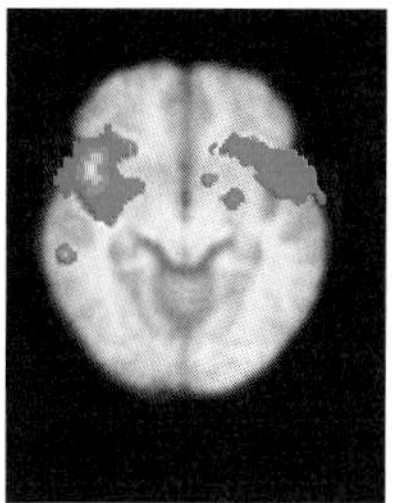

Figure 15 fMRI scan of endometriosis pain

Social organisms and altruism

Honeybees are one of the best known of social organisms. They live in colonies of thousands of bees, co-operating to give the colony as a whole the colony the best chance of survival. Individual bees cannot survive unless the colony survives. The colony is like a super-organism, with natural selection acting at this level rather than at the level of the individual bees in it.

Figure 16 Worker bees ventilate and cool the colony by fanning.

There are three castes of honeybee:

- Queens—fertile females, one only per colony, who lays eggs.
- Drones—fertile males, several hundred per colony, whose only role is to search for virgin females and mate with them.
- Workers—infertile females, thousands, who carry out all the tasks needed in the colony including foraging for food and rearing larvae.

Often worker bees give up their lives while defending the colony. They also help in rearing larvae, though they do not usually themselves breed. Both these actions can be regarded as altruistic. In studies of animal behaviour, altruism is defined as actions that increase another individual's lifetime number of offspring at a cost to one's own survival and reproduction.

The evolution of altruism is an interesting conundrum: if natural selection involves a struggle for existence in which the winners reproduce and pass on their alleles to their offspring, natural selection should always act against altruism. It is easier to understand the evolution of co-operation—a group of animals who help each other in some way may be more successful in the overall struggle for existence than animals that do not and all of them will increase their chance of reproduction. Of course there are cases where cheating or manipulation by some individuals disrupts these relationships. True altruism always seems to involve animals that are genetically related, with natural selection operating at the level of alleles. In a honeybee colony, the offspring that a worker helps to rear will usually be her siblings and share some of her alleles. In fact, because of the way in which gender is determined in honeybees, female siblings will on average share 75 per cent of their alleles. If the larvae that a worker rears become either queens or drones, they will have the chance of reproducing and passing on some of the worker's alleles to bees in the colony or another colony. This explains why workers toil so hard to ensure the survival of the colony—their alleles perish or survive with it. Over millions of years of evolution, any colonies where the worker bees did not act altruistically to defend the colony and rear the young will have died out and the alleles that made the workers act non-altruistically will have disappeared too.

Inquiry

There are many examples of social organisms apart from honeybees. Not all of them are insects. Choose one example of a social organism, other than honeybees or humans, in which there is altruistic behaviour. If possible, choose an organism that can be found in the area where you live—a local species.

1. Find out about the social organization of this species.
2. What examples of altruistic behaviour occur in this species?
3. How can the evolution of this altruistic behaviour be explained?

HL

Foraging behaviour

Foraging is searching for food. Animals must decide what type of prey to search for and how to find it. Studies have shown that animals tend to choose the prey that gives the highest rate of energy return. For example, the shore crab (*Carcinus moenas*) prefers to eat mussels of intermediate size, when presented in an aquarium with equal numbers of each size. As Figure 17 shows, these give the highest energy yield per second of time spent breaking open the shells. For most animals, the availability of food varies over both space and time. Foraging behaviour may therefore need to change, to optimize food intake at any moment. Evidence for this has come from studies of the bluegill sunfish (*Lepomis macrochirus*), which feed on small invertebrates, including *Daphnia*. Figure 18 shows how the consumption of *Daphnia* of different sizes changes as prey density increases.

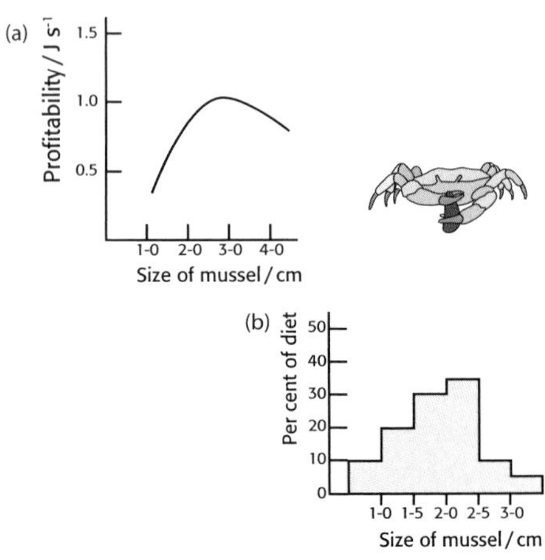

Figure 17 Crab diet investigation

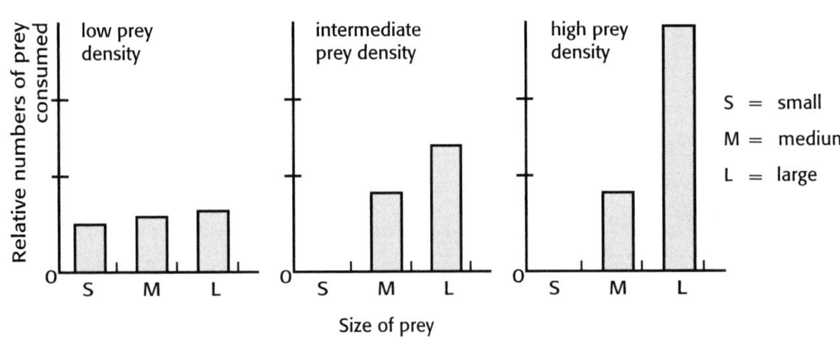

Figure 18 Sunfish prey size at different prey densities

Inquiry

Some animals have developed traits that seem highly exaggerated, e.g. the tail of the peacock.

1 What other examples of exaggerated traits can you find?
2 How could these types of trait have developed by natural selection? Hypotheses to explain these exaggerated traits usually involve mate selection.

Data-based question: rhythmical behaviour patterns

Many animal species show rhythmical variations in activity. These can be over an annual cycle. For example, mate selection and mating in red deer (*Cervus elaphus*) occurs in the autumn. As a result, calves are born in the spring, when most food is available. On the Great Barrier Reef most species spawn together over just one or two nights, soon after the full moon in November each year. This gives the highest possible concentration of sperm and eggs, raising the chance of fertilization and reducing the chance of predation. Rhythmical behaviour patterns can also follow daily cycles, for example, diurnal sleep and nocturnal activity in hamsters.

A study was done of the flight behaviour of serotine bats (*Eptesicus serotinus*), using radio transmitters attached to the bats. Each line indicates the timing and length of a flight.

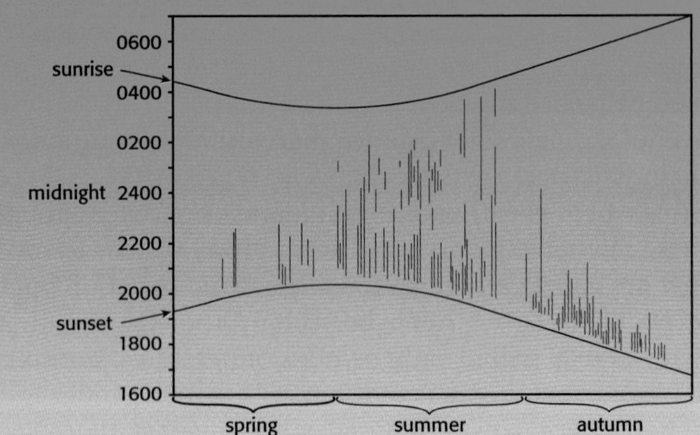

Figure 19 Flight behaviour in serotine bats

1 Outline the rhythmical behaviour pattern of the bats in autumn. [3]
2 (a) Compare the behaviour of the bats in summer and autumn. [3]
  (b) Suggest reasons for the differences in behaviour. [2]
3 Discuss whether the behaviour of the serotine bats shows any rhythmical variation. [2]

31 Option F: Microbes and biotechnology

Traditional classification systems have recognized two categories of organisms based on cell types: eukaryotes and prokaryotes. The inadequacy of this division is that the group categorized as prokaryotes is far more diverse than the eukaryotes in some ways. Based on the examination of the genes that are used to transcribe ribosomal RNA (rRNA), prokaryotes and eukaryotes can be divided differently, into three domains—bacteria (Eubacteria), Archaea and Eukaryota (Figure 1). Chapter 1 outlines some of the major differences between eukaryotes and prokaryotes. Table 1 shows some of the features that can be used to distinguish between the two groups of prokaryotes. For example, there are significant differences in the chemistry of the membranes (see Figure 2).

Archaea are found in a broad range of habitats such as the ocean surface, deep ocean sediments and even oil deposits deep below the surface of the Earth. They are also found in some fairly extreme habitats such as in water with very high salt concentrations (halophiles) and at temperatures close to boiling (thermophiles). The methanogens are obligate anaerobes and give off methane as a waste product of their metabolism. Methanogens live in the intestines of cattle and the guts of termites and are responsible for the production of "marsh gas" in marshes.

Inquiry

Microscopic eukaryotes are very diverse. This can be shown by comparing the following six types: *Saccharomyces*, *Amoeba*, *Plasmodium*, *Paramecium*, *Euglena*, and *Chlorella*.

1 Describe the mode of nutrition in each group.
2 Describe the methods of locomotion in each group, e.g. cilia or flagella.
3 Determine whether members of the group possess cell walls. Describe the composition of these cell walls.

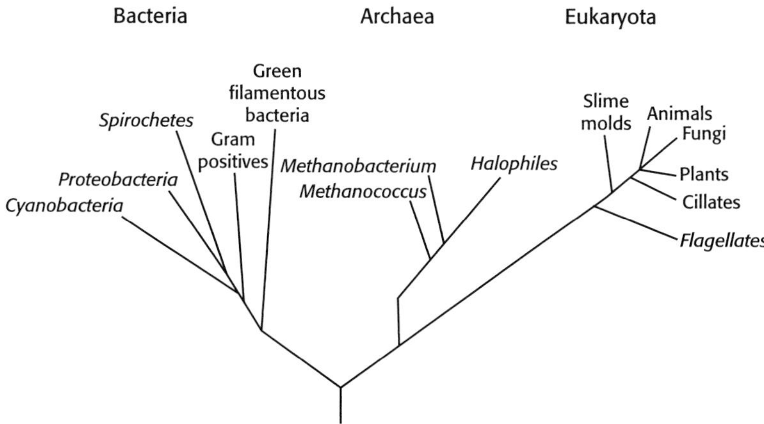

Figure 1 The evolutionary relationship between the three domains of life

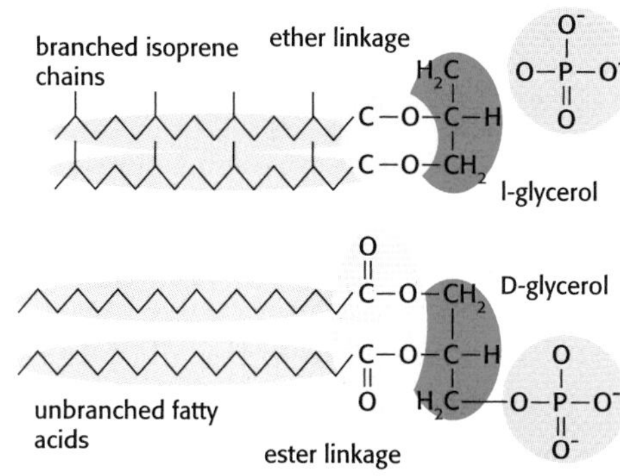

Figure 2 Lipids in the membrane bilayers of Archaea have ether linkages, while those in Eubacteria and Eukaryota have ester linkages.

Table 1

Feature	Domain		
	Bacteria	**Archaea**	**Eukaryota**
Histones associated with DNA	Absent	Proteins similar to histones bound to DNA	Present
Presence of introns	Rare or absent	Present in some genes	Frequent
Structure of cell walls	Made of chemical called peptidoglycan	Not made of peptidoglycan	Not made of peptidoglycan; not always present
Cell membrane differences	Glycerol-ester lipids; unbranched side-chains; D-form of glycerol	Glycerol-ether lipids; unbranched side chains; L-form of glycerol (see Figure 2)	Glycerol-ester lipids; unbranched side chains; D-form of glycerol

Data-based question: thermophiles

Some members of the cyanobacterial genus, *Synechococcus* can be found in hot springs. Five strains (*Synechococcus* strains 1 to 5) were isolated from a hot spring in Oregon, US, and tested for tolerance to high temperatures. Figure 3 shows how closely the five strains are related. The growth rate of each of these five strains was determined at different temperatures.

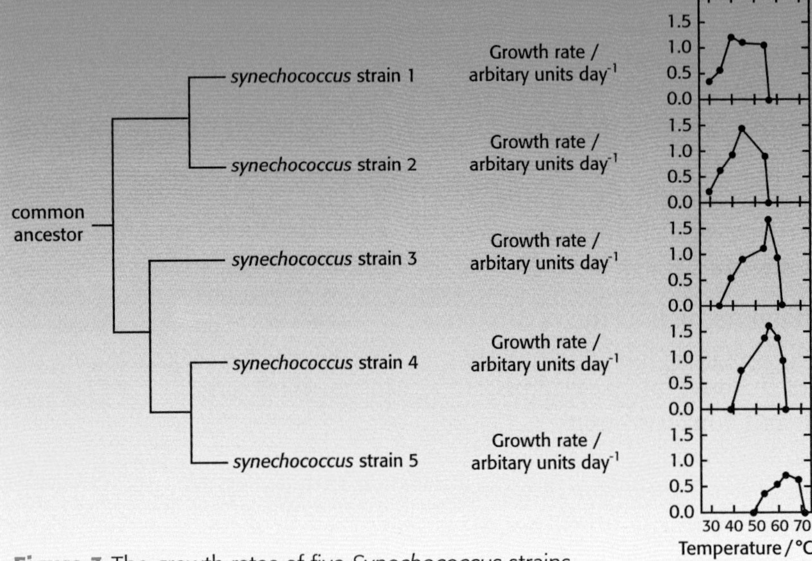

Figure 3 The growth rates of five *Synechococcus* strains as a function of temperature

1 State the domain in which cyanobacteria are placed. [1]
2 Identify, giving a reason, which strain would have the largest population after a growth period of 24 hours at a temperature of 45 °C. [1]
3 Compare the growth rate at high temperatures of strain 4 and strain 5. [2]
4 Suggest one advantage this genus of cyanobacteria has acquired by evolving to survive at higher optimum temperatures. [1]

Diversity of Eubacteria

A traditional test used to classify bacteria is whether they are Gram-negative or Gram-positive, based on how they react to Gram-staining (see below). The cell wall of Gram-positive bacteria consists of many layers of peptidoglycan (a polymer consisting of amino acids and sugars). These layers are connected by amino acid bridges. The greatest fraction of the Gram-positive cell wall is composed of peptidoglycan.

The cell wall of Gram-negative bacteria is much thinner—only about 20% peptidoglycan. Gram-negative bacteria have two unique regions that surround the outer plasma membrane—the periplasmic space and the lipopolysaccharide layer. The periplasmic space separates the outer plasma membrane from the peptidoglycan layer. The outer plasma membrane contains lipopolysaccharide and is attached to the peptidoglycan by lipoproteins.

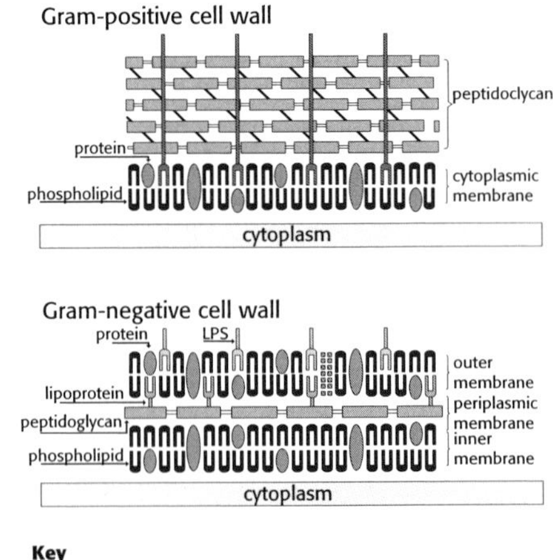

Figure 4 A comparison of the cell walls of Gram negative and Gram positive bacteria

Rediscovering biology: Gram-staining procedure

1 Prepare smears of *Bacillus cereus*, *Streptococcus fecalis*, *Escherichia coli* and *Micrococcus luteus*. Fix these preparations by heating over a bunsen burner.
2 Stain with crystal violet for about 30 seconds.
3 Rinse with water, then cover with Gram's iodine. Allow stain to act for about 30 s.
4 Rinse with water, then de-colourize with 95% alcohol for 10–20 s.
5 Rinse with water, then counterstain with safranin for 20–30 s.
6 Rinse with water and blot dry. Gram-negative bacteria will be pink. Gram-positive bacteria will be blue or violet.
7 Depending on local restrictions, you might choose to examine prepared slides of Gram-negative and Gram-positive bacteria.

Bacterial aggregates

Bacteria come in a diversity of sizes and shapes. The largest bacteria, *Thiomargarita namibiensis* can be 0.5 mm long. The smallest are *Mycoplasma* which lack cell walls and can be 0.2 μm. Some are intracellular parasites such as *Chlamydia* and *Rickettsia*. A coccus is a spherical bacteria, baccilus are rod-shaped bacteria, vibrio are comma shaped rods and spirilli are twisted.

Some bacteria can group together to form aggregates. Bacteria whose name has the prefix "strepto-" aggregate into filaments, while bacteria with the prefix "staphylo-" form clusters. Thus *Staphylococcus* is a genus of bacteria that form spherical clusters.

Some bacteria can release chemicals called autoinducers, which help coordinate the actions of a group of bacteria. *Vibrio fischeri* is a bacterium found in sea water that is able to bioluminesce (emit light). Individuals do not emit light unless they become part of a population of a certain density. *V. fischeri* releases an autoinducer into its surroundings. In a dense population, the concentration of the inducer becomes high enough to trigger bioluminescence. It happens, for example, when large numbers of *V. fischeri* are living together in the light organs of the Hawaiian bob-tailed squid (*Euprymna scolopes*).

Aggregates of the bacteria *Pseudomonas aeruginosa* can produce a biofilm—a surface-coating colony of microorganisms. This is of particular concern in burn patients and patients with cystic fibrosis, because antibiotic resistance can develop in the biofilm. The bacteria living on the biofilm usually have significantly different properties from free-floating bacteria of the same species, as the protection of the biofilm allows them to cooperate and interact in various ways.

Sewage treatment

The consequence of not treating sewage and allowing it to flow in to watercourses would be nutrient enrichment, or eutrophication. This favours algal blooms. When the mats of algae die, it leads to a loss of oxygen, because of bacterial activity on the dead organic matter. This is called biological oxygen demand.

Many sewage treatment plants make use of biofilms. A trickling filter system has a rock bed that can be up to 2 metres deep. The rocks are colonized by a biofilm of aerobic bacteria. Sewage water is sprayed onto the rocks. The process of spraying adds oxygen to the sewage, which is necessary for the aerobic bacteria to digest the sewage content.

Reed-bed systems make use of a range of ecological community members to treat sewage. Artificial oxygenation supports aerobic bacteria in lowering the biological oxygen demand of the sewage. Plants such as reeds extract nitrogen from the water; small animals such as rotifers and other filter-feeding invertebrates extract particulates.

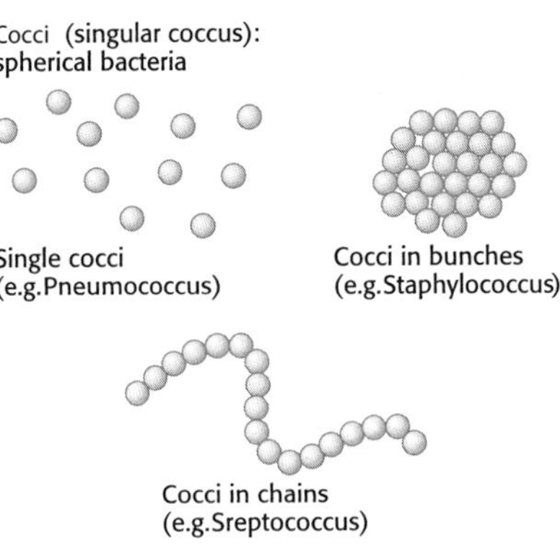

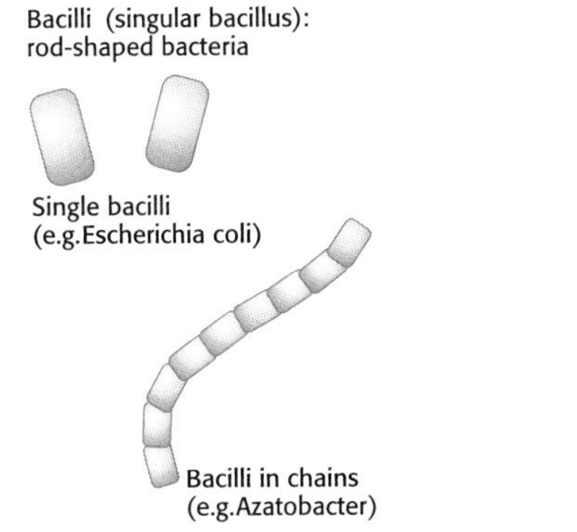

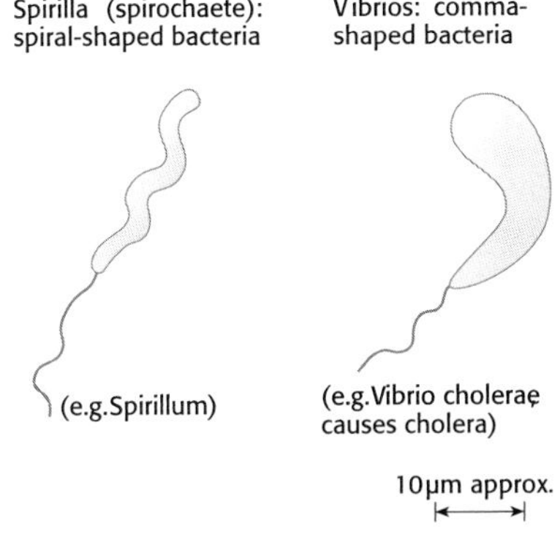

Figure 5 Shapes of bacterial cells and varieties of aggregates

Figure 6 A trickling filter bed

Data-based question: sewage release into a river

Figure 7 shows changes in the biotic and abiotic factors at increasing distances from an untreated sewage outfall into a river.

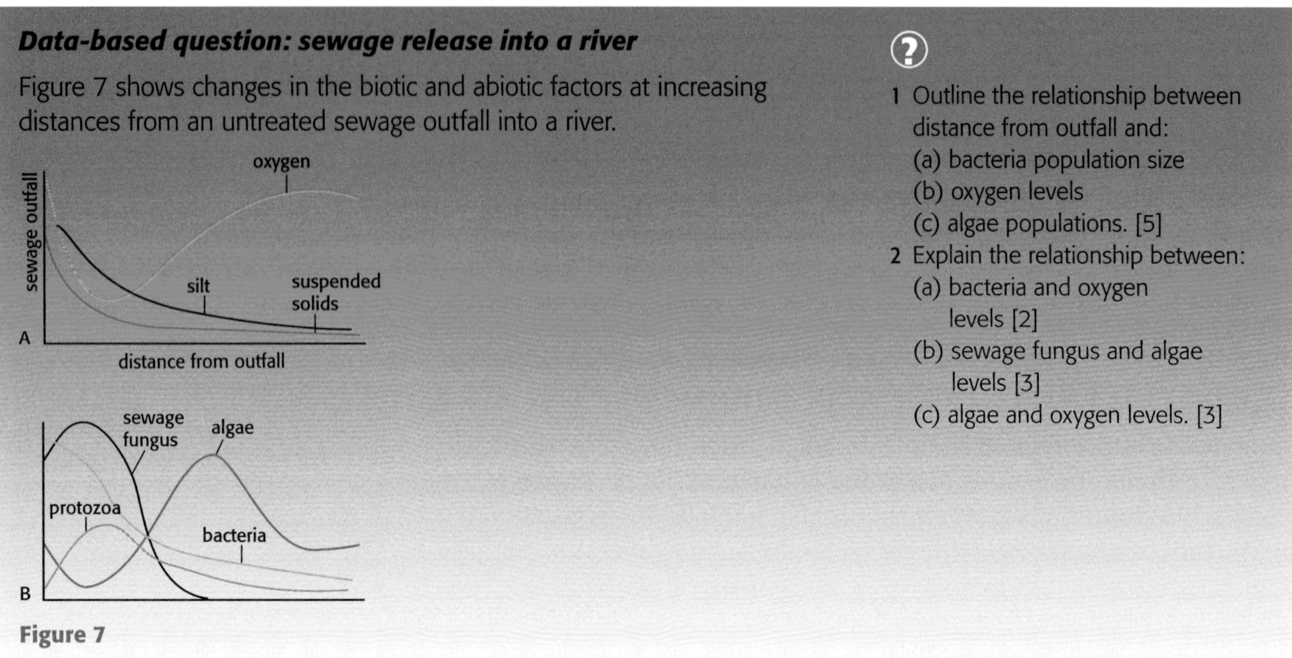

Figure 7

1 Outline the relationship between distance from outfall and:
   (a) bacteria population size
   (b) oxygen levels
   (c) algae populations. [5]
2 Explain the relationship between:
   (a) bacteria and oxygen levels [2]
   (b) sewage fungus and algae levels [3]
   (c) algae and oxygen levels. [3]

The nitrogen cycle

Microbes occupy a number of niches in ecosystems: as saprotrophs (decomposers) they release nutrients trapped in detritus and make it available to ecosystems. Cyanobacteria (blue-green algae) and protoctists such as algae and *Euglena* are photosynthetic and act as producers. The bacteria *Rhizobium* and *Azotobacter* can fix nitrogen and convert it to a form that living things can use. Bacteria such as *Nitrobacter* and *Nitrosomonas* can use inorganic chemicals as energy sources. They are known as chemoautotrophs.

Nitrogen would quickly become a limiting factor for ecosystems if it were not for the bacteria involved in the nitrogen cycle. In agriculture, soil nitrogen is often supplemented with fertilizer. Fertilizer can be industrially produced through the Haber process, which produces ammonia from atmospheric nitrogen. Alternatively, other sources such as manure from livestock production can be used.

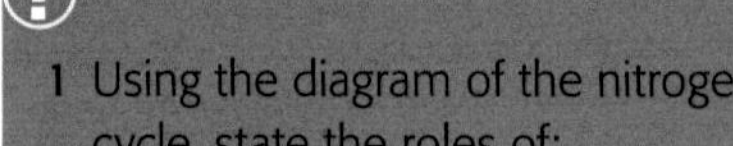

1 Using the diagram of the nitrogen cycle, state the roles of:
   (a) *Rhizobium*
   (b) *Azotobacter*
   (c) *Nitrosomonas*
   (d) *Nitrobacter*
   (e) *Pseudomonas denitrificans.* [5]
2 *Pseudomonas* usually uses O_2 as an electron acceptor, but under anaerobic conditions it will use nitrate instead. Explain why each of the following favours denitrification.
   (a) waterlogging due to poor drainage. [2]
   (b) compacted soil that has not been tilled. [2]

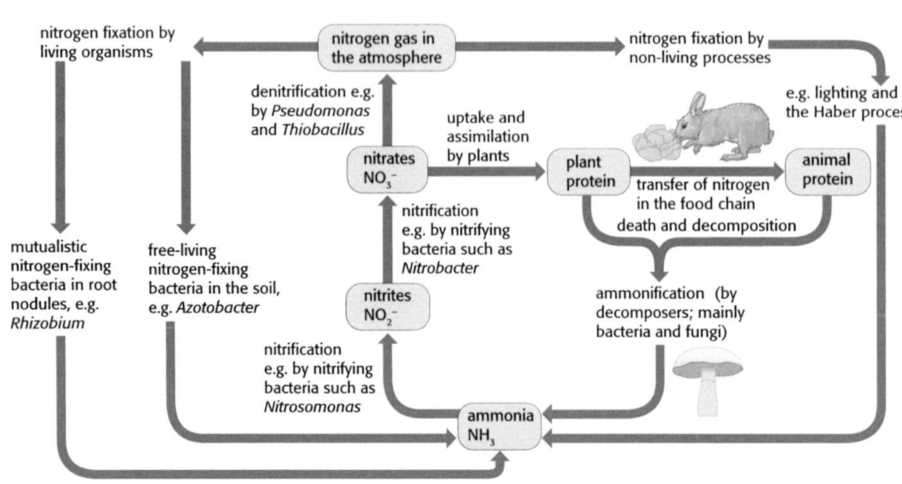

Figure 8 The nitrogen cycle

Rediscovering biology: ammonification and denitrification

1 Weigh out 0.1 g of soil.
2 Add the soil to a tube of peptone broth.
3 Incubate at room temperature for 1 week.
4 To test for the presence of ammonia, add a drop of Nessler's reagent on to a spot plate. Add a drop of the soil culture to the reagent. The presence of ammonia is shown by the development of a yellow, orange or brown coloured precipitate. It may occur after 24, 48 or 72 hours or 1 week. The length of time for a positive reaction or the strength of the positive reaction may indicate soil differences.
5 Inoculate a tube of nitrate broth with 0.1 g of soil and a second tube of nitrate broth with *Pseudomonas aeruginosa*. Inoculate a nitrate-free tube with *Pseudomonas aeruginosa*. Into all, place an inverted Durham tube (a small test-tube shaped tube) filled with fluid from the test tube. Gas formation indicates denitrification.

Data-based question: the nitrogen cycle and agriculture

Food production is linked to the nitrogen cycle because considerable quantities of nitrogen are added to soil as fertilizers. This extra nitrogen can become a contaminant of drinking water, causing eutrophication of rivers and lakes. Various chemical reactions in soil together with the burning of fossil fuels release nitrogen oxides into the air.

Nitrogen oxides are effective greenhouse gases and contribute to acid rain and smog. Figure 9 shows the release of nitrogen from fertilizer (millions of metric tonnes) in seven world regions between 1961 and 1996.

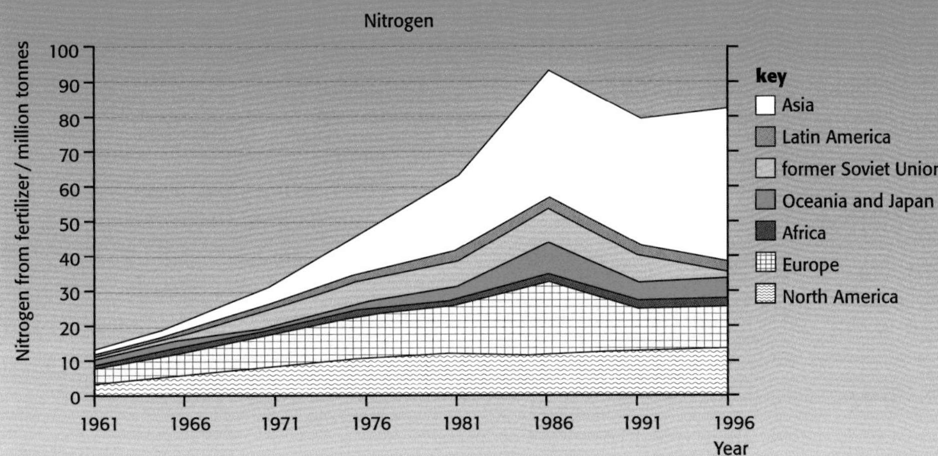

Figure 9 The release of nitrogen from fertilizer use in different regions of the world

1 Identify the year in which most nitrogen was released from fertilizers. [1]

2 Compare the nitrogen released in Latin America in 1986 with 1991. [1]
3 Suggest a reason for the difference in overall nitrogen release
   (a) between 1961 and 1986 [2]
   (b) between 1986 and 1996. [1]
4 State how nitrogen can be lost from the soil, other than by absorption in plants. [1]

Biomass fuel

Biomass is organic matter such as agricultural crops, wastes from agricultural production, wood, animal wastes, livestock farming wastes, aquatic plants such as macro- and micro-algae, fast-growing trees and plants, and sewage waste. Biomass can be used as raw material for the production of fuels such as methane gas and ethanol.

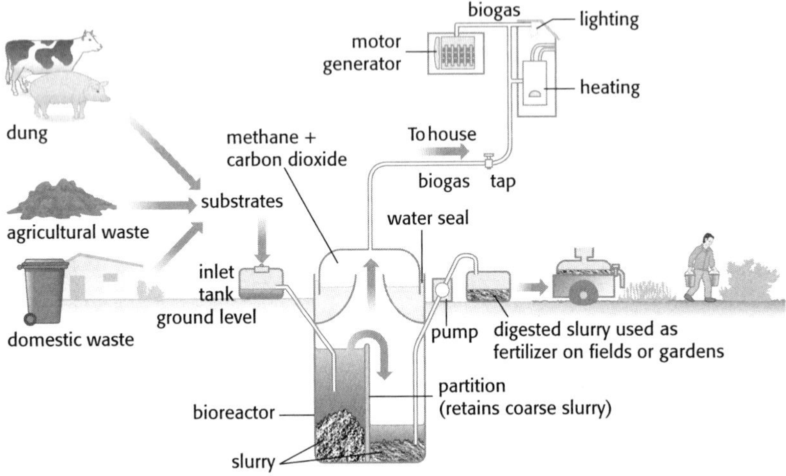

Figure 10 The process for producing biogas

Figure 10 shows the generation of biogas. Three different communities of anaerobic microbes are required. The first group convert the raw organic waste into a mixture of organic acids, alcohol, hydrogen and carbon dioxide. The second group use the organic acids and alcohol from the first stage to produce acetate, carbon dioxide and hydrogen. These first two communities are Eubacteria. The last group are Archaea called methanogens. The methanogens produce methane by one of the following two reactions:

(a) $$CO_2 + 4H_2 \rightarrow CH_4 + 2H_2O$$

reduction of carbon dioxide to methane

(b) $$CH_3COOH \rightarrow CH_4 + CO_2$$

splitting acetate to form methane and carbon dioxide.

Microbes and food production

Saccharomyces cerevisiae is a fungus that is widely used in biotechnology. Alcoholic fermentation produces alcohol and CO_2. Its fermentation reactions are used in the production of beer, wine and bread.

In wine production, *Saccharomyces* occurs naturally on the surface of grapes. Crushed grapes are allowed to ferment in the presence of the yeast. If the grape skins are left within the fermentation vessel during this time, the alcohol that develops will be a red wine.

In beer production (Figure 11), barley grains are allowed to germinate so that some starch is converted to maltose. The grains are then heated to denature the amylase and further processed so that the sugars are washed out of them. The resulting solution has *Saccharomyces* added and is allowed to ferment.

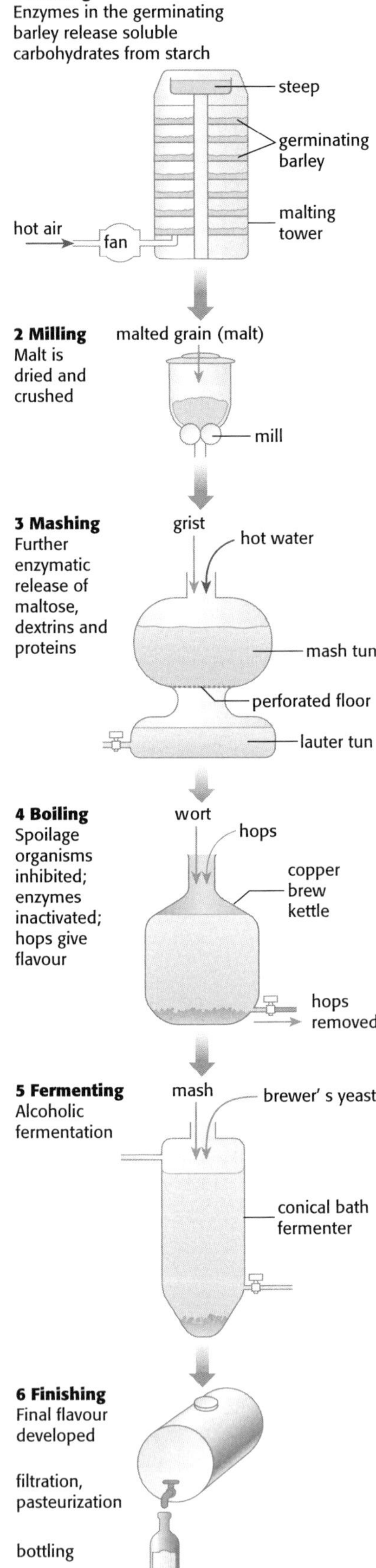

Figure 11 Beer production

Lastly, in the production of bread, *Saccharomyces* is added as an ingredient to the dough. The dough is left to rise, which means the yeast is allowed to operate on the sugars in the flour or on added sugars. The CO_2 produced as a result of fermentation causes the bread to rise. To make soy sauce, a salted mixture of crushed soybeans and wheat is treated with the fungus *Asperigillus oryzae,* which breaks down the starch in the substrate to glucose. This is then fermented, in the presence of salt, by other microbes to produce soy sauce.

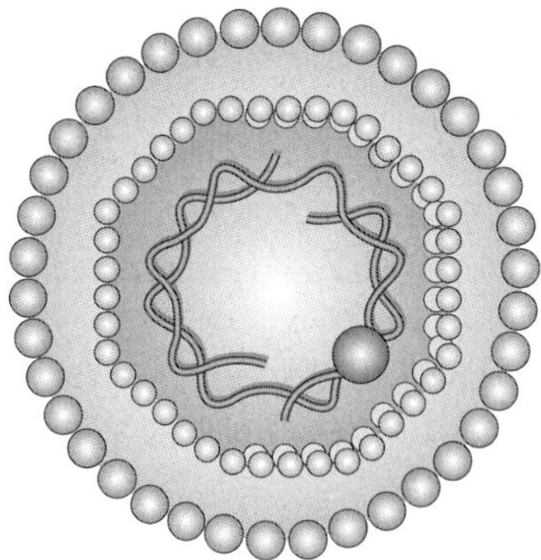

Figure 12 Hepatitis B is a double-stranded DNA virus with an envelope.

Food preservation involves controlling the growth of microorganisms. High salt or high sucrose concentrations provide osmotic inhibition. Pickling is a process that often involves storing food in a salt or vinegar solution and allowing anaerobic respiration to occur, The organic acids produced, such as lactic acid, inhibit growth of other kinds of microbes.

Food "poisoning" is a food-borne illness caused by toxins produced by pathogens. Poor preparation of foods such as poultry or other meat products is often the source. An example is α-toxin, produced by *Clostridium perfringens.* Symptom of infection include abdominal pain and diarrhea. *Salmonella* is one bacterium that causes food-borne infection. In the case of *Salmonella* the most common source is the consumption of contaminated poultry such as chicken or turkey. Normally a high dose of the organism must be ingested, meaning that it has been allowed to multiply for several hours in the food before consumption. The symptoms include vomiting, diarrhoea, fever and pain in the abdomen.

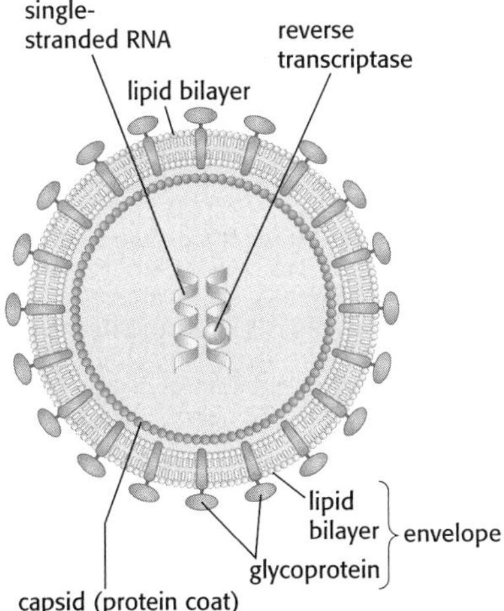

Figure 13 Human immunodeficiency virus (HIV) is a single-stranded RNA virus with an envelope.

Viral diversity

A virus is a biological structure that is organized but non-cellular. It can reproduce only within a host cell. Viruses consist of nucleic acid covered by a protein coat called a capsid. The nucleic acid can be RNA or DNA and it can be single-stranded or double-stranded. Some capsids are further covered in a membranous bilayer derived from the membranes of the host cells they infect.

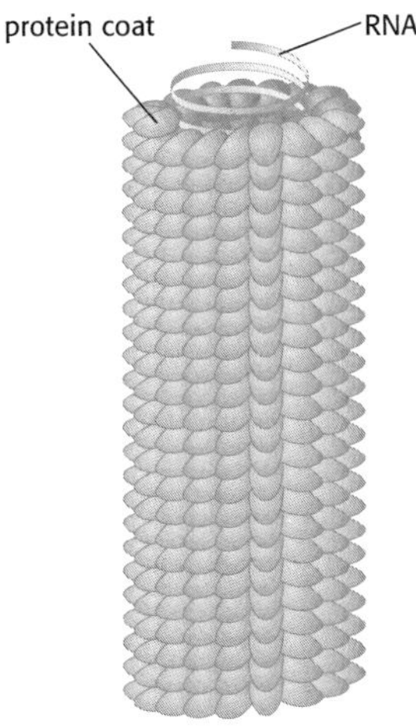

Figure 14 Tobacco mosaic virus is a single-stranded, naked, RNA capsid virus.

Some RNA viruses such as HIV rely on an enzyme called reverse transcriptase for reproduction. Reverse transcriptase converts the viral RNA into a complementary strand of DNA, digests away the original viral RNA template and then copies the new DNA to produce a double-stranded molecule of viral DNA. This DNA then becomes part of the host genome and directs the formation of new virus particles. Molecular biologists have made use of this property. Mature mRNA can be used to make DNA genes without the introns. The DNA can then be inserted into a bacterial host in genetic engineering applications. This is necessary because bacteria cannot carry out post-transcriptional modifications.

Data-based question: combination therapies for HIV infection

The incidence of simultaneous infection by human immunodeficiency virus (HIV) and *Mycobacterium tuberculosis* are steadily increasing and represent a major health crisis in many developing countries. Azidothymidine (AZT) is a drug that inhibits the activity of reverse transcriptase. LMP-420 is a drug that blocks the action of tumour necrosis factor (TNF), which is produced by white blood cells and plays a role in the proliferation of HIV particles and *M. tuberculosis* in an infected individual.

Figure 15 shows the quantity of HIV particles produced by infected blood cells treated with one or both drugs as a percentage of the number produced by untreated infected cells.

Figure 16 shows the growth rate of *M. tuberculosis cells* over a 120-hour period in the presence of TNF, with nothing added, with an anti-TNF agent added and with the drug LMP-420 added.

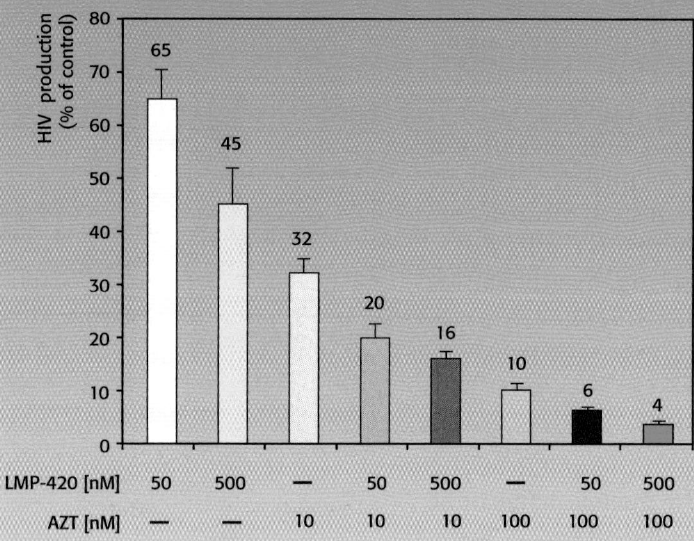

Figure 15 Effect of different combinations of the drugs LMP-420 and AZT on the ability of HIV to affect cells

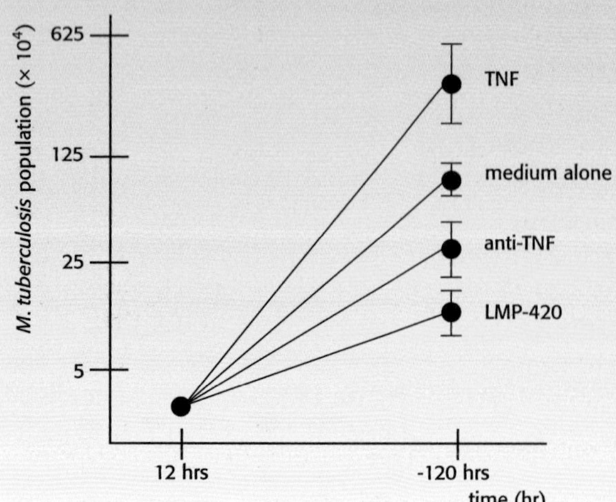

Figure 16 Tumour necrosis factor (TNF) is hypothesized to increase *M. tuberculosis* infection of cells in vitro. Anti-TNF and the drug LMP-420 block the action of TNF.

Table 2

Type of genetic material	Example
Double-stranded DNA (dsDNA)	Herpes virus
Single-stranded DNA (ssDNA)	*Enterobacter* phage X174
Double-stranded RNA (dsRNA)	Rotavirus
Single-stranded RNA (ssRNA)	Poliovirus

1 (a) Explain how AZT prevents HIV particle production. [2]
   (b) Determine the concentration of AZT alone that will decrease HIV particle production by 68 per cent. [1]
   (c) Compare the effectiveness of the different drug treatments in reducing HIV production. [3]

2 Explain the evidence from the data for TNF promoting *Mycobacterium tuberculosis* growth. [2]

Microbes and biotechnology

Some inherited diseases are caused by a defective gene, which results in the lack of a particular enzyme or protein. Cystic fibrosis is one such disease. It is caused by the lack of cystic fibrosis transmembrane protein (CFTP). This protein normally transports chloride ions out of cells and into mucus. The chloride ion draws water out of the cells and makes mucus watery. Cystic fibrosis patients suffer from thick mucus, which builds up in the airways.

Gene therapy may offer a cure for inherited diseases like cystic fibrosis. In gene therapy, working copies of the defective gene are inserted into a person's genome. To do this, a gene delivery system, or vector, is needed. Figure 17 shows how liposomes (artificial lipid vesicles) could be used as vectors to deliver working copies of the CFTP gene to a cystic fibrosis patient.

Though liposomes work in principle, in practice viruses are better vectors for gene therapy. Figure 18 shows two different ways of using viruses as vectors. The viral genome is altered so that the particles are not virulent. The therapeutic gene is then inserted into the virus. Gene therapy using retroviruses has successfully been used to treat severe combined immune deficiency syndrome (SCID). However two patients developed leukemia because the gene inserted itself into the genome at the wrong point, leading to uncontrolled cell division. This has led to concerns about using retroviruses as vectors.

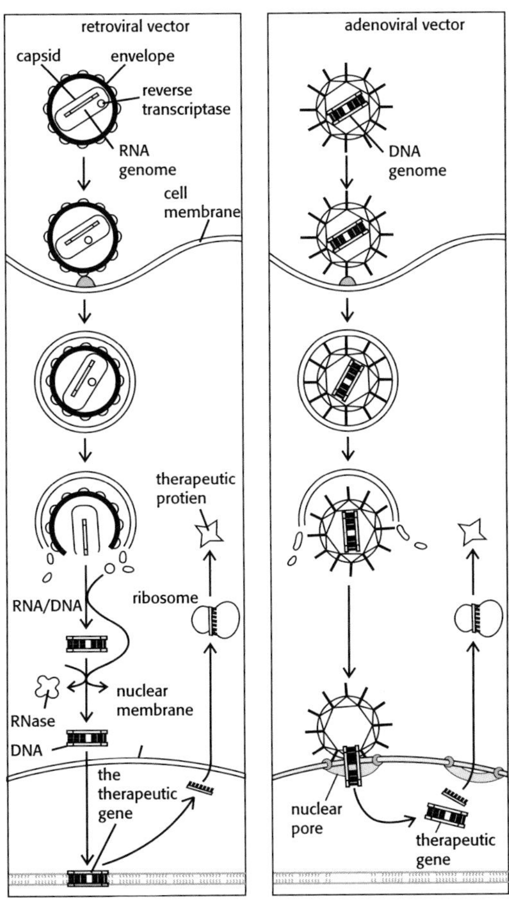

Figure 18 Two different gene therapy techniques involving viral vectors

Viruses that contain double-stranded (ds) DNA, such as adenovirus, cannot cause the problems found with retroviruses because the viral DNA is not inserted into the genome. However, the therapeutic gene is not passed on to the next generation of cells, so treatment has to be repeated more frequently. A challenge of using viruses as vectors is that the host may develop immunity to the virus.

The treatments described above are called somatic germ line therapy, because the cells being altered are somatic (body) cells. An alternative method would be to inject therapeutic genes into egg cells. The missing gene would be expressed in all cells of the organism.

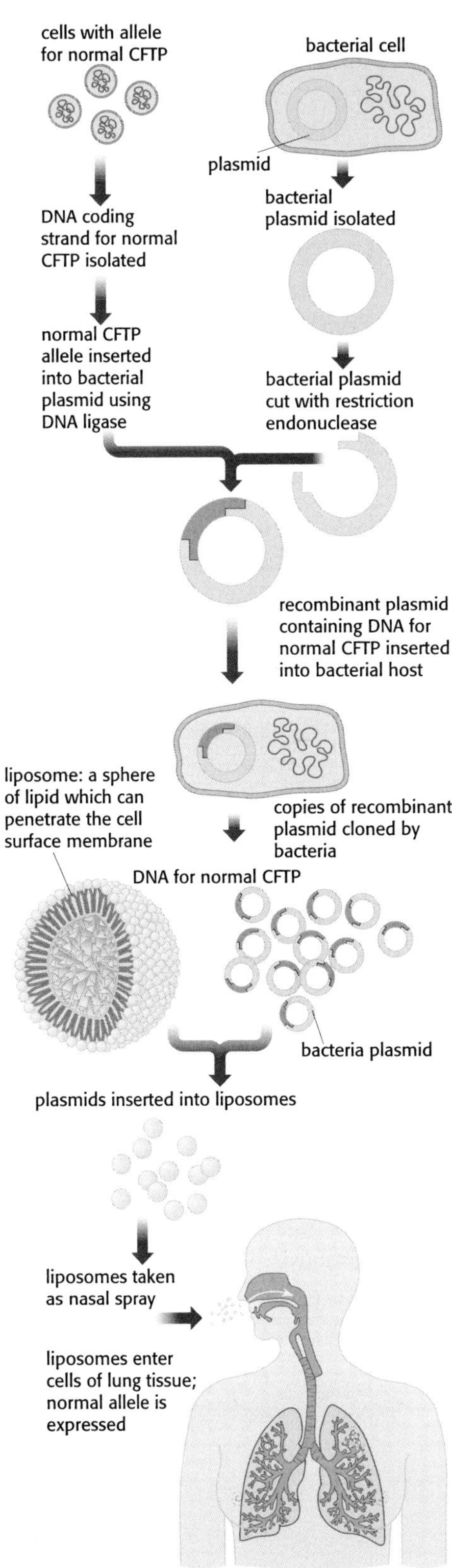

Figure 17 The procedure involved in using liposomes for gene therapy of cystic fibrosis

This is called germ line therapy. While such modifications have been successfully achieved in animals, it is currently illegal in many countries because of concerns over the risk of "designer babies", uncertainties over impacts later in life and the impact on the offspring of the treated individual.

Critical considerations: gene therapy

In 1999 a patient died as a result of participation in clinical trials for gene therapy. He suffered from ornithine transcarbamylase deficiency, or OTC, a liver disease marked by an inability to metabolize ammonia. Ammonia is a waste product of amino acid metabolism. He had been able to survive up to that point because of dietary modification and medication. The trial he participated in involved being injected with adenoviruses carrying the gene for transcarbamylase. He died within days due to a strong immune response to the viral vector. An investigation concluded that the scientists involved in the trial violated several rules of conduct.

- Four other patients who had received the treatment had reactions that were deemed so severe that the trial should have ended.
- The informed consent forms did not include information about primates that had died in similar trials.
- The patient had levels of ammonia that were so high he should have been excluded from the study.
- A principle investigator of the study had a major interest in the outcome of the trial as he held patents on the OTC treatment.

From *Welcome to the Genome* by Bob De Salle and Michael Yudell (Wiley).

1 Explain what is meant by informed consent.

2 (a) Suggest what policy instruments might be put in to place to prevent such occurrences.

(b) Who should administer these policies – governments, other scientists, or research institutions?

HL

Metabolism of microbes

Microbes can be classified by their mode of nutrition, including their sources of energy and of carbon (see Table 3). Phototrophs obtain their energy from light, while chemotrophs obtain their energy from organic compounds. Autotrophs require only CO_2 as a source of carbon. They can build their own organic molecules. Heterotrophs require organic molecules as carbon sources. The cyanobacterium *Anabaena* grows in a filamentous colony. Specialized cells in the filament called heterocysts fix nitrogen (see Figure 19). Intercellular connections allow the nitrogen to flow from the heterocysts and carbohydrates from photosynthetic colony members into the heterocysts.

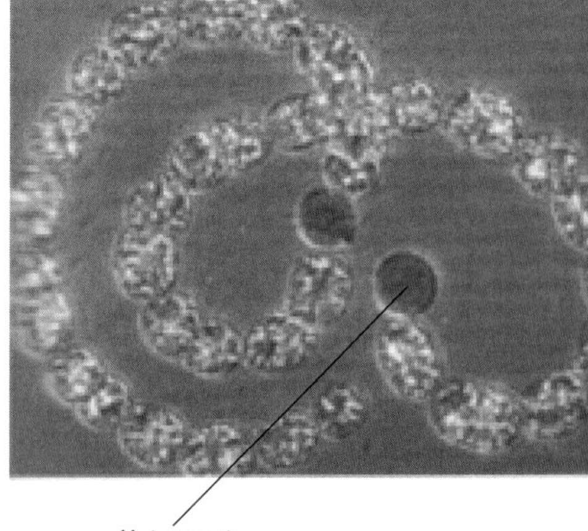

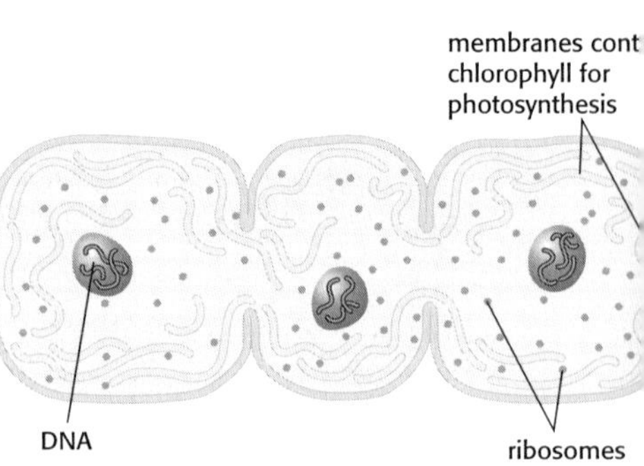

Figure 19 The cyanobacterium *Anabaena* fixes nitrogen and carries out photosynthesis.

Table 3

Category	Source of energy	Source of carbon	Example
photoautotroph	light	inorganic	cyanobacteria
chemoautotroph	inorganic chemicals; e.g. H_2S	inorganic	*Nitrobacter*
photoheterotroph	light	organic	*Rhodobacter*
chemoheterotroph	organic compounds	organic compounds	*Mycobacterium tuberculosis*

Bioremediation

Bioremediation is the use of microbes, fungi, plants or enzymes to remove environmental contaminants from water or soil. The bacterium *Dehalococcoides ethenogenes* has been used to break down chlorinated solvents in soil. The bacterium *Geobacter sulfurreducens* uses uranium as an electron acceptor converting it from a soluble to an insoluble form, which allows the uranium to settle out and be collected. Some members of the genus *Pseudomonas* can use crude oil for energy. They also require substances such as potassium and urea as nutrients. These are often sprayed on to an oil spill to aid the bacteria in their work.

Microbes and disease

Pathogenic microbes obtain nutrients through parasitism. To do this they need to gain entry to their hosts. Pathogens are transmitted in a number of ways, as summarized in Table 4.

After a pathogen enters the body, a disease can be caused if the microbe colonizes a tissue. Bacteria such as streptococci release digestive enzymes that allow them to invade tissues. *Streptococcus agalactiae*, also known as Group B streptococcus, can be a dangerous infection transmitted from mother to infant during childbirth. *Streptococcus pyogenes* is responsible for conditions such as "strep throat", scarlet fever and rheumatic fever.

Most bacterial infections are found around cells, but some actually infect cells intracellularly, such as *Rickettsia* and *Chlamydia*. The pathogen *Chlamydia trachomatis* is one of the leading causes of preventable blindness in the world.

In addition to the tissue damage due to colonization, bacterial toxins also contribute to the disease state. There are two categories of toxins. An exotoxin is a protein secreted by the bacterium. It may produce symptoms even in the absence of the bacterium. An endotoxin is a toxic component of the outer membrane of certain Gram-negative bacteria that is released only when they die. In Figure 4 on page 316, which shows the structure of the cell wall of Gram-negative bacteria, the label LPS refers to lipopolysaccharide. LPS is an endotoxin.

Exotoxins tend to be much stronger and are actively secreted by the living bacterium. For example, the exotoxin released by *Vibrio cholerae* causes intestinal cells to release chloride ions. Water follows the chloride ions, leading to severe diarrhea, one of the main symptoms of cholera. *Clostridium botulinum* produces a toxin that is a powerful inhibitor of neuromuscular junction activity. The botulinum toxin has been used in small doses as part of the commercial preparation termed "botox", which is used to help smooth out lines on the face.

Table 4

Method of transmission	Example
Direct contact between the infected and uninfected person or between a surface and the new host	Norwalk virus
Cuts	*Clostridium tetani*
Droplets: coming in contact with droplets from an infected person such as through a cough or sneezing	Influenza virus
Ingesting contaminated food or water	*Salmonella* infection
Insects act as vectors *	West Nile virus
Sexually transmitted diseases	*Chlamydia*

* see page 327.

Investigating biology: controlling microbial growth

A number of methods exist for controlling microbial growth. They include sterilization, pasteurization, antibiotics, antiseptics, disinfectants and irradiation. Depending on local restrictions students can make spread plates where nutrient agar is innolculated with a volume of pure culture spread out over the surface. Paper disks can be impregnated with the anti-microbial agent and placed on the spread plate. After a period of time, the plates can be examined to determine the size of the zone of exclusion of growth. The plate shown in Figure 20 shows the effectiveness of various antibiotics against the bacterium *Staphylococcus aureus*. Table 5 shows some of the mechanisms of actions of certain antibiotics.

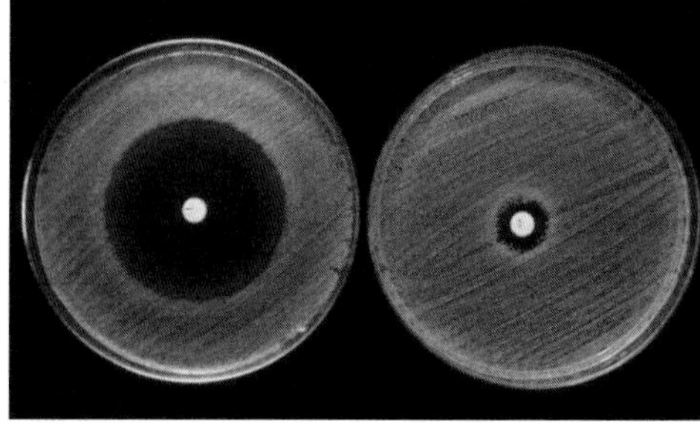

Figure 20 The zone of inhibition around an antibiotic disc is a measure of the effectiveness of that antibiotic against that particular microbe.

Table 5

Antibiotic	Source	Mode of action
Penicillin	Produced by the fungus *Penicillium notatum*, the green fungus that is commonly found on the surface of oranges	Prevents the cross-linking of proteins in the cell wall of Gram-positive bacteria
Erythromycin	*Saccharopolyspora erythraea*, a soil fungus	Binds to the bacterial ribosome and blocks the exit of the growing polypeptide, thus preventing protein synthesis
Rifamycin	*Amycolatopsis mediterranei*, a soil bacterium	Binds to RNA polymerase and prevents transcription
Polymyxin	*Bacillius polymyxa*, a bacterium	Acts as a detergent, dissolving bacterial cell membranes
Sulfanilamide	Chemically derived	Competitive inhibition of pathways leading to folic acid synthesis
Nalidixic acid	Chemically derived	Prevents DNA replication

Inquiry

1 (a) Outline the following methods of controlling microbial growth in food:
 (i) irradiation
 (ii) pasteurization.
 (b) Evaluate both methods.

2 Compare antiseptics and disinfectants.

HL

Influenza

Influenza (flu) is caused by a virus. The virus is spread by droplets, such as those released by an infected person when they cough or sneeze. When the viruses are inhaled, they bind to the surface of epithelial cells in the respiratory system. They are brought inside through the cell's own endocytosis mechanism.

Influenza is a single-stranded, enveloped RNA virus. Once inside the cell, the genetic material and accessory proteins are uncoated and the viral molecules enter the nucleus (see Figure 21). The viral enzyme RNA-dependent RNA transcriptase begins making complementary copies of the viral RNA. The viral RNA is either exported into the cytoplasm and translated, or it remains in the nucleus. Some newly synthesized viral proteins are inserted in the cell membrane and some are transported back into the nucleus to assemble the new viral particles.

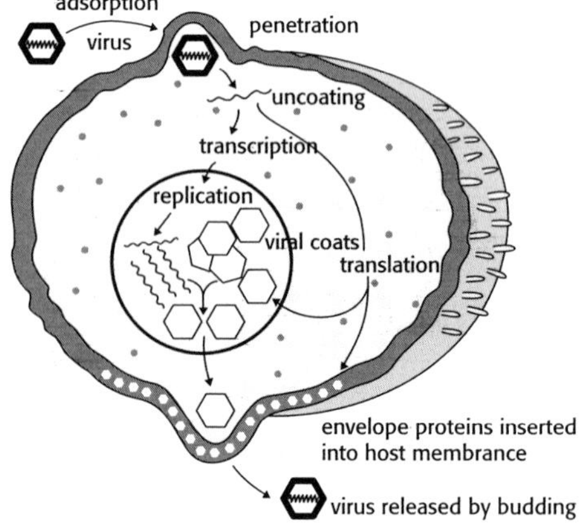

Figure 21 The influenza "life cycle"

Epidemiology

Epidemiology is the study of disease in human populations. It includes the occurrence, distribution and control of diseases. An epidemic is a widespread outbreak of an infectious disease, in which many people are infected at the same time. A pandemic is a very widespread epidemic that affects a large geographic area and crosses international boundaries. In 1918 and 1919, an influenza pandemic killed between 20 and 40 million people across the globe. Dubbed the "Spanish flu", evidence suggests that the virus was a bird zoonosis, i.e. a disease of birds that infected humans. The spread of the disease was aided by population movements due to world war I.

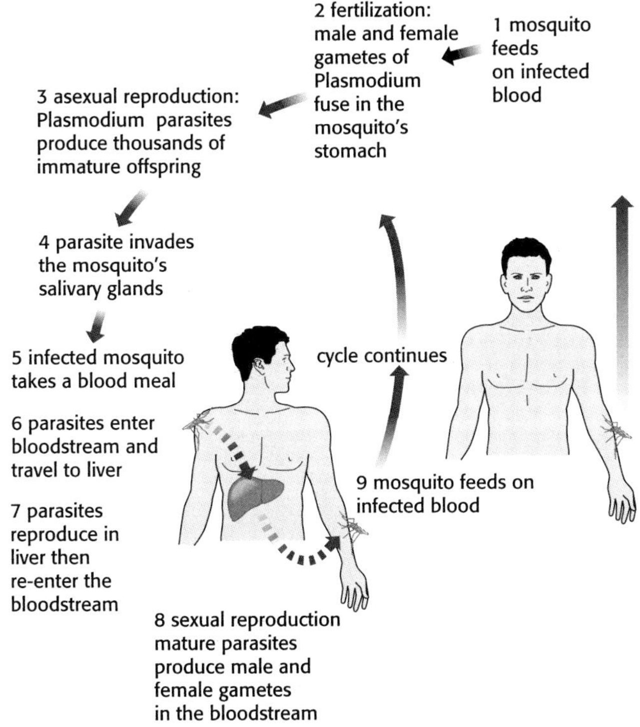

Figure 22 The life cycle of *Plasmodium*

Malaria

Figure 22 outlines the life cycle of the unicellular eukaryote *Plasmodium* sp., the pathogen that causes malaria. There are four different species of malarial parasite that infect humans, *P. falciparum*, *P. vivax*, *P. ovale* and *P. malariae*. The first two species are the most common causes of infection. *P. falciparum* often causes a severe, potentially fatal form of malaria. The *Plasmodium* parasite invades red blood cells, grows and multiplies. The blood cells burst and the released parasites infect additional red blood cells. The parasite is transmitted from person to person via the *Anopheles* mosquito. A carrier of an infectious agent that transmits infection from one host to another is called a vector. In this case, the *Anopheles* mosquito is the vector.

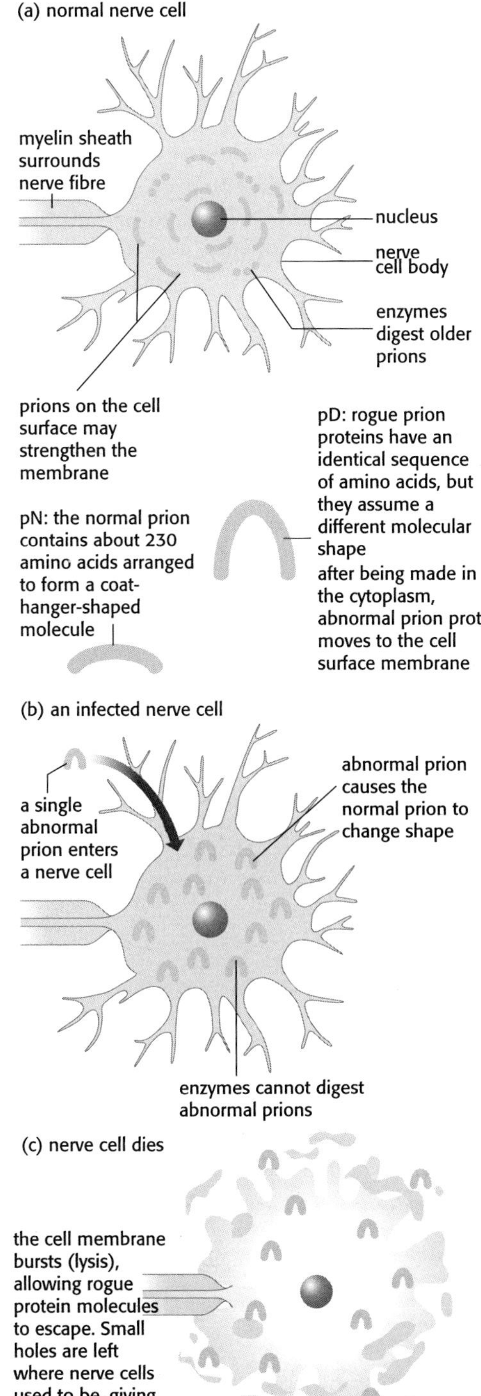

Figure 23 The prion hypothesis

Inquiry: spongiform encephalopathies

1 Creutzfeldt-Jakob disease (CJD) is the most common form of transmissible spongiform encephalopathies in humans. The prion hypothesis of transmission is outlined in Figure 23. Outline the effects of transmissible spongiform encephalopathies (TSE) on the brain.

2 (a) List three other examples of transmissible spongiform encephalopathies.
(b) Why is the explanation called the prion *hypothesis*? Is the hypothesis controversial?

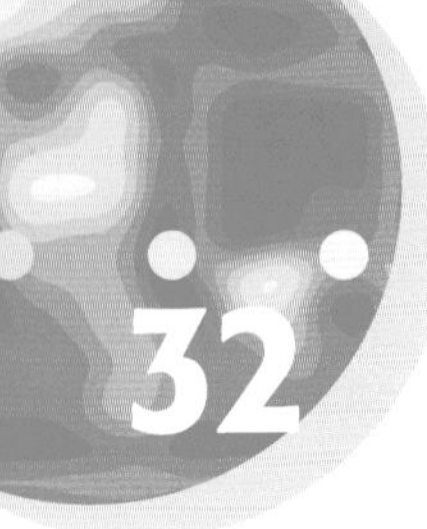

32 Option G: Ecology and conservation

Productivity refers to the quantity of energy fixed per unit area per unit time in an ecosystem by a particular trophic level. The net productivity is affected by the amount of energy that is used by the plant in respiration.

net productivity = gross productivity – energy used in respiration.

Within an ecosystem, each organism fulfills a unique role, called its ecological niche. All of the biotic and abiotic resources that the organism interacts with are components of its niche. If, in a certain habitat, the range of a variable falls outside what an organism can tolerate, then the organism will not be found there. For a plant, temperature, water, light, soil pH, soil salinity and mineral nutrients are examples of factors which limit distribution.

Inquiry

The distribution of animal species is affected by biotic and abiotic factors. Using examples of animals found in your local area, suggest how their distribution might be affected by

1 temperature
2 water
3 breeding sites
4 territory.

Inquiry

Transects

The distribution of organisms within a particular habitat is affected by a number of limiting factors. One method used to assess distribution is a transect. A tape is laid along the ground between two poles. In a line transect, sampling can be confined to describing all of the organisms that touch the line. In a belt transect, sampling is confined to a strip of fixed width such as 0.5 or 1 m.

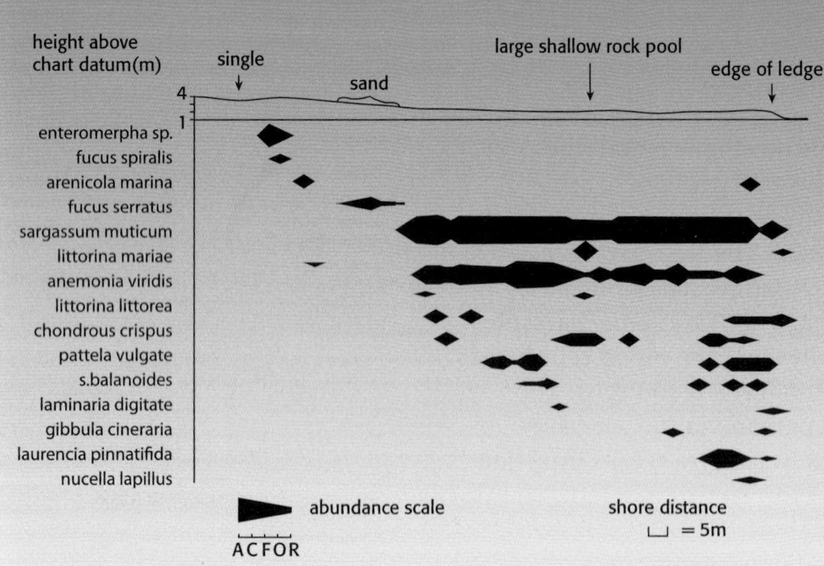

Figure 1 Species abundance as a function of distance from the shore

The kite diagram (Figure 1) illustrates the distribution of common intertidal species 300 m south of Bembridge Lifeboat Station on the Isle of Wight, UK. The thickness of the shaded region indicates whether the organism is abundant, common, frequent, occasional or rare according to the key (Abundance Scale).

?

1 (a) Examine the kite diagram and suggest a procedure by which such data could be collected.
(b) Through research, describe one method for determining the size of a population of organisms that make use of a quadrat.
2 Using the scale bar, determine the length of the large shallow rock pool.
3 Choose three species from the list shown. Conduct research on these organisms. Suggest factors that would affect the distribution of the species as shown.

The competitive exclusion principle

In the 1930s the Russian scientist Carl Friedrich Gauss investigated competition between two species of *Paramecium*: *P. caudatum* and *P. aurelia*. When cultured separately, under ideal laboratory conditions, both thrived. When cultured together, *P. caudatum* was eliminated.

The bay-breasted warbler and the yellow-rumped warbler are migratory birds that appear to occupy the same niche, as they feed on similar prey items and can be found foraging together on the same tree. Figure 2 illustrates observed parts of the tree where each warbler most commonly forages. Note that the birds feed in such a way as to avoid competition with one another.

It appears from these observations that two species cannot coexist in the same habitat if their niches overlap. This is known as the competitive exclusion principle. Either one species will lead to the decline and extinction of the other, or one or both of the competitors will narrow their niches to avoid competition, i.e. their actual or realized niche will be narrower than their potential or fundamental niche.

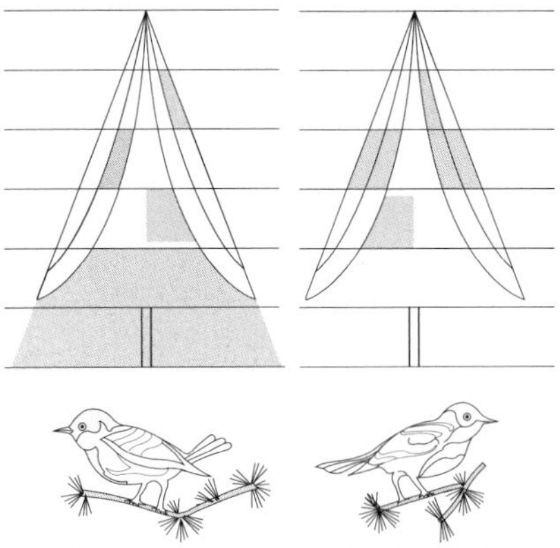

Yellow-rumped warbler

Bat-breasted warbler

Figure 2 The yellow-rumped warbler and bay-breasted warbler have similar prey yet they tend to forage in different parts of the same tree to avoid competition.

Investigating biology: competitive exclusion

Design a procedure to investigate a model of competitive exclusion using bird feeders. This can be done by examining species abundance and behavior at feeders hung at different heights, hung in different parts of a tree or containing different seeds. Alternatively, ant niches can be investigated by observing patterns of visitation timing, degree of aggression toward other ant species or preference for bait types.

Inquiry

Competition is only one possible type of interaction between two species. Referring to organisms found in your local area, find two examples of each of the following:

1 herbivory
2 predation
3 parasitism
4 mutualism.

Working with data: measuring biomass

Biomass is the dry weight of organic matter of a group of organisms found in a habitat. It may refer to the entire community or to a single trophic level or even to a single individual. There are a number of ways to measure plant biomass with varying degrees of destructiveness. In the area set aside for biomass sampling, randomly locate a quadrat.

1 Plant biomass can be assessed indirectly and completely nondestructively by
   (a) Counting the number of individuals of the target species.
   (b) Randomly selecting a sample of individuals.
   (c) Determining mean height within the sample (height will be an indirect measure of biomass).
   (d) Multiplying the mean height by the stem density.

2 A second, slightly more destructive method involves taking a sample of individuals of the target species and cutting them at the soil level. Tag each individual with a label, dry it to a stable weight and weigh it. Determine the mean mass of plants in the area and multiply by the stem density in the area.

3 A third method is to completely harvest all the target species in the quadrat, dry them to a stable weight and then determine the total biomass.

Regardless of the method, it is advised that organic matter should be returned to the harvest location.

To construct a pyramid of biomass, herbivorous insects can be collected from a plant to determine their mass and then the biomass of the plant being consumed can be determined.

Data-based question: calculating productivity values.

The energy flow diagram in Figure 3 is for a temperate ecosystem. It has been divided into two parts. One part shows autotrophic use of energy and the other shows heterotrophic use of energy. All values are kJ m^{-2} yr^{-1}.

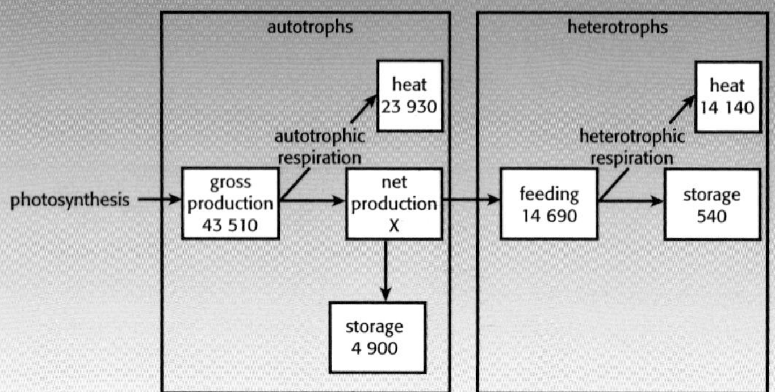

Figure 3 An energy flow diagram for a temperate ecosystem

1 Calculate the net production of the autotrophs. [1]
2 Compare the percentage of heat lost through respiration by the autotrophs with that lost by the heterotrophs. [1]
3 Most of the heterotrophs are animals. Suggest one reason for the difference in heat losses between the autotrophs and animal heterotrophs. [1]

Investigating biology: leaf area index

Leaf area index (LAI) is the ratio of the total area of all leaves on a plant to the area of ground covered by the plant (see Figure 4). For example, if a plant had a total leaf area equal to the ground it covered, then the leaf area index would be 1.0.

Design an experiment to assess a variable affecting leaf area index in a deciduous plant.

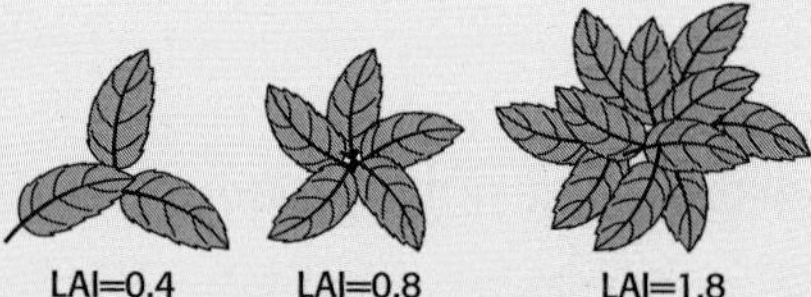

Figure 4 Leaf area index is the ratio of the total leaf surface area of a plant to the surface area of ground that is covered by the plant.

Succession

In an ecosystem, abiotic factors set limits to the distribution of biotic factors while biotic factors have an effect on abiotic factors. Consider a forest next to a field. Relative to the field, the forest will be darker, more humid and cooler, largely because of the presence of the trees. Leaf litter from trees will increase the rate of water infiltration in the soil, increase the nutrient concentration of the soil and will directly and indirectly affect the aeration of the soil.

Figure 5 Surtsey Island, off the coast of Iceland, was created in 1963.

Biotic factors may actually change the abiotic factors in the environment to such an extent that the environment becomes limiting to them and other species become better suited. This is known as succession. Figure 5 shows an image of a land mass being created because of volcanic activity. Primary succession is expected to occur on this island, i.e. a pioneering group of plant species will arrive on the island and contribute to the development of an ecosystem that will likely ultimately stabilize with different dominant species. Figure 6 summarizes a similar process, called secondary succession, where the change in dominant species with time again occurs as a result of the pioneering species altering the abiotic conditions such that other species become better suited. During the process of succession, initial changes are often much more rapid than subsequent changes.

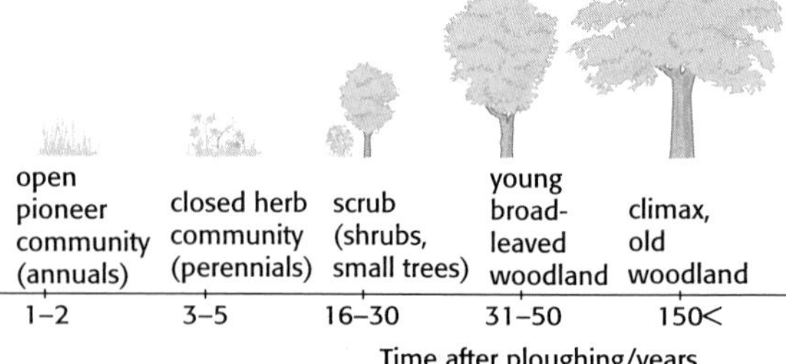

Figure 6 Secondary succession is the progression of communities where a pre-existing climax community has been disturbed but the soil is already developed.

Rediscovering biology: the Simpson index

The Simpson Index is used to assess plant species diversity. It is an index that takes into account the variety of species and the relative numbers of each of these individual species.

$$D = \frac{N(N-1)}{\Sigma n(n-1)}$$

Where N = total number of all individual organisms of all species found and n = number of individuals of a particular species.

The value of the index allows a comparison between two different areas. Locate an area undergoing secondary succession. Design an experiment to determine what happens to species diversity as secondary succession proceeds. You will need to compare a disrupted area or perhaps multiple disrupted areas with a stable community within the same ecosystem.

Data-based question: secondary succession

The three boxes in Figure 7 show a field undergoing secondary succession. It is shown 5 years, 25 years and 30 years after the original disruption. Each numbered shape represents a distinct plant species.

1 State what changes are occurring in:
   (a) leaf area index
   (b) number of species visible.
2 Deduce what changes might be occurring in:
   (a) productivity
   (b) diversity
   (c) soil quality
   (d) the water cycle.

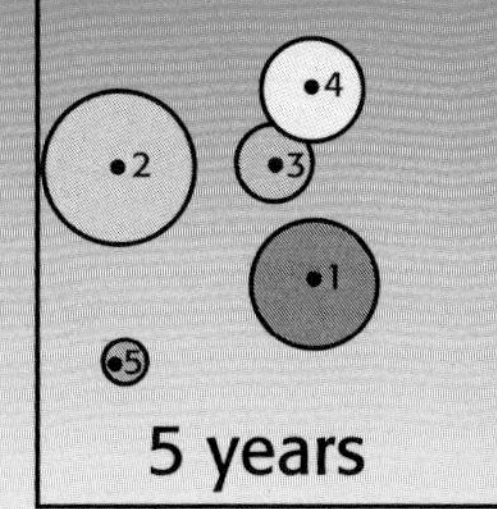

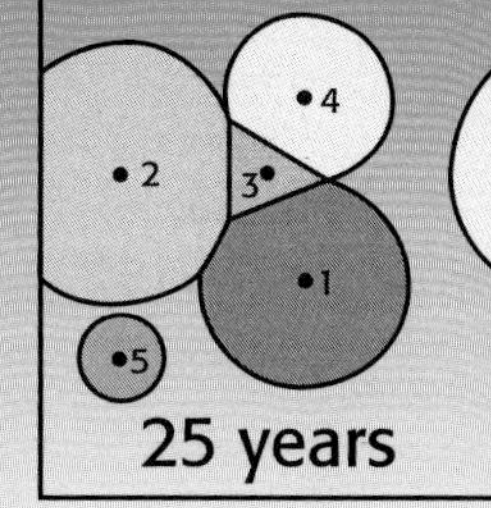

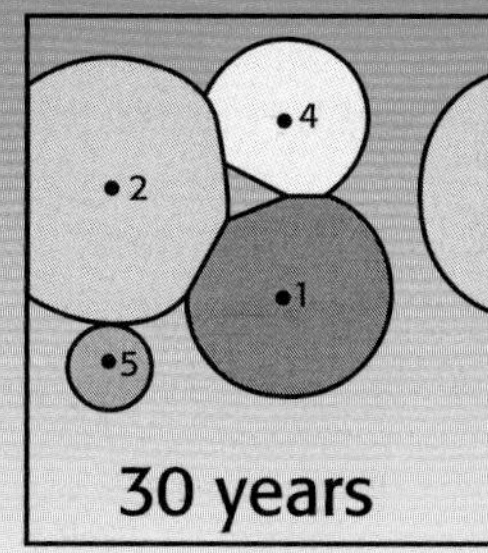

Figure 7 Community changes 5 years, 25 years and 30 years after an initial disturbance

Inquiry

Biomes

A biome is an ecological region dominated by a certain type of ecosystem characterized by certain precipitation and temperature conditions leading to a distinctive biological community adapted to those conditions.

Six of the word's major biomes are:
   (a) desert
   (b) grassland
   (c) shrubland
   (d) temperate deciduous forest
   (e) tropical rain forests
   (f) tundra.

1 Using the climograph in Figure 8, determine the range of temperature that exists for each biome.
2 Divide the biomes among your classmates and research the characteristics of vegetation in each of the biomes.

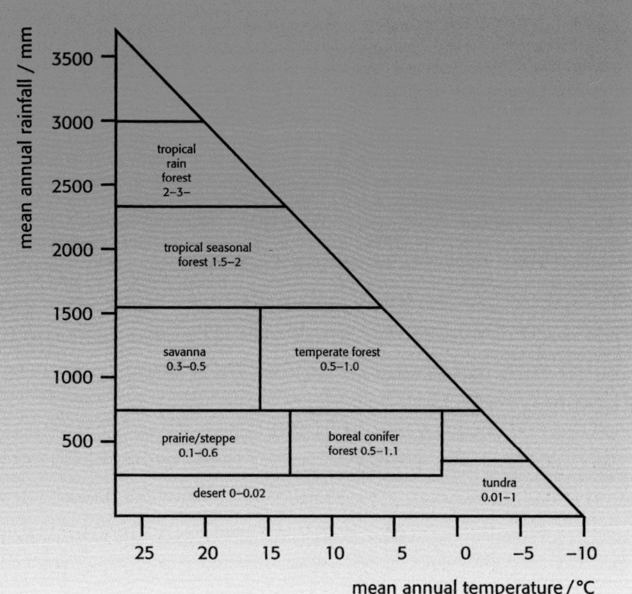

Figure 8

Invasive species

Human activity often results in an organism being introduced to an area where it did not previously occur naturally.

Because of niche overlap with native species and the absence of predators and other limiting factors present in its own habitat, the introduced species can become so reproductively successful and aggressive that it dominates the ecosystem. This can pose serious threats to the diversity of the habitat and so the introduced species is considered to be **invasive**.

In some cases, biological control can be used to limit an invasive species. This involves using natural predators of the invasive species to control its spread. For example, purple loosestrife is a plant that has become invasive in North America. A beetle (*Galerucella*) that eats the plant has been successfully used to limit its spread.

Sometimes biological control can go wrong. The cane toad (*Bufo marinus*) was introduced to Australia to control the cane beetle (*Dermolepida albohirtum*). Unfortunately the toad has become a generalist predator and is having a significant effect on Australian endemic species.

Inquiry

Choose one invasive species from Table 1 or elsewhere. Research the following aspects about the species:

1. its original habitat
2. predators in its original habitat
3. the vector or mechanism by which it became invasive
4. its ecological effect on its non-native habitat
5. other effects on its non-native habitat such as agricultural, tourism, public health etc.

Table 1 Some invasive species

Species	Location affected
Cane toad (*Bufo marinus*)	Australia
Japanese knotweed (*Fallopia japonica*)	Europe
Coconut rhinoceros beetle (*Oryctes rhinoceros*)	New Zealand
Mexican poppy (*Argemone mexicana*)	India
Zebra mussel (*Dreissena polymorpha*)	North America
Grey squirrel (*Sciurus carolinensis*)	UK

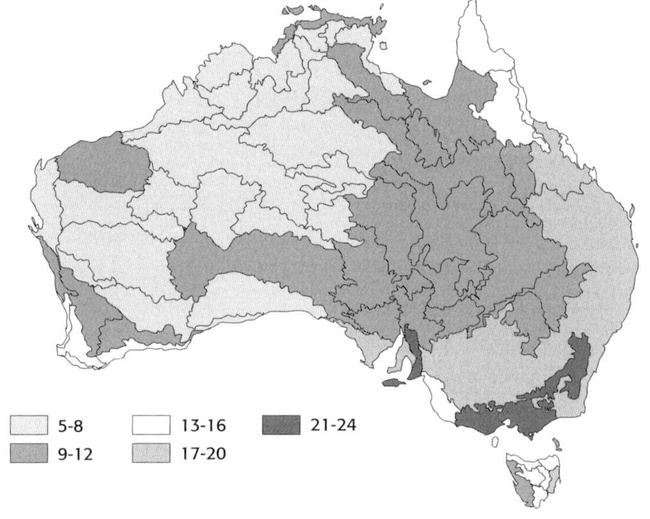

Figure 9 Number of invasive species in various geographic regions in Australia

Suggest reasons for the pattern of distribution of invasive species in Australia shown in Figure 9.

Biomagnification

Certain toxins can build up in the body of an organism, particularly if they are fat-soluble. This is known as bioaccumulation. The effects of the toxin can be magnified up the food chain, with top predators having increasing concentrations stored in their body. This is known as biomagnification.

Data-based question: biomagnification of caesium

In addition to nutrients, other atmospheric elements may also enter the ecosystem. Radioactive caesium-137 was released into the atmosphere by atomic bomb tests in 1961. The caesium-137 was deposited in the soil and on to plants. Figure 11 shows the amount of radioactivity found in the tissues of lichens (an alga and a fungus growing togather), caribou (a member of the deer family) and Inuit people in the Anaktuvuk Pass of Alaska

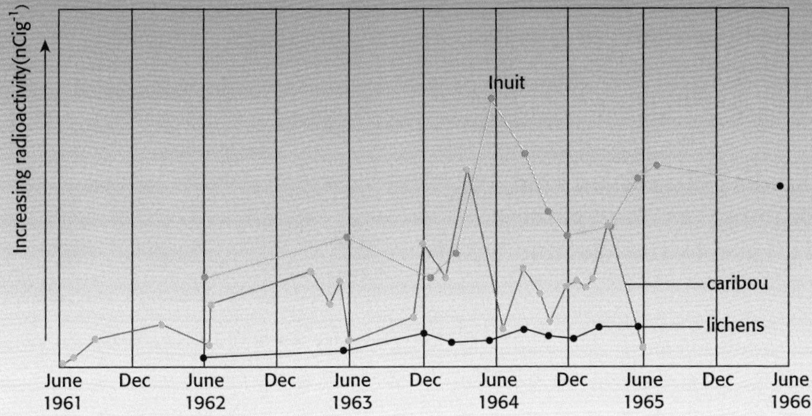

Figure 11 Radioactivity levels in various trophic levels in an Alaskan community after an atmospheric atomic bomb test

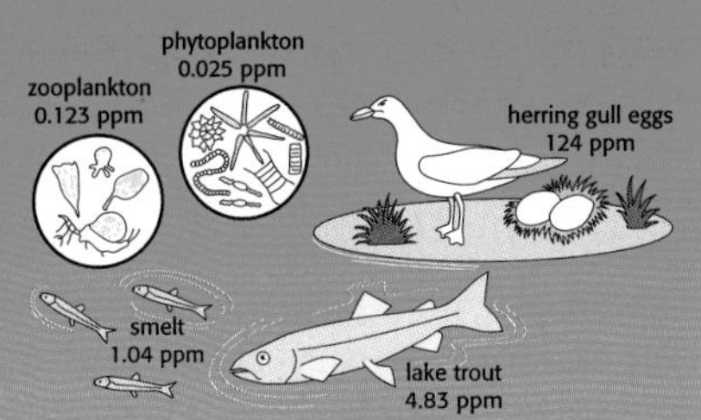

Figure 10 The concentration of pollutants called PCBs in each level of the North America Great Lakes aquatic food chain (in parts per million, ppm)

1. Describe the level of caesium-137 in the Inuit from 1962 through to 1965. [2]
2. The three organisms form a food chain. Deduce the trophic level of:
   (a) lichens
   (b) the Inuit. [2]
3. Suggest a reason for the difference in the amount of caesium-137 found in lichens, caribou and the Inuit.

Data-based question: ozone depletion

The ozone layer in the atmosphere is an effective barrier to ultraviolet A (UVA) and ultraviolet B (UVB) light. UVA and UVB light penetrate the surface waters of the ocean and cause damage to marine organisms. Destruction of the ozone has occurred because of CFCs (chlorofluorocarbons) in the atmosphere added by human activity. The most extensive destruction of the ozone layer occurs over Antarctica and the Southern Ocean. This causes increased amounts of UVA and UVB light to reach Antarctic marine environments.

The effect of UV light on the primary productivity of phytoplankton (microscopic photosynthetic organisms) at various depths was investigated in the Antarctic Ocean. Primary productivity was measured under three conditions:

- full sunlight
- full sunlight with UVB removed using a suitable filter
- full sunlight with UVB and UVA removed using a suitable filter.

Discuss the effect of a reduction in the ozone layer on marine food chains. [5]

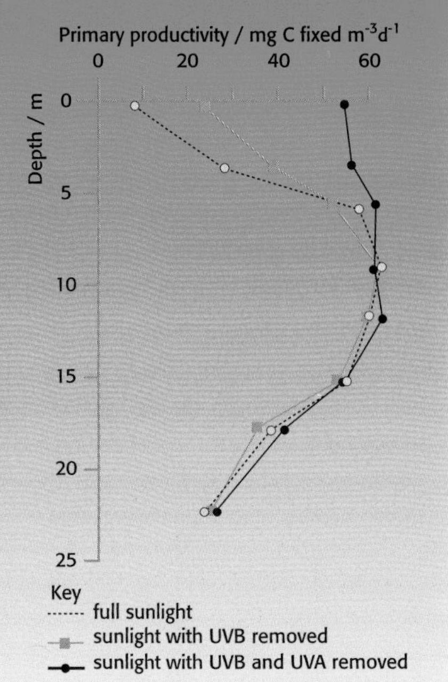

Figure 12 Primary productivity as a function of depth and amount of UV light exposure

Indicator species and biotic indices

An indicator species is an organism that occurs only when specific environmental conditions are present. The presence or absence of these species on a site is a good indicator of environmental conditions. For example, the distribution of understorey plants in a forest is a good indicator of such things as soil fertility or water drainage. Fruticose lichens are one group of lichens that are relatively pollution-intolerant, so their presence is an indicator of clean air.

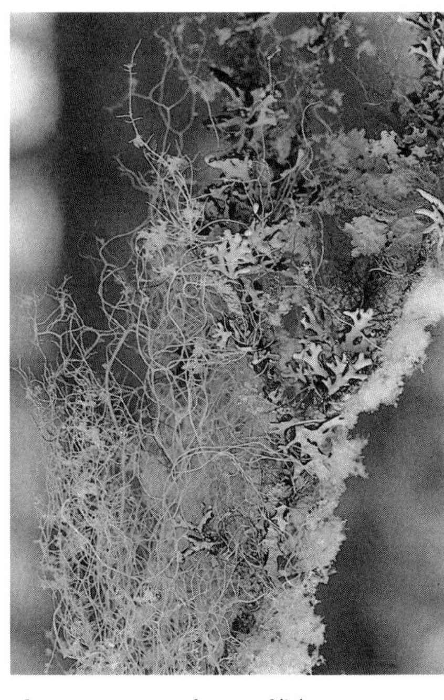

Figure 13 Some forms of lichen are indicators of clean air

A biotic index compares the relative frequency of indicator species. For example, the macroinvertebrate biotic index is a measure of stream health. The number of individuals of each indicator species in a sample is determined. Each number is multiplied by a pollution tolerance factor and a weighted average is determined.

One possible biotic index multiplies the number of a certain kind or organism (n_i) by its pollution tolerance rating (a_i). Each of these products is then added to the others and divided by the total number of organisms in the habitat.

$$BI = \frac{\Sigma n_i a_i}{N}$$

Investigating biology: biotic index

Figure 14 shows six different benthic (bottom-dwelling) macroinvertebrates that are found in rivers. Benthic macroinvertebrates are useful indicators of stream health for a number of reasons: they live in the water for the duration of the aquatic stage of their life cycle and so they reflect conditions in the water over a period of time; they are easy to capture; and most importantly, they vary in their response to chemical and physical changes in their habitat.

As pollution increases, oxygen levels in the water decrease. These low oxygen levels can only be tolerated by certain organisms.

In the kick-sampling procedure, a net is placed downstream of the collector with the edge of the net resting on the substrate of the stream. The collector walks backwards away from the net, kicking the substrate to disturb it. The net is kept close to the area being disturbed, to allow the current to carry disturbed material into the net.

If the collector is looking at overall stream health, then they should move across the stream and back in an upstream direction for a fixed period. This is necessary because certain species may be confined to a particular region of the stream. The net should then be emptied and relative numbers of each species determined. A local biotic index will provide pollution tolerance ratings for certain organisms. These ratings could then be considered along with the relative number of each kind of organism.

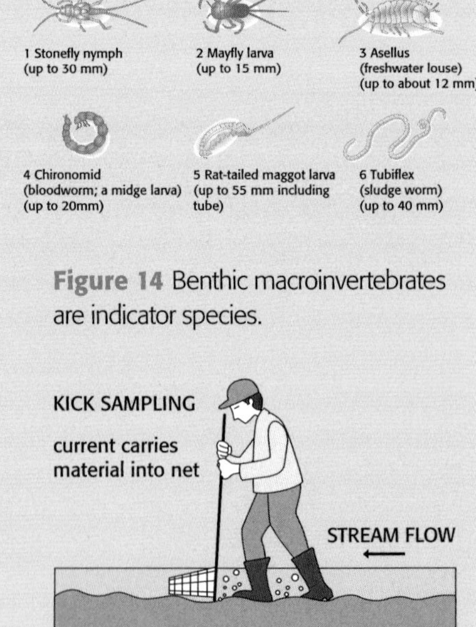

Figure 14 Benthic macroinvertebrates are indicator species.

Figure 15 Kick sampling is a method of collecting samples of benthic invertebrates.

Data-based question: spiders as indicator species

Old-growth forests are ecosystems that exhibit unique biological features. They typically contain large live trees, large dead trees and multiple layers of vegetation representing a variety of tree species. Old growth redwood (*Sequoia sempervirens*) forests have increasingly been logged in recent years.

Insects, litter spiders and other arthropods can be used to distinguish between old-growth forests, second-growth forests and tree farms. Second-growth redwood forests grow in an area following logging or other disturbances that removed the old trees. Tree farms are managed areas with uniformly aged trees of one species that are harvested by clear-cutting.

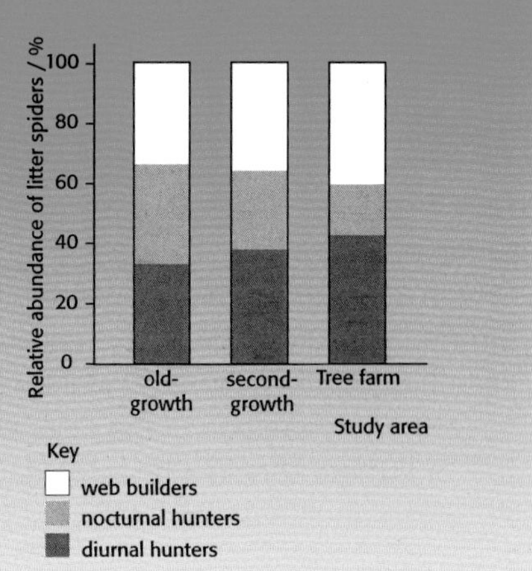

Figure 16 The relative abundance of litter spiders in a study area

1 Compare the relative abundance of types of litter spiders found in an old-growth forest to that of a second-growth forest. [2]
2 Suggest how the abundance of spiders can be used as a biotic index. [2]

Habitat loss is a significant variable in the extinction of species. Habitat fragmentation due to human activity often isolates small populations. Because of their size, these small populations are likely to go extinct. Human activities such as mining, agriculture, urban development, forestry and pollution can fragment ecosystems. When designing nature reserves it is critical to retain corridors between patches to maintain the connection between isolated populations (Figure 17).

At the boundary between two ecosystems there are unique conditions. Some organisms thrive because they are able to take advantage of resources in both ecosystems. Others who are adapted to narrow niches that are found only in the interior, suffer when fragments are below a critical size.

Figure 17 A viaduct for wildlife crossing in the Netherlands

Data-based question: cowbird parasitism

The brown-headed cowbird (*Molothrus ater*) is a parasitic bird that lays its eggs in the nests of other species. The parasitized hosts often raise the resulting cowbird offspring as their own. The true offspring may starve while the larger cowbird offspring consume most of the food brought by the parents.

The preferred habitat of the brown-headed cowbird is open agricultural areas. The results of a study into the effects of deforestation on cowbird parasitism of four different host species are shown in Figure 18. The vertical axis shows the percentage of nests parasitized.

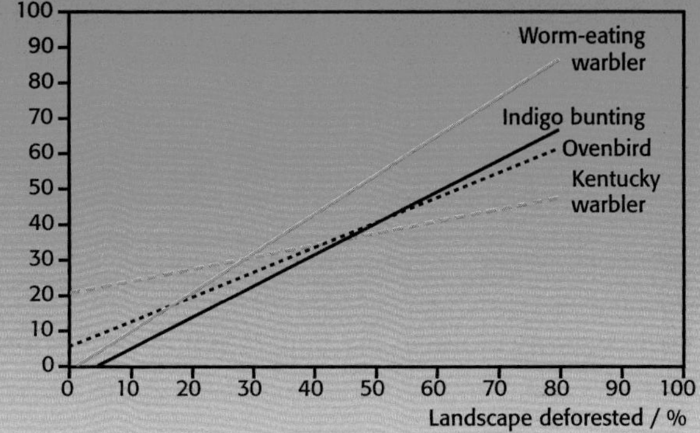

Figure 18 The effect of deforestation on cowbird parasitism

1 State the effect of deforestation on cowbird parasitism. [1]
2 Compare the effect of deforestation on cowbird parasitism of the worm-eating warbler and the Kentucky warbler. [2]
3 Determine the percentage of worm-eating warbler nests parasitized by cowbirds at a level of 60% deforestation. [2]

Critical considerations: **in situ *or* ex situ?**

Working in pairs, identify a species that belongs in Appendix 1 of CITES (the Convention on the International Trade in Endangered Species). Then carry out the following tasks.

Task 1

Group member A:

Design a nature reserve that would meet the needs of that species *in situ* (in its natural habitat) and encourage recovery. Design a "visitor's pamphlet" for the reserve, that includes:

- a short description of the reserve, including the kind of habitat it is and where should it be located
- a section that describes the species and gives reasons why it is endangered
- a list of the rules and regulations of the reserve. Outline what kinds of activities are allowed, what kinds of activities are prohibited (for example, "No hunting")
- an outline of the management challenges of the reserve.

Group member B:

With reference to the same species, imagine a comprehensive *ex situ* conservation programme that would avoid the extinction of that species and encourage recovery. Design a "funding proposal" for your programme aimed at preserving the species. Include the following in the proposal:

- a short description of the aspects of your conservation programme, including a list of the resources required, including artificial habitat and where should it be located
- an information section that describes the species and reasons for its endangerment
- an outline of the challenges facing the conservation programme.

Task 2

Discuss with your partner the strengths and limitations of *in situ* and *ex situ* conservation approaches. Make a choice about which approach you would recommend. Make a short presentation, or a written statement, or make a short post in the class forum defending your recommendation. Be explicit about your criteria for judgment.

Having reviewed the proposals of all of your classmates, be prepared to discuss in class the advantages and disadvantages of *in situ* and *ex situ* conservation measures.

Investigating biology: the population of your school

Design a procedure to determine the population of students in your school using sampling methods. One possibility is to use the Lincoln Index procedure. This makes use of the equation:

$$\text{population size} = \frac{n_1 \times n_2}{n_3}$$

where n_1 = number of individuals initially caught, marked and released; n_2 = total number of individuals caught in the second sample; and n_3 = number of marked individuals in the second sample.

The r–K spectrum

In a predictable environment, in order to maximize fitness, it pays to invest resources in long-term development and long life (K-strategy). In an unstable environment, it is better to produce as many offspring as quickly as possible (r-strategy).

A K-strategy involves investing more resources in development and long-term survival. This involves a longer lifespan, late maturity, and is more likely to involve parental care, the production of few offspring, and reproducing more than once. The "K" refers to the variable normally used to represent carrying capacity.

An r-strategy involves investing more resources into producing many offspring, having a short lifespan with early maturity, reproducing only once and having a small body size. Table 4 summarizes these strategies and gives examples of organisms with extreme r- or K-strategies.

Most organisms have life histories that are intermediate on the r–K continuum. Over time, some populations, such as *Drosophila* can switch strategies depending on environmental conditions.

Disruption to ecosystems, such as might be observed with climate change, favours r-strategists such as pest species. Table 5 summarizes some observations of plant pathogen population booms during warm weather conditions.

Table 4

r-strategists	K-strategists
short-lived	long-lived
small	large
fast maturation	slow maturation
opportunistic	consistent
large number of offspring	small number of offspring
small relative size at birth	large relative size at birth
little care for offspring	more care for offspring
variable population size	stable population size
examples: rabbits, pathogens	examples: elephants, parrots

Table 5

Pathogen	Effect
the fungus causing Mediterranean oak decline, *Phytophthora cinnamomi*	more severe root rot at higher temperatures than the current Mediterranean average
Dutch elm disease fungus (*Ophiostoma novoulmi*)	greater defoliation in warmer years
beech bark cankering by *Nectria spp.*	worse after mild winters or dry autumns
Wheat stripe rust (*Puccinia striiformis*)	less severe during years of low temperature and rainfall

Source: "Climate Warming and Disease Risks for Terrestrial and Marine Biota.", by Harvell, C. et al, *Science*, 2002

Fishing

Fish are an important food resource. Because they are an open access resource on the high seas, the incentives are limited for conservation.

An important component of managing fish is clear data about fish populations. The concept of maximum sustainable yield is related to the sigmoid growth curve. At point 2 in the graph in Figure 19 the population is growing at its maximum rate. If fish were harvested at this rate, then fishing would be able to continue indefinitely.

Figure 20 shows a graph of sustainable yield versus intensity of fishing. If there is no fishing, then the yield of fish would be zero. If there is very high-intensity fishing, it may be that the population of fish becomes

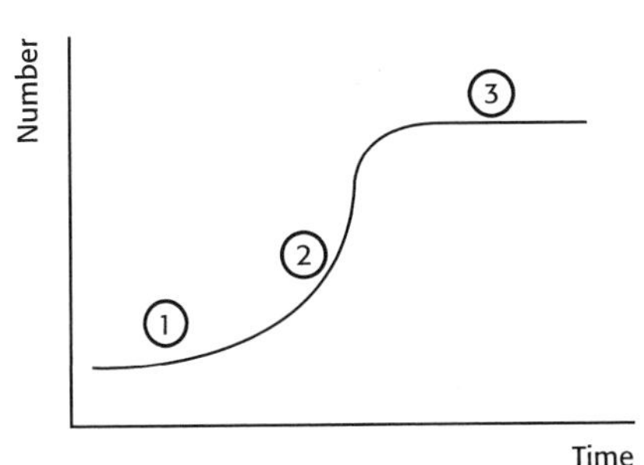

Figure 19 Population growth curve

extinct and there is no yield. The maximum of the curve in Figure 20 would correspond to point 2 on the S-curve in Figure 19.

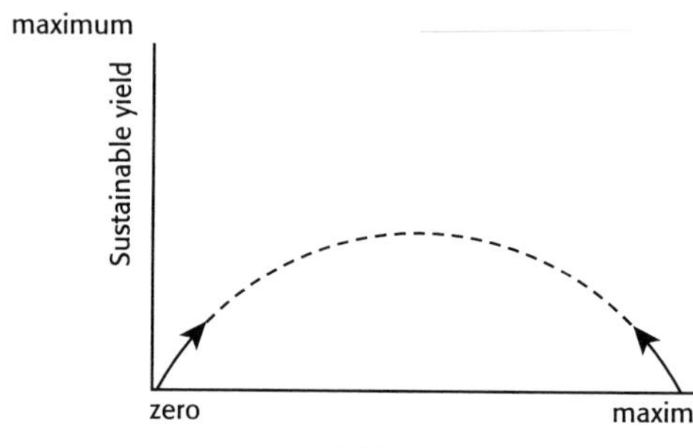

Figure 20 Intensity of fishing

Data-based question: fish harvest and ocean temperature

Environmental variables can alter fish populations so that regular monitoring of a population is necessary to be sure that yields are determined based on existing populations rather than historical populations.

The size of a population of fish can be estimated by determining the distribution of ages in landed fish. The greater the fraction of older fish, the healthier the population. One method of determining the age of fish is to extract and examine otoliths or ear bones. These have rings which have increasing numbers of rings with age.

Year class is the number of fish that are spawned and survived in a particular year. Figure 21 shows the relationship between the temperature of the water off the coast of California and the year class in various years.

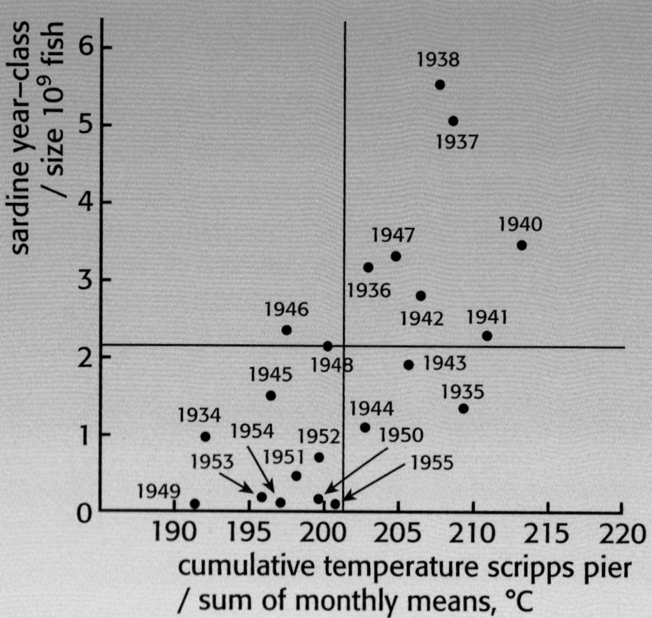

Figure 21

Figure 21 depicts the relationship between the year class or the number of sardines that were spawned and survived in a particular year and the temperature of the water off the coast of California.

1 Identify the two years with the smallest year classes. [1]
2 Identify the two years with the lowest cumulative temperature. [1]
3 Outline the relationship between year class and water temperature. [2]
4 Suggest why, in 1955, despite it being a relatively average year in terms of temperature, the year class is still low. [2]

33 Option H: Further human physiology

Endocrine glands are structures that secrete chemical messages, called hormones, directly into the blood. These messages are transported to specific target cells (Figure 1). Hormones can be steroids, proteins, glycoproteins, polypeptides, amines or tyrosine derivatives. Some important hormones are listed in Table 1.

Table 1

Hormone	Type	Source	Target	Action
leptin	protein	adipose tissue	hunger centre of hypothalamus	Signals to brain that body has eaten enough.
oxytocin	polypeptide	pituitary gland	uterus/mammary glands	Stimulates milk flow from mammary glands; uterine contractions.
FSH (follicle-stimulating hormone)	glycoprotein	pituitary gland	ovary and testes	Testosterone production; stimulates development of Graafian follicle.
adrenalin (epinephrine)	amine	adrenal gland	broad	Prepares body for "fight or flight" response.
progesterone	steroid	ovary/placenta	uterus	Maintains endometrium/pregnancy.
thyroid hormone	tyrosine derivative	thyroid gland	broad	Determines basal metabolic rate.

Peptide hormones and lipid hormones differ in their solubility. This leads to different mechanisms of action. Steroid hormones can cross directly through the plasma membrane and the nuclear membrane and bind to receptors (see Figure 2). An example is estrogen. The receptor–hormone complex then serves as a transcription factor, promoting transcription of a certain gene. For example, the steroid hormone vitamin D effects expression of the calcium transport protein calbindin in the small intestine. Some steroids, such as cortisol, bind to receptors in the cytoplasm and the hormone–receptor complex then passes into the nucleus to effect transcription. The hormone may have different effects in different cells.

Protein hormones, being hydrophilic, cannot pass through the membrane directly. Instead they bind to surface receptors that can trigger a cascade reaction mediated by chemicals called second messengers. The process is illustrated in Figure 3.

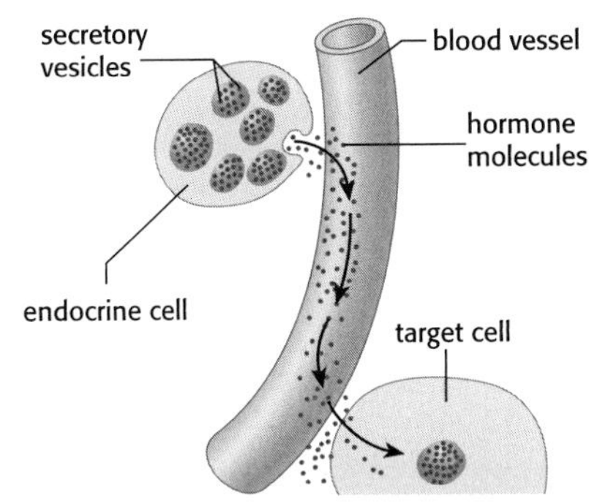

Figure 1 Endocrine glands secrete chemical messages directly into the blood.

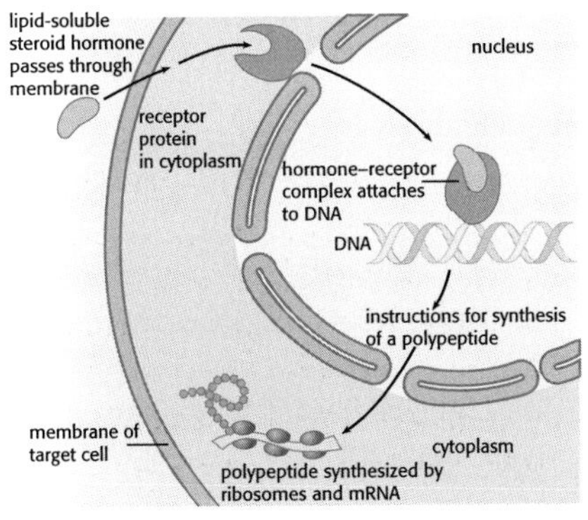

Figure 2 Mechanism of action of steroid hormones

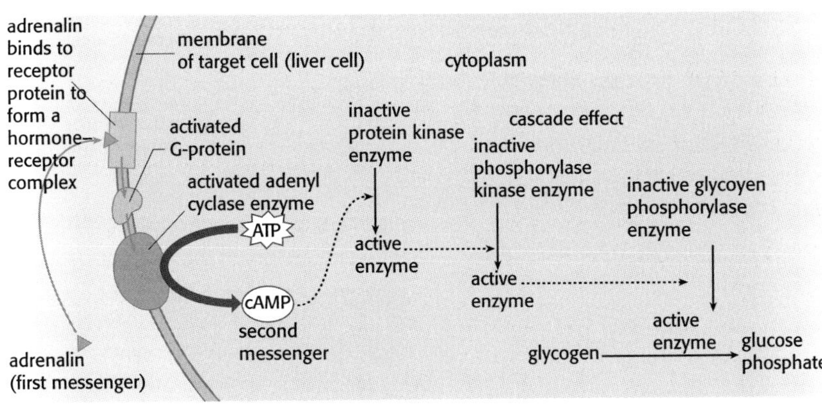

Figure 3 Mechanism of action of adrenaline on a liver cell

The hypothalamus and the pituitary gland

Both the nervous system and the endocrine system play a role in homeostasis. The hypothalamus links the nervous system to the endocrine system and the pituitary gland is a component of the endocrine system. The posterior pituitary gland stores oxytocin and ADH, two hormones produced in the hypothalamus. The anterior pituitary produces a number of hormones such as FSH and LH. Figure 5 shows the neurosecretory cells of the hypothalamus. These are a kind of neuron that releases chemicals into a blood vessel rather than into a synapse. ADH and oxytocin travel down the axons of the neurosecretory cells from the hypothalamus to the posterior pituitary for storage. The neurosecretory cells also carry releasing factors to the anterior pituitary, triggering the release of hormones (Figure 6).

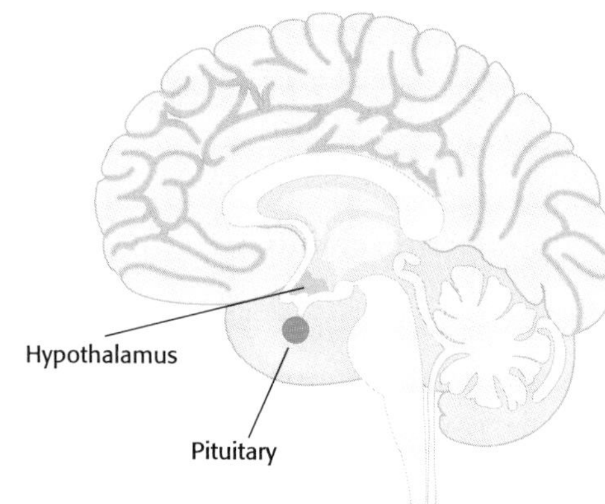

Figure 4

Data-based question: ADH and thirst

Blood solute concentration is monitored by osmoreceptors in the hypothalamus. Anti-diuretic hormone (ADH) acts on the collecting ducts of the kidney, making them more permeable to water. This allows the body to reabsorb more water and dilute blood solutes. The plasma solute concentration, plasma ADH concentration and feelings of thirst were tested in a group of volunteers. Figure 7 shows the relationship between intensity of thirst, plasma ADH concentration and plasma solute concentration.

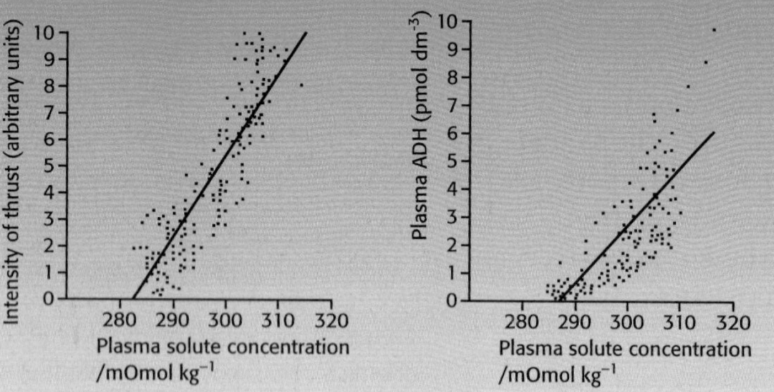

Figure 7 The relationship between intensity of thirst, ADH concentration and plasma solute concentration

1 Identify the plasma ADH concentration at a plasma solute concentration of 300 mOsmol kg^{-1} using the line of best fit. [1]
2 Compare the effect of plasma solute concentration on the intensity of thirst and plasma ADH concentration. [1]
3 Outline what would happen to plasma solute concentration and ADH concentration if a person were to drink water to satisfy their thirst. [2]
4 State two reasons why a person's plasma solute concentration may increase. [2]

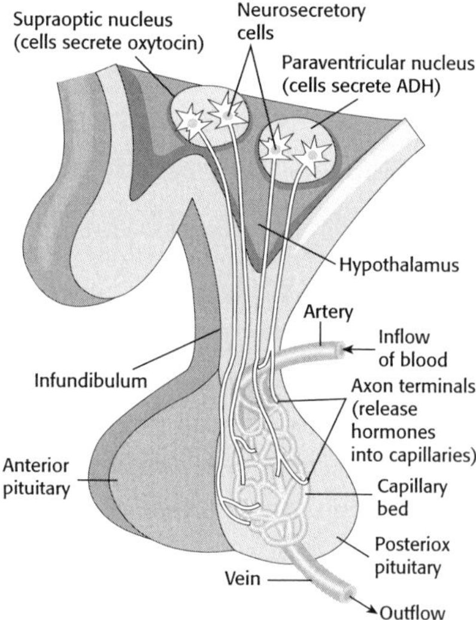

Figure 5 The neurosecretory cells of the hypothalamus

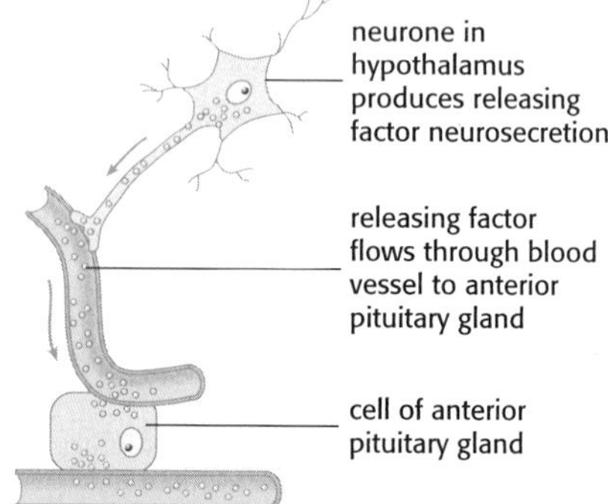

Figure 6 Neurosecretory cells can carry releasing factors to the anterior pituitary.

Inquiry

Outline the role of negative feedback in the control of ADH secretion.

Exocrine glands in digestion

The passage through which food passes from mouth to anus is called the alimentary canal. Digestive fluids are added into the alimentary canal at several points including the salivary glands, the gastric glands of the stomach wall, the pancreas, and the wall of the small intestine (see Table 2).

The structures responsible for the release of these fluids are exocrine glands. Unlike endocrine glands, which secrete directly into the bloodstream, exocrine glands secrete into ducts. Figure 8 shows that exocrine glands have a duct portion and a glandular portion. In the diagram of the salivary gland depicted, the ending of each branch is an acinus formed of secretory cells of two types – serous cells, which secrete proteins such as enzymes, and mucous cells, which secrete mucus.

Under the electron microscope, the exocrine cells show extensive endoplasmic reticulum, relatively numerous mitochondria and large numbers of secretory vesicles containing protein. In some exocrine glands, such as the mammary gland, the vesicles bud off, producing membrane-bound vesicles in the duct. In others, the vesicles release their contents by exocytosis.

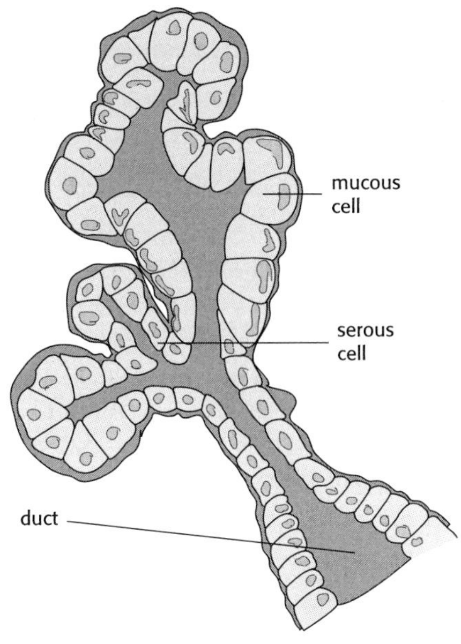

Figure 8 An exocrine gland

Table 2

Digestive fluid	Source	Composition
saliva	salivary glands	water; electrolytes, salivary amylase, mucus, lysozyme
gastric juice	stomach	water, mucus, enzymes including: pepsin, rennin and hydrochloric acid
pancreatic juice	pancreas	water, bicarbonate, enzymes including: amylase, lipase, carboxypeptidase, trypsinogen

Regulation of gastric juice production

The hormone gastrin is released into the bloodstream by cells of the duodenum as well as cells in the part of the stomach nearest the duodenum. When peptides are detected in the stomach by chemoreceptors, or when the distention of the stomach is detected by stretch receptors, gastrin is released under stimulus from the vagus nerve of the parasympathetic nervous system. Gastrin stimulates acid production from parietal cells and pepsinogen production from chief cells (Figure 9). When the pH in the stomach falls too low, gastrin release is inhibited by the hormones secretin and somatostatin.

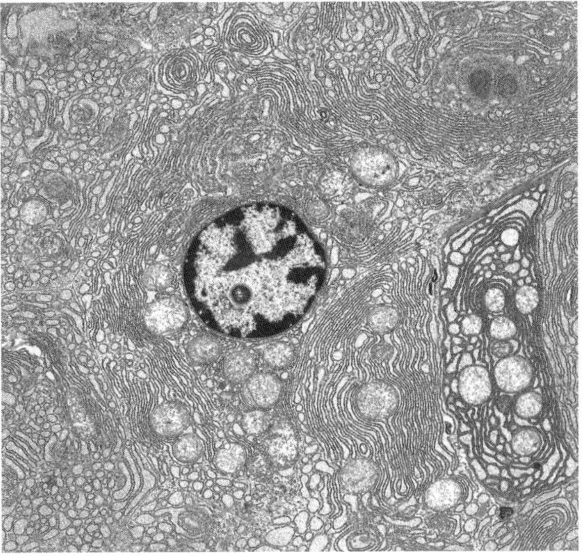

Figure 9 An exocrine cell

Thinking about science: the cause of ulcers

This excerpt is from an article in the *Financial Times*, "Settling the stomach, not lining the pocket" by Thomas Barlow (October 2000).

As a general rule, drug companies claim that they profit not from our misfortunes, but from their own ingenuity. Moreover, they stress the importance of profit to pay for future ingenuity through research. One story is often quoted as a counter-example. This centres on the stomach, or, more specifically on stomach ulcers.

A stomach ulcer is an open sore in the stomach wall, where digestive juices—mostly acid and the enzyme pepsin—have begun to eat away at the stomach lining. Twenty years ago, it was widely believed that emotional stress contributed to stomach ulcers. Too many curries, too many cigarettes and too much alcohol were considered compounding factors.

Yet it is now recognized about 80 per cent of ulcers are caused by infection from a bacterium called Helicobacter pylori *(the remaining 20 per cent is caused by over-use of anti-inflammatory drugs such as aspirin and ibuprofen).*

The corkscrew-shaped bacterium survives in the stomach by producing an enzyme called urease, which neutralizes stomach acid and allows the bacterium to colonize the stomach's mucous lining, opening up the stomach wall to attack form digestive fluids.

The transition from the earlier belief that stress caused ulcers to the modern one, that bacteria do, was not easy.

The theory that ulcers was the consequence of an infection was put forward in the early 1980's by Barry Marshall and Robin Warren, two little-known Australians working at the Royal Perth Hospital in Western Australia. By the mid-1980s, they worked out an inexpensive treatment that cured about 75 per cent of patients. (The success rate is even higher today.) By 1988, they had shown definitively that antibiotics which killed H. pylori *would cure ulcers for good. Yet it took until 1994 for a consensus to be reached. And there are wonderful conspiracy theories as to why the global medical community was so slow to believe the evidence.*

Marshall...attributes the slow take-up of his discovery to at least three different factors. The first problem seemed to lie, as it often does, with the inertia of existing beliefs. Doctors, drug companies and policy-makers had convinced themselves that they already knew the cause of ulcers: emotional stress. Marshall and Warren's infectious-agent theory had to displace the mindset.

Also the blockbuster drugs of the time, Smith Kline Beecham's Tagamet and Glaxo's Zantac were both very good at putting ulcers into remission. 'The effect was so dramatic, it was hard to believe there could be anything better,' says Marshall.

The second problem lay in the way funding was (and still is) allocated. Research grants are often awarded for three-year stints which can contribute to the delayed acceptance of any new theory. When, in 1988, Marshall and Warren demonstrated that antibiotics could cure ulcers, many medial researchers who might have confirmed their result were already locked into research on acid-lowering drugs.

Third, Marshall says that initially, they found it difficult for their publications to be noticed. Again this was an accident of research funding. Pharmaceutical companies fund an enormous amount of drug research in universities and hospitals. 'If pharmaceutical companies are funding, say, 50 other research programmes, there might be 50 times as many publications on other subjects, so the effect of a one-off work on H. pylori *is diluted,' says Marshall.*

This meant that, by the early 1990s, doctors might have heard a rumor, seen an editorial or read an occasional article about H. pylori *and ulcers by these were easily out-weighed in many minds by the cornucopia of papers published about stress and other factors.*

Pharmaceuticaal companies understandably tend to concentrate their efforts on conservative research that tends toward lucrative ongoing treatments rather than speculative ventures that might produce cheaper, permanent cures. This would mean the true cause of stomach ulcers would almost certainly never have been uncovered inside a pharmaceutical company.

1 Discuss the challenges a scientist faces when they discover an unexpected truth.
2 Discuss the relative merits of public versus private funding of medical research.
3 Investigate the role of *H. Pylori* in the development of stomach cancers.

Digestive enzymes

Pepsinogen, released in the stomach, and trypsinogen, released by the pancreas, are two examples of zymogens—inactive enzyme precursors that are chemically altered after secretion to produce the active form of the enzyme. This is done to prevent self-digestion of the cells that produce the zymogen.

Pepsinogen is converted to pepsin by the acid conditions of the stomach. Trypsinogen is converted to trypsin by the action of enteropeptidase. This is an enzyme bound to the membranes of the small intestine.

There are other enzymes immobilized on the membrances of the intestinal epithelium cells. Some other examples are epithelial maltase, lactase and sucrase. The active sites of the enzymes are oriented toward the lumen (middle) of the intestine. They remain functional even when epithelial cells have been sloughed off into the lumen.

Dietary plant matter contains cellulose. Because humans lack the digestive enzyme cellulase, then dietary cellulose remains undigested and is egested in the feces. Other animals, such as cows and sheep, also lack this enzyme. However, they have evolved symbiotic relationships with cellulose-digesting bacteria that allow the host to obtain energy from the cellulose. A similar symbiotic relationship exists between termites and an intestinal protozoan.

Data based question: Lipase and bile

Lipase is a water-soluble enzyme that acts on water-insoluble lipids. It converts triglycerides into free fatty acids and monoglycerides. In order to function the lipid must be accessible, despite being a hydrophobic substance in a hydrophilic medium. Bile molecules are released to support lipid digestion. Bile acts as a detergent, breaking fat into tiny droplets and forming an emulsion. This helps the hydrophilic lipase to attack the hydrophobic lipid substrate.

Bile is formed by the liver and drains through canaliculi (narrow tubes) into the gall bladder, where it is stored. During bile formation, liver cells secrete bile salts, which cause water to be drawn into the canaliculi by osmosis. The graph in Figure 10 shows the rate of bile flow into the gall bladder at different levels of bile salt secretion. The effect of a hormone, secretin, is also shown.

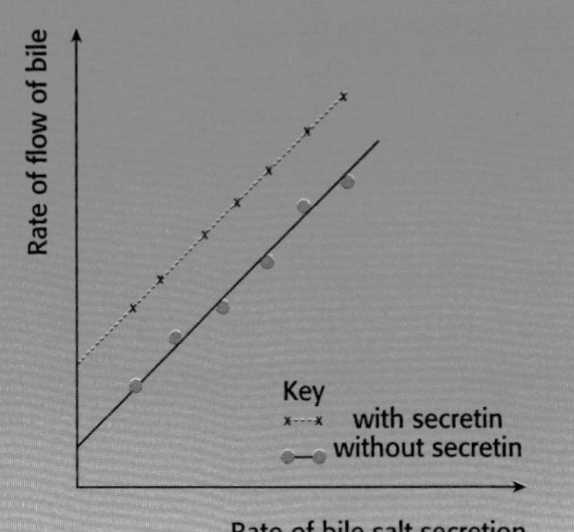

Homer Andrews, ***Liver*** p31. Edward Arnold 1979.

Figure 10 The effect of bile salts and secretin on the rate of flow of bile

Look at Figure 10.

1 (a) State the relationship between the rate of bile salt secretion and the rate of bile flow, without secretin. [1]
  (b) Suggest the cause of this relationship. [1]

2 Suggest when the rate of secretion of bile salt by liver cells needs to be highest. [1]

3 Secretin causes HCO_3^- (hydrogen carbonate) ions to be secreted into the bile. Using only the data in the graph, outline the effect of secretin on bile flow. [2]

4 The results in the graph show that in addition to HCO_3^- and bile salt, another solute is secreted into the bile. Explain how this conclusion can be drawn from the results in the graph. [2]

Absorption of digested foods

Figure 11 shows a cross-section through the ileum, the site of a significant amount of the absorption that takes place in the small intestine.

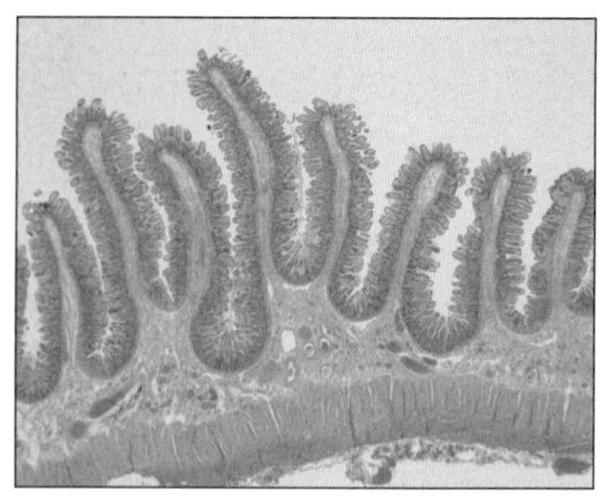

Figure 11 Cross-section through the ileum

The inner surface of the ileum has numerous circular folds (Figure 12a). Each of the folds is covered in tiny projections called villi (Figure 12b). Absorption takes place through the epithelial cells covering each villus.

Each epithelila cell covering the villus adheres to its neighbours through tight junctions, which ensure that most materials pass into the blood vessels lining the villi through the epithelial cell. The cell surface membrane on the intestinal lumen side has a number of extensions called microvilli. The collection of microvilli on the intestinal side of the epithelial cells is termed the brush border. The function of the brush border is to increase the surface area for absorption. Relatively high amounts of ATP are required to drive active transport processes such as endocytosis. Thus epithelial cells have large numbers of mitochondria.

Nutrients are absorbed into the intestine by four processes: simple diffusion, facilitated diffusion, active transport, and endocytosis.

Monoglycerides, bile salts, and fatty acids combine to form structures called micelles. These easily diffuse across the plasma membrane. Inside the epithelial cell, they reassemble into triglycerides and become coated with protein and cholesterol to form chylomicrons, which are also known as lipoproteins. The chylomicrons are then able to enter a structure called a lacteal. A lacteal is part of a system called the lymphatic system. The lacteals converge into lymph vessels that carry the chylomicrons to a junction point with the circulatory system.

Fructose is an example of a substance that passes into the epithelium by facilitated diffusion. Glucose, amino acids, and other molecules are pumped into the epithelium by active transport. One mechanism of active transport is for the epithelial cell to pump sodium ions out into the lumen of the intestine, which sets up a concentration gradient. The passive re-entry of sodium back into the epithelial cell is accompanied by nutrients such as glucose and amino acids through a co-transporter protein.

Unabsorbed material such as cellulose and other plant fibres such as lignin are eliminated through egestion of feces. Additional components of feces include bile pigments, bacteria, and intestinal cells.

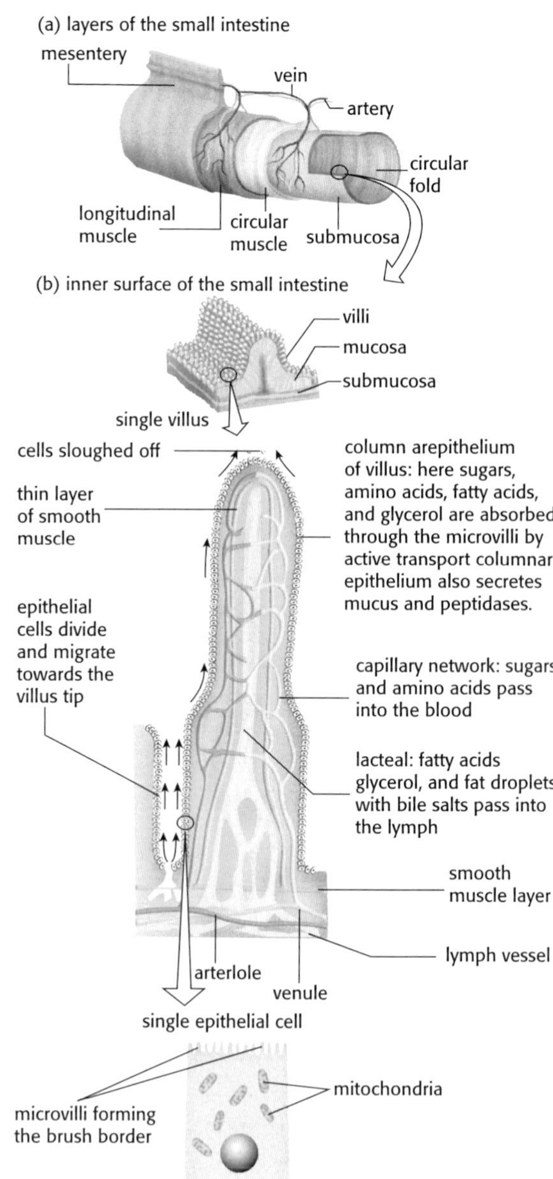

Figure 12 The small intestine

Data-based question: cholesterol

Cholesterol and lipids are not soluble in the blood because blood is water-based. To solve this problem, lipids are transported in the blood in the form of lipoproteins called chylomicrons. The concentration of cholesterol in the blood as lipoproteins is a determining factor in the onset of coronary heart disease.

In 1998 the blood cholesterol level of 70,000 people in Mexico was measured. The people were divided into two age groups: 1 to 19 (young people) and 20 to 98 (adults). Mean blood cholesterol levels were calculated for the two age groups in each of the different states of Mexico. Figure 13 shows the results. Each point on the graph shows the mean blood cholesterol level for the two age groups in one state.

1 State the relationship between cholesterol levels in young people and adults. [1]
2 Predict, using the data in the graph, how the blood cholesterol level usually changes over a lifetime. [2]
3 The maximum desirable blood cholesterol level is 200 mg 100cm^{-3} of blood. Suggest the implications of the survey of blood cholesterol levels for the population of Mexico. [3]

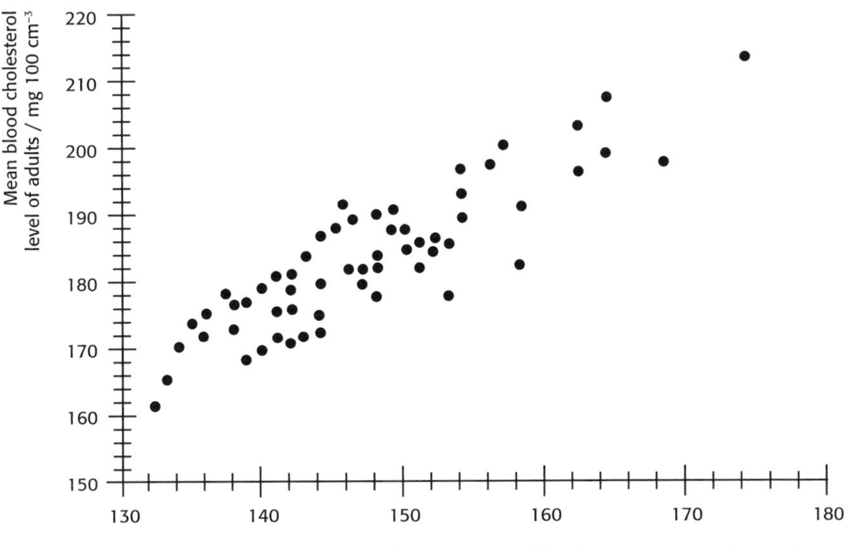

Figure 13 Relationship between blood cholesterol in adults and blood circulation in adolescents in different Mexican states

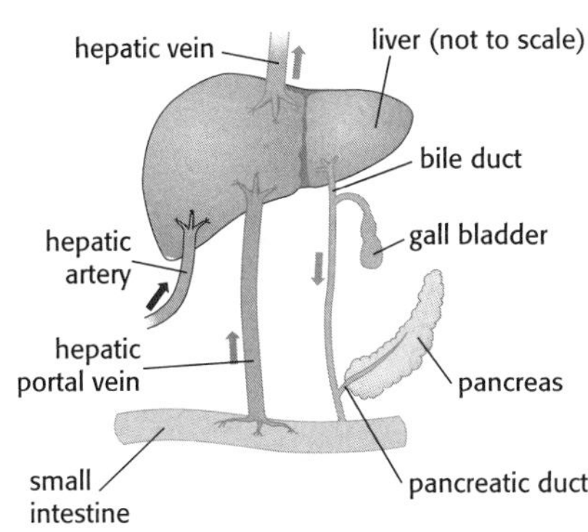

Figure 14 Hepatic circulation

Functions of the liver

Blood arrives at the liver from two sources. The hepatic artery from the aorta brings oxygen-rich blood to the liver from the heart. The hepatic portal vein brings blood from the gut to the liver (Figure 14). Within the liver, the vein subdivides into divisions called sinusoids. Sinusoids are like capillaries only they are wider and the walls are more porous, or fenestrated (Figure 15). This allows the blood flowing through to come in contact with the hepatocytes (liver cells). The fenestrations are large enough to allow proteins such as albumin to enter and leave the blood. The hepatic artery subdivides into arterioles, which merge with the sinusoids at various points, providing oxygenated blood. The sinusoids merge with venules that lead to the hepatic vein. This carries blood away from the liver to the vena cava.

One of the main functions of the liver is to regulate the quantity of nutrients circulating in the blood. It plays a key role in the regulation of circulating glucose by either storing glucose as glycogen

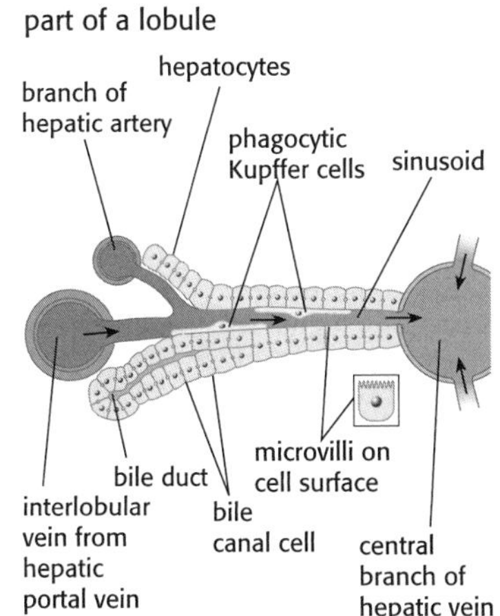

Figure 15 Circulation within the liver

or breaking glycogen down to glucose. Because the body cannot store proteins or amino acids, excess quantities of these in the diet are broken down in the liver to be utilized as energy sources. The liver processes the resulting nitrogenous waste. The liver also synthesizes plasma proteins such as albumin. It regulates the amount of circulating lipids such as cholesterol and lipoproteins, either synthesizing them as required, or breaking them down and secreting cholesterol and phospholipids in the bile. The liver also stores fat-soluble vitamins such as vitamin A and vitamin D.

Data-based question: diet and liver function

The liver plays an important role in detoxification. For example, the liver contains enzymes which degrade alcohol through several steps into fatty acids. One of the consequences of chronic excessive alcohol consumption is a fatty liver. Long-term alcohol consumption stimulates the liver's fat-storing cells to produce collagen, leading to the formation of scar tissue. This is a condition known as cirrhosis. Hepatitis is inflammation of the liver. Hepatitis can occur as a result of a reaction to the cells affected by fatty acid changes.

A number of chemicals have been shown to cause tissue damage due to the production of free radicals. Free radicals are highly reactive chemicals, such as superoxides and peroxides. Antioxidants produced by our body, such as reduced glutathione, combine with free radicals and decrease tissue damage. Reduced glutathione reacts with free radicals and in the process is converted to oxidized glutathione.

Recently dietary antioxidants such as lignins have also been shown to protect against tissue damage. Flax seed is known to contain lignins, but its antioxidant effects have yet to be evaluated. Research was done to see if flax seed could help prevent damage to the liver by tetrachloromethane. Metabolism of tetrachloromethane by the liver leads to the formation of free radicals. Rats were pre-treated by oral injection with flax seed extract (+) or corn oil (–) (control) for three days and then injected with buffered saline solution (control) or tetrachloromethane. The glutathione levels were then measured. The results are shown in Figure 16.

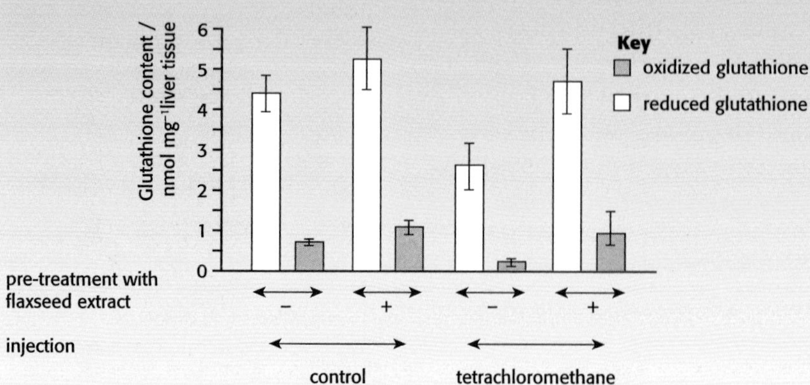

Figure 16 Effect of flaxseed oil on glutathione levels

1 State the reduced glutathione content of liver tissue injected with tetrachloromethane with no flaxseed pretreatment. [1]
2 Calculate the total glutathione content (oxidized + reduced) in liver tissue treated with flaxseed extract but not injected with tetrachloromethane.[1]
3 Describe the effect of tetrachloromethane injection on total glutathione and reduced glutathione content in liver tissue without flaxseed pretreatment. [2]
4 Predict, using the data, the effect of using flaxseed extract in protecting liver tissue from damage due to tetrachloromethane. [3]

Breakdown of erythrocytes

The typical lifespan of an erythrocyte (red blood cell) is about 120 days. As the cells age, they swell and are engulfed by Kuppfer cells, which line the sinusoids in the liver. Inside the Kupffer cell, the hemoglobin molecule is split into globin chains and a heme group. The globin chains are recycled, while the heme group is further broken down into iron and bilirubin, or bile pigment. Bile pigment is used in the production of bile and it gives feces its characteristic colour. A build-up of bile pigment in the blood can lead to a yellow appearance, known as jaundice.

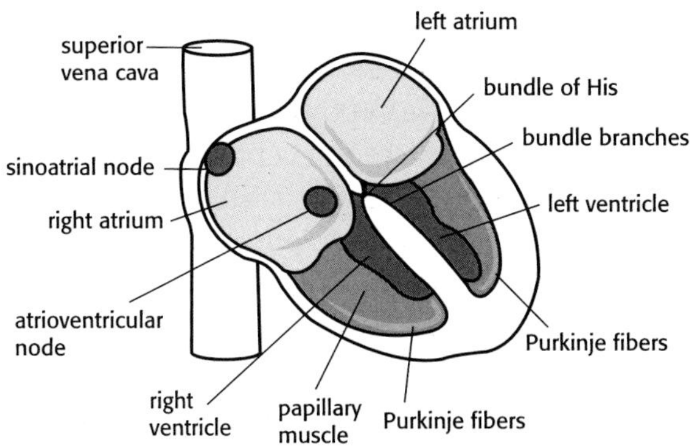

Figure 17 Wave of depolarization from the SA node

The cardiac cycle

The cardiac cycle is all the events from the beginning of one heartbeat to the beginning of the next. To start the cycle, the sino-atrial (SA) node fires an electrical signal throughout the walls of the atria. This causes the atria to undergo systole, i.e. they contract. The SA node is sometimes referred to as the pacemaker of the heart.

Data-based question: pressure changes in the heart

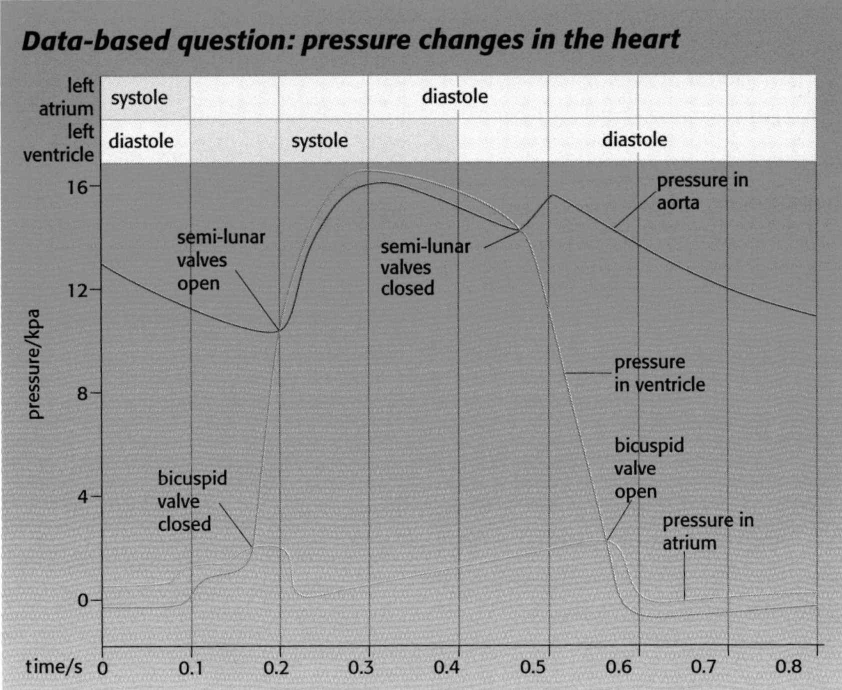

Figure 19 The pressure changes inside the heart during the cardiac cycle

1 Explain why the pressure in the aorta rises when the semilunar valves open. [2]
2 Outline the changes in the pressure in the ventricle. [3]
3 Determine the length of one cardiac cycle. [1]
4 Identify the chamber with the longest
  (a) diastole
  (b) systole. [2]
5 Explain the changes in the pressure in the atrium. [3]

(a) atrial systole

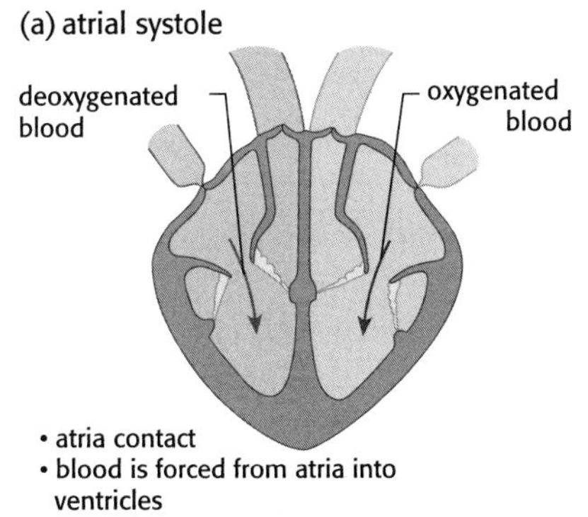

• atria contact
• blood is forced from atria into ventricles

(b) ventricular systole

bicuspid valve

tricuspid valve

• ventricles contract
• blood is forced into arteries
• bicuspid and tricuspid valves prevent blood flowing back into atria
• atria start to fill with blood again

(c) diastole

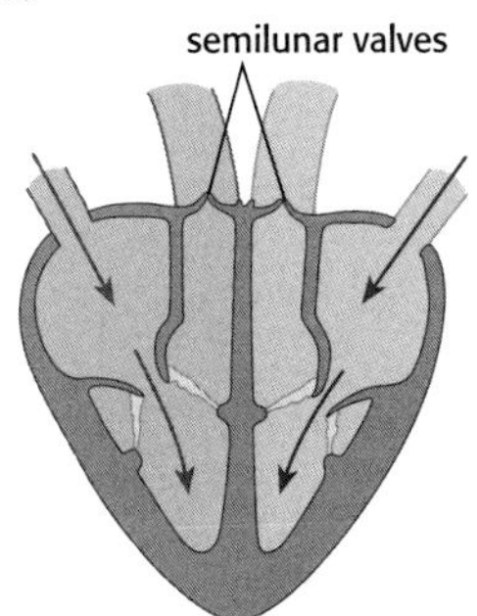

• heart relaxes and fills with blood from veins
• semi-lunar valves prevent blood entering through arteries

Figure 18 The cardiac cycle

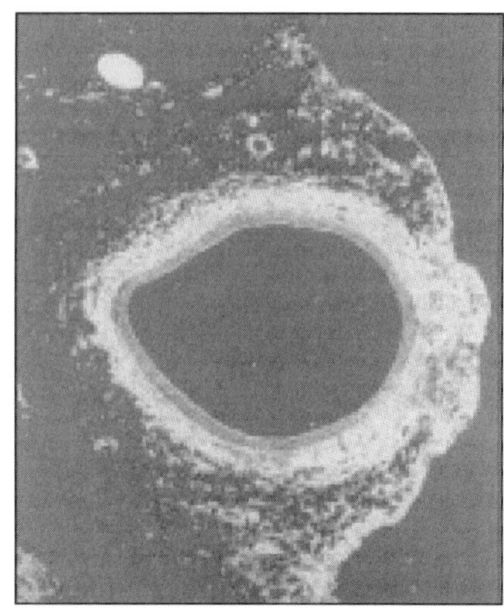

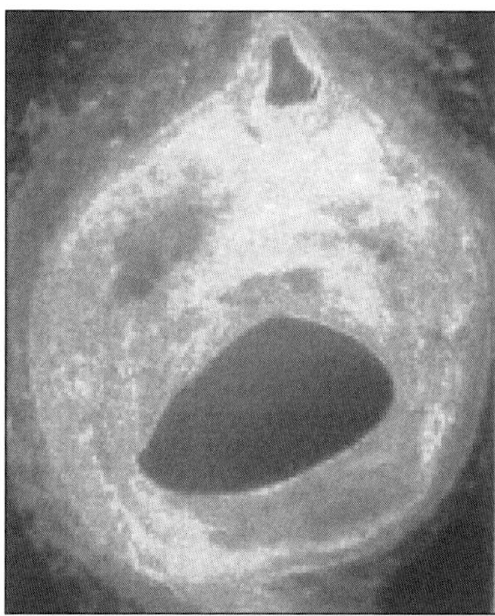

Figure 20 A normal artery (top) can be compared to an artery where a plaque has formed (bottom).

The signal from the SA node also reaches the atrio-ventricular (AV) node. From there the signal spreads throughout the heart via specialized heart muscle tissue called Purkinje fibres. This signal causes the ventricles to undergo systole. This snaps the atrio-ventricular valves shut, causing the "lub" sound of the heartbeat. After the ventricles are emptied the semilunar valves close, causing the "dub" sound.

The ventricles begin diastole, the atrio-ventricular valves open and the ventricles start filling with blood. Finally, all four chambers are in diastole and filling. When the atria are filled and the ventricles are 70 per cent filled, the cycle has ended.

Coronary heart disease

Coronary heart disease (CHD) refers to damage to the heart as a consequence of reduced blood supply. Atherosclerosis is hardening of the arteries caused by the formation of plaques, or atheromas, on the inner lining of arteries (Figure 20). Plaques are areas that are swollen and accumulate a diversity of debris. The plaques often develop because of high circulating levels of lipids and cholesterol. The plaques can reduce the speed at which blood moves through vessels. This can trigger a clot, or thrombosis, which can block the blood flow through the artery and deny the tissue access to oxygen. If this occurs on the surface of the heart, the consequence can be a myocardial infarction, or heart attack.

There are a number of factors that are correlated with a greater incidence of CHD.

- Having parents who have experienced heart attacks indicates a genetic precondition.
- Old age leads to less flexible blood vessels.

Data-based question: hypertension

High blood pressure (hypertension) is a major risk factor for coronary heart diseases. In a major study, more than 316,000 males were followed for 12 years to investigate the effects of high blood pressure (BP). Figure 21 shows the relationship between systolic and diastolic blood pressure and the effect on the death rate per 10 000 persons year^{-1}.

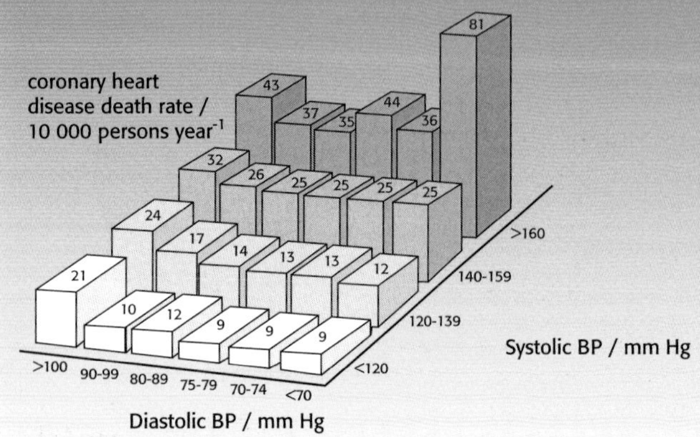

Figure 21 The effect of blood pressure on coronary heart disease

1 Determine the death rate for a systolic blood pressure between 140 and 159 mmHg and a diastolic blood pressure between 75 and 79 mmHg. [1]
2 Describe the effect of systolic blood pressure and diastolic blood pressure on the death rate. [2]
3 Calculate the minimum difference between systolic and diastolic blood pressure where the death rate is highest. [1]
4 Evaluate the impact of differences between systolic and diastolic pressure on death rate. [3]

- Risk in females increases post-menopause correlated with a fall in estrogen levels.
- Males are at greater risk compared with females correlated with lower levels of estrogen.
- Hypertension/high blood pressure is correlated with CHD.
- Smoking raises blood pressure because nicotine causes vasoconstriction.
- A high-salt diet, excessive amounts of alcohol and stress are also correlated with coronary heart disease.
- Eating too much saturated fat and cholesterol promotes plaque formation.
- Obesity strains the heart.
- Sedentary lifestyle, i.e. a lack of exercise is correlated with obesity.

Rediscovering biology: EKG analysis

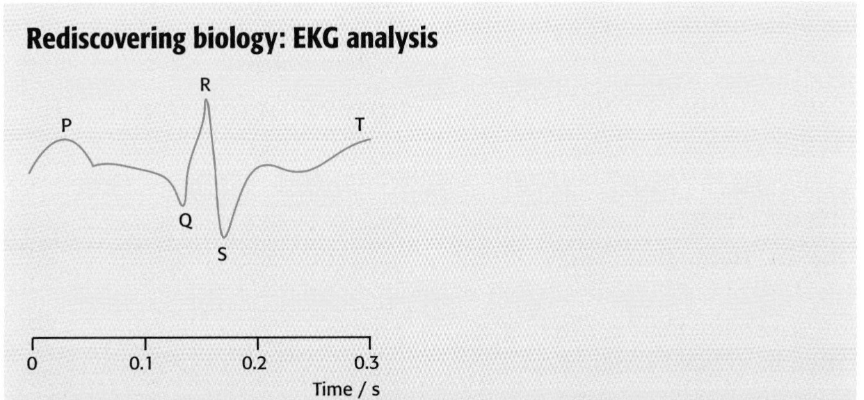

Figure 22 An EKG is a record of the electrical activity of the heart.

Cardiac muscle contracts because it receives electrical signals. These signals can be detected and quantified using an electrocardiogram (EKG). Data-logging EKG sensors can be used to produce a pattern as shown in Figure 22. The P-wave is caused by atrial systole, the QRS wave is caused by ventricular systole. The T-wave coincides with ventricular diastole.

Interval analysis can be performed on the EKG signal, for example on the times between the beginning of P and Q (P–Q), QRS, and Q to the end of T (Q–T) intervals. The height of the R-wave can be compared when the body changes position from standing to lying down. The overall pattern can be compared before and after mild exercise.

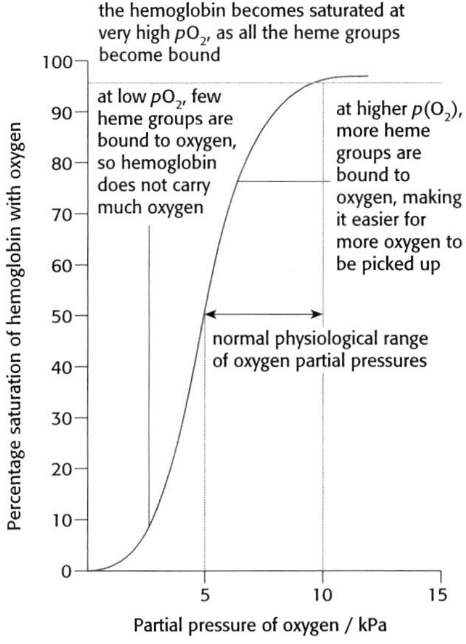

Figure 23 Saturation of hemoglobin by O_2 changes as the partial pressure of O_2 changes.

Gas exchange

In a mixture of gases, each gas has a partial pressure. This is the pressure of the gas if it occupied the volume without the other gases present.

Hemoglobin is an oxygen transport protein in the blood. The degree to which oxygen binds to hemoglobin is determined by the partial pressure of oxygen (pO_2). The oxygen dissociation curve shown in Figure 23 describes saturation of hemoglobin by oxygen at different partial pressures of oxygen. Note the significant change in saturation over a narrow range of oxygen partial pressure. This narrow range typifies oxygen pressures surrounding cells under normal metabolism. At low pO_2, such as might occur in the muscles, O_2 will dissociate from hemoglobin. At high pO_2, such as might occur in the lungs, the hemoglobin will have a high affinity for O_2. Increased

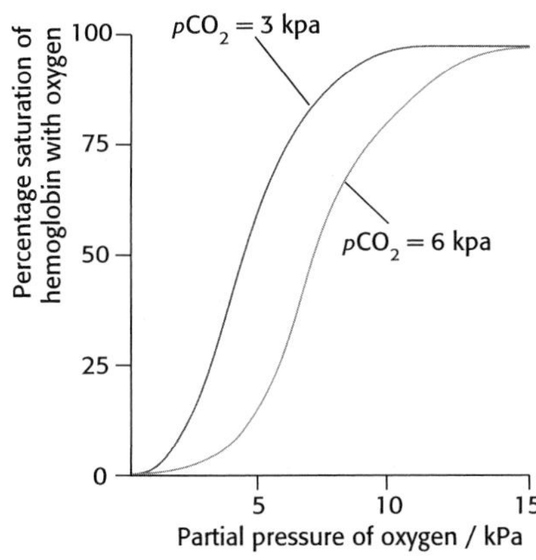

Figure 24 The Bohr shift

metabolism results in greater release of CO_2 into the blood, which lowers the pH of the blood. This increased acidity shifts the oxygen dissociation curve to the right, which results in a greater release of oxygen from hemoglobin at the same partial pressure of oxygen (see Figure 24). This is known as the Bohr shift. This ensures that respiring tissues have enough oxygen when their need for oxygen is greatest. Also, in the lungs, pCO_2 is lower, so saturation of hemoglobin can occur at lower partial pressures of oxygen.

Myoglobin is a specialized oxygen transport protein in muscles. It has a much higher affinity for oxygen and will only release its oxygen when the pO_2 is quite low, for example in the muscles during heavy exercise. The shapes of the two curves in Figure 25 are different because hemoglobin has four chains with four heme groups, whereas myoglobin has one. The release of each O_2 from hemoglobin triggers a conformational change, which causes the hemoglobin to more rapidly release subsequent O_2 molecules.

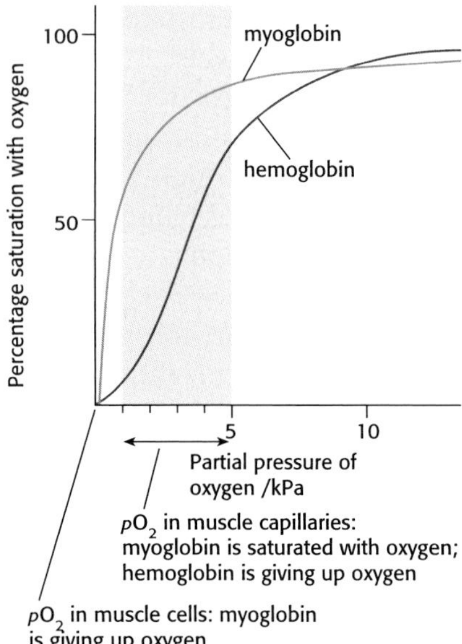

Figure 25 A comparison of the O_2 dissociation curves of hemoglobin and myoglobin

Figure 26 compares the oxygen dissociation curves of adult and fetal hemoglobin. Note that fetal hemoglobin has a higher affinity for O_2 at all partial pressures. This ensures that O_2 is transferred to the fetus from the maternal blood across the placenta.

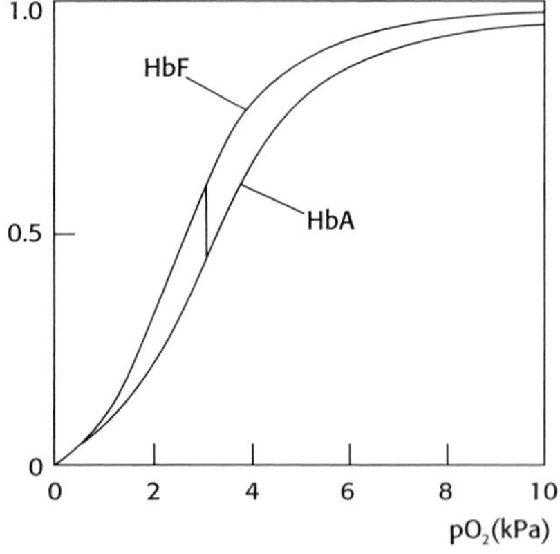

Figure 26 A comparison of the O_2 dissociation curves of fetal and adult hemoglobin

At high altitude there is a low pO_2 in the air. Hemoglobin may not become fully saturated and as a consequence, the tissues may not be adequately supplied with oxygen. To some degree, human physiology can adapt to high altitude. Red blood cell production can increase, which increases the total amount of circulating hemoglobin. Ventilation rate increases to increase gas exchange. Muscles produce more myoglobin to ensure delivery of oxygen to the tissues. Populations living permanently at high altitude have greater mean lung surface area and larger vital capacities than people living at sea level. Their oxygen dissociation curve shifts to the right, encouraging release of oxygen into the tissues.

Exercise

Exercise increases metabolism and leads to an increase in the production of CO_2 (see Figure 27). Increased CO_2 causes blood pH to decrease because CO_2 dissolves in water to form carbonic acid (H_2CO_3). This conversion is facilitated by the enzyme carbonic anhydrase. The carbonic acid then dissociates to form H^+ and HCO_3^-. The HCO_3^- ions move out of the red blood cell by facilitated diffusion. This movement is associated with a movement of Cl^- ions into the red blood cell. This ensures ionic balance within the red blood cell. This is known as the chloride shift. The hemoglobin is able to bind the H^+ ions that remain within the red blood cell, i.e. the hemoglobin is able to behave as a buffer. Other plasma proteins also play a role in regulating blood pH. Chemoreceptors in the aorta and the carotid artery are able to detect a change in blood pH. If a drop in pH is detected, the chemoreceptors send a message to the breathing centre in the pons of the medulla oblongata. Nerve impulses are sent from the medulla to the diaphragm and the intercostal muscles increase the ventilation rate. This leads to an increased rate of gas exchange. The long-term effect of exercise is increased lung surface area, i.e. a greater vital capacity.

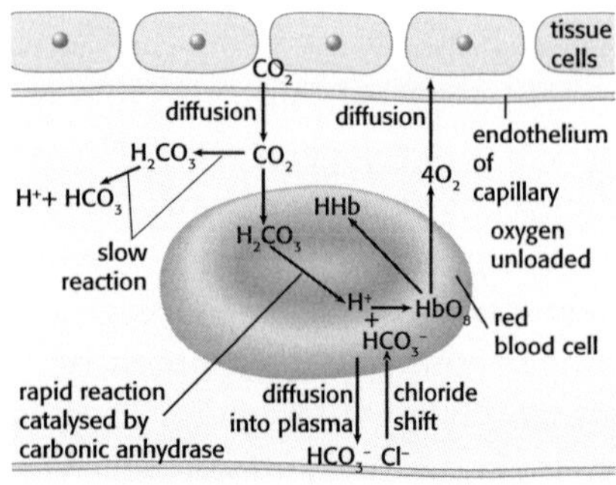

Figure 27 The chloride shift

Asthma

Asthma is a chronic lung condition that is characterized by inflammation of the air passages in the lungs. Often the bronchioles constrict in an over-reactive fashion. The consequence is breathing problems, coughing and shortness of breath. It is often triggered by environmental factors such as allergies, exercise, cold, viral illness, or strong responses to environmental triggers such as pollen. Asthma sufferers are more likely to live in industrial areas.

Data-based question: asthma

The symptoms of asthma vary according to the time of year. A study was carried out in New York to determine if an increase in the amount of pollen in the air caused an increase in the number of asthma attacks. Over a period of 270 days, the number of people admitted to New York hospitals with asthma attacks was recorded. Figure 28 shows this data together with the pollen count.

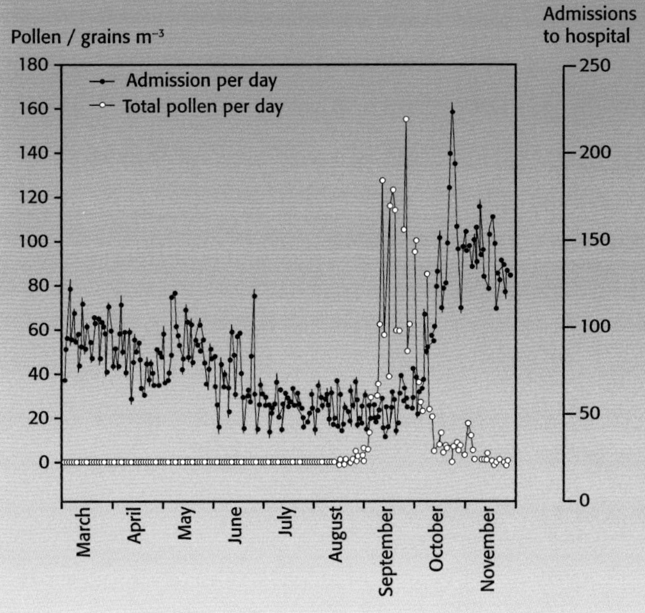

Figure 28

1 Identify the greatest number of hospital admissions in one day. [1]
2 Describe how the numbers of hospital admissions changed over the period of study. [3]
3 Evaluate whether the hypothesis that pollen in the air increases asthma attacks is supported by the data. [2]

Table 3

Form of transport	Arterial $mmol^{-1}$ blood	Rest Venous $mmol^{-1}$ blood	Exercise Venous $mmol^{-1}$ blood
dissolved CO_2	0.08	0.78	1.23
bicarbonate ion	13.52	14.51	14.66
CO_2 bound to protein	0.3	0.3	0.24
total CO_2 in plasma	14.50	15.59	16.22
pH of blood	7.4	7.37	7.14

CO_2 transport in blood plasma at rest and during excercise.

Use the data in the Table 3 to:

1 calculate the percentage of CO_2 found as bicarbonate ions in the plasma of venous blood at rest [1]
2 compare the changes in total CO_2 content in the venous plasma due to exercise [1]
3 identify which form of CO_2 transport shows the greatest increase due to exercise [1]
4 explain the pH differences shown in the data. [3]

34 Researching biology—extended essays

Getting started

Writing an extended essay in biology is a great opportunity to do the following things.

- Choose a topic that really interests you and adopt a personal approach to it.
- Be an independent learner, with the freedom to make the decisions yourself.
- Have the excitement of asking a question to which you don't know the answer.
- Have the time to look at a topic in depth and really get to grips with it.
- Do some real scientific research that may not have been done by anyone before you.
- Have the fun of gradually discovering the answer to your question.
- Show everyone, including the IB examiners, just what you can achieve.

Table 1 shows the sequence of tasks, with a space by each to add times by which they should be completed.

Table 1

Task	Deadline
Choosing a research question and approach.	
Preliminary experiments and literature searches.	
Main experiments and collecting data.	
Constructing results tables and graphs.	
Data analysis, including statistical hypothesis testing.	
Writing the main body of the essay.	
Adding an abstract, references and acknowledgments.	
Submitting the first draft of the extended essay.	
Revising the extended essay in response to comments.	
Proof-reading and final corrections.	
Submitting the final copy of the extended essay.	

The boxes at the foot of each page offer a summary of advice for writing your extended essay.

- As any research biologist will tell you, TTT, which stands for Things Take Time.
- Get started promptly and leave time for setbacks and delays. If you can do something today, don't leave it until tomorrow.
- Stick to any deadlines that your teacher or supervisor gives you.
- Your supervisor is there to help you—use them as your most valuable resource, by asking them for help when you need it and telling them promptly about difficulties.

Making observations, asking questions

Many research programs begin with an interesting observation. With sharp eyes and uncluttered minds, young people can be particularly good at making observations. For example, Frederick Sladen became interested in bumblebees at the age of 13 and started collecting colonies of them, keeping them in small home-made hives and making observations. In 1892, when he was just 16, he wrote a 40-page booklet on this group of insects and duplicated it on a stencil copying machine, selling copies at sixpence each. He included many observations that had never been made before and showed that some popular theories about bumblebees were wrong. He even corrected the distinguished French biologist Cuvier on one aspect of larval development. Frederick's booklet was, at the time, the most advanced work on bumblebees, which he called humblebees! Part of his "extended essay" is reproduced in Figure 1.

seed sooner than a healthy one.

80

VI.– Honey and Wax.

Honey gathered by Humble Bees, is deposited in the yellow cocoon-cells out of which the young bees have eaten their way. On very fine days these cells are lengthened, and almost sealed up with wax. The honey contained in them is extremely thick & rich, and, though much less in quantity is far superior to that of the Hive Bee. Each kind of Humble Bee has honey of a different flavour, but in addition to this, which is only stored in the cocoon cells, there is always a kind of sweet water, deposited in larger cells, which are entirely made of wax

This sweet water, for it can be called little more, seems to be gathered more especially for immediate use than the thick honey for it is generally all consumed by the next morning. Pollen is stored in cells of either kind. Humble Bees' wax is much darker & softer than Beeswax, & though more easily moulded, would not undergo the strain that the latter does.

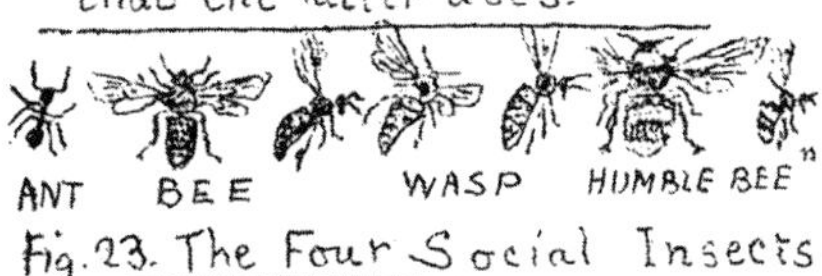

Fig. 23. The Four Social Insects

Figure 1

It would be easy to think that in the 21st century there are no more discoveries to be made. In fact, the natural world is so diverse and each species is has so many aspects that there are still plenty of unasked and unanswered questions and many discoveries waiting to be made!

- Choose a topic that interests enough to want to study it in real depth.
- Read about the topic that you are considering, in books or on the internet, but as soon as possible start to take a personal approach to investigating it.
- Make your own observations, and start asking questions.
- You need to find a research question that you can't immediately answer and is worth finding an answer to—this may take time but is vital to make a really good essay.

Where to look, how to choose a topic

You will have a supervisor for your extended essay, very likely a biology teacher. They might offer suggestions, but it is up to you to choose what you want to investigate. It obviously should be a biological topic if you have decided on an extended essay in biology, so it must involve one or more living organisms. Many students choose to investigate something about humans, often an aspect of human health. Be warned: this can make things very difficult and most essays dealing with human biology do not score highly, especially if the focus is on the diagnosis and treatment of disease. Ideas for how to find a non-human topic are given below. Life springs up almost everywhere, so it is *not* necessary to visit a tropical rainforest, though it would be nice!

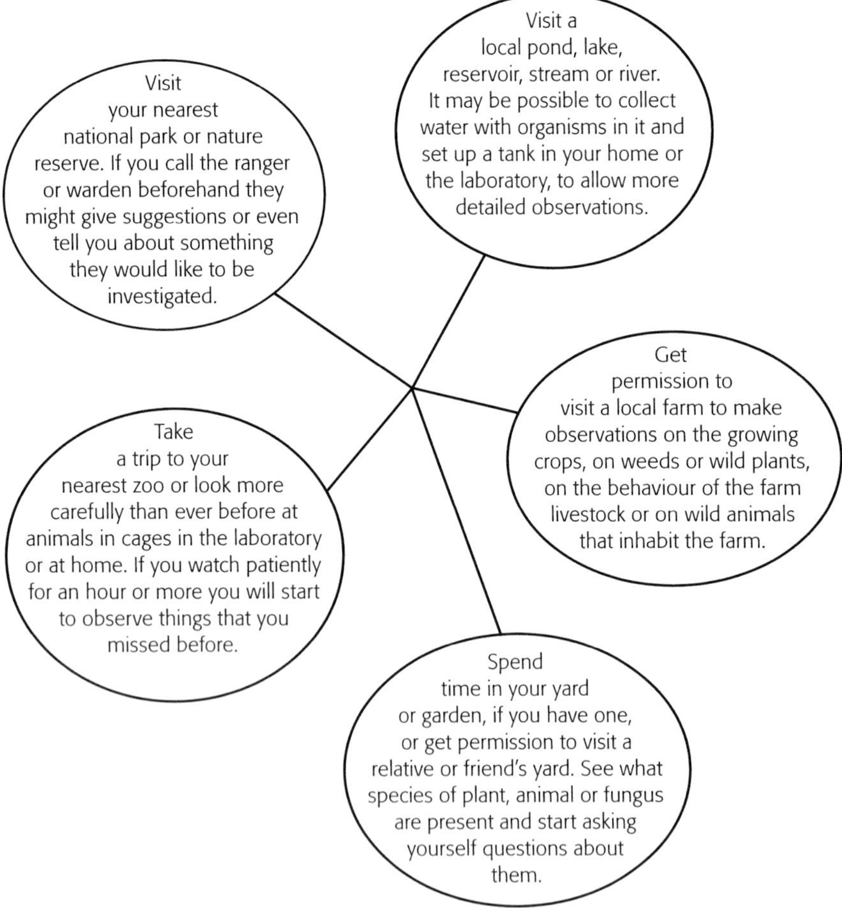

- Look for interactions—ways that living organisms affect each other, or affect their environment, or are affected by their environment.
- Look for explanations—reasons for things, causes of things, how things happen.
- Some of the simplest questions result in the most interesting answers, so don't overcomplicate things when you start.
- Avoid high-tech methods or it will be hard to show evidence of a personal approach.

Example

Observations and questions

Two gerbils were being kept in a biology laboratory and the person who cleaned the laboratory started giving them a peanut each when she cleaned the lab each morning. After a few weeks, she noticed that when she entered the lab the gerbils came over to the front of the cage, stood on their back legs and waited for their peanut. When other people came into the lab at other times they did not do this.
This observation suggests some interesting questions.

1. Were the gerbils able to recognize the cleaner and if they were, what were the recognition features?
2. If clothes of different colour, but the same design were worn, was she still recognizable?
3. Was the time of day critical to recognizing the cleaner?
4. Could the gerbils predict the arrival of the cleaner before she actually came into the lab?
5. Could sight, sound and/or smell be used for recognition?

A simple observation can lead to interesting and worthwhile questions.

Deciding on your research question

Making a final choice of research question is the most important stage in writing your extended essay. If you have made plenty of observations or preliminary experiments and have done some reading of relevant literature, you should have various possible questions from which to choose. You can use Figure 2 to help decide whether a question is suitable. You should also get help from your supervisor and may need to actively ask for this, rather than wait for your supervisor to come to ask you about it. You should avoid non-biological questions that are mainly concerned with ethics, religion, politics, economics or social issues. Many students choose to write about the diagnosis and treatment of diseases. These are rarely successful essays and it is very hard to demonstrate a personal approach in them as experimental work by the student isn't possible.

Your research question is best stated in the form of an actual question. Often the research question can be used to make a hypothesis or group of linked hypotheses. A hypothesis is a predictive statement. Scientists test hypotheses, usually by doing experiments. The Example gives an example of a research question and the hypothesis that can be formulated and tested, to try to answer the research question.

Example

Phrasing of research questions

A student visits an apiary and is shown inside some of the hives. In one of the hives the bees recently died out and the inside of this hive is much dirtier than the others, with signs of microbial growth.

The students reads about the duties of worker bees and finds out that they remove dirt from the hive and also that some of the substances produced by the workers have anti-microbial properties.

Research question: are there more saprotrophic fungi and bacteria in a beehive when there are no bees living in it to keep it clean?

Hypothesis: there more saprotrophic fungi and bacteria in a beehive when there are no bees living in it to keep it clean.

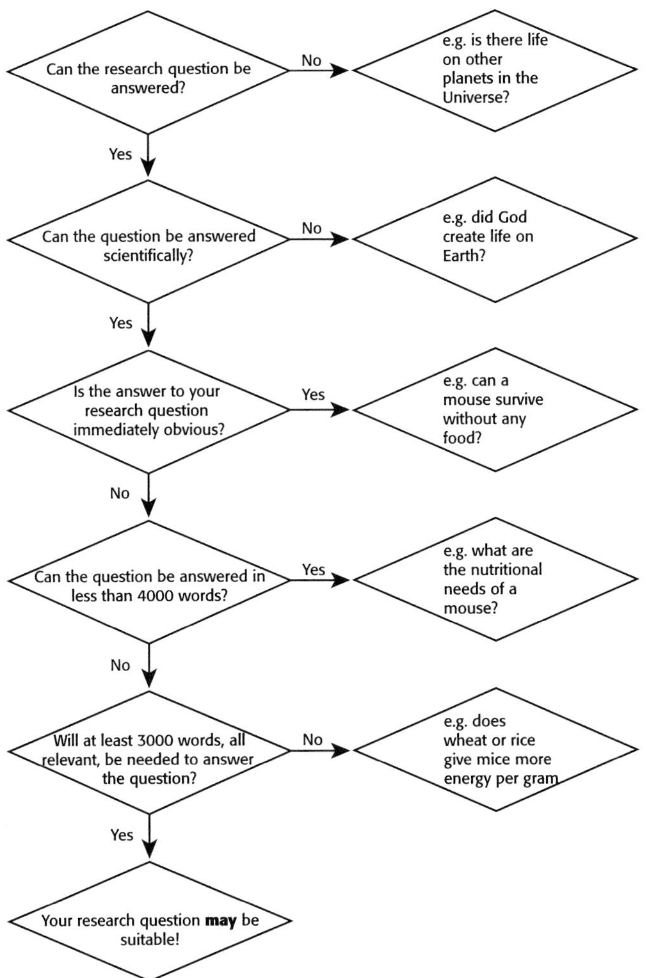

Figure 2

- Discuss your research question with your supervisor and don't change it without talking to them—they should be able to tell you if your idea is unsuitable.
- Don't just write an essay about a subject—the whole of your extended essay must form an answer to the research question, or a test of your hypothesis.
- State your research question clearly, in the first page or two of the essay.
- Don't be over-ambitious—you aren't expected to find a cure for AIDS, for example!

How to approach your research question

To answer your research question you need data or evidence. Biology is an experimental science so this almost always comes from the results of experiments. There are two main ways to get results.

1 Design and carry out some experiments yourself. The results could be numerical or in the form of microscope observations or biological drawings. Sometimes careful observation of the effects due to natural variation in a factor in a habitat is more suitable than artificial manipulation of a variable.
2 Do a literature search, in scientific books or journals, for experiments that give you the data or evidence that you need. You must then manipulate and analyse the data in an original way.

You will not find your extended essay in biology as rewarding if you do not carry out your own experiments. If you find an interesting biological topic where it would be impossible to do any experiments, perhaps you should read about this topic if it interests you, but choose a different one for your extended essay.

Whichever approach you use, you should include the actual experimental results in the main body of your essay, not just a description of the trends that they showed. If your experiments generated very large amounts of data, for example by data logging, it could go in an appendix, with a summary in the main body of the essay.

Advice if you do your own experiments

Everyone can do experiments if they choose a suitable research question, so don't be put off if you don't have access to superb laboratory facilities. An example of an experimental approach that needed very little equipment is given in the Example box.

Some students do high-tech experiments in research labs but it is often difficult for them to show that they had much personal input. Also, very complicated techniques are likely to go wrong the first time that you use them and students often reveal when they write up their experiments that they didn't really understand them.

More details about experimental design are given on the next page.

Advice if you do a literature search

You are unlikely to be successful if you try to write an essay without actually quoting data. If you merely copy out the conclusions of other biologists you will not be able to do any real analysis.

It is better to draw up your own results tables and graphs, rather than scan in published versions. In this way you can present the data in the best way to help answer your research question. Also you are following a more personal approach, which is always more impressive. However, you must include a reference, to state the source of the data. Indeed, wherever you use material from any source apart from your own experiments you must include references. This includes diagrams and quotations.

Example

Experimental work with minimal materials

A student wanted to show that two eyes gave more acute vision than one eye.

He designed and printed a chart for his experiments with these symbols, printed in five rows of ten, in a random sequence:

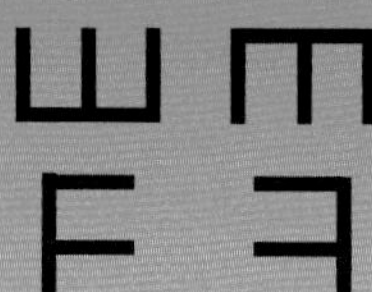

The student then found a quiet room to do the experiments, where lighting could be carefully controlled and there weren't any distractions. In the preliminary experiments, several volunteers tried to read the symbols as w, m, e and 3, sitting at different distances from the chart. A large number of volunteers were asked to read the symbols with both eyes open, with only the left eye open, and then with only the right eye open. Despite the original hypothesis, accuracy was greater with one eye than with two. A statistical test showed the trend to be significant.

- You should explain the reasons for your choice of approach in your essay.
- Although essays based on material obtained from published literature are acceptable, it is hard to score highly unless you do experiments yourself.
- Don't rely only on data from the internet, unless you can access journal articles.
- If you use any material from published sources, for example scientific journals, you must include text references in your essay to show who you are quoting.

A personal approach to experimental design

You should already know a lot about designing experiments from your biology lessons, but you will probably never have designed a large-scale experiment by yourself. Some IB students perform routine experiments that have been done many times before. There is little point in this and no excitement. You are far more likely to make a discovery if there is at least something novel about your experiment. This is one way to demonstrate a personal approach. Be sure to include photographs of your apparatus and experiments, if there is a digital camera that you can use. You can also include diagrams, especially if there was something unusual or complicated about your methods.

Remember all the advice about experimental design that your biology teacher has given you. A control may be needed and you must make sure that apart from the variable you are investigating, all other variables are kept constant. Decide in advance if you are going to do statistical hypothesis testing so that you can make sure you collect suitable data. Statistical hypothesis tests are described on later pages. Table 2 is a checklist for your experimental design, with spaces for your answers.

Table 2	
1 What is the factor that you will be varying in your experiments? This is the **independent** variable.	
2 How many levels of your independent variable will you have? The minimum is two.	
3 What factor will be affected by the independent variable? This is the **dependent** variable.	
4 Which other factors that could affect the dependent variable? These are the **control** variables.	
5 How will the control variables be kept constant, so results are due to the independent variable?	
6 How many repeats will you have, to increase reliability and help identify anomalous results?	
7 Will you use a statistical hypothesis test and if so, which would be most suitable?	

- It is hard to score highly unless you do some experiments yourself.
- Try to devise at least part of the method yourself, then make it clear that you did, with photographs if possible to give evidence that there was a personal approach.
- Describe methods clearly, but in less than one third of overall essay length.
- Put all the data that you use to help answer the research question in the main body of the essay, not in appendices at the end.

Example

A personal approach to design

A student is interested in diving and wants to investigate the slowing of the heart when a diver holds their breath underwater (bradycardia). She takes her pulse on the surface and then after holding her breath for 30 s at the bottom of a swimming pool. There is a reduction in the pulse rate. She takes her pulse using the simple method of feeling her radial artery and counting for 30 s. The student wants to find out how rapidly the pulse falls, whether it falls suddenly or gradually and whether it stabilizes at a lower rate. She needs to monitor the heart rate continuously for one minute or more underwater. For this she needs an electronic probe, but the equipment that her teacher offers her has to be kept dry. If the student designs her own method for this, she will certainly have demonstrated a personal approach.

Devise a method for monitoring heart rate underwater, so that second-by-second changes can be recorded.

Starting to write your essay

Once you have finished your experiments or completed your literature search for data you can start to write your essay. There are a few rules for how to do this and also some things to avoid. Almost all students use a word processor, so that they can revise and edit their essay as many times as are needed. Another advantage is that the word processor will count how many words you have used—you mustn't go above 4000!

However, software for plotting graphs or displaying data may not do quite what you want, and hand-drawn results tables and graphs are perfectly acceptable. Indeed, they would give your essay a more personal touch. They can be scanned in after drawing if you prefer electronic copies. The text of your essay should have the lines either double spaced or 1½ spaced, to make it easier to read. It also helps the reader if the essay is broken up into sections, with subheadings for each section. A typical sequence is shown below. Instructions for how to write each section are given on the following pages.

Title page
The title of your essay should say exactly what the essay is about.

Abstract
A brief summary in 300 or fewer words of your whole essay.

Contents
A list of the sections of your essay with the page on which each starts.

Introduction
A clear statement of the research question, with its biological context, reasons why it is significant, and the approach used to investigate it.

Methods
Description of experimental methods, with reasons for choosing them or an explanation of how published data was selected.

Results
Usually tables, but you can put in drawings or other types of results.

Analysis and evaluation
The main part of the essay—an in-depth look at the results, to find trends, patterns and significance and to assess the strength of the evidence in relation to your hypothesis or research question.

Conclusions
A brief overall summary of the answer to your research questions, including any unresolved questions.

References and bibliography
A numbered list of books, journal articles and websites that you used.

Appendix
Extra data, if there is any, which would be distracting in the main essay.

Example

Diagrams of apparatus

William Sladen's 1892 essay on bumblebees, mentioned on page 352, includes hand-drawn diagrams, with figure numbers so they can be referred to in the text. "Fig.12" in his essay shows the rims of the jars he used to collect bumblebees.

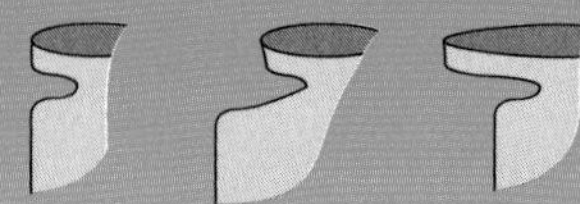

"Fig 13" in the essay shows how he used a glass and two cards to catch and transfer bees to the jar.

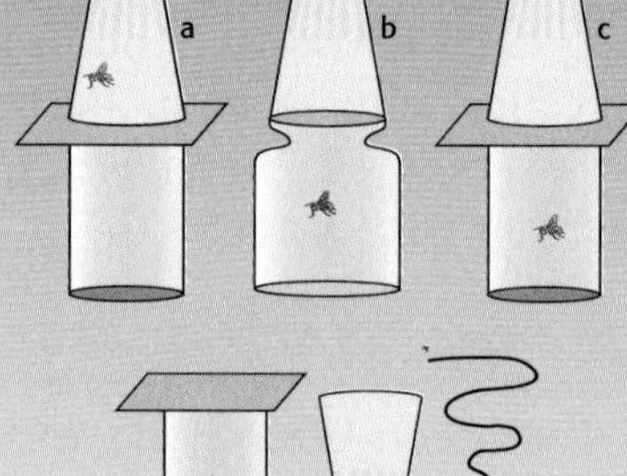

"Fig. 14" shows the end result.

- Divide your extended essay into sections with a heading for each section.
- Number all the pages and include a contents list at the start, with page numbers for each section.
- Try to include your own illustrations such as drawings, diagrams or photos. Place them, with a title for each, at the relevant point in your essay, not in an appendix.
- Decide how to show text references and put them in as you write the essay.

Analysis

The analysis is the section where you write in detail about your results, looking for any significant trends and patterns that they show and assessing the significance of the differences between the results. This needs to be a very thorough section of your extended essay.

To allow someone reading your essay to understand the analysis, your results or other data must presented as clearly as possible. Double-check your results tables to make sure that you have followed the basic rules that your teacher has told you. If you have put large amounts of raw data into an appendix, you should include summary results tables in the main body of your essay.

In your analysis, you should be asking a series of questions about the data. The pyramid below shows a hierarchy of questions, starting with basic ones at the bottom and rising up through increasingly sophisticated ones. These are suggestions and may not suit your results—every set of data presents its own challenges and opportunities.

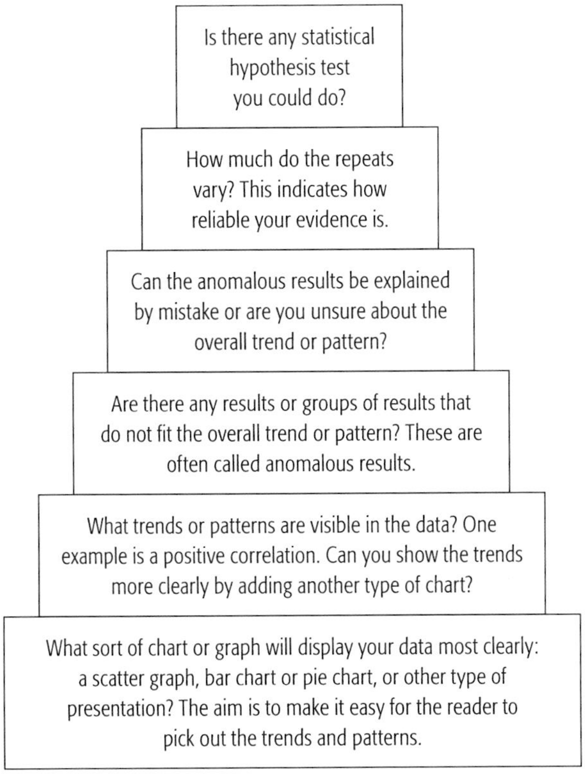

Is there any statistical hypothesis test you could do?

How much do the repeats vary? This indicates how reliable your evidence is.

Can the anomalous results be explained by mistake or are you unsure about the overall trend or pattern?

Are there any results or groups of results that do not fit the overall trend or pattern? These are often called anomalous results.

What trends or patterns are visible in the data? One example is a positive correlation. Can you show the trends more clearly by adding another type of chart?

What sort of chart or graph will display your data most clearly: a scatter graph, bar chart or pie chart, or other type of presentation? The aim is to make it easy for the reader to pick out the trends and patterns.

Start at the base of the pyramid

- Be careful how you use the words precision and accuracy: precise results do not vary much when they are repeated and accurate results are close to the true value.
- If results are both accurate and precise, then they are reliable (yet more jargon!)
- Get someone who doesn't know much about your work to read your analysis to see if it seems clear—if not then you should explain your reasoning more fully.
- It's fine to use a computer to do your calculations, but double-check the answers.

Example

Error bars

Error bars on a graph of average results give an indication of how widely results are spread above and below the average (mean).

A simple method is to show the range from the lowest to the highest values (see below).

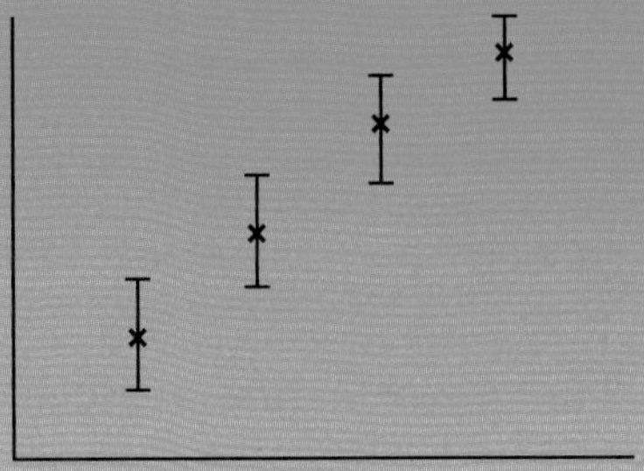

You can also calculate either the standard deviation or standard error and show plus and minus one SD/SE on graphs or plus one SD/SE on bar charts.

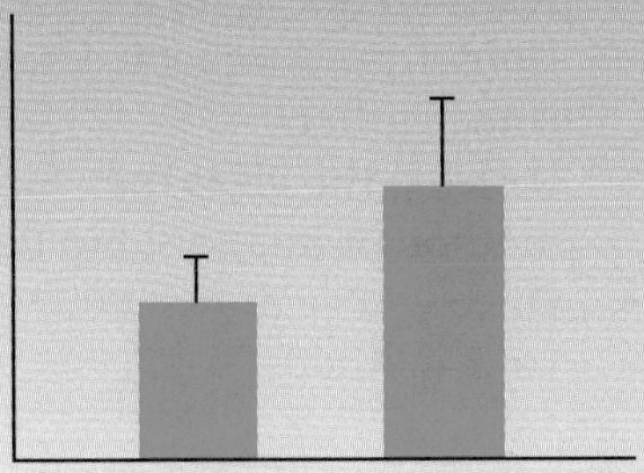

Evaluation and reasoned argument

Through your analysis, you should try to develop an overall answer to your research question. It is very unlikely that there will be an easy or simple answer. You will need to weigh up the evidence from your analysis and any data that you obtained from a literature search. There may well be two or more different possible views and you should look carefully at each of them in turn. This section of your essay is sometime called the **evaluation** and it can either be included with the analysis section or written as a separate section.

The analysis and evaluation are probably the sections of your essay that need the most careful thought. You are like a judge, weighing up the evidence as objectively as possible, to try to reach a final verdict. However, unlike a court trial there aren't just two possible verdicts—case proven or case dismissed. In fact, scientists should never claim to have proved something, as there will always be some doubt remaining. Your job is to assess the strength of the evidence and therefore how certain you can be in your answer to the research question. The whole of your essay should form a reasoned argument, with the analysis and evaluation the culmination of the argument where you explain your answer to the research question.

Table 3

Your evidence is strong if you answer "yes" to these questions	*Your evidence is weak if you answer "yes" to these questions*
Are your results consistent enough to give you reliable evidence to use to answer the research question?	Are your results variable or are there many anomalous results that can't easily be explained?
Was the design of your experiment successful so that only it gave precise and accurate results?	Were there faults in the experimental design which limited the precision or the accuracy?
Were all the variables controlled satisfactorily so that only the independent variable was varied?	Were there uncontrolled variables, which introduced uncertainties into your interpretation of the results?
Is there only one explanation that fits all the evidence and answers the research question?	Are there alternative explanations that would also fit the evidence and which you cannot refute?
Can you support each part of your answer to the research question with experimental evidence or by reference to other published data?	Are there parts of your answer to the research question which are unsubstantiated or uncertain and which need further investigation?

- All parts of your essay should help you to answer the research question. If part of your evaluation is irrelevant to the research question, it's irrelevant!
- Don't introduce new data in the evaluation—it should have come earlier.
- Develop the ideas that you wrote about in your introduction, method or analysis, but don't repeat them unchanged.
- Once you've drawn your conclusions, it is too late to evaluate your methods.

Example

Research question: do small grapes taste sweeter than large grapes?

Quantitative tests using Benedict's reagent showed a difference in the concentration of reducing sugars, with all of the 10 small grapes tested having a higher content of reducing sugars than the 10 large grapes. A statistical test gave evidence for the difference in reducing sugar content at the 5% significance level. However, only white grapes were tested, not red grapes. The large and small grapes were from different varieties and may have been grown under different conditions, so size may not have been the only variable.

Sucrose, a non-reducing sugar, may have been present in the grapes and is not a reducing sugar so is not detected by the test with Benedict's reagent. Also reducing sugars vary in how sweet they taste, with fructose tasting sweeter than glucose for example. The levels of these sugars may have varied.

Conclusion, abstract and references

If you are ready for the advice on this page, then the only bits of work that your extended essay needs are some "topping and tailing" tasks. You should add conclusions, an abstract and check that you have inserted all the necessary references. Instructions for this are given below. Writing your extended essay has probably seemed like a long race over rough terrain, but you are now near the finishing line!

Conclusions

This can be quite a brief section, where you summarize your answer to the research question, including what you decided in your discussion and what uncertainties and unresolved issues remain for further investigation.

Abstract

This goes immediately after the title page of your essay. In it you give a brief but clear summary of three things.

1 Your research question.
2 The scope of the study and the methodology.
3 The conclusions that you reached.

References and bibliography

If you followed the advice given on a previous page you will have been adding references as you wrote the essay. There are two main reasons for references.

1 It is a way of substantiating the information that you have given. If someone isn't sure whether the information is correct they can check it by looking up the references.
2 It is a way of acknowledging that you have used the words, graphics or ideas of another person. You have to sign a candidate's declaration that you've acknowledged wherever in the essay you have done this. If you don't put in a reference, it will look as though you are claiming that they are your words, graphics or ideas and this is known a plagiarism.

Usually a superscript number is given in text of the essay immediately after the place where the information from an outside source has been used. This is the **citation**. The number refers to a **reference**, which can be put at the bottom of the page or together with all the other references at the end of the essay, in a section called the **bibliography**. Each reference should include the author and date of publication, together with either for a book, the title, pages and publisher, or for a journal, the volume number and pages. The website of the Council of Biology Editors on the internet can be consulted for further advice.

- Proof-read your essay, to find and correct mistakes that would annoy readers.
- Complete and hand in the essay before the deadline, to give your supervisor as much time as possible to read it and make comments if necessary.
- If you revise your essay, check that page breaks are still in sensible places.
- Make sure that your supervisor has all the information that they need to write the supervisor's report on the front cover of your essay.

Example

Substantiation

If a statement is properly substantiated, it can be checked. In the following extract from an extended essay, the student has made some statements that can be checked, others that can't and some that were perhaps too obvious to be worth making.

Every person has his or her own biological clock. It's part of the circadian system. It gives signals to tell the body when to stay awake and when to go to sleep. Approximately one hour before a person normally goes to bed, the pineal gland in the brain will start producing melatonin. Melatonin is a hormone, which controls the circadian system. When it is dark, the release of melatonin is stimulated. Light suppresses its activity.[1]

Secretion starts after dark, peaks at midnight and is lower in the second half of the night[2]. It was thought for a long time that taking pills of melatonin easily solved people's sleeping disorders, but light is now thought to be a more powerful way of beating the problem.

References

1 http://www.umm.edu/altmed/ConsSupplements/Melatonincs.html.
2 de Leersneyder *et al., Journal of Medical Genetics* 2003; **40**:74–78.

Statistical hypothesis tests

Hypothesis tests

A hypothesis test is a statistical procedure for deciding whether or not to reject a "default belief", known as a null hypothesis. The decision is based on the result of a calculation, using the data that you collected as the results of your experiment. Hypothesis tests have the following key stages.

- **The hypotheses, H_0 and H_1**: H_0 stands for the null hypothesis, and is the belief that there is nothing out of the ordinary (e.g. the two means are equal, or there is no correlation). H_1 stands for the alternative hypothesis, and is the belief that there is something going on (e.g. the two means are different or there is a positive correlation).
- **The significance level and critical region**: the significance level is a somewhat arbitrary number, usually chosen to be 5 per cent. It is actually the probability of rejecting H_0 when H_0 is in fact true. Obviously, we want this probability to be small – but make it too small and you are unlikely to reject H_0 when it is false! In most cases, 5 per cent is fine. Most of the worked examples on the following pages use this value. The critical region depends on your significance level and the type of test. The critical region is found either by using statistics tables or by a statistics program on your computer. When you have done the calculation, you compare the calculated statistic with the critical region. If the result of your calculation lies in the critical region, H_0 is rejected.
- **The data**: it is good practice to include only a summary of the data where you show the hypothesis test in your extended essay. A suitable summary consists of as the number of observations (n), the mean (x), variance (S^2) and so on.
- **The calculations**: a formula is used to calculate a statistic. Obviously you must use the correct formula for the test that you are using, and do the calculation without any mistakes. The statistic is then compared with the critical region.
- **The conclusions**: in this last section you state whether there is evidence to reject H_0, and interpret the decision in the context of your experiment.

One- and two-tailed tests

In all hypothesis tests you are likely to come across (except ANOVA), you will have to decide whether a one-tailed or a two-tailed test would be more appropriate. If, according to your alternative hypothesis, one mean will be a larger than the other, or that the correlation will be positive, you should do a one-tailed test. This type of test is only concerned with extreme results on one side. If you do not know *before you collect your data* which mean will be larger than the other or whether the correlation will be positive or negative, you should do a two-tailed test. This test looks out for extreme values on both sides. Note that in a one-tailed test, H_1 has a $\neq$ sign, but in a two-tailed test, H_1 has a $<$ sign or a $>$ sign.

Basic calculations

Most hypothesis tests require the calculation of a mean and variance or standard deviation. In case you are not familiar with these, the formulae are given below.

The mean

$$\bar{x} = \frac{\Sigma x}{n}$$

where Σ is the sum of.

In Excel, this can be calculated with the AVERAGE function. For example, AVERAGE (A2:A21) calculates the mean of the values in the cells A2 to A21.

The variance

$$S^2 = \frac{\Sigma x^2 - \bar{x}^2 n}{n - 1}$$

In Excel, use the VARP function. For example, VARP(A2:A21).

The standard deviation

$$S = \sqrt{\frac{\Sigma x^2 - \bar{x}^2 n}{n - 1}}$$

In Excel, use the STDEVA function. For example, STDEVA (A2:A21).

Chi-squared tests

The chi-squared test is used with discrete or categoric variation, where the observed values are frequencies. For example, when Mendel found that pea plants were either tall or dwarf, he counted the frequency of each of these types.

The chi-squared test can be used to test whether an expected ratio fits the observed frequencies—this is called **goodness of fit**. We can use it to test whether a 3:1 ratio fits the numbers of tall and dwarf pea plants that Mendel counted. Another example would be using the test to see whether a 1:1 ratio fits the number of male and female births in a population.

The chi-squared test can also be used to see whether two variables are associated or independent. This use of the chi-squared test involves the use of a contingency table, which is a two-way table showing frequencies. The example box shows a contingency table.

Chi-squared tests can only be used where:
- all the expected frequencies are 5 or more
- the sample has been taken at random from the population.

Chi-squared is a statistic and is calculated using this equation:

$$\chi^2 = \Sigma \frac{(f_0 - f_e)^2}{f_e}$$

where f_o is the observed frequency
f_e is the expected frequency and
Σ is the sum of.

Each expected frequency is calculated from values on a contingency table using this equation:

$$\text{expected frequency} = \frac{\text{row total} \times \text{column total}}{\text{grand total}}$$

The number if degrees of freedom (ν) is

$\nu = (m - 1)(n - 1)$

where m and n are the number of rows and columns in the contingency table.

- The chi-squared test is just one of many statistical hypothesis tests. You must choose an appropriate one or the conclusions that you draw may not be valid.
- If the tests shown on the following pages aren't suitable, find out about others.
- Most statistics calculations can be done using a calculator or spreadsheet.
- Excel or other statistics software is useful for checking calculations, or for statistics such as ANOVA which are messy to do with a calculator.

Example

A holly tree (*Ilex aquifolium*) growing in shade seemed to have smaller berries than a holly tree in a sunny position. The number of seeds inside a holly berry is variable. It seems reasonable to predict that larger berries will contain more seeds than smaller berries. To investigate whether the number of seeds inside the berries on the two trees was different, a random sample of 60 berries was collected from each tree and the number of seeds inside each was counted. The two variables are the tree on which a berry grew and the number of seeds in the berry.

H_0: the two variables are independent (the numbers of seeds is not significantly different in the berries on the two trees)

H_1: the two variables are associated (there are more seeds in the berries on one tree than the other.

Significance level: $\alpha = 5\%$.

Number of degrees of freedom = $\nu = 2$.
Critical region: $\chi^2 > 5.99$.
Observed numbers:

	Seeds per berry			Totals
	2	3	4	
Tree in shade	12	40	8	60
Tree in sun	0	17	43	60
Totals	12	57	51	120

Expected numbers:

	Seeds per berry			Totals
	2	3	4	
Tree in shade	6	28.5	25.5	60
Tree in sun	6	28.5	25.5	60
Totals	12	57	51	120

$$\chi^2 = \frac{6^2}{6} + \frac{6^2}{6} + \frac{11.5^2}{28.5} + \frac{11.5^2}{28.5} + \frac{17.5^2}{25.5} + \frac{17.5^2}{25.5}$$

$\chi^2 = 6 + 6 + 4.64 + 4.64 + 12.01 + 12.01$

$\chi^2 = 45.3$.

This lies in the critical region, so we reject H_0.

There is evidence at the 5% level for an association between the variables, with the tree growing in the sun having more seeds per berry.

Two-sample *t*-tests

The type of *t*-test described here can be used to test for a difference between two means. For example, if two groups of mice are fed on different diets, and their growth rates are measured, the t-test can be used to find whether there is a significant difference between the growth rates with the two different diets. A two-tailed test is used if we have no idea which diet will produce significantly heavier mice. A one-tailed test (this is a statistical term and nothing to do with the mice!) is appropriate if we know which diet, if any will give the greater growth rate.

t-tests can only be used where:

- the data is not paired
- the two variables have independent normal distributions
- the two variables have the same unknown variance.

Although the test can be used for small samples, for example fewer than 30, large sample sizes are more likely to give enough evidence to reject H_0 where H_0 is untrue.

The statistic *t* is almost always calculated using a suitable calculator or computer, so the formula given here will rarely be needed:

$$t = \frac{\bar{x} - \bar{y}}{\sqrt{\dfrac{(n-1)s_x^2 + (m-1)s_y^2}{n+m-2}\left(\dfrac{1}{n} + \dfrac{1}{m}\right)}}$$

The values or results in the two samples are referred to as *x* and *y*.

$\bar{x}$ and $\bar{y}$ are the two means (the means of the two samples)

n and *m* are the numbers in each of the two samples

s_x and s_y are the standard deviations of the two samples

The number of degrees of freedom (ν) is

$$\nu = n + m - 2.$$

One-tailed *t*-tests can also be used, where it is known which mean will be the larger one.

Example

The population density of fish in an aquarium might affect their growth rate.

Two aquaria were set up which were identical in size and shape. Ten shubunkin fish were put in one and forty shubunkin were put in the other. These fish were taken from the same population and were assigned to the two aquaria randomly. The same amount of food and oxygen was supplied per fish to both tanks.

A two-tailed *t*-test is appropriate as it is not certain which population will show the more rapid growth. Crowding might inhibit or encourage growth.

H_0: *there is no difference in mean length between the denser and less dense populations.*

H_1: *there is a difference in mean length between the denser and less dense populations.*

Significance level: $\alpha = 5\%$
Number of degrees of freedom = 48
Critical region: $t \geq 2.011$

The lengths of the fish after two weeks were measured:

Length of fish (mm)				
10 fish aquarium	**40 fish aquarium**			
50	61	52	52	54
41	50	59	61	60
59	45	54	63	56
52	56	57	53	51
49	57	60	60	57
58	53	54	59	56
56	52	44	46	53
47	45	51	58	51
66	60	54	44	60
50	59	51	55	58

The mean length of the 10 fish was 52.8 mm and of the 40 fish was 54.5 mm—a higher mean with the denser population.

$t = 0.8903$

This does not lie in the critical region, so we do not reject H_0. There is insufficient evidence at the 5% level to suggest that the fish have a different length after two weeks of growth in denser and less dense populations.

After another six weeks of growth, the difference in the means was greater and *t* was 2.27, allowing H_0 to be rejected.

Spearman test

If a scattergraph suggests that there is a relationship between two variables, a statistical test can be used to test for a correlation. The Spearman test can be used to test for correlation in data that may or may not be normally distributed. For example, we might use it to test whether there is a correlation between the number of whales that are seen each day at a whale-watching site and the number of boats carrying whale-watching tourists at the site. The data may well not be normally distributed and it is not certain whether whales would be attracted or repelled by the presence of boats, so a two-tailed test would be appropriate. To test for either a positive or a negative correlation, a one-tailed test is used.

Spearman's test involves ranking the data for each of the variables. For tied ranks, the average rank is used. For example, two items in joint 4th place would each have rank 4.5.

Spearman's rank correlation test can only be used where:

- the data is bivariate – for each member of the sample two different quantities have been measured (x and y)
- there are at least 5 pairs of measurements, and preferably 10 or more
- the variables may have any distribution, since the data is ranked and ranks are used for calculating the statistic, not actual values.

For each pair of measurements the rank for the variable y is subtracted from the rank for the variable x. These differences are called d. All of the differences are squared and all of the squares are added together to give the sum of squares of differences of rank. The statistic r_s is then calculated using the equation:

$$r_s = 1 - \frac{6\sum d^2}{n(n^2 - 1)},$$

where d^2 is the square of a difference
Σd^2 is the sum of the squares of differences and
n is the number in the sample.

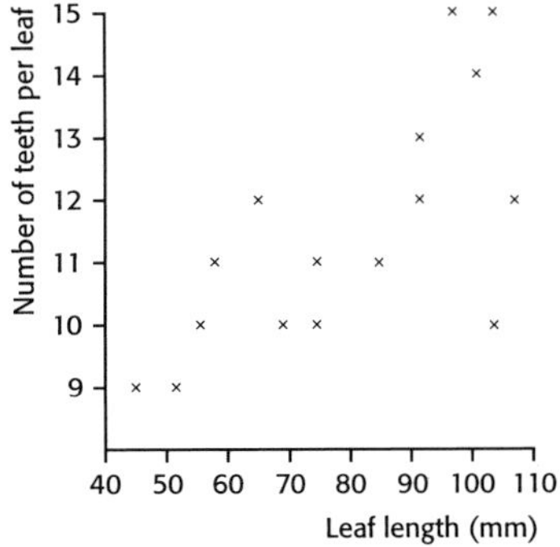

Figure 3

Example

The leaves on a Turkey oak tree (*Quercus cerris*), like most species of oak, have serrated edges, with a series of teeth. The leaves vary in size and also in the number of teeth on the edge of the leaf. There might be a positive correlation between the number of teeth and the size of the leaf, or the number of teeth might be independent of leaf size. A one-tailed test is appropriate. A random sample of sixteen leaves was collected from a Turkey oak tree. The length of each leaf was measured and the number of teeth counted.

H_0: *there is no correlation between leaf length and number of teeth.*
H_1: *there is a positive correlation between leaf length and the number of teeth.*

Significance level: $\alpha = 5\%$
$n = 16$
Critical region: $r_s > 0.425$

Length (mm)	Length rank	Teeth number	Teeth rank	d	d^2
107	16	12	11	5	25
69	6	10	4.5	1.5	2.25
92	10.5	12	11	−0.5	0.25
74	7	10	4.5	2.5	6.25
101	13	14	14	−1	1
85	9	11	8	1	1
45	1	9	1.5	−0.5	0.25
56	3	10	4.5	−1.5	2.25
77	8	11	8	0	0
104	15	10	4.5	10.5	110.25
92	10.5	13	13	−2.5	6.25
103	14	15	15.5	−1.5	2.25
97	12	15	15.5	−3.5	12.25
52	2	9	1.5	0.5	0.25
65	5	12	11	−6	36
58	4	11	8	−4	16

$\Sigma d^2 = 221.5$

$r_s = 1 - \frac{6 \times 221.5}{16(255)}$

$r_s = 1 - 0.325$

$r_s = 0.674.$

This lies in the critical region, so we reject H_0. There is evidence for a positive correlation between leaf length and number of teeth at the 5% level (see Figure 3).

Wilcoxon signed-ranks test

One use of the Wilcoxon signed-ranks test is for comparing the median of a set of data with another known or hypothesized median. For example, if the median height of children in a school in 1910 is known and the heights of children in the same school are measured in 2010, a Wilcoxon signed-rank test could be used to ascertain whether the children's height had changed.

However, the use that is described here is to test for a difference between the medians of two paired samples of data. The median is the middle value—the value that is half-way up the rank order from the lowest to the highest. An example of paired data would be measurements of the length of the thumb on the left and right hands of ten people. The null hypothesis in this case would be that left and right thumbs are equal in length.

The test involves subtracting one result in each pair of results from the other, and then arranging these differences in order, from the smallest or most negative to the largest or most positive. Each difference is then ranked, with 1 being the smallest or most negative difference. Differences of zero are not included in the ranking. For tied ranks, the average rank is used. For example, two items in joint 4th place would each have rank 4.5.

The ranks of all the positive differences are then added up to give S_+ and the ranks of all of the negative differences are added up to give S_-. The statistic S is whichever of S_+ and S_- is smaller. Comparing S with the critical value then completes the test.

Because the statistic is calculated from the ranks and not from the actual data values, the data does not have to be normally distributed. However, the distribution of differences must be symmetrical rather than skewed.

Wilcoxon signed-ranks (paired data) test can only be used where:

- the data is paired
- the difference has a symmetrical distribution, so the mean difference is equal or very nearly equal to the median difference
- the differences are independent
- the sample has been taken at random from the population.

Example

A narrow strip of willow trees is growing on Bettisfield Moss, a peat bog in Shropshire, England. Willows do not usually grow there. The warden of the reserve has suggested that water from a nearby canal might be leaking through the peat, making it less acid and allowing the willows to grow. Peat samples are collected from 10 random sites along the willow strip. Peat is also collected from sites 5 m south of each of the random sites, in an area with no willow trees growing. The pH of each of the 20 peat samples is measured. The pH scale is logarithmic, so the data is not normally distributed. A Wilcoxon test is therefore suitable. The hypothesis is that the pH in the area with willow trees is higher, not that there is a difference in pH. A one-tailed test is therefore appropriate.

H_0: *there is no difference in median pH.*
H_1: *the pH is greater in the willow area.*

Significance level: $\alpha = 5\%$

pH of peat		Difference in pH of peat	Rank
Non-willow area	Willow area		
4.5	4.5	0	-
4.5	4.6	0.1	1.5
4.4	4.6	0.2	3.5
4.4	4.6	0.2	3.5
4.3	4.6	0.3	5.5
4.3	4.6	0.3	5.5
4.8	5.2	0.4	7
4.7	4.6	−0.1	1.5
4.8	5.2	0.5	8
5.6	6.2	0.6	9

$n = 9$
Critical region: ≥ 8
S_+ (sum of ranks of positive differences) $= 1.5$
S_- (sum of ranks of negative differences) $= 43.5$
$S = 42.$
This lies in the critical region so we reject the null hypothesis. There is evidence at the 5% level to suggest that the pH of the peat in the willow area is higher than the pH of the peat in the non-willow area.

35 Internal assessment

The practical work component of the IB Diploma Programme biology course has number of aims. These include the following.

- Providing you with experience of methods and techniques used in biological investigations.
- Providing you with opportunity to explore biology-related questions that you genuinely wonder about.
- Developing your ability to analyse, evaluate and synthesize scientific information.
- Engendering your awareness of the need for, and the value of, effective collaboration and communication during scientific activities. This is one of the main aims of the group 4 project, described later in this section.
- Developing your ability to use and apply information and communication technology skills to the study of biology.

During the course, your teacher will assess your work in order to provide you with feedback for improvement and will also evaluate your work against five criteria.

The first three criteria—design (D), data collection and processing (DCP) and conclusion and evaluation (CE)—are each assessed a minimum of two times by your teacher.

The five assessment criteria

- *Design (D)*
- *Conclusion and Evaluation (CE)*
- *Data Collection and Processing (DCP)*
- *Manipulative Skills (MS)*
- *Personal Skills (PS)*

Manipulative skills (MS) are assessed over the whole course. The assessment is based on your performance on a wide range of manipulative skills. At the end of the course, your teacher will assign an overall MS level.

Personal skills (PS) are assessed once only and this will be during the group 4 project.

Each of the assessment criteria can be separated into three **aspects**. Descriptions are provided to indicate what is expected in order to meet the requirements of a given aspect **completely** (c) and **partially (p)**. A description is also given for circumstances in which the requirements are **not** satisfied **at all (n)**. A "complete" is awarded 2 marks, a **"partial"** 1 mark and a **"not at all"** 0 marks.

At the end of the course, your teacher is required to submit samples of written practical work to an external moderator. The external moderator checks for programme compliance and ensures that the standard of marking is not too harsh or too lenient. To help with this process, you will need to maintain a portfolio of assessed work from your lab activities.

You, or your teacher, will maintain a portfolio cover sheet called the Practical Scheme of Work (PSOW). The PSOW should indicate that higher level (HL) students have undertaken 60 hours of practical work and that standard level (SL) students have undertaken 40 hours of practical work.

Your teacher may offer you laboratory experiences that are primarily designed to reinforce content, verify well-known concepts or train you in a particular protocol. In such cases, it may not be appropriate for the teacher to assess the work using the IB internal assessment (IA) criteria, though it might form part of the practical scheme of work.

Independence in laboratory work

Where written products are assessed using the IA criteria, it is expected that the assessed portion will be your own work.

For the design criterion, this means that you should set the question, identify the variables and design or modify the method.

For the data presentation and processing criterion, this means that you should collect the data yourself (not in groups), decide on how to process the data and how to present the processed data.

For the conclusion and evaluation section, your teacher will not provide guiding questions to support the evaluation.

The design criterion

Ideally, a portion of your practical scheme of work will represent genuine inquiry, i.e. you will be exploring questions that you genuinely wonder about. Where the *design* assessment criterion is being applied, you are responsible for generating your own research question, though your teacher may specify a general aim or possibly even indicate the independent variable.

A clear, focused research question will state the dependent and independent variables. The dependent variable is the outcome variable. The independent variable is the cause variable. A student would usually set the level of the independent variable and then measure the resulting dependent variable. The variables stated in the research question should be those that are directly measured.

Aspect 1 of the design criterion requires a student to: *Formulate a focused problem/ research question and identify the relevant variables.*

For example, *The effect of blow-drying on the breaking force of human hair* is better than *The effect of hair treatment on the tensile strength of human hair.*

If an organism is part of the investigation, the scientific name should be included in the research question. You will then need to clearly indicate additional variables that could reasonably be anticipated to have an influence on the dependent variable if they are not held constant. The list could be infinite, so a suggestion would be to aim for the most significant three or four variables that need to be controlled. When describing the method, additional controls could be discussed. In some cases, variables are not controllable, such as in the case of field work. In this case, these variables should be monitored.

In carrying out the method, you must ensure, as far as possible, that the change in the dependent variable is due to changes in the independent variable only. You must change only one thing and attempt to hold everything else constant.

Aspect 2 of the design criterion requires a student to: *Design a method for the effective control of the variables.*

In the case of the hair experiment above, this would include whether the hair sample being tested was dry or wet, whether the hair was curly or straight, whether the hair had been chemically treated and the length of the hair being tested.

When designing a method, it is acceptable to modify an existing protocol that you learned to use earlier in the course. In this case, you should provide a citation for the handout, lab manual or website where the protocol was obtained. Your teacher should not provide you with the full method.

When describing your method, you should imagine a student who is perhaps two years younger than you following your method. Have you written down all the steps? A diagram of the apparatus could be included. Clear descriptions of procedural controls should be given to ensure that only the independent variable is changing, e.g. *Wrap each hair around the hook four times; Pull each hair approximately one centimeter from the bottom*, etc.

Aspect 3 of the design criterion reads:
Develops a method that allows for the collection of sufficient relevant data.

Biological systems, because of their complexity and variability, require replicate observations and multiple samples of material. As a rule, the lower limit is 5 measurements, or a sample size of 5. Very small samples run from 5 to 20, small samples run from 20 to 30, and big samples run from 30 upwards. Obviously this will vary within the limits of the time available for an investigation.

The data collection and processing criterion

Awareness of the errors associated with data collection is called for in the conclusion and evaluation, data collection and processing, and design criteria.

An appreciation of error should be apparent at all stages of a report on an investigation.

- **In the planning stages** where the limitations of the time and the materials should be assessed, and the potential sources of error should be controlled.
- **In the data collection and processing stages** where the degree of accuracy of a measuring device should be stated and a recognition of the effects of errors and uncertainties in processed data should be shown.
- **In the evaluation of the investigation** where the sources of error should be discussed, along with the possible ways of avoiding them.

?

1 A force meter is being used to measure breaking force of a strand of hair. With the hair attached, the force reading indicates a force of −0.12 N.
   (a) State whether this represents random error or systematic error.
   (b) Determine the effect that this error might have on the outcome of the experiment.
   (c) Describe how this error can be avoided.

2 In an investigation to measure residual lung volumes, repeat trials are performed. The same subject shows a residual volume of 1.0 dm^3 in the first trial and 1.2 dm^3 in the second trial.
   (a) State whether this variation represents random error or systematic error.
   (b) Explain how to overcome this type of error using the residual volume experiment as an example.

Systematic errors (equipment error)

Systematic errors can be reduced if equipment is regularly checked or calibrated to ensure that it is functioning correctly. For example, a thermometer should be placed in an electronic water bath to check that the thermostat is correctly adjusted, a colorimeter should be calibrated to a blank between measurements to compensate for the drift of the instrument or a force meter should be set to zero if there is nothing hanging from it.

Random errors (measurement error)

In biological investigations, errors can be caused by changes in the material used, or changes in the conditions under which the experiment is carried out. Biological material is notably variable. For example, three trials from the same subject show them having a vital capacity of 3.6 dm^3, 3.8 dm^3 and 3.9 dm^3 . Random errors can be kept to a minimum by careful selection of material and careful control of variables.

> Aspect 1 of the data collection and processing criterion reads: *Records appropriate quantitative and associated qualitative raw data, including units and uncertainties where relevant.*

Degrees of precision

All measured values have an associated uncertainty, therefore, all columns in data tables are required to have an accompanying uncertainty figure (a ± figure).

The following rules apply for *minimum* uncertainty figures.

1 When making a measurement, the last recorded digit should be an estimate (except in the case of a ruler).

For most measuring devices, the minimum uncertainty figure is ± half of the place value of the last measured value (as opposed to the estimated digit).

Examples: in the following quantities, the estimated digit is in italics and the last measured digit is underlined:

40.*7*
1*3*
3.*0*

The minimum uncertainty figures for the above quantities are shown:

40.*7* ± 0.5
1*3* ± 5
3.*0* ± 0.5

The ruler is a special case. For example, a leaf measured using a ruler whose smallest division is 1 mm would measure 4.5 cm ± 0.1 cm. This is due to the fact that there is a limit to the precision at both ends. This uncertainty figure indicates that if a number of students were to interpret the reading, most would place the value within 4.4 cm and 4.6 cm.

2 You must choose an appropriate instrument for measuring such things as length, volume, pH and light intensity. For example, very small lengths should be measured with calipers, not a ruler.

3 For an electronic instrument, the *minimum* uncertainty is ± one unit of the last decimal place; e.g. 13.7 g ± 0.1 g. Often the reference material that accompanies an electronic device will specify an uncertainty that is higher than the minimum uncertainty. For example, the reference material for a sensor used to measure the partial pressure of CO_2 indicates that the uncertainty of measurements made with this device is ± 100 ppm. The minimum uncertainty of a measurement of 505 ppm would be ± 1 ppm, but the absolute uncertainty of the measurement, according to the literature, is higher at ± 100 ppm.

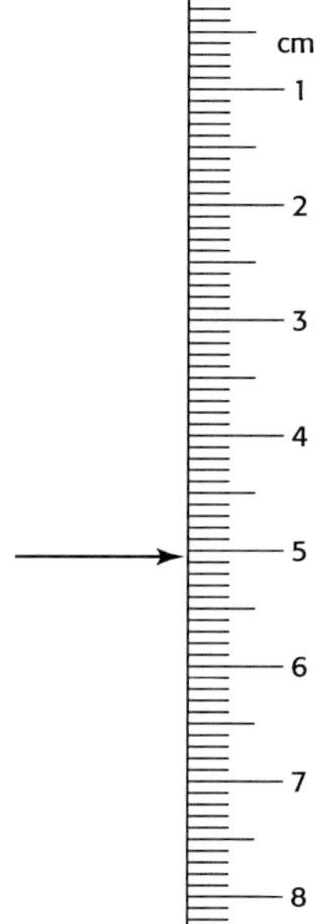

Figure 1 The minimum uncertainty in measurements using a mm ruler is ± 1 mm

> Using the precision rules for a ruler outlined, answer the following multiple choice questions, relating to Figure 1.
>
> 1 Which represents the measurement which shows the correct degree of precision for the ruler?
> (a) 4.95 cm.
> (b) 49 mm.
> (c) 4.9 cm.
>
> 2 What is uncertainty figure which should accompany the measurement?
> (a) ± 0.5 cm.
> (b) ± 1 mm.
> (c) ± 0.5 mm.

Errors and uncertainties when processing data

If a reading is particularly different from the others, it may be left out of the processing and analysis. However, you must justify why you have chosen to do this.

Means should be written with the same precision as the input data. Means could be written with uncertainty figures. You could write ± the standard deviation or you could write ± the standard error of the mean. You should indicate which you are writing.

You can write error bars on bar graphs representing the range or the standard deviation or the standard error. Error bars could also indicate ± half the range of measurements. You should indicate which you are writing.

For line graphs, biology students are not normally expected to write error bars representing minimum uncertainty. However, if the points on a graph represent means, then error bars are recommended.

Statistical tests are not expected in the internal assessment, but their use is recommended as it helps to reinforce the content of Topic 1.

Aspect 3 of the data collection and processing criterion requires a student to:
Present processed data appropriately and, where relevant, include errors and uncertainties.

Qualitative and quantitative data

Any experimental observation that is measured can be expressed as a quantity, i.e. as a number with an associated unit. Such data is called quantitative data. It is important that raw data is recorded with uniform precision. The working with data activity on page 29 provides guidance on the construction of data tables. Associated qualitative data are considered to be those observations that would play a role in helping you interpret or evaluate results. Such observations could be kept in a log book for later reference.

Aspect 1 of the data collection and processing criterion requires a student to:
Record appropriate quantitative and associated qualitative raw data, including units and uncertainties where relevant.

Example

Table 1

Time	pCO_2/ppm ± 100ppm
0	505
10	515
15	569
30	620
45	700

Changes in the partial pressure of CO_2 at floor level in the classroom from the time students enter the class until the end of the class period.

Associated qualitative data:

- when students moved around the room, there would be fluctuations in the reading
- readings had a certain inertia; one student breathed on the sensor and it stayed significantly elevated for a period of time.

?

1 Descriptive titles should contain enough information so that the table could "stand alone". Explain what this means using the example shown in Table 1.

2 (a) Compare the time column with the pCO_2 column in Table 1.

(b) Suggest improvements to the presentation of the time column in Table 1.

3 Discuss whether the data supports a conclusion that the pCO_2 is rising in the room with time.

Data processing

Data processing involves manipulating raw data. It could involve determining the mean of several measurements, constructing a histogram from frequency classes, determining a slope, drawing a line of best fit, conducting a statistical test, etc.

Aspect 2 of the data collection and processing criterion reads: *Processes the quantitative raw data correctly.*

Remember that titles should be descriptive so that the transformed data can stand alone. Axes of graphs should be fully labelled including quantity name and units. Graphs generated from software packages are acceptable as long as you have had a role in making decisions such as the selection of labels and scales for axes, the inclusion of appropriate units, a graph title, etc.

The working with data activity on page 113 provides guidance on how to decide which type of display is most appropriate for the transformed data, e.g. table, graph, kite diagram etc. You are expected to make this decision yourself.

Aspect 3 of the data collection and processing criterion reads: *Presents processed data appropriately and, where relevant, includes errors and uncertainties.*

You should present the data so that all the stages to the final result can be followed. For calculations, this could involve showing a sample calculation.

Inclusion of metric/SI units is expected for final derived quantities. Errors in the use of significant figures should be avoided. The propagation of uncertainties through calculations such as in the determination of per cent uncertainty is not expected.

The conclusion and evaluation criterion

Students should seek to explain the experimental results. The explanation should refer to observations, trends or patterns revealed by the data, i.e. conclusions should be supported by the data. Students should take care not to say that their results "prove" an explanation. The most that can be said about empirical evidence is that it "supports" an explanation.

Aspect 1 of the conclusion and evaluation criterion requires the student to: *State a conclusion, with justification, based on a reasonable interpretation of the data.*

When carrying out a procedure that should have lead to a well-known result, you should draw a conclusion as to your confidence in your result by comparing the experimental outcome with the published literature. The literature consulted should be cited.

The guiding question for Aspect 2 should be: *What makes me lack confidence in my conclusion?*

Aspect 2 of the conclusion and evaluation criterion requires the student to: *Evaluate weaknesses and limitations in the procedure.*

You should evaluate the design of the investigation as well as your performance of the procedure. You could also comment on the reliability of the data. You should go beyond identifying the weaknesses in design and performance. You should also evaluate the relative impact of the weaknesses.

The guiding question for Aspect 3 should be: *What modifications could I make to increase confidence in my conclusion?*

Suggestions for improvements should be based on the weaknesses identified in Aspect 2: they should cover both design aspects and aspects of the performance of the experiment. The modifications proposed should be reasonable in relation to the resources and time available in a standard secondary school classroom.

The manipulative skills criterion

Safety

A guiding question for students working in the lab is: *Can this investigation be conducted in a fashion that exposes me or my classmates to less risk?*

For example, a student sitting down at a bench when beakers of boiling water are being handled runs the risk of a lap burn if the beakers spill. The student could carry out the same investigation standing up without compromising their ability to complete the experiment.

A suggestion is to include a statement about relevant safety issues as part of the experiment design. The process of reflecting on possible threats to safety will decrease risk.

Aspect 3 of the conclusion and evaluation criterion requires the student to:
Suggest realistic improvements in respect of identified weaknesses and limitations.

Ethics

The International Baccalaureate Organization has issued a directive that laboratory or field experiments and investigations need to be undertaken in an ethical way. Following the directive is a good model of part of the professional life of a scientist and should be followed even when it seems trivial.

This directive covers three domains.

- Experiments involving human subjects should be undertaken with due consideration to informed consent.
- Experiments should not be undertaken that inflict pain or harm on humans or living animals.
- Experiments should be conducted in such a way that they have little or no negative impact on the environment.

A suggestion is to include a statement about ethical considerations as part of the experiment design.

Guiding questions for ethical investigations

- Are the subjects in my experiment willing participants? Do they have full understanding of the nature of the experiment?
- Do they have full access to the conclusions drawn from the experiment?
- Is it possible to investigate my question without the use of animals?
- Can this investigation be conducted in another way so that it has a lower negative impact on the environment?

Table 2 Manipulative shills

Levels/marks	Aspect 1: following instructions	Aspect 2: carrying out techniques	Aspect 3: working safely
Complete/2	Follows instructions accurately, adapting to new circumstances (seeking assistance when required).	Competent and methodical in the use of a range of techniques and equipment.	Pays attention to safety issues.
Partial/1	Follows instructions but requires assistance.	Usually competent and methodical in the use of a range of techniques and equipment.	Usually pays attention to safety issues.
Not at all/0	Rarely follows instructions or requires constant supervision.	Rarely competent and methodical in the use of a range of techniques and equipment.	Rarely pays attention to safety issues.

The group 4 project

Scientific publications often have long lists of authors. If one were to read the list of authors and the places where they work, it soon becomes apparent that research groups are often made up of specialists in different areas. For example, one research group looking at models of trace mineral absorption in the animal intestine consisted of a mathematician who developed the model that predicted absorption patterns, a chemist with expertise in mass spectroscopy, a radiation biologist with expertise in radioactive tracers as study tools and a biochemist with expertise in trace mineral absorption.

The group 4 project is a model of this kind of interdisciplinary team approach to research. The process is meant to be a collaborative experience where the emphasis is on the **processes** involved in scientific investigation rather than the **products** of such investigation.

The project is divided into three stages: planning, action and evaluation.

In the planning stages, students discuss the possibilities and narrow them down. This is best done in a rational way – the final focus should be the one where arguments demonstrated the greatest range of possible experiments for all of the sciences.

In programmes involving a small number of IB students, students are often all involved in the investigation of the same topic. For larger programmes, it is acceptable to divide up into several smaller groups containing representatives from each of the science subjects, where each group investigates a different topic.

The action phase often involves students working on a discipline-specific investigation that relates to the group's theme. This will involve communication within the group to ensure equitable work distribution and negotiation around timelines. For the group to be successful, individuals will need to complete their portion of the work.

The emphasis during the evaluation stage is on students sharing their conclusions as well as evaluation of their results and process with other students. This can be done as a website, a PowerPoint presentation or a poster. Evidence of an individual's contribution to the group project needs to be kept on file by your teacher in case your work needs to be submitted for external moderation.

The most successful topics focus on a genuine task or question rather than on a general theme, e.g. conduct an environmental impact assessment of building a dam on the River Dee or evaluation of the consequences of sequestering large quantities of CO_2 deep into the ocean.

Recommendations for PowerPoint© presentations

- Follow the six-by-six rule No more than six words per bullet; no more than six bullets per page
  Slides should be purposeful – plan on about 45 seconds to 1 minute for each slide.
- Empathise with the audience – try to be interesting.
- Animate bullets.
- Use a light background with a strong contrasting font.
- Be professional – no noise or sound files unless they are necessary to understand the content.
- Vary content: include animations, tables, movies, images, quotes and humour.
- Start from the general and move to the specific.

Table 3 Personal skills

Levels/marks	Aspect 1: self-motivation and perseverance	Aspect 2: working within a team	Aspect 3: self-reflection
Complete/2	Approaches the project with self-motivation and follows it through to completion.	Collaborates and communicates in a group situation and integrates the views of others.	Shows a thorough awareness of their own strengths and weaknesses and gives thoughtful consideration to their learning experience.
Partial/1	Completes the project but sometimes lacks self-motivation.	Exchanges some views but requires guidance to collaborate with others.	Shows limited awareness of their own strengths and weaknesses and gives some consideration to their learning experience.
Not at all/0	Lacks perseverance and motivation.	Makes little or no attempt to collaborate in a group situation.	Shows no awareness of their own strengths and weaknesses and gives no consideration to their learning experience.

Example experiment: the effect of blow-drying on the tensile strength of hair

The format of this experiment write-up is just one of many possibilities. You should consult with your teacher to determine how this format will differ from expectations in your class.

Introduction

Blow-drying hair not only speeds up the drying of hair, but many claim that it adds "volume" to hair appearance. Others claim that blow drying makes hair more likely to break and therefore actually lowers hair volume. This second claim will be the subject of this investigation

Research question *What is the effect of regular blow-drying on the breaking force of hair?*

Breaking force is a measure of tensile strength. The independent variable in this experiment is hair treatment (blow-dried or non-blow-dried) and the dependent variable is breaking force in newtons (N).

A number of other variables can affect the breaking force of hair: whether the hair sample being tested is dry or wet, whether the hair is curly or straight, whether the hair has been chemically treated, and the length of the hair being tested.

Materials

- Computer.
- Vernier force sensor.
- 50 blow-dried hairs.
- 50 non-blow-dried hairs.

Method

1 Hair from every member of the class was sampled. Gender was controlled as all students at the school are girls. Age was controlled as all students were either 17 or 18. All students had dry hair at the time–no one had just come from swimming.

2 Hair was divided into two groups depending on whether the student regularly blow-dried her hair or never blow-dried her hair. Occasional blow-driers were not found in the sample. No students who dyed their hair were found in the sample. Equal numbers of curly haired students were found in both groups though no one had pronounced curly hair. Between 5 and 10 hairs from each individual were sampled.

3 Data collection was set at 20 samples per second and time was left as the independent variable. The display was set as a graph.

4 The maximum force switch was set to 10 N.

5 The force meter was anchored to a ring stand.

6 Hair was wrapped around the force meter hook with 4 turns.

7 Hairs were pulled 15 cm below the hook.

8 When the hair snapped, data collection was stopped and the statistics tool of the software was used to determine the peak force.

9 The procedure was repeated 50 times for each treatment group, alternating each time to ensure that variations in technique affected both groups equally.

Results

The breaking force measurements of blow-dried and non-blow-dried hair are shown in Table 4. Additional observations made during the experiment are listed in Table 5. Figure 2 shows the results from Table 4 as a bar chart.

Conclusion

The data support the alternate hypothesis that blow-dried hair has a lower mean breaking force than non-blow-dried hair.

A possible explanation for the observations follows. Hair is primarily composed of the fibrous protein keratin. Fibrous proteins have tensile strength. In the case of keratin, part of this tensile strength comes from the physical structure of the hair, which can be damaged by too much manipulation. Heat can cause physical degradation of the hair surface. Heating also causes

Ethics

All hair donors were aware of the purpose of the experiment.

Hair was extracted by first running fingers through hair to find detached hairs. To obtain the balance of the sample, individual hairs were carefully cut out.

Safety

Care was taken to ensure that cables and cords were bound so that they were not a tripping hazard.

The ring stand was clamped down to avoid the apparatus falling over on the experimenter.

Table 4 Breaking force of non-blow-dried hair (NBD) vs blow-dried hair (BD)

NBD hair breaking force /N ± .01 N*			BD hair breaking force /N ± .01 N*		
1.794	0.854	1.779	0.634	2.434	1.104
1.027	1.784	2.503	1.811	1.403	1.227
3.472	2.787	1.166	2.142	1.84	0.856
2.884	2.454	1.196	1.967	1.932	0.859
2.111	1.963	3.213	0.155	1.994	0.061
1.956	1.819	2.872	2.024	1.626	0.031
1.413	2.607	2.616	1.563	1.932	1.072
1.782	2.055	3.421	2.076	1.714	2.001
2.186	1.411	3.011	2.065	0.061	1.721
1.738	2.362	2.704	3.881	2.454	1.231
1.765	2.276	1.986	2.399	2.856	2.172
1.897	2.546	2.721	2.481	2.515	1.982
2.261	2.669	3.002	2.434	2.853	2.21
1.309	2.393	2.947	2.445	1.012	1.70
1.165	1.334		2.787	1.258	
1.534	1.963		1.188	1.012	
1.251	1.503		2.474	1.258	
0.854	1.626		2.007	0.031	
$\overline{\chi}$ = 2.10 N s = 0.66 N			$\overline{\chi}$ = 1.70 N s = 0.84 N		

*Uncertainty figure is from force sensor literature.

keratin to soften. Blow-drying hair when it is excessively wet can lead to "bubble hair", i.e. water that has penetrated the hair can boil, and tiny bubbles of steam form inside the softened hair. This permanently weakens the hair (Gray, 2003) .

Limitations

The number of replicates per individual was satisfactory, but the overall sample size of donors was too small to yield a high degree of certainty in the results. Variation between individuals seemed to be greater than variation within an individual. Because of the small sample size, a single person with strong hair in that group could have skewed the results. The pulling method was controlled to some degree, but there was no method of regulating the uniformity of increasing the force. Hair varied in the degree to which it stretched before breaking suggesting that this was a variable that might have needed to be controlled. Variation in blow-drying times and heat settings was not attempted.

Modifications

The measurement was fairly easy to perform so doubling the number of donors would be possible without too much of an additional time burden. To achieve uniform increases in force, increments of fixed weight could be hung from the hair instead of pulling with the hand. Attempts could be made to encourage subjects to blow-dry their hair at a fixed setting for a fixed time before the experiment was carried out, though this is possibly not realistic.
A separate experiment should be performed to examine the effect of blow-drying on stretchiness of hair.

Table 5 Additional observations

• Hair varied in coarseness, which was a variable not controlled for.
• Coarser hair tended to require higher forces to break.
• Variation in coarseness appeared to be the same in both groups.
• Variation between individuals seemed to be greater than variation within an individual.
• Hairs vary in the degree to which they stretch before breaking.

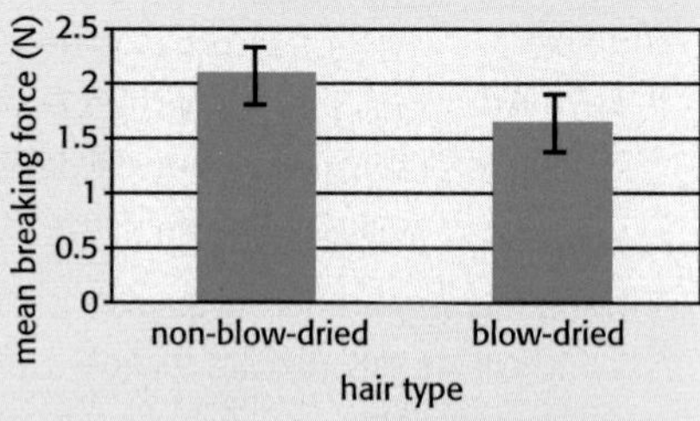

Figure 2 The mean breaking force of non-blow dried vs. blow-dried hair. Error bars represent +/– standard deviation.

Statistical test

H_0: *differences in the mean breaking force of non-blow dried hair and blow-dried hair are due to chance and are not significantly different.*

H_1: *differences in the mean are not due to chance but are statistically significant.*

Using Excel to calculate the *t*-value:
p=0.00093
p« 0.05.
Accept H_1.

Bibliography

Gray, John The Science of Styling. http://www.pg.com/science/haircare/hair_twh_140.htm

Summative assessment criteria

Table 6 Design

Levels/marks	Aspect 1: defining the problem and selecting variables	Aspect 2: controlling variables	Aspect 3: developing a method for collection of data
Complete/2	Formulates a focused problem/research question and identifies the relevant variables.	Designs a method for the effective control of the variables.	Develops a method that allows for the collection of sufficient relevant data.
Partial/1	Formulates a problem/research question that is incomplete or identifies only some relevant variables.	Designs a method that makes some attempt to control the variables.	Develops a method that allows for the collection of insufficient relevant data.
Not at all/0	Does not identify a problem/research question **and** does not identify any relevant variables.	Designs a method that does not control the variables.	Develops a method that does not allow for any relevant data to be collected.

Table 7 Data-collection and processing

Levels/marks	Aspect 1: recording raw data	Aspect 2: processing raw data	Aspect 3: presenting processed data
Complete/2	Records appropriate quantitative and associated qualitative raw data, including units and uncertainties where relevant.	Processes the quantitative raw data correctly.	Presents processed data appropriately and, where relevant, includes errors and uncertainties.
Partial/1	Records appropriate quantitative and associated qualitative raw data, but with some mistakes or omissions.	Processes quantitative raw data, but with some mistakes and/or omissions.	Presents processed data appropriately, but with some mistakes and/or omissions.
Not at all/0	Does not record any appropriate quantitative raw data or raw data is incomprehensible.	No processing of quantitative raw data is carried out or major mistakes are made in processing.	Presents processed data inappropriately or incomprehensibly.

Table 8 Conclusion and evaluation

Levels/marks	Aspect 1: concluding	Aspect 2: evaluating procedure(s)	Aspect 3: improving the investigation
Complete/2	States a conclusion, with justification, based on a reasonable interpretation of the data.	Evaluates weaknesses and limitations.	Suggests realistic improvements in respect of identified weaknesses and limitations.
Partial/1	States a conclusion based on a reasonable interpretation of the data.	Identifies some weaknesses and limitations, but the evaluation is weak or missing.	Suggests only superficial improvements.
Not at all/0	States no conclusion or the conclusion is based on an unreasonable	Identifies irrelevant weaknesses and limitations.	Suggests unrealistic improvements.

Answers to Questions

Chapter 1

1 a i eukaryote because there is a nucleus
ii root tip because it has a cell wall
iii interphase because chromosomes are not visible

b i 17 μm ii 12.5 mm

c water lost from cell by osmosis; volume of cytoplasm reduced; plasma membrane pulled away from cell wall;

2 a mitochondrion; nucleus; lyosome;

b measure the length of structure I in mm and multiply by 1000 to convert to micrometres; magnification is the length of I in micrometres divided by 1.5;

c multiply 10 μm by the magnification; divide by 1000 to obtain the length of the scale bar in millimetres

d measure the length of structure III in millimetres; multiply by 1000 to convert the length to micrometres; divide by the magnification to get the actual size in micrometres

3 a calculate length of smallest branch; measure length of image of smallest branch and length of scale bar in millimetres; actual length of branch is 600 μm × length of image of smallest branch, divided by length of scale bar

b measure diameter of siphon in millimetres; multiply by 1000 to obtain length in micrometres; divide by 180 to obtain actual length

c not an animal because it has a cell wall; not an animal because it has chloroplasts; not an animal because it has a large vacuole;

d not multicellular because no divisions between nuclei; could be unicellular if single cells can have many nuclei; could be acellular if we define a cell as small with only one nucleus;

e transport of material through the alga/support for whole alga; fluid emptied from whole vacuole if it is burst anywhere;

f entry of more water by osmosis/possible bursting; loss of mineral ions/inability to absorb mineral ions;

Chapter 2

1 a osmosis; b endocytosis;
c diffusion; d exocytosis;
e facilitated diffusion/active transport;

2 a i active transport; ii facilitated diffusion;
iii osmosis;

b contains secreted proteins; not enough water dilutes the solutes/proteins; because not enough chloride ions in it; so not enough osmosis happens;

3 a 98 130 square micrometres;

b plasma membrane area divided by the total area; multiplied by 100; 1.8%;

c outer membrane is smooth; inner membrane is invaginated; extra surface area of inner membrane needed for respiration;

d protein synthesis as there is much rough ER; ATP production as there is much mitochondrial membrane;

4 a diagram showing these clearly labelled structures: phospholipid bilayer; phosphate heads and hydrocarbon tails; cholesterol; glycoproteins; integral proteins; peripheral proteins;

b proteins allow cell to cell communication; hormone binding sites; when hormone binds a mesage passes into the cell; receptors for neurotransmitters at synapses; channel proteins allow nerve impulses to pass;

c vesicles can move around inside cells; vesicles can pinch off/fuse with membranes; vesicles carry proteins from the rough ER; vesicles carry proteins to the Golgi; vesicles carry substances from the Golgi; vesicles carry substances to the plasma membrane; vesicles are used to secrete/absorb substances; exocytosis/endocytosis;

Chapter 3

1 a The sequence is F, C, B, A, D, G, H, E

b A metaphase B prophase C prophase
D anaphase E telophase F prophase
G anaphase H anaphase

c in prophase they appear to be randomly arranged; initially telomeres are pointed away from the cell's equator at metaphase;

at anaphase, then are pointing away from poles/toward equator; at telophase, they are pointing toward the cleavage furrow/former equator of cell;

2 a spindle microtubule formation; the others are processes that occur during interphase;

b attachment of microtubules to centromeres; it is unrelated to prophase while the other two are related to prophase;

c formation of gametes; the others are roles of mitosis;

3 a I – G_1 or end of mitosis II – S III – G_2 or beginning of mitosis.

b i prophase – approximately 14 pg. nucleus^{-1}

ii telophase – approximately 7 pg. nucleus^{-1}

4 One mark per definition:

i diploid: one chromosome per parental set;

ii cytokinesis: division of the cytoplasm;

iii S phase: stage of the cell cycle/interphase when DNA replication occurs;

5 (early) prophase showing growing spindle microtubules; (early) prophase showing condensed chromosomes; (late) prophase showing spindle microtubules extending to the equator from the centromere/MTOC; (late) prophase showing even number of chromosomes with sister chromatids; metaphase showing spindle microtubules attached at centromere; metaphase showing chromosomes lined up at the cellular equator; anaphase with chromatids separated and moving to opposite poles; telophase with nuclear membrane re-forming and chromosomes uncoiling;

6 I – anaphase II – prophase

Chapter 4

1 a they can be the same thing if the protein is made up of only one polypeptide. Many proteins are composed of more than one polypeptide. For example insulin is formed from two polypeptides and hemoglobin is composed of four polypeptides

b both are triglycerides but fats are solid at room temperature and oils are liquid at room temperature/oils have more unsaturation/more double bonds per molecule than fats;

c both are polymers of glucose/starch is the storage form of carbohydrate in plants while glycogen is the storage form of carbohydrates in animals/fungus;

d condensation releases water while hydrolysis consumes water/condensation leads to polymerization while hydrolysis leads to polymer breakdown;

e hydrophobic is water hating while hydrophilic is water loving; (1 mark each)

2 a maltose → glucose + glucose + water

b 3 fatty acids + glycerol → triglyceride + $3H_2O$

c (starch composed of a polymer of n glucose molecules) + H_2O → (starch composed of a polymer of n-1 glucose molecules) + glucose

3 transparency is necessary for photosynthesis by aquatic plants/algae; aquatic plants/algae serve as base of aquatic food chains; enable aquatic organisms to see prey;

4 a epsilon and zeta globin

b gamma is expressed in the first 10 weeks while beta is not; while gamma declines, beta increases; gamma is not expressed at six weeks while beta is significant fraction of hemoglobin at six weeks; both are expressed in approximately equal amounts at 24 days;

c at 10 weeks, hemoglobin is composed of equal amounts of alpha and gamma globin; at 6 months, hemoglobin is primarily composed of alpha globin and beta globin with smaller amounts of delta globin;

d maternal blood;

e fetal blood and maternal blood will differ in affinity for oxygen; (difference in hemoglobin structure gives) fetal blood greater affinity so oxygen will move from mother to fetus; change (with development) necessary so fetus can prepare for independent gas exchange/transition from placental to pulmonary gas exchange;

5 a i wild birds 13.3 kg

ii captive birds 16.2 kg

b both groups lose most of their lipid; captive birds lose more of their lipids than wild ones; 11.2 kg versus 9.6 kg/ 93% lost versus 81%/other valid figures comparing change;

c insulation/source of waste heat when metabolized/source of metabolic water;

6 a α-helix/ β-pleated sheet;

b fibrous protein example, e.g., keratin; composed largely of non-polar amino acids/ tend to be insoluble in water; often serving a structural role; lacking complex tertiary structure; globular protein example, e.g. amylase; (*do not award mark for general term "enzyme" without specific example*) often with metabolic functions; tend to be water soluble;

c polar amino acids in a protein will orient toward water; help determine tertiary structure; help determine which domains of the protein will be found at the membrane's surfaces/lining membrane pores;

d Secondary structure is pattern of regular folding; For example α-helix/ β-pleated sheet; Stabilized by hydrogen bonds between carboxyl and amino groups of backbone; Tertiary structure is overall three dimensional shape; Position of secondary structures in overall protein is a component of tertiary structure; Tertiary structure determined by interaction of R-groups/hydrophobic interactions;

7 a as quantity of water goes up, perecentage of grain protein goes down

b optimum level of water; above optimum level may get reduction in soil air; might favour soil denitrification/prevent nitrogen fixation/reduce availability of soil nitrogen;

Chapter 5

1 nucleotide

2 a RNA primase b DNA polymerase I
c DNA ligase d DNA helicase

3 a deoxyribose b DNA
c III and V d III

4 a helix is unwound; two strands are separated; helicase (is the enzyme that unwinds the helix separating the two strands); by breaking hydrogen bonds between bases; new strands formed on each of the two single strands; nucleotides/nucleoside triphosphates bind; by complementary base pairing; A to T and G to C; DNA polymerase forms covalent bonds between bases (on the new complementary strands); replication is semi-conservative/each of the DNA molecules formed has one old and one new strand; occurs in the 5′ to 3′ direction;

b nucleosomes help condense DNA/help with DNA packing; serves a role in gene expression; complex of histone proteins and DNA; octet of histones; with DNA coiled around; additional histone stablizes structure;

c DNA is double stranded and RNA is single stranded; DNA has deoxyribose and RNA has ribose; DNA has thymine while RNA has uracil; Both are nucleic acids/polymers of nucleotides; Both are synthesized in the nucleus; Both possess the bases, G, C and A;

5 aids packing of DNA; affects gene expression;

6 a phosphate b H-bond
c covalent bond d nucleotide
e 3′

7 assuming same orientation, II and III would be radioactive

Chapter 6

1 D

2 D

3 B

4 C

5 Sample. Many other concept maps can be drawn. (organizer created using Inspiration software)

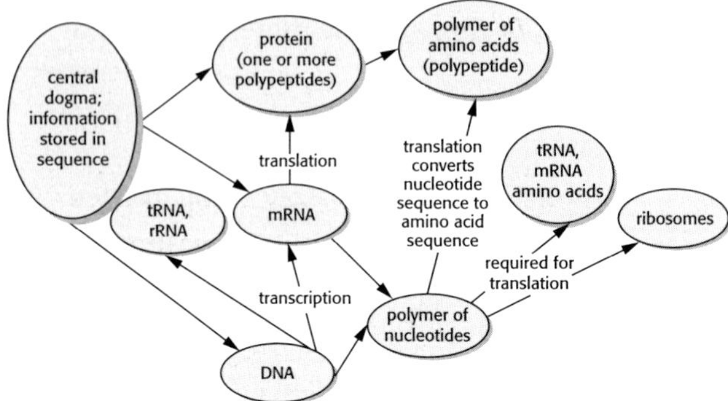

6 degenerate–more than one codon can code for the addition of a certain amino acid to a polypeptide; universal–in nearly all life, the same amino acids are coded for by the same codons;

7 a the genetic code is degenerate; therefore not possible to predict the exact sequence; would be possible to rule out wrong sequences;

b the genetic code is degenerate; therefore not possible to predict the exact sequence; would be possible to rule out wrong sequences; further complicated in eukaryotes by the presence of introns; (HL concept)

8 a (DNA) nucleotides;

b (mRNA) nucleotides;

c sugar, phosphate and nitrogenous base;

d amino acids;

9 a both in 5′ to 3′ direction; both require ATP; DNA is transcribed and mRNA is translated; transcription produces RNA and translation produces polypeptides/protein; transcription is nucleotide sequence to nucleotide sequence, translation is nucleotide sequence to amino acide sequence; RNA polymerase for transcription and ribosomes for translation/ ribosomes in translation only; transcription in the nucleus
(of eukaryotes) and translation in the cytoplasm/at ER; tRNA needed for translation but not transcription;

10 composed of mRNA base triplets; called codons; 64 different codons; each codes for the addition of an amino acid to a growing polypeptide chain; the genetic code is degenerate; meaning more than one codon can code for a particular amino acid; the genetic code is universal; meaning it is the same in almost all organisms; (AUG is the) start codon; some (nonsense) codons code for the end of translation;

11 many possible mRNA sequences. For example, UUUAUUGUUCUU; UUCAUCGUCCUG. For these two sequences, the corresponding DNA sequences are: AAATAACAAGAA and AAGTAGCAGGAC.

12 function of tRNA is transport of amino acids during translation; tRNA is composed of one chain of (RNA) nucleotides; tRNA has a position/end/site for attaching an amino acid; (*Reject tRNA contains an amino acid.*) at the 3′ terminal / consisting of CCA / ACC; tRNA has an anticodon complementary to codon; so that correct amino acid will be brought; anticodon of three bases which are not base paired/single stranded/forming part of a loop; tRNA has double stranded sections formed by base pairing; tRNA has (*three*) loops (sometimes with an extra small loop); tRNA has a distinctive three dimensional / clover leaf shape; ensures correct binding to specific enzymes;

13 both involve polymerases/RNA polymerase vs. DNA polymerase; polymerases both operate in 5′ to 3′ direction; RNA polymerase unwinds in transcription but helicase unwinds DNA in replication; RNA primer required by DNA polymerase but not RNA polymerase; no role for DNA ligase in transcription;

14 b/a/e/c/d

Chapter 7

1 a i no effect due to temperature between 20 °C and 40 °C; rate of activity falls above 40 °C;

ii higher temperatures cause changes to enzyme structure; active site no longer fits the substrate; denaturation;

b i also becomes less active at higher temperatures; activity is higher than soluble papain at temperatures above 40 °C;

ii immobilised papain is more heat stable; binding to a solid surface makes the enzyme molecule more stable;

iii maltase in cells lining the small intestine/other example;

2 a curve B;

b i enzyme activity was almost constant initially; then became less and less;

ii substrate concentration is reducing; as substrate is converted to product;

3 a chain or cycle of reaction; enzyme catalysed;

b i pyr; FDP; PEP;

c i pyruvate is accumulating; pyruvate is the product of the pathway;

ii enzyme that converts F6P to FDP; F6P accumulates; FDP concentration is lower;

4 a all reactions in the cytoplasm are catalysed by enzymes; there are many reactions occurring in the cytoplasm; enzymes are specific; each reaction is catalysed by a different enzyme; this allows control of individual reactions;

b enzyme/protein structure is very variable; active sites of different enzymes fit different substrates; different amino acids are found at the active site; amino acids (radicals) have different chemical properties; nucleus contains a gene/genes for making each enzyme type;

c inhibitor binds to active site versus at another site; inhibitor similar to substrate versus different; inhibitor blocks active site for substrate versus does not; inhibitor does not alter conformation/shape of active site versus does change it; increasing substrate concentration reduces effect of inhibitor versus does not; not used in end product inhibition versus can be used in end product inhibition; example of competitive inhibitor; example of non-competitive inhibitor;

5 a enzymes consist of polypeptides; sequence of amino acids in polypeptide must be correct; polypeptide must be folded up correctly; impossible artificially to link amino acids in long enough sequences correctly; many components/tRNA mRNA and ribosomes needed for polypeptide synthesis; cells have the entire polypeptide synthesis mechanism;

b proteins are very/more versatile in structure; they can form many/more different shapes; amino acids give many/more different chemical properties; genes can store the information for making many different proteins/enzymes;

c active site fits substrate precisely in lock and key model; active site has to change to fit substrate in induced fit model; substrate structure does not need to alter for binding in lock and key model; substrate structure altered during binding in induced fit model; induced fit model better explains how bonds are weakened in the substrate; induced fit model better explains how activation energy is reduced; induced fit model better explains the broad specificity of some enzymes; evidence for induced fit model is stronger/confirmed by studies of enzyme structure;

Chapter 8

1 D

2 C

3 B

4 D

5 B

6 D

7 a i glucose/fructose;
ii pyruvate/triose phosphate;
iii acetyl group/acetyl (CoA);
iv carbon dioxide;
v oxygen;
b i Krebs cycle; ii Krebs cycle;
c i NAD; ii ADP;

8 a oxidation is loss of electrons; reduction is gain of electrons; loss of hydrogen (atoms) is oxidation; gain of hydrogen (atoms) is reduction;
b hexose is phosphorylated (twice); lysis/splitting of hexose biphosphate; to form two triose phosphates; triose phosphate is oxidised; NADH + H^+ formed; ATP formed; pyruvate formed;
c ATP produced using energy from oxidations; respiratory substrates are oxidised by removal of hydrogen/electrons; NADH + H^+ formed; 2ATP produced per glucose in anaerobic respiration; substrates not fully oxidised in anaerobic respiration versus fully oxidised in aerobic; energy remaining in pyruvate/lactate/ethanol; more NADH + H^+ formed in aerobic respiration; energy from NADH + H^+ released in aerobic respiration; 3ATP formed per NADH + H^+; over 30 ATP per glucose formed by aerobic respiration;

Chapter 9

1 a i zero light intensity then sudden/instantaneous rise; slower rise to a maximum then these changes in reverse;
ii sudden rises and falls when light switches on or off; slower rise to a maximum and fall due to natural light changes;
b i same maximum/1200 lux/maximum beyond range of meter;
ii daily maximum falls; falls rapidly at first then more slowly;
iii no significant change in maximum intensity; fluctuation from day to day (due to natural light variation);
c *Chlorella* culture has multiplied/more cells per unit volume; more chlorophyll per unit volume; light absorbed by the chlorophyll/Chlorella cells;
d *Chlorella* culture has reached a maximum density; nutrients in water are used up; light intensity/carbon dioxide concentration is restricting growth;

2 a i light intensity;
ii carbon dioxide concentration;
iii temperature;
iv light intensity or carbon dioxide concentration or photosynthetic capacity;
b because carbon dioxide concentration is the limiting factor; they are at the same carbon dioxide concentration; the light intensity is not affecting the rate of photosynthesis;
c rate of photosynthesis is very low because of low light intensity; carbon dioxide is released in respiration; rate of respiration is greater than rate of photosynthesis;

3 a cell wall; plasma membrane; vacuole; nucleus; many chloroplasts;
b chloroplast envelope of two membranes/double membrane; thylakoids; small space inside thylakoids; stacks of thylakoids/grana; grainy stroma;
c light-dependent reactions happen in the thylakoids; light-independent reactions happen in the stroma; oxygen is produced in the light-dependent reactions; carbon dioxide is used in the light-independent reactions; ATP is produced in the light dependent and used in the light independent reactions; NADPH is produced in the light dependent and used in the light independent reactions; light independent reactions are affected by temperature but light dependent are not; light independent reactions are catalysed by enzymes but light dependent are not;

Chapter 10

1 a i active transport of sugar;
ii create high solute concentration; water drawn in by osmosis;
b i no oligosaccharides at sucrose concentrations below 0.25mol dm^{-3}; oligosaccharides concentration rises between 0.25 and 0.50mol dm^{-3}; no further increase above 0.50mol dm^{-3};
ii to reduce water loss from aphid/gut cells; by osmosis;
c i poor source of amino acids; many at a lower percentage in phloem sap that aphid proteins; especially essential amino acids;
ii plants synthesise amino acids for making plant proteins; plant and aphid proteins have different amino acid composition;

d i feed aphids on phloem sap containing antibiotics; test aphid growth rates/protein synthesis rates/amino acid contents;

ii physiological problems have to be overcome; problem of phloem sap dehydrating cells by osmosis; problem of lack of essential amino acids;

2 a stem/root growth;

b food storage;

c food storage in seeds/photosynthesis;

d controlling stomatal aperture;

e photosynthesis;

f stimulating/inhibiting flowering;

g gas exchange;

h support;

i reducing water loss;

j transport of water;

3 pollination is transfer of pollen from anther to stigma; by wind or insects or another agent; fertilisation is fusion of male and female gametes inside the ovule; involves pollen tube carrying male nuclei to ovule; seed dispersal is transfer of seeds from the parent plant to a new site; by wind or animals or another agent;

4 a roots are narrow cylindrical structures; that can grow between soil particles to reach water; roots can form lateral branches; root hairs can grow out from the root; surface area of root is increased by branching and root hairs;

b transpiration stream; water evaporates from the surface of mesophyll cells; water drawn through leaf cell walls to replace it; adhesion of water to cellulose in cell walls; cohesion of water molecules in cell walls/xylem vessels; water drawn out of xylem vessels in leaf; low pressure created in xylem; transpiration pull is enough to draw water up in the xylem from root to leaf;

c vascular bundles are distributed through monocot stem rather than in ring; no cambium/ no thickening of the stem; stem cannot grow into a thick trunk/cannot develop mass of xylem; no/limited branching possible without increasing support;

Chapter 11

1 For each, there is more than one possible answer.

a **non-disjunction** is linked to events that occur later in meiosis;

b random orientation, which occurs during metaphase I, results in independent assortment; **segregation** is a consequence of anaphase I

c **chemical composition** is not a method of sorting chromosomes for the purposes of making a karyotype

d **sickle cell anemia** is not a consequence of non-disjunction.

2 i IV (amniotic fluid)

ii I (chorionic villi of the placenta)

3 a the long arm of the chimp chromosome #12 and the short arm of the human chromosome appear to be identical; the entire length of the chimp chromosome #13 appears to be found on the long arm of the human chromosome; the final band on the end of the short arm of chimp chromosome #13 does not appear in the human chromosome; the human chromosome is longer than either of the chimp chromosomes;

b near the centromere on the long arm of the human chromosome, you would find a number of repeats that were more characteristic of telomeres than sequences normally found near the centromere;

4

Unique to meiosis	Shared features	Unique to mitosis
For sexual reproduction	Both can be used for reproduction	For growth maintenance, repair, asexual reproduction
Reduction in chromosome number; diploid to haploid;	Both involve nuclear division;	Maintains chromosome number; diploid to diploid or haploid to haploid;
Two divisions		
Four cells produced (though oogenesis results in one ovum and two polar bodies)		

5 C

6 B

Chapter 12

1 a genotype is the alleles of an organism; phenotype is the characteristics of an organism;

b dominant alleles have the same effect in heter zygotes and homozygotes; recessive alleles only have an effect in homozygotes;

c homozygous is having two of the same alleles of a gene; heterozygous is having two different alleles of a gene;

d gene is a heritable factor controlling a specific characteristic; genome is all the genetic information of an organism;

e haploid is having one chromosome of each type; diploid is having two chromosomes of each type;

2 a L is allele four leaflets and l is the allele for three leaflets; F1 genotype is Ll;

b i Punnett grid; 3 to 1 ratio of four leaved to three leaved; four possible genotypes are LL Ll lL and ll;

ii Not the expected ratio/most clovers have three leaflets; wild ratio is due to natural selection/allele frequencies in gene pool;

c three leaved clovers will produce more three leaved when crossed together; identify homozygous four-leaved plants by crossing with three leaved/test crossing; four leaved clovers that produce only four leaved offspring when crossed with three leaved clovers are homozygous

3 a AB individuals are all I^AI^B;

O individuals are ii;

A individuals are all I^Ai;

B individuals area all I^Bi except II 1 which may be I^BI^B;

b A or B or O or AB; 25% chance of each;

c i blood group O; 100%

ii 50% group A; 25% AB and 25% B;

4 a B is dominant; b is recessive; C^R is codominant;

b homozygous for A; heterozygous for B;

c homologous; have same genes in same sequence; but different alleles of these genes;

d have same position/locus on the chromosome; same letter used to denote them;

5 a allele for hemophilia is located on the X chromosome; mother was a carrier; son inherited mother's X chromosome carrying the hemophilia allele; and a Y chromosome from the father that does not have a copy of this gene;

b non-hemophiliac son does not have the hemophilia allele; he inherited the mother's X chromosome not carrying the hemophilia allele; his children will not have hemophilia unless his partner is a carrier; daughter has X chromosome from father carrying the non-hemophilia allele; daughter also has X chromosome from mother which might carry the hemophilia allele; 50% chance that she is a carrier; if she is a carrier there is a 50% chance of sons having hemophilia but not daughters;

c hemophilia involves periods of ill health and hospital visits; feelings of anxiety/guilt about passing on hemophilia to offspring; loss of employment/income/prejudice against AIDS patients; grief after death of relative that developed AIDS;

6 a light wing probably due to recessive allele; dark wing form due to dominant allele;

b Punnett grid showing DD Dd dD and dd genotypes or other suitable letters; 3 to 1 ratio of dark to light;

c sex-linkage; gene must be on a sex chromosome;

d males have two X chromosomes; females have one X and one Y;

e X^DY crossed with X^dX^d; all female offspring light winged and all male offspring dark winged;

Chapter 13

1 a autosomes are non-sex chromosomes; sex-chromosomes determine gender;

b genes in autosomal linkage are on the same autosome; genes in sex linkage are on the X chromosome/on a sex chromosome;

c continuous variation is due to one gene and discontinuous is polygenic; few discrete types with discontinuous but many/infinite with continuous;

d polygenic inheritance is a character controlled by two or more genes; multiple alleles is one gene with more than two alleles;

e skin colour; height;

2 a seven;

b i tall/dwarf and round/wrinkled or other example;

ii yellow/green cotyledons and purple/white flowers or other example;

c gametes FG Fg fG and fg; 4 × 4 Punnett grid constructed; sixteen possible genotypes shown on Punnett grid (FFGG, FFGg and so on); possible phenotypes are full green, full yellow, constricted green and constricted yellow; 9 3 3 1 ratio;

d crossing over occurs between the genes; large distance between genes so crossing over is very frequent/recombination rate is 50%;

3 a which allele of one gene is present does not influence which allele of another gene is present; in a gamete/a nucleus formed by meiosis; genes located on different types of chromosome/non-homologous chromosomes; homologous chromosomes pair up in meiosis; orientation of pairs of chromosomes is random; therefore movement of alleles to the poles on non-homologous chromosomes is independent;

b homologous chromosomes pair up; in prophase one of meiosis; breaks at same point in non-sister chromatids; non-sister chromatids join up; process is called crossing over; crossing over creates chiasmata;

c skin colour varies from very light to very dark; polygenic inheritance; several genes influence skin colour; offspring of dark and light skinned parents often have intermediate skin colour; environment also influences skin colour; sunlight/ultra violet causes skin to darken;

4 a rose parent is AA and single parent is aa; F1 hybrid is Aa;

b pea parent is BB and single parent is bb; F1 hybrid is Bb;

c AaBb;

d 4×4 Punnett square with sixteen dihybrid genotypes shown; walnut rose pea and single combed offspring 9 to 3 to 3 to 1 ratio;

e could be regarded as polygenic; more than one gene influences comb shape;

5 a same genotype; allows effects of environment and genes to be distinguished;

b same genotype; same/similar environment;

c smallest is 4; largest is 37;

d education affects IQ; the greater the difference in education the greater the IQ difference;

e other environmental factors affect intelligence; for example nutrition/other example;

Chapter 14

1 a zero cheetahs heterozygous and thirteen domestic cats

b one allele in cheetahs and three alleles in domestic cats

2 used to produce genetically modified organisms; host DNA and donor DNA are cut with same restriction enzyme; fragments are combined using DNA ligase; used in DNA profiling; sample DNA is cut with restriction enzymes; pattern created depends on frequency of recognition/restriction sites/two individuals will possess different recognitions sites/two individuals will produce a unique pattern of fragments; fragments separated by gel electrophoresis

3 therapeutic cloning requires eggs; limited supply of surplus eggs from IVF clinics; possibility that women will be commissioned to undergo fertility treatment to supply eggs; invasive procedure to extract eggs; "markets" in egg production might exploit vulnerable/poor populations; risk of unwanted multiple pregnancies; asexual reproduction/vegetative propagation in plants; binary fission in bacteria; monozygotic twins;

4 both processes are semi-conservative; primer required in both cases; primer is added artificially in PCR; primer is synthesized by RNA primase in natural process; replication normally proceeds at much lower temperature; in PCR process requires temperature that is only suitable to extremophile/thermophiles (such as *T. aquaticus*); PCR only amplifies fragments whereas replication copies whole genome; Unwinding done by heat in PCR but helicase in natural process; Both involve polymerase;

5 asexual reproduction/vegetative propogation in plants; binary fission in bacteria; monozygotic twins;

6 a IV is the smallest fragment

b each parent gives a child one allele; child 2 possesses alleles I and IV; father possesses alleles II and III; neither allele found in child 2 is found in the father;

7 A

Chapter 15

1 a natality is birth rate/ratio of number of live births per population unit (such as live births per 1000 population); mortality is death rate/ ratio of number of deaths per population unit;

b ecological community is all of the interacting populations in a certain area; ecological community is a component of ecosystems; ecosystem is all of the populations in a certain area combined with abiotic factors and all of the interactions between these elements;

c heterotroph and autotroph are terms which refer to the source of carbon for the organism; an autotroph is an organism that is able to utilize carbon dioxide as a carbon source; heterotrophs must ingest organic matter to obtain their carbon; autotrophs are producers in a food chain; heterotrophs are consumers in a food chain;

2 a Respiration loss = gross production – net production

$$= 1 \times 10^2 \text{ kJ m}^{-2} \text{ y}^{-1}$$

b answer presumes a student draws a pyramid of net production: base of pyramid is 50 units wide; second tier is 6 units wide; third tier is 0.6 units wide; (accept equivalent ratios) tiers labeled as producers, primary consumers, secondary consumers (accept equivalent terms)

3 a greater fraction of incident light energy lost in desert; deserts are less productive/less vegetation to fix energy;

b some plant matter falls as detritus and becomes trapped in this sink; undigested plant matter can be lost as feces from herbivores; energy lost as heat/respiration in herbivores;

4 a 1968;

b (max. 2 marks for any of the following) catch by fisherman fell before cormorants feeding was significant; fisherman's catch was low in the 1960s before cormorant numbers rose;

fisherman's catch fluctuated greatly without cormorant feeding; cormorants caught less than the fishermen in every year; (additional mark for one of the following) lowest catches by fisherman when cormorant numbers have risen; fishermens' catches remain low when cormorant numbers have risen;

5 a producer

b i two trophic levels/producer and primary consumer;

ii seven trophic levels; nanophytoplankton→ microzooplankton→macrozooplankton→ small fish/squid→emperor penguin→ leopard seal→toothed whale; bacteria→ microzooplankton→macrozooplankton→ small fish/squid→emperor penguin→ leopard seal→toothed whale;

iii small fish and squid/krill;

iv toothed whales;

Chapter 16

1 a all are in remote areas/areas uncontaminated by local pollution;

b both show linear increase; data for Baring Head gathered over a longer period; Alert shows seasonal variation not seen in Baring Head;

c greatest fraction of world's land mass is in northern hemisphere; seasonal variation due to photosynthesis by temperate forests/CO_2 uptake by northern hemisphere trees; causes drop in atmospheric CO_2 during northern hemisphere summer; effect is local/less pronounced in Southern hemisphere/New Zealand;

2 a with higher latitude, the daylength required to end hibernation is longer/direct relationship;

b 1996 requires shorter daylength to end hibernation; direct relationship the same in both data sets;

c two variables involved in ending hibernation/ temperature and daylength both contribute to ending hibernation; warmer temperatures mean prey species end hibernation early; adaptive advantage to mosquito to hatch earlier for a fixed daylength/earlier hatchers likely to enjoy more reproductive success;

3 a i 330 ppm ii 330 ppm- 310 ppm = 20 ppm

b i night;

ii CO_2 produced by cellular respiration; plants/ shrubs/forest floor plants produce the CO_2/ produced near the ground; CO_2 not as easily blown away on floor compared to canopy/ CO_2 may sink to floor because of density;

iii 7/8 am

Chapter 17

1 D

2 A

3 D

4 D

Data-based question

1 hybrids have genetic variation; genetic variation needed for natural selection;

2 variation in days to flowering time reduces; mean days to heading increases/flowering is later on average;

3 a the further north the earlier the flowering time;

b growth is later in the cooler north;

4 a less variation in flowering time at each latitude;

b initially they would flower at the same time as the F10 plants; hybridisation would create variety in flowering time; natural selection would favour the plants flowering at the intermediate times;

Chapter 18

1 I filicinophyta;

II bryophyta;

III filicinophyta;

IV angiospermophyta;

V coniferophyta;

VI angiospermophyta;

VII bryophyta;

VIII coniferophyta;

2 a 34;

b 24–30; and 31–33;

c 1–7; and 8–23;

d 1–23; and 24–33 and 34;

e more closely related to species 16; because they are in the same family and species 6 isn't;

f due to its different characteristics it is classified by itself; in its own genus family and order;

3 a mollusca have a mouth but porifera do not; mollusca have an anus but porifera do not; mollusca have a muscular foot but porifera do not; mollusca have a head but porifera do not; mollusca often have shell but porifera do not; porifera have many pores through their body by mollusca do not;

b both are segmented; both are bilaterally symmetric; both have a mouth and anus; arthropoda have jointed appendages but annelida do not; annelida have bristles/chaetae but arthropods do not; arthropoda have an exoskeleton but annelida do not;

c cnidaria are radially symmetric and platyhelminthes are bilaterally symmetric; cnidaria have a rounded body and platyhelminthes have a flattened body; both have a mouth; neither have an anus; both are unsegmented; cnidaria have stinging tentacles but platyhelminthes do not;

4 a Kingdom Animalia; phylum Chordata; class Mammalia; order Primate; family Hominidae; genus Homo; species sapiens;

b two names; first name is the genus name; second name is the species name; upper case letter for first letter of genus but lower case for first letter of species; international/Latin or Greek used/printed in italics;

c series of choices/options; usually two choices/ dichotomous; choices based on reliable characteristics found in all individuals; choices based on easily visible/external characterisitics; some choices give the identification; other choices refer to another stage in the key;

Chapter 19

1 B

2 A

3 C

4 D

5 A

6 A

7 a mouth to esophagus to stomach; stomach to small intestine to large intestine;

b salivary glands secrete into the mouth; wall of stomach secretes into the stomach; liver/ gall bladder and pancreas secrete into small intestine; wall of small intestine secretes into the small intestine;

8 a diagram clearly showing these labelled structures: mouth and esophagus; stomach; small intestine; large intestine and anus; liver with gall bladder; pancreas;

b food molecules are chemically stable/do not break down rapidly; enzymes increase the rate of digestion; would take too long at body temperature without enzymes; undigested food molecules cannot be absorbed;

c much digestion in the small intestine but little in the large intestine; fat/starch/protein digestion completed in the small intestine; enzymes secreted by small intestine wall but not large intestine wall; maltase/other named enzyme secreted; much absorption of digested food in the small intestine but little in the large intestine; glucose/amino acid absorption completed in small intestine; more absorption of water in the large intestine; storage of feces in the large intestine;

9 a i 3.7%; ii 96.3%;

b i digested; into nucleotides/subunits/by nucleases; in the small intestine;

ii absorbed in the small intestine; used to make human DNA/used in DNA replication in human cells;

c speed of movement of food through gut; amount of DNA digesting enzymes secreted;

d to check that DNA had time to pass to the end of the small intestine; to allow rate of digestion to be assessed;

e suggests that transgenes are mostly digested in the gut; therefore they are unlikely to cause harm; but in one individual less was digested/ more evidence needed;

Chapter 20

1 a i I is an erythrocyte; II is a leucocyte/ lymphocyte;

b II is a lymphocyte and III is a phagocyte; rounded nucleus versus lobed nucleus;

c i measure the diameter of a typical erythrocyte in millimetres; multiply by 1000 and divide by 7 to obtain the magnification

ii measure diameter of largest and smallest leucocytes in millimetres and multiply by 1000; divide by the magnification to obtain the actual sizes in millimetres

2 atria start to contract; blood is pumped from atria to ventricles; ventricles start to contract; atrioventricular valves close; semilunar valves open; blood is pumped from ventricles to artery; atria start to fill with blood; semilunar valves close; atrioventricular valves open; ventricles start to fill with blood;

3 a i lungs;

ii lungs;

iii respiring tissues;

iv small intestine;

v endocrine glands;

vi kidneys;

b i skin; lungs;

ii muscles;

c antibodies/leucocytes;

4 a T wave is repolarisation of ventricles /relaxation of ventricles;

b heart rate;

c no electrical activity between S and T waves in normal cardiac cycle; electrical activity continuing between S and T waves during heart attack; no electrical activity between T and P waves in normal cardiac cycle; electrical activity continuing between T and P waves during heart attack;

d less S-T and T-P electrical activity in the later stages of a heart attack; interval between and Q in the later stages of a heart attack;

e ambulance operatives/paramedics; doctors/ nurses/rescue workers;

Chapter 21

1 a Lysozyme-non-specific

b Phagocytes-non-specific

c Memory B-cells-specific

d Antibodies-specific

2 a 1986 (acceptable answers in the range from 1984 to 1990);

b i significant success in reducing number of people suffering;

ii early success in reducing number of new cases; resurgence in numbers in 1996; resurgence might represent better data gathering; rate of new cases was low even at the beginning of the study;

3 B

4 B

5 a T-cells are involved in activating specific immune response; such as B-cell proliferation; HIV targets T-cells specifically/reduces number of circulating T-cells; Leads to reduced ability to develop specific immune response;

b Example of disease; e.g., tuberculosis; Name of pathogen responsible; e.g., Mycobacterium tuberculosis; Method of transmission; e.g., droplet infection/exposure to coughs/sneezes from infected individual; Effects (up to two marks): e.g., development of tubercules in lungs/persistent cough/coughing up blood/lung collapse;

c macrophages ingest antigen / antigen presenting cells ingest antigen; T-helper cells bind to macrophages; T-helper cells stimulated / activated; antigen binds to B-cells; activated T-helper cells then bind to B-cells; activation of B-cells; which divide / undergo mitosis to form clones of (plasma) cells; B-cells / plasma cells produce antibodies; memory cells produced; with subsequent exposure more antibodies and faster response / graph to show this; antibodies are specific to antigen;

6 A clotting factors B thrombin

C fibrinogen D fibrin

Chapter 22

1 D

2 C

3 A

4 C

5 C

6 a smokers life expectancy is 7 years shorter at 35; smokers life expectancy is 5.2 years shorter at 70;

b more cancer in smokers; most lung cancer is caused by smoking; more heart disease/lung disease in smokers;

7 a i airways become blocked so ventilation stops; oxygen concentration of alveoli falls so saturation drops;

ii reduced oxygen saturation wakes the sleeper; airways reopened by moving the soft palate;

iii 55 cycles in one hour; 65 seconds per cycle;

b 65%; 5 hours 40 minutes;

c normal sleep initially; then apnoea for rest of night apart from two periods of normal sleep;

Chapter 23

1 a protein not filtered out in glomerulus; protein molecules too large to pass through basement membrane/filtration slits;

b glucose reabsorbed from filtrate; by active transport; in the proximal convoluted tubule;

c water reabsorbed; by osmosis; no/little urea reabsorbed;

d sodium ions are reabsorbed; by active transport; sodium ions are needed in the plasma whereas urea is a waste product;

e amount of water reabsorption varies; due to ADH secretion/concentration; variable need for reabsorption of sodium ions/variable reabsorption in the distal convoluted tubule;

2 a pump proteins for glucose; in cells in the proximal convoluted tubule; blood glucose concentration low enough for all glucose to by absorbed;

b limited capacity of pump proteins for glucose absorption; blood glucose concentration can be very high in diabetics; because insulin is not secreted; body cells not stimulated to absorb glucose;

3 a excretion and osmoregulation;

b neither;

c neither/small amount of excretion;

d excretion;

e osmoregulation;

f osmoregulation;

g excretion;

h osmoregulation;

4 a sodium ions/minerals/solutes reabsorbed by active transport; therefore water reabsorbed by osmosis;

b solute concentration in interstitial fluid in medulla becomes greater; water reabsorbed from the filtrate by osmosis;

c sodium ions diffuse out of the filtrate; active transport of sodium ions out of the filtrate;

d water reabsorbed by osmosis; due to solute concentration being higher outside the tubule;

e less absorption of water in collecting duct; due to absence of ADH/aquaporins;

5 a ADH secreted when the body is dehydrated; but does not cause increased reabsorption of water; large volumes of dilute urine produced; severe dehydration;

b due to a recessive allele; parent could be a carrier and not show symptoms;

c sex linked; gene for ADH receptor is located on the X chromosome;

d drink large volumes of water; increased intake to replace losses in urine;

Chapter 24

1 a glucose is stored in liver as glycogen when blood sugar is high; when blood sugar is low, glucose is released from liver stores; daily fluctuations in glycogen levels is linked to daily fluctuations in blood glucose; caused by eating; and activity;

b homeostasis is maintenance of variables at or near a set point; blood sugar regulated by the hormones insulin and glucagons; these hormones whether glucose is stored as glycogen; or released from glycogen stores; considerable daily variation in glycogen stores enables blood sugar homeostasis;

2 a depolarization of pre-synaptic membrane; triggers opening of voltage gated calcium channels to open; calcium influx triggers vesicles containing neurotransmitter to move to the pre-synaptic membrane; neurotransmitters released by pre-synaptic neurons (by exocytosis); diffuse across synapse; bind to specific receptors on post-synaptic membranes; some neurotransmitters increase permeability of post-synaptic membrane to ions; causing localized depolarization; which helps an action potential to form; neurotransmitter in the synaptic cleft is rapidly broken down to prevent continuous synaptic transmission;

b cell body with nucleus (nucleus not occupying more than one third the diameter of the cell body); dendrites (at least three shown with appropriate length); axon (with mark indicating length is longer than can be shown); axon terminus/synaptic knob;

c at rest, the neuron membrane is negative on the inside relative to the outside; action potential in one part of neuron triggers action potential to develop in adjacent region; due to diffusion of sodium ions from the region experiencing an action potential; to the region at rest; this reduces the membrane potential; if the reduction reaches a threshold potential, voltage gated sodium channels open; sodium ions diffuse down their concentration gradient; leading the inside of the membrane to develop a positive charge/depolarization; potassium channels then open; potassium diffuses out; a net internal negative charge re-develops/ repolarization; overshoot ensures the action potential moves in one direction; sodium-potassium pump re-establishes the resting potential;

3 More than one correct answer.

a Efferent carries impulses from the central nervous system and afferent carries a message to the CNS

b Action potential is achieved when the resting potential is reversed; after a threshold potential is achieved;

c Sodium-potassium pump is involved in active transport; ion channels are involved in facilitated diffusion

d Axon carries impulses away from the cell body and dendrite carries them to the cell body; in a motor neuron, the axon is much longer;

e Glucagon is released in response to low blood sugar and insulin is released in response to high blood sugar; Glucagon causes glucose to be released from glycogen stores and insulin causes glycose to be stored as glycogen;

f Negative feedback is inhibition caused by the presence of end-product; Positive feedback is amplification caused by the presence of end-product;

g Vasoconstriction is narrowing of skin surface arterioles and vasodilation is widening of skin surface arterioles; Vasoconstriction occurs to prevent core temperature from falling while vasodilation occurs to prevent core temperature from rising.

h Type I diabetes is due to low insulin production by pancreas and Type II diabetes is due to poor response of cells to insulin; There is a stronger dietary link to type II diabetes

4 a Vasoconstriction is narrowing of skin surface arterioles; preventing blood flow near surface; With exposure to cold, vasoconstriction occurs to prevent core temperature from falling; With prolonged exposure, some vasodilation must occur; Explaining temporary rise in skin surface temperature;

b negative feedback is end-product inhibition; immersion of hand in ice water lowers blood temperature near skin surface; vasoconstriction to reduce this effect; with temperature rising/ build-up of waste heat, vasodilation would occur; causing renewed temperature lowering near the surface; leading to vasoconstriction again;

c i 9 °C ± 1 °C

ii trade-off between preserving peripheral/ skin temperature; and preserving core body temperature; since core body temperature plays a critical role in survival; vasoconstriction threatening peripheral tissue health is preferable to vasodilation threatening core temperature;

5 an action potential normally triggers release of Ca^{2+} from the sarcoplasmic reticulum; the Ca^{2+} binds to the troponin complex causing muscle contraction; the Na^+/Ca^{2+} pump mimics the release of calcium from the sarcoplasmic reticulum; causing muscle contraction;

Chapter 25

1 B

2 A

3 D

4 B

5 a in a pivot joint the movement is limited to rotation. In a hinge joint, movement is restricted to a single plane.

b Radius is a bone in the lower arm and humerus is a bone in the upper arm

c hip offers greater range of motion than the knee joint in that it can flex and extend, rotate and move to the side of the body and back.

d actin is the protein to which myosin heads bind; myosin filaments are thicker than actin filaments;

e ADP has two phosphate bonded to it while ATP has three;

f Ligaments attach bone to bone while tendons attach muscle to bone;

g Extension is movement which increases the angle between two bones and flexion is movement which decreases the angle between two bones;

6 a the hip joint can flex and extend; Rotate; move to the side of the body and back/abduct and adduct

b calcium released from sarcoplasmic reticulum; calcium binds to troponin; troponin with calcium bound makes tropomyosin move; movement of tropomyosin exposes binding sites (for myosin) on actin; contraction of muscle fibres is due to the sliding of filaments (over each other); myosin heads bind to / form cross bridges with actin; ATP binds to the myosin heads causing them to detach from the binding sites; hydrolysis of ATP / conversion of ATP to ADP causes myosin heads to move; myosin heads reattach to actin further along; myosin pushes actin / actin pushed towards the centre of the sarcomere/ shortening of sarcomere;

c Draw a labeled diagram of the human elbow joint (4) humerus; radius; ulna (shown in correct position in relation to radius); cartilage (on ends of bones); ligaments (connecting humerus with radius/ulna); capsule (sealing joint); synovial fluid; biceps (attached to radius); triceps (attached to ulna); tendons (connecting muscle to bone correctly);

7 a between 3.5 and 3.6 ms^{-1}

b Group D (1500 m runners) and Group E (42 km marathon runners);

c both increase the amount of lactate in the blood / eventually respire anaerobically; 100 m runners increase lactate much more than marathon runner / lactate in 100 m runners starts off higher even at low speeds; marathon runner can run at greater speed aerobically;

d lactate produced as a consequence of anaerobic respiration; With greater distance trained for, the amount of lactate produced for a fixed speed decreases; Suggesting training for long distance decreases reliance on anaerobic respiration; Advantage is that more energy released from the same amount of stored energy/energy per glucose yield is higher; Possible improvements due to improved heart rates/more efficient muscles/ better oxygen delivery to muscles/changes in muscle composition.

Chapter 26

1 a diagram clearly showing these labelled structures: testes; sperm duct; penis; urethra; prostate gland/seminal vesicle/epididymis;

b stimulates fetal gonads to develop as testes; stimulates development of penis in fetus; stimulates development of secondary sexual characteristics during puberty; stimulates production of sperm; maintains sex drive;

c gender determined by sex chromosomes; XX is female; eggs carry an X chromosome; XY is male; X and Y chromosomes separate during first division of meiosis in males; sperm carries either an X or a Y chromosome; half of sperm carry an X and half carry a Y chromosome; chance of X or Y bearing sperm fertilising the egg is equal;

2 a diagram clearly showing these labelled structures: ovary; oviduct; uterus; cervix; vagina;

b lining of the uterus is shed during menstruation; rising levels of estrogen; cause the thickening of the uterus lining/ endometrium; in the days before ovulation; rising progesterone levels;maintain the thickened uterus lining/endometrium; endometrium is thickened at the time of implantation; falling progesterone levels cause the uterus lining/endometrium to break down;

c FSH; injected; high doses; stimulates many follicles to develop; down regulation of natural menstrual cycle using drugs;

3 a success rate increases as age of mother increases; success rate much lower above age 39; success rates slightly lower below 30 than 30–34;

b success rates increase as more embryos are transferred; but rate with two/three is not double/triple rate with one; rate with three embryos only slightly greater than rate with two;

c restricting the number of embryos transferred reduces chance of multiple birth; multiple births increase the health risks for mother/ child; restricting number of embryos to two would prevent (almost all) triplets; older mothers at less risk of multiple births so more embryos could be transferred;

4 a availability of contraceptives; cultural differences/whether women have careers; variation in infant mortality/life expectancy (and the need to ensure that some children survive);

b i Sub-Saharan Africa;

ii East Asia;

c decrease in infant mortality; expectation that children will survive/high costs of raising large numbers of children;

d death rate also affects population size; migration also affects population size; emigration and immigration;

5 a LH levels do not rise high enough; ovulation not stimulated; development of corpus luteum not stimulated;

b i FSH level much lower during pregnancy; FSH level more constant during pregnancy; higher and fluctuating levels of FSH at other times

ii HCG stimulates the maintenance of the corpus luteum/production of progesterone; progesterone levels rise during pregnancy; progesterone inhibits FSH production;

c i 110mIU versus 10mIU; 110 minus 10 = 100, divided by 10 = 10, multiplied by 100% = 1000%;

ii estrogen/progesterone levels remain low/ no follicles produce estrogen; FSH and LH production not inhibited;

iii ovary becomes insensitive to FSH; primary follicles used up/have degenerated/not available for development.

Index